PROTEIN CROSSLINKING
Biochemical and Molecular Aspects

ADVANCES IN EXPERIMENTAL MEDICINE AND BIOLOGY

Recent Volumes in this Series

PROTEIN CROSSLINKING

Biochemical and Molecular Aspects

Edited by

Mendel Friedman

Western Regional Research Laboratory
Agricultural Research Service
U.S. Department of Agriculture
Berkeley, California

PLENUM PRESS • NEW YORK AND LONDON

Library of Congress Cataloging in Publication Data

Symposium on Protein Crosslinking, San Francisco, 1976.
   Protein crosslinking.

  (Advances in experimental medicine and biology; v. 86A-86B)
  "Proceedings of the [1st and 2d halves] of a Symposium on Protein Crosslinking
held in San Francisco, California, August 30-September 3, 1976, with additional
invited contributions."
  Includes index.
  CONTENTS: v. 1. Biochemical and molecular aspects. — v. 2. Nutritional and
medical consequences.
  1. Proteins — Crosslinking — Congresses. I. Friedman, Mendel. II. Title. III.
Series. [DNLM: 1. Proteins — Congresses. W1 AD559 v. 86 pt. A etc. / QU55
S98644p 1976]
QP551.S939 1976   574.1'9245   77-23448

ISBN 978-1-4684-3284-8   ISBN 978-1-4684-3282-4 (eBook)
DOI 10.1007/978-1-4684-3282-4

Proceedings of the first half of a Symposium on Protein Crosslinking
held in San Francisco, California, August 30-September 3, 1976, with
additional invited contributions

PREFACE

The word <u>crosslinking</u> implies durable combination of (usually large) distinct elements at specific places to create a new entity that has different properties as a result of the union. In the case of proteins, such crosslinking often results in important changes in chemical, functional, nutritional, and biomedical properties, besides physical properties simply related to molecular size and shape. (Nucleic acids, carbohydrates, and other biopolymers are correspondingly affected.) Since proteins are ubiquitous, the consequences of their crosslinking are widespread and often profound. Scientists from many disciplines including organic chemistry, biochemistry, protein chemistry, food science, nutrition, radiation biology, pharmacology, physiology, medicine, and dentistry are, therefore, minutely interested in protein crosslinking reactions and their implications.

Because protein crosslinking encompasses so many disciplines, in organizing the Symposium on Nutritional and Biochemical Consequences of Protein Crosslinking sponsored by the Protein Subdivision of the Division of Agricultural and Food Chemistry of the American Chemical Society, I sought participants with the broadest possible range of interests, yet with a common concern for theoretical and practical aspects of protein crosslinking.

An important function of a symposium is to catalyze progress by bringing together ideas and experiences needed for interaction among different, yet related disciplines. To my pleasant surprise, nearly everyone invited came to San Francisco to participate. Furthermore, those that could not come usually agreed to contribute a paper for the Proceedings. Many participants told me privately that they had made a special effort to come to San Francisco to help celebrate the combined Centennial of the American Chemical Society and Bicentennial of the United States. I am grateful for this friendly gesture. To supplement the verbal presentations further, the Proceedings include several closely related, invited contributions. The distinguished international participation from at least nine countries increases the authority and usefulness of these Proceedings.

These papers are being published in two volumes in the series
Advances in Experimental Medicine and Biology under the following
titles: PROTEIN CROSSLINKING: BIOCHEMICAL AND MOLECULAR ASPECTS
(Part A) and PROTEIN CROSSLINKING: NUTRITIONAL AND MEDICAL CONSE-
QUENCES (Part B). The two volumes are intended to be complementa-
ry, but their interests necessarily overlap.

Part A, the first volume, encompasses detailed discussions of
natural crosslinks such as disulfide and peptide bonds, various
artificial crosslinks formed by means of bifunctional reagents,
radiation-induced crosslinks, and techniques to determine crosslinks.

Ultraviolet and gamma radiations are widely used to increase
vitamin D content of foods, to sterilize food and drug products,
and to treat diseases such as psoriasis and cancer. However, our
knowledge about the molecular and nutritional consequences of
irradiating food products and other proteins is still imperfect.
Such consequences include crosslink formation. Several contributions report recent results in this field. The results directly
concern those interested in radiation biology and cancer therapy
as well as food scientists and food technologists responsible for
balancing good and bad effects of radiation.

Part B, the second volume, includes detailed discussions of
crosslink formation in food proteins through lysinoalanine, isopeptide bonds, and products derived from protein-carbohydrate
reactions. Such crosslinks not only lower the nutritional quality
and digestibility of food products but sometimes introduce toxicity.
This volume discusses not only nutritional and pharmacological
consequences of crosslink formation in food proteins but also
factors that govern crosslink formation, effects of crosslinks on
protein structure, reactivity, and digestibility, and ways to
minimize crosslinking.

Part B also discusses structural and tissue proteins, such as
collagen and elastin, which contain many natural crosslinks derived
from lysine. Several papers report evidence that these crosslinks
are important in aging and connective tissue diseases. The chemistry and biochemistry of such natural crosslinks are thus important
to anyone concerned with the relation of nutrition, health, and
aging.

I want to emphasize considerations supporting the diversity of
the subject matter presented in these volumes and of contributors'
backgrounds and interests. The widest possible interaction of
viewpoints and ideas is needed to transcend present limitations
in our knowledge as expeditiously as possible and to catalyze
progress in the field of crosslinking. Scientists from related
disciplines need one another's results; results with different
biopolymers need to be compared; scientists and physicians

responsible for practical applications need to share experiences and problems with basic researchers. These volumes bring together many elements needed for such interactions. The range of material includes a great variety of specific and general topics. This scope should interest at least a similar range of readers, but it challenges all of us to think seriously about subjects beyond our primary interests. It is my hope, therefore, that the reader will look not only to those articles of primary interest to him but to others as well and so profit by a broad overview.

I am particularly grateful to all contributors and participants for excellent cooperation, to Dr. Wilfred H. Ward for constructive contributions to several manuscripts, to my son Alan D. Friedman for his help with the preparation of the index, to Roy Oliver of Pierce Chemical Company and Dr. Rao Makineni of Bachem Fine Chemicals for financial assistance, to Dawn M. Thorne for final typing of several manuscripts, and to the Protein Subdivision of the Division of Agricultural and Food Chemistry of the American Chemical Society for sponsoring the symposium. I hope that PROTEIN CROSSLINKING will be a valuable record and resource for further progress in this very active interdisciplinary field.

Finally, I dedicate this work to the late Professor S. Morris Kupchan, with whom I had the privilege of spending a post-doctoral year at the University of Wisconsin. His untimely death deprives us of a very great scientific benefactor whose twenty-year global search for natural anti-tumour protein (enzyme) alkylating compounds is just now beginning to bear fruit.

Mendel Friedman
Moraga, California
March, 1977

GENESIS 44:30...because the father's (Jacob's) life is
                crosslinked to his son's (Benjamin's)...

SAMUEL I 18:1...Jonathan's soul was crosslinked to David's...

CONTENTS OF PART A

CONTENTS OF PART B

BIOLOGICALLY IMPORTANT THIOL-DISULFIDE REACTIONS AND
THE ROLE OF CYST(E)INE IN PROTEINS: AN EVOLUTIONARY PERSPECTIVE

Robert C. Fahey

Department of Chemistry

University of California, San Diego

I.   ABSTRACT

Selected aspects of the reactions of thiols and disulfides
are reviewed in an evolutionary context with special emphasis on
the implications of the transition from a reducing to an oxidizing atmosphere on the earth. It is argued that thiols were important in prebiotic chemistry and in primitive metabolism but that
disulfides, owing to their general instability in a reducing environment, came to be of importance as structural links in proteins
only after the transition to an oxidizing atmosphere. The occurrence
of glutathione is reviewed and discussed in terms of the role of
glutathione in maintaining a reducing intracellular environment.
The occurrence of cysteine and cystine in intracellular and extracellular proteins of bacteria and of animals is examined in terms
of the redox state of the environment in which the protein functions.
Thiol-disulfide changes associated with the dormant state are described and the role of cellular water content in dormancy is discussed. The potential significance of reactions between thiols and
products of oxidative metabolism is discussed with special emphasis
upon thiol additions to carbonyl and α,β-unsaturated carbonyl groups
and their possible role in steroid-receptor interaction.

II.   INTRODUCTION

Of the options presented to me I have chosen to offer here a
review rather than a specialized technical paper. My intent is to
describe the broad rationale which has motivated our recent and

current research, to summarize much of our recent work, and to
raise what I feel are some important problems for future study.
Although I will be touching only briefly on disulfide links in proteins I hope my more general discussion of thiols and disulfides
will be helpful in understanding the significance of this important
functional group.

Biochemists and biologists have long been intrigued by the im-
portance of thiols -- especially of the most common thiol, glutath-
ione -- and to a lesser extent of disulfides in biological systems
(Hopkins, 1921; Rapkine, 1930; Barron, 1951; Colowick, et al, 1954;
Crook, 1959).   Interest in thiols as well as disulfides has been
renewed and intensified in recent years (Kosower and Kosower, 1969;
Jocelyn, 1972; Friedman, 1973; Torchinskii, 1974; Flohé, et al, 1974;
Fluharty, 1974; Meister, 1975; Benöhr and Waller, 1975; Arias and
Jakoby, 1976).   Although much of the early work on the biological
role of glutathione was inconclusive, recent advances in a variety
of fields are beginning to clarify not only the role of glutathione,
but the functions of many other thiol and disulfide components of
the cell.

In the very broadest sense, the fundamental problem of interest
is to understand the role played by thiols and disulfides in the
origin and the subsequent evolution of life.   Although this is clearly
an immense problem which touches on every aspect of biochemistry and
biology, I feel that we are now at a stage where it is useful to
begin organizing our knowledge and formulating questions in this
context.   My own interests have derived largely from considerations
of the consequences of the shift from a reducing to an oxidizing
atmosphere for the chemistry and biochemistry of thiols and disulfides.   These considerations, as well as relevant experimental results, are summarized in this review.

III.   THIOLS AND DISULFIDES ON THE PRIMITIVE EARTH

What were the abundances of thiols and disulfides on the primitive earth and what functions did they play in prebiotic syntheses?
This question has received very little attention but there are two
considerations which suggest that thiols were important on the primitive earth.

First, it is generally agreed that oxygen was absent on the
primitive earth (Miller and Orgel, 1974; Broda, 1975) and to the extent that the primitive atmosphere was reducing we might qualitatively expect thiols to be thermodynamically favored relative to dis-
sulfides.   A quantitative estimate would be possible if the pressure
of hydrogen on the primitive earth were known.   A value of about
10^{-3} atm has been estimated (Urey, 1952; Miller and Orgel, 1974) for
the early primitive atmosphere and the partial pressure in the

present atmosphere is about 5×10^{-7} atm. Using these values as limits the equilibrium thiol to disulfide ratios have been calculated for a number of thiols and dithiols (Table 1). The equilibria involving monothiols are concentration dependent, favoring the reduced form at lower concentrations, and a total concentration of 10^{-2} M has been used since this is the highest concentration of thiol or disulfide ordinarily found in cells (Jocelyn, 1972; Section VI).

TABLE 1

Equilibrium Data for the Reduction of Disulfides by Hydrogen

$$RSSR + H_2 \underset{}{\overset{K_{H_2}}{\rightleftharpoons}} 2\ RSH$$

Compound	K_{H_2} [a]	[RSH]:[RSSR] [b]	
		$P_{H_2} = 10^{-3}$	$P_{H_2} = 5 \times 10^{-7}$
Monothiols [b]			
Ergothioneine	1×10^{12}	1×10^{11}	5×10^{7}
Cysteine	5×10^{6}	5×10^{5}	300
Glutathione	1×10^{6}	1×10^{5}	50
Dithiols			
Thioredoxin	2×10^{7}	2×10^{4}	10
Dihydrolipoate	2×10^{3}	2	10^{-3}
Dithiothreitol	1×10^{3}	1	5×10^{-4}

a Calculated from the redox potentials compiled by Jocelyn (1972) and the value for dithiothreitol given by Cleland (1964); pH 7.

b Assuming a total concentration [RSH] + [RSSR] = 10^{-2} M.

The calculations indicate that the monothiols would have been stable relative to the corresponding disulfides at either hydrogen pressure. In the case of the dithiols, however, the disulfide form might have been an important or even dominant contributor at equilibrium. In the case of thioredoxin and dihydrolipoate the disulfides play a role in the biological function of the compounds and the intramolecular arrangement of the thiol groups makes such disulfide formation more favorable than in monothiols. It is possible to have pairs of thiol groups in proteins or other polymers located such that intramolecular disulfide formation would cause unfavorable distortion or strain in the structure and in such a case intramolecular disulfide formation would not be important. In such cases intermolecular disulfide formation could still occur unless prohibited by steric hinderence. In principle, the equilibrium constant for the reduction of proteins can range from a value similar to that for dithiothreitol to one markedly greater than any given in Table 1. Thus, with the exception of some special cyclic disulfides, thiol should have been favored thermodynamically relative to disulfides on the primitive earth. There is, of course, no assurance that equilibrium was achieved.

The second consideration is that a number of prebiotic syntheses depend upon hydrogen sulfide or other thiol and thereby imply the presence of thiols on the primitive earth. Friedmann and Miller (1969) have described syntheses of phenylalanine and tyrosine from phenylacetylene using hydrogen sulfide. Methane thiol and hydrogen sulfide have been employed to produce methionine and homocysteine, respectively (Van Trump and Miller, 1972). Hydrogen sulfide may also have been important in enhancing photoproduction of amino acids by long-wavelength ultraviolet light (Sagan and Khare, 1971).

A systematic study of the role of sulfur in prebiotic syntheses has been undertaken by Raulin and Toupance (1975). Such studies should serve to elaborate in greater detail the possible roles played by thiols and other sulfur compounds on the primitive earth.

IV. THIOLS IN PRIMITIVE METABOLISM

What roles did thiols play in the first living cells? Here again we are forced to speculate, but we can do so on a somewhat firmer basis if we assume that the essential features of the enzymes and coenzymes involved in fundamental biochemical processes have been preserved in modern cells. Table 2 lists some fundamental cell components which either require thiols to retain activity or are inhibited by disulfides or thiol reagents. For nucleic acid and protein synthesis it is not yet clear whether thiols play an essential mechanistic role or whether their modification merely leads to changes in protein conformation or aggregation.

In the case of Coenzyme A and acyl carrier protein (ACP) the central role of the thiol group is clear and Abiko (1975) lists over seventy metabolic reactions which are known to depend upon one or the other of these cofactors. At least some of these must have been important to the first primitive organisms so that CoA or a related compound must logically have been a component of primitive metabolism.

TABLE 2

The Role of Thiols in Fundamental Metabolic Processes

Process	Thiol dependent component	References
DNA Synthesis	DNA Polymerase (II, III)	Kornberg (1974)
RNA Synthesis	RNA Polymerase	Krakow (1975)
Protein Synthesis	Cell Free System	Nolan and Hoagland (1971); Kosower, Vanderhoff, and Kosower (1972); Ondrejickova, et al (1974)
	Translocation	Siler and Moldave (1971)
	Aminacyl-tRNA Synthetase	Loftfield (1971)
Fatty Acid Synthesis	CoASH and ACP	Abiko (1975)
Glycolysis	Glyceraldehyde-3-phosphate Dehydrogenase	Segal and Boyer (1953)

Glyceraldhyde-3-phosphate dehydrogenase is considered to be present in all organisms and is a key enzyme in sugar metabolism and fermentation. The enzyme contains an SH group in the active site which forms a hemithioacetal upon combination with the substrate and is then converted to a thioester upon oxidation (Segal and Boyer, 1953, Boyer and Segal, 1954). Thus, a thiol group is essential for activity in this enzyme as isolated from contemporary organisms and presumably also in the corresponding enzyme from the primitive ancestors of present cells.

Thiol dependence is not restricted to the processes listed in Table 2. In fact, most intracellular enzymes can be classified as sulfhydryl proteins (those containing cysteine rather than cystine) and exhibit some kind of sulfhydryl dependence (Jocelyn, 1972; Friedman, 1973; Torchinskii, 1974). The importance of sulfhydryl dependent enzymes has been recognized for many years (Hopkins and Morgan, 1938; Rapkine, 1938; Barron, 1951).

Iron-sulfur proteins such as the ferredoxins depend upon cysteine thiol groups to bind iron. The ferredoxins from *Clostridia* and other anaerobes have a potential near that of the hydrogen electrode (Yasunobu and Tanaka, 1973; Hall, Cammack, and Rao, 1974). Since such anaerobes are considered to be primitive organisms it can be inferred that another function of thiol groups in primitive cells was to provide sites for metal binding.

The role played by disulfides in primitive metabolism is less clear. There appears to be no reason to postulate that disulfides played a structural role in primitive proteins and their general thermodynamic instability in a reducing environment mitigates against such a role. To the extent that disulfides were important their occurrence may have been limited to redox proteins such as thioredoxin and their corresponding reductases where dithiol-disulfide interconverion is part of the redox mechanism.

V. THIOLS AND DISULFIDES IN AN OXIDIZING ATMOSPHERE

Although uncertainties exist in our knowledge of events on the primitive earth we can be quite certain about the status of thiols and disulfides in our present atmosphere. It is well known that thiols are oxidized by molecular oxygen under catalysis by a variety of common agents (Capozzi and Modena, 1974). The equilibrium for disulfide formation strongly favors disulfide for thiols of biological interest (Table 3). The mechanisms for oxidation of thiols by oxygen are not clearly understood, but are believed to involve sulfenic acids (RSOH) as intermediates (Capozzi and Modena, 1974). Thiols are also quite reactive toward superoxide radical, hydrogen peroxide, and hydroxyl radical, these being the intermediates produced in the successive one electron reductions of molecular oxygen

TABLE 3

Equilibrium Data for the Oxidation of Thiols by Oxygen

$$2RSH + \tfrac{1}{2}O_2 \xrightleftharpoons{K_{O_2}} RSSR + H_2O$$

Compound	K_{O_2} [a]	[RSSR]:[RSH] (P_{O_2} = 0.2 atm)
monothiols [b]		
Ergothioneine	2×10^{29}	3×10^{13}
Cysteine	3×10^{34}	1×10^{16}
Glutathione	2×10^{35}	3×10^{16}
dithiols		
Thioredoxin	7×10^{33}	3×10^{33}
Dihydrolipoate	7×10^{37}	3×10^{37}
Dithiothreitol	2×10^{38}	9×10^{37}

a Calculated from the redox potentials compiled by Jocelyn (1972) and the value for dithiothreitol given by Cleland (1964); pH 7.

b Assuming a total concentration [RSH] + [RSSR] = 10^{-2}M.

(Fridovich, 1974). Given the overwhelming instability of thiols relative to disulfides in the presence of oxygen it is understandable that most extracellular enzymes and proteins are disulfide proteins, that is proteins which contain cystine rather than cysteine.

Because of the instability and relatively facile reactivity of thiols toward oxygen the accumulation of oxygen in the atmosphere as a consequence of photosynthesis must have had adverse consequences for the thiol dependent components of cells. The toxic effects of oxygen upon microorganisms, plants and animals have been extensively studied (Gerschman, 1964; Haugaard, 1968; 1974). Many sulfhydryl enzymes have been shown to be inactivated by oxygen either in extracts or in purified form, and in some cases the oxidation of thiol groups has been demonstrated to accompany inactivation (Haugaard, 1974). Various substances protect against the hazards produced by oxygen. These include the enzymes superoxide dismutase and catalase which promote the decomposition of superoxide and

hydrogen peroxide, respectively (Fridovich, 1974). Vitamin E is
thought to have an antioxidant function in cells, but glutathione
is the substance considered to provide the primary protection of
intracellular sulfhydryl enzymes.

VI. GLUTATHIONE

It has long been considered that glutathione is ubiquitous and
plays some essential function in all living cells (Baron, 1951;
Wieland, 1954; Meister, 1975). It is now clear that this is not the
case. Mutants of *E. coli* have been isolated which lack glutathione
by virtue of being deficient in the biosynthesis of glutathione
(Apontoweil and Berends, 1975; Fuchs and Warner, 1975). These
mutants grow normally under laboratory conditions, but exhibit increased sensitivity toward a wide variety of chemical agents
(Apontoweil and Berends, 1975). These observations support the view
advanced by Barron and Singer (1943) that the primary function of
glutathione is to protect sulfhydryl enzymes. It would be of interest to learn to what extent the glutathione deficient mutants
exhibit greater sensitivity to oxygen and peroxides than do the
parent strains.

If glutathione metabolism evolved as a protective mechanism
against the adverse effects of oxygen it might be expected that the
more primitive bacteria, especially obligate anaerobes, would lack
the ability to make glutathione. To explore this possibility we
undertook in conjunction with Willie Brown a survey of the occurrence of glutathione in bacteria. The obligate anaerobe *Clostridium
pasteurianum* appears to lack glutathione, but surprisingly, so do
many other gram positive bacteria including some obligate aerobes
(Table 4). By contrast all of the gram negative bacteria examined
contained at least some glutathione (Table 5). In both gram positive
and gram negative bacteria the thiol level is highly variable, being
quite low in most species, but very high in *E. coli*. The low thiol
content of bacilli and cocci was first noted in 1938 by Miller and
Stone.

A number of important questions can be formulated on the basis
of these results. What are the thiols which occur in those bacteria
which lack glutathione? To what extent are those bacteria having
low thiol contents more sensitive to oxygen and to thiol-reactive
antiseptics and antibiotics? Can the occurrence of glutathione and
related thiols be used to establish the evolutionary relationships
in prokaryotes? Definitive answers to these questions are not yet
available, but in the latter context we can at least note that the
tendency for gram positive bacteria to lack glutathione and for
gram negative bacteria to have it is consistent with the view that
the divergence between these classes of organism occurred prior to
the accumulation of oxygen in the atmosphere (Margulis, 1970).

TABLE 4

Ethanol Soluble Thiol and Glutathione in Gram Positive Bacteria

species	oxygen[a] requirement	umoles per g[b]	
		SH	GSH
Clostridium pasteurianum	AN	1	< 0.03
Lactobacillus casei	M	2	< 0.05
Bacillus subtilus	F	1.5	< 0.01
Bacillus cereus	F	1.3	< 0.01
Staphylococcus auereus	F	3	0.5
Staphylococcus epidermidis	F	2	< 0.02
Streptococcus agalactiae	F	3	3
Streptococcus salivaris	F	3	< 0.05
Streptomyces griseus	AE	3	< 0.02
Micrococcus lysodeikticus	AE	3	< 0.05
Micrococcus roseus	AE	2	< 0.02

a   AN, obligate anaeorbe; M, microaerophile; F, faculatative
    anaerobe; AE, obligate aerobe.
b   g, grams residue dry weight from 80% ethanol extraction.
    Fahey, Brown, Adams and Worsham (1977).

In those cells which have glutathione it is maintained in the reduced state by the enzyme glutathione reductase and it is of interest to examine the intracellular redox state as reflected in the oxidation state of glutathione. Table 6 summarizes some GSH:GSSG ratios measured for a variety of organisms and tissues. Owing to difficulties involved in preparing extracts without accompanying oxidation of GSH to GSSG these ratios probably represent lower limits. With this in mind it is evident that the GSH:GSSG ratio is at least about 10^2 and may exceed 10^3 in some systems. The total glutathione concentrations are not far from 10^{-2} M so that these values are within the range of the equilibrium ratios estimated for glutathione at the hydrogen pressures which may have prevailed on the primitive earth (Table 1). Thus, it appears that cells having glutathione maintain

TABLE 5

Ethanol Soluble Thiol and Glutathione in Gram Negative Bacteria

species	oxygen[a] requirement	umoles per g[b]	
		SH	GSH
Escherichia coli	F	29	27
Alcaligenes faecalis	F	23	25
Chromobacter violaseum	F	3	3
Enterobacter aerogenes	F	1.5	1.2
Serratia Marcescens	F	1.6	0.2
Acinetobacter calcoaceticus	AE	7	6
Azotobacter vinelandii	AE	5-7	7-9
Myxococcus xanthus	AE	3	1
Pseudomonas fluorescens	AE	1.6	1.3

a,b See Table 4.

Fahey, Brown, Adams, and Worsham (1977)

their thiol-disulfide state at about the equilibrium level that might have existed at the time life first evolved.

It now appears rather certain that the function of glutathione in prokaryotes is primarily that of a protective agent and there is much evidence that its role as a detoxifying agent has been retained and refined in higher organisms (Arias and Jakoby, 1976). It may also be involved in other more specialized functions in higher organisms such as those suggested by Meister (1974) and by Kosower and Kosower (1974, 1976).

VII CYST(E)INE IN BACTERIAL PROTEINS

It has been suggested that if the bases of DNA are considered to occur randomly then the occurrence of amino acids in proteins should be predictable from the genetic code, and for many amino acids

TABLE 6

The Cellular Glutathione Thiol-Disulfide Status

Organism	GSH μmoles per g[a]	$\frac{[GSH]}{[GSSG]}$	Reference
Eschericia coli	20[b]	~1000	Fahey, Brown, Adams and Worsham (1977)
Saccharomyces cerevisiae	20[b]	~200	Fahey, Brody, Worsham and Mikolajczyk (1976)
Neurospora crassa (mycelia)	15	~1000	Fahey, Brody and Mikolajczyk (1975)
Sea urchin embryo (*Litechinus pictus*)	10	~50	Fahey, Mikolajczyk, Meier, Epel and Carroll (1976)
Sheep erythrocytes high GSH	2.5[c]	~300	Young, Nimmo and Hall (1975)
low GSH	0.7[c]	~500	
Human erythrocytes	~2[c]	~500	Srivastava and Beutler (1968) Güntherberg and Rost (1966)
L-5178Y lymphoma cells	2.3[c]	~100	Bump (1975)
Rat heart	~9	~80	Wendell (1970)
Rat liver	~6[d]	~60	Tietz (1969)
Rat kidney	~3[d]	~50	Tietz (1969)

a Based upon total dry weight except as noted.

b Near the end of exponential growth phase.

c As μmoles per ml of packed cell volume.

d As μmoles per g wet weight.

such predictions correspond roughly with the observed amino acid
content (King and Jukes, 1969). Two codons code for cysteine and
on this basis we expect the cyst(e)ine content of proteins to be
around 3.3%. The values for bacterial proteins are well below this
value and there are two features of the occurrence of cyst(e)ine in
bacterial proteins which may have evolutionary significance.

First, many of the extracellular proteins produced by bacteria
are cyst(e)ine-free proteins, being lacking in both cystine and
cysteine (M. R. Pollock and M. H. Richmond, 1962). Probably the
most thoroughly studied example is subtilisin which has the same
specificity as the mammalian enzyme chymotrypsin. X-ray cystal-
lographic studies in the laboratory of Joe Kraut have shown that
subtilisin has essentially identical catalytic and binding sites
to chymotrypsin but that the two enzymes differ completely in other
structural features; these two enzymes have therefore been considered
to represent an example of convergent evolution (Robertus, et al,
1972). Chymotrypsin contains five disulfide bonds, corresponding to
about four percent of the amino residues being half-cystine, whereas
subtilisin is a cyst(e)ine-free protein.

What is the reason for this difference? Pollock and Richmond
(1962) advanced the hypothesis that the lack of disulfides in extracellular bacterial proteins is associated with the mechanism by which
they are released from the cell, flexibility being required "to pass
more freely through a rigid cell-wall meshwork". An alternative, or
possibly additional, explanation merits consideration. Many of the
organisms which produce cyst(e)ine-free extracellular proteins are
strict or facultative anaerobes. If these extracellular proteins
evolved their function in an anaerobic, reducing environment it is
possible that cystine was excluded because of the instability of
disulfides under reducing conditions. Disulfide containing extracellular proteins would by this argument have evolved under aerobic
conditions. The fact that the extracellular proteases from the gram-
positive aerobe *Streptomyces griseus* (Jurasek, et al, 1974; Olafson,
Jurasek, Carpenter, and Smillie, 1975) and from the gram-negative
aerobe *Sorangium sp.* (Olson, et al, 1970) all contain disulfide
links is consistent with this argument. Moreover it refutes the
hypothesis of Pollock and Richmond (1962) unless different mechanisms
for the release of extracellular proteins occur in aerobic bacteria.

The second unusual feature is that intracellular bacterial proteins have generally low cyst(e)ine contents. Reeck and Fisher (1973)
found in analyzing data for purified proteins that eucaryotic enzymes
have twice the cyst(e)ine content of prokaryotic enzymes, the latter
containing about 1.1% cyst(e)ine residues. Table 7 summarizes the
data obtained by Pollack and Richmond (1962) for the cyst(e)ine content of the total cellular protein from various bacteria. The mean
value, 1.1 ± 0.3 %, agrees well with that of Reeck and Fisher (1973).

TABLE 7

The Cyst(e)ine Content of Bacterial Protein

Organism	Strain	Per Cent Cyst(e)ine[a]
Bacillus subtillis	6346, H	0.67, 1.2
Bacillus cereus	569, 5/B	0.86, 1.1
Corynebacterium diphtheriae	PW 8	1.2
Escherichia coli		1.4
Pseudomonas fluorescens	KV 2	0.77
Pseudomonas saccharophila		0.84
Staphylococcus aureus	524	1.6
Streptococcus haemolyticus	C 748, D 58	1.3, 1.1
Vibrio cholerae	4 Z	0.97

a   Calculated from the data of Pollock and Richmond (1962).

Differences among bacteria, as well as between bacteria and higher organisms, appear to be significant, and the effect can be seen in specific purified proteins. Thus, malate dehydrogenase from *Bacillus subtilis* contains no cyst(e)ine whereas that from *E. coli* contains 0.85 % cyst(e)ine, that from *Neurospora crassa* contains 1.25% cyst(e)ine, and those from chicken heart mitochondria and supernatant contain 2.1 and 1.24 % cyst(e)ine, respectively (Murphey, Barnaby, Lin and Kaplan, 1967). Since intracellular proteins are predominantly sulfhydryl proteins and since sulfhydryl groups are reactive toward oxygen and toward many oxidation by-products, one plausible explanation for the low cysteine content of intracellular bacterial proteins is that natural selection has limited the occurrence of cysteine in such proteins to those residues essential to their function. Differences between organisms might then be attributable in part to differences in the effectiveness of glutathione or other protective thiols in ameliorating the adverse effects of oxygen. Additional data on the nature of protective thiols in bacteria and

on the cysteine content of bacterial proteins are needed to provide
a more rigorous test of this hypothesis.

VIII INTRACELLULAR AND EXTRACELLULAR MAMMALIAN PROTEINS

From the standpoint of evolution it is interesting to compare
the occurrence of cyst(e)ine in bacterial proteins with that in mammalian proteins. The average cyst(e)ine content of eukaryotic proteins
generally is greater than that of bacterial proteins and can approach
or exceed the value of 3.3% predicted from the code depending upon
the proteins averaged (King and Jukes, 1969; Reeck and Fisher, 1973).
However, the standard deviation for cyst(e)ine is nearly as large
as the mean value itself suggesting nonumiformity in the distribution (Reeck and Fisher, 1973). Such surveys have not distinguished
between the occurrence of cystine and that of cysteine in proteins.
When this is done for mammalian proteins an interesting difference
is mean content and in distribution pattern is found.

Since intracellular mammalian proteins are generally sulfhydryl
proteins and extracellular mammalian proteins primarily disulfide
proteins, a comparison of the cysteine content with the half-cystine
content can be made by comparing intracellular and extracellular
proteins. Extracellular proteins are taken here to include those
proteins which are stored intracellularly in lysozomes or similar
granules prior to release from the cell. Figure 1 shows a comparison
for 32 intracellular mammalian proteins and 34 extracellular mammalian proteins having at least fifty amino acid residues per chain.
These were taken from the compilations by Jocelyn (1972) and by Reeck
(1970) with identical proteins from different species and closely
related proteins being deleted.

The first feature to note in Figure 1 is that the cysteine contents of intracellular proteins are distributed over a relatively
narrow range with a mean value similar to the higher values found
for intracellular bacterial proteins (Table 7). Thus, mammalian
intracellular proteins, like their bacterial counterparts, have a
mean cysteine content well below the value of 3.3% expected from
the genetic code.

Extracellular mammalian proteins differ markedly from extracellular bacterial proteins as well as from intracellular proteins.
The half-cystine content of these disulfide proteins is distributed
over a wide range of values and the mean value is greater than the
3.3% predicted from the code. Thus, natural selection does not
appear to have generally restricted the occurence of half-cystine
to either high or low values. These disulfide links presumably
serve to stabilize the protein structure. In this context it is

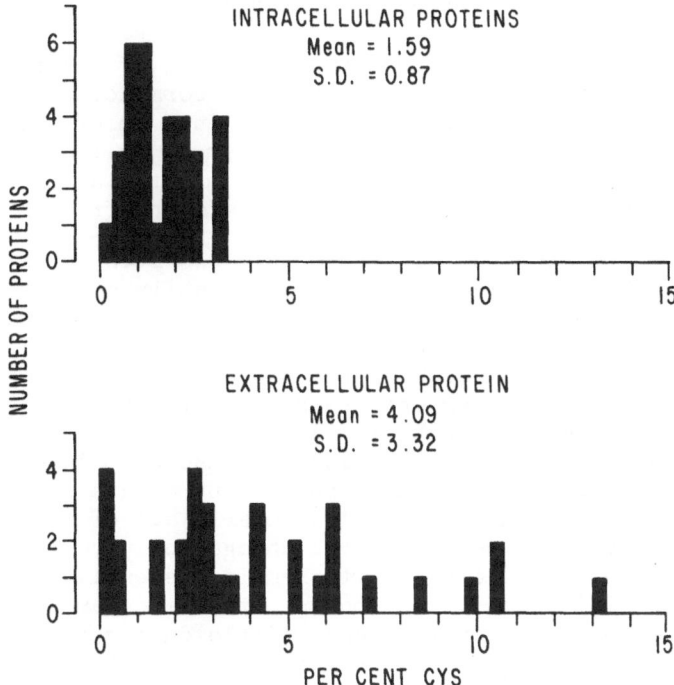

Figure 1.   Distribution of cyst(e)ine in intracellular and extra-
             cellular mammalian proteins. Fahey, Hunt, and Windham
             (1977).

interesting to note that protease inhibitors are among the most
cystine rich proteins. Thus, inhibitors of serine proteases have
about 10% half-cystine residues while the sulfhydryl protease inhibitors from pineapple stem contain some 20% half-cystine (Reddy,
et al, 1975).

From the discussion in the present and preceding sections it
should be clear that the occurence of cyst(e)ine in proteins exhibits
a far wider variation than that of any other amino acid. Moreover,
the general variations appear to be understandable in terms of the
oxidation-reduction state of the environment in which the protein
functions and the role played by thiol or disulfide groups in the
protein. Proteins which function in an aerobic environment should
lack thiol groups, except when these are essential to their function,
and may contain disulfide groups in low or high amounts depending
upon the requirements for stabilization of the protein structure.

IX. DORMANCY

How are the thiol dependent components of cells protected in an oxidizing environment during periods of dormancy when cellular metabolism is greatly slowed or stopped? In a joint study with Stuart Brody we have been studying this problem as well as trying to ascertain whether thiol-disulfide reactions might play a role in the mechanisms controlling the transitions to and from the dormant state.

The system studied most thoroughly has been the asexual life cycle of the fungus *Neurospora crassa*. While glutathione is maintained in a highly reduced state during vegetative growth, the formation of conidia (spores) is accompanied by a significant increase in GSSG content. The GSSG content of the spores drops again within a few minutes after the spores are transferred to germination media (Fahey, Brody, Mikolajczyk, 1975). When the spores are allowed to age following harvesting the viability and GSSG level is influenced by the relative humidity in which the spores are stored. At 0% relative humidity the conidia lose nearly all of their water content in a few hours and the GSSG level changes only very slowly over a period of months while viability remains generally high. At 100% relative humidity the water content stabilizes around 30-40% by weight and the GSSG level changes little during the first few days after which viability falls dramatically and GSSG levels increase. At 50% relative humidity the water content stabilizes around 5-7% and the GSSG level increases steadily over a two month period until essentially all of the glutathione has been oxidized (Figure 2). Protein bound glutathione (PSSG) levels also increase and there is some decline in viability after the first month. When such aged spores are germinated there is still a rapid reduction of GSSG with a corresponding rise in the GSH content (Fahey, Brody, and Mikolajczyk, 1977).

In fact, merely increasing the water content of the spores by storage at 100% relative humidity suffices to cause reduction of GSSG. This is seen in Figure 3 where the results of storing conidia alternately at 50% relative humidity and at 100% relative humidity on the GSSG content are shown. The oxidation-reduction cycle can be repeated many times. Thus, increasing the water content of the cell appears to mobilize the metabolism necessary to generate NADPH and also to activate the utilization of these reducing equivalents by glutathione reductase to reduce GSSG.

Other eucaryotic systems having anhydrobiotic dormant states exhibit qualitatively similar behavior. Table 8 summarizes data obtained with fungal, plant, and animal systems. All exhibit an increase in GSSG level in the dormant state and a decrease upon

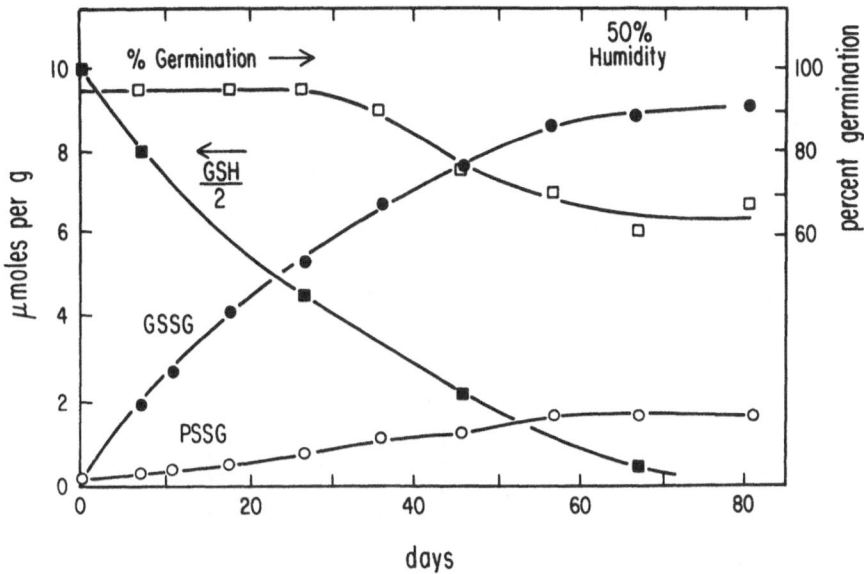

Figure 2.   Changes in GSH, GSSG, protein-glutathione disulfide
             (PSSG), and percent germination during storage of
             Neurospora crassa conidia at 50% relative humidity.
             Fahey, Brody, and Mikolajczyk (1977).

hydration or activation in growth media. These results suggest that
the thiols of dormant systems are protected against oxidation only
to the extent that dehydration slows the oxidative process and to the
degree that the reduction system activated by hydration is capable
of reversing the oxidation process.

     The observed changes in thiol-disulfide status accompanying the
transitions to and from the dormant state are consistent with the
view that thiol-disulfide reactions involving proteins serve as a
control mechanism in these systems but the details of such control
are not yet clear. One possibility which has been examined in our
studies is that proteins are converted to disulfide forms in the
dormant state and thereby stabilized against thermal inactivation.
We find however that the thermal stability of conidia is independent
of the thiol-disulfide state. It depends instead upon the water
content, dehydrated spores showing high thermal stability while moist
spores are rapidly inactivated upon heating. Paul Price (personal
communication) finds that purified enzymes when lyophilized will
withstand extended heating at temperatures above 100° C in the

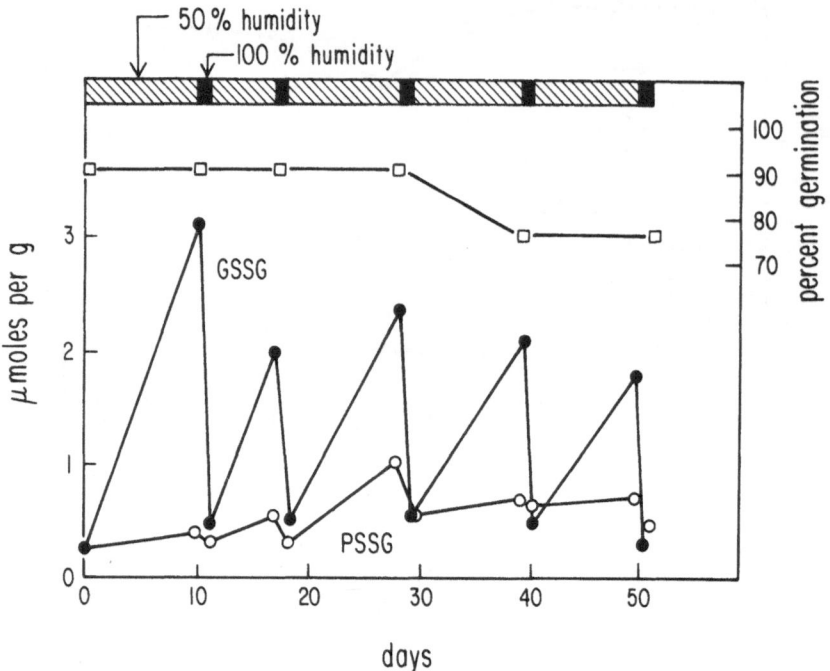

Figure 3.   Changes in GSSG, PSSG, and percent germination during
             alternate storage of conidia at 50% and 100% relative
             humidity. Fahey, Brody, and Mikolajczyk (1977).

absence of oxygen.   Thus the evidence indicates that dehydration
is a major factor in determining the thermal stability of macromolecules and thereby of dormant systems.   What, if any, other properties
are subject to thiol-disulfide control in the transition to and from
the dormant state is the subject of continuing study in our laboratories.

TABLE 8

The Glutathione Thiol-Disulfide Status and Dormancy

| Organism | State or Treatment | umoles per g[a] | |
		GSH	GSSG
Neurospora crassa	Freshly Harvested	13	0.1
conidia[b]	aged 10 days	10	1.4
	germinated 15 min	10	0.03
Saccharomyces	dry	5	4
cerevisiae[c]	18 hours at 100% RH	15	0.3
	10 min in glucose growth medium	11	0.2
Wheat embryo[d]	dry	7	0.4
	imbibed 15 min.	6	0.05
Barley embryo[d]	dry	2	0.6
	imbibed 60 min.	1.4	0.13
Artemia salina[e]	dry	0.3	0.13
	39 h at 100% RH	0.6	0.05

RH, relative humidity

a Total cell dry weight. Residue dry weight values have been adjusted to total dry weight basis.

b Fahey, Brody and Mikolajczyk (1975).

c Fahey, Brody, Worsham and Mikolajczyk (1976).

d Fahey, Meier and DiStefano (1977).

e Fahey and Windham (1976).

X.   ADDITION REACTIONS OF THIOLS

Addition reactions of thiols play an important role in some mechanisms for oxidation of thiols and also in the interaction of thiols with various compounds derived from oxidative metabolism. Two addition reactions of major importance are the addition of thiols to carbonyl groups and the addition of thiols to α,β-unsaturated carbonyl groups (Equation 1 and 2).

$$\text{RSH} \; + \; \text{>C=O} \; \rightleftharpoons \; \text{>C} \overset{SR}{\underset{OH}{<}} \tag{1}$$

$$\text{RSH} \; + \; \text{>C=C-C=O} \; \rightleftharpoons \; \text{RS-C-CH-C=O} \tag{2}$$

An example involving oxidation via thiol addition to a carbonyl group is the oxidation of GSH to GSSG by dehydroascorbic acid. This reaction is believed to occur spontaneously in animal tissues but is catalyzed by dehydroascorbate reductase in plants (Vennesland and Conn, 1954). The mechanism of the uncatalyzed reaction does not appear to have been studied but almost certainly involves the steps shown in Figure 4. Thiol addition to the central carbonyl group of dehydroascorbic acid gives rise to a hemithioketal (two geometric isomers possible). A nucleophilic displacement by thiolate ion at sulfur then gives the products directly. The leaving group is formally a carbanion and these are ordinarily poor leaving groups in nucleophilic displacements, but in this case the "carbanion" is the highly stabilized ascorbate ion.

If the carbon attached to sulfur in the adduct is not a good leaving group the reaction may stop at the stage of the hemithioketal or hemithioacetal when these are stable. Such reactions have been carefully studied by Lienhard and Jencks (1966) and by Barnett and Jencks (1969) who find the equilibrium constants corresponding to equation 1 to be largely independent of the pK_a of the thiol and to be favorable for simple aldehydes. Thiol addition is more favorable than hydration and carbonyl compounds which are hydrated to any extent in water can be expected to have a favorable equilibrium constant for thiol addition. Acetone is not hydrated in water and hemithioketal formation from acetone is unfavorable, but the presence of electron-withdrawing groups adjacent to the carbonyl, as in pyruvate, shifts the balance toward favorable formation of the hemithioketal (Lienhard and Jencks, 1966). Similarly, the presence of electron-withdrawing substituents in dihydroxyacetone phosphate causes appreciable hydration of the carbonyl to occur in water (Reynolds, Yates, and Pogson, 1971) implying that hemithioketal formation would be favorable in this case. Since aldehydes, β-keto acids, and α-hydroxy carbonyl compounds are common biochemical funct-

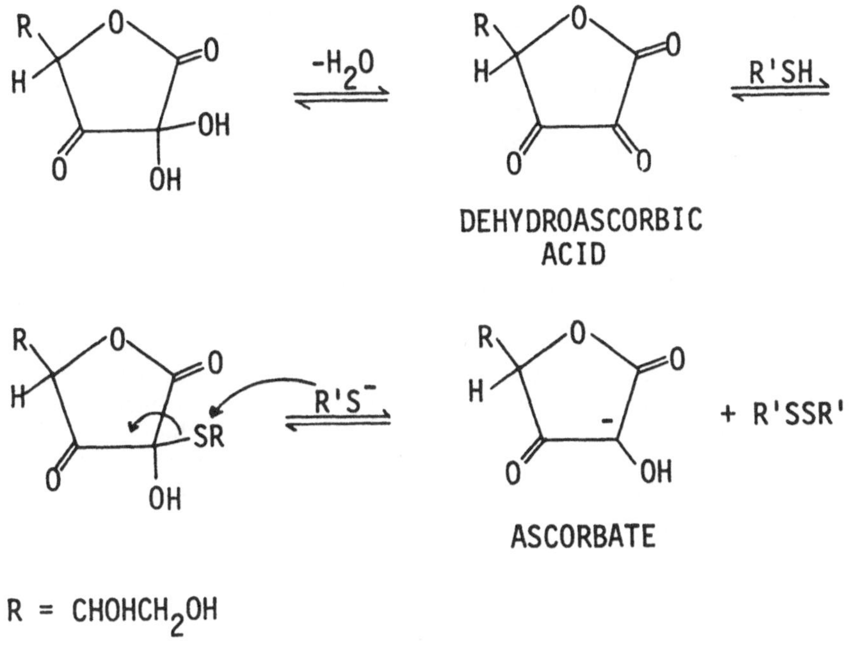

DEHYDROASCORBIC
ACID

ASCORBATE

R = CHOHCH$_2$OH

Figure 4.   Postulated mechanism for the oxidation of glutathione
by dehydroascorbic acid.

ionalities the possibilities for thiol additions to carbonyl compounds
are numerous.   Such reactions are considered to play a key role in the
mechanism of action of glyceraldehyde-3-phosphate dehydrogenase
(Segal and Boyer, 1953) and also in the glyoxalase and related reactions (Flohé and Günzler, 1976).   More examples can be anticipated
as we expand our understanding of how the thiol groups of sulfhydryl
proteins interact with biologically important carbonyl compounds.

Biologically important examples of thiol addition to α,β-un-
saturated carbonyl compounds include the maleylacetoacetate isomerase
reaction in which cis-trans isomerization is considered to be accomplished by an addition elimination sequence involving glutathione
(Selzer, 1972).   Other examples are the additions of GSH to a wide
range of olefins activated by the presence of electron-withdrawing
groups attached to the double bond.   These reactions are catalyzed
by glutathione S-transferases and are thought to comprise a key component of the process by which potentially hazardous compounds such
as ethyl acrylate, crotonaldehyde, and ethyl vinyl ketone are de-
toxified in animals (Chasseaud, 1976).   Higher plants produce a

variety of compounds which are thought to have growth regulatory properties and which contain α,β-unsaturated carbonyl groups (Gross, 1975). Of these certain unsaturated lactones, which have been of special interest because of their anti-tumor activity, have been actively studied by Kupchan and coworkers (1970, 1974, 1976). These authors associate their facile reactivity toward thiols and protein sulfhydryl groups with their **anti-tumor** properties.

The steroid hormones constitute a large class of compounds whose biosynthesis depends upon oxygen and many steroids possess carbonyl or α,β-unsaturated carbonyl groups which are potential sites for the reversible addition of thiol groups. Thus, the dihydroxyacetone functionality at C-20 in cortisol, the aldehyde group of aldosterone, and possibly even the C-20 monohydroxyacetone functionality in corticosterone and aldosterone can be expected on the basis of the chemistry discussed above to exhibit reactivity toward a thiol group (Figure 5). A β,β-disubstituted α,β-unsaturated carbonyl group is

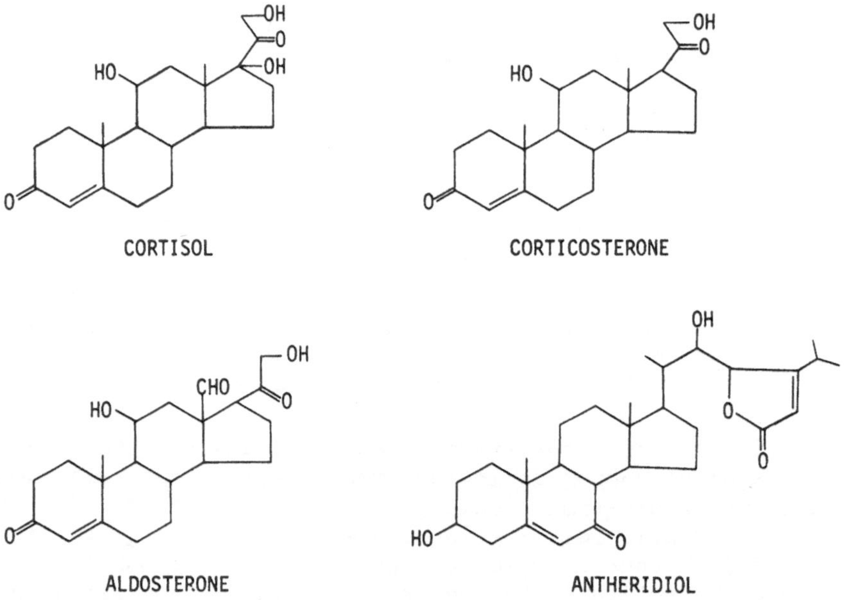

Figure 5.   Structures of some steroid hormones.

present in the A ring of all three of these steroids, in the B ring
and the side chain butenolide group of the fungal hormone antheridiol
(Arsenault, Biemann, Barksdale and McMorris, 1968), and in the B ring
of the insect hormones, the ecdysones and phytoecdysones (Berkoff,
1969). Nonsteroid hormones such as the plant hormone abscisic acid
(Milborrow, 1974) and the insect juvenile hormone (Berkoff, 1969)
also possess this type of group.

The presence of two alkyl groups at the β-position is expected
to markedly decrease the reactivity of the α,β-unsaturated carbonyl
system (Kupchan, et al, 1970; Miyadera and Kosower, 1972) but
thiol addition might still be favorable, especially in the binding
site of a protein receptor or enzyme. Thus, it appears reasonable
to postulate that carbonyl and α,β-unsaturated carbonyl groups in
hormones can serve as sites for reversible thiol addition reactions
which contribute to the binding of the hormone to receptors or other
proteins. Sulfhydryl reagents have been found to inhibit the binding
of steroids to their receptor proteins (Rees and Bell, 1975; and
references therein) which is consistent with this hypothesis and
indicates that it merits specific testing.

We see then that, in addition to peroxides and epoxides, certain
kinds of naturally occurring α,β-unsaturated carbonyl compounds
produced by oxygen dependent metabolism undergo irreversible reaction
with thiols. Other such α,β-unsaturated compounds undergo potentially
reversible reaction and thiol reactive carbonyl groups are also
found in compounds derived from biosynthetic or degradative pathways
which depend upon oxygen. Thus, oxygen dependent metabolism adds a
new dimension to the biological role of thiols by producing a wide
range of compounds that can react either reversibly or irreversibly
with thiol groups of sulfhydryl proteins.

XI. CONCLUSION

By way of concluding this review I would like to summarize the
main themes which have been developed and to add a few additional
comments.

From our present perspective it appears likely that thiols played
important roles in prebiotic chemistry and it seems virtually certain
that they were essential components of the first living cells. Disulfides, on the other hand, probably had a more restricted role in
prebiotic chemistry and in primitive metabolism owing to their instability in a reducing environment relative to the corresponding
thiol. According to this view most of the enzymes of primitive cells
would have been sulfhydryl proteins, i.e. those containing cysteine
rather than cystine. Cyst(e)ine-free proteins may also have been
important but the occurrence of cystine in proteins was likely

restricted to instances were the disulfide link was functionally
involved in redox reactions.

The accumulation of oxygen in the atmosphere reversed the balance in the thiol-disulfide equilibrium and introduced a serious
challenge to the essential thiol components of cells. The ability
of cells to produce glutathione in substantial quantities and to
maintain it in a redox state reflecting that of the primitive reducing environment appears to have evolved in Gram-negative bacteria
as a protective response to the toxic effects of oxygen in the environment. Superoxide dismutase and catalase serve similar functions
by destroying reduction products of oxygen which are highly reactive
toward thiols. But even with these protective mechanisms present
the cysteine content of intracellular proteins is found to be only
about half that expected on the basis of the genetic code and this
may be, at least in part, still another consequence of the instability of thiols in an oxidizing environment.

The process of adaptation to an oxidizing environment was not
restricted to the evolution of protective mechanisms for essential
thiols but also involved the incorporation of oxidized forms of
sulfur into new functional roles in proteins. The most clearly
defined of these is the structural role played by the disulfide
link in the extracellular disulfide proteins of aerobic organism.
The cystine content of mammalian disulfide proteins exhibits a
wide variation in values, presumably reflecting diverse requirements of individual proteins for structural stabilization. The
mean cyst(e)ine content of such proteins (4.1%) is much greater
that that of the corrsponding intracellular sulfhydryl proteins
(1.6%) but the average of these values (2.9%) approaches the value
expected on the basis of the genetic code if the bases are assumed
to occur randomely.

In addition to the structural role of disulfides and their
catalytic function in dithiol-disulfide dependent redox processes,
oxidized forms of sulfur might reasonably be expected to have
evolved new catalytic roles in aerobic organisms. Probably the
best prospective example of this is the sulfenic acid group which
has been extensively studied by Allison (1976) and has been
postulated by him to be involved in the nonflavin amine oxidase
reaction. As we learn more about the mechanisms of such reactions
it will be interesting to see to what extent sulfenic acids and
other sulfenyl derivatives are involved in oxidative processes.

In plants and animals oxygen dependent degradative and biosynthetic processes are responsible for the production of an almost
endless array of organic compounds, many of which incorporate
functionalities which are reactive toward thiols. When such reactions are irreversible, as in the case of epoxides and some α,β-un-
saturated carbonyl compounds, the compounds may have toxic proper-

ties, as in the case of the anti-tumor agent Jatrophone (Kupchan, et al, 1976), and presumably can function for the producing organism as a defense mechanism against predators or competing organisms. Where such reactions are reversible they provide a mechanism for interaction with thiol components of the cell in a fashion having potential significance in regulatory processes. An example in the latter category is the postulated involvement of thiol addition to carbonyl and α,β-unsaturated carbonyl groups in the mechanism of action of steroids.

To the extent that these general views are correct it becomes clear that reactions of thiols and disulfides play central roles in many of the most important developments involved in the evolution of aerobic organism. However, some of these views represent little more than speculation at present and much additional data is needed to test them in detail. I am told that Daniel Mazia applied the term "Thiology" to the study of glutathione and thiols as it developed through the 1950's. Borrowing from Mazia and recognizing the advances of the past two decades, it seems appropriate to dub the current efforts in this area as studies of "Molecular Thiology".

XI.   ACKNOWLEDGMENTS

I am pleased to acknowledge stimulating and useful discussions with my collaborators, Stuart Brody and Willie Brown, and also with my colleagues William Allison, F. Thomas Bond, Russell Doolittle, Jack Kyte, Stanley Miller, and Leslie Orgel. Financal support from the National Institutes of Health under grant GM 22122 is also gratefully acknowledged.

REFERENCES

Abiko, Y. (1975). in "Metabolism of Sulfur Compounds", D. M. Greenberg, Ed., Academic Press, New York, p. 1.

Allison, W. S. (1976). Accounts Chem. Res., 9, 293.

Apontoweil, P. and Berends, W. (1975). Biochim. Biophys. Acta, 399, 10.

Arias, I. M. and Jakoby, W. B. (1976). "Glutathione: Metabolism and function", Raven Press, New York.

Arsenault, G. P., Biemann, K., Barksdale, A. W., and McMorris, T. C. (1968). J. Amer. Chem. Soc., 90, 5635.

Barnett, R. E. and Jencks, W. P. (1969). J. Amer. Chem. Soc., 91, 6758.

Barron, E. S. G. (1951). Adv. Enzymology, 11, 201.

Barron, E. S. G. and Singer, T. P. (1943). Science, 97, 356

Benöhr, H. Ch. and Waller, H. D. (1975). Klin. Wschr., 53, 789.

Berkoff, C. E. (1969). Quart. Rev., 23, 372.

Boyer, P. D. and Segal, H. L. (1954). in "A Symposium on the Mechanism of Enzyme Action", W. D. McElroy and B. Glass, Eds., Johns Hopkins University Press, Baltimore, p. 520.

Broda, E. (1975). "The Evolution of the Bioenergetic Process", Pergamon Press, New York.

Bump, E. A. (1976). Thesis, Oregon State University.

Capozzi, G. and Modena, G. (1974). in "The Chemistry of the Thiol Group", S. Patai, Ed., Wiley, New York.

Chasseaud, L. F. (1976). in "Glutathione: Metabolism and Function", I. M. Arias and W. B. Jakoby, Eds., Raven Press, New York, p. 77.

Cleland, W. W. (1964). Biochem., 3, 480.

Colowick, S., Lazarow, A., Racker, E., Schwarz, D. R., Stadtman, E., and Waelsch, H. (1954). "Glutathione: A Symposium", Academic Press, New York.

Crook, E. M. (1950). "Glutathione", Biochem. Soc. Symp. No. 17, Cambridge Univ. Press, London and New York.

Fahey, R. C., Brody, S., and Mikolajczyk, S. D. (1975). J. Bacteriol., 121, 144.

Fahey, R. C., Brody, S. and Mikolajczyk, S. D. (1977). Submitted.

Fahey, R. C., Brody, S., Worsham, M. B., and Mikolajczyk, S. D. (1976). Unpublished results.

Fahey, R. C., Brown, W. C., Adams, W. B., and Worsham, M. B. (1977). Submitted.

Fahey, R. C., Hunt, J. S. and Windham, G. C. (1977). Submitted.

Fahey, R. C., Meier, G. P. and DiStefano, D. (1977). Submitted.

Fahey, R. C., Mikolajczyk, S. D., Meier, G. P., Epel, D. and Carroll, E. J. (1976). Biochim. Biophys. Acta, 437, 445.

Fahey, R. C. and Windham, G. C. (1976). Unpublished results.

Flohe', L., Benöhr, H. Ch., Sies, H., Waller, H. D. and Wendel, A. (1974). "Glutathione", Academic Press, New York.

Flohe', L. and Gunzler, W. A. (1976). in "Glutathione: Metabolism and Function", I. M. Arias and W. B. Jakoby, Eds., Raven Press, New York, p. 17.

Fluharty, A. L. (1974). in "The Chemistry of the Thiol Group", S. Patai, Ed., John Wiley, New York, Part 2, p. 589.

Fridovich, I. (1974). in "Horizons in Biochemistry and Biophysics", E. Quagliariello, F. Palmieri, and T. P. Singer, Eds., Addison-Wesley, Reading, Massachusetts, Vol. 1, p. 1.

Friedman, M. (1973). "The Chemistry and Biochemistry of the Sulfhydryl Group in Amino Acids, Peptides and Proteins", Pergamon, New York.

Friedmann, N. and Miller, S. L. (1969). Science, 166, 766.

Fuchs, J. A. and Warner, H. R. (1975). J. Bacteriol., 124, 140.

Gerschman, R. (1964). in "Oxygen in the Animal Organism", F. Dickens and E. Neil, Eds., MacMIllan, New York, p. 475.

Gross, D. (1975). Phytochem., 14, 2105.

Guntherberg, H. and Rost, J. (1966). Anal. Biochem., 15, 205.

Hall, D. O., Cammack, R. and Rao, K. K. (1974). Origin of Life
    5, 363.

Haugaard, N. (1968). Physiol. Rev., 48, 311.

Haugaard, N. (1974). in "Molecular Oxygen in Biology", O. Hayaishi,
    Ed., North-Holland Publishing Co., Amsterdam, p. 163.

Hopkins, F. G. (1921). Biochem. J., 15, 286.

Hopkins, F. G. and Morgan, E. J. (1938). Biochem. J. 32, 611.

Jocelyn, P. C. (1972). "Biochemistry of the SH Group", Academic
    Press, London.

Jurasek, L., Fackre, D. and Smillie, L. B. (1969). BBRC, 37, 99.

King, J. L. and Jukes, T. H. (1969). Science, 164, 788.

Kornberg, A. (1974). "DNA Synthesis", Freeman, San Francisco.

Kosower, N. S. and Kosower, E. M. (1974). in "Glutathione: Metabo-
    lism and Function", I. M. Arias and W. B. Jakoby, Eds.,
    Raven Press, New York, p. 159.

Kosower, N. S. and Kosower, E. M. (1969). Nature, 224, 117.

Kosower, N. S., Vanderhoff, G. A. and Kosower, E. M. (1972). Biochim.
    Biophys. Acta, 272, 623.

Krakow, J. S. (1975). Biochemistry, 14, 4522.

Kupchan, S. M., Giacobbe, T. J., Krull, I. S., Tbomas, A. M., Eakin,
    M. A. and Fessler, D. C. (1970). J. Org. Chem., 35, 3539.

Kupchan, S. M. (1974). Fed. Proceed., 33, 2288.

Kupchan, S. M., Sigel, C. W., Matz, M. J., Gilmore, C. J. and Bryan,
    R. F. (1976). J. Amer. Chem. Soc., 98, 2295.

Lienhard, G. E. and Jencks, W. P. (1966). J. Amer. Chem. Soc., 88,
    3982.

Loftfield, R. B. (1971). in "Proteins Synthesis", E. H. McConkey,
    Ed., Marcel Dekker, New York, p. 1.

Margulis, L. (1970). "Origin of Eukaryotic Cells", Yale University
    Press, New Haven.

Meister, A. (1974). in "Glutathione", L.Flohe, et al, Eds, Academic
    Press, New York, p. 56.

Meister, A. (1975). in "Metabolism of Sulfur Compounds", D. M.
    Greenberg, Ed., Academic Press, New York, p. 101.

Milborrow, B. V. (1974). Recent Adv. in Phytochem., 7, 57.

Miller, S. L. and Orgel, L. E. (1974). "The Origins of Life
    on Earth", Prentice-Hall, Englewood Cliffs, New Jersey.

Miller, T. E. and Stone, R. W. (1938). J. Bacteriol., 36, 248.

Miyadera, T. and Kosower, E. M. (1972). J. Med. Chem., 15, 534.

Murphey, W. H., Barnaby, C., Lin, F. J. and Kaplan, N. O. (1967).
    J. Biol. Chem., 242, 1548.

Nolan, R. D. and Hoagland, M. B. (1971). Biochim. Biophys. Acta,
    247, 609.

Olafson, R. W., Jurasek, L., Carpenter, M. R. and Smillie, L. B.
    (1975). Biochemistry, 14, 1168.

Olson, M. O. J., Nagabhushan, N., Dzwinill, M., Smillie, L. B.
    and Whitaker, D. R. (1970). Nature, 228, 488.

Ondrejickova, O., Drobnica, L., Sedlacek, J., and Rychlik, I. (1974).
    Biochem. Pharmacol., 23, 2751.

Pollock, M. R. and Richmond, M. H. (1962). Nature, 194, 446.

Rapkine, L. (1930). Compt. rend., 191, 871.

Rapkin, L. (1938). Biochem. J., 32, 1729.

Rauline, F. and Toupance, G. (1975). Origins of Life, 6, 91, 507.

Reeck, G. R. (1970). in "Handbook of Biochemistry", H. A. Sober,
    Ed., 2nd edn., Chemical Rubber Co., Cleveland.

Reeck, G. R. and Fisher, L. (1973). Int. J. Peptide Protein Res.,
    5, 109.

Reddy, M. N., Keim, P. S., Heinrikson, R. L. and Kezdy, R. J. (1975).
    J. Biol. Chem., 250, 1741.

Rees, A. M. and Bell, P. A. (1975). Biochim. Biophys. Acta,
    411, 121.

Reynolds, S. J., Yates, D. W. and Pogson, C. I. (1971). Biochem.
    J., *122*, 285.

Robertus, J. D., Alden, R. A., Birktoft, J. J., Kraut, J. C.,
    Powers, J. C. and Wilcox, P. E. (1972). Biochemistry, *11*, 2439.

Sagan, C. and Khare, B. N. (1971). Science, *125*, 417.

Segal, H. L. and Boyer, P. D. (1953). J. Biol. Chem., *204*, 265.

Selzer, S. (1972). in "The Enzymes", Vol. VII, 3rd edition, P. D.
    Boyer, Ed., Academic Press, New York, p. 381.

Siler, J. G. and Moldave, K. (1971). in "Protein Synthesis", Vol. 1,
    E. H. McConkey, Ed., Marcel Dekker, New York, p. 121.

Srivastava, S. K. and Beutler, E. (1968). Anal. Biochem., *25*, 70.

Tietz, F. (1969). Anal. Biochem., *27*, 502.

Torchinskii, Yr. M. (1974). "Sulfhydryl and Disulfide Groups of
    Proteins", H. B. F. Dixon, Translator, Consultants Bureau,
    New York.

Urey, H. C. (1952). Proc. Nat. Acad. Sci., *38*, 349.

Van Trump, J. E. and Miller, S. L. (1972). Science, 178, 859.

Vennesland, B. and Conn, E. E. (1954). in "Glutathione", S.
    Colowick, et al, Eds., Academic Press, New York, p. 105.

Wendell, P. L. (1970). Biochem. J., *117*, 661.

Wieland, T. (1954). in "Glutathione". S. Colowick, et al, Eds.,
    Academic Press, New York, p. 45.

Yasunobu, K. T. and Tanaka, M. (1973). in "Iron-Sulfur Proteins",
    Vol. II, W. Lovenberg, Ed., Academic Press, New York, p. 29.

Young, J. D., Nimmo, I. A. and Hall, J. G. (1975). Biochim. Biophys.
    Acta, *404*, 124.

2

DISULFIDE CROSSLINKS AND THE SPECIFICITY OF PROTEIN TURNOVER IN
PLANTS

Gary Gustafson and Clarence A. Ryan

Department of Agricultural Chemistry

Washington State University, Pullman, WA 99163

ABSTRACT

Studies of the protein metabolism of detached tomato leaves,
hormonally induced to accumulate proteinase inhibitors, have indicated that the state of oxidation of protein-bound half-cystine
residues may be a principal parameter affecting *in vivo* and
in vitro stability of leaf proteins. Induced leaves exhibited a
general specificity of intracellular protein degradation directed
towards the preferential hydrolysis of proteins having free-
sulfhydryl residues. Proteins having disulfide cross-linkages,
including the proteinase inhibitors, were markedly stable to *in vivo*
degradation, and as a result, accumulated. These results provide a
precedence for a cellular protein selection process, resulting from
a directed specificity of intracellular protein degradation, which
is focused on a particular protein structural parameter.

INTRODUCTION

Potato tubers store high concentrations of several proteins
which are potent inhibitors of serine-type proteinases (Ryan, 1973).
Many of these proteins are distinguished from other tuber proteins
by their marked resistance to heat-denaturation and their high content of disulfide cross-linkages (Bryant et al., in press; Iwazaka
et al., 1971; Melville and Ryan, 1972). Two of the heat-stable
inhibitors, called Inhibitors I and II, also accumulate in the
leaves of potato and tomato plants when their leaves are wounded by
insect attack (Green and Ryan, 1972). This wound response is mediated by a hormone-like substance, termed PIIF (proteinase inhibitor
inducing factor), which is released from wounded tissues and

transported through the plants' vascular systems (Green and Ryan, 1973; Ryan, 1974). The wound response can be mimicked by treating detached leaves with a solution of the isolated hormone. This detached leaf system has proved useful in developing a biological assay for the hormone (Ryan, 1974).

In the present study we have utilized the detached tomato leaf system as a model for determining whether PIIF elicits a major alteration in the overall protein metabolism of tomato leaves. In these investigations we have focused on the unique structural characteristics of the inhibitors (i.e., their heat-stability and disulfide content) in distinguishing between alterations in their cellular levels relative to changes in the levels of other leaf proteins. Because the inhibitors are rich in disulfide residues, it was anticipated that their accumulation in detached leaves might result in a marked increase in the quantity of total protein disulfide residues relative to total protein sulfhydryl residues, and that measurements of changes in the quantities of these two protein structural parameters might provide insights regarding the specificity of the hormone's activity.

Experimental Procedure

The general approach to each experiment is given in the text below. More detailed descriptions of the experimental conditions have been presented elsewhere (Gustafson and Ryan, in press).

Results and Discussion

Accumulation of proteinase inhibitors and heat-soluble protein in induced leaves. The following procedure was used to induce inhibitor accumulation. Leaves were excised from young tomato plants with a razor blade and supplied with a crude solution of PIIF for 15 min through their cut petioles. The hormone treated leaves were then transferred to water and incubated under constant light. Protein was extracted from induced leaves after various intervals of incubation with Tris-HCl buffer, pH 8.5. These extracts were then adjusted to pH 4.0, incubated in a 60 C water bath for 30 min, and centrifuged. The supernatant fluid obtained from this treatment is referred to as heat-soluble protein. Inhibitors present in the heat-soluble fraction were quantified using immunological assays. Preliminary experiments indicated that Inhibitors I and II were quantitatively recovered in the heat-soluble fraction. Fig. 1 illustrates the relationship between the increase in inhibitor levels and the increase in total heat-soluble protein in PIIF-induced leaves. Approximately 70% of the heat-soluble proteins present in induced leaves after 60 h of incubation was accounted for in Inhibitors I and II.

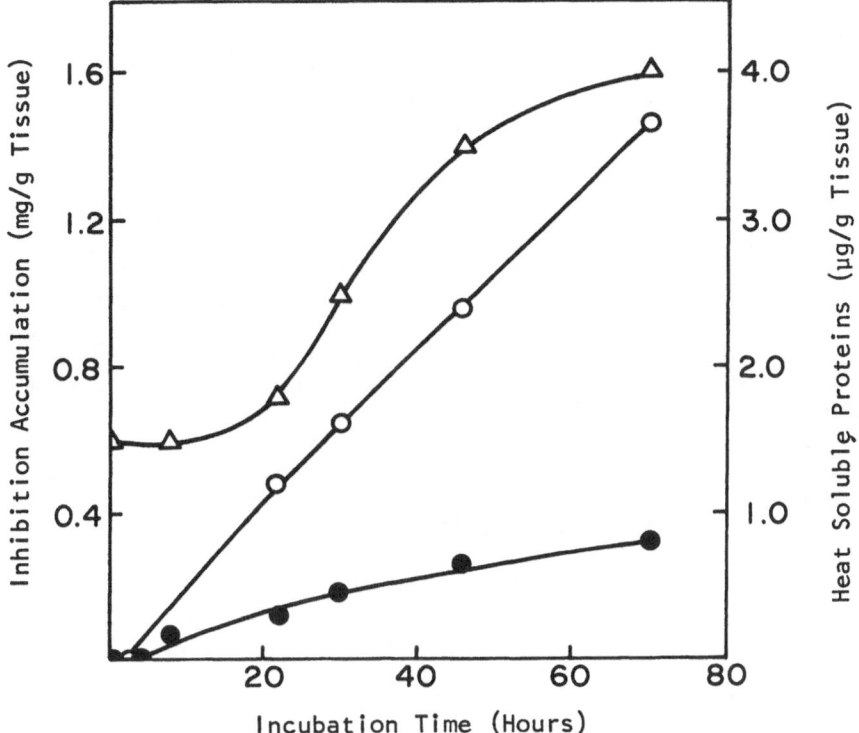

Fig. 1. Time-course for the accumulation of Inhibitor I and Inhibitor II, and heat-soluble proteins, in tomato leaves treated with the wound hormone, PIIF. Total soluble leaf protein was extracted as described in the text. Inhibitors were assayed immuno-logically. Inhibitor I -o-o-; Inhibitor II -•-•-; heat-soluble proteins -△-△-.

Changes in protein sulfhydryl and disulfide residues during induction. Table I illustrates the results of an experiment in which the quantities of cysteine (-SH) and cystine (-SS-) present in the proteins of uninduced and induced leaves were compared. Leaf proteins, separated into heat-soluble and heat-precipitable frac-tions as described above, were assayed for total half-cystine resi-dues according to the following procedure. Protein in the heat-soluble and heat-precipitated fractions were collected and washed with 15% TCA. The washed TCA pellets were then redissolved in 0.5 M Tris-HCl, pH 8.5 buffer containing 1% SDS. An excess of dithio-threitol was added to reduce all (-SS-) residues. After 30 min

incubation the mixtures were adjusted to pH 2. The dithiothreitol
was then removed by chromatography on a G25 Sephadex column with
0.01 N HCl containing 0.1% SDS. The void volume fractions were
assayed for total reduced half-cystine using Ellman's reagent. In
a parallel experiment the proportions of half-cystine present originally as cystine in the two protein fractions were also determined.
For these determinations leaf extracts were prepared and fractionated
in the presence of an excess of iodoacetamide in order to block all
protein (-SH) residues. The unmodified (-SS-) residues, present in
these alkylated protein fractions, were reduced and quantified using
the same procedure employed for determining total half-cystine. The
proportions of half-cystine present originally as (-SH) residues in
the heat-soluble and heat-precipitated protein fractions were determined as the difference between total half-cystine and total (-SS-)
residues.

     As shown in Table I, essentially all proteins which contained
(-SH) residues were localized in the heat-precipitated fractions.
Although most (-SS-) proteins initially present in uninduced leaves
were also precipitated by the heat treatment, (-SS-) proteins that
accumulated in 48 h induced leaves (including Inhibitors I and II)
were recovered in the heat-soluble fraction. The decrease in heat-
precipitable protein (-SH) residues, occurring during induction, was
paralleled by a decrease in total heat-precipitable protein. The

TABLE I

Quantitation of Cysteine and Cystine in the Heat-Soluble and
Heat-Precipitable Soluble Proteins of Tomato Leaves

Fraction	Induction Time (hr)*	mg Protein / g Tissue	μmoles Half-Cystine/g Tissue		
			Total	As Cysteine	As Cystine
Heat Precipitated	0	22	1.52	1.20	0.32
	48	14	0.94	0.55	0.39
Heat Soluble	0	1.6	0.12	0.07	0.05
	48	4.0	0.73	0.08	0.65

*After supplying leaves with PIIF for 15 min. Leaves were
incubated in 1000 ft c light at 31 C.

increase in heat-soluble protein (-SS-) residues was paralleled by
an increase in heat-soluble protein. Although the quantity of heat-
precipitable protein lost was greater than the quantity of heat-
soluble protein gained, total protein half-cystine remained constant
during the 48 h induction period because the (-SS-) proteins accumulated had a considerably higher concentration of half-cystine. Sulfhydryl proteins lost from the heat-precipitable pool had an average
half-cystine content of 0.07 µmoles/mg protein, whereas the heat-
soluble disulfide proteins accumulated had an average of 0.3 µmoles/
mg protein.

Fig. 2 shows a comparison of the change in half-cystine residues present in inhibitor proteins and total heat-soluble protein as
a function of leaf incubation time. The proportion of heat-soluble
protein half-cystine present in the inhibitors was calculated based
on measurements of inhibitor protein levels and known values for
their cystine content. Approximately 50% of the total half-cystine
in heat-soluble proteins after 71 h of incubation was associated
with the inhibitors.

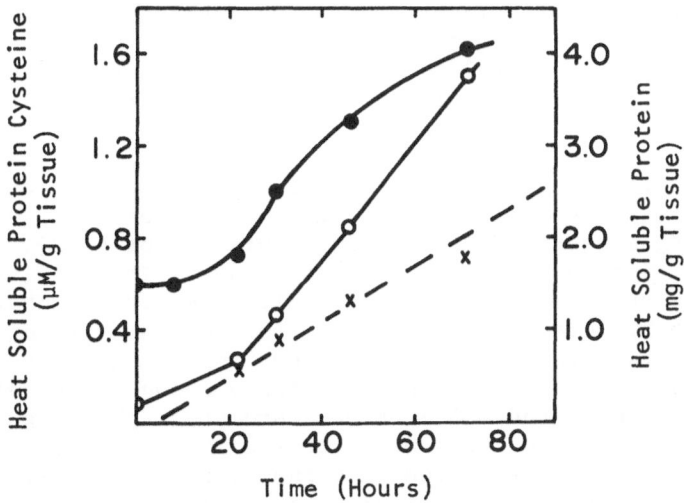

Fig. 2. Time-course of the accumulation of heat-soluble proteins
and heat-stable soluble protein cysteine during PIIF-induced accumulation of proteinase inhibitors in tomato leaves. Heat-soluble proteins -•-•-; heat-soluble protein cysteine -o-o-; Inhibitor I --x--x--.

Purification of inhibitors from induced leaves. Isotope incorporation experiments were performed in order to characterize factors contributing to the induced alterations in the levels of (-SS-) and (-SH) proteins in hormone treated leaves. These studies focused on comparing the rate of incorporation of [^{35}S] cysteine into Inhibitors I and II relative to its incorporation into total non-inhibitor protein. In performing these studies it was necessary to develop procedures for isolating the inhibitors from labeled leaves. The procedure adopted for this purpose has also proved useful as a general method for preparing large quantities of the inhibitors.

Table II summarizes the results of a large scale purification. In the first step of the purification, leaves were homogenized in 20 mM EDTA, pH 5.6. Under these conditions the inhibitors were quantitatively bound to the particulate cell debris, whereas the majority of other leaf proteins were soluble. The particulate material in the homogenates was collected and washed several times with fresh aliquots of 20 mM EDTA. The inhibitors were then quantitatively extracted from this washed material using 1 M KCl. The KCl extract was next adjusted to pH 4.0 and incubated in a 60 C water bath for 30 min. Following this incubation, the heat-precipitated protein was removed by centrifugation. The remaining protein solution was concentrated by precipitation with ammonium sulfate at 70% saturation and redissolved in a minimal volume of 0.05 M KCl. In the final step of the purification the protein concentrate was chromatographed on a G75 Sephadex column. As shown in Fig. 3, the inhibitors were resolved into two separate peaks with this chromatography. Fractions containing Inhibitor I and Inhibitor II were collected separately and evaluated for homogeneity by electrophoresis in polyacrylamide gels. Only one stainable protein band, corresponding to inhibitor protein, was observed with each of the inhibitor fractions (Fig. 4).

Isotope incorporation studies. The rate of total protein synthesis relative to the rates of syntheses of Inhibitors I and II was examined by measuring rates of incorporation of [^{35}S], derived from $Na_2{}^{35}SO_4$, into proteins of induced leaves. The results of this experiment are shown in Fig. 5. Induced leaves, pre-incubated in water for 23.5 h, were exposed to $^{35}SO_4$ for periods varying from 2 to 6 h. Inhibitor proteins were isolated from labeled leaves as described above. The center panel of Fig. 5 shows a typical elution profile obtained with the G75 Sephadex chromatography step of inhibitor purification. The lower panel of Fig. 5 illustrates plots of total radioactivity incorporated into Inhibitors I and II, and into total soluble leaf protein (excluding inhibitors) as a function of isotope exposure time.

Utilizing the data in Fig. 5, it was possible to calculate the proportion of total protein synthesis in induced leaves which was

TABLE II

Purification of Proteinase Inhibitors I and II from Tomato Leaves

Step	Total Protein	Inhibitor I	Inhibitor II
	mg	*mg*	*mg*
1 M KCl extraction of lyophilized tomato leaves (EDTA extracted)	271	73	27
Heat treatment, 60°, pH 4, 30 minutes	210	63	25
Ammonium sulfate precipitation (0-70%)	88	56	17
G75 Sephadex (Peak 2)	31.6	32	--
G75 Sephadex (Peak 3)	14.9	--	12

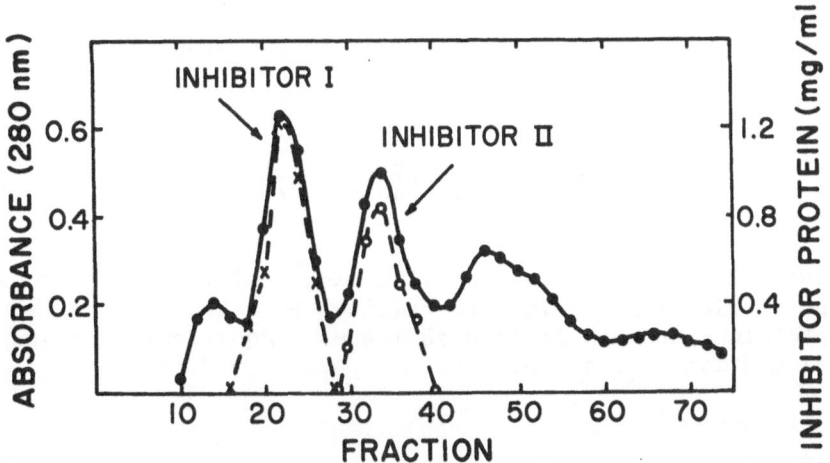

Fig. 3. Gel filtration of Inhibitor I and Inhibitor II proteins. The concentrated protein solution obtained from the ammonium sulfate precipitation step of the purification procedure was applied to a column (2.5 × 95 cm) of G75 Sephadex previously equilibrated with elution media (0.05 M KCl). Six ml fractions were collected. Absorbancy -•-•-; Inhibitor I -x-x-; Inhibitor II -o-o-. Inhibitor concentrations were determined immunologically.

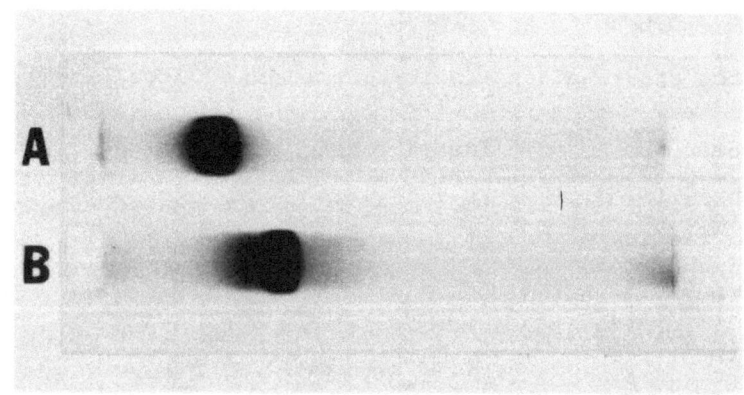

Fig. 4. Disc gel electrophoresis of purified proteinase Inhibitor I and Inhibitor II from tomato leaves. A, Inhibitor I; B, Inhibitor II. Migration is from left (anode) to right (cathode).

directed towards the synthesis of inhibitors. The specific radioactivity of intracellular [^{35}S] cysteine utilized for protein synthesis was first calculated according to the following equations:

Specific Radioactivity of Inhibitor Synthesized $= \dfrac{\text{(Total CPM into Inhibitor/h)}}{\text{(Total μg Inhibitor Accumulated/h)}}$

Specific Radioactivity of Intracellular ^{35}S Cysteine $= \dfrac{\text{(Specific Radioactivity Inhibitor Synthesized)}}{\text{(Total μg Inhibitor/μmole Half-cystine)}}$

The value calculated for the specific radioactivity of intracellular cysteine (Table III) was in turn utilized to calculate the rate of total protein synthesis in induced leaves. Equations utilized for this calculation were as follows:

Rate of Incorporation of Half-Cystine into Total Protein $= \dfrac{\text{(Total CPM into Total Protein/h)}}{\text{(Specific Radioactivity } ^{35}\text{S Cysteine Pool)}}$

Rate of Total Protein Synthesis $= \dfrac{\text{(Rate Incorporation of Half-Cystine into Protein)}}{\text{(μmole Half-Cystine/μg Protein)}}$

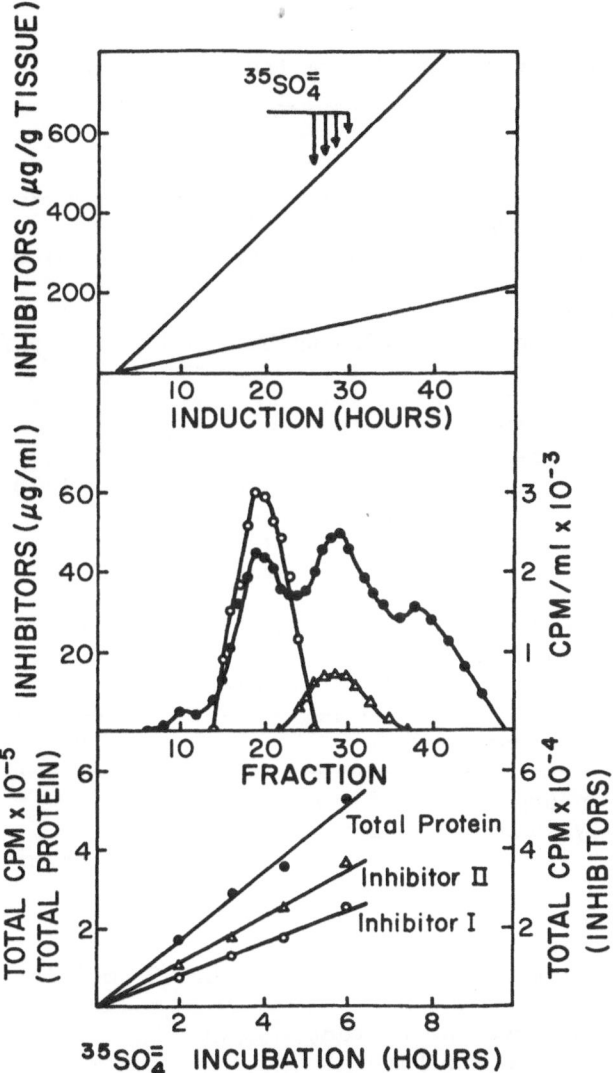

Fig. 5. Summary of the incorporation of $[^{35}S]$ from $[^{35}SO_4^=]$ into proteinase Inhibitor I, Inhibitor II, and soluble proteins during a time-course experiment of the PIIF-induced accumulation of proteinase inhibitors in tomato leaves. The upper figure shows the times during which $[^{35}SO_4^=]$ was supplied to sets of leaves. The middle figure shows a typical isolation of Inhibitor I and Inhibitor II, as performed for each set of leaves. Inhibitor I -o-o-; Inhibitor II -Δ-Δ-; assayed immunologically. Counts per minute -●-●-●-. The bottom figure illustrates the time-course of radioactivity incorporation into Inhibitor I, Inhibitor II, and total non-inhibitor proteins.

TABLE III

Summary of Values Calculated from Isotope Incorporation Experiment

Parameter	From Inhibitor I Data	From Inhibitor II Data
Rate of incorporation of [^{35}S] cysteine	4300 cpm/hr	6100 cpm/hr
Rate of inhibitor protein accumulation	20.5 µg/hr	5 µg/hr
Specific radioactivity of inhibitor synthesized during labeling period	210 cpm/µg	1220 cpm/µg
Half-cystine content of inhibitor	2×10^{-4} µmole/µg	1.2×10^{-3} µmole/µg
Specific radioactivity of intracellular [^{35}S] cysteine	1.05×10^6 cpm/µmole	1.02×10^6 cpm/µmole

Parameter	Value
Rate of incorporation of radioactivity into total, non-inhibitor protein	9.2×10^4 cpm/hr/g tissue
Rate of incorporation of half-cystine into total, non-inhibitor protein	0.088 µmole/hr
Average half-cystine content of total, non-inhibitor protein	7×10^{-5} µmole/µg
Rate of total, non-inhibitor protein synthesis	1250 µg/hr/g tissue
Average concentration of total, non-inhibitor proteins in leaves	26.5 mg/g tissue
Rate of turnover of total, non-inhibitor proteins	5%/hr

In comparing the values obtained for the rate of inhibitor synthesis and total protein synthesis (Table III), it can be seen that only about 2% of total protein synthesis in induced leaves was attributable to inhibitor synthesis. That is, about 98% of protein synthesis appeared to be directed towards the large pool of non-inhibitor protein comprised primarily of (-SH) proteins. Because there is a net loss in total (-SH) protein during induction, it must be concluded that these proteins are subject to rapid turnover in induced leaves--the rate of degradation being more rapid than the rate of synthesis. The turnover rate calculated for these proteins was 5%/h.

Based upon the above results it is concluded that the reciprocal alterations in levels of (-SS-) and (-SH) proteins in induced tomato leaves is attributable more to the specificity of protein turnover than to the specificity of protein synthesis. In a general way the specificity of protein turnover appeared to be the preferential degradation of (-SH) proteins and the preferential stability of (-SS-) proteins.

Studies of protein turnover in a variety of other systems have also suggested that the structural properties of proteins can influence their susceptibilities to degradation *in vivo* (Bond, 1975; Dice et al., 1973; Goldberg et al., 1974; Pine, 1975). In general it has been found that proteins which exist in more tightly folded conformations or are more resistant to denaturation *in vitro*, are also degraded more slowly *in vivo*. The observation that disulfide cross-linked proteins in tomato leaves are both more stable to heat treatment *in vitro* and degradation *in vivo* than sulfhydryl proteins agrees with this general concept.

ACKNOWLEDGMENTS

This research was supported in part by NSF grants GB-37972 and PCM-75-23629, and United States Department of Agriculture, Cooperative States Research Service grant 316-15-60. The authors thank Mr. Charles Oldenberg for growing the plants.

REFERENCES

Bond, J. (1975) *in Intracellular Protein Turnover* (Schimke, R. T., and Katunuma, N., Eds.) p. 281, Academic Press, New York.
Bryant, J., Green, T. R., Gurusaddaiah, T., and Ryan, C. A. (in press) *Biochemistry*.
Dice, J. F., Dehlinger, P. J., and Schimke, R. T. (1973) *J. Biol. Chem.* 248, 4220.
Green, T. R., and Ryan, C. A. (1972) *Science* 175, 776.
Green, T. R., and Ryan, C. A. (1973) *Plant Physiol.* 51, 19.

Goldberg, A. L., Howell, E. M., Li, J. B., Martel, S. B., and Prouty,
    W. F. (1974) *Fed. Proc.* <u>33</u>, 1112.
Gustafson, G., and Ryan, C. A. (in press) *J. Biol. Chem.*
Iwazaka, T., Kiyohara, T., and Yoshikawa (1971) *J. Biochem.* <u>70</u>, 817.
Melville, C. J., and Ryan, C. A. (1972) *J. Biol. Chem.* <u>247</u>, 3445.
Pine, M. J. (1975) *in Intracellular Protein Turnover* (Schimke, R. T.,
    and Katunuma, N., Eds.) p. 65, Academic Press, New York.
Ryan, C. A. (1973) *Ann. Rev. Plant Physiol.* <u>24</u>, 173.
Ryan, C. A. (1974) *Plant Physiol.* <u>54</u>, 328.

3

PROTEIN THIOL-DISULFIDE INTERCHANGE AND INTERFACING

WITH BIOLOGICAL SYSTEMS

D.B. Wetlaufer, V.P. Saxena, A.K. Ahmed, S.W. Schaffer,
P.W. Pick, K.-J. Oh, and J.D. Peterson
Department of Chemistry, University of Delaware
Newark, Deleware 19711 (present address) and
Department of Biochemistry, University of Minnesota
Minneapolis, Minnesota 55455

ABSTRACT

Disulfide-containing proteins offer unique advantages for
mechanistic studies of the formation of native three-dimensional
structure from unordered, reduced precursors. The main advantage
is that covalent intermediates are formed; by characterizing these
intermediates, one obtains substantial information about the
reaction pathway. Thiol- disulfide interchange is a major component of most oxidative mechanisms carrying thiol to disulfide;
thus, it required some attention in its own right. Anfinsen's
descriptions of a "shuffle-ase" enzyme led us to examine the rates
of the uncatalyzed exchange under physiologically plausible conditions. Somewhat surprisingly, we found that the rates for formation of several native proteins in uncatalyzed systems containing
GSSG and GSH are as great as with the "shuffle-ase" enzyme, suggesting that a substantial portion of biological thiol oxidations
proceed by uncatalyzed exchange. While thiol-disulfide exchange
of course results in no net change in the oxidation level of a
system, catalytic linkage of thiol or disulfide to other redox
systems provides a mechanism for achieving net changes.

We became interested in thiol-disulfide interchange reactions
some years ago in connection with our studies on protein folding.
Disulfide proteins offer unique advantages for studying folding
mechanisms in that discrete covalent intermediates are formed; by
characterizing the intermediates one obtains substantial information on the folding mechanism. The results of several such

studies have been reported elsewhere (refer to Anderson and
Wetlaufer, (1976), for references), and will not further concern
us here. Since thiol-disulfide interchange is a major component
of most oxidative mechanisms carrying thiol to disulfide, it
deserves some attention in its own right. The reactions in question are represented in the following equations:

$$R'S-SR'' + R'''S^- \rightleftharpoons \begin{array}{l} R'S-SR''' + R''S^- \\ R''S-SR''' + R'S^- \end{array}$$

These equations summarize the familiar relevant facts: The
reaction is reversible, the thiolate anion is the reactive species
in a physiologically accessible pH range, and one obtains both of
the possible mixed disulfides.

To know whether a reaction is feasible on a given time scale
we also need to know something of the kinetics of the reaction.
Our reading of the experiments from Anfinsen's laboratory (Goldberger et. al., 1964; Epstein, 1970) on the "shuffle-ase" enzyme
and its catalysis of thiol- disulfide interchange left us less
than convinced of the physiological role assigned to the protein.
Several kinetic studies of exchange (Eldjarn and Pihl, 1957;
Lamfrom and Nielsen, 1958; Bradshaw et. al., 1967; Jocelyn, 1967)
showed that uncatalyzed exchange is quite rapid. It appeared to
us that uncatalyzed exchange could be rapid enough to provide the
mechanism for rapid regeneration of disulfide-reduced proteins.

In our studies with hen eggwhite lysozyme (Saxena and Wetlaufer, 1970), these predictions were borne out. To a first
approximation, a suitable regeneration system would require
oxidizing capabilities and shuffling (i.e., reducing) capabilities.
We therefore reasoned that a mixture of low molecular weight
disulfide and thiol compounds would be required. Our first experiments with mixtures of reduced and oxidized glutathione were
successful in producing a rapid regeneration; other studies
(Saxena and Wetlaufer, 1970; Creigton, 1975) have shown that
regenerations can be carried out with a number of low molecular
weight thiols and disulfides. Optimization experiments showed
that a broad range of concentrations and ratios of GSH and GSSG
gave nearly maximal rates and yields (Saxena and Wetlaufer, 1970,
Wetlaufer et. al., 1974); interestingly, anlytical determinations
of GSH and GSSG in tissues (Tietze, 1969) showed concentrations in
the same broad range.

While the mixtures of GSH and GSSG used for regenerations
constitute redox buffers, there is no net change in the total
number of thiols or disulfides in the system over the course of a
regeneration. There is only a redistribution of disulfide bonds

among the various thiol groups present. The concept of biologically functioning redox buffers, and the use of such buffer systems _in vitro_ to simulate a biochemical milieu, do not have the degree of acceptance they deserve. This is surprising, considering the elements in common between acid—base and redox buffers and the universal acceptance of the former.

Since our original experiments, several disulfide proteins have been shown, in our laboratory and elsewhere, to undergo regeneration from their reduced forms in systems containing a GSH—GSSG redox buffer. These findings are summarized in Table 1.

Table 1

Reduced Proteins Regenerated with a Glutathione Oxido-Reduction Buffer

Bovine Plasma Albumin and BPA Fragments	Teale & Benjamin, 1976
Bowman-Birk Protease Inhibitor	Hogle & Liener, 1973
Hen Egg Lysozyme	Saxena & Wetlaufer, 1970
Human Lysozyme	Wetlaufer, et. al., 1974
Immunoglobulin G	Petersen & Dorrington, 1974
Pancreatic Typsin Inhibitor	Creighton, 1974
RNase A	Ahmed, et. al., 1975
Trypsinogen and Trypsin	Sinha & Light, 1975

An inspection of Table I makes it apparent that there is considerable generality in the non-enzymic reactivation of reduced proteins. Which mechanism is responsible for the physiological formation of native disulfides in proteins? The data available do not permit a clear answer. On the one hand, the demonstrated existence of a plausible non-enzymic system abolishes the need to invoke an enzymic system to account for the formation of protein disulfides _in vivo_. Furthermore, the enzymic system does not seem to be widespread (DeLorenzo and Molea, 1967), and the heat-stable components of the enzymic system have not been fully characterized (Anfinsen, 1967); thus, the possibility exists that the observed "shuffle-ase" activity is a fortuitous and nonspecific property of an exposed thiol group on a protein whose physiological role is unrelated to protein folding. On the other hand, there are numerous enzymes known which catalyze reactions that have a

moderately rapid uncatalyzed rate: CO_2 hydration and hexose
phosphate anomerizations quickly come to mind. Since glutathione
is ubiquitous and generally occurs in effective concentrations,
we think it likely that it provides a general mechanism for
forming native disulfides in proteins. The less widely distributed
"shuffle-ase" enzymes, if they are physiologically significant,
are probably employed in situations with special requirements.

Glutathione, although not unique in promoting regeneration,
is present in most tissues at many times the concentration of all
other low molecular weight thiols and disulfides combined (Jocelyn,
1972). We will therefore limit our discussion to glutathione, on
the assumption that it is the only generally significant low
molecular weight thiol and disulfide. We should begin by noting
that the normal physiological ratios of GSH to GSSG are quite
high in actively metabolizing tissues -- ranging from 20 to 100
(Tietze, 1969). In some physiological situations that ratio may
be considerably lower than 20 (i.e., a higher fraction in the
oxidized form). In such cases one would expect an appreciable
extent of formation of mixed disulfides between proteins and
gluathione. Observations in Neurospora (Fahey et. al., 1975) and
rat liver (Isaacs, 1976) are consistent with this expectation.
The possibility exists that this provides a general route for
control of intracellular enzymes, since a number of enzymes are
inhibited by mixed disulfide formation (Birkett, 1973; Nowak &
Himes, 1971; Frankfater & Fridovich, 1970; Ernest & Kim, 1973;
Wang & Volini, 1968; Sumegi et. al., 1971). On the other side of
the coin, two enzymes have been shown to be activated by mixed
disulfide formation (Mize et. al., 1962; Pontremoli & Horecker,
1970).

Protein thiols with ordinary reactivity should form mixed
disulfides with gluatathione with about the same reduction potential as GSSG. Therefore an intracellular pool of readily reactive
disulfides should be viewed as containing Protein-S-S-G mixed
disulfides in addition to GSSG.

We expect that GSH and GSSG pool sizes should depend on the
rates of biosynthesis from and degradation to the constituent
amino acids, and on rates of transfer among other sources and
sinks. While mechanisms of biosynthesis and degradation of
glutathione are reasonably well understood, we do not yet understand how these processes are controlled. As for "other sources
and sinks," NADPH linked glutathione reductase (Black, 1963),
glutathione S-transferase (Habig et. al., 1974) and perhaps
gluathione peroxidase (Mills, 1960) will be important.

We ought not end these brief considerations of GSH and GSSG
pools, sources, and sinks without examining oxidation of GSH by
O_2 and by various activated species of oxygen, notably peroxide
and superoxide. The later two species of oxygen do, of course,

oxidize thiols, and this presumably is a basis of superoxide toxicity (Gregory & Fridovich, 1973). O_2 oxidation of thiols has long been known, but in general requires some catalytic metal ion (Bernheim & Bernheim, 1938). It is unclear whether O_2 oxidation will occur in a completely metal-free system. The relative stability of thiols in cereal grains (Bloksma, 1971), and in the wool of sheep with a dietary deficiency of copper (Lee, 1956), suggests that uncatalyzed air oxidation of thiols is, at most, very slow.

REFERENCES

Ahmed, A. K., Schaffer, S. W. and Wetlaufer, D. B. (1975). Non-enzymic reactivation of reduced bovine pancreatic ribonuclease by air oxidation and by gluathione oxido-reduction buffers. J. Biol. Chem. 250, 8477-82.

Anderson, W. L. and Wetlaufer, D. B. (1976). The folding pathway of reduced lysozyme. J. Biol. Chem. 251, 3147-53.

Anfinsen, C. B. (1967). The formation of the tertiary structure of proteins. The Harvey Lectures 61, 95-116.

Bernheim, F. and Bernheim, M. L. C. (1938). The effects of various metals and metal complexes on the oxidation of sulfhydryl groups. Cold Spring Harbor Sympos. on Quant. Biol. 7, 174-183.

Birkett, D. J. (1973). Mechanism of inactivation of rabbit muscle glyceraldehyde 3-phosphate dehydrogenase by ethacrynic acid. Mol. Pharmacol. 9, 209-218.

Black, S. and Colman, R. F. (1963). The biochemistry of sulfur-containing compounds. Ann. Rev. Biochem. 32, 399-418.

Bloksma, A. H. (1971) in Pomeranz, Y. ed. "Wheat Chemistry and Technology." Amer. Assn. of Cereal Chemists, St. Paul, 1971, pp. 523-584.

Bradshaw, R. A., Kanarek, L. and Hill, R. L. (1967). The preparation, properties, and reactivation of the mixed disulfide derivative of egg white lysozyme and L-crystine. J. Biol. Chem. 242, 3789-3798.

Creighton, T. E. (1975). The two-disulfide intermediates and the folding pathway of reduced pancreatic trypsin inhibitor. J. Mol. Biol. 95, 167-99.

DeLorenzo, F. and Molea, G. (1967). Relative levels of the disulfide-interchange enzyme in the microsomes of bovine tissues. Biochem. Biophys. Acta 146, 593–595.

Eldjarn, L. and Pihl, A. (1957). On the mode of action of X-ray protective agents II. The interaction between biologically important thiols and disulfides. J. Biol. Chem. 225, 499–510.

Eldjarn, L. and Pihl, A. (1957). On the mode of action of X-ray protective agents III. The enzymatic reduction of disulfides. J. Biol. Chem. 227, 339–345.

Epstein, C. J. (1970) in Anfinsen, C. B. ed. "Aspects of Protein Biosynthesis." Part A, Acad. Press, N.Y. pp. 367–431.

Ernest, M. J. and Kim, K. (1973). Regulation of rat liver glycogen synthetase; reversible inactivation of glycogen synthetase D by sulfhydryl–disulfide exchange. J. Biol. Chem. 248, 1550–1555.

Fahey, R. C., Brody, S. and Mikolajczyk, S. D. (1975). Change in the glutathione thiol-disulfide status of Neurospora crassa conidia during germination and aging. J. Bacteriol. 121, 144–51.

Frankfater, A. and Fridovich, I. (1970). The purification and properties of oxidized derivatives of L-histidine ammonia-lyase. Biochem. Biophys. Acta 206, 457–472.

Givol, D., DeLorenzo, F., Goldberger, R. F., and Anfinsen, C. B. (1965). Disulfide interchange and the three dimensional structure of proteins. Proc. Nat'l Acad. Sci. U.S. 53, 676–684.

Givol, D., Goldberger, R. F., and Anfinsen, C. B. (1963). Oxidation and disulfide interchange in the reactivation of reduced ribonuclease. J. Biol. Chem. 239, 3114–3116.

Goldberger, R. F., Epstein, C. J. and Anfinsen, C. B. (1964). Purification and properties of a microsomal enzyme, system catalyzing the reactivation of reduced ribonuclease and lysozyme. J. Biol. Chem. 239, 1406–1410.

Gregory, E. M. and Fridovich, I. (1973). Oxygen toxicity and superoxide dismutase. J. Bacteriol. 114, 1193–1197.

Habig, W. H., Pabst, M. J. and Jakoby, W. B. (1974). Glutathione S-transferases; the first enzymatic step in mercapturic acid formation. J. Biol. Chem. 249, 7130–39.

Isaacs, J. T. (1976). Glutathione systems and the control of the protein SS/SH ratio. Fed. Proc. 35, Abstract No. 605.

Jocelyn, P. C. (1967). The standard redox potential of cysteine-cystine from the thiol-disulfide exchange reaction with glutathione and lipoic acid. Eur. J. Biochem. 2, 327-331.

Jocelyn, P. C. (1972). Biochemistry of the SH group. Acad. Press, London, p. 10.

Lamfrom, H. and Nielsen, S. O. (1958). Mercaptolysis of cystine--synthesis of an asmymetric disulfide containing half-cystine. Compt. Rend. Trav. du Lab. Carlsberg, Ser. Chim. 30, 349-359.

Lee, J. H. (1956). The influence of copper deficiency on the fleece of British breeds of sheep. J. Agr. Sci. 47, 218-24.

Liener, I. E. and Hogle, J. (1973). Reduction and reactivation of the Bowman-Birk soybean inhibitor. Canad. J. Biochem. 51, 1014-1020.

Mills, G. C. (1960). Glutathione peroxidase and the destruction of hydrogen peroxide in animal tissues. Arch. Biochem. Biophys. 86, 1-5.

Mize, C. E. Langdon, R. and Thompson, T. E. (1962). Hepatic gutathione reductase II. Physical properties and mechanism of action. J. Biol. Chem. 237, 1596-1600.

Nowak T. and Himes, R. H. (1971). Formyltetrahydrofolate synthetase; the role of the sulfhydryl groups. J. Biol. Chem. 246, 1285-1293.

Peterson, J. G. L. and Dorrington, K. J. (1974). An in vitro system for studying the kinetics of interchain disulfide bond formation in immunoglobulin G. J. Biol. Chem. 249, 5633-5641.

Pontremoli, S. and Horecker, B. L. (1970) Current topics in cellular regulation. Acad. Press, N.Y., Vol. 2, 188-195.

Saxena, V. P. and Wetlaufer, D. B. (1970). Formation of three-dimensional structure in proteins. I. Rapid nonenzymic reactivation of reduced lysozyme. Biochemistry 9, 5015-5023.

Sinha, N. K. and Light, A. (1975). Refolding of reduced, denatured trypsinogen and trypsin immobilized on agarose beads. J. Biol. Chem. 250, 8624-8629.

Sumegi, J., Sanner, T. and Pihl, A. (1971). Involvement of Highly reactive sulfhydryl groups in the action of RNA polymerase from E. Coli. FEBS Letters 16, 125.

Teale, J. M. and Benjamin, D. C. (1976). Antibody as an immunological probe for studying the refolding of bovine serum albumin. I. The catalysis of reoxidation of reduced bovine serum albumin by glutathione and a disulfide interchange enzyme. II. Evidence for the independent refolding of the domains of the molecule. J. Biol. Chem. 251, 4603-4608, 4609-4615.

Tietze, F. (1969). Enzymic method for quantitative determination of nanogram amounts of total and oxidized glutathione: applications to mammalian blood and other tissues. Anal. Bioch. 27, 502-522.

Wang, S. and Volini, M. (1968). Studies on the active side of rhodanase. J. Biol. Chem. 243, 5465-5470.

Wetlaufer, D. B., Johnson, E. R. and Clauss, L. M. (1974) in Osserman, E., Beychok, S. and Canfield R. eds. Lysozyme. Acad. Press, N.Y., pp. 269-280.

4

ON THE MECHANISM OF RENATURATION OF PROTEINS CONTAINING DISULFIDE
BONDS

Hiroshi Taniuchi, A. Seetharama Acharya, Generoso Andria

and Diana S. Parker

Laboratory of Chemical Biology, National Institute of

Arthritis, Metabolism, and Digestive Diseases, National

Institutes of Health, Bethesda, Maryland 20014

The renaturation of bovine pancreatic ribonuclease A and
hen egg white lysozyme from their reduced forms involves two
statistical processes, pairing of half-cystine residues by
oxidation and rearrangement of disulfide bonds by enzyme or
thiol catalyzed sulfhydryl-disulfide interchange. The stability
against sulfhydryl-disulfide interchange of the native or native-
like conformation thus attained, which could be a form containing
three native disulfide bonds and one open disulfide bond,
causes the system to accumulate the renatured enzyme. Thus,
the native-like conformation is associated with the lowest free
energy only in the late phase of folding.

INTRODUCTION

Anfinsen and his colleagues (Anfinsen, 1967) have established
that the information contained in the amino acid sequence is
sufficient for determination of the native three-dimensional
structure of proteins including disulfide bonds. They (Epstein,
Goldberger and Anfinsen, 1963) have further proposed the thermodynamic hypothesis that the native fold of proteins is the
three-dimensional configuration that is associated with the
lowest configurational free energy.

On the basis of this groundwork, however, many mechanisms
can be considered for protein folding. For example, Phillips

(1967) speculated that the hen egg white lysozyme chain may
fold from the NH$_2$-terminus during biosynthesis. Evidence
accumulated during the past several years has given insight
into the mechanism which we wish to discuss here.

CONFORMATION OF REDUCED PROTEINS

Reduced RNase A contains only 8 more hydrogen atoms than
native RNase which is composed of more than 2000 atoms. Therefore, the simplest application of the thermodynamic hypothesis
would be to assume that reduced RNase folds to a native-like
conformation by guidance of interactions between residues and
thereby most, if not all, of the population of reduced RNase A
directly form the native disulfide bonds upon oxidation (Watson,
1976). However, measurements of viscosity and optical rotation
by Harrington and Sela (1959) show that cleavage of the disulfide
bonds of RNase A by reduction-alkylation disrupts the ordered
structure, which is consistent with the earlier observation of
performic acid-oxidized RNase A (Anfinsen, 1955). Givol et al.
(Givol, Goldberger and Anfinsen, 1964) and Venetianer and Straub
(1964) independently showed that fast oxidation by dehydroascorbic
acid of reduced RNase A yields an inactive species which is
rapidly converted to the active form by enzyme catalyzed sulfhydryl-
disulfide interchange. It is confirmed by gel filtration and
ultracentrifugation (Andria and Taniuchi, 1971) that this
inactive species is not a high molecular weight material artificially formed by interchain disulfide bonds but a monomeric
form presumably containing incorrect intrachain disulfide
bonds. Recently Hantgan et al. (Hantgan, Hammes and Scheraga,
1974) showed that the native disulfide bonds are not favorably
formed in the early phase of oxidation of reduced RNase A.
Cupric ion catalyzed fast oxidation of reduced hen egg white
lysozyme also yields monomeric species containing non-native
disulfide bonds (Acharya and Taniuchi, 1976a and b). Thus,
these reduced enzymes do not possess a compactly folded native-
like conformation but an enlarged, flexible conformation.

The intrinsic viscosity of reduced lysozyme (12.5 ml per g
in 0.1 M acetic acid at 25°) (Acharya and Taniuchi, 1976) is
greater than that of the native enzyme but smaller than that
expected for a random coil (Tanford, 1968). Thus, the reduced
enzyme does not sample all sterically allowable conformations
but only a limited number of conformations loosely constrained
presumably due to interactions between residues. Nonetheless,
the frequency of the native format (Anfinsen and Scheraga,
1975) sampled by these reduced enzymes does not appear to be
greater than that of the non-native formats as judged by the
frequency of the native set of four disulfide bonds formed by
oxidation without sulfhydryl-disulfide interchange (Acharya and
Taniuchi, 1976a). The formation of some incorrect disulfide bonds

in the early phase of oxidation of reduced proteins is also
reported with pancreatic trypsin inhibitor (Creighton, 1974) and
hen egg white lysozyme (Ristow and Wetlaufer, 1973).

THE DECREASE IN ENERGY IN THE LATE PHASE OF FOLDING

Haber and Anfinsen (1961) showed that RNase S-protein
(Richards and Vithayathil, 1959) can yield the "native" derivative
in a quantity 15 times greater than that expected for random
combination of half-cystine residues after reduction and reoxidation and suggested that information determining the native
structure is contained in RNase S-protein. In the later study
by Kato and Anfinsen (1969), the importance of cooperative
interaction between S-protein and S-peptide in stabilization of
RNase-S (Richards and Wyckoff, 1971) is emphasized. Thus,
these experiments, though significant, did not permit a conclusion
about whether the conformational energy progressively decreases
during folding or whether the conformational energy distinctly
decreases in a certain phase of folding.

Neither des-(121-124)-RNase (Anfinsen, 1956) nor des-(119-
124)-RNase (Lin, 1970) can form the native set of 4 disulfide
bonds in a significantly large population (for example, 1% or
less in the case of the latter derivative) but instead form
non-native bonds after reduction and oxidation (Taniuchi, 1970;
Andria and Taniuchi, 1971) in contrast with intact RNase A
(Anfinsen, 1967). The "non-native" des-(119-124)-RNase thus
formed can be converted at least in part (approximately 33%) to
the native form by enzyme catalyzed sulfhydryl-disulfide interchange in the presence of a complementing fragment containing
residues 105 to 124 (Andria and Taniuchi, 1971) (Fig. 1). In
the absence of this fragment there is no detectable accumulation
of the native derivative in this system (Andria and Taniuchi,
1971). Thus, the conformational energy of the native derivative
is equivalent to the energy of a derivative having non-native
disulfide bonds. In other words, native des-(119-124)-RNase is
not associated with the lowest configurational free energy.

Removal of the 6 residue segment from the carboxyl-end of
RNase A may be considered in effect equivalent to dislocation or
unfolding of the segment from the native three-dimensional
structure (Karta, Bello and Harker, 1967; Wyckof et al., 1967;
Richards and Wyckoff, 1971). Then, it would follow that unfolding
of a small portion of the native three-dimensional structure
renders the conformational energy equivalent to that of non-
native RNase A containing incorrect disulfide bonds. Further,
it would follow that a distinct decrease in conformational
energy would occur only after completion of folding (Taniuchi,
1973; Taniuchi and Bohnert, 1975) and thereby the initial phase
of folding would involve little change in energy, as any small

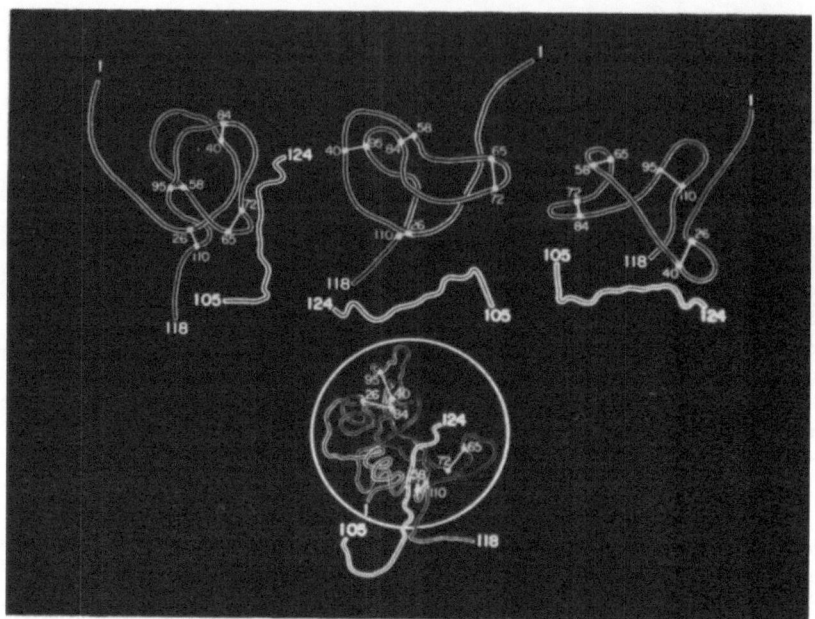

Figure 1. A schematic illustration of trapping by complementation of the native form of des-(119-124)-RNase formed from
non-native des-(119-124)-RNase by statistical sulfhydryl-
disulfide interchange. The non-native derivatives containing
incorrect disulfide bonds do not bind to the complementing
fragments of residues 105 to 124 (top). Only the native form of
the derivative binds to the fragment and then the conformation is
stabilized (bottom). The drawing of the ordered polypeptide
chain backbone (circled) is based on the three-dimensional structure of RNase-S (Wyckoff et al., 1967). The studies of Gutte et
al. (1972) using various lengths of synthetic complementing fragments indicate that the discontinuity of the polypeptide chain in
the ordered structure of the active complex occurs between residues 116 and 124, assuming that the ordered structure has no
overlap, as is the case with the complex of the two overlapping
nuclease fragments (Taniuchi and Anfinsen, 1971). The conversion
by enzyme catalyzed sulfhydryl-disulfide interchange from the
non-native to the native form is assumed not to involve a decrease
in energy until the latter form is bound with the fragment.

perturbation may move the molecule from one conformation to
another in the ensemble with the exception that disulfide bonds
require catalysis for their rearrangement.

STABILIZING ENERGY

Although the conformational energy of native des-(121-
124)-RNase is close to that of the non-native forms of the
derivative containing incorrect disulfide bonds, measurements
of viscosity and optical rotation by Anfinsen and Sela (Anfinsen,
1956; Sela and Anfinsen, 1957) showed that the native derivative
retains some native features of packing of the polypeptide
chain, presumably due to the restriction by the native disulfide
bonds. This is confirmed by gel filtration and CD studies of
the derivative (Taniuchi, 1970; Andria and Taniuchi, 1971). On
the other hand, Ottesen and Stracher's deuterium exchange
experiments (Ottesen and Stracher, 1960) demonstrated that the
motility of the native derivative is increased, indicating
weakened interatomic interactions throughout the molecule.
These observations suggest that the increase in the energy of
RNase A by removal of the 4 carboxyterminal residues occurs by
weakening interatomic interactions throughout the three-dimensional
structure without a drastic spatial rearrangement of the polypeptide chain (flexing nearly within the native domain).

The cooperative change in motility within an ordered structure
has been known in the ligand induced stabilization of proteins.
For example, Markus et al. (1968) showed that binding of 2'-
cytidylate to RNase A increases resistance against proteolytic
cleavage in regions distributed over the entire molecule. Two
alternative explanations are possible for this. In the first
mechanism, binding of the ligand shifts the conformational
equilibrium in favor of one conformation having a higher free
energy than the native one and a reduced motility and capable
of binding to the ligand (Markus et al., 1968; Koshland, 1970;
Anfinsen and Scheraga, 1975). In the second, the ligand-ligand
binding site interaction is linked to the cooperative interactions operative throughout the molecule and these two linked
interaction systems strengthen each other without a large change
in conformation (Taniuchi and Bohnert, 1975). This second
mechanism would provide the protein molecule with means by
which the interactomic interactions could be strengthened without a large change in conformation. Since this hypothetical
mechanism is important for protein folding, the concept is
discussed below.

Nuclease contains 149 residues and is devoid of sulfhydryl
groups and disulfide bonds (Cone et al., 1971; Bohnert and
Taniuchi, 1972). An enzymically active derivative of nuclease,
Nuclease-T, is composed of two fragments, Nuclease-T-(6-48) and

Nuclease-T-(50-149) (or Nuclease-T-(49-149)), held together
by non-covalent interactions (Taniuchi and Anfinsen, 1968;
Taniuchi, Davies and Anfinsen, 1972). The studies of exchange
between ^{14}C-labelled Nuclease-T-(50-149) incorporated in Nuclease-
T and free non-labelled Nuclease-T-(50-149) (Taniuchi, 1973)
showed that Nuclease-T is maintained in the equilibrium of
folding and unfolding under physiological conditions. The
presence of the ligands, deoxythymidine-3',5'-diphosphate and
calcium ion, does not accelerate folding (Light, Taniuchi and
Chen, 1974) but strongly suppresses unfolding (Taniuchi and
Bohnert, 1975).

     Using this system, it is observed (Taniuchi and Bohnert,
1975) that binding of Nuclease-T to either ligand shows apparently
one association constant when measured in the absence of or in
near saturation with a second ligand and that liganded Nuclease-
T unfolds without prior dissociation of ligands as indicated by
the increase in activation free energy of unfolding due to binding
of ligands being fractional to the magnitude of the decrease in
Gibbs standard free energy by binding of ligands (Table 1).
The greater the decrease in Gibbs standard free energy by binding
of ligands, the greater the increase in conformational stabilizing
energy (Table 1). It is also shown (Taniuchi and Bohnert,
1975) that the two ligands cooperatively bind to Nuclease-T
(Table 1) which is consistent with earlier qualitative observations
(Cuatrecasas, Fuchs and Anfinsen, 1967).

     The cooperative binding of two ligands is generally explained
as shifting the conformational equilibrium by binding of one
ligand in favor of one perturbed conformation capable of binding
to a second ligand without another energy consuming conformational
change (Koshland, 1970). Since no further change in conformation
is assumed upon binding of a second ligand and liganded Nuclease-
T unfolds without dissociation of ligands and thereby the
energy barrier of unfolding would not involve the ligand binding
interactions, the rate of unfolding should remain unchanged
upon binding of a second ligand. However, the increase in the
activation free energy of unfolding is not smaller when Nuclease-
T containing one ligand is bound with a second ligand than when
unliganded Nuclease-T is bound with the same ligand (Table 1).
In order to explain this observation on the basis of conformational
equilibrium, one may assume that one ligand, when present
alone, binds to one conformation incapable of binding to a
second ligand and that a third perturbed conformation, having
a further decreased rate of unfolding and incapable of binding
to either ligand alone, binds to the two ligands in combination
with a high affinity. However, in the three-dimensional structure
of liganded nuclease (Arnone et al., 1971) the two ligands are
shown to be too far from each other to have any significant
interaction between them. Furthermore, if this assumption of

TABLE 1

Relationship between changes of activation free energy ($\Delta F^{\ddagger}$) of unfolding of Nuclease-T and Gibbs standard free energy ($\Delta G°$) by binding of ligands to Nuclease-T at 20°C.[a]

Ligands	R^{b} s^{-1}	$\Delta F^{\ddagger}$ kcal/mol	K^{c} M^{-1}	$\Delta G°$ kcal/mol
none	4.6×10^{-4}	21.6		
pdTp	9.0×10^{-5}	22.6	1.0×10^{4}	-5.4
Ca^{2+}	1.6×10^{-4}	22.2	2.0×10^{2}	-3.1
pdTp in presence of Ca^{2+}	2.1×10^{-5}	23.4	4.0×10^{5}	-7.5
Ca^{2+} in presence of pdTp	2.3×10^{-5}	23.4	1.4×10^{4}	-5.6

[a] Data are taken from the previous report (Taniuchi and Bohnert, 1975).
[b] Rate constant of unfolding of Nuclease-T at 20°.
[c] Association constant of ligands at 20°.

simultaneous binding of the two ligands is correct, the association constant of one of the two ligands measured in the presence of the other ligand should be equal to each other in contrast to the observations (Table 1).

An alternative explanation which we propose is to assume the ligand binding interaction involving one ligand is linked with that involving a second ligand through cooperative interactions operating in the protein molecule and that this hypothetical interaction linkage strengthens all the interactions involved in the linkage, resulting in additional stabilization of the molecule without a large change in conformation (Taniuchi and Bohnert, 1975). Thus, the stabilizing energy due to this hypothetical interaction linkage is not localized but distributed throughout the three-dimensional structure. By analogy, we speculate that only after completion of folding this hypothetical interaction linkage, the physical reality of which is unknown, would be operational in the molecule and should strengthen otherwise weak interatomic interactions throughout the

three-dimensional structure (Taniuchi and Bohnert, 1975). In
this view the hypothetical interaction linkage is assumed to
provide the energetic constraint underlying the conformational
cooperativity (Taniuchi, Parker and Bohnert, 1976). In other
words, without this linkage there would be no conformational
cooperativity. In this connection it should be mentioned that
to explain cooperativity in hemoglobin Hopfield (1973) has
proposed the distribution or delocalization of strain energy
which Wyman (1964) treated using linked functions.

ENERGY BARRIER OF FOLDING

If protein folding is initiated by assembly of the native
formats of several segments, "nucleation" (Rose, Winter and
Wetlaufer, 1976), the folding rate would be related to the
frequency of the native formats and thereby their thermodynamic
stability (Anfinsen and Scheraga, 1975; Karplus and Weaver,
1976). For example, if the native formats are heat labile, the
folding rate would decrease with increasing temperature.
However, Epstein et al. (1971) have shown that the initial
renaturation rate of acid denatured nuclease is independent of
temperature from 13 to 38°. The folding rate of Nuclease-T
from its two component fragments follows apparent first order
kinetics and is also independent of temperature from 5 to 45°
(Light, Taniuchi and Chen, 1974).

On the other hand, if disordered nuclease would go from
one conformation to another in the ensemble with little change
in energy, driven by any small perturbation, until the native
conformation would be reached, then, the energy barrier of
folding would be entropic and independent of temperature provided
that the number of conformations in the ensemble does not
change with temperature. If this hypothesis is correct, the
rate of folding would not be related to the decrease in energy
occurring after completion of folding by development of the
hypothetical linkage of interatomic interactions. This prediction has been tested as follows (Taniuchi and Bohnert, 1973;
Parker, Davis and Taniuchi, 1975; Taniuchi, Parker and Bohnert,
1976).

The interaction of Nuclease-(1-126) and Nuclease-T-(50-
149) simultaneously forms two alternative, enzymically active,
complementing structures, type I and II, in 1 min (Taniuchi and
Anfinsen, 1971) (Fig. 2). The complementation is quantitative
and the equilibrium between type I and II complementing structures
is attained through unfolding and refolding after the initial
complementation. Type II complex folds 3 times faster than type I
complex and the folding rate is independent of temperature and
ligands (Taniuchi, Parker and Bohnert, 1976). The ratio of type
I to type II complex at the equilibrium state increases with

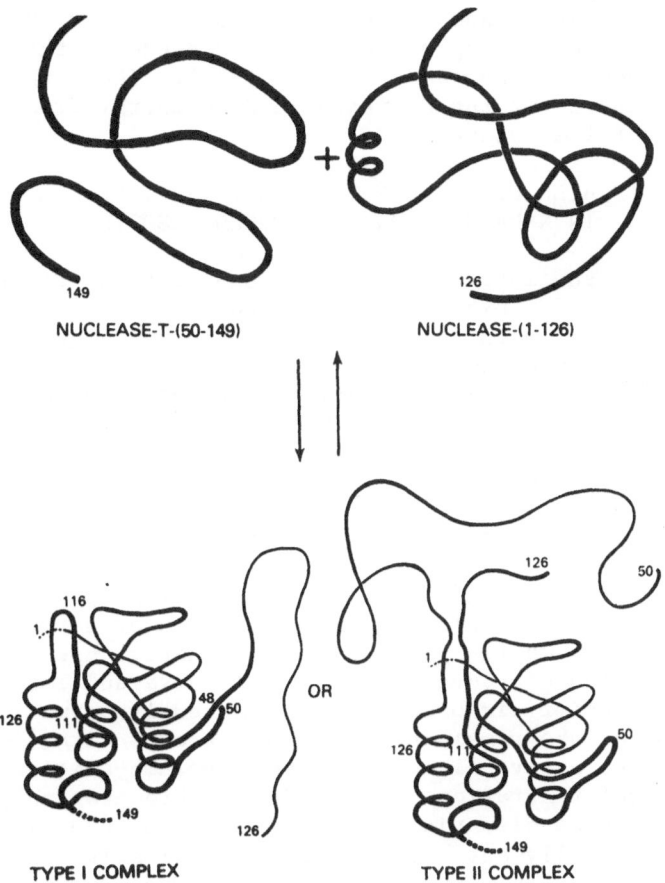

NUCLEASE-T-(50-149)   NUCLEASE-(1-126)

OR

TYPE I COMPLEX   TYPE II COMPLEX

Figure 2.   The simultaneous formation of type I and II
complexes by complementation of Nuclease-(1-126) and Nuclease-
T-(50-149).   In type I structure, Nuclease-T-(50-149) binds to
the portion of Nuclease-(1-126) that contains residues 6 to 48.
In type II, Nuclease-(1-126) combines with the portion of
Nuclease-T-(50-149) that contains residues 111 to 149.   The
redundant amino acid sequences flexibly protrude from the
ordered portions of the complexes without interfering with the
ordered structure (Taniuchi and Anfinsen, 1971).   The ordered
structure of type I has no overlap.   Similarly, the ordered
structure of type II is assumed to have no overlap and the
discontinuity of the polypeptide chain in the ordered structure
may occur between residues 114 and 124 (Taniuchi and Anfinsen,
1971; Parikh, Corley and Anfinsen, 1971).   The simultaneous
formation of type I and II complexes from the two fragments
follows first order kinetics (Light, Taniuchi and Chen, 1974).
Presumably a disordered complex of the two fragments is formed
as an intermediate.

increasing temperature and its value at 23° (greater than two) indicates that type I complex has lower free energy than type II complex (Taniuchi, Parker and Bohnert, 1976). Thus, the folding rate is independent of the decrease in free energy from the un-folded to the folded state.

INTERMEDIATES CONTAINING THREE NATIVE DISULFIDE BONDS IN RENATURATION OF REDUCED LYSOZYME

If only in the late phase of the renaturation process of reduced lysozyme, for example, the native-like conformation would be associated with the lowest free energy, a question would arise as to how many native disulfide bonds would have to be formed before the native or native-like molecule could be energetically differentiated from the non-native forms. This problem has been studied as follows (Acharya and Taniuchi, 1976a and b).

Initially it was found that monomeric RNase containing non-native disulfide bonds elutes sooner than native RNase by gel filtration (Andria and Taniuchi, 1971). This procedure is also found to be applicable for the separation of native-like species of hen egg white lysozyme from non-native species formed during reoxidation of reduced lysozyme (Acharya and Taniuchi, 1976a). After brief air oxidation of reduced lysozyme followed by alkylation of the remaining sulfhydryl-groups with $[1-^{14}C]$iodo-acetic acid, the native-like hydrodynamic volume species was separated by gel filtration. Ion exchange chromatography of this native-like species yielded completely renatured lysozyme and three enzymically active, radioactive derivatives containing one or two mol of S-carboxymethylcysteine. The examination of radioactive tryptic peptides from each radioactive derivative indicated that these three radioactive species contain one open disulfide bond in the native pair of half-cystine residues between Cys 6 and Cys 127, Cys 76 and Cys 94 or Cys 64 and 180 (Acharya and Taniuchi, 1976a) (Fig. 3). Preliminary observations showed that the two active derivatives containing one open disulfide bond between Cys 6 and Cys 127 or Cys 76 and 94 can be partially renatured (for example, approximately 50% in the case of the former derivative) after reduction and oxidation in the presence of β-mercaptoethanol (Acharya and Taniuchi, 1976a and b). Since it is reasonable to assume that the three disulfide bonds in each of these radioactive derivatives are all native pairs of half cystine residues, it appears that the mechanism responsible for the distinct lowering of the energy of the native conformation is operational in the molecule containing three native disulfide bonds and one open disulfide bond. The fourth derivative containing one open disulfide bond between Cys 30 and Cys 115 (Fig. 3) has not been found in this experiment.

STRUCTURE I   STRUCTURE II

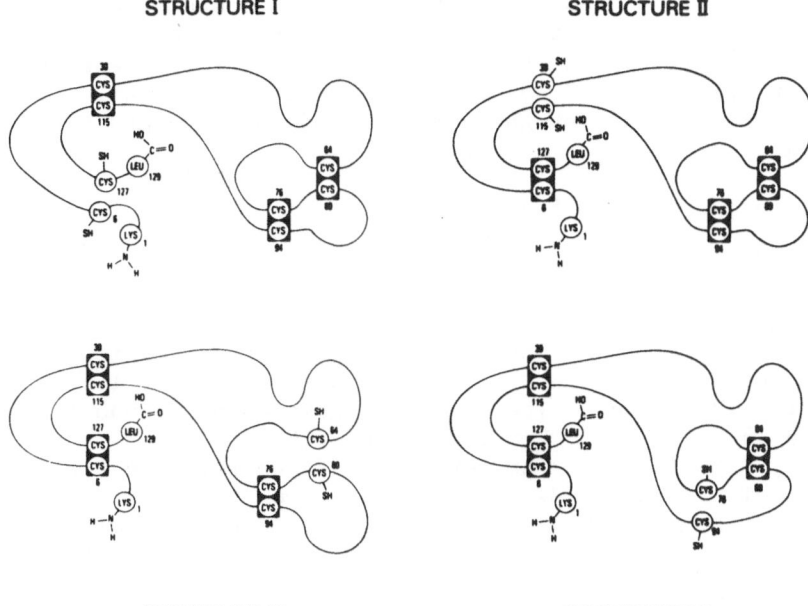

STRUCTURE III   STRUCTURE IV

Figure 3. Illustration of the four isomers of native lysozyme containing three native disulfide bonds and one open disulfide bond on the basis of Canfield's presentation of the amino acid sequence (Canfield and Liu, 1965).

In the native three-dimensional structure (Blake et al., 1965) the Cys 30–Cys 115 bonds is the only disulfide bridge having a helix on both sides. However, the formation of this fourth derivative has been indicated (Acharya and Taniuchi, 1976b) as follows.

Reduced lysozyme partially carboxymethylated with less than one mol [1-^{14}C]iodoacetic acid was completely air oxidized in the presence of β-mercaptoethanol. The native-like hydrodynamic volume derivatives containing less than one mol radioactive S-carboxymethyl cysteine were separated by gel filtration from the non-native higher hydrodynamic volume species of the oxidation product. The examination of radioactive tryptic peptides from the native-like derivatives showed that all the eight isomeric forms of monocarboxymethylated lysozyme having three disulfide bonds and one free sulfhydryl group are present in this sample. Clearly all the four isomeric forms of lysozyme with one open disulfide bond possess the lower hydrodynamic volume structure (Fig. 3). The three native disulfide bonds are formed in any reduced derivatives having one of the eight cysteinyl residues carboxymethylated. Thus, it would follow that none of the four native disulfide bonds are obligatory in the formation of the

other three native disulfide bonds. This is consistent with
the hypothesis that in the early phase of renaturation the
native format is not associated with the lowest free energy and
thereby pairing of half cystine residue is statistical. It is
possible that only after the formation of the intermediates
containing three native disulfide bonds and two sulfhydryl
groups the native-like conformation may be associated with the
lowest configurational free energy.

REFERENCES

Anfinsen, C.B. (1955). Studies on the structural basis of ribo-
    nuclease activity. Biochim. Biophys. Acta, 17, 141-142.
Anfinsen, C.B. (1956). The limited digestion of ribonuclease with
    pepsin. J. Biol. Chem., 221, 405-412.
Anfinsen, C.B. (1967). The formation of the tertiary structure of
    proteins. Harvey Lectures, 61, 95-116.
Anfinsen, C.B. and Scheraga, H.A. (1975). Experimental and
    theoretical aspects of protein folding. Advan Prot. Chem.,
    29, 205-300.
Andria, G. and Taniuchi, H. (1971). Renaturation of randomly
    oxidized des-(119-124)-ribonuclease by disulfide interchange
    in the presence of the fragment, 105-124. Fed Proc, 30, 1288.
    The full account is in preparation.
Acharya, S.A. and Taniuchi, H. (1976a). A study of renaturation
    of reduced hen egg white lysozyme: enzymically active inter-
    mediates formed during oxidation of the reduced protein. J.
    Biol. Chem., in press.
Acharya, S.A. and Taniuchi, H. (1976b). The manuscript is in
    preparation.
Arnone, A., Bier, C.J., Cotton, F.A., Day, V.W., Hazen, E.E., Jr.,
    Richardson, D.C., Richardson, J.S. and, in part, A. Yonath
    (1971). A high resolution structure of an inhibitor complex
    of the extracellular nuclease of Staphylococcus aureus. I.
    Experimental procedures and chain tracing. J. Biol. Chem.,
    246, 2302-2316.
Blake, C.C.F., Koenig, D.F., Mair, G.A., North, A.C.T., Phillips,
    D.C. and Sarma, V.R. (1965). Structure of hen egg white
    lysozyme. A three-dimensional Fourier synthesis at 2 Å
    resolution. Nature, 22, 757-761.
Bohnert, J.L. and Taniuchi, H. (1972). The examination of the
    presence of amide groups in glutamic acid and aspartic acid
    residues of staphylococcal nuclease (Foggi strain). J. Biol.
    Chem., 247, 4557-4560.
Canfield, R.E. and Liu, A.K. (1965). The disulfide bonds of egg
    white lysozyme (Muramidase). J. Biol. Chem., 240, 1997-2002.
Cone, J.L., Cusumano, C.L., Taniuchi, H. and Anfinsen, C.B. (1971).
    Staphylococcal nuclease (Foggi strain). II. The amino acid
    sequence. J. Biol. Chem., 246, 3103-3110.

Creighton, T.E. (1974). The single-disulfide intermediates in the refolding of reduced pancreatic trypsin inhibitor. J. Mol. Biol. 87, 603-624.

Cuatrecasas, P., Fuchs, S., and Anfinsen, C.B. (1967). The binding of nucleotides and calcium to the extracellular nuclease of Staphylococcus aureus: studies by gel filtration. J. Biol. Chem., 242, 3063-3067.

Epstein, C.J., Goldberger, R.F. and Anfinsen, C.B. (1963). Genetic control of tertiary protein structure: studies with model systems. Cold Spring Harbor Symposia on Quant. Biol., 28, 439-449.

Epstein, H.F., Schechter, A.N., Chen, R.F. and Anfinsen, C.B. (1971). Folding of staphylococcal nuclease: kinetic studies of two processes in acid denaturation. J. Mol. Biol., 60, 499-508.

Givol, D., Goldberger, R.F. and Anfinsen, C.B. (1964). Oxidation and disulfide interchange in the reactivation of reduced ribonuclease. J. Biol. Chem., 239, 3114-3116.

Gutte, B., Lin, M.C., Caldi, D.G. and Merrifield, R.B. (1972). Reactivation of des(119-, 120-, or 121-124) ribonuclease A by mixture with synthetic COOH-terminal peptides of varying lengths. J. Biol. Chem., 247, 4763-4767.

Harrington, W.F. and Sela, M. (1959). A comparison of the physical chemical properties of oxidized and reduced alkylated ribonuclease. Biochim. Biophys. Acta, 31, 427-434.

Hantgan, R.R., Hammes, G.G. and Scheraga, H.A. (1974). Pathways of folding of reduced bovine pancreatic ribonuclease. Biochemistry, 13, 3421-3431.

Habor, E. and Anfinsen, C.B. (1961). Regeneration of enzymic activity by air oxidation of reduced subtilisin-modified ribonuclease. J. Biol. Chem., 236, 422-424.

Hopfield, J.J. (1973). Relation between structure, co-operativity and spectra in a model of hemoglobin action. J. Mol. Biol., 77, 207-222.

Karplus, M. and Weaver, D.L. (1976). Protein-folding dynamics. Nature, 260, 404-406.

Kartha, G., Bello, J. and Harker, D. (1967). Tertiary structure of ribonuclease. Nature, 213, 862-865.

Kato, I. and Anfinsen, C.B. (1969). On the stabilization of ribonuclease S-protein by ribonuclease S-peptide. J. Biol. Chem. 244, 1004-1007.

Koshland, D.E. (1970). The molecular basis for enzyme regulation. "The Enzymes" P.D. Boyer, ed., Academic Press, New York, Vol. 1, pp. 341-396.

Light, A., Taniuchi, H. and Chen, R.F. (1974). A kinetic study of the complementation of fragments of staphylococcal nuclease. J. Biol. Chem., 249, 2285-2293.

Lin, M.C. (1970). The structural roles of amino acid residues near the carboxyl terminus of bovine pancreatic ribonuclease A. J. Biol. Chem., 245, 6726-6731.

Markus, G., Barnard, E.A., Castellani, B.A. and Saunders, D. (1968). Ligand-induced conformational changes in ribonuclease. J. Biol. Chem., 243, 4070-4076.

Ottesen, M. and Stracher, A. (1960). Deuterium exchange of sub-tilisin-modified ribonuclease and pepsin-inactivated ribo-nuclease. Compt.-rend. Lab. Carlsberg. Ser. Chim., 31, 457-467.

Parikh, I., Corley, L., and Anfinsen, C.B. (1971). Semisynthetic analogues of an enzymically active complex formed between two overlapping fragments of staphylococcal nuclease. J. Biol. Chem., 246, 7392-7397.

Parker, D.S., Davis, A. and Taniuchi, H. (1975). Determination of the difference in energy between two alternative enzymically active structures formed by two overlapping fragments of staphylococcal nuclease. Fed. Proc. 34, 597.

Phillips, D.C. (1967). The hen egg-white lysozyme molecule. Proc. Nat. Acad. Sci. U.S.A., 57, 484-495.

Richards, F.M. and Vithayathil, P.J. (1959). The preparation of the subtilisin-modified ribonuclease and the separation of the peptide and protein components. J. Biol. Chem., 234, 1459-1465.

Richards, F.M. and Wyckoff, H.W. (1971). Bovine pancreatic ribo-nuclease. "The Enzymes", P.D. Boyer, ed., Academic Press, New York, Vol. 4, pp. 647-806.

Ristow, S.S. and Wetlaufer, D.B. (1973). Evidence for nucleation in the folding of reduced hen egg lysozyme. Biochem. Biophys. Research Commun., 39, 544-550.

Rose, G.D., Winters, R.H. and Wetlaufer, D.B. (1976) A testable model for protein folding. FEBS Letters, 63, 10-16.

Sela, M. and Anfinsen, C.B. (1957). Some spectrophotometric and polarimetric experiments with ribonuclease. Biochim. Biophys. Acta, 24, 229-235.

Tanford, C. (1968). Protein denaturation. Advan. Protein. Chem., 23, 122-282.

Taniuchi, H. and Anfinsen, C.B. (1968). Steps in the formation of active derivatives of staphylococcal nuclease during trypsin digestion. J. Biol. Chem., 243, 4778-4786.

Taniuchi, H. (1970). Formation of randomly paired disulfide bonds in des-(121-124)-ribonuclease after reduction and reoxidation. J. Biol. Chem., 245, 5459-5468.

Taniuchi, H. and Anfinsen, C.B. (1971). Simultaneous formation of two alternative enzymically active structures by complementa-tion of two overlapping fragments of staphylococcal nuclease. J. Biol. Chem., 246, 2291-2301.

Taniuchi, H., Davies, D.R. and Anfinsen, C.B. (1972). A comparison of the X-ray diffraction patterns of crystals of reconstituted Nuclease-T and of native staphylococcal nuclease. J. Biol. Chem., 247, 3362-3364.

Taniuchi, H. and Bohnert, J.L. (1973). Regulation of the comple-
    mentation of two overlapping fragments of stpahylococcal
    nuclease. Fed. Proc., 32, 458.
Taniuchi, H. (1973). The dynamic equilibrium of folding and un-
    folding of Nuclease-T'. J. Biol. Chem., 248, 5164-5174.
Taniuchi, H. and Bohnert, J.L. (1975). The mechanism of stabili-
    zation of the structure of Nuclease-T' by binding of ligands.
    J. Biol. Chem., 250, 2388-2394.
Taniuchi, H., Parker, D.S. and Bohnert, J.L. (1976). A study of
    equilibration of the system involving two alternative, enzy-
    mically active complementing structures simultaneously formed
    from two overlapping fragments of staphylococcal nuclease.
    J. Biol. Chem., in press.
Venetianer, P., and Straub, F.B. (1964). The mechanism of action
    of the ribonuclease-reactivating enzyme. Biochim. Biophys.
    Acta, 89, 189-190.
Watson, J.D. (1976). "Molecular Biology of the Gene." 3rd Edition,
    W.A. Benjamin, Inc., New York.
Wyckoff, H.W., Hardman, K.D., Allewell, N.M., Inagami, T., John-
    son, L.N. and Richards, F.M. (1967). The structure of ribo-
    nuclease-S at 3.5 A resolution. J. Biol. Chem., 242, 3984-
    3988.
Wyman, J., Jr. (1964). Linked functions and reciprocal effects in
    hemoglobin: a second look. Advan. Prot. Chem., 19, 224-286.

5

DISULFIDE BONDS: KEY TO WHEAT PROTEIN FUNCTIONALITY

F. R. Huebner, J. A. Bietz, and J. S. Wall

Northern Regional Research Center, Agricultural
Research Service, U.S. Department of Agriculture,
Peoria, Illinois 61604

ABSTRACT

Disulfide bonds in wheat proteins are major factors
that determine the properties of the proteins and their
functionality in wheat flour. The gliadin proteins contain
mostly intramolecular disulfide bonds. In contrast, the
high-molecular-weight glutenins are formed by disulfide
linkages of several diverse polypeptide chains which have
been separated and characterized. The linkage of these
proteins in a fairly linear array contributes to the unique
viscoelastic properties of glutenin. The glutenin has been
separated into two fractions differing in molecular weight.
The amount of highest molecular weight component is correlated
with the rheological behavior of the flours from different
wheat varieties. Various oxidizing and reducing agents are
widely used to alter the functional behavior of wheat proteins
by the action on sulfhydryl and disulfide groups.

INTRODUCTION

Wheat has been valued as a food for thousands of years because of its ability to form doughs and produce leavened bread products. The unique cohesive-elastic properties of wheat flour doughs are due primarily to the water-insoluble gluten protein complex which can be concentrated from dough by washing out the starch. Other components in flour such as lipids, polysaccharides, and low-molecular-weight (MW) compounds also influence dough properties. During the early 1900's, it was discovered that adding small quantities of oxidizing or reducing agents to flour could often change the physical characteristics of doughs to make them more suitable for bread making (Kent-Jones, 1939). The pioneering work of Sullivan and coworkers (1940) suggested that disulfide bonds in cystine residues of wheat gluten proteins were essential to dough functionality and that the flour additives effected modification of these bonds during dough mixing.

Pence and Olcott (1952) found that addition of reducing agents to solutions of isolated wheat gluten proteins reduced the viscosity of the solutions. They concluded that reducing agents cleaved disulfide bonds which produced smaller protein subunits. Research in the past 2 decades has revealed that disulfide bonds link certain smaller proteins into a fraction of gluten protein called glutenin. Intermolecular and intramolecular disulfide bonds determine the molecular conformation of these and other proteins in wheat. All of the proteins of gluten tend to associate strongly in neutral aqueous solutions by noncovalent forces due to their high levels of nonpolar and amide-containing residues. Current evidence indicates that the higher MW components of glutenin are most active in such associations and therefore have a major effect on dough quality and strength. In this paper, we shall seek to characterize the proteins of wheat gluten and to indicate how disulfide bonds contribute to their structure and properties.

GLUTEN PROTEINS

Fractionation. As early as 1907, Osborne recognized that wheat gluten protein was heterogeneous and devised a method of fractionating it based on differences in solubility. Wheat proteins are now routinely extracted with saline solutions to obtain albumins and globulins; then stirring with 70% ethanol removes gliadin proteins and, finally, vigorous extraction with dilute acetic acid yields glutenin. Of the protein in flour 10-15% is saline soluble, 35% is gliadin, 35% is glutenin, and the remainder is designated residue protein. These amounts vary by up to 8% depending on the variety and extraction method used. Some cross contamination of proteins in these extracts is common, so it is essential that reprecipitation or re-extraction of fractions be undertaken to improve their purity (Bietz and Wall, 1975).

Gel filtration chromatography separates the gluten proteins according to MW differences. The gluten extract is resolved into four distinct peaks by Sephadex G-100 filtration using a dilute acetic acid solvent (Fig. 1). The high MW components which elute first are the glutenins, the second peak contains gliadins, the third consists of albumins and globulins, and the last peak consists of nonprotein low MW material. Analysis of the 70% ethanol (gliadin) extract by gel filtration also reveals that it is not homogeneous relative to MW (Beckwith et al., 1966). Thus gel filtration serves as an additional means of fractionating gluten, and also establishes the existence of new protein classes in gliadin extracts.

Gliadins. Gliadin is a very heterogenous mixture, but individual proteins have been isolated and characterized. Moving boundary electrophoresis resolved gliadin into four fractions: α, β, γ, and ω (Jones et al., 1959). Later use of starch-gel electrophoresis demonstrated that each of these components could be further separated into subfractions of different mobilities designated as α_1, α_2, etc. (Woychik et al., 1964). When chromatographed on Sephadex G-100, gliadin separated into two major peaks differing in MW (Beckwith et al., 1966). The first fraction was designated high MW gliadin. Upon starch-gel electrophoresis, it caused streaking near the origin rather than a discrete band and has, therefore, also been called low MW glutenin. A small fraction consisting of ω gliadins eluted between high MW gliadin and the following peak which contained the α, β, and γ components.

Removal of high MW gliadin and ω gliadin facilitated chromatographic separation of the remaining gliadins on columns of sulfoethyl cellulose (SEC) (Huebner and Wall, 1966). Using dimethyl formamide as solvent and a salt gradient for elution, an extensive fractionation was obtained which led to isolation of the first purified gliadin proteins

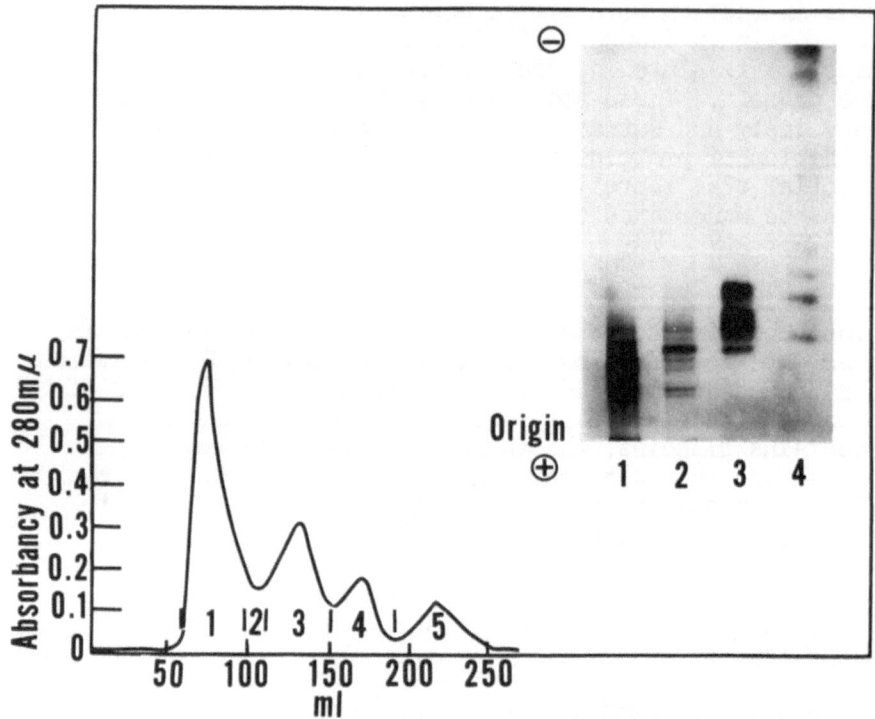

Fig. 1. Left: Sephadex G-100 filtration of Wells
gluten. Buffer: 0.2M acetic acid; fractions combined from
area between vertical lines and numbered 1 to 5. Right:
Starch-gel electrophoresis of fractions.

(Fig. 2) (Huebner et al., 1967; Huebner and Rothfus, 1968).
Most gliadins have very similar amino acid composition; they
are rich in glutamine and proline, but contain little lysine.
Molecular weights of γ_1 and γ_3 gliadin in 3M urea pH 3.1
were 30,300 and 34,700, respectively (Sexson and Wu, 1972);
and most other gliadins have similar MWs (Bietz and Wall,
1972).

Omega gliadins, however, have MWs of about 73,000; some
contain no cystine or methionine (Booth and Ewart, 1969)
while other ω gliadins (Charbonnier, 1971) isolated by ion-
exchange chromatography contained cystine but too few residues

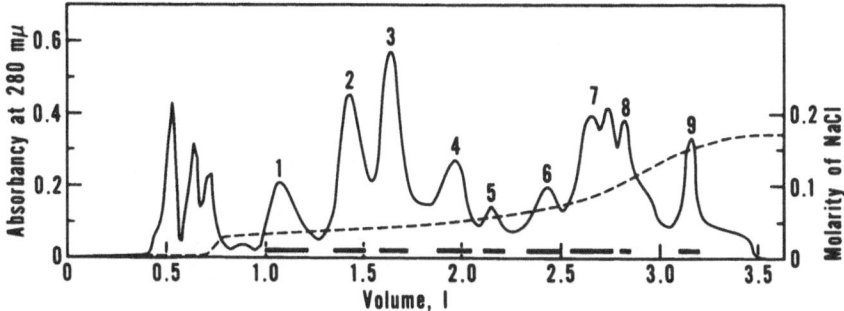

Fig. 2.   Chromatographic separation of gliadin on
sulfoethyl cellulose (SEC) Column, 3.9 X 50 cm, temp, 30 °C.
Buffer: 2M dimethylformamide (DMF), 0.03N acetic acid,
0.015N HCl, pH 2.2.   Broken line shows salt gradient.   Straight,
dark lines along the abscissa indicate fractions that were
pooled.

to permit intra- or intermolecular disulfide linkages.   High
MW gliadin was shown to have a MW of 104,000-125,000 (Beckwith
et al., 1966; Nielsen et al., 1968) and more cystine than
the other gliadins thereby permitting intermolecular disulfide
bonds.

Glutenin.   Glutenin is the high MW component of wheat
gluten and has been reviewed in detail by Bietz et al.
(1973) and Bushuk (1974).   Rothfus and Crow (1968) first
used gel filtration on polyacrylamide to try to fractionate
glutenin into components of different MW, but obtained only
partial separation based on size.   Bushuk and Wrigley (1971)
obtained fractions differing in MW by filtration on Sephadex
G-150, but resolution was restricted by the gel's limited
pore size relative to protein MW's.   By using 4% agarose gel
columns, Huebner and Wall (1976) have been able to fractionate
glutenin into two distinctly different fractions; one
having a MW of as high as 20 million and the other ranging
from 100,000 to 15 million (peaks I and II, Fig. 3).   Except
for the MW and physical properties, little difference in
amino acid composition was observed between these two fractions.
Glutenin was also separated into fractions by gel filtration
on Bio-Gel A-5m (Huebner and Rothfus, 1971) and by differential
solubility (Bietz and Wall, 1975).

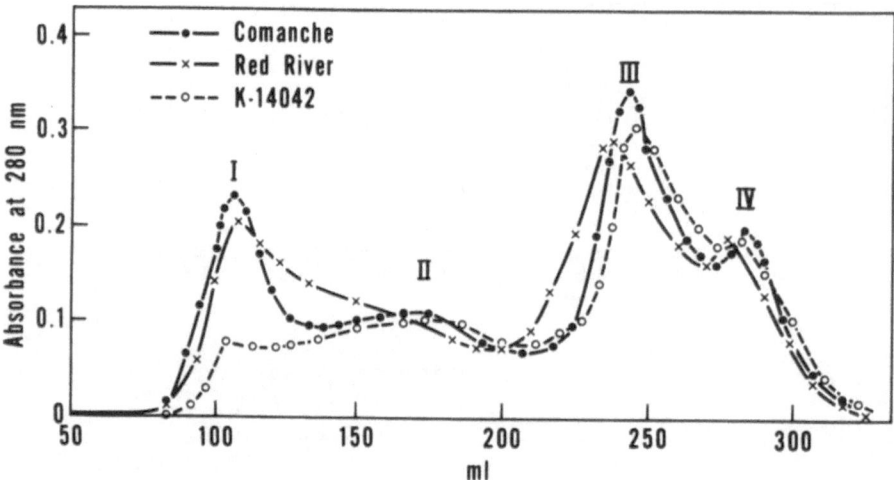

Fig. 3.   Gel filtration of 0.1N acetic acid, 3M urea, and 0.01N cetyltrimethylammonium bromide (AUC) extracts from three wheat flours on Sepharose 4B.   Solvent: 5.5M guanidine hydrochloride.

Residue Protein.   After acetic acid extraction of flour, 20 to 40% of the protein remains in the residue (Orth and Bushuk, 1972) (Bietz and Wall, 1975).   Meredith and Wren (1966) reported a solvent consisting of acetic acid, urea, and cetyltrimethylammonium bromide (AUC) extracted protein from flour more effectively than acetic acid.   Huebner and Wall (1976) found that only 9 to 23% of the flour protein, depending on variety, was not extracted by AUC.   The additional protein extracted from the residue by AUC was mainly high MW glutenin (Glut-I, Table I).

CLEAVING DISULFIDE BONDS

Reduction and Alkylation.   While oxidation of cystine to yield cysteic acid or reduction by sodium borohydride to cysteine disrupt disulfide linkages, these methods are generally not used because of their destructive action on other amino acids.   The preferred method of reducing protein disulfides employs disulfide interchange with excess amounts of sulfhydryl agents such as 2-mercaptoethanol or dithiothreitol. Reduction is conducted at pH 7.5 to 8.0 in a strong dissociating

TABLE I

Relative Contents of Different Protein Fractions in Wheat Flour Differing in Functionality

Wheat	Mixing[a]	Loaf volume	% Total protein	% of total protein in fractions				
				Glut-I	Glut-II	Gliadin	Albumin + globulin	Insoluble
HRW[b]								
Ponca	Med. Long	Good	10.5	16.2	22.8	33.3	14.3	15.2
Comanche	Long	Good	11.0	15.5	18.0	32.0	13.0	21.0
K-14042	Very Short	Very poor	11.7	6.8	25.0	38.4	15.4	9.4
K-501099	Short	Poor	13.6	6.6	26.4	50.0		9.6
66-2558	Med. Long	Good	12.3	16.3	17.9	37.4	12.2	13.8
66-2560	Medium	Good	14.0	15.0	19.3	43.6	9.3	12.2
HRS								
Red River	Very Long	Poor	12.7	17.3	13.4	33.0	11.8	22.8
Rescue	Long	Good	16.5	12.1	18.8	36.4	12.1	18.2
SRS								
Chinese Spring	Short	Poor	21.2	10.5	21.7	42.0		10.9
Durum								
Wells	Medium	Poor	14.6	17.1	19.2	41.2	11.0	9.6

[a]Mixing times. Long = 5.5-7.5 min, medium long = 4-5 min, medium = 3-3.5 min, short = 2-2.5 min, and very short = 1-1.5 min.
[b]HRW = hard red winter; HRS = hard red spring; SRS = soft red spring.

solvent to insure complete unfolding and deaggregation of
proteins to permit quantitative reaction. Because cyanates
formed in 8M urea may cause carbamylation of proteins (Cole
and Mecham, 1966), 6M guanidine hydrochloride is the preferred
solvent. To prevent reoxidation, sulfhydryls are blocked by
alkylation with compounds such as iodoacetamide, acrylonitrile
(Woychik et al., 1964), ethyleneimine (Rothfus and Crow,
1968) (now used infrequently because of its carcinogenicity),
or 4-vinyl pyridine (Friedman et al., 1970). Choice of
alkylating agent varies with subsequent research objectives.
For example, pyridylethyl derivatives of reduced glutenin
will have electrophoretic mobilities in an acid buffer
greater than derivatives prepared from acrylamide (Huebner
and Wall, 1974). Also, pyridylethyl derivatives separate
better on cation exchange columns and pyridylethyl cystine
may be accurately determined upon amino-acid analysis after
hydrolysis of the protein.

    Mecham et al. (1972) observed that mercuric chloride
had a solubilizing effect on the residue or "gel" protein of
wheat flour. Danno et al. (1975) have established that the
mercuric chloride cleaved disulfide bonds. Thus, the solubilizing
action of mercuric chloride appears to be due to its ability
to split disulfide bonds.

    Effect on Proteins. The MWs of α, β, and γ gliadins
are not appreciably altered by reduction and alkylation.
Nielsen et al. (1968) reported that the weight average MW
determined by ultracentrifuge of this gliadin fraction was
26,900 and after reduction and alkylation, 22,300. Woychik
and Huebner (1966) reported that a purified γ gliadin had MW
of 37,000 before and after reduction. These data establish
that most gliadins contain only intramolecular disulfide
bonds. However, following reduction of disulfides in gliadin,
Nielsen et al. (1968) observed significant increase in
gliadin's axial ratio determined from viscosity data in
8M urea solutions. Value of f/f_0 increased from 2.33 to
3.26 after reduction. This difference indicates an unfolding
of the molecule. Evidently intramolecular disulfide bonds
maintain gliadin proteins in a more compact structure even
in 8M urea. Starch gel electrophoresis also indicates a
change in conformation after reduction. The alkylated
reduced protein exhibits lower mobilities, indicative of a
more random coil structure which hinders penetration of the
starch gel matrix (Beckwith et al., 1965).

In contrast, high MW gliadin decreases in MW from 125,600 to 36,800 after reduction and alkylation of disulfides (Nielsen et al., 1968). Intermolecular disulfide bonds must be involved in the structure of these proteins. The native protein chains may also have intramolecular disulfide bonds, since the frictional ratio in 8M urea shows a slight increase after reduction. Figure 4 compares changes in protein structure of purified gliadin and glutenin, or high MW gliadin, before and after reduction (Krull and Wall, 1969). Prior to reduction, high MW gliadin migrates in starch gel as indistinct streaks, but after reduction its pattern consists of several bands with mobilities similar to the gliadins (Nielsen et al., 1968).

Glutenin MW is drastically diminished when its disulfide bonds are cleaved, from values ranging in the millions to values of 11,000 to 133,000 (Bietz and Wall, 1972). The disulfide bonds that link the polypeptides of glutenin must be limited in number since viscosity studies indicate that glutenin molecules in denaturing solvents are highly extended and have high axial ratios (Wu et al., 1967). Since more extensive crosslinking would be expected to give rise to a less extended, less soluble protein network, residue protein may owe its properties to a greater degree of intermolecular disulfide bonding than that present in glutenin.

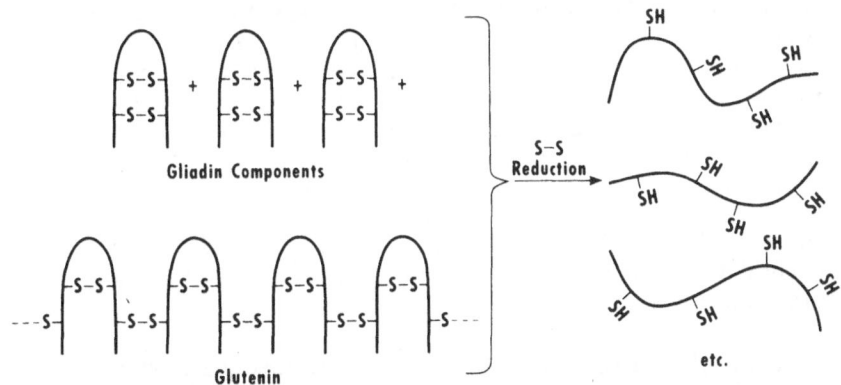

Fig. 4. Reduction of disulfide bonds of gliadin and glutenin.

Glutenin appears fibrous or rodlike by scanning electron microscopy (Orth et al., 1973). When its disulfide bonds are reduced and alkylated, the glutenin loses this fibrous structure and appears amorphous. These structural changes offer further evidence that disulfide bonds link the subunits of glutenin in long chains which would likely associate to form fibers.

Glutenin does not migrate from the origin upon starch gel electrophoresis (Woychik et al., 1961) due to its high MW. After reduction and alkylation, the resulting polypeptide chains migrate in the gel with mobilities resembling those of gliadins. It was then postulated that glutenin was a disulfide-linked polymer consisting of subunits of gliadin; however, later evidence showed this to be incorrect.

Subunits. Polyacrylamide gel electrophoresis (PAGE) in the presence of sodium dodecylsulfate (SDS) is an alternative and powerful tool for analyzing subunits of reduced glutenin and gliadins by MW differences (Fig. 5). Bietz and Wall (1972) first compared glutenin with low and high MW gliadin by this technique, and determined subunit MWs by comparison to reduced globular proteins. Most low MW gliadins had MWs near 37,000, while high MW gliadin consisted mainly of 44,000 MW subunits. However, glutenin migrated as many components with variation in MW from 11,600 to approximately 133,000. Interestingly, 44,000 MW subunits are prominent both in glutenin and in high MW gliadin. Recently, Hamauzu et al. (1975) reported that the MW of wheat protein subunits obtained by SDS PAGE are too high relative to values obtained by sedimentation equilibrium analysis. Probably the high proline content, which causes the molecule to be kinked or twisted, gave a false higher MW. However, the SDS PAGE values do provide good relative values of MW and can be corrected by reference to MW of these proteins determined by other methods. SDS PAGE, along with starch gel electrophoresis, indicates that the low MW gliadin and glutenin subunits are not identical.

To further characterize subunits of reduced glutenin, Huebner and Wall (1974) isolated them by a combination of gel filtration and ion exchange chromatography. S-Pyridylethyl derivatives of glutenin subunits were first separated on Sephadex G-200 with 4M urea, 0.03N acetic acid as solvent into three distinct fractions (A, B, and C) differing in MW (Fig. 6). Fractions B and C were further separated on columns of sulfoethyl cellulose to yield pure protein components. Huebner et al. (1974) determined considerable differences in amino acid content between the components of the three main

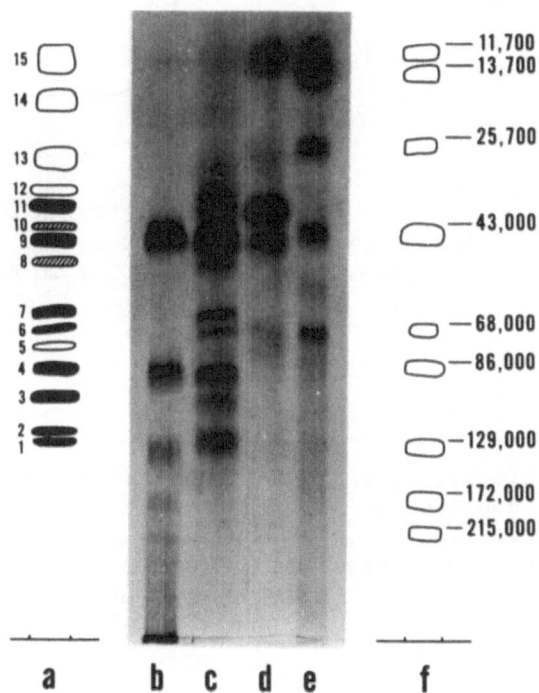

Fig. 5. Comparison of SDS-electrophoresis patterns of glutenin and gliadin with those of proteins of known MW's: (a) diagram and numbering system for glutenin subunits, with prominent bands shaded; (b) polymeric ovalbumin; (c) glutenin; (d) gliadin; (e) standard protein calibration mixture; and (f) composite diagram and MW's of components of calibration mixtures.

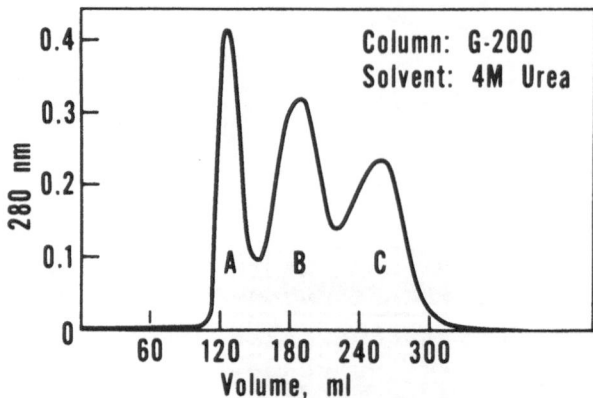

Fig. 6. Gel filtration of pyridylethyl glutenin (PE-Glu) from Ponca wheat. Column: Sephadex G-200, 2.5 X 72 cm; Solvent: 4M urea, 0.03N acetic acid; 50-mg sample.

fractions (Table II). Fraction A proteins were low in glutamic acid and proline but higher in basic amino acids. Of the B protein fraction, 72% consisted of glutamic acid, proline, and glycine. Fraction C protein resembled gliadin in amino acid content, but had more cystine. These data demonstrate further that α, β, and γ gliadins do not contribute to glutenin structure.

Reduced and alkylated glutenin may also be partially fractionated on the basis of solubility in neutral 70% ethanol (Bietz and Wall, 1973). The soluble fraction (62%) consists mainly of 44,000 MW subunits, similar to reduced high MW gliadin. The 44,000 MW subunits were separated from both fractions by gel filtration on Sephadex G-200 (Fig. 7) and compared. The amino acid analysis of the two 44,000 MW fractions were similar. This finding also supports the concept that a major subunit in reduced high MW gliadin is similar or identical to a subunit of glutenin.

The amino acid content of the insoluble residue protein (Cluskey and Dimler, 1967) is similar to fraction A of reduced glutenin (Huebner et al., 1974). These data suggest that besides comprising part of glutenin, components in addition to those of 44,000 MW may exist as separate entities in wheat flour.

TABLE II

Amino Acid Composition of Fractions from
Reduced and Alkylated Ponca Wheat Glutenin[a]

Amino acid	A-1	A-2	B-1	B-2	B-6	HC-3	Whole PE-Glu[b]
Tryptophan	1.2	0.6	0.8	0.9	0.5	0.2	0.5
Lysine	3.3	2.4	0.8	0.9	1.0	0.7	1.5
Histidine	2.0	1.9	0.5	0.6	1.6	1.5	1.6
Arginine	3.8	3.1	1.2	1.5	2.1	1.9	2.4
Aspartic acid	6.8	5.2	0.7	0.7	1.1	1.5	2.7
Threonine	4.1	3.8	2.9	3.2	3.3	2.7	3.2
Serine	7.0	8.3	6.2	7.2	7.4	8.9	7.4
Glutamic acid	20.9	25.2	37.8	37.2	36.9	37.7	33.3
Proline	9.3	10.3	14.4	12.7	11.7	14.9	13.0
Glycine	10.6	6.2	19.4	19.2	14.2	3.6	8.4
Alanine	6.5	5.2	2.0	2.4	2.6	2.3	3.4
Half-cystine	0.9	1.7	0.4	0.5	1.3	2.2	1.8
Valine	5.4	5.6	1.8	1.8	2.8	4.3	4.0
Methionine	1.7	1.8	0.6	0.5	0.9	1.7	1.4
Isoleucine	3.6	3.8	0.6	0.7	2.0	3.6	3.0
Leucine	3.8	8.0	4.3	4.1	5.0	7.3	6.5
Tyrosine	3.9	2.9	5.4	5.8	4.4	1.4	3.0
Phenylalanine	3.6	4.5	0.4	0.5	1.6	4.6	3.6

[a] Values shown are as molar percent of the total protein.
[b] PE-Glu = S-pyridylethyl-glutenin.

FORMATION OF DISULFIDE BONDS

Reoxidation. Anfinsen (1962) concluded that the amino
acid sequence of reduced proteins determines their most
stable conformation in aqueous buffers and, therefore, the
specific intramolecular disulfide bonds that form on reoxidation
are similar to those of the native protein. To achieve
reoxidation of the reduced gliadins, the protein solution
was dialyzed to remove denaturing and reducing agents and
then air bubbled through it (Beckwith et al., 1965). For
restoration to the native state, the protein was reoxidized
in low concentration, 0.1% in dilute acetic acid. Under
these conditions, reoxidation yielded a product that contained
mainly the native protein, as evidenced by measurement of
MW, viscosity, and optical rotatory dispersion. Reoxidation
in high concentrations of urea which overcomes the natural
association of side chain residues yielded a product with
properties that indicated random disulfide reformation
occurred.

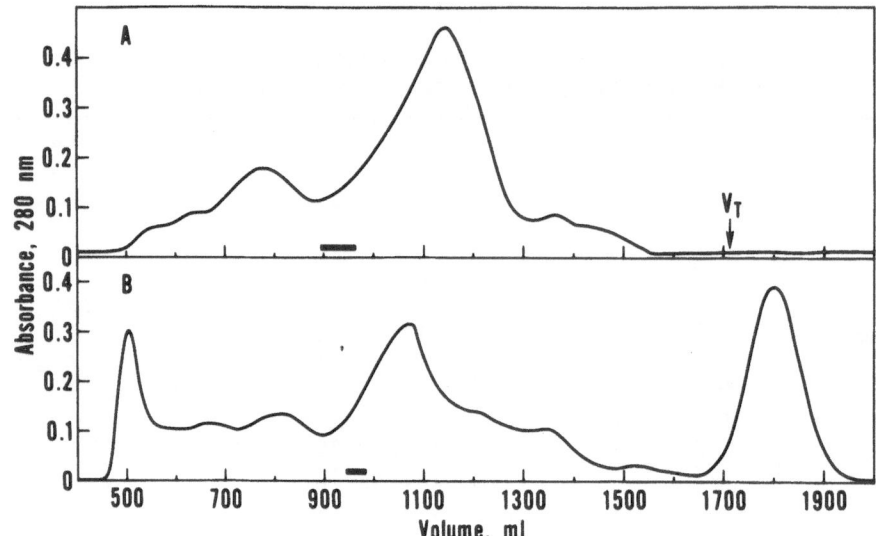

Fig. 7. Sephadex G-200 elution patterns of A,
312 mg of ethanol-soluble AE-glutenin; and B, 235 mg of
reduced high-MW gliadin. Tubes which were combined into
44,000 MW fractions are indicated on the abscissa. Total
column volume (V_T) is indicated. The peak in B eluting after
V_T contained only nonproteinaceous material.

In laboratory studies, protein concentration determines
whether intra- or intermolecular disulfide bonds are formed
upon reoxidation of reduced proteins. Reduction of intermolecular
disulfide bonds in glutenin decreases the sedimentation
constant (Beckwith and Wall, 1966). When the reduced glutenin
is reoxidized at 5.0% protein concentration, the resulting
products contain intermolecular disulfide bonds, as evidenced
by the sedimentation pattern, which is fairly broad but
indicates components of high MW. The viscoelastic properties
of this product resemble native glutenin. When reduced
glutenin is reoxidized at a concentration of 20%, the protein
is highly insoluble, inelastic, and extensively crosslinked.
Reduced glutenin can also be reoxidized at 0.1% concentration
to yield protein of low MW, but its reoxidation rate was
slow (11 days) compared to that of gliadin (3 days). Differences
between amino acid sequences and chain lengths of gliadin
and glutenin polypeptides may be responsible for the preferential
development of intermolecular bonds in glutenin.

Biosynthesis of Gliadin and Glutenin. Jennings et al. (1963) observed the formation of protein bodies in immature wheat endosperm by accumulation of gliadins within a lipoprotein membrane adjacent to ribosomes on the endoplasmic reticulum. After the grain matures, protein bodies cannot be distinguished. Simmonds (1971) has examined mature wheat storage protein with the scanning electron microscope and observed that it contains remnants of membranes, endoplasmic reticulum and protein bodies. Since glutenin consists of diverse protein subunits differing in solubility, amino acid composition, and molecular size, it is probable that they were synthesized at different sites.

The question remains as to when and how these varied proteins become linked by disulfide bonds. Bushuk and Wrigley (1971) analyzed the grain proteins of three wheat varieties by gel filtration at various stages of development. They showed that high MW proteins are laid down at all stages of kernel development. Of special interest was a low MW glutenin which appeared at a late stage of kernel development in bread wheats but not in a durum which is unsatisfactory for breadmaking. Thus, some subunit polymerization through intermolecular disulfide bonding occurs promptly after synthesis of the primary protein chains, but, as Simmonds (1971) states, in the mature kernel the proximity of various protein groups also provides ample opportunity for additional interactions as soon as water is introduced.

FLOUR PROTEINS AND BAKING PERFORMANCE

Varietal Differences. Hard wheats generally have more protein than soft wheats, and consequently form stronger doughs which perform better in baking. However, some high-protein wheats such as durums do not make good bread. Huebner (1970) found that glutenins of various wheats differed in their tendency to associate. Glutenins from better quality bread wheats tended to be salted out from solution more readily than others. However, glutenins from durum wheats and some soft wheats were also very salt sensitive.

Tsen (1967) first showed by gel filtration that weak flours have more acetic acid extractable glutenin than do flours from strong varieties. Orth and Bushuk (1972) compared the relative amounts of extracted protein fractions from 26 wheat varieties with varied baking qualities. They found a definite negative correlation between baking quality and percentage of acetic acid extractable glutenin but a positive correlation to percent of residue protein. Huebner and Wall

(1976) extracted more glutenin protein by using the AUC buffer, and by using agarose gel filtration determined that the ratio of high MW glutenin (Glut-I plus insolubles) to lower MW glutenin (Glut-II) was greater in the stronger wheats (Fig. 8). These findings were consistent with those of Orth and Bushuk (1972). Huebner and Wall (1976) also observed (Table I) a correlation between mixing time and content of high MW glutenins plus residue proteins. These studies, as well as those on the salt susceptibility of glutenin (Huebner, 1970), agree that baking quality and strength are more dependent on high MW glutenin (including residue protein) than on lower MW acetic acid soluble glutenin.

Since the amino acid sequence in the polypeptide chains determines the folding and accessibility of sulfhydryls to oxidation to form disulfides, differences in these sequences must govern the tendency to form high MW glutenin proteins. Huebner (1970) compared glutenin subunits from a number of wheat varieties by starch gel electrophoresis. While there were differences among bread wheats, no definite relationship to dough quality was apparent. In durum wheats, two subunits were absent that were prominent in all bread wheats. Bietz and Wall (1972) used SDS PAGE to establish that these missing components in durums were three of the highest MW subunits of bread wheat (Fig. 9). An extensive survey, by SDS PAGE, of bread wheats of different origins and qualities, however, revealed little difference in subunit number and MW between most varieties (Bietz et al., 1975). However, according to most results, it appears that genetic factors governing the biosynthesis of wheat proteins are important determining factors in wheat quality.

Changes in Disulfides During Mixing. The quality and strength of wheat varieties are also influenced by cystine content and reactivity. As first suggested by Goldstein (1957), disulfide-sulfhydryl interchange is a mechanism for transferring the attachment of disulfide bonds from one protein chain to another as shown below.

$$P_1S\text{-}SP_2 + P_3SH \longrightarrow P_1S\text{-}SP_3 + P_2SH$$

This interchange permits either a strengthening of the protein networks by chain elongation or a relaxation of a too rigid structure. Tsen and Bushuk (1968) related the sulfhydryl and disulfide content of various wheat flours to their rheological behavior. Flours yielding strong doughs having long development times had a higher disulfide content and fewer sulfhydryl groups than did weak flours. Similarly,

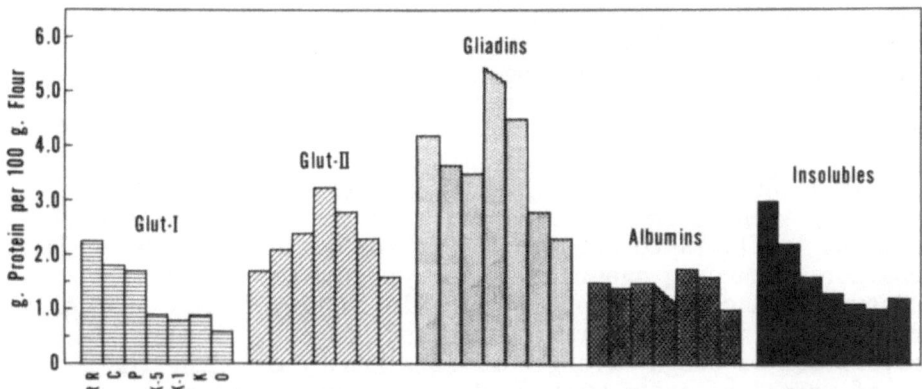

Fig. 8. Amounts of protein fractions for flours of selected varieties of wheats. RR = Red River, C = Comanche, P = Ponca, K-5 = K-501099, K-1 = K-14042, K = Knox, O = Omar.

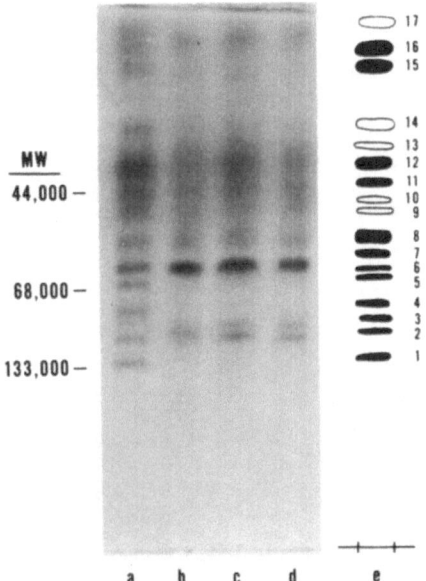

Fig. 9. SDS-PAGE at pH 8.9 of total protein extracts of (a) Chinese Spring wheat; (b) Wells durum; (c) Steward durum; (d) Mindum durum; and (e) diagram of Chinese Spring glutenin subunits.

Archer (1972) found that mixing time was negatively correlated
with the reduced glutathione level of flour.   Glutathione
and other low MW compounds may serve to catalyze disulfide
interchange, thereby promoting partial breakdown of large
glutenin molecules and rapid rearrangement into a gluten
network.

It is now widely accepted that maturing of flour and
oxidizing agents, such as bromate, serve to reduce sulfhydryl
contents in flours to optimum levels for suitable dough
behavior.   Reducing agents such as cysteine are used in
industry to accelerate dough mixing in continuous mix processes
or very long mixing flours (Henika and Rodgers, 1965).
Restoration of dough strength is obtained by coupling the
use of reducing agents with oxidizing agents.   Further
knowledge about the disulfide links in wheat proteins can
lead to new processing methods to facilitate optimum performance
of poor quality wheats, or will provide the means to genetically
improve new varieties.

REFERENCES

Anfinsen, C. (1962). The tertiary structure of ribonuclease. Brookhaven Symp. Biol., 15, 184.

Archer, M. J. (1972). Relationship between free glutathione content and quality assessment parameters of wheat cultivars. J. Sci. Food Agric., 23, 485-491.

Beckwith, A. C., Wall, J. S. and Jordan, R. W. (1965). Reversible reduction and reoxidation of the disulfide bonds in wheat gliadin. Arch. Biochem. Biophys., 112, 16-24.

Beckwith, A. C. and Wall, J. S. (1966). Reduction and reoxidation of wheat glutenin. Biochim. Biophys. Acta, 130, 155-162.

Beckwith, A. C., Nielsen, H. C., Wall, J. S. and Huebner, F. R. (1966). Isolation and characterization of a high-molecular weight protein from wheat gliadin. Cereal Chem., 43, 14-28.

Bietz, J. A. and Wall, J. S. (1972). Wheat gluten subunits: Molecular weights determined by sodium dodecyl sulfate-polyacrylamide gel electrophoresis. Cereal Chem., 49, 416-430.

Bietz, J. A. and Wall, J. S. (1973). Isolation and characterization of gliadin-like subunits from glutenin. Cereal Chem., 50, 537-547.

Bietz, J. A., Huebner, F. R. and Wall, J. S. (1973). Glutenin: The strength protein of wheat flour. Bakers Dig., 47, 26-35, 67.

Bietz, J. A. and Wall, J. S. (1975). The effect of various extractants on the subunit composition and associations of wheat glutenin. Cereal Chem., 52, 145-155.

Bietz, J. A., Shepherd, K. W. and Wall, J. S. (1975). Single-kernel analysis of glutenin: Use in wheat genetics and breeding. Cereal Chem., 52, 513-532.

Booth, M. R. and Ewart, J. A. D. (1969). Studies on four components of wheat gliadins. Biochim. Biophys. Acta, 181, 226-233.

Bushuk, W. and Wrigley, C. W. (1971). Glutenin in developing wheat grain. Cereal Chem., 48, 448-455.

Bushuk, W. (1974). Glutenin: Functions, properties and genetics. Bakers Dig., 48(4), 14-22.

Charbonnier, L. (1971). Fractionnement de l'ω-gliadine sur sulfoethylcellulose. C. R. Acad. Sci. Paris, t.272, 709-712.

Cluskey, J. E. and Dimler, R. J. (1967). Characterization of the acetic acid-insoluble fraction of wheat gluten protein. Cereal Chem., 44, 611-619.

Cole, E. G. and Mecham, D. K. (1966). Cyanate formation and electrophoretic behavior of proteins in gels containing urea. Anal. Biochem., 14, 215-222.

Danno, G., Kanazawa, K. and Natake, M. (1975). Cleavage of disulfide bonds of wheat glutenin by mercuric chloride. Agric. Biol. Chem., 39, 1379-1384.

Friedman, M., Krull, L. H. and Cavins, J. F. (1970). The chromatographic determination of cystine and cysteine residues in proteins as S-β-(4 pyridylethyl) cysteine. J. Biol. Chem., 245, 3868-3871.

Goldstein, S. (1957). Sulfhydryl- und disulfidgruppen der klebereiweisse und ihre beziehung zur backfahigkeit der brotmehle. Mitt. Lebensm. Hyg. Bern., 48, 87.

Hamauzu, Z., Kamazuka, Y., Kanazawa, H. and Yonezawa, D. (1975). Molecular weight determination of component polypeptides of glutenin after fractionation by gel filtration. Agric. Biol. Chem., 39, 1527-1531.

Henika, R. G. and Rodgers, N. E. (1965). Reactions of cysteine, bromate and whey in a rapid breadmaking process. Cereal Chem., 42, 397-408.

Huebner, F. R. and Wall, J. S. (1966). Improved chromatographic separation of gliadin proteins on sulfoethyl cellulose. Cereal Chem., 43, 325-335.

Huebner, F. R., Rothfus, J. A. and Wall, J. S. (1967). Isolation and chemical comparison of different γ-gliadins from hard red winter wheat flour. Cereal Chem., 44, 221-229.

Huebner, F. R. and Rothfus, J. A. (1968). Gliadin proteins from different varieties of wheats. Cereal Chem., 45, 242-253.

Huebner, F. R. (1970). Comparative studies on glutenins from different classes of wheat. Agric. Food Chem., 18, 256-259.

Huebner, F. R. and Rothfus, J. A. (1971). Evidence for glutenin in wheat: Stability toward dissociating forces. Cereal Chem., 48, 469-478.

Huebner, F. R. and Wall, J. S. (1974). Wheat glutenin subunits. I. Preparative separation by gel-filtration and ion-exchange chromatography. Cereal Chem., 51, 228-239.

Huebner, F. R., Donaldson, G. L. and Wall, J. S. (1974). Wheat glutenin subunits. II. Compositional differences. Cereal Chem., 51, 240-249.

Huebner, F. R. and Wall, J. S. (1976). Fractionation and quantitative differences of glutenin from wheat varieties varying in baking quality. Cereal Chem., 53, 258-269.

Jennings, A. C., Morton, R. K. and Polk, B. A. (1963).
    Cytological studies of protein bodies of developing
    wheat endosperm. Aust. J. Biol. Sci., 16, 366-374.
Jones, R. W., Taylor, N. W. and Senti, F. R. (1959).
    Electrophoresis and fractionation of wheat gluten.
    Arch. Biochem. Biophys., 84, 363-367.
Kent-Jones, D. W. (1939). Modern Cereal Chemistry. 3rd
    ed. The Northern Publishing Co.
Krull, L. H. and Wall, J. S. (1969). Relationship of amino
    acid composition and wheat protein properties. Bakers
    Dig., 43(4), 30-39.
Mecham, D. K., Cole, E. W. and Ng, H. (1972). Solubilizing
    effect of mercuric chloride on the "gel" protein of
    wheat flour. Cereal Chem., 49, 62-67.
Meredith, O. B. and Wren, J. J. (1966). Determination of
    molecular weight distribution in wheat flour proteins
    by extraction and gel filtration in a dissociating
    medium. Cereal Chem., 43, 169-186.
Nielsen, H. C., Beckwith, A. C. and Wall, J. S. (1968).
    Effect of disulfide-bond cleavage on wheat gliadin
    fractions obtained by gel filtration. Cereal Chem.,
    45, 37-47.
Orth, R. A. and Bushuk, W. (1972). A comparative study of
    the proteins of wheats of diverse baking qualities.
    Cereal Chem., 49, 268-275.
Orth, R. A., Dronzek, B. L. and Bushuk, W. (1973). Studies
    of glutenin. IV. Microscopic structure and its relations
    to breadmaking quality. Cereal Chem., 50, 688-696.
Osborne, T. B. (1907). The proteins of the wheat kernel.
    Carnegie Institution, Washington, D.C.
Pence, J. W. and Olcott, H. S. (1952). Effect of reducing
    agents on gluten proteins. Cereal Chem., 29, 292-298.
Rothfus, J. A. and Crow, M. J. A. (1968). Aminoethylation
    and fractionation of glutenin. Evidence of differences
    from gliadin. Biochim. Biophys. Acta, 160, 404-412.
Sexson, K. R. and Wu, Y. V. (1972). Molecular weights of
    wheat γ_1- and γ_3-gliadins in various solvents. Biochim.
    Biophys. Acta, 263, 651-657.
Simmonds, D. H. (1971). Morphological and molecular aspects
    of wheat quality. Wallerstein Lab. Commun., 34, 17-
    31.
Sullivan, B., Howe, M., Schmalz, F. D. and Astleford, G. R.
    (1940). The action of oxidizing and reducing agents on
    flour. Cereal Chem., 17, 507-528.
Tsen, C. C. (1967). Changes in flour proteins during dough
    mixing. Cereal Chem., 44, 308-317.

Tsen, C. C. and Bushuk, W.   (1968).   Reactive and total
     sulfhydryl and disulfide contents of flours of different
     mixing properties. Cereal Chem., 45, 58-65.
Woychik, J. H., Boundy, J. A. and Dimler, R. J.   (1961).
     Starch gel electrophoresis of wheat gluten proteins
     with concentrated urea. Arch. Biochem. Biophys., 94,
     477-482.
Woychik, J. H., Huebner, F. R. and Dimler, R. J.   (1964).
     Reduction and starch-gel electrophoresis of wheat
     gliadin and glutenin. Arch. Biochem. Biophys., 105,
     151-155.
Woychik, J. H. and Huebner, F. R.   (1966).   Isolation and
     partial characterization of wheat γ-gliadin. Biochim.
     Biophys. Acta, 127, 88-93.
Wu, Y. V., Cluskey, J. E. and Sexson, K. R.   (1967).   Effect
     of ionic strength on the molecular weight and conformation
     of wheat gluten proteins in 3M urea solutions. Biochim.
     Biophys. Acta, 133, 83-90.

6

CHEMICAL STRATEGY FOR STUDYING THE ANTIGENIC STRUCTURES OF
DISULFIDE-CONTAINING PROTEINS: HEN EGG-WHITE LYSOZYME AS A MODEL*

M.Z. ATASSI

Department of Immunology, Mayo Medical School,
Rochester, Minnesota 55901, Department of Biochemistry,
University of Minnesota, Minneapolis, Minnesota 55455

ABSTRACT

A critical approach in studying the immunochemistry of proteins
relies on the preparation of a variety of long and overlapping peptides from the parent molecule. Disulfide-containing proteins are
usually inaccessible to proteolytic attack and therefore it is not
possible to prepare from such proteins a variety of peptides without rupturing the disulfides. Unfortunately, the preparations with
broken disulfides bear no immunochemical relationship to the native
proteins. These have until recently virtually eluded investigation.
To break the deadlock, we introduced a novel cleavage approach for
obtaining peptides from the native protein without rupturing the
disulfide bonds. Reversible citraconylation of the amino groups
induced conformational changes in the protein that were sufficient
to render it completely accessible to tryptic hydrolysis at the arginyl bonds. Complete cleavage at the lysyl bonds may be effected,
if desired, after deprotection of the amino groups which proceeds
quantitatively. By this approach we obtained peptides from lysozyme
which accounted for almost all (90%) of the immune reaction of native lysozyme. From these and the immunochemistry of numerous chemical derivatives of lysozyme, it was possible to derive the location of the antigenic sites of lysozyme. Two antigenic sites were
then accurately delineated. One was achieved by organic synthesis
of 9 disulfide peptides corresponding to various overlaps around
the disulfide 6-127. For the delineation of the second site (around
the disulfides 64-80 and 76-94), we devised (Atassi, Lee and Pai,
1976, Biochim. Biophys. Acta, 427, 745) an entirely novel "Surface-
Simulation" synthetic approach. In this, the conformationally-
adjacent residues constructing the site were linked directly via

peptide bonds into a single peptide which does not exist in native
lysozyme but simulates a surface region of it.   This unorthodox approach provides a novel and powerful technique for final delineation
of antigenic reactive sites (and perhaps other types of binding
sites) in native proteins.

I.   INTRODUCTION

Delineation of the antigenic structure of myoglobin, which
constituted the first antigenic structure of a protein to be completed, was recently achieved in our laboratories (see Fig.1) (Atassi,
1975).   However, myoglobin (although a globular protein) will serve
as an appropriate model only for proteins that do not contain disulfide bonds.   In the case of disulfide-containing proteins (e.g.
lysozyme, ribonuclease, albumin, etc.) their antigenic structures
have been extremely difficult to study.   The covalent cross-linking
of the protein by the disulfide bonds imparts on the molecule a
high structural stability and a 'tight' mode of folding which renders it almost completely inaccessible to cleavage procedures.   It
has already been shown (Atassi, 1972 and 1975) that one of the useful approaches in the delineation of a protein antigenic structure
depends on the isolation of a large variety of overlapping peptide
fragments representing various parts of the protein molecule.   However, because of the inaccessibility of these tight proteins it has
not been possible to apply the cleavage approach in a systematic
and effective manner.   Although it has been possible to obtain some
limited information from peptic and similar fragments, these accounted only for a very small portion of the reactivity of the intact
protein (see Section IV-A).   It became clear that these proteins
presented a major technical challenge.   Accordingly, for lack of a
better alternative, many investigators resorted to studying the
immunochemistry of protein derivatives with broken disulfide bonds
since such derivatives are completely accessible to cleavage procedures.   Examples of such studies are those on performic acid-
oxidized ribonuclease (Brown et al., 1959; Brown, 1962) and reduced
S-carboxymethylated lysozyme (Gerwing and Thompson, 1968; Young and
Leung, 1970).   However, these unfolded preparations are unfortunately
quite distinct immunochemically from the native parent protein. In
fact, the two proteins are entirely unrelated immunochemically
(Brown et al., 1959; Brown, 1962; Gerwing and Thompson, 1968; Young
and Leung, 1970; Lee and Atassi, 1973; Atassi et al., 1973).

---

* Abbreviations and terminology: ORD, optical rotatory dispersion; CD, circular dichroism. Lysozyme derivatives are abbreviated
for convenience and these abbreviations are described in Table 1 and
Section III.   In the present terminology, previously defined by
Atassi and Saplin (1968), an antigenic reactive site could incorporate one or more antigenic reactive regions which are close in
three-dimensional structure but might be distant in sequence.

In order to understand the complex primary and three-dimensional structural features of protein antigenic sites, only the native protein is the appropriate model for investigation, even though it posed a seemingly unsurmountable obstacle.

   Lysozyme represents a typical member of this class of disulfide-containing tight proteins. Its antigenic structure should be highly revealing as to the manner in which such proteins interact with their antibodies. When this work started early in 1967, the covalent structure of hen egg-white lysozyme had been previously determined (Canfield, 1963; Jollès et al., 1963, 1964 and 1965; Canfield and Liu, 1965). Also, its three-dimensional structure had been elucidated (Blake et al., 1965). Lysozyme is a single polypeptide chain of 129 amino acid residues and is internally cross-linked by

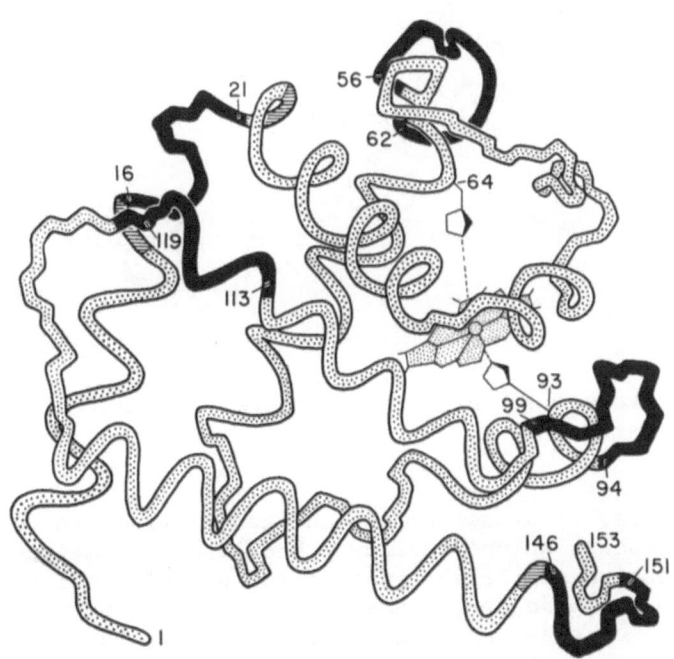

Fig.1. A schematic diagram showing the mode of folding of sperm-whale myoglobin and its antigenic structure. The solid black portions represent segments which have been shown to comprise accurately entire antigenic reactive sites . The striped parts, each corresponding to one amino acid residue only, can be part of the antigenic reactive site with some antisera. The dotted portions represent parts of the molecule which have been exhaustively shown to reside outside reactive sites (from Atassi, 1975).

four disulfide bonds (Fig.2), which stabilize the various folds of
the globular protein into a 'tight' proteolytically-inaccessible
mode of folding. Unlike myoglobin, this protein does not carry a
prosthetic group and has a much lower helical content.

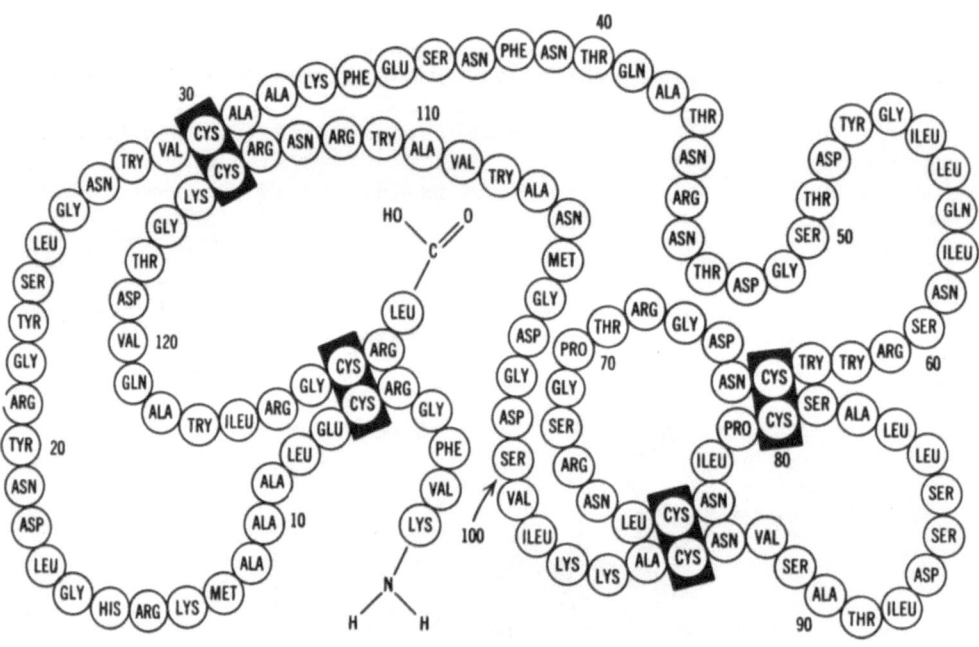

Fig.2. The covalent structure of lysozyme. (From Canfield and Liu,
1965, by permission).

     Our strategy of attack at the antigenic structure relied on
the five approaches which previously had been extremely effective
in the delineation of the antigenic structure of myoglobin (Atassi,
1972 and 1975). These approaches were : (1) to study the effect of
conformational changes on the immunochemistry of the protein; (2)
to isolate and characterize immunochemically-reactive fragments that
can quantitatively account for the total reaction of the native protein; (3) to study the immunochemistry and conformation of chemical
derivatives of lysozyme specifically modified at appropriate amino
acid locations; (4) to study the effect of chemical modification at
selected amino acid locations on the immunochemistry and conformation of immunochemically-reactive peptides; (5) after hopefully narrowing down each of the antigenic sites by approaches (1-4) to a
conveniently small region, the final delineation would rely on studying the immunochemistry of synthetic peptides corresponding to many

overlaps around this region. The application, usefulness and shortcomings of these approaches to protein immunochemistry have recently
been discussed in considerable detail (Atassi, 1976). It is also
necessary to mention here that none of these approaches by itself
is capable of yielding the full antigenic structure. We invariably
used the results from one approach to confirm and correct those from
the others. The complete structure is a composite logical synthesis
of all the information.

Initially, it was felt that the experience we have gained with
myoglobin should be quite valuable in our attack at the lysozyme
antigenic structure. Whereas this subsequently proved to be useful,
it soon became clear that not only is the lysozyme antigenic structure different in character from that of myoglobin, but it also demanded different concepts and necessitated the introduction of several novel and often unorthodox approaches to protein chemistry.
In the following sections, the findings derived from each of these
approaches will first be presented briefly, following which the information will be coordinated to derive our present position in delineation of the antigenic structure of the native protein.

II. IMMUNOCHEMISTRY AND CONFORMATION OF LYSOZYME
     DERIVATIVES WITH BROKEN DISULFIDE BONDS

Very recent studies from our laboratories (Atassi et al., 1973,
1976b, 1976c and 1976d) have shown that the disulfides are extremely
important in bringing into conformational proximity various parts
of each antigenic site from otherwise distant (in sequence) parts
of the molecule. In this section, therefore, some of the effects
of disulfide bond cleavage on the immunochemical properties of the
protein will be considered.

The complete cleavage of the disulfide bonds in lysozyme gives
rise (immunochemically speaking) to a new protein antigen that is
entirely unrelated to the parent native enzyme (Gerwing and Thompson,
1968; Young and Leung, 1970; Lee and Atassi, 1973). Antisera to
lysozyme will not react with the reduced S-carboxymethylated derivative (SCM-lysozyme) (Table 2). Similarly, antisera to SCM-lysozyme
will not react with lysozyme. Occasionally, small cross-reactions
have been reported by some workers, but these are most likely due
to the presence of some residual unmodified lysozyme in SCM-lysozyme.
Complete reduction of the disulfides in lysozyme is not always attained. The reaction of lysozyme with its antisera is not inhibited
by the tryptic fragments (Gerwing and Thompson, 1968; Young and
Leung, 1970; Atassi et al., 1973) or by the chymotryptic or cyanogen
bromide fragments (Young and Leung, 1970) of SCM-lysozyme.

The foregoing studies show that cleavage of the disulfides effects a complete disruption of the conformation and immunochemical properties of the protein in spite of the directive force of long-range interactions. Clearly, a satisfactory approach of the previously disulfide-linked regions is prevented by like-charge repulsion or by steric obstruction by the substituent or by both. Recently, we investigated (Lee and Atassi, 1973) the possibility of improving the reapproach of regions previously linked by disulfide bonds hoping to achieve a better approximation of the native three-dimensional structure of lysozyme without disulfide bond formation. This was shown to be feasible (Lee and Atassi, 1973) by eliminating the like-charge repulsion and minimizing the effect of steric hindrance by bulky side chains. Two lysozyme derivatives were prepared (Lee and Atassi, 1973) one in which the disulfides were reduced and then the resultant thiol grups carboxymethylated (SCM-lysozyme), and in the other reduction was followed by methylation (SM-lysozyme). ORD and CD measurements in water showed that both derivatives were greatly unfolded relative to native lysozyme, although the CD results indicated that SM-lysozyme was somewhat more folded than SCM-lysozyme. Conformational studies in increasing concentrations of methanol suggested that SM-lysozyme assumed, around 35% methanol, some stabilized structure whose ORD parameters approximated those of native lysozyme (Fig.3). In contrast, SCM-lysozyme showed no discretely stabilized structure in the range 0-60% methanol (Fig.3). This was further confirmed by immunochemical studies. SCM-lysozyme showed no reaction (0%) with antisera to lysozyme, while SM-lysozyme had appreciable (35-38%) cross-reaction with these antisera (Table 2). However, neither derivative had any enzymatic activity, suggesting that more rigid structural requirements are needed for this property than for immunochemical cross-reaction. These findings indicated that it was indeed feasible, at least to a limited extent, to effect a stabilized structure in SM-lysozyme due to the ability of the S-methyl groups to participate in non-polar interactions. In SCM-lysozyme, the directive effect of long-range interactions is ineffective because a refolded, stabilized structure is prevented by the like-charge repulsion and steric hindrance of the carboxy-methyl anions as they approach one another (Lee and Atassi, 1973).

It is relevant to emphasize that the complete loss of immunochemical reactivity upon cleavage of the disulfides is not a purely conformational effect resulting from the unfolding of the protein. It has been shown (see Sections VII and VIII) to arise from an actual scission of the reactive sites into two or more parts previously held together by the disulfide bonds (Atassi et al., 1973; Lee and Atassi, 1975; Atassi et al., 1976b, 1976c and 1976d).

III. IMMUNOCHEMISTRY AND CONFORMATION OF SPECIFIC CHEMICAL
      DERIVATIVES OF LYSOZYME

      Determination of the structural features responsible for the
antigenicity of the native protein and correlation of these with
the three-dimensional structure was our prime goal, and it was apparent from the foregoing that only the intact protein can be studied.
Critical information for the delineation of the antigenic structure

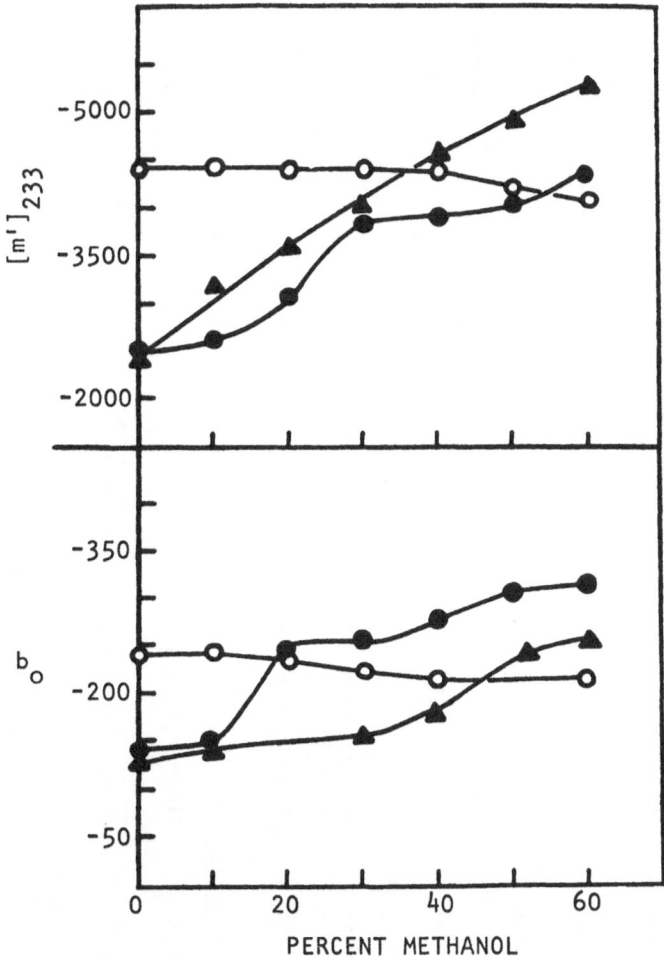

Fig.3. Effect of methanol on the rotatory behavior of (o) lysozyme;
(▲) SCM-lysozyme and (●) SM-lysozyme. The upper part shows change
of $[m']_{233}$ and the lower part shows change of b_o with increasing
methanol concentration. For description of the derivatives, see
Table 1. (Figure is from Lee and Atassi, 1975).

TABLE 1. Summary of Results from Some Chemical Derivatives of Lysozyme

Derivative	Residues modified	Conformational change	Immunochemical change +	Conclusion
A. Derivatives with broken disulfides				
1) SCM-lysozyme[a]	The 4 disulfides, reduced and carboxy-methylated	large	total	none made because of the large conformational change
2) SM-lysozyme[a]	The 4 disulfides, reduced and methylated	large	total	
B. Tyrosyl derivatives				
1) NT_2-lysozyme[b,c]	Tyr-20 & -23, nitrated	present	present	Tyr -20 and/or -23 at or near an antigenic site
2) AT_2-lysozyme[b,c]	Tyr-20 & -23, to aminotyrosine	present	none	
C. Tryptophan derivatives				
1) NPS_6-lysozyme[d,c]	6 tryptophans, with 2-nitrophenylsulfenyl chloride	large	very large	none made because of large conform. change
2) DISA-lysozyme[e]	Trp-123, with 2,3-dioxo-5-indolinesulfonic acid	none	none	Trp-123 not in antigenic site
D. Methionine derivatives				
1) CNBr-lysozyme[f]	Cleavage at Met-12 & -105, with CNBr	large	present	none made due to large conform. change
2) CE-lysozyme[g]	Met-12 & -105, carboxyethylated	none	none	Met-12 & -105 not in antigenic site
E. Arginine derivatives				
1) CHD-lysozyme I[h]	10 arginines, with cyclohexanedione in 0.1N NaOH	large	very large	none made because of large conform. change
2) CHD-lysozyme II[h]	10 arginines, with cyclohexanedione in 0.1M triethylamine	large	very large	
3) PG-lysozyme[h]	Arg-61, with phenylglyoxal	minor	none	Arg-61 not in antigenic site

F. __Amino group derivatives__

1) Gu_5-lysozyme[i]	5 amino groups, guanidinated	none	none	none, modification does not alter charge
2) Ac_7-lysozyme[i]	7 amino groups, acetylated	large	large	none made because of
3) ML_7-lysozyme[i,j]	7 amino groups, maleylated	large	large	the large
4) Su_7-lysozyme[i]	7 amino groups, succinylated	large	large	conformational
5) Su-lysozyme I[k]	Lys-1(α-$\delta\epsilon$-), 13, 97 & 116; -OH at 43 (or 36 or 40), succinylated	large	large	change
6) Su-lysozyme II[k]	Lys-1(α-$\delta\epsilon$-), 13, 96 & 116, succinylated	considerable	large	
7) Su-lysozyme III[k]	Lys-1(α-$\delta\epsilon$-), 13, 97 & 116, succinylated	considerable	large	
8) Su-lysozyme IV[k]	Lys-1(α-NH_2), 33, 96 & 116, succinylated	minor or none	present	one or more of Lys-33, -96 & -116 in antigenic sites
9) Su-lysozyme V[k]	Lys-1(α-NH_2), 33, 96, succinylated	minor or none	present	Lys-33 & -96 in antigenic sites
10) Su-lysozyme VI[k]	Lys-33 and 116, succinylated	minor or none	present	Lys-33 & -116 in antigenic sites

G. __Carboxyl group derivatives__

1) BH_3-lysozyme[l]	Asp-119 & Leu-129, reduced by BH_3	minor	none	Asp-119 & Leu-129 not in antigenic sites
2) CME_2-lysozyme[m]	Asp-119 & Leu-129, coupled to gly-methyl ester	minor	none	
3) HME_2-lysozyme[m]	Asp-119 & Leu-129, coupled to his-methyl ester	large	large	none made because of large conform. change

+ Some of the immunochemical results with antisera to native lysozyme are given in Tables 1 and 2.

For the immunochemical results with antisera to the derivatives, see references cited below.

References: (a) Lee and Atassi, 1973; (b) Atassi and Habeeb, 1969; (c) Atassi et al., 1971; (d) Habeeb and Atassi, 1969; (e) Atassi and Zablocki, 1976; (f) Johnson et al., 1976; (g) Atassi et al., 1976b; (h) Atassi et al., 1972; (i) Habeeb and Atassi, 1971b; (j) Habeeb and Atassi, 1971b; (k) Lee et al., 1975; (l) Atassi et al., 1975a; (m) Atassi et al., 1974; Atassi and Rosemblatt, 1974.

of lysozyme has been obtained (see Table 1 for summary) from the
immunochemical results of pure and well-characterized chemical derivatives of lysozyme (see Tables 1, 2 and 3), and which suffered
no conformational changes from the modification. The advantages
and shortcomings of this approach have been critically analyzed and
discussed and the chemistry of chemical modification and cleavage
reactions has been reviewed in detail (Atassi, 1976). In this section, the findings from the chemical derivatives of lysozyme (see
Table 1) will be briefly discussed.

A. Amino Group Derivatives

The amino groups in lysozyme were modified by guanidation,
acetylation, succinylation or maleylation (Habeeb and Atassi, 1971b)
and the specificity of each reaction was carefully determined. A
homogeneous guanidinated derivative, in which five amino groups were
modified (Habeeb and Atassi, 1971b), exhibited no conformational
changes and suffered no loss in enzymatic activity. Two homogeneous
acetylated derivatives were prepared and each was acetylated at all
seven amino groups, but one had seven and the other had fifteen hydroxy amino acids esterified. Succinylation or maleylation each
gave electrophoretically heterogeneous preparations, even when all
amino groups were modified. The heterogeneity was attributed to
non-uniform esterification of aliphatic hydroxy amino acids. The
acetylated, succinylated, or maleylated derivatives showed small,
but measurable, conformational changes. The enzymatic activity was
abolished upon modification of four or more amino groups and this
loss was attributed to changes in electrostatic interactions between the modified enzyme and the negatively charged bacterial cell
wall. Guanidination of five amino groups did not alter the antigenic reactivity with antisera to lysozyme, suggesting that an intact surface-charge distribution is essential for full antigenic
reactivity. However, the results from the succinylated and maleylated derivatives showed that this charge distribution may be partially disturbed with little effect on the immunochemical reactivity
whereas it was critical for enzymatic activity (Habeeb and Atassi,
1971b). Significantly, the two acetyl derivatives showed identical
immunochemical reactivity (78% relative to lysozyme) despite the
esterification of seven and fifteen aliphatic hydroxyl groups (Habeeb
and Atassi, 1971b). Partially or completely succinylated lysozymes
were antigenic in rabbits and antibodies to the completely succinylated derivative exhibited (a) a specificity against parts of lysozyme, (b) a specificity against antigenic regions containing succinyl groups and showing carrier specificity and (c) a haptenic
specificity against succinyl groups which reacted with unrelated
carrier proteins (Habeeb and Atassi, 1971b). Of interest was the
finding that fully maleylated lysozyme showed a lower (72%) reaction
with antisera to fully succinylated lysozyme indicating that the

TABLE 2. Immunochemical Reactivity of Some Chemical Derivatives of Lysozyme

Values represent per cent precipitation at equivalence relative to reaction of native lysozyme as 100 per cent. Results were obtained from three or more independent determinations which varied $\pm$ 1.3 per cent or less.

Protein or[+] derivative	Reaction with antiserum[*] (%)					
	G9	G10	HM	L1	L2	L7
Lysozyme	100	100	100	100	100	100
A. Derivatives with broken disulfides						
1) SCM-lysozyme[a,b]	0	0	0	0	0	-
2) SM-lysozyme[a]	35	38	-	36	-	-
B. Tyrosyl derivatives						
1) NT$_2$-lysozyme[c]	78.8	76.8	90.0	-	-	-
2) AT$_2$-lysozyme[c]	94.8	98.6	97.6	-	-	-
C. Tryptophan derivatives						
1) NPS$_6$-lysozyme[d]	10.5	9.8	12.0	8.6	-	-
2) DISA-lysozyme	97.5	99.1	-	100	-	99.0
D. Arginine derivatives						
1) CHD-lysozyme I[f]	8	19	20	-	-	-
2) CHD-lysozyme II[f]	9	21	22	-	-	-
3) PG-lysozyme[f]	100	99	99	-	-	-
E. Amino group derivatives						
1) Su-lysozyme I[g]	-	12.5	-	18.6	15.3	
2) Su-lysozyme II[g]	-	24.5	-	20.7	20.4	
3) Su-lysozyme III[g]	-	24.4	-	24.6	29.4	
4) Su-lysozyme IV[g]	42.5	42.8	-	41.1	40.0	
5) Su-lysozyme V[g]	53.7	56.4	-	55.5	57.4	
6) Su-lysozyme VI[g]	64.0	62.4	-	60.7	68.5	
F. Carboxyl group derivatives						
1) BH$_3$-lysozyme[h]	96.2	99.5	-	96.4	98.5	-
2) HME$_2$-lysozyme[i]	41	58	-	91	72	-
3) GME$_2$-lysozyme[i]	98.7	99.1	-	97.0	95.9	-

[+] For structural descriptions of the abbreviations see Table 1.
[*] These antisera are against native lysozyme. G9 and G10 are goat antisera and the other four are rabbit antisera. Antisera were also raised against the derivatives. For reactions with antisera to lysozyme derivatives, original references may be consulted.

References: (a) Lee and Atassi, 1973; (b) Atassi et al.,1973; (c) Atassi and Habeeb, 1969; (d) Habeeb and Atassi, 1969; (e) Atassi and Zablocki, 1976; (f) Atassi et al., 1972; (g) Lee et al., 1975; (h) Atassi et al., 1975a; (i) Atassi et al., 1974.

antibodies were capable of differentiating between the succinyl and
the maleyl group. On the other hand, the citraconyl group did not
react with antibodies to the succinyl substituent (Habeeb and Atassi,
1971b). From the foregoing results, it was concluded that at least
eight aliphatic hydroxyl groups and about three amino groups are
not parts of antigenic sites in native lysozyme (Habeeb and Atassi,
1971b).

      Succinylation of lysozyme in the presence of 7 molar excess of
$[1,4-^{14}C_2]$-succinic anhydride yielded a heterogeneous reaction product from which six homogeneous derivatives were isolated by column
chromatography (Lee, Atassi, and Habeeb, 1975). The locations of
the modifications in each derivative are shown in Table 1. Significantly, these findings indicated some differences in the side-
chain reactivity in solution and their expected accessibility from
the conformation of the protein in the crystalline state. Conformational changes were detectable only in derivative I by ORD and CD
measurements, and in each of derivatives I, II and III by accessibility of their disulfide bonds to reduction. On the other hand,
derivatives IV, V and VI showed little or no conformational changes.
Of the six succinyl derivatives, only derivative VI possessed some
(10%) enzymatic activity. The reactivity of each of the derivatives
with antisera to lysozyme was lower than the homologous reaction
(Table 2). Since conformational changes in succinyl derivatives
IV, V and VI were virtually absent, correlation of the extent of
decrease in their immunochemical reaction with the locations of
modifications led to the conclusion (Lee et al., 1975)that lysines
33, 96 and 116 are parts of antigenic reactive sites in lysozyme.

      In another study, the reaction of lysozyme with diketene and
tetrafluorosuccinic, maleic and citraconic anhydrides was investigated (Habeeb and Atassi, 1970). The results have been reviewed
elsewhere (Atassi and Habeeb, 1972) and will be mentioned only briefly here. Complete unmasking of the amino groups was not achieved
with acetoacetylated (by diketene), tetrafluorosuccinylated or maleylated derivatives. The deblocked preparations were highly heterogeneous in disc electrophoresis with partial recovery of enzymatic
activity, immunochemical properties and native conformation. In
contrast, the corresponding citraconyl derivative gave upon unmasking which proceeds readily (see Fig.4), homogeneous preparations
with full (100%) recovery of amino groups, enzymatic activity, immunochemical properties and native conformation. Citraconylation
of lysozyme introduced into the protein conformational changes (Habeeb
and Atassi, 1970) which formed the basis for a novel approach (Atassi
et al., 1973) to obtain all the tryptic peptides from lysozyme with
intact disulfide bonds (see Section IV-B).

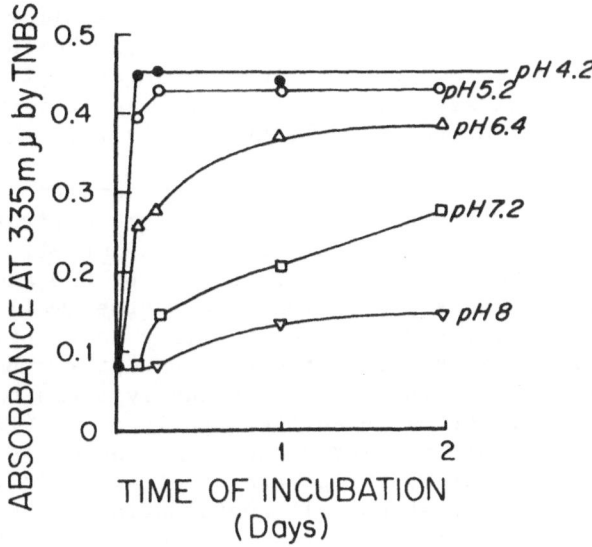

Fig.4. Removal of the citraconyl masking group from citraconyl-lysozyme at different pH values. (Figure is from Habeeb and Atassi, 1970).

B. Arginine Derivatives

The arginine residues in lysozyme were modified by reaction with 1,2-cyclohexanedione under two different conditions. One derivative was prepared by reaction in 0.1 N NaOH and the second by reaction in 0.1 M triethylamine (Atassi et al., 1972). Each derivative was modified at 10 arginine residues. The two derivatives had identical conformations indicated from their ORD and CD parameters, disulfide reducibility and susceptibility to chymotryptic attack, but were greatly unfolded relative to native lysozyme. They had no enzymatic activity and very little antigenic reactivity with antisera to lysozyme (Table 2). Conversely with antisera to one of the derivatives, lysozyme reacted only partially (60%), and these antisera showed appreciable (28%) cross-reaction with a de-

rivative of human serum albumin that had been similarly modified at
the arginine residues. Because of the extensive nature of the modifications and the large conformational changes that accompanied
them, no unequivocal assignment of residues to antigenic (and enzymatic) sites could be made (Atassi et al., 1972).

By investigating a variety of reaction conditions with phenylglyoxal, it was possible to prepare a lysozyme derivative that was
modified at one arginine (residue 61) (Atassi et al., 1972). The
derivative showed no conformational changes by ORD and CD measurements, but minor changes were detectable by the small increases in
disulfide reducibility and in susceptibility to tryptic attack.
Its enzymatic activity retained the same pH optimum but was slightly decreased (to 83%). The derivative and lysozyme had equal antigenic reactivities, both with antisera to lysozyme (Table 2) or to
the derivative. It was concluded that arginine 61 is not located
in an antigenic site of lysozyme (Atassi et al., 1972). Also, the
findings demonstrated that not all conformational changes will influence antigenic reactivity. This phenomenon was subsequently observed with other derivatives of lysozyme.

C. Carboxyl Group Derivatives

Two derivatives of lysozyme were prepared (Atassi et al., 1974)
by reaction with carbodiimide followed by coupling with glycine
methyl ester (GME) or with histidine methyl ester (HME). After
column chromatography, the two derivatives were homogeneous by
starch gel electrophoresis. Both derivatives were modified at aspartic acid 119 and at the C-terminal carboxyl group. However, in
one derivative the two carboxyl groups were coupled to GME (i.e.
GME_2-lysozyme) and in the other, they were coupled to HME (HME_2-
lysozyme). No changes were observed in the spectral and sedimentation behaviors of the two derivatives relative to lysozyme. The
ORD and CD parameters of lysozyme and GME_2-lysozyme were identical
throughout the pH range 2-11 (Atassi and Rosemblatt, 1974). At a
given pH, the corresponding parameters of HME_2-lysozyme were lower.
However, some conformational differences between lysozyme and GME_2-
lysozyme were detectable by the increase in accessibility to tryptic
hydrolysis and in disulfide reducibility. These parameters were
even more greatly increased in the HME_2-lysozyme derivative relative
to lysozyme, and the enzymatic activities of both derivatives were
drastically decreased (Atassi and Rosemblatt,1974). Lysozyme and
GME_2-lysozyme had equal antigenic reactivities, both with antisera
to the native protein (Table 2) or with antisera to the derivative.
On the other hand, HME_2-lysozyme had much lower reactions with these
antisera (Table 2). From these studies it was concluded that aspartic acid 119 and the terminal carboxyl group of leucine 129 are not
essential parts of an antigenic site in lysozyme (Atassi et al.,

1974). This conclusion was further confirmed by studies on yet
another derivative of lysozyme.

Diborane reduction followed by air oxidation of the reduced
disulfides and chromatography on CM-cellulose yielded a homogeneous
lysozyme derivative in which the carboxyl groups of aspartic acid
119 and the end-chain leucine residue were reduced to their corresponding alcohols (Atassi et al., 1975a). Correct disulfide pairing
was demonstrated by peptide mapping of the tryptic hydrolysates of
the derivative without breaking the disulfide bonds (see Section
IV-B), followed by chemical and immunochemical characterization of
the disulfide-containing peptides. Conformational differences between the derivative and lysozyme were almost undetectable by ORD
and CD measurements, but were readily detected by chemical monitoring of the conformation and appeared to be small. The lytic activity of the derivative decreased (to 52%) but retained the same pH
optimum. Lysozyme and the derivative possessed identical antigenic
reactivities, with antisera to either protein (see Table 2). These
findings further confirmed that aspartic acid 119 and the C-terminal
leucine are not part of an antigenic site in lysozyme (Atassi et al.,
1975a). Again it may be noted here that the slight conformational
change had no effect on the immunochemical properties.

D. Methionine Derivatives

Reaction of CNBr with lysozyme effected scissions at methionines
12 and 105 and because of the disulfide bonds the molecule remained
held together (Bonavida et al., 1969). Large conformational changes
were recently detected in the derivative by ORD and CD measurements
(Johnson et al., 1976). It is remarkable that despite the large
conformational changes the immunochemical reactivity of the derivative remained very high (Bonavida et al., 1969; Johnson et al., 1976)
indicating essentially the preservation of the intactness of the
antigenic sites. However, because of the large conformational changes, conclusions from this derivative concerning the role of the
methionines in antigenic sites are inadvisable.

It had been suggested (Strosberg and Kanarek, 1970) from carboxymethylation studies that one or both of the two methionine residues is involved in reaction of lysozyme with its antibodies. However, the two methionines in lysozyme are totally buried with regard
to all the atoms of their respective side chains (Shrake and Rupley,
1973). Even allowing for solution perturbation, dynamic fluctuation
and rearrangements induced upon the protein by antibody (Atassi, 1975)
and because of the rigidity of the lysozyme mode of folding, it remained difficult to visualize how either one of the two methionines
could participate in binding with antibody. The immune response to
native protein antigens is directed against their native, three-

TABLE 3. Antigenic Reactivities of Lysozyme and Derivative Carboxyethylated at the Two Methionines

Values are given in percent immune precipitation with antiserum at equivalence relative to reaction of native lysozyme. Results represent the average of 4 or 6 replicate determinations and varied ± 1.1% or less.

Antiserum	G9	G10	G531	G528	L1	L2	L110	HM
Species	Goat	Goat	Goat	Goat	Rabbit	Rabbit	Rabbit	Rabbit
Bleeding (No. of wks after 1st injection)	3	3	3	5	3	3	5	12
Protein:								
Lysozyme	100	100	100	100	100	100	100	100
Control Lysozyme*	100	99.2	n.d.	98.9	100	99.6	99.5	100
Derivative	98.6	100	100	97.7	99	100	101	101

*This preparation has been subjected to the same reaction conditions as those used for the preparation of the derivative, except in this case β-propiolactone was not added.

(Table is from Atassi et al., 1976b)

dimensional structure (Atassi and Thomas, 1969). A reinvestigation of the role of the methionines in native lysozyme was quite necessary to resolve the conflict.

The two methionine residues in lysozyme were specifically carboxyethylated by reaction with β-propiolactone (Atassi et al., 1976b), a reagent of high specificity for methionine (Taubman and Atassi, 1968; Atassi, 1969). The electrophoretically homogeneous derivative showed no conformational changes by ORD and CD measurements, but exhibited a slight increase in disulfide reducibility relative to native lysozyme. Its lytic activity was about half that of native lysozyme, probably as a result of the slight conformational change. The derivative and native lysozyme had identical antigenic reactivities (Table 3) with eight different rabbit and goat (both early-course and late-course) lysozyme antisera and overwhelmingly demonstrated that methionines 12 and 105 are not involved in interaction of lysozyme with its antibodies. The early conclusions of other workers (Strosberg and Kanarek, 1970) derived from carboxymethylation with iodoacetic acid were probably the result of nonspecificity problems inherent in this reaction. This is evidenced from the complete loss of enzymatic activity of the carboxymethyl derivative (Strosberg and Kanarek, 1970) which cannot otherwise be explained solely on the basis of methionine modification. Our conclusion derived from the carboxyethyl lysozyme derivative was further confirmed by evidence obtained from immunochemical study of synthetic peptides (see Section VII-A).

E. Tryptophan Derivatives

The six tryptophan residues in lysozyme were modified by reaction with 2-nitrophenylsulfenyl chloride in 98% formic acid (Habeeb and Atassi, 1969). This solvent effected the esterification of 12 (out of 17) hydroxy amino acids. The derivative suffered large conformational changes demonstrated by ORD and CD parameters, by increased susceptibility of its disulfides to reduction and by increased accessibility to proteolysis (Atassi et al., 1971). Moreover, binding of SDS to the derivative resulted in a constrained conformation accompanied by decreased availability of the disulfide bonds to reduction (from 2.6 to 1.3 disulfide bonds). Reactivity of the derivative with lysozyme antisera was minor (9-12%) (Table 2). Because of the large conformational changes, an unequivocal conclusion cannot be made from this derivative regarding the involvement of the tryptophans in antigenic sites. It was shown (Habeeb and Atassi, 1969) that a lysozyme control treated with 98% formic acid was esterified at 12 aliphatic hydroxyls. This control exhibited conformational changes (detected by availability of 0.7 disulfide bond to reduction and susceptibility to tryptic digestion) but retained 90% of its reactivity with lysozyme antisera and only 70%

of the enzymic activity. It was concluded that the decrease in both
the immunochemical reactivity and enzymic activity is due to conformational changes and that at least 12 hydroxy amino acid residues
are not involved in antigenic sites (Habeeb and Atassi, 1969).

Reaction of lysozyme with 2,3-dioxo-5-indolinesulfonic acid,
a reagent highly specific for tryptophan (Atassi and Zablocki, 1975),
yielded a homogeneous derivative which was modified at tryptophan
123 (Atassi and Zablocki, 1976). No conformational changes were
observed in the derivative by ORD and CD measurements, but some
slight changes were detectable by increases in accessibility to
tryptic hydrolysis and in disulfide reducibility relative to lysozyme. The lytic activity of the derivative was greatly decreased
(50%), probably as a result of the conformational change. However,
with several antisera to lysozyme the derivative and native protein
possessed equal reactivities (Table 2), indicating that the small
conformational change had no detrimental effect on the antigenic
reactivity. This is not unusual since previous studies on other
derivatives of lysozyme (Atassi and Habeeb, 1969; Atassi et al.,
1971, 1972 and 1974) had shown that not every conformational change
will be expected to influence the antigenic reactivity. The effect
will depend on the protein and on the nature of the conformational
change (Atassi and Habeeb, 1969). From this derivative it was concluded that tryptophan 123 is not part of an antigenic site in native lysozyme (Atassi and Zablocki, 1976).

F. Tyrosine Derivatives

Two derivatives of lysozyme modified at tyrosines 20 and 23 in
more than one way were prepared (Atassi and Habeeb, 1969). In one
derivative, tyrosines 20 and 23 were nitrated (NT_2-lysozyme) and in
the other the nitrotyrosine residues were reduced to aminotyrosine
(AT_2-lysozyme). Detailed conformational studies were performed
(Atassi et al., 1971) on these two derivatives. In ORD and CD
measurements, NT_2-lysozyme and AT_2-lysozyme had identical parameters
that were higher than the corresponding parameters of lysozyme.
The rotatory behavior of lysozyme and the derivatives showed no
change in the pH range 7 to 3. The conformational changes revealed
by ORD and CD measurements were also shown by increase in disulfide
reducibility and in accessibility to tryptic attack. Also, the effect of sodium dodecyl sulfate (SDS) on the availability of the disulfide bonds to reduction was determined. Both derivatives exhibited appreciable but equal disulfide reducibility relative to lysozyme. Changes in conformation obtained on binding of each of these
proteins with SDS revealed that lysozyme assumed a relaxed conformation. In the two tyrosyl derivatives, a relaxed conformation was
also observed in each case, and the SDS complexes of the derivatives
showed identical disulfide accessibility (Atassi et al., 1971).

Whereas lysozyme is completely inaccessible to tryptic hydrolysis, the two tyrosyl derivatives showed   appreciable and again equal degrees of accessibility. The foregoing information clearly indicated that the tyrosyl derivatives had closely similar, if not identical, conformations (Atassi et al., 1971). It is, of course, likely that minor differences in conformation may exist that cannot be detected by the present methods. Comparable enzymic activities were observed in NT_2-lysozyme (50%) and AT_2-lysozyme (56%), which may be explained by the similar conformational changes.

The antigenic reactivity decreased slightly in nitrated lysozyme (77-90% relative to homologous reaction) but was entirely recovered upon reduction of the nitrotyrosine residues to aminotyrosine (Table 2) despite the fact that conformational changes still existed (Atassi and Habeeb, 1969; Atassi et al., 1971). Conversely, lysozyme and AT_2-lysozyme reacted equally but less efficiently with antisera to NT_2-lysozyme than the homologous antigen (Atassi and Habeeb, 1969).

On nitration of the two tyrosyl residues ortho to the phenolic hydroxyl, the electron-withdrawing nitro group effects an increase in the acidity of the phenolic hydroxyl (cf. Pka values: tyrosine, 10.1, Edelhoch, 1962; 3-nitrotyrosine, 7.2, Sokolovsky et al., 1967). Hence, nitration would alter the properties of the tyrosyl residues sufficiently to influence their involvement, if any, in the biological activity of a protein (Atassi, 1968). This would explain the complete recovery of the antigenic reactivity (Atassi and Habeeb, 1969) upon restoration of the pK_a to its original value by reduction to 3-aminotyrosine (pK_a=10.0; Sokolovsky et al., 1967). From the foregoing results it was concluded (Atassi and Habeeb, 1969) that one or both of tyrosines 20 and 23 is located in or very close to an antigenic reactive site.

IV. IMMUNOCHEMISTRY OF PEPTIDE FRAGMENTS

The immunochemistry of a large number of peptide fragments with a variety of overlaps and representing various parts of the protein molecule affords a very effective approach for narrowing down of the antigenic reactive sites of the protein. The shortcomings of this approach and precautions to be observed in its application have been discussed in detail, together with a review and critical analysis of chemical cleavage reactions for proteins at given amino acid locations (Atassi, 1976).

A. Peptides Obtained by Peptic and
Other Cleavage Procedures

Even though lysozyme is not accessible to tryptic hydrolysis, it can be digested by pepsin (Canfield and Liu, 1965). Many investigators have exploited this susceptibility to isolate fragments from peptic digests of lysozyme and study their immunochemistry (Shinka et al., 1967; Fujio et al., 1968 a,b; Komatsu et al., 1975; Ha et al., 1975; Arnon and Sela, 1969; Maron et al., 1971). Some synthetic parts of such peptides have been studied (Arnon et al., 1971; Geiger and Arnon, 1974). Also, peptides prepared from digestion of lysozyme with thermolysin have been studied (Sakato et al., 1972). The results of these investigations, which have recently been reviewed elsewhere in detail (Atassi and Habeeb, 1977), show that they have been troubled by work with impure peptides, often yielding confusing and contradictory results. Also, the broad selectivity of the peptic digestion produced many intermediates which varied considerably with the conditions and made reproducibility difficult. Furthermore, these peptides accounted additively for only a small part (38%) of the lysozyme immune reaction (Fujio et al., 1968b). This situation has been frustrating in the search for immunochemically-reactive peptides from lysozyme.

B. A Novel Cleavage Approach that Yielded Fragments
Accounting for the Full Antigenic Reactivity

To break the aforementioned deadlock, a reproducible cleavage procedure of high specificity was needed that can yield a variety of peptides directly from the native protein without rupturing the disulfide bonds. To achieve this, we introduced a novel cleavage approach (Atassi et al., 1973). It has already been discussed (see Section III-A) that examination of several amino group reversible blocking reagents showed (Habeeb and Atassi, 1970; Singhal and Atassi, 1971) that citraconic anhydride was the most satisfactory. Based on the observation (Habeeb and Atassi, 1970) that reversible masking of the amino groups by citraconylation induced in the protein conformational changes which rendered it accessible to tryptic attack at the arginyl peptide bonds we introduced a novel cleavage approach for obtaining fragments with intact disulfide bonds from "tight" (i.e. disulfide-containing, proteolytically inaccessible) proteins (Atassi et al., 1973). The tryptic cleavage may be terminated, after scission of the arginyl bonds, by adding trypsin inhibitor before removal of the citraconyl masking groups at pH 4 (Fig. 4). If no trypsin inhibitor is added, then, following the removal of the protecting groups, cleavage of the lysyl bonds may be continued, if desired. By this approach, it was possible to effect the complete tryptic hydrolysis of lysozyme without rupturing the disulfide bonds (Atassi et al., 1973). The total tryptic hydrolysate showed substantial inhibitory activity (85-89%) of the reaction of lysozyme with its antibodies. The fragments responsible for this inhibition were identified mainly as the three disulfide-containing tryptic peptides: 22-33 (Cys 30-Cys 115) 115-116; 62-68 (Cys 64-Cys 80) 74-96 or 97 (Cys 76-Cys 94); and 6-13 (Cys 6-Cys 127) 126-128 (see Fig.5). This remarkably high inhibitory activity of the three peptides enabled us to account for the first time for almost all the immune reaction of native lysozyme.

Sequence and location of peptide in primary structure

22   33
Gly – Tyr – Ser – Leu – Gly – Asn – Trp – Val – |Cys| – Ala – Ala – Lys
                                              115 | 116
                                                  |Cys| – Lys

                        62   68
                        Trp – Trp – |Cys| – Asn – Asp – Gly – Arg
74
Asn – Leu – |Cys| – Asn – Ile – Pro – |Cys| – Ser – Ala – Leu – Leu – Ser – Ser

Lys – Ala – |Cys| – Asn – Val – Ser – Ala – Thr – Ile – Asp ————————
96

         6   13
         |Cys| – Glu – Leu – Ala – Ala – Ala – Met – Lys
126   128
Gly – |Cys| – Arg

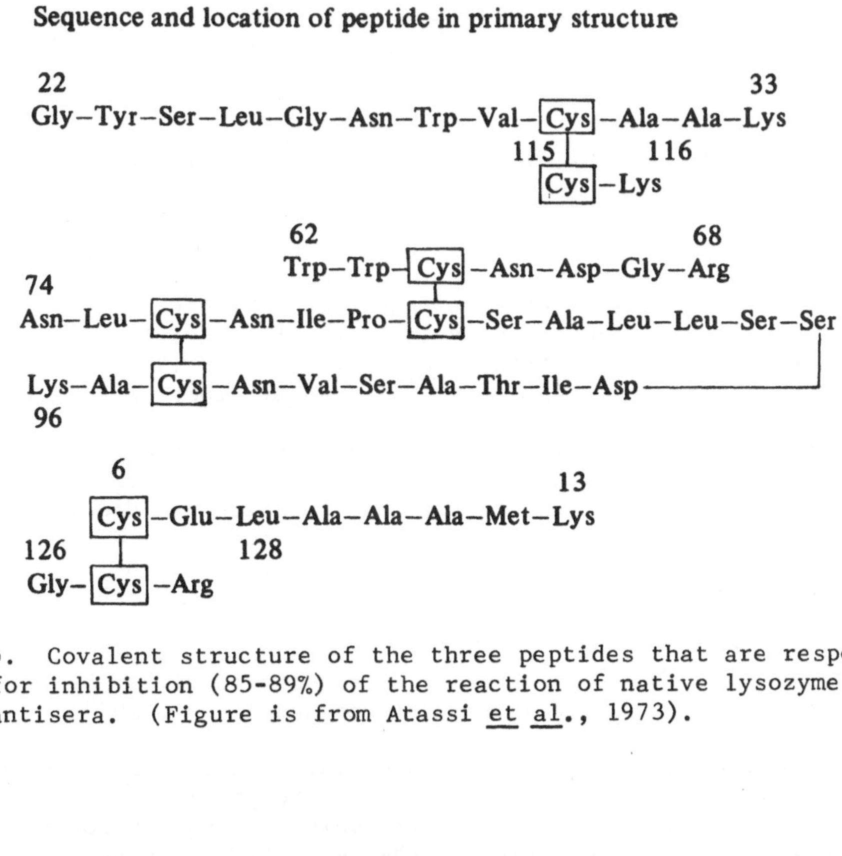

Fig.5. Covalent structure of the three peptides that are responsible for inhibition (85-89%) of the reaction of native lysozyme with its antisera. (Figure is from Atassi et al., 1973).

The approach is not limited to lysozyme and has proved to be of general applicability. Fragments have been obtained from cleavage of bovine serum albumin at the arginyl peptide bonds which accounted for all (97%) of the antigenic reactivity of the native protein (Habeeb et al., 1974; Atassi et al., 1976a). Application of the approach to bovine ribonuclease A has afforded fragments which also reacted with antibodies to the native protein (Habeeb and Atassi, unpublished results). Also, the preparation of peptides with intact

disulfide bonds was introduced and applied as a procedure to determine the correct disulfide pairing in proteins (Atassi et al., 1973
and 1975a).

V.   SPECIFIC CHEMICAL DERIVATIVES OF
IMMUNOCHEMICALLY-REACTIVE PEPTIDES

Identification of the residues involved in binding with antibody and further narrowing down of antigenic reactive sites in an
immunochemically-reactive peptide is best achieved (Atassi, 1968)
by immunochemical studies of chemical derivatives of the peptide
modified at specific amino acid locations. The advantages and
shortcomings of the approach have previously been outlined (Atassi,
1972 and 1975).

A.   Derivatives of the Two-Disulfide Peptide

The immunochemical reactivity of the peptide corresponding to
sequence 62-68 (Cys 64-Cys 80) 74-96 (Cys 76-Cys 94), isolated from
the tryptic hydrolysate by the approach described in Section IV-B,
was quite strong and accounted for about one-third (see Fig.6 and
Table 5) of the total antigenic reactivity of native lysozyme with
its early-course antisera (Atassi et al., 1973 and 1975a). The reactivity of the peptide was later confirmed by other workers (Fujio
et al., 1974) using a peptide immunoabsorbent and a pool of late-
course rabbit antisera. The peptide was also isolated with lysine
97 attached to it (Lee and Atassi, 1975). ORD measurements on this
two-disulfide peptide showed that it was greatly unfolded in solution (see Table 4) relative to its expected mode of folding within
the intact lysozyme molecule. Despite its unfolding the peptide
exhibited a surprising binding efficiency which can be seen from
the relatively low molar excess required to achieve half the maximum inhibition (Table 5). This was attributed to the presence of
the relevant parts of the antigenic site anchored firmly by the two
disulfide bonds (Atassi et al., 1973; Lee and Atassi, 1975).

For identification of the residues involved in binding with
antibody, several chemical derivatives of the peptide were prepared,
purified, and characterized and their conformations studied (Table
4). The antigenic reactivities of the derivatives are shown in
Table 5. The derivatives suffered no conformational change relative
to the unmodified peptide (Table 4). However, modification of the
two typtophans by reaction with 2,3-dioxo-5-indolinesulfonic acid
provided a derivative which had only about half the immune reaction
of the peptide (Lee and Atassi, 1975). Also, succinylation of the
amino groups caused the loss of about half of immunochemical reactivity (Lee and Atassi, 1975). Modification of the two tryptophans

TABLE 4. ORD Parameters of the Disulfide Peptide and Its Derivatives

Measurements were carried out on the peptide solutions in water

Protein or Peptide*	$[m']_{233}$	b_o
Lysozyme[a]	-4,150	-205
Unmodified $(SS)_2$-peptide[a]	-1,480	- 31
DISA-peptide[a]	-1,400	- 29
Succinyl-peptide[a]	-1,470	- 31
Succinyl-DISA-peptide[a]	-1,450	- 31
HME_4-peptide[b]	-1,520	- 32
CHD-peptide[b]	-1,470	- 31
PG-peptide[b]	-1,450	- 31
Pre-protected PG-peptide[b]	-1,460	- 31

*Abbreviations: $(SS)_2$-peptide, the two-disulfide peptide corresponding to the sequence 62-68 (Cys 64-Cys 80) 74-97 (Cys 76-Cys 94); DISA-peptide, derivative of the $(SS)_2$-peptide modified at tryptophans 62 and 63 by reaction with 2,3-dioxo-5-indoline-sulfonic acid; Succinyl-peptide, derivatives succinylated at the amino groups and succinyl-DISA-peptide, a succinylated DISA-peptide; HME_4-peptide, derivative coupled to histidine methyl ester at the four carboxyl groups; CHD-peptide, derivative modified with cyclohexanedione at arginine 68, the two tryptophans, the two lysine and the N-terminal asparagine 74; PG-lysozyme, derivative modified by phenylglyoxal at Arg-68, Trp-62, Asn-74 and one ϵ-amino group; Pre-protected PG-peptide, derivative in which the NH_2-groups were first protected by citraconylation and after reaction with phenylglyoxal the protecting groups were removed giving a derivative modified only at Arg-68.

References: (a) Lee and Atassi, 1975; (b) Atassi et al., 1976c

TABLE 5. Inhibitory Activities of the Two Disulfide Peptide and Its Derivatives

Results are expressed in maximum percent inhibition of the precipitin reaction of native lysozyme with its antisera. Each value represents the average of six or more replicate determinations which varied ± 0.8% or less. Antisera G9 and G10 are goat antisera against native lysozyme.

Peptide or derivative*	Antiserum G9		Antiserum G10	
	Max. inhibitory activity	Molar ratio at ½ max. inhibn**	Max. inhibitory activity	Molar ratio at ½ max. inhibn**
$(SS)_2$-peptide[a,b]	26.4	14.2	32.7	4.5
DISA-peptide[a]	12.0	14.5	14.5	7.0
Succinyl-peptide[a]	12.2	14.3	14.7	4.5
Succinyl-DISA-peptide[a]	2	52	2.5	40
HME_4-peptide[b]	11.6	13.7	16.1	6.5
CHD-peptide[b]	3.2	43	4.6	40
PG-peptide[b]	13.2	19	15.2	4.8
Pre-protected PG-peptide[b]	26.5	14.2	32.6	4.5
Chymotrypsin-cleaved peptide[a]	3	40	3.5	35
S-Carboxymethyl-peptide[a+]	0	310‡	0	300‡

*Abbreviations: are as in the footnotes to Table 4.

**These values represent peptide/antigen molar ratio at 50% of the maximum inhibition. +These values represent the maximum molar excess of peptide, relative to lysozyme, which was employed in the inhibition experiment.

+Similar results were obtained either with the total chymotryptic hydrolysate or with peptide 64-68 (Cys 64-Cys 80) 76-83 (Cys 76-Cys 94) 94-97.

‡These values represent the maximum molar excess of peptide relative to lysozyme in the inhibition reaction.

References: (a) Lee and Atassi, 1975; (b) Atassi et al., 1976c.

followed by succinylation of the amino groups abolished the antigenic
reactivity almost completely (Lee and Atassi, 1975). From these
results it was concluded that the antigenic site in this part of
lysozyme incorporates one or both of tryptophans 62 and 63 as well
as one or both of lysines 96 and 97. This agreed with previous

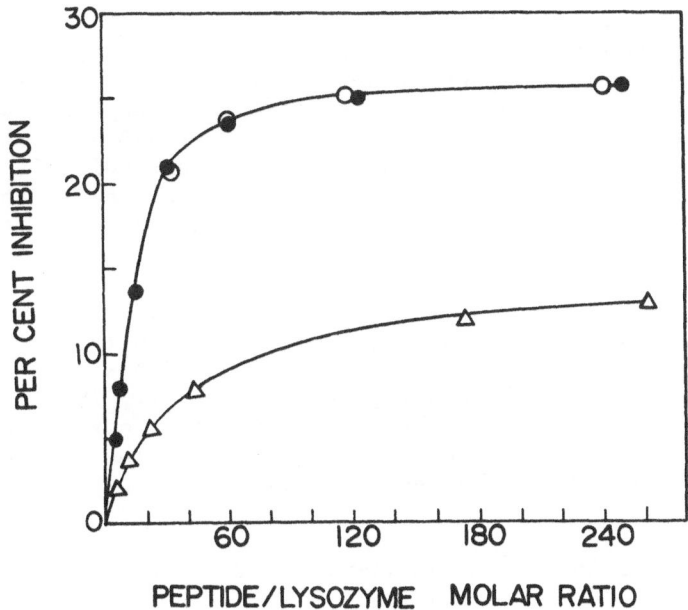

PEPTIDE / LYSOZYME   MOLAR RATIO

Fig.6. Inhibition of the precipitin reaction of native lysozyme
with an antiserum (G9) to lysozyme by: (●) $(SS)_2$-peptide; (Δ) PG-
peptide and (○) pre-protected PG-peptide. Each point represents
the average of six or more determinations which varied $\pm$ 0.8% or
less. For other results, see Table 5. (Figure is from Atassi et
al., 1976c).

results (Lee et al., 1975) derived from modification of these lysines in intact lysozyme (see Section III-A). Since reduction and
carboxymethylation of the disulfides abolished the antigenic reactivity of the peptide (Atassi et al., 1973; Lee and Atassi, 1975),
it was concluded that the two disulfides 64-80 and 76-94 bring these
two parts of the lysozyme molecule into a single reactive site. The
intactness of the disulfides is essential for maintenance and reactivity of the site.

Recently we investigated (Atassi et al., 1976c) the role of arginine 68 and aspartic acids 66 and 87. Activation of carboxyl groups with carbodiimide followed by coupling with histidine methyl ester yielded a homogeneous derivative which was modified at all carboxyl groups and suffered no conformational changes (Table 4) but retained only less than half of the immunochemical reactivity of the peptide (Table 5). Reactions with phenylglyoxal or 1,2-cyclohexanedione were not specific for arginine 68 but modified other functional groups on the peptide (see Table 4). The derivatives showed no changes in conformational parameters (Table 4) but lost most or all of the inhibitory activity (Table 5). Reaction with phenylglyoxal became specific for arginine after protection of the free amino groups by citraconylation followed by removal of the protecting group, and yielded a homogeneous derivative which was modified at arginine 68 only. This derivative suffered no change in conformation (Table 4) or in immunochemical behavior (Fig.6 and Table 5) relative to the unmodified peptide. From these studies it was concluded that one (or both) of aspartic acids 66 and 87 is part of the antigenic site, whereas arginine 68 is not located in this site on the peptide (Atassi et al., 1976c).

Studies on the chemical derivatives of the peptide, which are summarized diagramatically in Fig.7, enabled us in fact to define the complex boundaries of the antigenic site in this part of the molecule with considerable accuracy (Atassi et al., 1976c). This definition paved the way for the formulation of a novel synthetic strategy for the final delineation of the site. These results are described and coordinated in Section VIII.

VI. PEPTIDE SYNTHESIS FOR FINAL DELINEATION OF ANTIGENIC SITES

Following the accurate narrowing down of antigenic sites in our laboratories by the foregoing chemical approaches, the final delineation of the sites was accomplished by the organic synthesis and immunochemistry of peptides representing different parts of each site. The precautionary measures that must be employed in the application of this approach to problems of protein antigenic structures have been outlined elsewhere (Atassi, 1977).

The foregoing chemical approaches demonstrated that the antigenic sites were located around disulfide bonds. The strategy, therefore, required the synthesis of disulfide-containing peptides. Such a synthetic scheme was employed in the delineation of the reactive site around the disulfide 6-127 (i.e. site 1) and for which nine disulfide peptides were synthesized by Atassi et al., (1976b). For this, two groups of peptides A and B were first synthesized, S-sulfonated (Fig.8) and purified by ion-exchange chromatography to

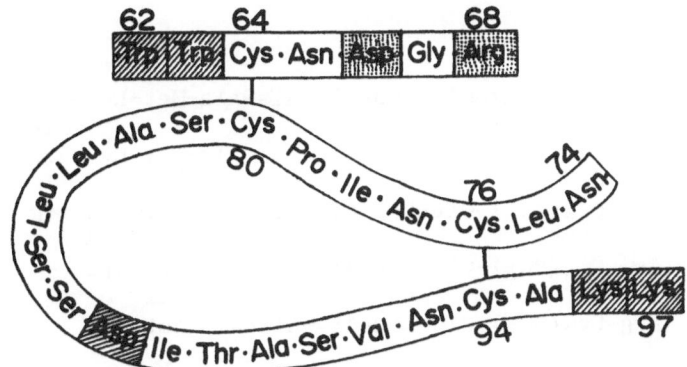

Fig.7. Schematic diagram showing the primary structure of the $(SS)_2$-peptide studied here. The marked residues are those whose involvement or otherwise in antigenic reactivity has been derived from their chemical modification. Striped residues (Asp 87; one or both of Trp 62 and Trp 63; Lys 96 and, to a lesser extent, Lys 97) are part of the antigenic site while the dotted residues (Arg 68 and Asp 66) are outside the antigenic site. (Figure is from Atassi et al., 1976c).

homogeneity before being used for the preparation of the disulfide peptides (Fig. 9). Following disulfide bond formation between two homogeneous S-sulfonyl peptides, the heterogeneous product was resolved by ion-exchange chromatography to obtain the correct disulfide-containing peptide. This approach proved to be extremely time-consuming and laborious. Thus, in the synthesis of the nine disulfide peptides by Atassi et al., (1976b) a total of 73 chromatographic experiments were required to prepare peptides with a purity level in excess of 98%. The approach was subsequently not employed in the delineation of the other two antigenic sites of lysozyme. For these, an entirely novel and unorthodox concept was introduced (Atassi et al., 1976d) and an extremely powerful and unique synthetic approach was devised. The various peptides synthesized in our laboratory for the delineation of the antigenic site around the disulfides 64-80 and 76-94 (site 2) are shown in Figs. 12 and 14. The rationale behind the choice of these peptides is best handled in the following section which deals with the accurate assignment of the antigenic sites.

GROUP A

```
3                                                                          14
Phe-Gly-Arg-Cys(SO3)-Glu-Leu-Ala-Ala-Ala-Met-Lys-Arg
        5                                                                  14
        Arg-Cys(SO3)-Glu-Leu-Ala-Ala-Ala-Met-Lys-Arg
            6                                                              14
            Cys(SO3)-Glu-Leu-Ala-Ala-Ala-Met-Lys-Arg
        5                                                              13
        Arg-Cys(SO3)-Glu-Leu-Ala-Ala-Ala-Met-Lys
        5                                                          12
        Arg-Cys(SO3)-Glu-Leu-Ala-Ala-Ala-Met
```

```
Analog: 5                                                              14
        Arg-Cys(SO3)-Glu-Leu-Ala-Ala-Ala-Gly-Lys-Arg
```

GROUP B

```
            125                           129
            Arg-Gly-Cys(SO3)-Arg-Leu
            125                     128
            Arg-Gly-Cys(SO3)-Arg
                126                 128
                Gly-Cys(SO3)-Arg
                    127             128
                    Cys(SO3)-Arg
            125     127
            Arg-Gly-Cys(SO3)
```

Fig.8. Amino acid sequence of the synthetic peptides representing various overlaps on both sides of the disulfide 6-127. (Figure is from Atassi _et al_., 1976b).

VII. ACCURATE ASSIGNMENT OF THE ANTIGENIC SITE
      AROUND THE DISULFIDE 6-127 (SITE 1)

In this and Sections VIII and IX, the information obtained from our chemical and synthetic approaches will be coordinated to derive the accurate location of the antigenic reactive sites of native lysozyme. It has recently been shown (Atassi _et al_., 1976c) that lysozyme has only three major antigenic sites.

```
          3                                        14
          Phe-Gly-Arg-Cys-Glu-Leu-Ala-Ala-Ala-Met-Lys-Arg
             125          |         129
   I         Arg-Gly-Cys-Arg-Leu


               5                                      14
               Arg-Cys-Glu-Leu-Ala-Ala-Ala-Met-Lys-Arg
          125          |    128
  II      Arg-Gly-Cys-Arg


               5                                      14
               Arg-Cys-Glu-Leu-Ala-Ala-Ala-Met-Lys-Arg
             126    |   128
 III         Gly-Cys-Arg


                 6                                      14
                 Cys-Glu-Leu-Ala-Ala-Ala-Met-Lys-Arg
             126    |   128
  IV         Gly-Cys-Arg


               5                                          13
               Arg-Cys-Glu-Leu-Ala-Ala-Ala-Met-Lys
          125          |    128
   V      Arg-Gly-Cys-Arg


               5                             12
               Arg-Cys-Glu-Leu-Ala-Ala-Ala-Met
             126    |   128
  VI         Gly-Cys-Arg


               5                                      14
               Arg-Cys-Glu-Leu-Ala-Ala-Ala-Met-Lys-Arg
                 127|    128
 VII              Cys-Arg


               5                                      14
               Arg-Cys-Glu-Leu-Ala-Ala-Ala-Met-Lys-Arg
          125          |
VIII      Arg-Gly-Cys


               5                                      14
               Arg-Cys-Glu-Leu-Ala-Ala-Ala-Gly-Lys-Arg
             126    |   128
  IX         Gly-Cys-Arg
```

Fig.9. Structure of the disulfide peptides synthesized from group A and group B peptides (shown in Fig.8). Peptide IX is an analogue containing glycine at position 12 instead of methionine. (From Atassi et al., 1976b).

A.   Assignment of the Reactive Site and Its Synthesis

From the immunochemical and conformational studies on derivatives of the intact protein, it was shown (see Section III) that
aspartic acid 119 and leucine 129 were not parts of an antigenic
site in native lysozyme (Atassi et al., 1974; Atassi and Rosemblatt,
1974; Atassi et al., 1975a). Also, modification of Trp-123 (Atassi
and Zablocki, 1976) or Met-12 (as well as Met-105) (Atassi et al.,
1976b) demonstrated that these residues were not located in an antigenic site. The peptide 6-13 (Cys 6-Cys 127) 126-128 carried substantial antigenic reactivity which, with two other disulfide-
containing peptides (see Fig.5), jointly accounted for almost all
(90%) of the antigenic reactivity of native lysozyme (Atassi et al.,
1973). These results indicated the presence of an antigenic site
around the disulfide bond 6-127. On one side of the disulfide,
the antigenic site clearly begins after Trp-123 and ends at or
before Arg-128. On the other side of the disulfide, the second
part of the reactive site must end at or close to methionine 12.

Having accomplished this degree of delineation chemically, the
final narrowing down was achieved by immunochemical studies of synthetic peptides corresponding to various parts of the site. The
aforementioned results were critical in planning the correct synthetic strategy for the final delineation of the site. Synthesis
of nine disulfide peptides comprising various overlaps of the sequences 3-14 and 125-129 (see Fig.9) was recently carried out and
their immunochemistry with antisera to native lysozyme studied
(Atassi et al., 1976b).

None of the peptides 3-14, 5-14, or 125-129 (representing
either one side or the other of the disulfide-linked antigenic site)
had an inhibitory effect on the lysozyme immune reaction (Table 6).
However, each of the disulfide peptides exhibited an inhibitory
activity towards the reaction of lysozyme with its antisera (Atassi
et al., 1976b). These results confirmed our previous findings that
the integrity of the disulfide bond is essential for bringing the
two distant (in sequence) parts of the site together. The behaviors of each of the peptides with three different antisera are
given in Table 6. With antisera L1 and G9, the presence or absence
of Phe-3, Gly-4, Arg-5, Arg-125 or Leu-129 did not make any contribution to the inhibitory activity. On the other hand, absence of
Arg-14, Lys-13, Gly-126 or Arg-128 had detrimental effects on the
inhibitory activity. Therefore, for antisera L1 and G9, the antigenic site comprises the two regions (6-14) + (126-128). With antiserum G10, residues 3, 4, 5, 125 and 129 again do not contribute to
the inhibitory activity, whereas residues 13, 14 and 126 are also
critical for full inhibitory ability. However, the presence or absence of Arg-128 did not change the inhibitory ability of the two
respective peptides towards this antiserum. Therefore, for antiserum G10, the reactive site comprised the two regions (6-14) +

TABLE 6. Inhibitory Activities of the Pure Synthetic Peptides Around the Disulfide 6–127

Results are expressed in maximum percent inhibition of the precipitin reaction of native lysozyme with various antisera. Each value represents the average of four or more replicate determinations which varied ± 0.8% or less. Peptides denoted by Roman numerals refer to the disulfide peptides in Fig. 9. Linear peptides denoted by their locations in the primary structure are in the S-sulfonated form. L1 is a rabbit antiserum and G9 and G10 are goat antisera, each against native lysozyme.

Peptide	Antiserum L1		Antiserum G9		Antiserum G10	
	Max. inhib. act. (%)	Molar ratio, peptide/antigen	Max. inhib. act. (%)	Molar ratio, peptide/antigen	Max. inhib. act. (%)	Molar ratio, peptide/antigen
I	28.0	19*	26.8	17*	32.1	10*
II	27.4	18*	27.2	18*	31.9	11*
III	28.3	20*	26.5	16*	31.6	10*
IV	27.5	18*	27.0	17*	31.5	11*
V	14.6	28*	9.9	26*	21.7	23*
VI	6.3	39*	7.7	43*	8.9	38*
VII	7.5	45*	7.3	40*	8.5	38*
VIII	17.6	23*	15.8	25*	31.7	21*
IX	27.9	18*	27.1	18*	31.8	11*
3-14	0	730**	0	650**	0	610**
5-14	0	740**	0	655**	0	620**
125-129	0	735**	0	645**	0	625**

*These values represent peptide/lysozyme molar ratio at 50% of the maximum inhibition.

**These values represent the maximum molar excess of peptide, relative to lysozyme, which was employed in the inhibition experiment.

(Table is from Atassi et al., 1976b)

(126-127). The finding that Arg-128 is not part of the antigenic
site only with one antiserum of those studied is analogous to the
one-residue shift observed (Koketsu and Atassi, 1973 and 1974a;
Atassi, 1975) with some antigenic sites of sperm-whale myoglobin.
Finally, substitution of Met-12 by glycine (peptide IX, Fig.9) did
not cause any change in the inhibitory ability of the resultant
peptide (cf. peptide IX and peptide III in Table 6). Therefore,
Met-12 does not actively participate in binding with antibody.
This confirms the foregoing conclusion which was independently
derived from modification of lysozyme at the two methionines (Section
III-D).

B. Description of the Reactive Site

The covalent structure of the antigenic site is shown in Fig.
10. A schematic diagram of the relative positions of the residues
comprising the site is shown in Fig. 11. It can be seen that Arg-
14, Lys-13, Glu-7, Gly-126 and Arg-128 have the spatial possibility
to construct the antigenic site. Also, the sulfur of Cys-6 may come
in contact with the antibody binding site. It must be emphasized
that the mere occurrence of favorable spatial orientation of a few
residues on the surface of the protein does not by itself indicate
the presence of an antigenic site in that location. The three-
dimensional structure is _not_ used here to predict but to explain
strong experimental results. The structure described in Fig. 11
shows very clearly why Met-12 does not participate in interaction
of native lysozyme with antibody. In the three-dimensional structure, Met-12 (as well as Met-105) is completely buried within the
interior of the molecule (Imoto _et al._, 1972). In fact, not a single
atom in the side-chain of either methionine residues makes any contact with the surface (Shrake and Rupley, 1973). Since the immune
response to a native protein antigen is directed against its native
three-dimensional structure (Atassi and Thomas, 1969), at least with
early-course antisera, it is not possible for the two methionines
in lysozyme to constitute part of an antigenic site.

VIII. ACCURATE DELINEATION OF THE ANTIGENIC SITE AROUND THE DISULFIDES 64-80 AND 76-94 (SITE 2)

A. Assignment of the Antigenic Site

Of our immunochemical studies on specific chemical derivatives
of native lysozyme that are of direct relevance here are the findings
that arginine 61 is not part of an antigenic site (Atassi _et al._,
1972), whereas lysine 96 is located within an antigenic site in
native lysozyme (Lee _et al._, 1975). Subsequently, we showed that
the two-disulfide peptide 62-68 (Cys 64-Cys 80) 74-97 (Cys 76-Cys 94)

```
        6                                    14
      Cys-Glu-Leu-Ala-Ala-Ala-Met-Lys-Arg
 126  ┆   128
 Gly-Cys-(Arg)
```

Fig.10. The structure of the antigenic reactive site that we have
delineated around the disulfide 6-127 (site 1). Residues underlined by a solid line are the most likely to be directly involved
in the binding with antibody. The sulfur of Cys-6 (underlined by
a dashed line) may also come in contact with the antibody combining
site. Arg-128 may or may not be part of the antigenic site, depending on the antiserum. For details, see text. (From Atassi et
al., 1976b).

[i.e. (SS)$_2$-peptide] accounted for about a third of the total antigenic reactivity of native lysozyme (Atassi et al., 1973; Atassi et
al., 1975a). Immunochemical studies (Lee and Atassi, 1975) of
derivatives of the (SS)$_2$-peptide (see Section V-A) showed that Arg-
68 and Asp-66 are not part of the antigenic site but that either one
(or both) of tryptophans 62 and 63 and that either one (or both) of
lysines 96 and 97 are located in the antigenic site (Lee and Atassi,
1975), in agreement with the aforementioned results on derivatives
of the intact protein (Lee et al., 1975). Furthermore, it was
demonstrated that the tryptophan (s) and the lysine (s) are parts
of the same antigenic site (Lee and Atassi, 1975), which requires
intactness of the disulfide bonds to effect its three-dimensional
construction (Lee and Atassi, 1975). Recently, it was shown that
the region around Asp-87 was essential for the full reactivity of
the site (Atassi et al., 1976c). After modification of Asp-87,
the antigenic site, however, retained about half of its reactivity
(Atassi et al., 1976c). It became clear, therefore that this site
begins (or ends) at tryptophan 62 or 63, requires lysine 96 or 97
(or both) and some or all of the region 84-93 (Lee and Atassi, 1975;
Atassi et al., 1976c). Since the (SS)$_2$-peptide carries a single
antigenic site and the total of its reactivity together with the two
smaller single-disulfide peptides (Atassi et al., 1973) accounts for
90% of the entire antigenic reaction of native lysozyme, it became
clear at this stage that lysozyme has only three major antigenic
sites (Atassi et al., 1976c).

    B. Novel Synthetic Peptides with Diglycyl
        Bridges Instead of Disulfides

    Chemical modification and cleavage studies on this region will
not yield any further information on the roles of tryptophans 62
and 63 or of lysines 96 and 97 in the antigenic site. Accordingly,

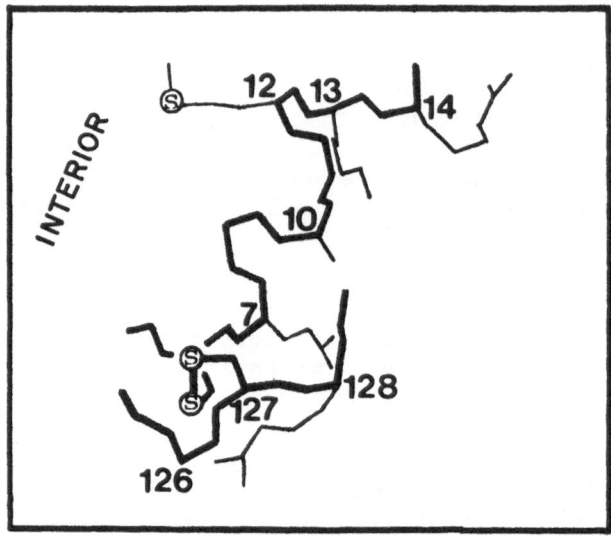

Fig.11. A schematic diagram showing the mode of folding of the
antigenic site (1) within the intact protein. The backbone is
presented by a heavy line and the thinner line represents the side-
chains of some selected residues. Only the side-chains of residues 7, 10, 13, 14 and 128 are shown here. Side-chains of the
other residues are buried and are omitted in order not to over-
crowd the diagram. In addition, the orientation of the side-chain
of Met-12 is given here to point out that it is completely buried
within the interior (left part of the diagram) of the molecule.
The side-chain of Arg-128 is presented here at variance with its
orientation in the crystalline structure of lysozyme (Imoto et al.,
1972). This is done in order to achieve a more favorable align-
ment of side-chains. This slight readjustment of surface side-
chains may be easily induced by the antibody (Atassi, 1975). The
immunochemical structure shown represents a true example of an
antigenic reactive site (Atassi and Saplin, 1968). For details,
see text. (Figure is from Atassi et al., 1976b).

we then focused our effort on the organic synthesis and immunochemical
studies of peptides related to the $(SS)_2$-peptide. However, synthesis was complicated by the fact that this peptide is composed of
two segments, one (sequence 74-97) forming a loop closed by an in-
trapeptide disulfide (Cys 76-Cys 94), and the other (sequence 62-
68) connected to this loop by a second disulfide (Cys 64-Cys 80).
The synthetic efforts were almost entirely unsuccessful and therefore a different approach was sought. We then pointed out (Lee et
al., 1976) that the problem may be studied in part by synthetic
peptides carrying the sequence 62-68 linked by a disulfide to sequence 76-80 which is also linked in turn by a second disulfide to
sequence 94-97 (see Figs. 5 and 7). This constitutes only part of
the site and does not include the essential region around Asp-87
and therefore should be expected, by analogy with peptide derivatives (Lee and Atassi, 1975; Atassi et al., 1976c), to have only
about half of the reactivity of the site (i.e. a decrease from
about 33% down to about 16% inhibition of the immune reaction of
native lysozyme). The lower (16%) inhibition is still a substantial amount of activity and should be extremely useful to
determine the independent roles of the two tryptophans (residues
62 and 63) and the two lysines (residues 96 and 97). However,
attempts at synthesis of these three-segment, two-disulfide
peptides were not too fruitful. Accordingly, we devised (Lee
et al., 1976) a completely different approach by synthesis of
peptides that did not contain disulfide bonds and in which the
disulfides were substituted by diglycyl segments. Such synthesis
will produce peptides in which the central four peptide bonds will
be in the opposite direction to those in the natural peptide (see
Fig. 12). However, since in binding with antibody only amino acid
chains will most likely participate, it was felt that the direction
of the peptide bond may not be too critical and that, furthermore,
reorientation of the side-chains by antibody (Atassi, 1975) may be
possible through rotation about certain peptide bonds. The outcome
of this reasoning was not entirely certain because no precedents of
such an undertaking have been reported. The results, however, showed
that this unorthodox reasoning proved to be sound.

Several peptides were synthesized (Lee et al., 1976) and are
shown in Fig. 12. Table 7 summarizes the reactions of the peptides
with two antisera and compares these with the maximum expected reactivity of the intact site as well as of the derivative lacking the
contribution of Asp-87.

With the antisera studied, peptide III exhibited the highest
inhibitory activity which approximated the expected value. Peptide
III carries two phenylalanine residues instead of two tryptophan
residues (Fig.12). Peptides II and IV (which carry one tryptophan
and one phenylalanine respectively at the amino end) showed equal
but somewhat lower inhibitory activity and lower binding efficiency
than peptide III. Therefore, the presence of the second phenylalanine at the amino end changed very little the final maximum of the

A.

```
               62        →    64
               TRP - TRP - CYS
                            |          ←          76
                           CYS - PRO - ILE - ASN - CYS
                                                    |         →
                                                   CYS - ALA - LYS - LYS
                                                   94                97
```

B.

```
                    →                        →                        →
I      TRP - TRP -|GLY - GLY|- PRO - ILE - ASN -|GLY - GLY|- ALA - LYS - LYS
II           TRP -|GLY - GLY|- PRO - ILE - ASN -|GLY - GLY|- ALA - LYS - LYS
III    PHE - PHE -|GLY - GLY|- PRO - ILE - ASN -|GLY - GLY|- ALA - LYS - LYS
IV           PHE -|GLY - GLY|- PRO - ILE - ASN -|GLY - GLY|- ALA - LYS - LYS
V                 |GLY - GLY|- PRO - ILE - ASN -|GLY - GLY|- ALA - LYS - LYS
VI     PHE - PHE -|GLY - GLY|- PRO - ILE - ASN -|GLY - ALY|- ALA - LYS
VII    PHE - PHE -|GLY - GLY|- PRO - ILE - ASN -|GLY - GLY|- ALA
```

Fig.12. Amino acid sequence of: (A) The disulfide-linked sequences (62-64) (76-80) (94-97) of native lysozyme. (B) The
peptides synthesized in our laboratory. The arrows indicate the
direction (N to C) of the peptide chains. The vertical dashed
lines are used to outline the diglycyl segments which were used to
substitute for the disulfides in peptide A. (Figure is from Lee
et al., 1976).

inhibition but improved the binding efficiency of the peptide.
Removal of the two phenylalanine residues from the amino terminal
(peptide V) had a drastic effect on the inhibitory activity as well
as the binding efficiency of the peptide. On the other end of the
peptides, removal of one lysine from the C-terminal (peptide VI)
resulted in substantial loss of the inhibitory activity and binding
efficiency and when the second lysine was removed (peptide VII), both
parameters were further decreased. The immunochemical differences
were not due to effects of peptide size (Lee et al., 1976). Clearly,
both lysine residues are important parts of the antigenic site and
on the other end only one tryptophan (or one phenylalanine) is needed
(Lee et al., 1976).

The ability of phenylalanine to substitute effectively in
immunochemical interaction for tryptophan is not unexpected. However, the lack of immunochemical reaction of peptide I (two trypto-

TABLE 7. Inhibitory Activities of the Pure Synthetic Peptides with Diglycyl Bridges*

Results are expressed in maximum percent inhibition by the peptide of the precipitin reaction of lysozyme with various antisera. Each value represents the average of at least three determinations which varied $\pm$ 0.8% or less.

Peptide	Antiserum G9[a]		Antiserum G10[a]	
	Max. inhibit. act. (%)	Molar ratio at ½ max. inhibn.[b]	Max. inhibit. act. (%)	Molar ratio at ½ max. inhibn.
(SS)$_2$-Peptide[c]	26.40	14.2	32.70	4.5
(SS)$_2$-Peptide Deriv.[d]	11.60	13.7	16.10	6.5
I	2.56	220	0	-
II	11.50	240	10.10	440
III	13.83	220	12.73	380
IV	10.68	540	10.82	540
V	4.74	420	2.86	400
VI	7.89	360	5.86	420
VII	5.53	520	3.44	530

*These peptides are shown in Fig. 12.

[a]G9 and G10 are goat antisera.

[b]These values represent peptide/antigen molar ratio at 50% of the maximum inhibition.

[c]Obtained from Lee and Atassi (1975).

[d]The decrease in inhibitory activity of this (SS)$_2$-peptide derivative is due to modification of Asp 87 which we have previously reported (Atassi et al., 1976c). This reactivity is the maximum expected extent of reaction for a synthetic peptide. For details, see text.

(Table is from Lee et al., 1976)

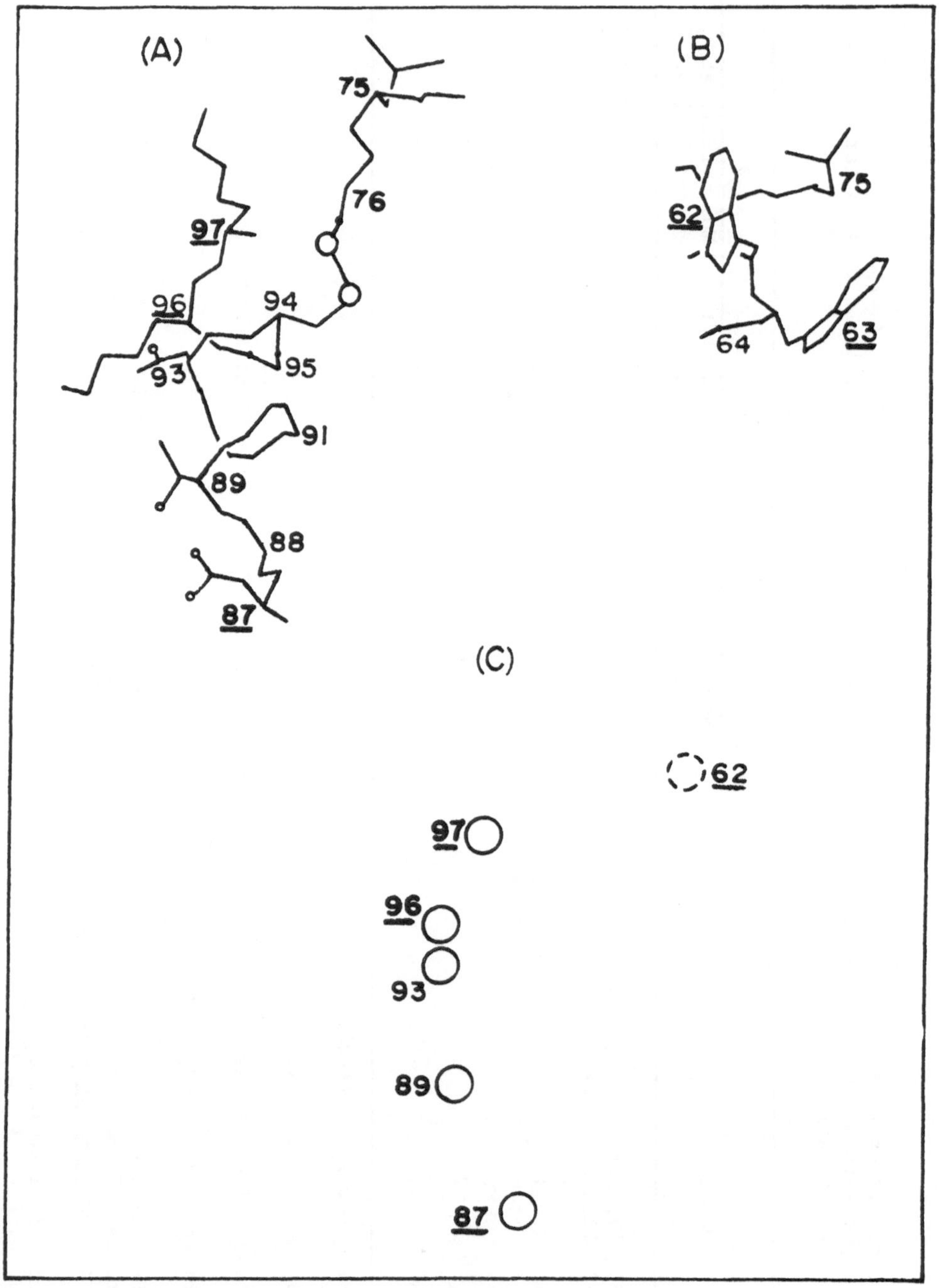
(A)
75
76
97
96
94
93
95
91
89
88
87
(B)
75
62
63
64
(C)
62
97
96
93
89
87

Fig. 13. A schematic diagram showing the relative conformational arrangement of the residues in antigenic site (2). Residues that we have shown by chemical modification to be part of the antigenic site are underlined. (A) shows the mode of folding and the relative arrangement of the residues in the antigenic site. To avoid over-crowding the diagram, only the side chains of the residues that are part of the antigenic site are shown. Aspartic 87 is closest to the observer and the residues 89, 93, 96, 97 and 75 steadily recede away from the observer so much that Trp-62 is behind the plane of the paper and is not shown in (A) and is continued in (B). The view in (A) is from the opposite face to that in which the enzyme's cleft is located. In going from 87 to 62, the antigenic reactive site moves from one side of the cleft (from behind it though) to the other. For elegant diagrams of the three-dimensional structure of lysozyme, see Imoto et al., (1972). (B) shows only the relative positions and proximity of Leu-75, Trp-62 and -63 and is obtained by looking at the molecule from the surface of the enzyme's cleft (i.e. the opposite surface to that in (A). Residues 62, 63 and 75, in that order, gradually recede away from the observer. (C) is a simplified diagram showing the relative positions of the α-carbons, only of the residues constituting the antigenic reactive site. This is quite useful since the positions of the side-chains can fluctuate to adjust themselves to the antibody combining site. From 87 to 62, the residues are receding steadily from the observer so that in fact Trp-62 is behind the plane of the paper and is therefore shown as a broken circle. Residues 87, 96, 97 and 62 have been shown by specific chemical derivatives of the $(SS)_2$-peptide (see text) to be part of the antigenic site. Residues 89 and 93 are intervening residues that must, because of their three-dimensional location, constitute part of the site. However, the boundaries of the site are well defined since the residues at the two ends of the site have been well characterized. The dimensions of the antigenic site are given in Figure 14. (Figure is from Atassi et al., 1976c).

phans at the amino end) was attributed to steric hindrance (Lee et al., 1976). In native lysozyme, Trp-63 is mostly buried whereas Trp-62 is appreciably exposed (Imoto et al., 1972). It is extremely unlikely for free peptide I to fold in solution in such a way so as to orient the indole nucleus equivalent to Trp-63 (Tryptophan-2 in peptide I) out of the way so that Trp-62 can effect binding with antibody. This steric hindrance would, of course, be considerably less pronounced with the two adjacent smaller phenyl side-chains (peptide III). The steric hindrance observed here with peptide I is not unusual and has previously been reported for three reactive sites of myoglobin which suffered decreases in antigenic reactivity when amino acids with bulky side-chains were incorporated on either end of the synthetic sites (Koketsu and Atassi, 1974a, 1974b; Atassi and Pai, 1975). The anomaly was explained (Koketsu and Atassi, 1974a) that the extraneous amino acids may exert, with certain antisera, steric obstruction due to an unsatisfactory mode of folding of the longer peptides in solution.

The remarkable effectiveness of the substitution by diglycyl segments for the disulfide bonds may be applicable to the solution of other similar problems in proteins and should merit consideration in certain studies. The foregoing analysis clearly shows that both lysine residues are required for the reactivity of the antigenic site with one being more critical than the other. On the other end of the site only one tryptophan is required and this, in the native protein, will be Trp-62 because Trp-63 is buried. Therefore, the residues Asp-87, Lys-96, Lys-97, and Trp-62 are essential parts of the antigenic site in this part of the molecule.

   C. "Surface-Simulation" Synthesis: a Novel Synthetic
      Approach Directly Linking the Conformationally-
      Adjacent Residues Forming the Site

Examination of the three-dimensional structure of lysozyme enabled explanation of the manner in which the aforementioned four residues construct the antigenic site, and we proposed (Atassi et al., 1976c) that it comprised the residues: Asp-87, Thr-89, Asn-93, Lys-96, Lys-97 and Trp-62 (Fig.13). From the foregoing description it can be seen that residues 87, 96, 97 and 62 were directly implicated in the active interaction with the antibody. No chemical evidence had really yet been obtained to implicate Thr-89 and Asn-93 and their possible involvement in the site was concluded because in the three-dimensional structure they lie reasonably well in an imaginary plane (or line) bearing the other residues. Since the two end and two middle residues were unambiguously assigned, the boundaries of the site were therefore clearly defined. Of course, the mere occurrence of favorable spatial proximity is not a sufficient factor for the residues in question to form an antigenic site in a protein. The three-dimensional model explains how this is possible and is <u>not</u> used to predict the location of the antigenic site.

At the time these studies were completed, this was the first antigenic reactive site (i.e. spanning residues that are close in three-dimensional structure but distant in sequence) to be described. The antigenic structure of only one protein (i.e. sperm-whale myoglobin) had been completed (Atassi, 1975) and we have shown and cautioned that the sequence and three-dimensional features that confer antigenicity on certain parts of a protein molecule are not too clear (Atassi, 1975). Accordingly, we felt it necessary to provide a more direct and completely unambiguous evidence for the boundaries and the participating residues of the reactive site.

To accomplish this goal, an entirely novel and powerful approach was devised (Atassi <u>et al.</u>, 1976d) for the study of antigenic (and perhaps other binding) sites of proteins. From examination of the three-dimensional structure of lysozyme, the distances between the contiguous residues of the site were measured and are shown in Fig. 14. It occurred to us (Atassi <u>et al.</u>, 1976d) that a very direct and totally unambiguous proof of the structure of the antigenic site would be through synthesis and immunochemistry of a peptide in which the relevant residues are <u>directly</u> linked by peptide bonds. However, we estimated that it would be necessary to introduce a glycine spacer between tryptophan and lysine in order to obtain the correct separation between their side-chains (Fig. 14). The peptide: Phe-Gly-Lys-Asn-Thr-Asp (Fig. 14) was synthesized and its immunochemistry studied (Atassi <u>et al.</u>, 1976d). It was gratifying that the approach worked so well. This peptide (which does not exist in native lysozyme but simulates a surface region of the protein:(hence named (Lee and Atassi, 1976) "surface-simulation" synthesis) was shown (Atassi <u>et al.</u>, 1976d) to possess an inhibitory activity (see Table 8) which was almost equal to the maximum expected reactivity of the site (i.e. a third of the total antigenic reactivity of lysozyme.).

These findings defined conclusively and accurately the reactive site (Figs. 13 and 14). Also, the results of this approach provided (Atassi <u>et al.</u>, 1976d) probably the strongest and most direct chemical verification for the correctness of the three-dimensional structure of lysozyme derived from the x-ray studies of the crystalline protein. The demonstration that spatially-adjacent residues can act as a single antigenic site, as if in direct linkage constituted a powerful approach to protein immunochemistry. Even though the presence of such an immunochemical feature was proposed relatively early (Atassi and Saplin, 1968), it was subsequently found <u>not</u> to exist in myoglobin (Atassi and Koketsu, 1975), where all five sites comprise residues that are directly linked to one another (see Fig.1)

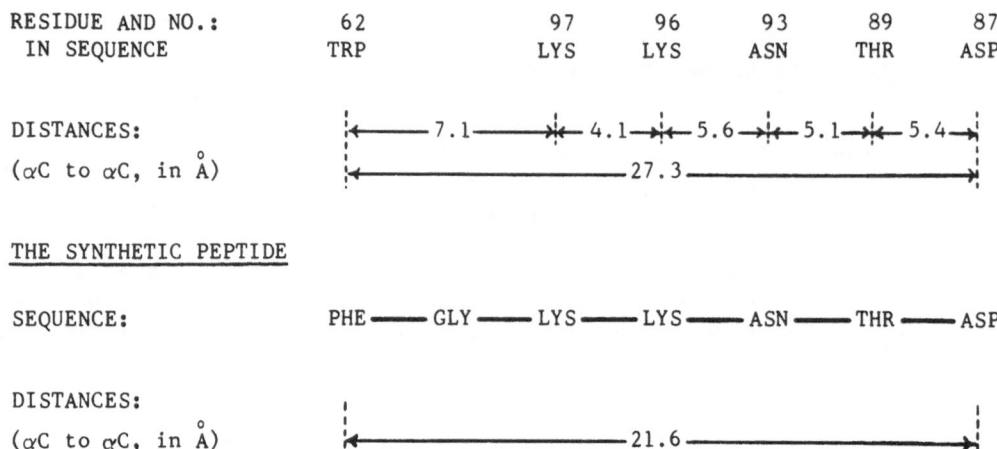

Fig.14. A diagram showing the spatially-contiguous surface residues
which we had previously established (Fig.13) to comprise antigenic
site (2). The distances (in Å) separating the consecutive residues
of the reactive site are given as C^{α} to C^{α} distances, together with
the overall "linear" dimension of the site. Below, the primary
structure of the "surface-simulation" synthetic peptide is given.
Previous studies had demonstrated (Fig. 12 and Table 7) that
phenylalanine can substitute for tryptophan with equal efficiency.
The total length of the synthetic peptide shown above assumes an
ideal C^{α}- C^{α} distance of 3.6 Å. Note that C^{α}- C^{α} distances are not
indicative of the separations between side-chains. The latter can
move over a wide range and need not be oriented in one line on
binding with antibody. (Figure is from Atassi et al., 1976d).

in the sequence (Atassi, 1975). The unequivocal establishment
of the concept described in this section indicated for the first
time that antigenic sites need not always be composed of residues
in direct peptide linkage in the sequence (Atassi et al., 1976d).
The unorthodox attack at the problem provides a novel and powerful
approach for final delineation of the antigenic sites (and perhaps
other types of binding sites) in native proteins. However, in view
of the numerous possibilities of surface-residues line-ups, it can
and indeed should only be applied in the final stages of the en-
deavor after all the exhaustive and accurate narrowing down by
chemical methods had been achieved (Atassi et al., 1976d).

TABLE 8. Inhibitory Activity of the Pure "Surface-Simulation" Synthetic Site 2[a]

Results are expressed in maximum percent inhibition by the peptide of the precipitin reaction of native lysozyme with two goat antisera. Each value is the average of at least three determinations which varied $\pm$ 0.7% or less. The immunochemical reactivity of the $(SS)_2$-peptide is given here for comparison.

Peptide	Antiserum G9		Antiserum G10	
	Max. inhibition activity (%)	Molar ratio at $\frac{1}{2}$ max. inhibn.*	Max. inhibition activity (%)	Molar ratio at $\frac{1}{2}$ max. inhibn.*
Synthetic peptide	35.6	170	27.7	530
$(SS)_2$-peptide**	26.4	14.2	32.7	4.5

[a]The structure of the synthetic site is shown in Fig. 14.

*These values represent peptide/antigen molar ratio at 50% of the maximum inhibition.

**Obtained from Lee and Atassi (1975).

(Table is from Atassi et al., 1976d)

IX.   THE ANTIGENIC SITE AROUND THE DISULFIDE 30-115 (SITE 3)

1.   Assignment of the reactive site

From the immunochemical and conformational studies of specific chemical derivatives of lysozyme it was shown (see Section III-A) that both lysines 33 and 116 are parts of an antigenic site (Lee et al., 1975). Similarly from derivatives modified at tyrosyl residues (Section III-F), it was concluded that one or both of tyrosines 20 and 23 is located in, or very close to, an antigenic site in lysozyme (Atassi and Habeeb, 1969; Atassi et al., 1971). A disulfide peptide corresponding to sequence 22-33 (Cys 30-Cys 115) 115-116 (see Section IV-B) possessed a substantial inhibitory activity towards the immune reaction of lysozyme (Atassi et al., 1973). These findings indicated that the approximate location of the antigenic site will be around the disulfide 30-115 and that the lysines 33 and 116 are essential residues of the site which could also incorporate one or both of tyrosines 20 and 23. This antigenic site is now being delineated by application of the novel 'surface-simulation' synthetic approach described above.

X.   CONCLUDING REMARKS

The number of antigenic sites in lysozyme is very limited despite the numerous possibilities of surface-residues line-ups. Obviously, the manner of arrangement of the side-chains in the imaginary surface-encircling line is of critical importance. But why some particular orders are antigenic is not entirely apparent at this stage. It is well to emphasize again here the caution that the sequence and three-dimensional features that confer immunogenicity on certain parts of a protein molecule are not too clear (Atassi, 1975).

Some general conclusions relating to antigenic structures of proteins which we previously formulated at various stages during the accurate mapping out of the antigenic structure of myoglobin have been reported elsewhere (Atassi, 1972, 1975 and 1977). All these conclusions are quite applicable here as well. One important aspect of the antigenic structure of lysozyme was the nature of its antigenic sites which were both fascinating and unexpected. These are the first such sites to be defined in proteins. The complete delineation of the lysozyme antigenic structure clearly shows that the spatially-adjacent residues of the site can act as if in direct linkage. These sites are quite different in character from the reactive sites in sperm-whale myoglobin (Fig. 1) (Atassi, 1975). We had earlier suggested (Atassi and Saplin, 1968) that antigenic reactive sites can be composed of regions or residues that are close in three-dimensional structure but distant in sequence.

Following this early suggestion, it is somewhat interesting that we have not found (Atassi and Koketsu, 1975) this immunochemical feature to exist in myoglobin, at least in early-course antisera. As previously stated (Atassi et al., 1976d), "in effect, therefore, an antigenic site is like a specific short ribbon of residues poised upon the surface of the protein molecule. The residues may either be directly linked to one another like in myoglobin (Atassi, 1975) or be so spatially arranged on the surface so that they behave functionally (towards the antibody) as if they are directly linked". Whether one type of antigenic site or the other or a mixture of both will exist in a given protein will obviously depend on the protein. An important question here is what are the factors which determine the type of the site? This question may not be answered unequivocally at this stage. Lysozyme differs from myoglobin in that its structure is stabilized by four disulfide bonds. Perhaps it can be tentatively concluded that an important factor in determining the type of the antigenic site may be dependent on the stabilization or otherwise of the structure by internal disulfide cross-links (Lee and Atassi, 1976). It is likely that in proteins which are not reinforced by covalent cross-links, the mode of folding is not sufficiently rigid to permit a stable conformationally-adjacent arrangement of residues in construction of antigenic sites. A more definite understanding of this subject must await knowledge of the antigenic structures of several proteins from both classes (i.e. with and without internal cross-links).

The rapid accumulation of completed protein antigenic structures is not likely at the present time. The chemical strategy set up from the approaches to myoglobin and lysozyme antigenic structures, may simplify but will not remove the element of challenge and fascination of future such studies on proteins. For example, although the chemical strategy of approach that we have set forth for myoglobin proved to be valuable for lysozyme it would not have been, by itself, sufficient. Our studies with lysozyme presented different chemical challenges for which new chemical reactions had to be devised. Some of these reactions were first developed in connection with our studies of myoglobin and then evolved to suit lysozyme. However, many new approaches were devised to solve obstacles that were encountered only in lysozyme. Ultimately, our studies with lysozyme confronted us with a different concept of a protein antigenic site that demanded to be tackled by a novel imaginative approach which linked the conformationally-adjacent residues of the site directly (Atassi et al., 1976d). However, it is well to caution here against taking any short cuts. The final structure is based on the coordination of enormous amounts of evidence which must, like a jig-saw puzzle, fit when put together. Care, thoroughness and the necessary numerous double-checks should not be compromised. The complexity of the task will continue to demand a unique blend of imagination, chemical expertise and sustained long-term commitment.

XI.   ACKNOWLEDGEMENTS

The work was supported by grants AI 11973, AI 11974 and AI 13181 from the National Institute of Allergy and Infectious Diseases, National Institutes of Health, U.S. Public Health Service. In the early part of our studies, the work was sponsored by the American Heart Association (Grant No. 71-910). A considerable part of the work was completed during the tenure to MZA of an Established Investigatorship of the American Heart Association.

REFERENCES

Arnon, R., and Sela, M. (1969) Proc. Natl. Acad. Sci. USA 62:163

Arnon, R., Maron,E., Sela, M., and Anfinsen, C.B. (1971) Proc. Natl. Acad. Sci. USA 63:1450

Atassi, M.Z. (1968) Biochemistry 7:3078

Atassi, M.Z. (1969) Immunochemistry 6:801

Atassi, M.Z. (1972) Specific Receptors of Antibodies, Antigens and Cells, 118-136. 3rd International Convocation of Immunology, June 12-15, S. Karger

Atassi, M.Z. (1975) Immunochemistry 12:423

Atassi, M.Z. (1976) Immunochemistry of Proteins, p.1-161, Plenum.

Atassi, M.Z. (1977) Immunochemistry of Proteins, Volume 2, in press, Plenum.

Atassi, M.Z., and Habeeb, A.F.S.A. (1969) Biochemistry 8:1385

Atassi, M.Z., and Habeeb, A.F.S.A. (1972) Methods in Enzymology 25B:546

Atassi, M.Z., and Habeeb, A.F.S.A. (1977) Immunochemistry of Proteins Volume 2, in press, Plenum.

Atassi, M.Z. and Koketsu,J. (1975) Immunochemistry 12:741

Atassi, M.Z., and Pai, R.C. (1975) Immunochemistry 12:735

Atassi, M.Z. and Rosemblatt, M.C. (1974) J. Biol. Chem. 249:482

Atassi, M.Z. and Saplin, B.J. (1968) Biochemistry 7:688

Atassi, M.Z., and Thomas, A.V. (1969) Biochemistry 8:3385

Atassi, M.Z., and Zablocki, W. (1975) Biochim. Biophys. Acta 386:233

Atassi, M.Z., and Zablocki, W. (1976) J. Biol. Chem. 251:1653

Atassi, M.Z., Perlstein, M.T. and Habeeb, A.F.S.A. (1971) J. Biol. Chem. 246:4291

Atassi, M.Z., Suliman, A.M., and Habeeb, A.F.S.A. (1972) Immunochemistry 9:907

Atassi, M.Z., Habeeb, A.F.S.A., and Ando, K. (1973) Biochim. Biophys. Acta 303:203

Atassi, M.Z., Rosemblatt, M.C., and Habeeb, A.F.S.A. (1974) Immunochemistry 11:495

Atassi, M.Z., Suliman, A.M., and Habeeb, A.F.S.A. (1975a) Biochim. Biophys. Acta 405:452

Atassi, M. Z., Habeeb, A.F.S.A., and Lee, C.L. (1976a) Immunochemistry, accepted

Atassi, M.Z., Koketsu, J., and Habeeb, A.F.S.A. (1976b) Biochim. Biophys. Acta 420:358

Atassi, M.Z., Lee, C.L., and Habeeb, A.F.S.A. (1976c) Immunochemistry 13:7

Atassi, M.Z., Lee, C.L., and Pai, R.C. (1976d) Biochim. Biophys. Acta 427:745

Blake, C.C.F., Koenig, D.F., Mair, G.A., North, A.C.T., Phillips, D.C., and Sarma, V.R. (1965) Nature (Lond) 206:757

Bonavida, B., Miller, A., and Sercarz, E.E. (1969) Biochemistry 8:968

Brown, R.K. (1962) J. Biol. Chem. 238:1162

Brown, R.K., Durieux, J., Delaney, R., Leikem, E., and Clark, B.J. (1959) Ann. N.Y. Acad. Sci. 81:524

Canfield, R.E. (1963a) J. Biol. Chem. 238:2691

Canfield, R.E. (1963b) J. Biol. Chem. 238:2698

Canfield, R.E., and Liu, A.K. (1965) J. Biol. Chem. 240:1997

Edelhoch, H. (1962) J. Biol. Chem. 237:2778

Fujio, H., Imanishi, M., Nishioka, K., and Amano, T. (1968a)
    Biken J. 11:207

Fujio, H., Imanishi, M., Nishioka, K., and Amano, T. (1968b)
    Biken J. 11:219

Fujio, H., Martin, R.E., Ha, Y.M., Sakato, N., and Amano, T. (1974)
    Biken J. 17:73

Geiger, B., and Arnon, R. (1974) Eur. J. Immunol. 4:632

Gerwing, J., and Thompson, K. (1968) Biochemistry 7:3888

Ha,Y.M., Fujio, H., Sakato, N., and Amano, T. (1975)
    Biken J. 18:47

Habeeb, A.F.S.A., and Atassi, M.Z. (1969) Immunochemistry 6:555

Habeeb, A.F.S.A., and Atassi, M.Z. (1970) Biochemistry 9:4939

Habeeb, A.F.S.A., and Atassi, M.Z. (1971b) Immunochemistry 8:1047

Habeeb, A.F.S.A., Atassi, M.Z., and Lee, C.L. (1974) Biochim. Biophys.
    Acta 342:389

Imoto, T., Johnson, L.N., North, A.C.T., Phillips, D.C., and Rupley,
    J. A. (1972) The Enzymes (Boyer, P.D., Ed.) 7:665, Academic
    Press, N.Y.

Johnson, E.R., Anderson, W.L., Wetlaufer, D.B., Lee, C.L., and
    Atassi, M.Z. (1976) J. Biol. Chem., submitted

Jollès, J., Jauregui-Adell, J., Bernier, I., and Jollès, P. (1963)
    Biochim. Biophys. Acta 78:668

Jollès, J., Sportono, G., and Jollès, P. (1965) Nature (Lond)
    208:1204

Jollès, P., Jauregui-Adell, J., and Jollès, J. (1964) Compt. Rend.
    Acad. Sci. 258:3926

Koketsu, J., and Atassi, M.Z. (1973) Biochim. Biophys. Acta
    328:289

Koketsu, J., and Atassi, M.Z. (1974a) Immunochemistry 11:1

Koketsu, J., and Atassi, M.Z. (1974b) Biochim. Biophys. Acta
    342:21

Komatsu, T., Shinka, S., Dohi, Y., and Amano, T. (1975) _Biken J._ 18:61

Lee, C.L., and Atassi, M.Z. (1973) _Biochemistry_ 12:2690

Lee, C.L., and Atassi, M.Z. (1975) _Biochim. Biophys. Acta_ 405:464

Lee, C.L., and Atassi, M.Z. (1976) _Biochem. J._ 159:89

Lee, C.L., Atassi, M.Z., and Habeeb, A.F.S.A. (1975) _Biochim. Biophys. Acta_ 400:423

Lee, C.L., Pai, R.C., and Atassi, M.Z. (1976) _Immunochemistry_ 13:681

Maron, E., Shiozawa, C., Arnon, R., and Sela, M. (1971) _Biochemistry_ 10:763

Sakato, N., Fujio, H., and Amano, T. (1972) _Biken J._ 15:135

Shinka, S., Imanishi, M., Miyagawa, N., Amano, T., Inouye, M., and Tsugita, A. (1967) _Biken J._ 10:89

Shrake, A., and Rupley, J.A. (1973) _J. Mol. Biol._ 79:351

Singhal, R.P., and Atassi, M.Z. (1971) _Biochemistry_ 10:1756

Sokolovsky, M., Riordan, J.F., and Vallee, B.L. (1967) _Biochem. Biophys. Res. Commun._ 27:20

Strosberg, A.D., and Kanarek, L. (1970) _Eur. J. Biochem._ 14:161

Taubman, M.T., and Atassi, M.Z. (1968) _Biochem. J._ 106:829

Young, J.D., and Leung, C.Y. (1970) _Biochemistry_ 9:2755

7

CROSSLINKING OF ANTIBODY MOLECULES BY BIFUNCTIONAL ANTIGENS

Danute E. Nitecki, Virgil Woods and Joel W. Goodman

Department of Microbiology, University of California,

San Francisco, San Francisco, California 94143

ABSTRACT

A requirement for at least two antigenic determinants to induce humoral antibody responses has been demonstrated using derivatives of a small molecule, L-tyrosine-p-azobenzenearsonate (RAT).
This molecule itself induces only cellular immunity in guinea pigs.
Assymetric bifunctional antigens composed of one RAT moiety and one
haptenic determinant, such as DNP, with or without a spacer, induce
cellular immunity to RAT and anti-DNP antibody. A symmetrical bifunctional antigen comprised of two RAT determinants separated by
a rigid spacer, $(PRO)_{10}$, induces cellular and humoral responses,
but the same two functions separated by a flexible spacer (6-amino-
caproyl) gives cellular responses only. The bifunctionality of the
latter antigen is probably compromised by intramolecular stacking
of azoarsonate groups, since this molecule exhibits extensive hypochromism in physiological solution. The simplest hypothetical model
of cell cooperation leading to a humoral response, i.e., bridging
of T and B cells by antigen, requires the congregation of two lymphocytes on a molecule as small as DNP-RAT. Since the means of direct
demonstration of such an interaction are not available, the capacity
of such bifunctional molecules to bridge the receptors of specific
antibody molecules was examined here. Thin layer gel chromatography
was used to assess the crosslinking of the specific antibodies. It
was found that all the bifunctional molecules examined were able to
polymerize specific antibodies, regardless of whether they were effective mediators of cell cooperation in vivo.

The immune response in vertebrate animals can be roughly divided into two compartments: namely, the cellular immune response
and the humoral immune response (circulating antibody). The cells
reacting in both of these responses are lymphocytes. Somewhere in
the maturation pathway of these cells, those involved in cellular
immune reactions are influenced by the thymus and are hence designated as T cells. The cells involved in the humoral response are
designated B cells and are ultimately responsible for synthesis and
secretion of circulating antibody (Greaves, Owen, and Raff, 1974).

In order to produce a full immune response, that is, humoral
as well as cellular immunity, the antigen must possess at least two
functional structures or determinants (Alkan et al., 1972). One
determinant provides the immunogenicity and stimulates the T cell
compartment. The second determinant, frequently called a hapten,
provides the structure against which the antibody specificity is
directed.

Complex protein molecules usually contain multiple determinants
and even small proteins, such as lysozyme, has been shown to present
several antigenic sites (Atassi, Lee, and Pai, 1976). It has been
possible to show in our laboratory that a peptide hormone, glucagon,
composed of twenty-nine amino acids (MW 3647), contains the immuno-
genic determinant in its C-terminal dodecapeptide and the haptenic
determinant (against which most of the antibody was produced) in
its N-terminal heptadecapeptide chain (Senyk, Nitecki, and Goodman,
1971; Nitecki et al., 1971).

The sequence of events leading to cellular and antibody responses is unknown and subject to much speculation. It is possible,
although by no means certain, that a bifunctional antigen such as
glucagon bridges the T and the B cells at some point in time through
their appropriate receptors.

We have shown earlier that the small molecule, L-tyrosine-p-
azobenzenearsonate   (RAT), induces in guinea pigs cellular immunity
only, i.e., no significant levels of humoral antibody can be demonstrated (Alkan, Nitecki, and Goodman, 1971). Thus, under these
conditions the RAT molecule is a monofunctional immunogen and does
not carry a haptenic determinant.

Haptenic determinants, such as the dinitrophenyl moiety, can
be attached to the α-amino group of RAT with or without a spacer.
Immunization with such bifunctional assymetric antigens, e.g., di-
nitrophenyl-(6-aminocaproyl)$_{0-3}$-RAT [DNP-(SAC)$_{0-3}$-RAT], resulted in
cellular immunity to tyrosine-azobenzenearsonate group and antibody
with anti-dinitrophenyl specificity (Alkan et al., 1972).

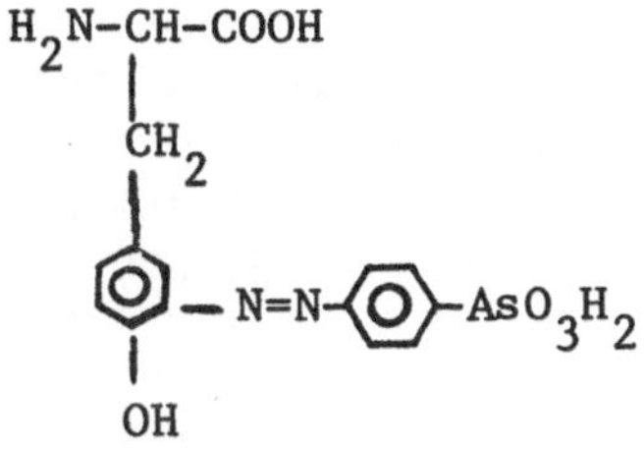

$$H_2N-CH-COOH$$

RAT

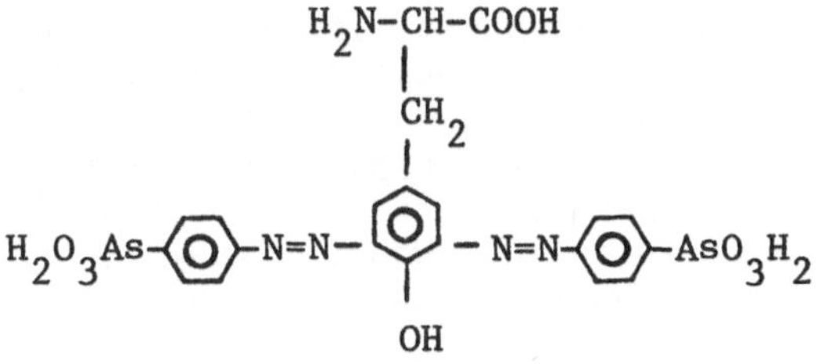

Bis-RAT

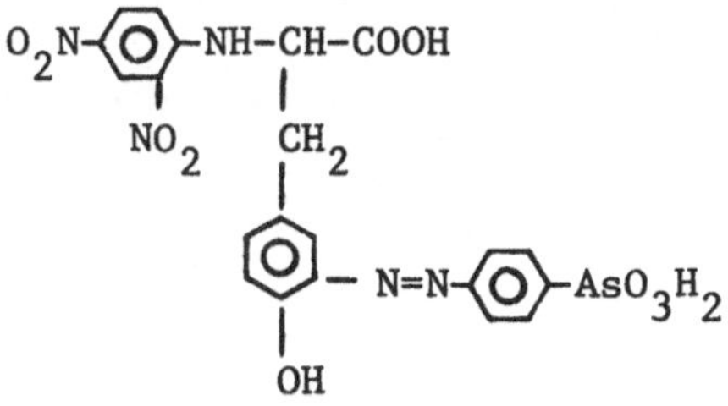

DNP-RAT

Symmetrical bifunctional antigens composed of two identical RAT determinants, i.e., molecules containing two azobenzenearsonate tyrosines separated by various spacers, were synthesized and used as antigens (Bush et al., 1972). These were found to induce only cellular immunity if the two determinants were separated by flexible spacers; that is, they behaved as monofunctional antigens. However, replacement of the flexible spacers by a rigid decaproline chain provided an antigen which was able to provoke cellular as well as humoral anti-RAT responses.

We have hypothesized that the inability of, for example, flexible RAT-6-aminocaproyl-RAT molecule, to behave like a bifunctional
antigen and to induce antibody responses is due to intramolecular
stacking of the two azobenzenearsonate groups, which could compro-
mise its bifunctionality as an antigen. This was supported by an
observed hypochromic decrease in the extinction coefficient of the
RAT-6-aminocaproyl-RAT compound as compared with rigid RAT-decaprolyl-
RAT compound (in physiological solution).

These observations raise intriguing questions in terms of simple models of T and B cell cooperation leading to the humoral antibody response. The requirement for at least two antigenic determinants is well established; mechanisms involving bridging of T and
B cells have been postulated (Goodman, 1975). It is difficult,
however, to visualize how two lymphocytes could congregate on a molecule as small as dinitrophenyl-RAT. Since direct binding between
T and B cells is impossible to test experimentally at present, we
attempted to assess the capacity of these bifunctional antigens to
bridge specific antibody molecules (Woods, Nitecki, and Goodman,
1975). While antibodies cannot be equated with cells, the antigen
receptors on cell surfaces are probably akin to antibody, at least
in the case of the B cell, and this approach represents a first approximation. Anti-dinitrophenyl and anti-RAT antibodies were obtained in rabbits by conventional immunization methods with appro-
priately conjugated proteins. These antibodies were purified by
affinity chromatography on cyanogen bromide treated Sepharose columns
(Porath et al., 1973) conjugated with either 6-aminocaproyl-RAT or
DNP-ovalbumin; the purified antibodies were quantitated by specific
precipitin reactions. The crosslinking of the antibodies by various
antigens was investigated by thin layer gel chromatography (TLG) on
glass plates coated with Sephadex G-200 superfine (Klaus, Nitecki,
and Goodman, 1972). The antigens used were: RAT; DNP-RAT; acetyl-
RAT-(6-aminocaproyl)$_{1-3}$-RAT; cyclo-L-RAT-L-RAT (L-tyrosine-diketo-
piperazine conjugated with one azobenzenearsonate group on each
phenol ring); cyclo-L-RAT-D-RAT; acetyl-RAT-(Prolyl)$_{10}$-RAT; Bis-RAT
(L-tyrosine conjugated with two azobenzenearsonate groups on the
same phenol ring); N,N'-bis-DNP-1,5 pentanediamine. Preparations
of antibody and bifunctional molecules were mixed in equimolar ratios
considered to favor crosslinking, based on the bivalency of antibodies. After brief incubation at room temperature, the mixtures
were assayed by TLG, the developed plates printed off on paper and
the paper stained by Coomassie BB R250 dye for visualization. TLG
on Sephadex G-200 allows rapid and simultaneous analysis of antibody
polymerization in multiple samples. Monomer rabbit IgG and a human
dimer IgA myeloma protein (provided by Dr. A.-C. Wang) were readily
resolved by this technique, but pentameric IgM and dimeric IgA were
not distinguishable from each other. Hence, in these experiments
dimers and higher order polymers were resolved from monomeric IgG
but not from each other.

The results are shown in Figures 1 and 2. The purified antibody preparations migrated as single components on Sephadex G-200
TLG (Fig. 1) with mobility identical to that of an authentic sample
of rabbit IgG (not shown). $N-DNP-C_5H_{10}-N-DNP$ did not change the chromatographic pattern of anti-RAT but led to the appearance of a spot
of more rapid mobility with anti-DNP, indicating the formation of
antibody oligomers. Conversely, the cyclic bifunctional RAT molecule, cyclo-L-RAT-D-RAT, in which the arsonate groups extend from
opposite sides of the diketopiperazine ring plane, had no effect on
the pattern of anti-DNP but produced an oligomeric component with
anti-RAT. Hence, specific cross-linking of antibody molecules took
place in the presence of an appropriate symmetrical bifunctional
antigen.

The asymmetric bifunctional antigen, DNP-RAT, in which the DNP
group was substituted directly on the amino group in the side chain
of tyrosine, did not crosslink either anti-DNP or anti-RAT antibodies
but did produce polymers with a 1:1 mixture of the two preparations
(Fig. 1). As seen in Figure 1, the two antibodies did not interact
in the absence of the crosslinking antigen.

Symmetrical bifunctional RAT antigens with flexible SAC spacers
or rigid decaproline spacers crosslinked anti-RAT antibodies to
varying degrees, based on relative intensities of the monomeric and
oligomeric spots (Fig. 2). Increasing the size of the flexible
spacer beyond a single SAC unit appeared to enhance polymerization
(compare G, I and J in Fig. 2), although quantitative comparisons
may be somewhat questionable. Even Bis-RAT, with two azobenzenearsonate groups on a single phenol ring, produced some polymerization, whereas RAT itself did not. Bifunctional RAT did not polymerize anti-DNP antibody.

The present study attempts to determine if a correlation exists
between the bridging capability and the capacity of bifunctional
molecules to implement the (presumed) cooperation between T and B
cells in the in vivo induction of a humoral antibody response. The
rationale is based on the premise that a molecule which is unable
to bridge antibodies should likewise be incapable of mediating cell
cooperation, although the converse does not necessarily follow. A
comparison of the capacities of bifunctional antigens to bridge antibody molecules and to induce antibody responses in vivo, presumably through the mediation of cell cooperation, is shown in Table 1.
All these compounds induce a cellular type of response (delayed hy-
persensitivity) in guinea pigs. All the bifunctional molecules were
able to crosslink antibody although only DNP-RAT and $Ac-RAT-(PRO)_{10}-$
RAT have been effective inducers of humoral immune responses and,
therefore, agents presumably able to mediate cell cooperation.

Figure 1. TLG patterns of A: anti-RAT + N-DNP-C$_5$H$_{10}$-N-DNP; B: anti-DNP + cyclo-L-RAT-D-RAT; C: anti-RAT + cyclo-L-RAT-D-RAT; D: anti-DNP + N-DNP-C$_5$H$_{10}$-N-DNP; E: anti-DNP + DNP-RAT; F: anti-RAT + DNP-RAT; G: anti-DNP + anti-RAT + DNP-RAT; H: anti-DNP + anti-RAT.

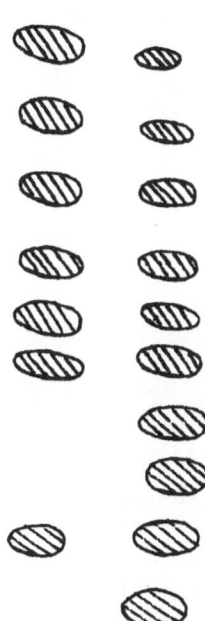

Figure 2. TLG patterns of A: anti-RAT; B: anti-RAT + Bis-RAT; C: anti-RAT + arsanilic acid; D: anti-RAT + RAT; E: anti-RAT + Ac-RAT-(PRO)$_{10}$-RAT; F: anti-RAT + cyclo-L-RAT-L-RAT; G: anti-RAT + Ac-RAT-SAC-RAT; H: anti-RAT + cyclo-L-RAT-D-RAT; I: anti-RAT + Ac-RAT-(SAC)$_3$-RAT; J: anti-RAT + Ac-RAT-(SAC)$_2$-RAT.

TABLE 1

Capacity of Mono- and Bifunctional Antigens to
Bridge Antibody Molecules and to Induce
Humoral Antibody Responses

Antigen	Bridging capacity	Humoral response
		Ppting antibody µg/ml
RAT	—	< 2
DNP-RAT	+	52 ± 17
cyclo-L-RAT-D-RAT	+	< 2
Ac-RAT-SAC-RAT	+	< 2
Ac-RAT-$(SAC)_2$-RAT	+	< 2
Ac-RAT-$(SAC)_3$-RAT	+	< 2
Ac-RAT-$(PRO)_{10}$-RAT	+	135 ± 18

Monofunctional RAT induces a pure cellular immune response, apparently does not mediate T cell-B cell cooperation, and does not crosslink anti-RAT antibodies, as expected. The smallest asymmetric bifunctional antigen, DNP-RAT, induces anti-DNP antibody responses in guinea pigs and crosslinks a mixture of anti-DNP and anti-RAT antibodies. The bond distances between the one carbon of DNP and the carbon ortho to the hydroxyl group of the phenol ring in DNP-RAT total 8.18Å, a maximum value since it does not take bond angles or conformational folding into account. The fact that antibody molecules can be bridged by this bifunctional antigen does not, of course, prove that it can serve as a cellular bridge, but the observation that it induces an antibody response suggests that this may be so.

Most interesting and surprising is the finding that Bis-RAT, a molecule in which the two functional groups (azobenzenearsonates) are separated by one carbon atom, is also able to crosslink two antibody molecules and, moreover, crosslink them firmly enough to survive the rigors of thin layer gel filtration. The antibody combining sites have been generally estimated to accomodate 5-7 hexose

units or 3-6 amino acids (Goodman, 1975). The crosslinking of antibody by Bis-RAT is tantamount to a paperclip holding two cars
together.

Symmetrical bifunctional RAT molecules with flexible spacers
do not induce significant humoral responses but polymerize anti-RAT
antibody at least as effectively as the rigidly spaced RAT bifunctional (Fig. 2), which does induce anti-RAT responses. Intramolecular stacking has been proposed as an explanation of the inability
of RAT-(SAC)$_{1-3}$-RAT compounds to implement cell cooperation. This
hypothesis was supported by hypochromism in the absorption spectra
of the compounds, a predictable consequence of stacking. If this
is the correct explanation, then stacking is an effective barrier
for cell cooperation but not for antibody crosslinking. In any
event, it is clear that the capacity of bifunctional antigens to
crosslink antibody molecules is not a reliable measure of their
ability to implement cell cooperation.

ACKNOWLEDGMENTS

We thank Miss Inge Stoltenberg for her skillful technical assistance. This work was supported by a grant from the United States
Public Health Service (AI 05664).

REFERENCES

Alkan, S.S., Nitecki, D.E. and Goodman, J.W. (1971). Antigen-recog-
    nition and the immune response. The capacity of L-tyrosine-
    azobenzenearsonate to serve as a carrier for a macromolecular
    hapten. J. Immunol., 107, 353.
Alkan, S.S., Williams, E.B., Nitecki, D.E. and Goodman, J.W. (1972).
    Antigen recognition and the immune response. Humoral and cel-
    lular immune responses to small mono- and bifunctional antigen
    molecules. J. Exp. Med., 135, 1228.
Atassi, M.Z., Lee, C.L. and Pai, R.C. (1976). Enzymic and Immuno-
    chemical properties of lysozyme. XVI. A novel synthetic approach
    to an antigenic reactive site by direct linkage of the relevant
    conformationally adjacent residues constituting the site.
    Biochim. Biophys. Acta., 427, 745.
Bush, M.E., Alkan, S.S., Nitecki, D.E. and Goodman, J.W. (1972).
    Antigen recognition and the immune response: "Self-help" with
    symmetrical bifunctional antigen molecules. J. Exp. Med., 136,
    1478.
Goodman, J.W. (1975). "Antigenic Determinants and Antibody Combin-
    ing Sites," In The Antigens, M. Sela, Ed., Vol. 3, p. 127,
    Academic Press, New York, N.Y.
Greaves, M.F., Owen, J.J.T. and Raff, M.C. (1974). T and B lympho-
    cytes. Excerpta Med., Amsterdam.

Klaus, G.G.B., Nitecki, D.E. and Goodman, J.W. (1972). Estimation
    of the molecular weights and Stokes' radii of proteins of thin-
    layer gel filtration in guanidine hydrochloride. Anal. Biochem.,
    45, 286.
Nitecki, D.E., Senyk, G., Williams, E.B. and Goodman, J.W. (1971).
    Immunologically active peptides of glucagon. Intrasci. Chem.
    Rep., 5, 295.
Porath, J., Aspberg, K., Drevin, H. and Axen, R. (1973). Prepara-
    tion of cyanogen bromide-activated agarose gels. J. Chromatog.,
    86, 53.
Senyk, G., Nitecki, D.E. and Goodman, J.W. (1971). The functional
    dissection of an antigen molecule: Specificity of humoral and
    cellular immune responses to glucagon. J. Exp. Med., 133, 1294.
Woods, V., Nitecki, D.E., Goodman, J.W. (1975). The capacity of
    bifunctional antigens to bridge antibody molecules and to me-
    diate cell cooperation. Immunochemistry, 12, 379.

8

MODIFICATION OF THE BIOLOGICAL PROPERTIES OF PLANT LECTINS BY

CHEMICAL CROSSLINKING

Reuben Lotan[1] and Nathan Sharon

Department of Biophysics, The Weizmann Institute

of Science, Rehovot, Israel

ABSTRACT

Soybean agglutinin, a glycoprotein lectin (m.w. 120,000, two
N-acetyl-D-galactosamine binding sites/mole) was crosslinked with
glutaraldehyde. Polymers (m.w. $\geqslant 240,000$) thus obtained were separated from uncrosslinked molecules by gel filtration and found to be
polyvalent with respect to sugar binding ($\geqslant 4$ binding sites/mole).
Soybean agglutinin polymers exhibited high agglutinating activity
toward erythrocytes and lymphocytes and triggered mitogenic stimulation of the latter cells, while the uncrosslinked divalent soybean agglutinin was a weaker agglutinin and failed to stimulate
lymphocytes. Similar activities were found with aggregates of soybean agglutinin which formed upon lyophilization and storage of the
lectin.

The effect of crosslinking of soybean agglutinin on its activities is assumed to be the result of increase in valency of the
divalent molecule by conversion into multivalent polymers. Polyvalent soybean agglutinin is more efficient in forming multiple
crossbridges between adjacent cells which lead to cell agglutination.
Stimulation of lymphocytes may require crosslinkage and/or topo-
graphic redistribution of cell surface receptors; this requirement
is satisfied by the poly- but not by the divalent lectin. The
effect of crosslinking of other lectins, such as concanavalin A,
phytohemagglutinin and the pokeweek mitogens, on their biological
properties is discussed.

[1] Present address: Dept. of Cancer Biology, The Salk Institute for
Biological Studies, San Diego, California 92112, U.S.A.

INTRODUCTION

Lectins are sugar-binding and cell agglutinating proteins widely distributed in nature, predominantly in the seeds of plants (Sharon and Lis, 1972, 1975; Lis and Sharon, 1973, 1976). In addition to cell agglutination, some lectins exhibit a remarkable ability to induce cell growth and division in lymphocytes, an effect known as "mitogenic stimulation".

The mechanism of action of lectins on cells is poorly understood. Even the seemingly simple cell agglutination is a complex phenomenon depending on the properties of both the lectin and the cell (Nicolson, 1974; Sharon and Lis, 1975). Although it is generally accepted that the initial step in the interaction of lectins with cells is their binding to specific receptor sites on the cell surface, binding of lectins to cells without subsequent agglutination is frequently observed. There is no satisfactory explanation for the ability of many lectins to preferentially agglutinate malignant cells and the enhanced susceptibility of erythrocytes and normal somatic cells to agglutination by lectins following mild treatment of the cells with proteolytic enzymes (Lis and Sharon, 1973).

The mechanism by which lectins stimulate lymphocytes is even more complex and less well understood than the agglutination phenomenon. Lymphocyte stimulation also requires binding of lectin molecules to cell surface glycoproteins (or glycolipids). The binding apparently generates some changes in the plasma membrane which serve as a signal to initiate a series of biochemical events that culminate in cell growth. It has been proposed that some degree of crosslinkage of lectin receptors is required for stimulation (Greaves and Janossy, 1972; Edelman et al., 1973).

One approach which may provide an insight into the mechanism of cell agglutination and lymphocyte stimulation by lectins is to study the effect of structural modifications of these proteins on their biological activity. Here we describe the effect of crosslinking of soybean agglutinin (SBA)(Fig. 1) with glutaraldehyde on its cell agglutinating and mitogenic activities. This modification was aimed at producing high molecular weight, multivalent derivatives which may be useful in the elucidation of the effect of molecular size and the number of saccharide binding sites in a lectin molecule on cell agglutination and lymphocyte stimulation.

CROSSLINKING OF SOYBEAN AGGLUTININ WITH GLUTARALDEHYDE

Glutaraldehyde has been used to crosslink antibodies (Avrameas and Ternynck, 1969) and concanavalin A (Donnelly and Goldstein, 1970). In both studies the purpose of crosslinking was to obtain insoluble

Molecular weight	120,000 ± 5,000
Subunits	4 x 30,000
Carbohydrate	D-Man, 4.5% D-GlcNAc, 1.5%
S-S, SH groups	0
NH₂ terminal	4 Ala
Binding sites	2
K_a , GalNAc	3×10^4

Figure 1.   Molecular properties of soybean agglutinin (Lotan et al., 1974).

polymers. To obtain water-soluble polymers of SBA we used mild reaction conditions which are described below.

SBA (200 mg) was dissolved in 15 ml of phosphate buffered saline, pH 7.4 (PBS) containing 2 M D-galactose. Glutaraldehyde was added dropwise to a final concentration of 0.03% (w/v), and the solution was stirred at 23° for 1 hr (Donnelly and Goldstein, 1970, used a final concentration of glutaraldehyde of about 0.2% (w/v) which insolubilized concanavalin A within 15 min.). The clear SBA solution was dialyzed extensively at 4°, first against glycine (2 mg/ml) in PBS and then against PBS alone. Insoluble material which precipitated during dialysis (about 25% of the starting amount of SBA) was removed by centrifugation. The soluble fraction was applied to an affinity chromatography column (Sepharose N-ε-aminocaproyl-β-D-galactopyranosylamine, Gordon et al., 1972). Unbound material was washed off the column with saline; this fraction contained about 15% of the starting amount of SBA and was devoid of hemagglutinating activity. Specific elution with D-galactose, 0.5% (w/v) in saline, afforded 110 mg (55% yield) of SBA capable of saccharide binding.

ISOLATION OF SBA POLYMERS

Gel electrophoresis of the glutaraldehyde-treated, affinity purified SBA preparation demonstrated the presence of several molecular species with varying degrees of polymerization as well as of uncrosslinked SBA molecules (compare gels I and II; inset, Fig. 2). To separate the polymers from the latter uncrosslinked material the preparation was subjected to gel filtration on a column of Sephadex G-150. Two peaks of protein were obtained, the first peak consisted

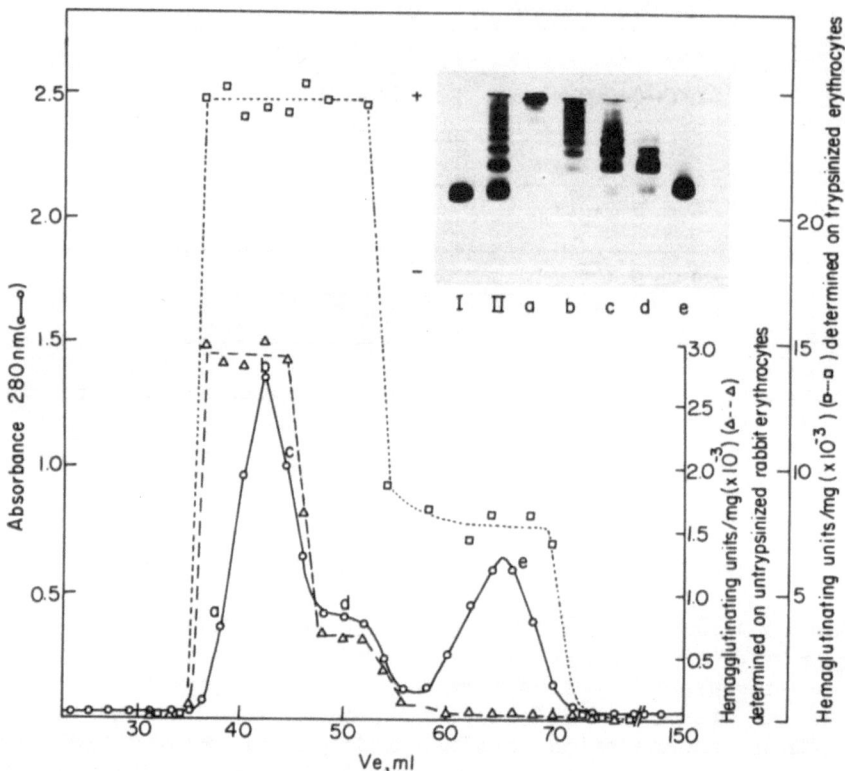

Figure 2. Fractionation of SBA crosslinked with glutaraldehyde. Subsequent to treatment with glutaraldehyde and affinity
chromatography, SBA (10 mg in 1 ml of saline) was applied to a
column (1.9 x 50 cm) of Sephadex G-150 superfine, pre-equilibrated
in saline at 4°. Elution with saline was carried out at a flow
rate of 4 ml/hr and fractions (2 ml) were collected, their absorbance at 280 nm and the hemagglutinating activity were measured.
Inset: Gel electrophoresis patterns of: I - native SBA; II - crosslinked SBA after affinity chromatography; a-e - selected fractions
eluted from the Sephadex column at points indicated on the elution
profile. Electrophoresis was performed on a gradient of acrylamide
from 4% (top) to 10% (bottom), prepared according to Kamm and Mes
(1971). The discontinuous buffer system, pH 4.3 (Reisfeld et al.,
1962) was used. Samples (100 μg) were applied to cylindrical gels
(0.5 x 8 cm) and electrophoresis was conducted at 2 mA per gel for
5 hr. The gels were stained for protein with Coomassie Brilliant
Blue R-250 (Lotan et al., 1973).

of several types of crosslinked SBA polymers which differed in size
(Fig. 2, inset gels a-d), and the second peak was uncrosslinked SBA
which comigrated with native SBA both on the Sephadex column and on
the gels (compare gels 1 and e, inset Fig. 2). The molecular weights
of the polymers were outside the fractionation range of the column
used except for fraction d which was not completely excluded. This
fraction was eluted at a volume corresponding to a molecular weight
of about 240,000 suggesting that it consisted mainly of a SBA dimer
(Fig. 2, inset gel d). The molecular weights of the other polymers
were assumed to be equal to higher multiples of 120,000 daltons,
and when the logarithms of the assumed molecular weights were plotted
against the distance of migration of the corresponding polymers
into the gel (Fig. 3), a linear correlation was obtained indicating
that the polymers were indeed of the type $(120,000)_n$ where
$n = 2,3,4$ .... The range of molecular weights of the SBA polymers
eluted in the first peak of the Sephadex column (Fig. 2) was between
360,000 and 1,200,000 (Fig. 3).

SELF-AGGREGATION OF SBA

SBA preparations stored in the lyophilized state at room temperature ($ca.$ 23°) for prolonged periods of time have been found to
contain high molecular weight aggregates (Lotan et al., 1975).
These aggregates resembled the glutaraldehyde-crosslinked SBA in
their electrophoretic migration and pattern of elution from the
Sephadex G-150 column; however, their quantities were much smaller
than those obtained by chemical crosslinking, and their degree of
polymerization was lower. The chemical nature of the self-aggregation process is not known. Treatment of SBA with 0.1% sodium
dodecyl sulphate for 3 min at 100° - conditions leading to almost
complete dissociation of native SBA into its subunits (Fig. 4, I) -
resulted in incomplete dissociation of the aggregated form (Fig. 4,
II). This finding suggests that the bonds holding the aggregates
together are stronger than those responsible for the association of
the subunits into the tetrameric structure of native SBA. Since
SBA is devoid of SH groups and S-S bonds (Lotan et al., 1974), self
aggregation must be the result of an interaction of other groups.
We are not aware of any report on similar aggregation of proteins
in dry state. There was, however, a report on somewhat similar
phenomenon: lyophilization of bovine pancreatic ribonuclease in 50%
acetic acid afforded aggregates of two, four or more molecules
(Crestfield et al., 1962).

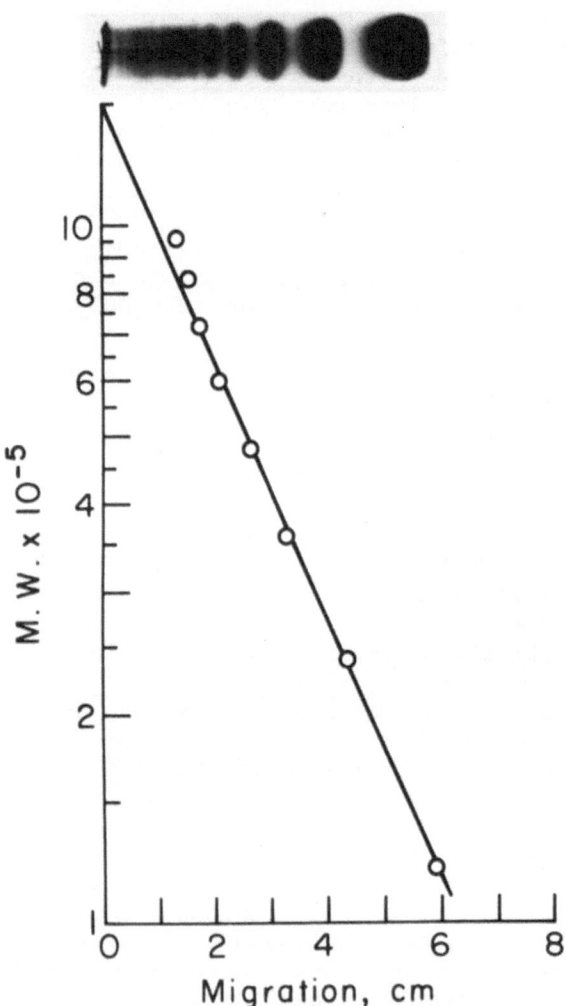

Figure 3. Estimation of the molecular weight of SBA polymers from their electrophoretic migration. Crosslinked SBA after affinity chromatography (100 µg) was subjected to gel electrophoresis under conditions identical to those described in the legend to Fig. 2. The m.w. was plotted on a logarithmic scale against the distance of migration (in cm) of the protein polymers.

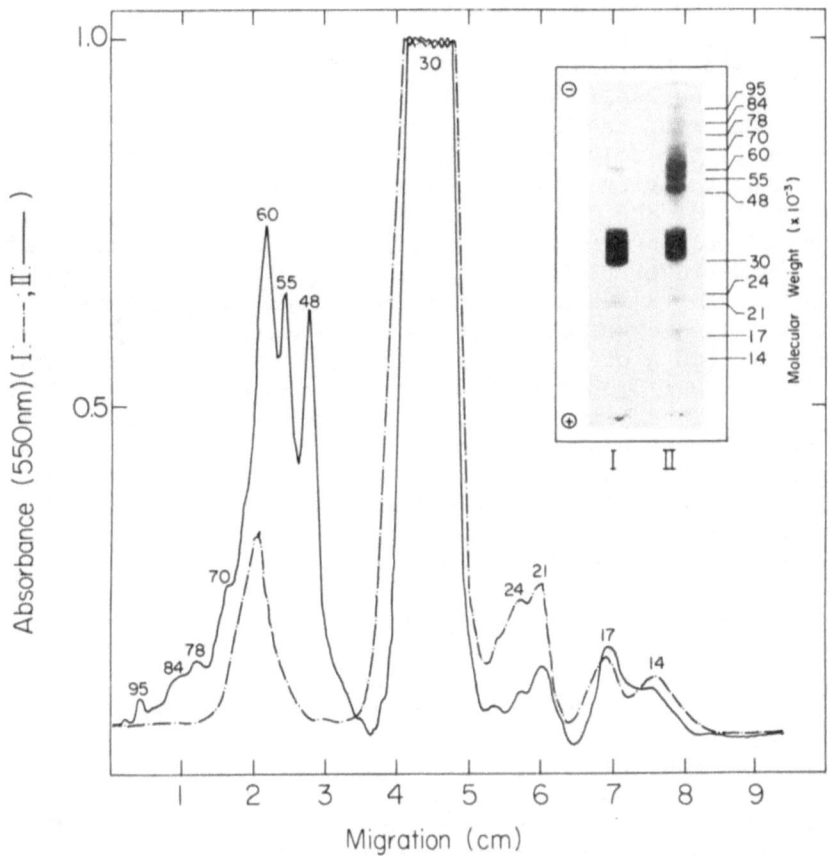

Figure 4. Gel electrophoresis of SBA in the presence of 0.1% SDS at pH 7.2, according to Weber and Osborn (1969). I - native, unaggregated SBA (200 µg); II - self-aggregated SBA (200 µg). After electrophoresis the gels were fixed and stained in 12.5% (w/v) CCl₃COOH. Molecular weights were calculated from the migration of standard proteins under the same conditions. Migration was from the top. Densitometric tracing was performed with a Gilford Model 2400-S scanner at 550 nm, (Lotan et al., 1975).

EFFECT OF CROSSLINKING ON SOME CHEMICAL AND
BIOLOGICAL PROPERTIES OF SBA

Saccharide Binding

Crosslinking by glutaraldehyde may affect the sugar binding
sites of SBA either by sterically blocking access of saccharides
or glycoproteins to these sites or by modifying essential amino
groups. Likewise, self-aggregation of SBA may also decrease the
effective number of saccharide binding sites per mole. To determine
the valence of the polymeric forms of SBA we have analyzed the
sugar-binding capacity of native SBA (m.w. 120,000), a fraction consisting mainly of SBA dimer (m.w. 240,000) and a fraction containing
a mixture of polymers (m.w. >360,000) by the gel filtration method
of Hummel and Dreyer (1962) on a column of Sephadex G-25 equilibrated with 0.1 mM [^{3}H]N-acetyl-D-galactosamine. The results of such
experiments, when expressed as moles of N-acetyl-D-galactosamine
bound per 120,000 g of SBA, gave values of 1.99 ± 0.04 for the
native SBA, 1.99 ± 0.08 for the dimers of SBA and 1.93 ± 0.08 for
the polymers, Thus, it seems as if polymerization by self aggregation did not cause a significant decrease in the number of N-acetyl-
D-galactosamine binding sites per 120,000 g SBA, implying that the
240,000 m.w. species is tetravalent and that the polymers of higher
m.w. are polyvalent (>6 binding sites per mole).

Cell Agglutination

The highly crosslinked SBA fractions which were eluted in the
first peak of the Sephadex column exhibited higher specific hemagglutinating activity than the dimeric and the uncrosslinked SBA
(Fig. 2). The enhanced activity was most noticeable when untreated
erythrocytes were used for agglutination, whereas a considerably
lower increase in the activity was observed when the assay was
carried out with trypsinized or with neuraminidase-treated erythrocytes (Table I). These results indicate that much lower concentrations of polyvalent SBA are required to agglutinate untreated
cells. However, treatment of the cells with neuraminidase or with
trypsin, both of which increase the susceptibility of cells to
agglutination by uncrosslinked SBA, decreased the difference in the
activities of polyvalent and divalent (uncrosslinked) SBA. The
activity of the tetravalent dimer was lower than that of the polyvalent lectin but higher than that of the uncrosslinked SBA. When
the rate of agglutination of untreated erythrocytes by polyvalent
and divalent SBA preparations was compared (Fig. 5), it was found
that cells were agglutinated faster by the polyvalent lectin. The
results given in Fig. 5 were obtained with the self-aggregated SBA,
but similar rates were observed with the glutaraldehyde-crosslinked
SBA.

TABLE 1

ENHANCEMENT OF THE CELL AGGLUTINATING ACTIVITY OF
SBA BY CROSSLINKING WITH GLUTARALDEHYDE

Cell Type	Treatment	Relative agglutinating activity of SBA[a]		
		Uncrosslinked (Fraction e)	Dimer (Fraction d)	Polymer (Fraction b)
Mouse spleen lymphocytes	—	1	10	30
	Neuraminidase[b]	30	60	120
Human erythrocytes	—	1	25	200
	Neuraminidase[c]	125	300	600
	Trypsin[d]	150	350	600
Rabbit erythrocytes	—	1	20	120
	Neuraminidase[c]	200	700	600
	Trypsin[d]	220	700	700

[a] The agglutinating activity of uncrosslinked SBA obtained with each of the untreated type of cell used was taken as 1. The SBA fractions tested are those eluted from the Sephadex column (Fig. 2). The Table does not express the differences in the agglutinability of the different cell types, but it demonstrates the enhanced activity within each type of cell.

[b] Mouse spleen lymphocytes (10^8/ml) in PBS were treated with 50 U/ml of neuraminidase for 30 min at 37°, then washed twice with PBS and resuspended in PBS to a concentration of 5×10^6 cells/ml used for agglutination.

[c] Erythrocytes (10^8/ml) in PBS were treated with neuraminidase as in b, then washed and resuspended in PBS to about 10^6/ml.

[d] Erythrocytes (10^8/ml) in PBS were treated with trypsin (100 µg/ml) for 1 hr at 37°, then washed and resuspended to a concentration of 10^6/ml.

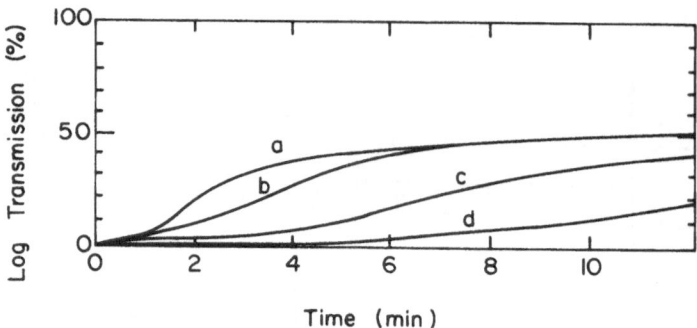

Figure 5. Rate of agglutination of human erythrocytes by SBA.
To a 2 ml suspension of erythrocytes, Type A (0.2% cells in PBS)
was added 0.4 ml of a solution containing: 1200 µg of aggregated
SBA (a); 1200 µg of unaggregated SBA (b); 150 µg of aggregated
SBA (c); 150 µg of unaggregated SBA (d). The increase in light
transmission resulting from agglutination was recorded automatically
in a Fragiligraph (Lotan et al., 1975).

    Cell agglutination by lectins is the result of the formation
of multiple lectin crossbridges between cells. The factors affecting the formation of such bridges include the structure and number
of sugar binding sites in the lectin molecule, cell rigidity and
deformability, the metabolic state of the cell, the electrical
charge of the cell surface, the number of lectin receptors on the
cell, their distribution and their mobility in the membrane
(Nicolson, 1974).

    The enhancement in the agglutinating activity of SBA upon
polymerization with glutaraldehyde is probably the result of the
increase in the valency and in the size of the lectin. Polyvalent
SBA may be expected to form multiple crossbridges between cells at
lower concentrations than the divalent lectin. The larger size of
SBA polymers may allow them to form bridges between cells without
bringing them as close to one another as would the smaller unaggregated lectin; therefore, the electrostatic repulsion between the
cells will be less pronounced with the polymerized SBA and agglutination will be facilitated. Another factor which should be considered for the understanding of the enhanced activity of crosslinked SBA is the effect of the number of sugar binding sites of
a lectin on its binding affinity to cell surfaces. Hammarström
(1973) has investigated the binding of native *Helix pomatia*
anti-A lectin which is hexavalent, and the binding of a partially

reduced lectin derivative which is divalent, to human erythrocytes and to tumor cell lines. He found that the apparent association constant of the multivalent lectin was 1,000-fold higher than that of the divalent derivative.

The rate of agglutination of erythrocytes by polyvalent SBA, which is faster than that by divalent SBA, also reflects the difference in their efficiency in crossbridging cells (Fig. 5).

Recently the effect of valence on binding affinity has been demonstrated numerically (Sawyer et al., 1975). As an example, if the occupation of a binding site involves a decrease in free energy of 5 Kcal/mol, then the binding of a monovalent monomeric lectin to a single membrane receptor would express an association constant of 4.6×10^3 M^{-1}. This is based on the following relation between the free energy of binding, $\Delta G°$ and the association constant K.

$$K = S \cdot e^{-p\Delta G°/RT}$$

where p is the number of bonds formed between the lectin and the receptors, S is the statistical factor which represents the number of binding configurations possible, R is the gas constant and T the absolute temperature. If a divalent dimer cross-links two receptor sites, the effective association constant would increase to 2.1×10^7 M^{-1}. Such a calculation illustrates the very large change in affinity which can result from a change of valence.

The importance of lectin valence for agglutination was noticed with other lectins. Galbraith and Goldstein (1972) have isolated two active components of lima bean lectin: component II (m.w. 247,000) was four times more active in agglutination of erythrocytes than component III (m.w. 124,000). It was later shown that component II was tetravalent and component III divalent (Bessler and Goldstein, 1974). Gunther et al. (1973) have found that succinyl-Con A, a divalent derivative of concanavalin A, was 500 times less active than native tetravalent Con A in agglutination of sheep erythrocytes. Pokeweed mitogens exist in five distinct molecular species differing in their molecular weight and with the exception of one (Pa-4), they show increased agglutinating activity with increased molecular weight; presumably the higher the molecular weight, the more sugar binding sites are present (Waxdal, 1974).

Following our report on the crosslinking of SBA, Hořejší and Kocourek (1974) used glutaraldehyde to prepare oligomers of lysozyme which have sugar binding properties similar to those of the lectin from wheat germ. They also separated the crosslinked lysozyme preparation on a Sephadex column and found that starting from the tetramer, their "semisynthetic lectin" had hemagglutinating activity. This activity increased with the molecular weight of the lysozyme

oligomers. It is not clear, however, whether binding of the
lysozyme oligomers to the cells was via the sugar binding site of
the modified enzyme, since no evidence for inhibition of agglutination by lysozyme inhibitors, such as GlcNAc-β(1→4)-GlcNAc-β(1→4)-
GlcNAc, was presented.

The finding that crosslinking of SBA has a much smaller effect
on its ability to agglutinate trypsinized erythrocytes than on the
agglutination of untreated ones may be a reflection of basic differences in the mobility of the receptors in these cells. In untreated erythrocytes, the distribution of receptors is presumably
random and membrane fluidity low, so that only an agglutinin with
high valency may be expected to achieve the degree of crosslinking
of receptors necessary for agglutination. It is assumed that in
trypsinized cells the fluidity of the membrane is high and permits
the rearrangement of receptor sites into clusters even with lectins
of low valency. Another reason could be the decrease in the charge
on the surface of cells treated with trypsin or with neuraminidase.
The removal of sialic acid residues by neuraminidase treatment and
the cleavage of sialoglycopeptides by trypsin (Cook and Eylar,
1965; Winzler et al., 1967) decrease the electrostatic repulstion
between the cells, thus facilitating their agglutination. Therefore, with neuraminidase- or trypsin-treated cells, the difference
between the agglutinating activity of polymerized SBA and that of
the native lectin will be smaller than with untreated cells.

Lymphocyte Stimulation

SBA was not mitogenic when tested with lymphocytes from mouse,
guinea pig, rat or man. However, these cells were stimulation by
the lectin following their treatment with neuraminidase (Novogrodsky
and Katchalski, 1973; Novogrodsky et al., 1975). Maino et al. (1975)
found that with pig lymphocytes the mitogenic activity of SBA was
high with both untreated and neuraminidase-treated lymphocytes.
The above experiments were carried out with SBA preparations that
had been stored in the lyophilized state. When it was found that
such preparations usually contain aggregates, the mitogenic properties of SBA were reinvestigated (Schechter et al., 1976). It was
found that both aggregated and crosslinked SBA preparations were
potent mitogens whereas native divalent SBA, when carefully separated from aggregates, completely lacked mitogenic capacity (Fig. 6).
The tetravalent SBA was as potent a mitogen as the polyvalent
lectin, both exhibiting maximal stimulation at around 10 μg/ml.
These observations were made both with untreated pig lymphocytes
and with neuraminidase-treated mouse lymphocytes.

A similar effect of lectin valency on its mitogenic activity
has been reported for the lima bean lectins: Ruddon et al. (1974)
have compared the mitogenic properties of lima bean component II,

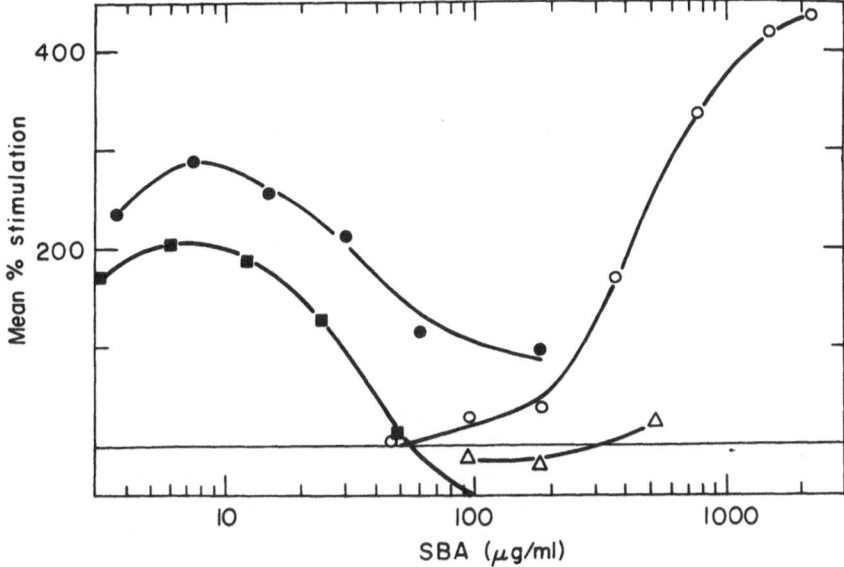

Figure 6.   Stimulation of pig lymphocytes by unfractionated (o), polyvalent (●), tetravalent (▪), and divalent (Δ) SBA (Schechter et al., 1976).

which is tetravalent, with those of component III, which is divalent, and found that the tetravalent lectin was several fold more potent than the divalent component III.   The latter studies were with human normal and leukemic lymphocytes.   Recent studies with lymphocytes from bovine, rabbit, rat and mouse have demonstrated that the tetravalent lima bean lectin is a far better mitogen than the divalent lectin; e.g., with bovine lymph node lymphocytes the uptake of thymidine into cells stimulated by the tetravalent lectin was almost 300 times higher than into the control cells, whereas cells treated with the divalent lima bean were not stimulated at all (Bessler et al., 1976, and personal communication).

The latter results, as well as our findings, demonstrate that a minimum of four binding sites per lectin molecule is required for stimulation of lymphocytes by SBA and by lima bean lectin.   Although the mechanism(s) by which lectins exert their stimulating effect is at present unknown, it is undisputable that the initial step in the sequence of events leading to mitogenesis is binding of the lectin through its sugar binding sites to sugar-containing cell surface receptors.   Binding *per se* is not sufficient for stimulation, and

many lectins bind to lymphocytes but fail to stimulate them (e.g.
Helix pomatia agglutinin [Dillner et al., 1975], peanut agglutinin
when tested with neuraminidase-treated mouse or guinea pig lymphocytes [Novogrodsky et al., 1975]). It was suggested, therefore,
that some degree of crosslinking of cell surface receptors may be
required for stimulation (Greaves and Janossy, 1972; Edelman et al.,
1973). If this assumption is correct, it is plausible to suggest
that the tetravalent SBA or lima bean component II are capable of
crosslinking receptors to an extent which triggers mitogenic stimulation, while divalent lectin (unaggregated SBA or lima bean component III) cannot achieve the required degree of crosslinking and
therefore are incapable of stimulating lymphocytes. Both SBA and
the lima bean lectin have a similar sugar specificity as they both
bind *N*-acetyl-D-galactosamine. It seems as if other lectins with
different sugar specificities may behave in a different way.
Gunther et al. (1973) reported that the divalent succinyl-Con A
was as potent a mitogen as was the native, tetravalent lectin.
Likewise the divalent garden pea lectin, which has a sugar specificity similar to that of Con A (methyl α-D-mannoside, methyl α-D-
glucoside) is a potent mitogen. Recently it has been reported that
even monovalent Con A is capable of stimulating lymphocytes (Beppu
et al., 1975; Fraser et al., 1976). These preparations are monovalent dimers that may undergo subunit exchange to give inactive
dimers and divalent dimers; therefore, it is difficult to interpret
the results, especially since a monovalent, monomeric Con A derivative prepared by Wands et al. (1976) was not mitogenic.

Sela et al. (1976) prepared monovalent succinylated anti-
carbohydrate antibody and demonstrated its ability to stimulate
lymphocytes. This result would suggest that the univalent antibody
may bypass the requirement for receptor crosslinking on the cell to
which it binds. However, as shown by Rosenstreich and his coworkers
(Rosenstreich and Wilton, 1975; Rosenstreich et al., 1976) there is
an absolute requirement of macrophages for lymphocyte stimulation by
lectins. It is possible that the receptor crosslinkage necessary
for lymphocyte stimulation may be achieved by indirect mechanisms
of mitogen presentation in concentrated areas on the surface of
macrophages as well as by the multivalent properties of the mitogenic agent itself (Sela et al., 1976). The latter hypothesis may
explain both our findings on the requirement of multivalency of
SBA for stimulation and the findings on stimulation by divalent or
monovalent mitogens.

CROSSLINKING OF LECTINS OTHER THAN SBA

Thus far, SBA is the only lectin which was crosslinked to yield soluble derivatives. There are, however, many reports on chemical crosslinking of other lectins to afford insoluble derivatives. Donnelly and Goldstein (1970) have crosslinked Con A with glutaraldehyde under conditions that converted the lectin into an insoluble form. The crosslinked lectin was capable of binding saccharides and was used for the isolation by affinity chromatography of polysaccharides (e.g. dextran, glycogen) or glycoproteins (e.g., immunoglobulin M). Recently such a preparation was used for the purification of carcinoembryonic antigen (Brattain et al., 1975). The properties of glutaraldehyde-crosslinked Con A are different from those of Con A bound to cyanogen bromide activated Sepharose; Brattain et al. (1976) found that Con A insolubilized with glutaraldehyde was able to bind carcinoembryonic antigen and horseradish peroxidase even in the presence of 8 M urea whereas the Con A-Sepharose conjugate was not active under the same conditions.

Several mitogens, when covalently attached to insoluble matrices, have been reported to exhibit altered biological activities. T cell mitogens such as Con A and phytohemagglutinin (PHA) are able to stimulate B cells when the lectins are bound to Sepharose beads (Greaves and Bauminger, 1972; Andersson and Melchers, 1973). Likewise, Con A crosslinked to plastic culture plates with water soluble carbodiimide stimulated B lymphocytes; however, it lost T-stimulating activity (Andersson et al., 1972). (Interestingly, however, aggregated soybean agglutinin appears to be mainly a T cell mitogen [Novogrodsky, 1974]). To explain these modified activities it was suggested that the binding of mitogens to insoluble matrices results in a local concentration effect which leads to B cell stimulation by a mechanism analogous to the concentration of antigens on the surface of T cells and their presentation to B cells in a form that leads to stimulation of the latter cells (Andersson et al., 1972). In an attempt to explore further the altered activities of insolubilized mitogens, Basham and Waxdal (Waxdal, 1975 and personal communication) crosslinked several mitogens by a bifunctional imidoester, dimethyl suberimidate. The insoluble aggregates of pokeweed mitogen Pa-2 and Pa-4, which are T-mitogens in their soluble form, were able to stimulate B cells without losing their T cell activity. The dose-response curves of these insoluble aggregates were markedly changed, and between 10 and 100 times more insolubilized   Pa-2 was required for maximal stimulation of T cells than by soluble Pa-2.

Unlike the soluble Pa-2, the insoluble derivative stimulated immunoglobulin synthesis by spleen cells from both BALB/c mice (possessing both T and B cells) and nude mice (possessing only B cells). Con A aggregated with dimethyl suberimidate maintained its

ability to stimulate T cells, though greater mitogen concentrations
were required for maximal stimulation. Unlike Con A attached to
plastic plates, the aggregated Con A did not acquire B cell stimulating capacity. The differences between the activities of soluble,
matrix-bound or crosslinked mitogen derivatives may be a reflection
of difference in the number of functional binding sites (e.g.
aggregation or binding to a matrix may sterically block many sites)
and the distribution of the binding sites in a unit of volume or
area (insoluble as well as soluble derivatives possess multiple
binding sites which may participate in multivalent interactions
with cell surface receptors). One of the pokeweed mitogens (Pa-1)
exhibits both T and B cell stimulation in its soluble form. This
dual activity may be due to its natural polymeric state (Waxdal,
1974, 1975).

CONCLUDING REMARKS

Crosslinking of lectins under mild conditions affords soluble
high molecular weight polymers which possess many sugar binding
sites. Extensive crosslinking, either with glutaraldehyde or with
dimethyl suberimidate, converts soluble lectins into insoluble
derivatives which conserve many of their sugar binding sites.
Soluble and insoluble polyvalent lectin derivatives are useful
tools in studies on the effect of multivalent interactions at the
cell surface of lymphocytes. Insolubilized lectin derivatives
have also been used to purify glycoproteins and polysaccharides.

ACKNOWLEDGMENTS

We thank Drs. H. Lis, A. Novogrodsky and B. Schechter and Mr.
A. Ravid for their collaboration in most of our investigations
reported here, and for stimulating discussions. Thanks are also
due to Drs. Bessler and Waxdal for communicating to us results of
their studies prior to publication. This work was supported in
part by grants from the United States-Israel Binational Science
Foundation, Jerusalem, a contribution from a friend of the Weizmann
Institute in Buenos Aires, Argentina, and the Foundation for
Advancement of Mankind, Jerusalem.

REFERENCES

Andersson, J. and Melchers, F. (1973). Induction of immunoglobulin
    M synthesis and secretion in bone-marrow derived lymphocytes
    by locally concentrated concanavalin A. *Proc. Natl. Acad. Sci.
    U.S.A., 70, 416-420.
Andersson, J., Edelman, G.M., Möller, G. and Sjöberg, O. (1972).
    Activation of B lymphocytes by locally concentrated concanavalin
    A. *Eur. J. Immunol., 2, 233-235.
Avrameas, S. and Ternynck, T. (1969). The cross-linking of proteins
    with glutaraldehyde and its use for the preparation of immuno-
    adsorbants. *Immunochemistry, 6, 53-66.
Beppu, M., Terao, T. and Osawa, T. (1975). Photoaffinity labeling
    of concanavalin A. Preparation of concanavalin A derivative
    with reduced valence. *J. Biochem. 78, 1013-1019.
Bessler, W. and Goldstein, I.J. (1974). Equilibrium dialysis studies
    on two lima bean lectins. *Arch. Biochem. Biophys. 165, 444-445.
Bessler, W., Resch, K. and Ferber, E. (1976). Valency-dependent
    stimulating effects of lima bean lectins on lymphocytes of dif-
    ferent species. *Biochem. Biophys. Res. Commun. 69, 578-585.
Brattain, M.G., Jones, C.M., Pittman, J.M. and Pretlow, T.G. (1975).
    The Purification of carcinoembryonic antigen by glutaraldehyde
    crosslinked concanavalin A. *Biochem. Biophys. Res. Commun.
    65, 63-67.
Brattain, M.G., Geer, J.C. and Pretlow, T.G. (1976). The binding
    of insolubilized concanavalin A to glycoproteins in the presence
    of 8 M urea. *Fed. Proc., 35, p. 754* (abst. 3027).
Cook, G.M.W. and Eylar, E.H. (1965). Separation of the M and N
    blood group antigens of the human erythrocyte. *Biochim.
    Biophys. Acta, 101, 57-66.
Crestfield, A.M., Stein, W.H. and Moore, S. (1962). On the aggreg-
    ation of bovine pancreatic ribonuclease. *Arch. Biochem.
    Biophys. Suppl. 1, 217-222.
Dillner, M.L., Hammarström, S. and Perlmann, P. (1975). The lack of
    mitogenic response of neuraminidase-treated and untreated human
    blood lymphocytes to divalent, hexavalent, or insoluble *Helix
    pomatia* A hemagglutinin. *Exptl. Cell Res., 96, 374-382.
Donnelly, E.H. and Goldstein, I.J. (1970). Glutaraldehyde-insolubil-
    ized concanavalin A: an adsorbant for the specific isolation of
    polysaccharides and glycoproteins. *Biochem. J., 118, 679-680.
Edelman, G.M., Yahara, I. and Wang, J.L. (1973). Receptor mobility
    and receptor-cytoplasmic interactions in lymphocytes. *Proc.
    Natl. Acad. Sci. U.S.A., 70, 1442-1446.
Fraser, A.R., Hemperly, T.T., Wang, J.L. and Edelman, G.M. (1976).
    Monovalent derivatives of concanavalin A. *Proc. Natl. Acad.
    Sci. U.S.A., 73, 790-794.
Galbraith, W. and Goldstein, I.J. (1972). Phytohemagglutinin of the
    lima bean (*Phaseolus lunatus*): Isolation, characterization and
    interaction with type A blood group substance. *Biochemistry,
    11, 3976-3984.*

Gordon, J.A., Blumberg, S., Lis, H. and Sharon, N. (1972). Purification of soybean agglutinin by affinity chromatography on Sepharose-*N*-ε-aminocaproyl-β-D-galactopyranosylamine. *FEBS Lett.*, *24*, 193-196.

Greaves, M.F. and Bauminger, S. (1972). Activation of T and B lymphocytes by insoluble phytomitogens. *Nature*, *235*, 67-70.

Greaves, M.F. and Janossy, G. (1972). Elicitation of selective T and B lymphocyte responses by cell surface ligands. *Transplant. Rev.*, *11*, 87-130.

Gunther, G.R., Wang, J.L., Yahara, I., Cunningham, B.A. and Edelman, G.M. (1973). Concanavalin A derivatives with altered biological activities. *Proc. Natl. Acad. Sci. U.S.A.*, *70*, 1012-1016.

Hammarström, S. (1973). Binding of *Helix pomatia* A hemagglutinin to human erythrocytes and other cells: Influence of multivalent interaction on affinity. *Scand. J. Immunol.*, *2*, 53-66.

Hořejší, V. and Kocourek, J. (1974). Studies on phytohemagglutinins. XXI. The covalent oligomers of lysozyme-first case of semisynthetic hemagglutinin. *Experientia*, *30*, 1348-1349.

Hummel, J.P. and Dreyer, W.J. (1962). Measurement of protein-binding phenomena by gel filtration. *Biochim. Biophys. Acta*, *63*, 530-532.

Kamm, L. and Mes, J. (1971) An apparatus to facilitate the preparation of continuous and gradient disc acrylamide gels. *J. Chromatogr.* *62*, 383-390

Lis, H. and Sharon N. (1973). The biochemistry of plant lectins (Phytohemagglutinins). *Ann. Rev. Biochem.* *42*, 541-576.

Lis, H. and Sharon, N. (1976). Lectins: their chemistry and applications to immunology. In:"The Antigens", ed. M. Sela, Vol. 4, Academic Press, in press.

Lotan, R., Lis, H., Rosenwasser, A., Novogrodsky, A. and Sharon, N. (1973). Enhancement of the biological activities of soybean agglutinin by crosslinking with glutaraldehyde. *Biochem. Biophys. Res. Commun.*, *55*, 1347-1355.

Lotan, R., Siegelman, H.W., Lis, H. and Sharon, N. (1974). Subunit structure of soybean agglutinin. *J. Biol. Chem.* *249*, 1219-1224.

Lotan, R., Lis, H. and Sharon, N. (1975). Aggregation and fragmentation of soybean agglutinin. *Biochem. Biophys. Res. Commun.*, *62*, 144-150.

Maino, V.C., Hayman, M.J. and Crumpton, M.J. (1975). Relationship between enhanced turnover of phosphatidylinositol and lymphocyte activation by mitogens. *Biochem. J.*, *146*, 247-252.

Nicolson, G.L. (1974). The interaction of lectins with animal cell surfaces. *Int. Rev. Cytol.*, *39*, 89-190.

Novogrodsky, A. (1974). Selective activation of mouse T and B lymphocytes by periodate, galactose oxidase and soybean agglutinin. *Eur. J. Immunol.* *4*, 646-648.

Novogrodsky, A. and Katchalski, E. (1973). Transformation of neuraminidase-treated lymphocytes by soybean agglutinin. *Proc. Natl. Acad. Sci. U.S.A., 70,* 2515-2518.

Novogrodsky, A., Lotan, R., Ravid, A. and Sharon, N. (1975). Peanut agglutinin, a new mitogen that binds to galactosyl sites exposed after neuraminidase treatment. *J. Immunol., 115,* 1243-1248.

Reisfeld, R.A., Lewis, U.J. and Williams, D.E. (1962). Disk electrophoresis of basic proteins and peptides in polyacrylamide gels. *Nature, 195,* 281-283.

Rosenstreich, D.L. and Wilton, J.M. (1975). The mechanism of action of macrophages in the activation of T-lymphocytes *in vitro* by antigens and mitogens. In: "Immune Recognition", ed. A.S. Rosenthal, Academic Press, New York, *pp. 113-132.*

Rosenstreich, D.L., Farrar, J.J. and Dougherty, S. (1976). Absolute macrophage requirement of T lymphocyte activation by mitogens. *J. Immunol. 116,* 131-139.

Ruddon, R.W., Weisenthal, L.M., Lundeen, D.E., Bessler, W. and Goldstein, J.J. (1974). Stimulation of mitogenesis in normal and leukemic human lymphocytes by divalent and tetravalent lima bean lectins. *Proc. Natl. Acad. Sci. U.S.A., 71,* 1848-1851.

Sawyer, W.H., Hammarström, S., Möller, G. and Goldstein, I.J. (1975). Precipitin and mitogenic behavior of dimeric and tetrameric concanavalin A. *Eur. J. Immunol., 5,* 507-510.

Schechter, B., Lis, H., Lotan, R., Novogrodsky, A. and Sharon, N. (1976). The requirement for tetravalency of soybean agglutinin for induction of mitogenic stimulation of lymphocytes. *Eur. J. Immunol., 6,* 145-149.

Sela, B.A., Wang, J.L. and Edelman, G.M. (1976). Lymphocyte activation by monovalent fragments of antibodies reactive with cell surface carbohydrate. *J. Exp. Med., 143,* 665-671.

Sharon, N. and Lis, H. (1972). Lectins: Cell agglutinating and sugar-specific proteins. *Science, 177,* 943-959.

Sharon, N. and Lis, H. (1975). Use of lectins for the study of membranes. *Method. Membrane Biol., 3,* 147-200.

Wands, J.R., Podolsky, D.K. and Isselbacher, K.J. (1976). Mechanism of human lymphocyte stimulation by concanavalin A: Role of valence and surface binding sites. *Proc. Natl. Acad. Sci. U.S.A., 73,* 2118-2122.

Waxdal, M.J. (1974). Isolation, characterization and biological activities of five mitogens from pokeweed. *Biochemistry, 13,* 3671-3677.

Waxdal, M.J. (1975). Differential stimulation of murine T and B cell populations by purified mitogens from pokeweed. In: "Immune Recognition", ed. A.S. Rosenthal, Academic Press, New York, *pp. 85-101.*

Weber, K. and Osborn, M. (1969). The Reliability of molecular weight
    determinations by dodecyl sulfate-polyacrylamide gel electro-
    phoresis. *J. Biol. Chem.*, *244*, 4406-4412.
Winzler, R.J., Harris, E.D., Pekas, D.J., Johnson, C.A. and Weber,
    P. (1967). Studies on glycopeptides released by trypsin from
    intact human erythrocytes. *Biochemistry*, *6*, 2195-2202.

9

INTRODUCTION OF ARTIFICIAL CROSSLINKS INTO PROTEINS

Rosa Uy and Finn Wold

Department of Biochemistry
University of Minnesota
St. Paul, MN 55108

ABSTRACT

In this chapter, we present a brief overview of the current
status of protein crosslinking technology. An attempt is made to
compare the natural crosslinks and natural crosslinking agents to
the artificial ones, and a brief section is devoted to the potential use of enzymes (transglutaminase and peroxidase) as crosslinking agents in vitro. Homobifunctional (x-R-x) and heterobifunctional (x-R-y) reagents are considered in terms of the kinds of
functional groups and R-groups that have been used in protein crosslinking, and some examples of reagents and applications from the
recent literature are tabulated.

INTRODUCTION

It was clear from the earliest studies on the covalent structure of proteins that the crosslinks contributed unique properties
to both the structure and function of the protein, and that the
elucidation of the location of the natural crosslinks could contribute unique information about the folding of the polypeptide
chain. It was natural that artificially produced crosslinks should
become one of the tools used by protein chemists to explore the
structural stability and establish the three-dimensional structure
of proteins. A number of bifunctional protein reagents were developed; initially they were applied mostly to induce increased
strength to structural proteins and stability to globular ones
(Zahn, 1955; Wold, 1961), or as "yardsticks" to measure interresidue distances in crosslinked protein derivatives (Zahn and

Meyenhofer, 1958; Fasold, 1965; Hartman and Wold, 1967), but their use was rapidly expanded to the exploration of the spatial arrangements of subunits in oligomeric proteins (Davies and Stark, 1970) and of the organization of protein molecules in more complex structures such as ribosomes (Slobin, 1972; Sun et al., 1974) and membranes (Steck, 1972; Wang and Richards, 1974; Ji and Ji, 1974). The development of protein immobilization techniques (Zaborsky, 1976) and also to some extent affinity chromatography gave a tremendous impetus to the search for new crosslinking technology with the main emphasis on intermolecular covalent linkages between proteins and various insoluble matrices, and more recently some new crosslinking methodology has been added for the specific purpose of establishing the molecular arrangement in nucleic acid-protein complexes (Sperling and Havron, 1976). The latter area, which is considered in other chapters of this volume, will not be included here.

The purpose of this chapter is to present a summary of current crosslinking technology, specifically as it applies to proteins. Since several reviews of different aspects of crosslinking have been published in the last 5 years (Fasold et al., 1971; Wold, 1972; Uy and Wold, 1976; Zaborsky, 1976) it was deemed desirable not to use the normal review format in the present context. Thus, rather than to survey in detail all the individual reagents and systems that have been studied to date, the discussion will be focussed on general considerations of protein crosslinking and the specific applications and techniques will only be identified through references to the recent literature.

CHOICE OF CROSSLINKING CONDITIONS

The Protein. Scheme 1 illustrates the most obvious states of the protein to be considered in selecting conditions favoring a given type of crosslinked product. Under standard conditions the protein may exist primarily as a simple monomer of conformation P. If a ligand A (substrate for example) binds to or induces a different conformation, P*, saturation with ligand will shift the equilibrium represented by K_1 to the left giving crosslinked P* rather than crosslinked P, which would be the expected product in the absence of A. The protein P may have a tendency to aggregate (the K_2-step) and high protein concentration would then favor formation of intermolecular crosslinks, which would yield stable dimers as the reaction product. Changes in the standard conditions such as pH, ionic strength, temperature, dielectric constant would also be likely to affect the K_2-step and therefore the relative yield of intra- and intermolecular crosslinks. The same considerations apply to the multiprotein complex on the right hand side of the scheme. Specific ligands and the solvent could affect the K_3-step or perturb the conformation of individual components and of the complex itself

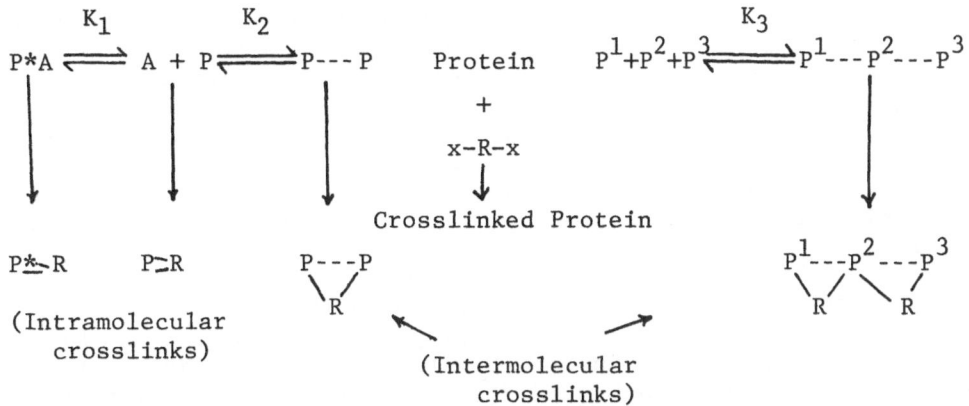

Scheme 1

and thus determine the nature of the crosslinking reaction.

The point of scheme 1 is simply that any knowledge of the behavior of the protein(s) to be crosslinked is important for the experimental design, and in fact, makes all the difference between a completely empirical "hit or miss" approach to a given experimental goal and a rationally planned experiment. The one unknown that will always remain in the consideration of a new protein system is whether reactive amino acid residues are present in the proper juxtaposition for a given reagent to form crosslinks, and this question will probably always have to be answered empirically.

Scheme 1 can also be used to outline the kind of information sought in crosslinking experiments. With the intramolecular crosslinks formed in P and P* the most obvious information provided after establishing the crosslink location is the interresidue distances in the respective derivatives and the conclucions about three-dimensional folding that this entails. If the derivative retains its biological activity (the presence of the specific ligand A would favor activity in P*) such information would have direct bearing on the relation of structure and activity. If the presence of A leads to different crosslinking patterns and/or different activity recoveries, the results can be interpreted in terms of the conformational changes that accompany ligand binding. In the case of the multisubunit (P···P) or multiprotein complexes ($P^1 \cdots P^2 \cdots P^3$) the main information from crosslinking experiments relates to the geometrical arrangements of the monomers in the complex and to the role of molecular associations in biological activity. As already mentioned this is becoming a very important application of crosslinking techniques in the study of structure and function of membranes, ribosomes and other subcellular organelles.

The crosslinking agents. A good deal is known about the chemical reactivity of the amino acid sidechains with a number of chemical reagents (Means and Feeney, 1971), so with the knowledge of the properties of the protein dictating optimal reaction conditions the selection of suitable reagents can also generally be based on rational planning. The persistent unknowns in selecting crosslinking agents are the nature of and the distance between the residues to be crosslinked, and again some trial and error will undoubtedly be required.

The conclusion of these brief considerations is that based on preliminary studies of the solution properties, stability and biological activity of the protein and with the available information on chemical reactivity and specificity of a large number of protein reagents, conditions and reagents can be selected with a good deal of confidence that a given desired product should be produced. Based on the limited information available, it appears that the probability of crosslink formation increases with the length of the crosslinking agent, and this may well be the main guideline to follow in this regard.

TYPES OF CROSSLINKS

It may be useful to attempt to classify all types of crosslinks, both natural, artificial and synthetic ones in one section, and then elaborate on specific ones in subsequent sections. The classification given in Table 1 based on crosslink length, is obviously quite

Table 1

Classification of Crosslinks

A. "Zero Length" Crosslinks
   1. Natural Crosslinks (disulfides, lysinonorleucine, ε-(β-aspartyl)-lysine, ε-(γ-glutamyl)-lysine)
   2. Artifacts of Protein Processing (lanthionine, lysinoalanine)
   3. Synthetic Crosslinks (ε-(β-aspartyl)-lysine, ε-(γ-glutamyl)-lysine, lysinonorleucine)

B. ">Zero Length" Crosslinks (introduction of new R-groups)
   1. Crosslinks Formed with Homobifunctional Reagents (x-R-x)
   2. Crosslinks Formed with Heterobifunctional Reagents (x-R-y)
   3. Artificial disulfides

---

arbitrary but may have some merit in juxtaposing in a brief discussion widely different aspects of this symposium volume.

"Zero Length" Crosslinks. This term is apparently catching on in describing crosslinks formed by the direct fusion of two amino acid sidechains in a protein. (The proper term should actually be "less than zero length" crosslinks, since atoms are often eliminated in the process of crosslink formation; even in the absence of any elimination, by definition the covalent link between the two crosslinked atoms must be shorter than the sum of the van der Waal's radii of the same two atoms in the precursor). This group of crosslinks includes all the natural ones such as the disulfide bonds, the whole family of crosslinks in collagen and elastin derived from lysine, hydroxylysine and histidine and the γ-glutamyl-lysine crosslinks which are all discussed elsewhere in this volume. It also includes crosslinks formed as artifacts of in vitro protein processing at high temperature or at extremes of pH. Examples of this type of crosslinks, lanthionine and lysinoalanine have also been discussed elsewhere in this volume. Finally the group includes the synthetic crosslinks such as those formed by chemically induced condensation between the carboxylic acid groups of aspartic and glutamic acids and the primary amino group of lysine. These derivatives will be discussed together with the natural γ-glutamyl lysine crosslinks below.

The >Zero Length Crosslinks are the crosslinks produced with the introduction of a new chemical structure into the protein. The new chemical entity is the R-group in different kinds of bifunctional reagents of the general type x-R-y, where x and y represent reactive or potentially reactive groups. The bifunctional reagents will be considered in some detail below.

SOME NATURAL AND SYNTHETIC "ZERO LENGTH" CROSSLINKS

The natural amide crosslink between glutamate and lysine is well established in a number of mammalian systems. Its formation is catalyzed by the enzyme transglutaminase acting on glutamine and lysine sidechains as substrates:

$$\text{Protein} \Big\langle \begin{array}{l} (CH_2)_2\text{-COOH} \\ (CH_2)_4\text{-NH}_2 \end{array} \quad \xrightarrow{\text{Transglutaminase}} \quad \text{Protein} \Big\langle \begin{array}{l} (CH_2)_2 \overset{CONH_2}{\underset{|}{}} \\ (CH_2)_4\text{-NH} \end{array}$$

According to Folk and Chung (1973) it is useful to consider three distinct mammalian transglutaminase types: a tissue type represented by the guinea pig liver enzyme; a platelet type, specifically

Factor XIIIa of the blood clotting cascade; and a hair follicle
type. In addition a similar enzyme has been found in lobster blood.
We have recently pursued a report by Chung and Folk (1975) that
transglutaminase also is present in red cells and have purified the
enzyme from human erythrocytes in good yield (Brenner, 1976). The
purification is summarized in Table 2 to illustrate the ease of
preparing the enzyme from the readily available starting material,

Table 2

Purification of Human Erythrocyte Transglutaminase (Brenner, 1976)

Purification Step	Protein	Specific Activity[a]	Yield[b]
Hemolysate	55 g		
DEAE batch elution	570 mg	29 u/mg	100%
DEAE chromatography	106 mg	111 u/mg	71%
Gel filtration	32 mg	123 u/mg	24%[c]
Preparative Gel Electrophoresis 6% Gel	2.3 mg	860 u/mg	12%
Preparative Gel Electrophoresis 8.5% Gel	1.3 mg	610 u/mg	5%

[a]Activity determined by routine fluorescent assay; protein determined by the Biuret-Phenol method.

[b]Transglutaminase activity was unassayable in the hemolysate, therefore activity recovered in the batch elution was defined as 100%.

[c]Recovery of activity after gel filtration and subsequent steps
requires 1 mM ATP (or, less effectively EDTA) in the buffers. The
example shown here used only EDTA to facilitate the protein determinations. The enzyme obtained after the gel filtration step is of
sufficiently high purity to be used in crosslinking experiments.

outdated human blood. Brenner (1976) explored the use of this enzyme as a general protein crosslinking reagent, but concluded that
its use in the formation of "zero length" crosslinks is limited
indeed. It appears that the erythrocyte enzyme is a typical tissue
transglutaminase and that it like the other transglutaminases must
have a high degree of specificity for the glutamine substrate in
native proteins. Denatured proteins are better sources of glutamine
substrates and carbobenzoxyglutamylglycine will act as a substrate
as well. Based on exhaustive methylation studies before and after
incubation with the enzyme, only a few native proteins have given
evidence of crosslink formation: fibrin and fibrinogen (McKee
et al., 1972), hair (Wajda et al., 1971), casein (Cooke and Holbrook,
1974) and hemoglobin (Brenner, 1976). However, the specificity for
the amine donor is much less stringent, and transglutaminase has
been used to introduce a number of primary amine-containing reagents
into proteins (Clark et al., 1959; Wajda et al., 1971; Lorand et al.,
1972; Brewer and Singer, 1975; Dutton and Singer, 1975; Toda and
Folk, 1969; Brenner, 1976). This suggests a potential use of transglutaminase in introducing crosslinks of the greater than zero length
type. Compounds of the general type x-R-NH$_2$ can perhaps be introduced into proteins through the enzyme catalyzed transamidation of
specific glutamine residues, and depending on the nature of the
second reactive group, the crosslink could form with different residues in the vicinity of the labeled glutamine residue. The ideal
make-up of such a reagent is one in which x is inert and must be
activated before a crosslink can be formed. Thus a reagent containing a photoactivatable x-group (see below) could be incorporated
enzymatically in the dark and then activated by illumination to
induce crosslink formation. Since carbobenzoxyglutaminylglycine is
a substrate for some of the transglutaminases, one can also visualize
a series of compounds with an R-x component added to this glutamine
derivative as two-step crosslinking agents. In this case there
would presumably be a relatively large number of reactive lysine
amino groups in a given protein providing a higher probability for
the first step to be successful. Because of the enzyme's low specificity for the amino donor, it is even possible that a compound like
carbobenzoxyglutaminyl-glycyl-glutaminyl-glycine could act as a
homobifunctional reagent with which long crosslinks could be formed
in two enzyme-catalyzed transamidations with two protein lysine
groups.

There is good evidence that the transglutaminase catalyzed reactions proceed through an acyl-enzyme intermediate in which an
active site sulfhydryl group forms a thioester with the glutamyl
group of the substrate (Folk et al., 1967). The enzyme-catalyzed
reaction is thus completely analogous to the chemical methods used
to form "zero length" amide crosslinks between carboxyl groups and
amino groups in proteins. The reactive acyl intermediate in this
case is produced chemically with reagents such as carbodiimides

(Goodfriend et al., 1965), ethyl chloroformate (Patramani et al., 1969), Woodward's reagent K_1 (Patel and Price, 1967) and in non-aqueous solution, carbonyldiimidazole (Barthing et al., 1974), and the active acyl group then reacts with adjacent amino groups.

$$
P\underset{(CH_2)_4-NH_2}{\overset{COO^-}{<}} \xrightarrow{\text{Activation}} P\underset{(CH_2)_4-NH_2}{\overset{COOX}{<}} \longrightarrow P\underset{(CH_2)_4}{\overset{\overset{O}{\overset{\|}{C}}}{<}}NH
$$

The side chain amide bond formation which appears to occur when proteins are treated with diethylpyrocarbonate is in a different category in that the intermediate in this case is more likely to be an activated amine function. As discussed by Wolf et al. (1970), the main product of the reaction, the N_ε-carbethoxy lysine derivative should be too stable to give further reaction with carboxyl groups, whereas a putative N-carboxy anhydride might undergo transamidation with carboxyl groups to yield amide crosslinks. Since diethylpyrocarbonate is generally used as a monofunctional protein reagent (Means and Feeney, 1971), its role in crosslink formation has not been explored to any significant degree.

A recent and important addition to synthetic "zero length" crosslinks which mimic both natural crosslinks and their mechanism of biosynthesis is the peroxidase-induced formation of lysinonorleucine reported by Dr. M. A. Stahmann in this volume. The possibility of using an enzyme as the crosslinking agent and thus avoiding the more drastic reaction conditions generally required for conventional reagents is a most exciting aspect of the potential applications of both peroxidase and transglutaminase in synthetic crosslink formation.

BIFUNCTIONAL REAGENTS

The bifunctional reagents are by far the most common crosslinking agents in protein chemistry. A large number of reagents have been prepared over the last 20 years, and there is virtually no limit to the possible permutations of reactive functional groups and associated R-groups that could be combined in bifunctional reagents. To simplify the discussion of the reagents that have been used to date, the functional groups and the R-groups will be considered separately in general terms in the following. Some specific examples of reagents will be given together with references to some recent additions in both reagents and applications. The reviews referred to in the introduction contain quite complete literature surveys of known reagents and applications.

The Functional Groups. A survey of functional groups in bi-functional reagents is tabulated in Table 3. It is useful to consider homobifunctional reagents (two identical functional groups; x-R-x) and heterobifunctional reagents (two different functional groups; x-R-y) separately. In principle the homofunctional reagents should give a very nearly simultaneous reaction of both groups

$$P\begin{matrix}\diagup BH\\\diagdown BH\end{matrix} + x-R-x \longrightarrow P\begin{matrix}\diagup B\diagdown\\\diagdown B\diagup\end{matrix}R\ ,$$

while the heterofunctional reagents are designed to give a step-wise reaction:

$$P\begin{matrix}\diagup BH\\\diagdown BH\end{matrix} + y-R-x \longrightarrow P\begin{matrix}\diagup B-R\\\diagdown BH\ y\end{matrix}\Big| \longrightarrow P\begin{matrix}\diagup B\diagdown\\\diagdown B\diagup\end{matrix}R$$

In practice, many of the homofunctional reagents exhibit some degree of interaction between the functional groups, however, and even in model reactions with free BH a difference in the rate of reaction of the two groups will be apparent. In reactions with protein further rate differences could be caused by different reactivity of the protein-bound BH-groups, and most of the homofunctional reagents that react slowly enough to allow monitoring of the reaction rate will show a two-step reaction pattern. As an example of this, consider a typical homofunctional reagent, p,p'-difluoro-m,m'-dinitrodiphenylsulfone. Since the interaction between the two parapositions should be disconnected by the sulfone group, this reagent was thought to have two identically reactive functional groups. However, a kinetic analysis of its reaction with butyl amine in dioxan or acetone revealed a 6-fold decrease in the second order rate constant for the second step compared to that for the first step (Jen, 1967), and this differential reactivity is undoubtedly reflected in its reaction with proteins.

The distinction between homofunctional and heterofunctional reagents is still a useful one, in that the latter ones are designed specifically to give a clearly differentiated sequential reaction. In the ideal case this could be an all-and-none situation, in which conditions can be defined for group x to react rapidly and completely while group y remains inert until a change in conditions or an activation process renders it reactive. The most obvious example of this type of reagent is one in which x undergoes a normal thermochemical and y a photochemical reaction. The first step can be carried out to completion in the dark (excess reagent

Table 3

Functional Groups (x,y) in Bifunctional Reagents (x-R-x, x-R-y)

	Functional Group	Reacts with	Examples[a]
1.	Aromatic $\diagup\!\!\!\!\diagdown$C-F	Lys, (Tyr, Cys)	A. p,p'-difluoro-m,m'-di-nitrodiphenylsulfone (1); 1,5-difluoro-2,4-dinitrobenzene (2). B. 1-Fluoro-2-nitro-4-Azidobenzene (3).
2.	$-SO_2Cl$	Lys	A. Phenol-2,4-disulfonyl chloride; α-naphthol-2,4-disulfonyl chloride (4).
3.	$-COOC_6H_4NO_2$	Lys	A. Adipate bis-(p-nitrophenyl ester (5); carbonyl bis(methionine p-nitrophenyl ester (6).
4.	$-CON_3$	Lys	A. Tartaryl diazide; Tartryl bis-(glycylazide) (7).
5.	$-COO-N\diagup\diagdown$	Lys	A. Succinate bis-(hydroxysuccinimide ester) (8). B. N-(Azidonitrophenyl)γ-aminobutyrate hydroxysuccinimide ester (3).
6.	$-COCH_2$ Br(I)	Cys, Met, His, Lys	A. 1,3-dibromoacetone (9) B. p-azidophenacyl bromide (10) See also 7-B and 10-B.
7.	$-CO-CHN_2$	Asp, Glu (Cys)	A. 1,1-bis-(diazo acetyl)-2-phenylethane (11). B. 1-diazoacetyl-1-bromo-2-phenylethane (11).
8.	Aromatic $\diagup\!\!\!\!\diagdown$C-N_2	Tyr, His	A. Bis diazo benzidine (12).
9.	$-CHO$	Lys (Cys,His,Tyr)	A. Glutaraldehyde (13).
10.	$-\overset{\oplus NH_2}{\underset{}{C}}$ - OR	Lys	A. Polymethylene (n=3-12) di-imidates (14); Di-

Table 3 (continued)

			methylsuberimidate (15).
			B. Ethyl (chloroacetimidate (16).
11.	$-N = C = O$	Lys	A. Hexamethylene diisocyanate (17).
			B. Toluene 2-isocyanate, 4-isothiocyanate (18).
12.	$-N = C = S$	Lys	B. See above (11B).
13.		Cys (Lys)	A. Bis(maleidomethyl)ether (19); N,N'-phenylenedimaleimides (20).
			B. (See Trommer et al., this volume).
14.	Aromatic $\diagdown$C-N$_3$ [b]	Nonspecific	A. N,N'-Bis(p-Azido-o-nitrophenyl)1,3 diamino-2-propanol (21).
			B. See above (1B, 6B).

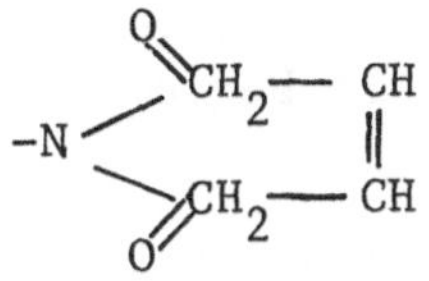

[a]The examples have been selected to illustrate homofunctional reagents (A) and heterofunctional reagents (B) with recent references either to reagent preparations or applications. For a more complete survey of reagents and applications see Uy and Wold (1976).

[b]Photoactivated: $-N_3 \xrightarrow{h\nu} -\ddot{N}$ nitrene is the reactive species.

References to Table 3:

1. Macleod and Hill (1970)
2. Grow and Fried (1975)
3. Yagub and Guire (1974)
4. Herzig et al. (1964)
5. Brandenburg (1972)
6. Busse and Carpenter (1974)
7. Lutter et al. (1974)
    See also Wetz et al. (1974)
8. Lindsay (1971)
9. Husain and Lowe (1970)
10. Hixson and Hixson (1975)
11. Husain et al. (1971)
12. Silman et al. (1966)
13. Josephs et al. (1973)
14. Hucho et al. (1975)
15. Tinberg et al. (1976)
    See also Wang and Richards (1975)
16. Olomucki and Diopoh (1972)
17. Snyder et al. (1974)
18. Schick and Singer (1961)
19. Freeberg and Hardman (1971)
20. Chang and Flaks (1972)
21. Guire (1976)

can be removed and the modified protein can be fractionated under
these conditions if necessary) and the second step is subsequently
initiated by illumination. In most cases, the heterofunctional reagents do not represent all-or-none, two-step reagents; as illustrated by a reagent containing an isocyanate and an isothiocyanate
group, one group (the isocyanate in this case) is considerably more
reactive than the other, and conditions can be established under
which its reaction is greatly favored. Thus at a pH around neu-
trality the isocyanate should react quite readily with amino groups,
while the isothiocyanate only will react after the pH has been
raised one or two pH units. The heterofunctional reagents have
heretofore had their major application in the preparation of immobilized protein derivatives through intermolecular crosslinks between proteins and non-protein matrices, but should be ideal protein crosslinking reagents.

A special type of heterofunctional reagent, leading to artificial disulfide crosslinks should be included in this summary. Their
unique feature is that the second step in the crosslinking process
is highly specific (or can be made specific) and will only take
place between the two incorporated reagent molecules. Two examples
of such reagents, N-acetyl homocysteine thiolactone (Benesch and
Benesch, 1956) and methyl(4-thiobutyrimidate) (Kenny et al., 1975)
are illustrated below.

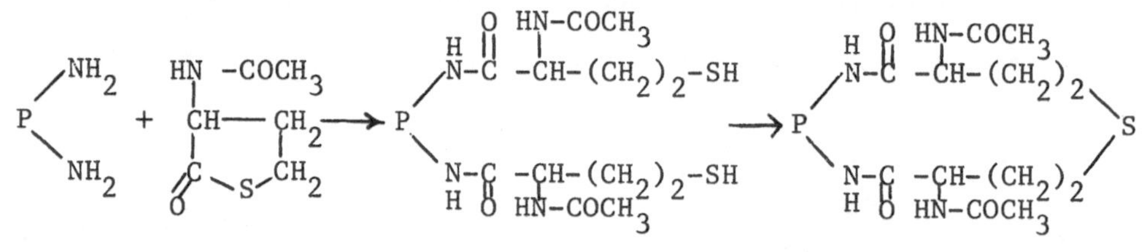

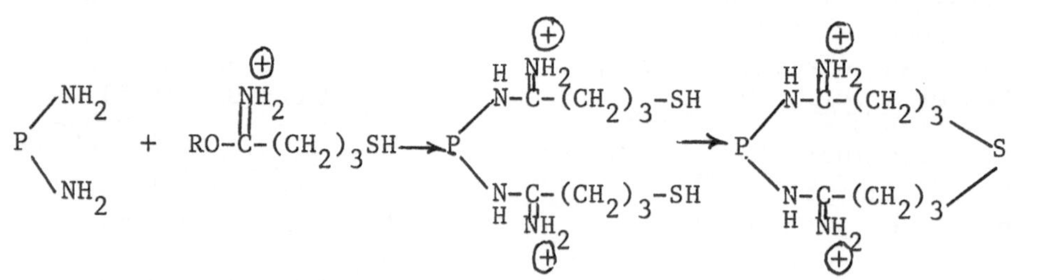

Table 4

Bifunctional Reagent Properties Determined by the R-Groups

1. Length of the crosslink
2. Hydrophobic-hydrophilic properties of the reagent
3. Cleavable crosslinks
4. Specificity of reagent: bifunctional affinity labels
5. Introduction of "reporter groups" with crosslink.

PROPERTIES OF THE R-GROUPS IN BIFUNCTIONAL REAGENTS

Several points worthy of consideration in selecting or de-
signing bifunctional reagents are outlined in Table 4. What should
be the optimal reagent length, and how can the reagent's solubility
properties (hydrophobic or hydrophilic nature) help direct it to
the proper region of the protein or the protein complex? How can
certain groups be included to simplify subsequent characterization
of the crosslinked product? How can the R-group be modified to induce specificity for a given site in the protein? All of these are
obviously important considerations, and will be discussed briefly
in the following. As mentioned above it is generally impossible to
predict the optimal reagent length for a given experiment, and one
is consequently forced to compare a series of reagents for optimal
yield of a desired product. Unfortunately, there are not many
families of bifunctional reagents that present one with a wide
choice of reagent lengths associated with the same reactive groups.
The polymethylene diisocyanates, the polymethylene diimidoesters
and the polymethylene dicarboxylic acids (activated as nitrophenyl
or hydroxysuccinimide esters for example) are the most obvious
homologous series of reagents available at this stage. Based on
both solubility and reaction properties, the diimidoesters are the
most attractive of these three, and there can be little doubt that
the wide choice of diimidoester reagent lengths has contributed
significantly to the popularity of this family of reagents in recent years. As to guidelines in the selection of reagent length,
it appears to be a generally accepted rule of thumb to use short
reagents for intramolecular and long reagents for intermolecular
crosslinks. The experimental bases for this rule are rather meager,
and exceptions to the rule should not be met with consternation and
surprise.

The possibility of directing a given reagent to the hydrophobic
interior of a protein molecule or a membrane complex, or restricting
its reaction to the hydrophilic surfaces of the same systems by the
polarity of the R-group, has received a good deal of attention in
recent years. As is the case for the whole field of crosslinking
technology the basic ideas and reagent properties have been explored

with monofunctional reagents and it is clear that the reaction
pattern of a given reagent R-x with a given protein or protein
complex can be altered quite drastically by simply changing the
polarity of the R-group (Staros and Richards, 1974; Staros et al.,
1975).

There are two reasons why it is very useful to cleave a crosslink once it has been incorporated and its effects have been studied.
The first reason has to do with proper controls for sound interpretation of the results. In all experiments involving chemical modification with bifunctional reagents, changes in protein properties
after the reaction can be due either to chemical derivatization of
certain amino acid residues or to the crosslink formation. By far
the cleanest way in which to distinguish between the two effects is
to cleave the crosslink, while leaving the chemical derivatives on
the participating residues intact. The other reason for seeking
cleavable crosslinking reagents has to do with experimental procedures involved in establishing the crosslink position. The
natural procedure is to degrade the crosslinked protein down to
short peptides and to isolate the crosslinked peptide pair from the
peptide mixture. Once this pair has been obtained, the next step
is to characterize each peptide separately, and again this requires
cleavage of the crosslink. Several crosslinks with readily cleavable groups have been used:

$$-C_6H_4-N=N-C_6H_4- \xrightarrow{S_2O_4^=} 2-C_6H_4NH_2$$ (Fasold et al., 1964)

$$-CH_2-S-S-CH_2- \xrightarrow{SO_3^=} 2-CH_2-S-SO_3^-$$ (Wang and Richards, 1975;
$$\xrightarrow{reduction} 2-CH_2-SH$$ Ruoho et al., 1975;
Kenney et al., 1975;
Lomant and Fairbanks,
1976).

$$\overset{OH}{\underset{|}{-CH}}-\overset{OH}{\underset{|}{CH}}- \xrightarrow{IO_4^-} 2-CHO (\xrightarrow{NaBH_4} 2-CH_2OH)$$ Lutter et al., 1974;
Coggins et al., 1976)

In addition to these general procedures, some special cleavable
reagents have been used. Thus, with carbonyl bis(L-methionine p-
nitrophenyl ester) as the bifunctional reagent, the crosslink can
be severed through cyanogen bromide cleavage of the methionine
amide bonds (Busse and Carpenter, 1974), and with a reagent such as
a di-α-amino (blocked) di-α-carboxylic acid (activated) derivative,
the crosslink can be cleaved through a regular Edman degradation
after unblocking the α amino groups (Geiger and Obermeyer, 1973).
An attractive area to explore for cleavable crosslinking reagents
is the introduction of groups (carboxyl and phosphate esters, for
example) which may be readily susceptible to enzymatic cleavage.

The last categories to consider under R-group variations are specialized types of refinements which may have only limited application. Nevertheless, the inclusion of "reporter groups", either unique chromophores (Benesch et al., 1976), fluorescent groups (Wu et al., 1976) or spin labels may provide useful probes of structure in the vicinity of the crosslink and also aid in any subsequent isolation of the crosslinked peptides. For this latter purpose a radioactive isotope is obviously by far the most powerful tool. A bifunctional reagent with a substrate (ligand) analog as part of the R-group could also have unique desirable features primarily in terms of specificity, although the incorporation of such an affinity label by design would lead to inactivation of the crosslinked protein. Several such bifunctional affinity labels have been explored (Bechet et al., 1973; Husain and Lowe, 1970; Husain et al., 1971).

CROSSLINKS FORMED AS SIDE PRODUCTS

It may be relevant in closing to recall that several reagents designed as monofunctional protein reagents and even impurities in common chemicals may lead to crosslinks where they are not expected and not sought. The classical example of "monofunctional" reagents which cause extensive crosslink formation are the aldehydes. As illustrated with formaldehyde, strongly electrophilic intermediates may form from the initial Schiff base and these can react with a number

of nucleophiles in the protein (Means and Feeney, 1971). Recently, there have been several reports of unexpected results with different protein reagents, results that have subsequently been established to reflect crosslink formation. The recent report that impurities in high purity grade of ethylene glycol and glycerol will crosslink proteins (Bello and Bello, 1976) emphasizes the need for caution in this area. The impurities have been established to be aldehydes and peroxide and the crosslinking reaction should thus not be surprising. Reagents such as tetranitromethane (Boesel and Carpenter, 1970) and diethyl pyrocarbonate (see above, Wolf et al., 1970) have also been shown to give crosslinks as side products in the intended monofunctional modification reactions.

REFERENCES

Barthing, G. J., Chattopadhyay, S. K., Barker, C. W., Forrester, L. J. and Brown, H. D. (1974) Int. J. Peptide Protein Res. $\underline{6}$,287.

Bechet, J. J., Dupaix, A. and Yon, J. (1973) Eur. J. Biochem. 35, 527.

Bello, J. and Bello, H. R. (1976) Arch. Biochem. Biophys. 172, 608.

Benesch, R. and Benesch, R. E. (1956) J. Am. Chem. Soc. 78, 1597.

Benesch, R., Ikeda, T. and Benesch, R. E. (1976) J. Biol. Chem. 251, 465.

Boesel, R. W. and Carpenter, F. H. (1970) Biochem. Biophys. Res. Commun. 38, 678.

Brandenburg, D. (1972) Hoppe-Seyler's Z. Physiol. Chem. 353, 869.

Brenner, S. C. (1976) Ph.D. Thesis, University of Minnesota.

Brewer, G. J. and Singer, S. J. (1975) Biochemistry 13, 3580.

Busse, W. D. and Carpenter, F. H. (1974) J. Am. Chem. Soc. 96, 5947.

Chang, F. N. and Flaks, J. G. (1972) J. Mol. Biol. 68, 177.

Chung, S. I. and Folk, J. E. (1975) Fed. Proc. 34, 259.

Clarke, D. D., Mycek, M. J., Neidle, A. and Waelsch, H. (1959) Arch. Biochem. Biophys. 79, 338.

Coggins, J. R., Hooper, E. A. and Perham, R. N. (1976) Biochemistry 15, 2527.

Cooke, R. D. and Holbrook, J. J. (1974) Biochem. J. 141, 71.

Davies, G. E. and Stark, G. R. (1970) Proc. Nat. Acad. Sci. U.S. 66, 651.

Dutton, A. and Singer, S. J. (1975) Proc. Nat. Acad. Sci. U.S. 72, 2568.

Fasold, H. (1965) Biochem. Z. 342, 295.

Fasold, H., Groschel-Stewart, U. and Turba, F. (1964) Biochem. Z. 339, 487.

Fasold, H., Klappenberger, J., Meyer, C. and Remold, H. (1971) Angew. Chem. Int. Ed. Engl. 10, 795.

Folk, J. E. and Chung, S. I. (1973) Adv. Enzymol. 38, 109.

Folk, J. E., Cole, P. W. and Mullooly, J. P. (1967) J. Biol. Chem. 242, 4329.

Freeberg, W. B. and Hardman, J. K. (1971) J. Biol. Chem. 246, 1439.

Geiger, R. and Obermeier, R. (1973) Biochem. Biophys. Res. Commun. 55, 60.

Goodfriend, T. L., Levine, L. and Fasman, G. D. (1964) Science 144, 1344.

Grow, T. E. and Fried, M. (1975) Biochem. Biophys. Res. Commun. 66, 352.

Guire, P. (1976) Fed. Proc. 35, 1632.

Hartman, F. C. and Wold, F. (1967) Biochemistry 6, 2439.

Herzig, D. J., Rees, A. W. and Day, R. A. (1964) Biopolymers 2, 349.

Hixson, S. H. and Hixson, S. S. (1975) Biochemistry 14, 4251.

Hucho, F., Müllner, H. and Sund, H. (1975) Eur. J. Biochem. 59, 79.

Husain, S. S., Ferguson, J. B. and Fruton, J. S. (1971) Proc. Nat. Acad. Sci. (U.S.) 68, 2765.

Husain, S. S. and Lowe, G. (1970) Biochemistry 117, 341.

Jen, L. C. (1967) Ph.D. Thesis, University of Illinois.

Ji, T. H. and Ji, I. (1974) J. Mol. Biol. 86, 129.

Josephs, R., Eisenberg, H. and Reisler, E. (1973) Biochemistry 12, 4060.

Kenny, J. W., Sommer, A. and Traut, R. R. (1975) J. Biol. Chem. 250, 9434.

Lindsay, D. G. (1971) FEBS Letters 21, 105.

Lomant, A. J. and Fairbanks, G. (1976) J. Mol. Biol. 104, 243.

Lorand, L., Campbell, L. K. and Robertson, B. J. (1972) Biochemistry 11, 434.

Lutter, L. C., Ortanderl, F. and Fasold, H. (1974) FEBS Letters 48, 288.

Macleod, R. M. and Hill, R. J. (1970) J. Biol. Chem. 245, 4875.

McKee, R. A., Schwartz, M. L., Pizzo, S. V. and Hill, R. L. (1972) Ann. N.Y. Acad. Sci. 202, 127.

Means, G. E. and Feeney, R. E. (1971) "Chemical Modification of Proteins", Holden-Day, Inc. San Francisco, Cambridge, London, Amsterdam.

Olomucki, M. and Diopoh, J. (1972) Biochim. Biophys. Acta 263, 213.

Patel, R. P. and Price, S. (1967) Biopolymers 5, 583.

Patramani, I., Katsiri, K., Pistevou, E., Kalogerakos, T., Pawlatos, M. and Evangelopoulos, A. E. (1969) Eur. J. Biochem. 11, 28.

Ruoho, A., Bartlett, P. A., Dutton, A. and Singer, S. J. (1975) Biochem. Biophys. Res. Commun. 63, 417.

Schick, A. F. and Singer, S. J. (1961) J. Biol. Chem. 236, 2477.

Silman, H. I., Albu-Weissenberg, M. and Katchalski, E. (1966) Biopolymers 4, 441.

Slobin, L. I. (1972) J. Mol. Biol. 64, 297.

Snyder, P. D., Wold, F., Bernlohr, R. W., Dollum, C., Desnick, R.J. Krivit, W. and Condie, R. M. (1974) Biochim. Biophys. Acta 350, 432.

Sperling, J. and Havron, A. (1976) Biochemistry 15, 1489.

Staros, J. V. and Richards, F. M. (1974) Biochemistry 13, 2720.

Staros, J. V., Richards, F. M. and Haley, B. E. (1975) J. Biol. Chem. 250, 8174.

Steck, T. L. (1972) J. Mol. Biol. 66, 295.

Sun, T. T., Traut, R. R. and Kahan, L. (1974) J. Mol. Biol. 87, 509.

Tinberg, H. M., Nayndu, P.R.V. and Packer, L. (1976) Arch. Biochem. Biophys. 172, 734.

Toda, H. and Folk, J. E. (1969) Biochim. Biophys. Acta 175, 427.

Uy, R. and Wold, F. (1976) in "Biomedical Applications of Immobilized Enzymes and Proteins", T.M.S. Chang, ed., Plenum Publishing Corporation, New York, N.Y.

Wang, K. and Richards, F. M. (1974) J. Biol. Chem. 249, 8005.

Wang, K. and Richards, F. M. (1975) J. Biol. Chem. 250, 6622.

Wajda, I. J., Hanbauer, I., Manigault, I. and Lajtha, A. (1971) Biochem. Pharmacol. 20, 3197.

Wetz, K., Fasold, H. and Meyer, C. (1974) Anal. Biochem. 58, 347.

Wold, F. (1961) J. Biol. Chem. 236, 106.

Wold, F. (1972) in "Meth. in Enzymol.", C.H.W. Hirs and S. N. Timasheff, eds., Vol. 25B, p. 623.

Wolf, B., Lesnaw, J. A. and Reichman, M. E. (1970) Eur. J. Biochem. 13, 519.

Wu, C.-W., Yarbrough, L. R. and Wu, F.Y.H. (1976) Biochemistry 15,
     2863.
Yaqub, M. and Guire, P. (1974) J. Biomed. Mater. Res. 8, 291.
Zaborsky, O. R. (1976) in "Biomedical Applications of Immobilized
     Enzymes and Proteins", T.M.S. Chang, ed., Plenum Publishing
     Corporation, New York, N.Y.
Zahn, H. (1955) Proc. International Wool Textile Research Conf.,
     Australia C-425.
Zahn, H. and Meienhofer, J. (1958) Makromol. Chem. 26, 153.

10

SYNTHESIS AND APPLICATION OF NEW BIFUNCTIONAL REAGENTS

Wolfgang E. Trommer, Klaus Friebel, Hans-Hermann Kiltz
and Hans-Jörg Kolkenbrock

Lehrstuhl für Biochemie, Abteilung Chemie
Ruhr-Universität Bochum
Postfach 2148, 4630 Bochum/West Germany

SUMMARY

Two new bifunctional reagents suited for the step-wise cross-linking of cysteine and lysine residues in proteins are described. Application to lactate dehydrogenase yields a cross-link between cysteine-165 and lysine-179, which suggests an alternative mechanism by which the "essential" cysteine reacts. For the mapping of the environment of a known and well defined amino acid the use of semireversible bifunctional reagents is suggested.

INTRODUCTION

In a preceding paper of these proceedings Finn Wold has given a thorough account of the many applications of bifunctional cross-linking reagents. Ever since the first introduction of this method by Zahn (Zahn, 1955) more and more reagents of various types have been proposed to avoid disadvantages very often encountered with the commonly used compounds. Symmetrical reagents are particularly limited in their application. Since both groups may react simultaneously, their incorporation into the macromolecule can not really be controlled. A complex mixture of mono- and disubstituted derivatives will be formed. With asymmetric reagents the situation may be even worse when the groups exhibit similar reactivities. However, a similar reactivity may be necessary because almost all functional groups in amino acid side chains require electrophilic reagents.

What should be the feature of an "ideal" cross-linking reagent when the main objective is the mapping of the environment of a certain amino acid or even of a certain protein:

1. The reagent should be rigid, i.e. its length should be well defined.
2. It should react selectively with a particular amino acid side chain. This applies to both functional groups which should be selective for different amino acids.
3. The reaction should take place in a step-wise manner. After completion of the reaction at one side, uncoupled excess reagent can be removed and only then the second group should be activated and brought to reaction.
4. The reagent should be cleavable under mild conditions, under which peptide bonds are not attacked in order to allow for a separate sequencing and identification of both halves of the cross-linked peptides after isolation.
5. To help isolation and to determine the degree of incorporation the reagent should be colored or, preferably, due to the higher sensitivity, radioactively labeled.

RESULTS AND DISCUSSION

Among the functional groups in amino acid side chains the SH-group of cysteine often plays a special role. Fairly selective SH-reagents are N-substituted maleimides, at least when the pH of the solution is carefully kept below 7 to keep primary amines fully protonated. A group suited for a step-wise reaction is the azide function. However, the nitrene formed upon irradiation is completely unspecific, which may sometimes be advantageous. Such a reagent, p-azidophenyl maleimide, which has recently been synthesized in our laboratory (Trommer and Hendrick, 1973), is shown in figure 1.

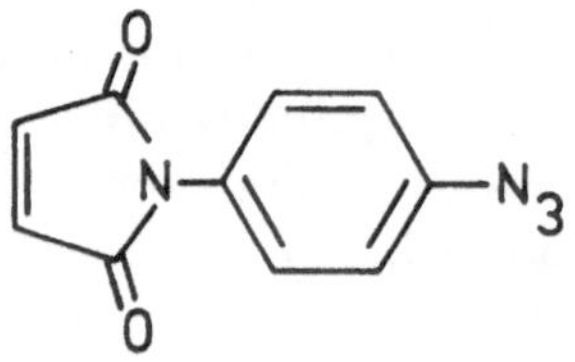

Figure 1.   4-Azidophenyl maleimide (I)

Since phenylsuccinimides, the products from SH-addition, are cleavable under mildly acidic conditions to the corresponding aniline derivatives this reagent would seem to meet most of the initial postulates. However, the photolysis of the azide requires irradiation at wavelengths below 300 nm, conditions under which tryptophan residues are destroyed.

A superior reagent is shown in figure 2, 4-azidocarbonyl-3-hydroxyphenyl maleimide (Trommer and Kolkenbrock, 1975).

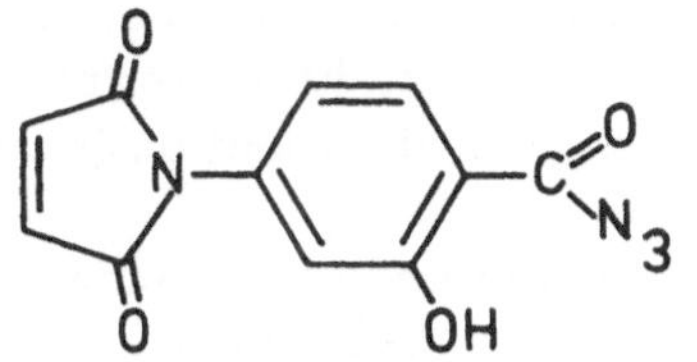

Figure 2.   4-Azidocarbonyl-3-hydroxyphenyl maleimide (II)

Again the maleimide function serves as the first anchor point. The hydroxyl group is to increase the water solubility of the reagent. Figure 3 shows two model compounds used to determine its specificity and to elaborate optimal reaction conditions separately for both functional groups.

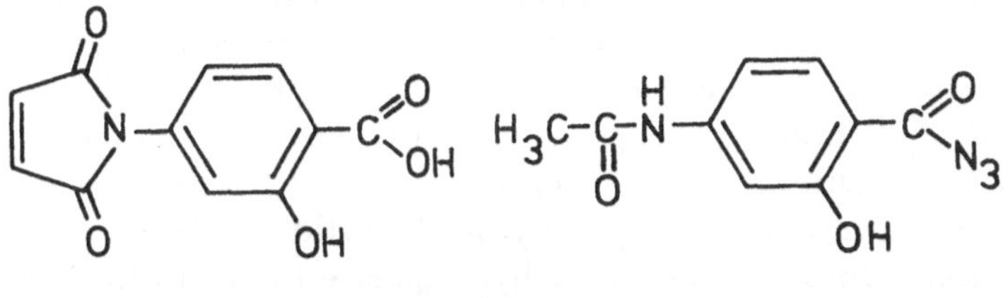

Figure 3.   4-Carboxy-3-hydroxyphenyl maleimide and 4-acetamido-2-hydroxybenzoyl azide

The azidocarbonyl function was originally conceived to be photochemically activated, however it is highly activated in itself and reacts nicely with nucleophiles.

At pH-values below 7 the only nucleophiles in proteins are the mercapto-group of cysteine and the imidazole ring of histidine. The latter, however, forms a labile N-acyl derivative upon reaction which is cleaved readily during work up. The reaction rates of the azidocarbonyl function and of the maleimide ring with the SH-group of cysteine in glutathion were determined by means of the model compounds. The bimolecular rate constants differ by at least a factor of 2000 in favor of the maleimide (Trommer and Kolkenbrock, 1975). In the bifunctional reagent this side will therefore react nearly ex-clusively and a step-wise cross-linking is feasible. After attachment to a cysteine residue excess reagent may be removed. Subsequent raising of the pH of the solution to about 8.5 allows for a smooth reaction of the azidocar-bonyl group with primary amines, e.g. lysine residues under amide formation. Hydrolysis may be disregarded when the reaction is carried out at $0^{\circ}C$.

Lactate dehydrogenase from pig heart was chosen as a test system for the usefulness of this reagent. This NAD-dependent enzyme catalyzes the interconversion of lactate and pyruvate. It contains one highly reactive SH-group per subunit, the modification of which leads to concomitant loss of activity (Holbrook et al., 1975). Although each subunit contains four additional SH-groups, only cysteine-165 is modified by monovalent male-imides as demonstrated by tryptic digestion which yields exclusively the "essential cysteine peptide" (Holbrook et al., 1966).

For the cross-linking experiment 4-azidocarbonyl-3-hydroxyphenyl maleimide had been radioactively labeled (0.08 Ci/mol) by starting its synthesis from $\left[2,3 - {}^{14}C_2\right]$ maleic anhydride. A sixfold excess of the reagent in dimethylformamide was added in small portions over a period of 3 hours to the enzyme solution (10 mg/ml) in 20 mM phosphate buffer pH = 7. Excess reagent was then removed by gel chromatography on Sephadex G-25. After vacuum dialysis against pyrophosphate buffer pH 8.5, any denatured protein likely to contain cross-links impossible in the native state was re-moved prior to the enzymic hydrolysis by gel filtration on Sephadex G-200 in the same buffer. Unfolded lactate dehydrogenase elutes in the void volume under these conditions (Jeckel, 1976). The incorporation of reagent into this denatured fraction actually corresponded to 4 moles per mole sub-unit, while the native portion which showed a residual activity of 35 % contained the reagent in a ratio of 0.6 to 1, which corresponds well to the residual activity (Trommer and Becker, 1976).

Before the enzymic degradation we checked whether any intersubunit cross-links had been formed. However, this was not the case as demonstra-ted by disc electrophoresis under denaturing conditions (Laemmli, 1970). No dimers or even higher aggregates could be detected. Tryptic cleavage of the modified enzyme (Holbrook et al., 1966) yielded only the non-crosslinked essential cysteine peptide, while about 50 % of the radioactivity remained in larger peptides not migrating when a fingerprint on silica was carried out (Wieland et al., 1964). However, these peptides could be cleaved by additional extensive chymotryptic digestion.

One major product was isolated by a series of purification steps as outlined in table 1. It corresponds to a cross-link between cysteine-165 and lysine-179 located both in peptides resulting from chymotryptic subcleavage at two leucine residues and methionine (Figure 4).

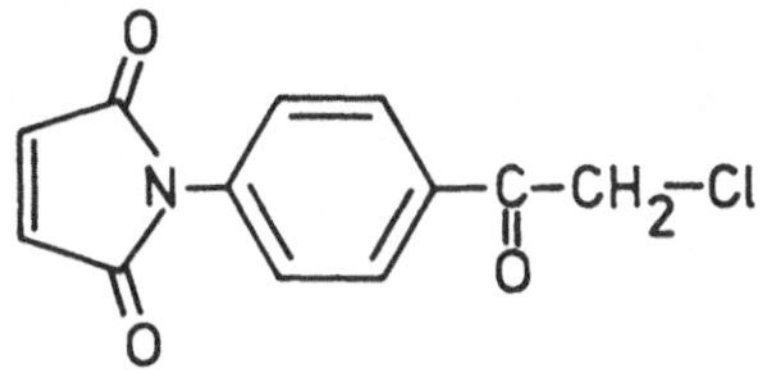

Figure 4.   Partial sequence of lactate dehydrogenase from pig heart indicating the cross-link obtained with the reagents II and III.

The structural assignment is unequivocal on the basis of the amino acid analysis and the determination of the two N-termini, valine and alanine by dansylation (Woods and Wang, 1967). The modified cysteine was identified by comparison with a synthetic model compound from cysteine with our reagent (Kolkenbrock, 1976).

The same cross-link was observed when a different reagent, 4-chloro-acetylphenyl maleimide (figure 5) (Trommer and Hendrick, 1973) was used.

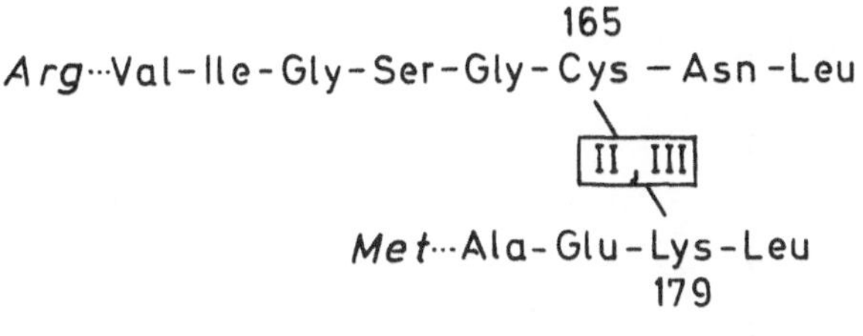

Figure 5.   4-Chloroacetylphenyl maleimide (III)

TABLE 1

Purification scheme of pig heart lactate dehydrogenase peptides cross-linked by 4-azidocarbonyl-3-hydroxyphenyl maleimide

Step	Method	Procedure
1	gel chromatography	Sephadex G-25, 30 mM NH_4HCO_3
2	ion exchange chromatography	DEAE-cellulose, 0.02 - 0.5 M NH_4HCO_3
3	thin layer chromatography	silica: a) pyridine / acetic acid / butanol-1 / water (40 : 14 : 68 : 25), pH = 6.5 b) pyridine / acetic acid / water (1 : 10 : 589), pH = 3.1
4	high voltage electrophoresis	silica plates, pyridine / acetic acid / water (25 : 1 : 225), pH = 6.5; 50 V/cm
5	gel chromatography	Sephadex G-50, 30 mM NH_4 (CH_3COO)

Although the chloroketone shows similar specifity as maleimides, its reactivity is so low that steric approximation effects caused by attachment to the first anchor point are necessary for satisfactory reaction rates. Its incorporation into the protein therefore takes place in a nearly step-wise manner. To prevent unspecific cross-linking during work up the reaction was quenched after 1 hour by addition of mercaptoethanol. The reagent was radioactively labeled after its attachment to the protein by reduction of its carbonyl function with tritiated sodium borohydride. In this case the residual activity in the native fraction was only 11 % with a corresponding incorporation of 0.9 moles of reagent per mole subunit. Isolation of the cross-linked peptide followed the same scheme as outlined before in table 1. However, chromatographic and electrophoretic separations were carried out on paper instead of silica plates.

We have correlated these findings with the newly revised sequence and tertiary structure of the pig heart isozyme of lactate dehydrogenase (Kiltz et al., 1976; Eventoff et al., 1976). Both reagents have about the right length for a cross-link between cysteine-165 and lysine-179. Moreover, the two anchor points are connected by something like a hydrophobic channel, a channel formed by hydrophobic amino acids. Lysine-179 is situated at its beginning on the surface of the molecule. One could therefore speculate that cysteine-165 is also initially attacked via this channel and not via the active center thus explaining its wellknown increased reactivity against hydrophobic maleimides. Modification via the active center seemed to be well established because cysteine-165 is not reactive in ternary complexes. There is, however, an alternative explanation. Cysteine-165 does not actually play a role in catalysis (Holbrook et al., 1975). When modified with bulky reagents it pushes aside the chain containing the essential histidine-195 thus preventing ternary complex formation. Let us reverse this argument. Once a ternary complex is formed, cysteine-165 can not be modified any more due to steric hindrance, regardless from where the reagent is coming.

SEMIREVERSIBLE BIFUNCTIONAL CROSS-LINKING REAGENTS

Regardless of the possible success with reagents of the type just discussed, the isolation of cross-linked and therefore rather large peptides, is very laborious and one has to deal always with non-cross-linked byproducts. We should like to suggest a different approach (Friebel, 1975). A reagent which can undergo a step-wise reaction and which can easily be cleaved at the first anchor point after cross-linking would yield exclusively labeled peptides which had definitely been cross-linked before. Non-cross-linked reagent would simply be cleaved off at the last step. For these reagents, which we should like to name semireversible bifunctional cross-linking reagents, the first anchor point has to be a known and well defined amino acid, which, however, quite often is the case. It will then give information on amino acid side chains in a certain distance from this known position as defined by its length. Such a reagent (IV) is shown in figure 6.

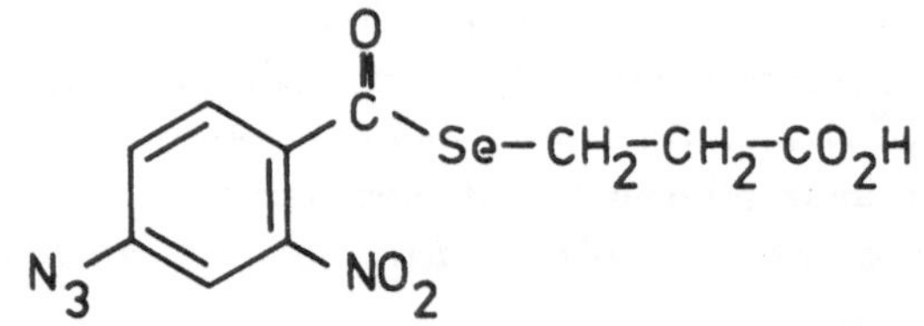

Figure 6.   2-Carboxyethyl selenol-4-azido-2-nitrobenzoate (IV)

It is composed of a selenol ester and an activated phenylazide. Due to the nitrogroup phenylnitrene may be generated by irradiation at a wavelength of 350 nm and even above. Selenol esters have recently been introduced as selective SH-reagents (Makriyannis et al., 1973). The compound used by these authors is shown in figure 7.

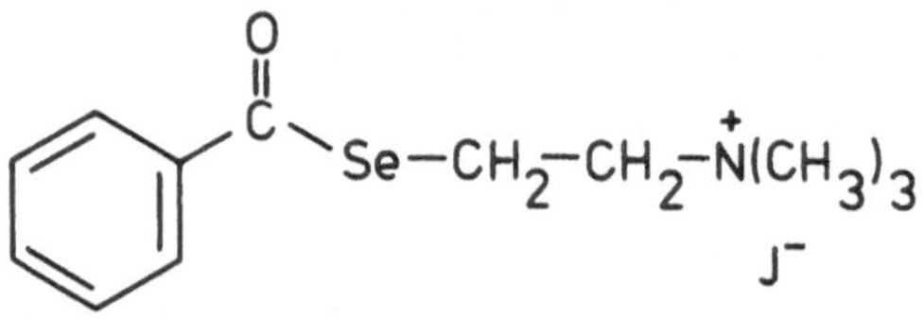

Figure 7.   Selenocholine benzoate

Upon reaction with thiols, thiolesters are formed. The liberated selenols readily form diselenides causing a very favorable equilibrium of this reaction. With Mautner's reagent we have been able to modify lactate dehydrogenase efficiently and then to reactivate the enzyme by treatment with thiols or amines (Friebel, 1975) proving the reversibility of this attachment. The bifunctional reagent IV (Figure 6) should therefore be quite useful for mapping the environment of the essential cysteine residue in lactate dehydrogenase and in similar systems including work on the relative arrangement of ribosomal proteins.

REFERENCES

Eventoff, W., Rossmann, M.G., Taylor, S.S., Torff, H.-J., Meyer, H., Keil, W. and Kiltz, H.-H. (1976). Structural adaptations of lactate dehydrogenase isozymes. Proc. Natl. Acad. Sci. US, in press.

Friebel, K. (1975). Semireversible Reagentien zur Quervernetzung von Proteinen. Diplomarbeit, Ruhr-Universität Bochum, West Germany.

Holbrook, J.J., Pfleiderer, G., Schnetger, J. and Diemair, S. (1966). The importance of SH-groups for enzyme activity. IV. Preparation of the tryptic peptides containing the essential cysteine residue of lactate dehydrogenase, Isozymes I and V. Biochem. Z., 344, 1.

Holbrook, J.J., Liljas, A., Steindel, S.J. and Rossmann, M.G. (1975). Lactate dehydrogenase. The Enzymes, (Boyer, P.D., ed.), 3rd edition, 191, Academic Press, New York.

Jeckel, D. (1976). Beteiligung einzelner Aminosäureseitenketten an der Ligandenbindung und an der Stabilisierung der nativen Konformation der Lactatdehydrogenase. Habilitationsschrift, Ruhr-Universität Bochum, West Germany.

Kiltz, H.-H., Keil, W., Meyer, H., Poth, E. and Trost, G. (1976). The structure of porcine lactate dehydrogenase. Amino acid sequence of both isoenzymes, M_4 and H_4, presented at the IUB-Meeting, Hamburg, West Germany.

Kolkenbrock, H. (1976). Dissertation, Ruhr-Universität Bochum, West Germany.

Laemmli, U.K. (1970). Cleavage of structural proteins during the assembly of the head of bacteriophage T4. Nature, 227, 680.

Makriyannis, A., Günther, W.H. and Mautner, H.G. (1973). Selenol esters as specific reagents for the acylation of thiol groups. J. Am. Chem. Soc., 95, 8403.

Trommer, W.E. and Hendrick, M. (1973). The formation of maleimides by a new mild cyclization procedure. Synthesis, 484.

Trommer, W.E., Kolkenbrock, H. and Pfleiderer, G. (1975). Synthesis and properties of a new selective bifunctional cross-linking reagent. Hoppe-Seyler's Z. Physiol. Chem., 356, 1455.

Trommer, W.E. and Becker, G. (1976). The separation of partially modified lactate dehydrogenase by affinity chromatography. The specific activity of protomers? BBA, 422, 1.

Wieland, T. and Georgopoulos, D. (1964). Zweidimensionale Auftrennung von Peptiden (Fingerprinttechnik) auf Dünnschichtplatten. Biochem. Z. 340, 476 and 483.

Woods, K.R. and Wang, K.T. (1967). Separation of dansyl amino acids by polyamide layer chromatography. BBA, 133, 369.

Zahn, H. (1955). Brückenreaktionen an Aminosäuren und Faserproteinen. Angew. Chem. 67, 561.

11

SYNTHESIS AND APPLICATION OF CLEAVABLE AND HYDRO-PHILIC CROSSLINKING REAGENTS

Hans J. Schramm and Thomas Dülffer

Abteilung für Strukturforschung I
Max-Planck-Institut für Biochemie
D-8033 Martinsried/W. Germany

ABSTRACT

Bifunctional imidoesters are due to their mild reaction especially suitable for the crosslinking of proteins. Most often used are diimidates with a medium span (4 or 6 CH_2-groups). Reagents with a wider span might be of interest. In such a case, however, the bridge should be more hydrophilic. Bifunctional imidoesters have, therefore, been prepared from dinitriles $NC-(CH_2)_2-X-(CH_2)_2-CN$; X being $-O-$, $-O-(CH_2)_n-O-$ or $-O-(CH_2)_m-O-(CH_2)_n-O-$. The bridge of diimidoesters can also be labelled by coloured or fluorescent groups. Diimidoesters containing $-S-S-$bonds in the bridge can be cleaved more easily after their reaction with proteins. Dimethyl-3.3'-(γ,δ-dithiahexamethylenedioxy)-dipropionimidate, a "long", cleavable and more hydrophilic diimidate, is a promising new reagent. In cases, where $HS-(CH_2)_3-C(=NH)-OCH_3 \cdot HCl$ is used for crosslinking of proteins, this reagent can possibly be replaced by 2-iminothiolane hydrochloride, a cyclic thioimidate, which is easier to prepare and to handle.

INTRODUCTION

It is widely accepted that bifunctional imidoesters are well suited for the crosslinking of proteins. They are especially valuable for investigations on the subunit structure of protein complexes (Davies & Stark, 1970). The most often used diimidoesters dimethyl-suberimidate $(CH_3O(HN=)C-(CH_2)_n-C(=NH)OCH_3$; n=6) and dimethyl-adipimidate (n=4) contain hydrophobic polymethylene bridges. It is evident that by structural alterations of the bridge crosslinking reagents with interesting new properties might be produced. For example, in two laboratories (Wang & Richards, 1974; Ruoho et al., 1975) the mildly cleavable -S-S-group has been introduced into the bridge, while Traut et al. (1973) have used a monofunctional, HS-group carrying, imidate with successive oxidation of the modified protein complexes to form the crosslinks. The vic-glycol group, cleavable by periodate, has also been introduced (Lutter et al., 1974; Coggins et al., 1976).

We have attempted (in different reagents) the following modifications: a) labeling of the bridge by attachment of a coloured or fluorescent group, b) introduction of ether-oxygen-groups into the bridge, which render the reagent more hydrophilic, c) elongation of the span of the bridge.

A. BRIDGE-LABELLED REAGENTS

The coloured groups introduced first into the bridge of crosslinking reagents were dinitrophenyl-derivatives (I) (Schramm, 1967). At this time it was intended to use the reagents for studies on the tertiary structure of proteins. The suitability of these compounds for investigations on quaternary structures has not yet been tested.

Fluorescent diimidoesters may very well prove to be valuable crosslinking reagents. For example, reagent II can crosslink oligomeric proteins like lactate dehydrogenase (Schramm, 1975).

I   II

B. LONG AND HYDROPHILIC REAGENTS

A second possibility to change or improve the properties of crosslinking reagents is to synthetize compounds with a wider span (Wetz et al., 1974). However, the presence of hydrophobic "patches" (Kauzman, 1959) on the surface of proteins might prevent long hydrophobic reagents from fully extending. So it seemed likely that all attempts to encrease the span of a reagent would only succede if the bridge is rendered more hydrophilic at the same time. This has been tried by reacting acrylonitrile with dihydroxy compounds (or water) (Bruson, 1949). The resulting dinitriles of the type $NC-CH_2-CH_2-X-CH_2-CH_2-CN$, X being $-O-$, $-O-(CH_2)_n-O-$, $-O-(CH_2)_m-O-(CH_2)_n-O-$ etc. can be transformed to diimidoesters using the PINNER-reaction (McElvain & Schroeder, 1948). Some of these more hydrophilic reagents have also a wider maximal span (e.g. $\sim 16\ \overset{o}{A}$ for VI) than dimethyl-suberimidate ($\sim 11\ \overset{o}{A}$).

The table gives a summary of the compounds synthetized up to now. Some of the diimidates have not yet been obtained in solid form. Since the maximal span and the chemical constitution of the reagents are not too different, not all of the reagents seem to be necessary. The use of the reagents III and VI, from the commercially obtainable dinitriles, could be a good choice for the study of oligomeric proteins (Schramm & Dülffer, 1976; Luduena et al., 1975).

It seems that the ether-bridge containing diimidates favour intermolecular crosslinking , especially at higher protein concentrations. The reagents should therefore be applied at low protein concentrations. They can also be applied in cases where intermolecular reaction is desired. We have used, for example, VI to stabilize protein crystals for embedding and sectioning in electron microscopy (Langer et al., 1975).

C. CLEAVABLE REAGENTS

Although imidoesters (bound as amidines to the protein lysine-ϵ-amino-groups) can in fact be removed again by treatment with rather concentrated solutions of NH_3 (Hunter & Ludwig, 1972), the presence of the more mildly cleavable -S-S-bond is frequently of great value. It allows identification of the protein subunits by 2-dimensional electrophoresis with 2-mercaptoethanol-cleavage between the first and the second run.

TABLE

$$X(-CH_2-CH_2-CN)_2 \xrightarrow[CH_3OH]{HCl} X\left(-CH_2-CH_2-C\diagdown^{NH\cdot HCl}_{OCH_3}\right)_2$$

No.	X	dinitrile			diimidate	
		b.p.(°C)$_{mm}$	m.p.(°C)	commerc. available	m.p.(°C)	max. span(Å)
III	O	174–178$_{10}$	–	+	101–102	10
IV	O–CH$_2$–O	155–160$_2$ a)	–	–	d)	13
V	O–(CH$_2$)$_2$–O	158$_2$ b)	–	–	d)	14
VI	O–(CH$_2$)$_4$–O	–	45	+	98–99	16
VII	O–(CH$_2$)$_2$–O–(CH$_2$)$_2$–O	190$_1$ b)	–	–	d)	17
VIII	O–(CH$_2$)$_2$–S–S–(CH$_2$)$_2$–O	210–215$_4$ c)	–	–	81–85	18
IX	HN–$\overset{O}{C}$–(CH$_2$)$_2$–S–S–(CH$_2$)$_2$–$\overset{O}{C}$–NH	–	170–171	–	56–64	20

a) (Walker, 1945)
b) (Bruson & Riener, 1943)
c) (Schramm & Dülffer, 1975)
d) viscous oil

Two of the diimidoesters mentioned in the table contain this disulfide-group. Only the first one (VIII; dimethyl-3, 3' -(ɣ·ɗ - dithiahexamethylenedioxy)-dipropionimidate) has been tested so far (Schramm & Dülffer, 1976). It crosslinks protein complexes (like DNA-dependant RNA-polymerase). After -S-S-bond reduction the free subunits turn up again.

The preparation of the second -S-S-diimidoester needs a slight modification of the PINNER-method: The solid dinitrile (50 mg), which is insoluble in most solvents, is added to 5 ml of dry methanol saturated with HCl gas. The slurry is stirred till solution is complete. Then, after standing overnight, the usual ether-precipitation is performed. The preparation of the dinitrile from dithiobis-(succinimidyl propionate) (Lomant & Fairbanks, 1976) and 3-aminopropionitrile is performed similar to the synthesis of biocytin by Bayer & Wilchek (1974) (yield: 90 %; elementary analysis: calc.: C = 45.8, H = 5.8, N = 17.8, S = 20.4; found: C = 45.96, H = 5.77, N = 17.76, S = 20.50). Both -S-S-group containing diimidates are hygroscopic.

For different protein complexes different crosslinking reagents may give the best results. Fig. 1 shows distinctly different crosslinking patterns of four different diimidoesters with the same protein. The higher tendency for intermolecular crosslinking of the more hydrophilic diimidates (gel No. 2, 4, 5) is evident. There is also a higher overall-reaction in comparison to suberimidate (gel No. 3). The "incomplete -S-S-bond cleavage" (gel No. 6) may be due to partial crosslinking of subunits by one imidoester group. This reaction should be possible, if the mechanism of Browne & Kent (1975) is correct (There are also reports of crosslinkings by monofunctional imidoesters (e. g. Wood et al., 1975)).

D. 2-IMINOTHIOLANE

For some investigations of protein complexes monofunctional imidoesters carrying a HS-group have been used. After modification the thiolgroups are mildly oxidized to form the -S-S-crosslinks in situ (Traut et al., 1973). This method has shown advantages in the case of ribosomes. Here some additional crosslinks are formed, which do not turn up if the corresponding -S-S-diimidoesters are used (Peretz et al., 1976). The mercaptanes, however, are unpleasant to handle and to synthesize. We propose

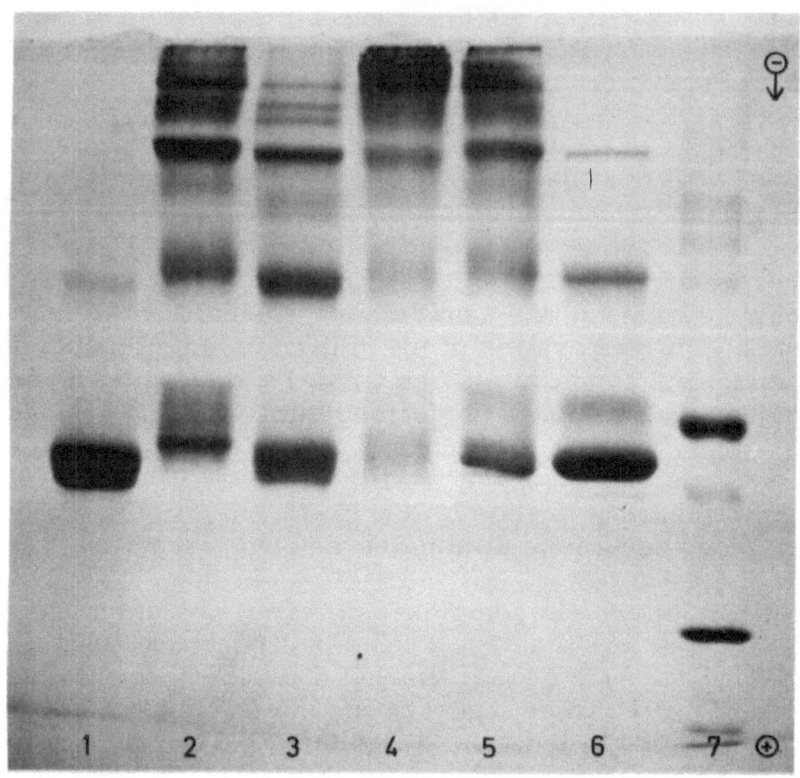

Fig. 1. SDS-electrophoresis (logarithmic gel 5-15 %) of glutamate dehydrogenase (1 = blank) crosslinked with dimethyl-3, 3' -
oxydipropionimidate III (= 2), dimethyl suberimidate (= 3), di-
methyl-3, 3' -(tetramethylenedioxy)-dipropionimidate VI (= 4),
dimethyl-3, 3' -(γ.δ -dithiahexamethylenedioxy)-dipropionimidate
VIII (= 5 + 6); 6 = reduced with 2-mercaptoethanol after crosslinking; 7 = chymotrypsin + serum albumin. Crosslinking conditions: 0. 1 mg enzyme (EC 1. 4. 1. 3; Boehringer Mannheim) in
0. 01 ml 50 % glycerol solution, 0. 3 mg reagent in 0. 04 ml 0. 2 M
triethanolamine/HCl - 0. 2 M NaCl - buffer pH 8. 7; 4 [h] at room
temperature; for gel No. 6 sample buffer containing 5 % 2-mer-
captoethanol; electrophoresis according to Laemmli (1970).

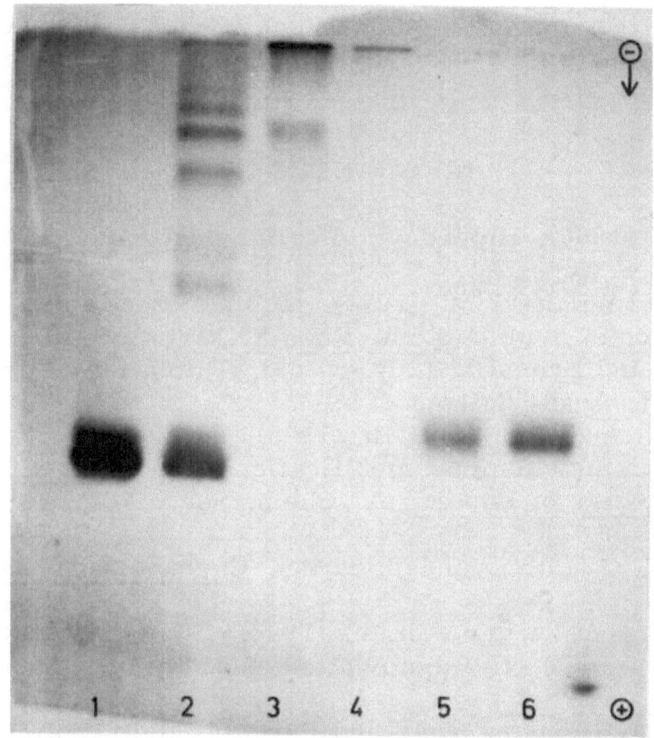

Fig. 2. SDS-electrophoresis of lactate dehydrogenase (1 = blank) crosslinked with 2-iminothiolane (= 2-6; reagent/protein (w/w) = 1:1 for 2, 3:1 for 3 and 5, 5:1 for 4 and 6; 5 and 6 = reduced with 2-mercaptoethanol). Crosslinking conditions: 0.05 mg enzyme (EC 1.1.1.27, from hog muscle, Boehringer Mannheim) in 0.005 ml of 50 % glycerol solution, 0.05 (- 0.25) mg reagent in 0.038 ml 0.2 M triethanolamine/HCl buffer pH 9.0; 2 [h] at room temperature; oxidation by addition of 0.007 ml of 3 % H_2O_2 for 2 [h] at room temperature. Other conditions as in Fig. 1.

the use of 2-iminothiolane (2-iminotetrahydrothiophen), a cyclic thioimidoester, as a substitute for 4-mercaptobutyrimidate. 2-Iminothiolane hydrochloride is easy to prepare (R. W. Addor, 1966), to store and to handle. It is an efficient crosslinking reagent, as can be seen from Fig. 2. Further studies are in progress. We are also investigating whether the reagent can be used to replace N-acetyl-homocystein-thiolactone (Benesch & Benesch, 1958), or similar amide-bond forming reagents, to introduce thiol groups for heavy atom labelling studies.

REFERENCES

Addor, R. W. (1966). Heterocyclic sulfur pesticides. C. A. 64, 5048 h (U. S. patent).

Bayer, E. and Wilchek, M. (1974). Insolubilized biotin for the purification of avidin. Meth. Enz. XXXIV, 265-267.

Benesch, R. and Benesch, R. E. (1958). Thiolation of proteins. Proc. Nat. Acad. Sci. 44, 848-853.

Browne, D. T. and Kent, S. B. H. (1975). Formation of non-amidine products in the chemical modification of proteins with imido-esters. Abstr. of Papers, A. Chem. Soc. 170th National Meeting, 156.

Bruson, H. A. (1949). Cyanoethylation. Chem. Reactions 5, 79-135.

Bruson, H. A. and Riener, Th. W. (1943). The chemistry of acrylonitrile. IV. Cyanoethylation of active hydrogen groups. J. Am. Chem. Soc. 65, 23-27.

Coggins, J. R. , Hopper, E. A. and Perham, R. N. (1976). Use of dimethyl suberimidate and novel periodate-cleavable bis(imido esters) to study the quaternary structure of pyruvate dehydrogenase multienzyme complex of Escherichia coli. Biochemistry 15, 2527-2533.

Davies, G. E. and Stark, G. R. (1970). Use of dimethyl suber-imidate, a crosslinking reagent, in studying the subunit structure of oligomeric proteins. Proc. Nat. Acad. Sci. 66, 651-656.

Hunter, M. J. and Ludwig, M. L. (1972). Amidination. Meth. Enz. XXV, 585-596.

Kauzman, W. (1959). Some factors in the interpretation of protein denaturation. Adv. Protein Chem. 14, 1-63.

Laemmli, U. K. (1970). Cleavage of structural proteins during the assembly of the head of bacteriophage T 4. Nature (London) 227, 680-685.

Langer, R., Poppe, Ch., Schramm, H.J. and Hoppe, W. (1975). Electron microscopy of thin protein crystal sections. J. Mol. Biol. 93, 159-165.

Lomant, A.J. and Fairbanks, G. (1976). Chemical probes of extended biological structures: synthesis and properties of the cleavable protein cross-linking reagent (^{35}S)dithiobis(succinimidyl propionate). J. Mol. Biol. 104, 243-261.

Luduena, R., Wilson, W. and Shooter, E.M. (1974). Cross-linking of tubulin: evidence for the heterodimer model. J. Cell. Biol. 63, 202 a (Abstract).

Lutter, L.C., Ortanderl, F. and Fasold, H. (1974) The use of a new series of cleavable protein-crosslinkers on the Escherichia coli ribosome. FEBS Letters 48, 288-292.

McElvain, S.M. and Schroeder, J.P. (1949). Orthoesters and related compounds from malono- and succinonitriles. J. Am. Chem. Soc. 71, 40-46.

Peretz, H., Towbin, H. and Elson, C. (1976). The use of a cleavable crosslinking reagent to identify neighboring proteins in the 30-S ribosomal subunit of Escherichia coli. Eu. J. Biochem. 63, 83-92.

Ruoho, A., Bartlett, P.A., Dutton, A. and Singer, S.J. (1975). A disulfide-bridge bifunctional imidoester as a reversible cross-linking reagent. Biochem. Biophys. Res. Comm. 63, 417-423.

Schramm, H.J. (1967). Synthese von farbigen Nitrilen, Dinitrilen und bifunktionellen Imidsäureestern. Hoppe-Seyler's Z. Physiol. Chem. 348, 289-292.

Schramm, H.J. (1975). The synthesis of mono- and bifunctional nitriles and imidoesters carrying a fluorescent group. Hoppe-Seyler's Z. Physiol. Chem. 356, 1375-1379.

Schramm, H.J. and Dülffer, Th. (1976). The synthesis of a cleavable and hydrophilic cross-linking reagent. Hoppe-Seyler's Z. Physiol. Chem. 357, 477-479.

Traut, R.R., Bollen, A., Sun, T.-T., Hershey, J.W.B., Sundberg, J. and Pierce, L.R. (1973). Methyl 4-mercaptobutyrimidate as a cleavable cross-linking reagent and its application to the Escherichia coli 30 S ribosome. Biochemistry 12, 3266-3273.

Walker, J.F. (1945). Reaction of CH_2O with acrylonitrile. C. A. 39, 223 (U.S. patent).

Wang, K. and Richards, F.M. (1974). The behavior of cleavable crosslinking reagents based on the disulfide group. Israel J. of Chem. 12, 375-389.

Wetz, K., Fasold, H. and Meyer, Ch. (1974). Synthesis of
    "long", hydrophilic, protein-crosslinking reagents. Analyt.
    Biochem. 58, 347-360.
Wood, F.T., Wu, M.M. and Gerhart, J.C. (1975). The radio-
    active labeling of proteins with an iodinated amidination
    reagent. Analyt. Biochem. 69, 339-349.

12

COMPARISON OF HYDROPHOBIC AND STRONGLY HYDROPHILIC
CLEAVABLE CROSSLINKING REAGENTS IN INTERMOLECULAR BOND
FORMATION IN AGGREGATES OF PROTEINS OR PROTEIN-RNA.

H. Fasold, H. Bäumert, and G. Fink

Institut für Biochemie, University of Frankfurt
Sandhofstrasse, Gebäude 75A
D 6000 Frankfurt/Main-Niederrad (GFR)

ABSTRACT

Most of the bifunctional reagents in protein
chemistry possess a strongly hydrophobic backbone, derived from aliphatic or aromatic hydrocarbons. Even
bifunctionals of more than 30 $\overset{o}{A}$ in length of this sort
form intramolecular bridges preferentially. In recent
years, the intermolecular crosslinking of physiological
protein aggregates has gained in importance. As shown
in the crosslinking of hemoglobin with two sets of hydrophobic and strongly hydrophilic reagents, derived
from azo dyes and tartaric acid, respectively, in this
case it is not primarily the length of the bifunctional,
but the hydrophilic structure that will enhance intermolecular crosslinking. Artificial dimers of native
structure may be obtained. For the crosslinking of RNA
to protein, we have synthesized a new reagent, 3-(2-bromo-
3-oxobutane-1-sulphonyl)-propionic acid p-nitrophenyl
ester. In a two step reaction, it is attached to adenine
and cytosine moieties at pH 6 first, and to lysine side
chains at pH 7,5. The reagent has been applied to the
poly-A sequence of globin messenger RNA nucleoprotein.

INTRODUCTION

This report is divided into two distinctly separate
parts. On one hand, we have tried to study the influence
of length and hydrophilic character of bifunctional reagents upon the average length of the effected crosslink. In a second project we synthesized a bifunctional
that should permit a preferential cross-linking reaction
between the adenine and cytosine moieties of nucleic
acids or single nucleotides and attached proteins.

I. Hydrophilic cross-linkers. The first problem started
to interest us, when we tried to determine distances and
nearest neighbourhoods in protein aggregates as in the
contractile system of striated muscle or in bacterial
ribosomes. Evidently, most of the bifunctional reagents
that have been proposed were designed for the purpose of
short-length cross-linking within one protein molecule,
and most of them possess a strongly hydrophobic backbone,
an aliphatic or aromatic hydrocarbon moiety. In this
class belong all of the most frequently used reagents,
like suberimidate, difluoro-dinitrobenzene, or phenylene-
diisocyanate. To this day, a certain lack of hydrophilic
protein cross-linking reagents persists, although the
gap has begun to be filled by some of the substances
described here and by those of Schramm (1).

In order to effect detectable amounts of intermolecular cross-links, these short hydrophobic reagents
have to be used in high concentrations, as in the determination of subunit number in protein aggregates (2).
As a simple way out, longer reagents were proposed
simply extending the hydrophobic backbone by further
elements (3):

However, this sort of compound tends to denature proteins even at low concentrations. Thus, while hemoglobin will bear the covalent attachment of up to five residues of compound I per molecule without losing its normal oxygen binding curve, the initial phase of denaturation is signalled already after coupling of 3.5 residues of Ia by loss of full cooperativity and a high content of methemoglobin, as well as a change from the spectrum of native hemoglobins in the region between 380 and 440 nm.

In consequence, several reagents were devised that combined a fairly long and rigid structure with the prevalent use of hydrophilic building elements, such as polyproline(II)(4, 5):

$$N_3^-OC\text{-}(Pro)_n^{\geq}N\text{-}CO\text{-}NH\text{-}C_6H_4\text{-}N= \quad | \text{ symm.} \qquad \text{II}$$

$$JCH_2CO\text{-}NH\text{-}(CH_2)_z^-NH\text{-}CO\text{-}(Pro)_n^{\geq}N\text{-}CO\text{-}NH\text{-}C_6H_4\text{-}N= \quad | \text{ symm} \qquad \text{III}$$

Suitable centerpieces, in this case the azo bond, permit the cleavage of the cross-link under mild conditions (3).

As expected, these fairly hydrophilic reagents effected a satisfactory amount of intermolecular covalent bridges when coupled to a model system, as hemoglobin (5). However, the comparison of reagents as I and III could not decide the question, whether bifunctional reagents of similar length, but differing in hydrophobicity, should give different yields in intermolecular cross-links, when tested on the same protein model system.

More recently, very strongly hydrophilic compounds were synthesized, primarily for the cross-linking of ribosomal proteins, and of subunits of F-actin (6):

$$|\text{-CHOH-CO-N}_3$$

$$|\text{-CHOH-CO-NH-CH}_2\text{CO-N}_3$$

symm. | IV,V,VI,VII

$$|\text{-CHOH-CO-NH-(CH}_2)_4\text{CO-N}_3$$

$$|\text{-CHOH-CHOH-CH}_2\text{C}\underset{\text{OCH}_3}{\overset{\text{NH}}{\diagdown}}$$

$$|$$

In the case of VII: 3,4,5,6-tetrahydroxy-suberimidate
(THS), the synthesis started from 1,6 Didesoxy-dibromo-
mannite a commercial product. The two halogen atoms
were substituted by two nitrile groups in a reaction
with potassium bromide in 50% ethanol solution at $60^{\circ}C$,
and the dinitrile was transformed into the imido ester
in methanol saturated with dry HCl in the usual manner.
The application of $K^{14}CN$ provided for an easily available radioactive label of the reagent.

All of the compounds of these series are almost
insoluble in organic solvents, but extremely well
soluble in water. Without taking accurate measurements
of the relative solubilities, we think we may state that
these reagents are all very strongly hydrophilic, and
should not show any attachment to the hydrophobic core
of many globular proteins, when coupled to their ex-
terior covalently.

It then became possible to compare the intermolecular cross-linking yield on two suitable protein aggregate models, hemoglobin and F-actin, between reagents of
similar length, but different hydrophobicity. Both compounds III_{Pro1} and V will span distances of 30 Å on the
average,but while the first, due to the prevalent central azobenzene part, is of moderate hydrophobicity, the
second is distinctly more hydrophilic.

In the hemoglobin experiments, the tetramer was
cross-linked at pH 7.55 in 0.1 M phosphate buffer,
using varying molar ratios of protein and reagents. A
weak radioactive label in the proline or glycine building blocks permitted the precise determination of the
amount of covalently bound molecules, after the excess
reagent had been removed by gel filtration. The dialyzed crosslinked protein was then treated with sodium
dithionite in the case of reagent III, and sodium periodate in the case of reagent VI. After renewed dialysis

and recounting of the specific radioactivity of the protein, this gave the amount of the reagent that had been
coupled to hemoglobin by only one of its two reactive
groups. Thus, in the end, the number of true crosslinks
on the tetramer was calculated.

To determine the fraction among them that had
formed bridges between subunits of the tetramer, the
crosslinked hemoglobin was now dissociated irreversibly
by the method of Fasold et al. (5). The mixture of
monomer subunits and artificial dimers was then separated on Sephadex G 150 columns. Control runs were performed starting from native hemoglobin, and showed only
very small amounts of material at 32000 daltons molecular weight. Integration of the monomer and dimer peaks
from the elution curves then permitted the calculation
of the amount of true intersubunit crosslinks per
original hemoglobin tetramer.

The results are summarized in Table 1. It could
be shown that length of a bifunctional reagent is not,
at least, the primary factor determining the yield of
long cross-links between diferent tertiary structures
in protein aggregates, utilizing the full distance
between the two protein-reactive groups. Although the
bis-(ureido)-prolylazobenzene (III1) is more rigid than
tartryl-bisglycineazide (V), and its two azide groups
are farther apart, TGA reached a higher percentage of
interchain bonds, and was only equalled by bis-(ureido)
-tetraprolyl-azobenzene. The latter, in addition to its
greater length also had decreased in hydrophobicity
against III1.

It seems, therefore, that distinctly hydrophilic
bifunctionals like the tartaric acid derivatives and
related compounds described here, tend to stick out
from the surface of the protein after attachment by
one arm, thereby enhancing reaction with a second protein molecule. Even minor hydrophobic portions of the
reagent molecule, in contrast, tend to flatten the
reagent against the protein, and favour internal crosslinks.

II. Cross-linking between RNA and Protein. A bifunctional reagent, coupling with one arm to protein, and with
the other to nucleic acids exclusively,in one reaction
mixture,would have wide application in structural studies on the many biologically relevant aggregates of
these two macromolecules. The difficulty in the planning

Table 1

Reaction of Bifunctional Reagents with Hemoglobin[a]

Reagent	Final concentration mM	Reagent residues introduced/ molecule of protein	Bifunctional coupling	Artificial dimers, % of total residues introduced
PAPA	1.2	2.6	1.6	36
	2.0	4.1	2.0	30
	2.5	4.8	2.2	25
	3.0	5.4	2.5	20
	4.0	6.5	2.9	15
TGA	1.2	2.6	1.8	70
	2.0	4.2	2.4	54
	2.5	4.8	2.6	48
	3.0	5.6	2.9	44
	4.0	6.5	2.9	32

of such a reagent of course lies in the similarity of
the functional groups available on proteins and nucleic
acids for covalent modification. Usually, many of the
aliphatic amino or sulfhydryl groups will compete
successfully with a substance applied for the substitution of amino- or hydroxyl groups on the nucleic acid
bases; except for periodate cleavage at terminal nucleotides, the ribose hydroxyls are not amenable to
specific modification, and carboxyls will prevent a
specific modification of phosphate groups.

We have tried to circumvent these difficulties by choosing two successive reaction media of different pH value for the modification of nucleic acid and protein in a two-step procedure. It then becomes possible to utilize the modification of the adenine and cytosine amino groups by α-halogen carbonyl compounds at pH 4 - 6.5 selectively, the rate of reaction with protein amino groups and also with sulfhydryls remaining very much slower at this stage. In the second step, modification of the protein partners was achieved by p-nitrophenyl or N-hydroxy succinimide ester groups.

The synthesis of the reagent (VIII) started from methyl vinylketone and 3-mercapto propionic acid. The acetonyl-propionic acid mixed sulfide was coupled to p-nitrophenole or N-hydroxysuccinimide with the aid of dicyclohexyl carbodiimide, and the resulting activated ester was cautiously oxidized to the corresponding sulfone. This last step proved necessary to avoid recleavage of the sulfide during the subsequent addition of bromine to obtain the halogenoketone.

$$CH_3CO-CHBr-CH_2 SO_2-CH_2 CO-X$$

$$X = O-\langle\bigcirc\rangle-NO_2$$

VIII

Several model reactions were tried out for the new reagent. In a typical experiment, equal amounts of yeast tRNA mixture and polylysine of average molecular weight of 100000 daltons were incubated with the p-nitrophenylester reagent at pH 6.3 at 30°C. The amount of reagent was an approximately double molar excess of the adenine and cytosine moieties of the tRNA. After 3 hours it was removed by chloroform extraction or gel filtration. The pH of the reaction mixture was then raised to 7.5 - 8.0, the second incubation lasted for 2 - 5 hours. The reaction was then stopped by lowering the pH to 3.0, and the solution was dialyzed for a prolonged time.

Control experiments were carried out ommitting either the addition of reagent, or of tRNA. For evaluation of the cross-link yield, the final mixtures were treated with trypsin at conditions suitable for an optimal digestion of the polylysine. The residual tRNA was dialyzed thoroughly against dilute hydrochloric acid, hydrolyzed in 6 N HCl, and subjected to a quantitative determination of lysine in an amino acid analyzer. Controls contained only traces of the amino acid. From the amount of lysine in the cross-linked tRNA preparation, an estimate of the amount of nucleic acid - proteins links was made under the assumption that trypsin would leave approximately three lysine residues at every point of covalent intermolecular attachment.

In this favourable example 30% of the adenines and cytosines were found to be linked to polylysine. But even when bovine serum albumin was used in the place of polylysine, more than 10% of the two bases participated in cross-links. In the actual experiments on E.coli ribosomes that have been performed since, therefore, the amount of reagent could be distinctly reduced, and this has led to fewer, but well reproducible links between ribosomal RNA and adjacent proteins (7).

REFERENCES

1) H.J. Schramm and Th. Dülffer (1977). This volume.

2) F.C. Hartman, and F. Wold, J.Am.Chem.Soc. 88, 3890(1966).

3) H. Fasold, J. Klappenberger, Ch. Meyer, and H. Remold, Ang. Chem. 83, 875 (1971).

4) H. Fasold, Biochem.Zschr. 339, 482 (1964).

5) K. Wetz, H. Fasold, and Ch. Meyer, Anal.Biochem. 58, 347 (1974).

6) L.C. Lutter, F. Ortanderl, and H. Fasold, FEBS Letters 48, 288 (1974).

7) J. Rinke, K. Möller, A. Ross, and R. Brimacombe, 10th Int.Congr. of Biochemistry, Hamburg (1976), Abstr. 03-3-115.

13

CROSSLINKING OF RIBOSOMES BY CLEAVABLE BIFUNCTIONAL MERCAPTOIMIDATES

Robert R. Traut and James W. Kenny

Department of Biological Chemistry, School of Medicine,

University of California, Davis, California 95616

ABSTRACT

Methods are described for determining protein:protein proximity relationships in complex cellular structures containing multiple protein components. Studies on the 50S ribosomal subunit of Escherichia coli illustrate the procedures employed. The intact ribosomal subunit is first incubated with methyl 4-mercaptobutyrimidate. Lysine amino groups become modified and thus converted to amidine derivatives which contain sulfhydryl groups. The modified ribosomal subunits are oxidized to promote crosslinking by formation of intermolecular disulfide bonds. The proteins are extracted and subjected to polyacrylamide/dodecyl sulfate disc gel electrophoresis under non-reducing conditions. The gel is immersed in a reducing solution and then embedded in a second polyacrylamide/dodecyl sulfate gel slab for electrophoresis in a second dimension. Non-crosslinked proteins retain their relative mobility in the two electrophoretic steps and fall on a diagonal line. Proteins crosslinked by disulfide bonds migrate slowly in the first electrophoretic separation but, following reduction, give rise to faster migrating monomeric proteins which appear beneath the diagonal in the second electrophoretic separation. Crosslinked proteins can be identified by their position on the gel pattern and by analysis of the apparent molecular weights of crosslinked species and monomeric proteins derived from them upon reduction. The use of the reversible protein crosslinking procedure and two-dimensional "diagonal" gel electrophoresis provides a characteristic fingerprint of the protein:protein interactions in ribosomal subunits. These techniques, developed to study the ribosome, should be valuable in investigations of other

biological ultrastructures in which it is useful to obtain information
concerning the arrangement of multiple protein components.

INTRODUCTION

The results of intensive investigations of the ribosome of
Escherichia coli during the 1960's led to the conclusion that knowledge of the spatial arrangement or topography of its protein constituents would be invaluable in understanding the function of the
ribosome in protein synthesis. The ribosome was shown to contain
fifty-five distinct proteins, most of which were present in quantities equimolar with the ribosome particle. The smaller 30S ribosomal
subunit was reconstituted from purified proteins and ribosomal RNA;
omission of single proteins in such reconstitution experiments
indicated that ribosomal proteins acted cooperatively to determine
functional properties of the ribosome; genetic studies provided
independent evidence for cooperative interactions among ribosomal
proteins (for review, see Nomura, Tissières and Lengyel, 1974).
It was also shown that the 50S ribosomal subunit itself catalyzed
peptide bond formation and hence could be considered as a complex
enzyme containing thirty-four polypeptide and two RNA components
(Traut and Monro, 1964).

The accumulated evidence strongly suggested that groups of
ribosomal proteins were required to form active sites and that there
were cooperative interactions among them. In no case has it been
shown that a single protein is uniquely associated with a single
functional property. Thus many laboratories undertook investigations
of ribosomal protein topography.

One of the first experimental approaches was that of treating
the ribosomes with protein specific bifunctional reagents (Bickle,
Hershey and Traut, 1972; Chang and Flaks, 1972; Lutter, Zeichardt,
Kurland and Stöffler, 1972; Shih and Craven, 1973) in order to crosslink adjacent proteins. These early experiments were hindered by
the fact that the reagents employed were not cleavable, or cleavable
only in low yield. This necessitated the identification of the
components of crosslinked species by time-consuming techniques
involving the use of antibodies against each purified ribosomal
protein (Sun, Bollen, Kahan and Traut, 1974; Lutter, Bode, Kurland
and Stöffler, 1974), peptide mapping (Shih and Craven, 1973) or the
analysis of cleavage products formed in low yield (Clegg and Hayes,
1972; Bickle, Hershey and Traut, 1972).

Since the aim of such research efforts was to reconstruct models
placing each protein relative to the others, the need for new techniques capable of detecting and identifying large numbers of crosslinked proteins became apparent. Imidates are mild protein modifying
reagents and a thioimidate, 3-mercaptopropionimidate, had previously

been described by Perham and Thomas (1971). We reasoned that by introducing new sulfhydryl groups into the intact ribosome with mercaptoimidates it would be possible to form intermolecular protein: protein crosslinkes by disulfide bond formation. The analysis of the composition of such crosslinked products would be immensely simplified by virtue of the fact that the disulfide bond linking two proteins could be readily cleaved by reduction to regenerate monomeric proteins which would have the same electrophoretic mobilities as the native proteins. The reversibility of the crosslinking procedure suggested the use of two-dimensional diagonal polyacrylamide/dodecyl sulfate gel electrophoresis to simplify further the identification of crosslinked proteins. The application of this crosslinking technique, using the reagent methyl 4-mercaptobutyrimidate, to the 50S ribosomal subunit of Escherichia coli is described here. The object of this brief review, however, is not so much to present an analysis of ribosomal structure and function, as it is to illustrate the general applicability of reversible crosslinking for the study of protein assemblies in which it is of interest to determine the relative spatial arrangement of different polypeptide components.

THE TWO-STEP CROSSLINKING PROCEDURE

Figure 1 outlines the two-step crosslinking procedures: modification of ribosomal subunits with methyl 4-mercaptobutyrimidate followed by oxidation. In the first step, ribosomal subunits are incubated with methyl 4-mercaptobutyrimidate to permit formation of amidine linkages between lysine amino groups and imidate group of the reagent. The ribosome solution is buffered at pH 8 with triethanolamine which does not react with the imidate. A reducing agent, mercaptoethanol, is included in the incubation mixture. At pH 8, the integrity of the ribosome is preserved and a sufficient number of lysine amino groups are present in the unionized form to permit appreciable reaction with the imidate. After incubation with 10 mM methyl 4-mercaptoimidate for 20 min at 0° followed by dialysis for three hours (see below), it has been found that approximately 42 sulfhydryl groups are added per 30S ribosomal subunit and 70 per 50S subunit. The imidate reagent appears relatively stable in solution since modification, as assayed by titration of ribosomal sulfhydryl groups, continues for at least three hours. Under the conditions employed approximately two sulfhydryl groups are added per ribosomal protein. The reagent appears to react preferentially with "exposed" or more reactive lysines since all of the added sulfhydryl groups react rapidly with sulfhydryl reagents as compared to the majority of endogenous sulfhydryl groups present as half-cysteine residues in the native ribosomal proteins which react slowly. With respect to the 50S ribosomal subunit, there are approximately 24 endogenous half-cystine residues (Acharya and Moore, 1973);

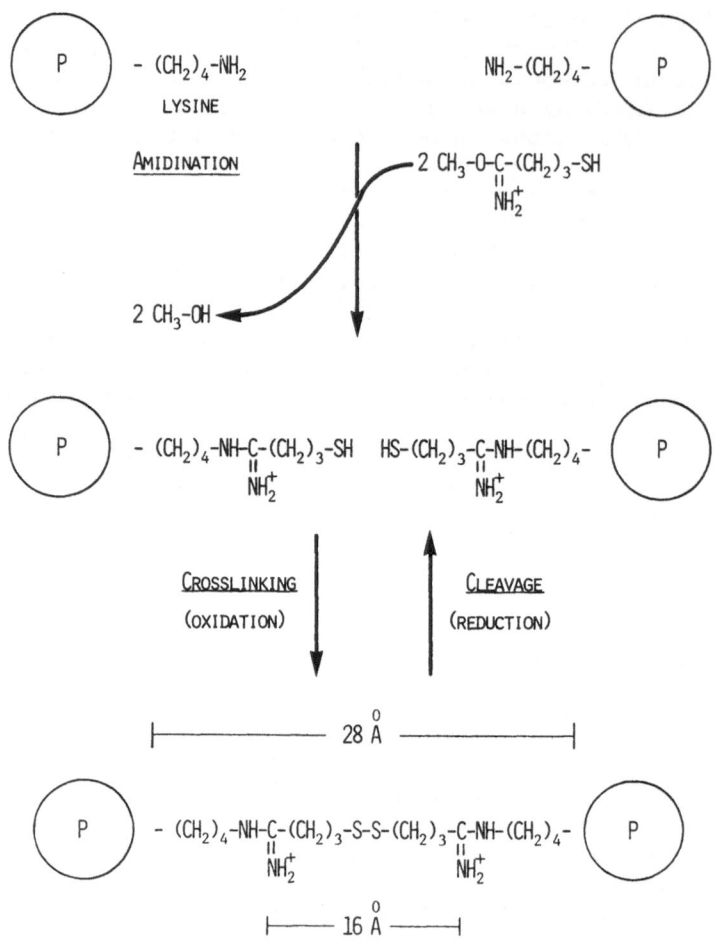

Figure 1.   Reversible crosslinking with methyl 4-mercapto-butyrimidate.

of these 5 to 6 react rapidly. Following modification with methyl 4-mercaptobutyrimidate, there are 94 sulfhydryl groups associated with the particle of which 74 react rapidly (R. Jue, unpublished results).

After incubation with methyl 4-mercaptobutyrimidate the solution of ribosomal subunits is dialyzed for three hours against buffers free of sulfhydryl reagents and then incubated for 30 min at 4° with 40 mM H_2O_2 to promote the formation of disulfide bonds. As a result of oxidation, it can be predicted that disulfide bonds of three types

are formed: intermolecular protein:protein, intramolecular, and
protein:mercaptan (mercaptoimidate or mercaptoethanol, present in
the solution during modification).   Only the first are of interest
in the studies described here and they can be detected by molecular
weight analysis upon electrophoresis in polyacrylamide/dodecyl
sulfate gels.   In summary, the first step of the procedure consists
of the addition of extra sulfhydryl groups to the ribosomal subunits
by amidination with a mercaptoimidate; the second "crosslinking"
step consists of the formation of intermolecular protein:protein
disulfide bonds by oxidation.

YIELD AND REVERSIBILITY OF INTERMOLECULAR CROSSLINKING

The polyacrylamide/dodecyl sulfate gels in Figure 2 illustrate
the formation of protein complexes of elevated molecular weight due
to oxidation of the particles previously treated with methyl 4-mercap-
tobutyrimidate.   It is apparent (panels a and b) that modification of
30S ribosomal subunits with mercaptoimidate under reducing conditions
results in no appreciable change in the gel pattern.   However, oxidation (panel c) leads to the formation of numerous stained bands of
lower mobility, or higher molecular weights, than the monomeric
ribosomal proteins.   If the modified and oxidized ribosomal subunits

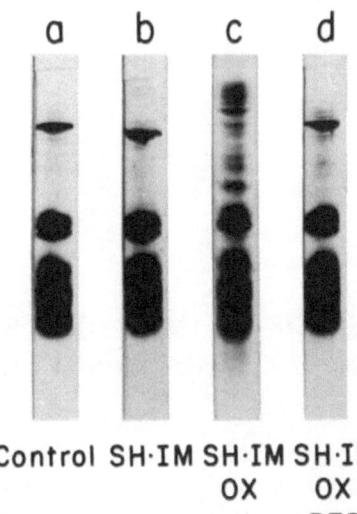

Figure 2.   Polyacrylamide/dodecyl gel electrophoresis showing
reversibility of crosslinking (Traut et al., 1973).

are incubated with a reducing agent (e.g., β-mercaptoethanol) prior to gel electrophoresis, the control pattern of monomeric proteins is restored (panel d). The results indicate the oxidation-dependent formation of protein products of molecular weight greater than single ribosomal proteins and that these products are removed by reduction. The results are interpreted to demonstrate that inter-protein disulfide bonds are formed in the oxidative step and that they can be essentially completely cleaved by reduction. These early experiments (Traut et al., 1973) showed that reversible crosslinking with the mercaptoimidate could be applied to the study of ribosomal protein topography. The major problems remaining were the resolution and identification of the multiple crosslinked products illustrated in Figure 2, panel c.

Before describing the method employed for analysis of cross-linked ribosomal proteins the yield of crosslinked products will be discussed briefly. Figure 3 shows polyacrylamide/dodecyl sulfate gel patterns of 30S ribosomal proteins oxidized after incubation with varying concentrations of methyl 4-mercaptobutyrimidate. In this

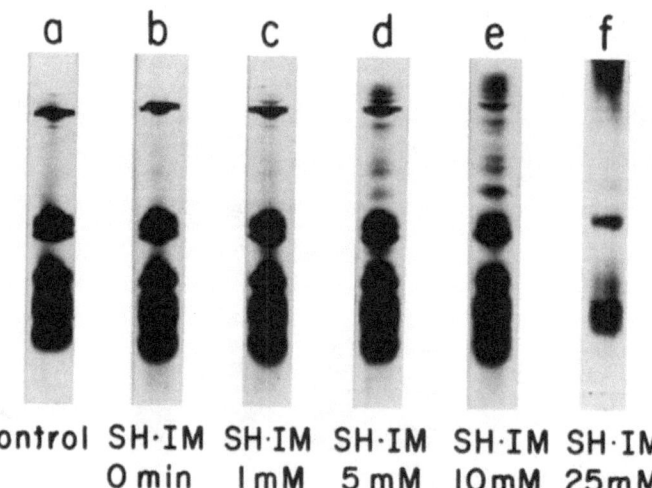

Figure 3. Effect of mercaptobutyrimidate concentration on yield and molecular weight distribution of crosslinks. 30S ribosomal subunits were incubated at 3 mg/ml in triethanolamine for 20 min at 0° with the concentrations of methyl 4-mercaptobutyrimidate (SH·IM) indicated (Traut et al., 1973). The band of lowest mobility in panel a represents the monomeric protein S1, molecular weight approximately 70,000.

experiment the time of incubation was constant and only the initial
concentration of the imidate was varied. The reaction was stopped
by addition of sodium dodecyl sulfate. The stained protein patterns
show that at 25 mM mercaptoimidate (panel f vs. panels a and b) there
is a substantial disappearance of monomeric ribosomal proteins coupled
with the formation of material of molecular weight so great as to not
migrate appreciable in the 10% gels used for analysis. Large complexes of this kind have molecular weights considerably in excess of
70,000 and therefore are composed of three or more ribosomal proteins
(average molecular weight 16-17,000). The complexes of high molecular weight in panel f therefore are not of value in a topographical
analysis based on the identification of pairs of neighboring proteins.
The experiment in panel f does show that crosslinking by disulfide
bond formation can take place in high yield with the formation of
products containing more than two protein components. However, in
order to obtain a variety of protein dimers for further analysis, we
reduced the concentration of mercaptoimidate used in the modification
of the ribosomal subunits. Panels d and e in Figure 3 show the formation of numerous crosslinked components with molecular weights
between 30,000 and 70,000 at lower mercaptoimidate concentrations.
Many of these must represent protein:protein dimers; accordingly, in
subsequent experiments ribosomal subunits were incubated for 20 min
with 5 to 10 mM methyl 4-mercaptobutyrimidate at 1-3°. The conditions used to optimize the formation of dimers produce a lower overall yield of crosslinking.

The use of intermolecular disulfide bonds to produce crosslinks
between proteins suggested two-dimensional polyacrylamide/dodecyl
sulfate (diagonal) gel electrophoresis for the separation of crosslinked from non-crosslinked proteins. The method is shown schematically in Figure 4. In the first dimension, carried out in tube gels,
the oxidized ribosomal proteins are subjected to electrophoresis
under non-reducing conditions. A second electrophoresis, with the
tube gel embedded in a slab gel, run under the same non-reducing
conditions would result in a pattern in which both monomeric and
crosslinked proteins would fall on a diagonal line: the relative
mobility of each component in the mixture would be the same in both
dimensions. On the other hand, if the protein mixture contains
disulfide linked complexes and if the first tube gel is soaked in
reducing agent prior to the second electrophoresis, then each crosslinked complex will be converted to monomeric proteins of lower
molecular weight. The previously crosslinked proteins will have
greater electrophoretic mobility in the second dimension and will
appear below the diagonal line of monomeric non-crosslinked ribosomal proteins on the stained diagonal gel pattern. An illustration
of the diagonal pattern for the proteins crosslinked in the intact
50S ribosomal subunit of Escherichia coli is shown in Figure 5. The
pattern of stained protein spots beneath the diagonal line represents a characteristic fingerprint of protein:protein interactions

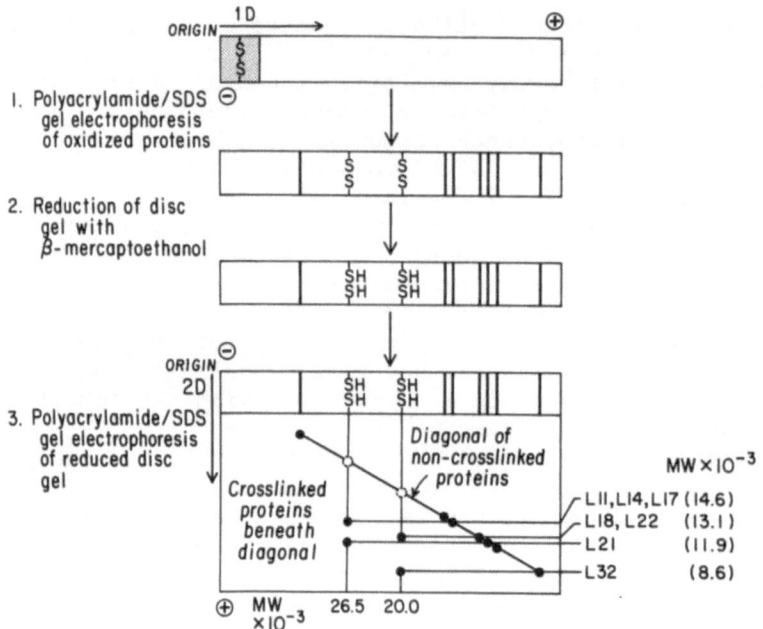

Figure 4. Diagonal polyacrylamide/dodecyl sulfate (SDS) gel electrophoresis of crosslinked ribosomal proteins. The molecular weight scales in both dimensions were established both with standard proteins and purified ribosomal proteins. The "L" numbers at the lower right refer to specific proteins of the large 50S ribosomal subunit.

in the native ribosomal particle (Sommer and Traut, 1974).

The next problem was the analysis of the complex pattern: the determination of the individual proteins which had been members of specific dimers crosslinked by disulfide bonds. Two preliminary criteria were employed: first, any monomeric components of an originally crosslinked species must necessarily fall on the same vertical line below the diagonal; second, the sum of the apparent molecular weights of the monomeric proteins arising by reduction of the disulfide crosslinked species must equal that of the crosslinked complex. The first criterion is obvious; the second depends upon the finding (Sommer and Traut, 1975) that crosslinked ribosomal

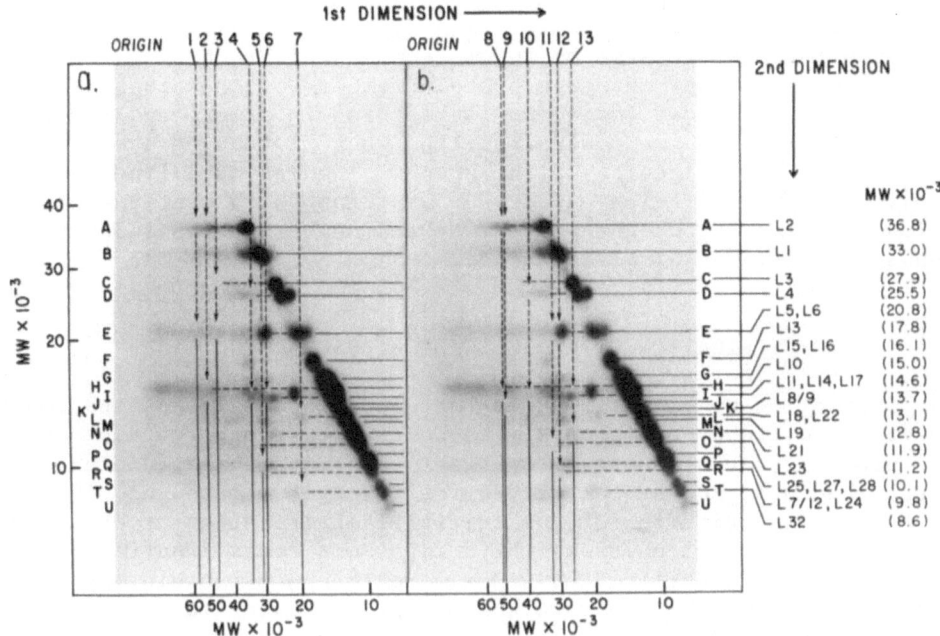

Figure 5. Diagonal gel electrophoresis of total proteins cross-linked with methyl 4-mercaptobutyrimidate in intact 50S ribosomal subunits. Lyophilized ribosomal protein (700 to 800 μg) derived from ^{35}S-labeled 50S subunits which had been modified with 10 mM methyl 4-mercaptobutyrimidate and oxidized with 40 mM mydrogen peroxide at 0° was resuspended in 25 μl of a buffer containing 4% sodium dodecyl sulfate, 80 mM Tris·HCl (pH 7.8), 40 mM iodoacetamide, and 10% glycerol. The solution was incubated at 65° for 15 min and then applied to a discontinuous polyacrylamide/sodium dodecyl sulfate disc gel (0.4 x 10 cm) containing 17.5% acrylamide, 0.35% methylenebis-acrylamide, 330 mm Tris·HCl (pH 8.7), and 0.1% sodium dodecyl sulfate. The sample was subjected to electrophoresis toward the cathode at 2 mM/gel for 3 hours. The electrophoresis buffer contained 30 g/liter of Tris base, 5 g/liter of sodium dodecyl sulfate, 144 g/liter of glycine. The pH was 8.7. Following electrophoresis the gel was removed from the tube and soaked first at 65° for 15 min in 50 ml of electrophoresis buffer containing 3% 2-mercaptoethanol and then at room temperature for 30 min in 50 ml of electrophoresis buffer, adjusted to pH 6.8 with HCl, without 2-mercaptoethanol. The gel·was embedded at the origin of a second dimension polyacrylamide/sodium dodecyl sulfate slab gel of length 24 cm whose composition was identical with that of the first dimension gel. Electrophoresis proceeded toward the cathode at 50 volts for 1 hour, then at 100 volts for 30 hours. The thirteen vertical ordinates identify pairs of monomeric proteins found in crosslinked protein dimers.

protein dimers have apparent molecular weights, determined by electrophoretic mobilities in polyacrylamide/dodecyl sulfate gels, which
are approximately the same as the molecular weight obtained by
summing those of the monomeric protein members of the crosslinked
complex. It was shown that crosslinked protein dimers and monomeric
proteins fall on the same straight line when plotted on a standard
mobility vs. log molecular weight graph. The vertical lines and
arrows in Figure 5 indicate pairs of proteins which meet the criterion of molecular weight additivity and hence are concluded to have
been crosslinked by disulfide bonds.

IDENTIFICATION OF CROSSLINKED RIBOSOMAL PROTEINS

Inspection of Figure 5 shows that 50S ribosomal proteins can be
divided into twenty-one groups. Each lettered group contains one or
more proteins having approximately the same apparent molecular
weight when analyzed by polyacrylamide/dodecyl sulfate gel electrophoresis. The criteria for the initial identification of crosslinked
protein pairs by molecular weight are in some cases insufficient for
their absolute identification since more than one protein may be
found in the same molecular weight group. In order to facilitate
the unambiguous identification of the proteins crosslinked in
ribosomal subunits, we employed subunits prepared from cells grown
in the presence of [^{35}S]sulfate for the crosslinking and diagonal
gel analysis just described. A stained spot on the diagonal gel
pattern is cut out and eluted with dodecyl sulfate and urea from the
macerated polyacrylamide gel. The radioactive material is mixed
with a total nonradioactive 50S ribosomal protein and passed through
a short column of Dowex AG 1 x 8 to remove both amido black and
dodecyl sulfate. The resulting protein solution is analyzed by
standard two-dimensional polyacrylamide/urea gel electrophoresis
(Howard and Traut, 1973; Knopf et al., 1975) which separates all the
50S (or 30S) ribosomal proteins as distinct stained spots. Gel
slabs of 2-3 mm are employed; these are stained with Coomassie brilliant blue, dried and exposed to X-ray film. Superimposition of the
autoradiograph on the stained gel shows directly the identity of the
radioactive crosslinked protein first characterized on the diagonal
gel. Many crosslinked protein pairs in both ribosomal subunits have
been identified by the techniques described in the preceding para-
graphs (Sommer and Traut, 1976; Kenny, Sommer and Traut, 1975; Kenny
and Traut, unpublished results).

FRACTIONATION OF PROTEIN MIXTURES EXTRACTED FROM RIBOSOMAL
SUBUNITS PRIOR TO DIAGONAL GEL ELECTROPHORESIS

Figure 5 shows the complexity of the diagonal gel pattern
resulting from crosslinking 50S ribosomal subunits. The vertical
lines and arrows indicate the stained spots which had been members

of protein dimers prior to reduction. The complexity of the pattern, the partial overlapping of certain spots containing monomeric proteins or formed from crosslinks of nearly the same molecular weight, makes it difficult to cut out precisely the desired regions on closely spaced vertical lines for final identification of the proteins. Therefore we turned to fractionation of the proteins from crosslinked ribosomal subunits prior to diagonal gel electrophoresis. A typical experimental procedure consists of the following steps: 1) reaction with methyl 4-mercaptobutyrimidate; 2) oxidation with hydrogen peroxide; 3) addition of catalase to remove unreacted hydrogen peroxide; 4) addition of iodoacetamide to block unoxidized sulfhydryl groups; 5) successive extraction of ribosomal subunits and protein-deficient ribonucleoprotein "cores" with increasing concentrations of LiCl; 6) electrophoresis of each salt-extracted protein fraction on polyacrylamide/urea gels at pH 5.5; 7) slicing of the gels into 5 mm pieces; 8) diagonal gel electrophoresis of the gel slices from 6). Iodoacetamide (40 mM) is included in all buffers used in the successive salt extractions and in the dodecyl sulfate solution used for preparing the sample for diagonal gel electrophoresis to avoid the possibility of disulfide interchange between proteins containing free sulfhydryl groups and crosslinked dimers. However, diagonal gel proteins of crosslinked ribosomal proteins prepared by the procedures described above are identical to those obtained with earlier procedures which involved extraction of protein with sixty-six percent acetic acid, dialysis against 7.5% propionic acid, lyophilization and suspension in the dodecyl sulfate buffer for diagonal gel electrophoresis.

Figures 6 and 7 show diagonal patterns of purified fractions of crosslinked 50S ribosomal protein. Comparison with Figure 5 (total protein from crosslinked 50S subunits) shows the simplification of the diagonal patterns. In many instances only two protein spots fall on the same vertical line: the molecular weight additivity relationship is still used to confirm that such monomeric proteins were part of the same single crosslinked dimer. As in the final analysis of protein spots eluted from diagonal gels of total 50S crosslinked protein, [35S]labelled subunits are used and final identification is made by two-dimensional electrophoresis in polyacrylamide/urea gels. This is illustrated in panel 2, Figures 6 and 7. These illustrations show the identification of the protein pairs L2-L5, (Figure 6) and L17-L21, L17-L32, L18-L32 and L22-L32 (Figure 7). All of the protein pairs unambiguously identified by fractionation of the crosslinked protein mixture can be found on the diagonal gel pattern of total protein. The advantage of the fractionation procedures is obvious in sorting out specific pairs from the complex diagonal pattern of total protein. At this time 33 protein pairs from the 30S subunit have been identified (Sommer and Traut, 1976) and 18 from the 50S subunit (Kenny and Traut, unpublished results). The latter are summarized in Table 1.

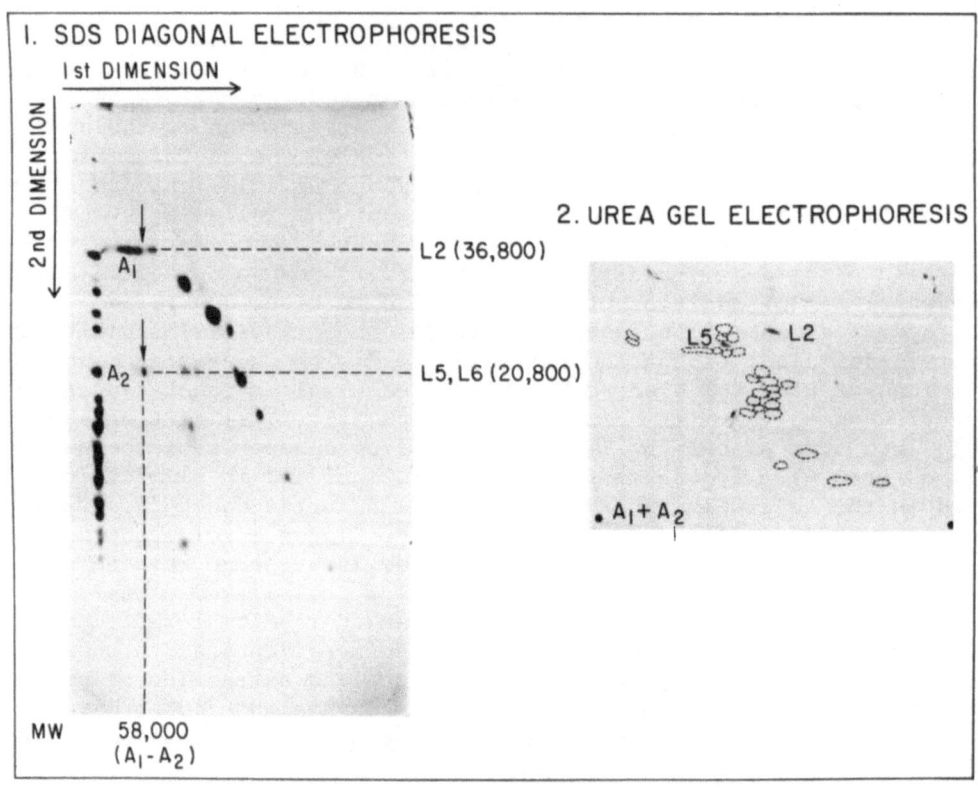

Figure 6. Diagonal gel electrophoresis and identification of partially purified proteins extracted from crosslinked 50S ribosomal subunits. Subunits were extracted with 1 M LiCl and the remaining "core" proteins were further fractionated by electrophoresis in a polyacrylamide/urea gel at pH 5.5. One 0.5 cm slice of the gel was analyzed. The proteins indicated by arrows (panel 1) are also observed in Figure 5. Panel 2 shows autoradiographs of polyacryla-mide/urea gels used for the identification of proteins indicated on the diagonal gel. Spots A_1 and A_2 were eluted from the diagonal gel mixed with nonradioactive total 50S protein prior to electrophoresis. The gels were stained, dried and exposed to X-ray film. The dotted lines show the positions of stained proteins in the region of the radioactive crosslinked protein. Identification of L2 and L5 from spots A_1 and A_2 shows the formation of the dimer L2-L5.

DISCUSSION

Studies on ribosomes have been described here to illustrate the value of reversible crosslinking by intermolecular disulfide bonds for the study of the topography of multiprotein ultrastructures. The

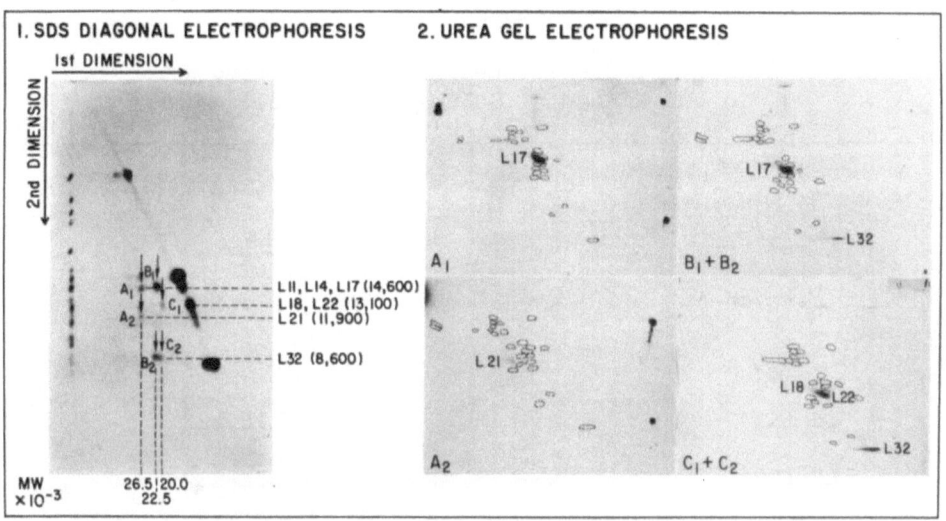

Figure 7. Diagonal gel electrophoresis and identification of partially purified proteins extracted from crosslinked 50S ribosomal subunits. Subunits were incubated with 0.5 M LiCl and the extracted proteins separated by electrophoresis at pH 5.5. One slice of the gel was analyzed by diagonal electrophoresis. Three conspicuous protein pairs beneath the diagonal are indicated by arrows (panel 1). Each of the radioactive spots was analyzed further as described in the legend to Figure 6. Spots A_1, A_2, B_1 and B_2 each contain single proteins, L17, L21, L17, and L32, respectively. Spot C_1 contains two proteins L18 and L22, while C_2 contains L32. Therefore L32 is crosslinked to both L18 and L22. The figure shows the identification of crosslinked dimers L17-L21, L17-L32, L18-L32 and L22-L32.

schematic diagram shown in Figure 8 summarizes the crosslinking results obtained for the 50S ribosomal subunit of <u>Escherichia coli</u> (Table 1). The crosslinking data alone do not permit an unambiguous arrangement of the proteins represented. The model based largely on the results of crosslinking is in agreement with additional experimental evidence from affinity labelling, partial reconstitution and binding of specific proteins to ribosomal RNA. We are still far from a three-dimensional molecular structure for either of the procaryotic ribosomal subunits. However, it is likely that the use of a variety of techniques such as immunoelectron microscopy, fluorescence transfer, neutron diffraction, reconstitution, genetics, RNA and protein sequence determination and crosslinking (both protein: protein and protein:RNA) will lead to a molecular understanding of the structure of this highly complex cellular organelle in the foreseeable future.

TABLE 1

Summary of Crosslinked Pairs Identified
in 50S Ribosomal Subunit of E. coli

L2–L5	L3–L5	L5–L25
L2–L7	L4–L14	L11–L14
L2–L8/9	L5–L7	L17–L21
L2–L10	L5–L12	L17–L32
L2–L11	L5–L17	L18–L32
L2–L12	L5–L23	L22–L32
L2–L17	L5–L24	

Certain qualifications concerning the crosslinking technique described here, as applied to the ribosome, should be noted. The rate of crosslinking may be slow relative to the rate of slight conformational changes in ribosomal subunits: this could lead to a situation in which even a single sulfhydryl group in one protein could be crosslinked to more than one other neighboring protein. Ribosomes may, as isolated, represent a heterogeneous population with respect to protein composition as well as conformation: this also could result in the formation of crosslinks not representative of the population as a whole. There is a variety of evidence indicating both conformational and compositional heterogeneity in ribosomes. Nevertheless, there is a striking consistency between the results obtained from crosslinking and those obtained with other structural and functional investigations on the ribosome. A major advantage of the crosslinking technique is its simplicity and the amount of information which can be obtained in a relatively brief period of time.

Inspection of Figure 5 indicates that the relative yields of crosslinked protein pairs varies considerably. There are many possible explanations for this variability: differences in reactivity of lysine residues and of the sulfhydryl groups introduced; steric properties of both the endogenous cysteine residues and the sulfhydryl containing amidine derivatives; possible conformational and compositional heterogeneity; conformational changes occurring or induced during the crosslinking reactions; competition between intermolecular

crosslinking with intramolecular disulfide bond formation and protein:
mercaptan disulfide formation; and perhaps most important, competition between dimer formation and the formation of higher crosslinked
oligomers.

What does it mean to identify a protein dimer in the ribosome?
It does not mean necessarily that the two proteins are in close
physical contact. The length of the reagent itself (see Figure 1)
even without taking into account the lengthy lysine side chains,
suggests that intermolecular crosslinks could be formed between two
proteins separated by RNA. The formation of an intermolecular crosslink does indicate that certain regions of the two proteins involved
are within 16-28 Å of each other. This does not imply that their
centers of mass are equally close. There is more and more evidence
from immunoelectron microscopy, neutron scattering and studies on
single proteins in solution that many ribosomal proteins are not
globular as represented in Figure 8, but are elongated with axial
ratios of up to 10:1. The next step in the use of crosslinking for
the mapping of ribosomal proteins will require identification of the

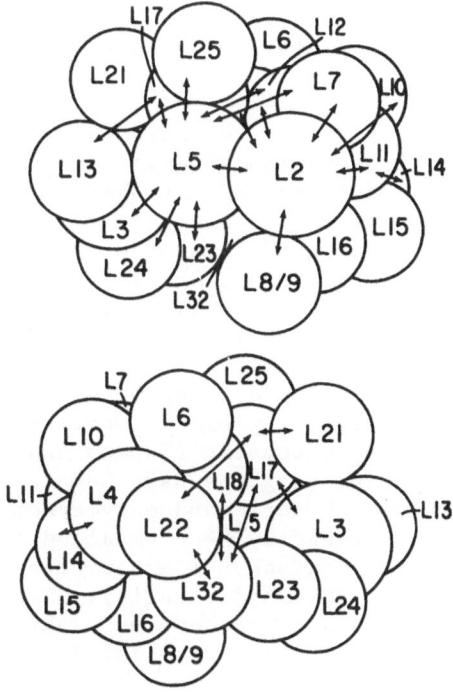

Figure 8. Schematic representation of crosslinks identified
in the 50S ribosomal subunit. "Top" and "bottom" views are shown
in order to depict all the crosslinks.

crosslinked amino acid residues in each protein dimer. Such an
investigation is fraught with difficulties since it must first be
established that two proteins are crosslinked at unique positions.
Fortunately, primarily through the efforts of Dr. H. G. Wittmann and
his collaborators in Berlin, there is extensive information on the
amino acid sequences of many of the ribosomal proteins.

In addition to its usefulness in determining the topography of
ribosomal proteins as described here, crosslinking can be employed
to investigate the binding sites for protein ligands which interact
to form stable complexes with the ribosome or ribosomal subunits.
Dr. J. W. B. Hershey of this University has prepared purified radioactively labelled initiation factors from Escherichia coli. Specific
factor : ribosome complexes have been formed and subjected to crosslinking with both cleavable and noncleavable bifunctional reagents.
Crosslinked complexes containing the radioactive factor have been
analyzed by immunochemical techniques using antibodies against single
ribosomal proteins by Dr. L. Kahan of the University of Wisconsin.
It has been possible to identify a small number of 30S ribosomal
proteins, antibodies to which precipitate radioactive crosslinked
factors (Bollen et al., 1975; Heimark et al., 1976). Collectively,
these proteins are considered to comprise the "initiation site" on
the 30S ribosomal subunit. The results show that crosslinking can
be used to investigate functional sites on the ribosome as well as
its overall structure.

Other cleavable crosslinking reagents have been employed for
the study of ribosomal protein topography: dimethyl 3,3'-dithiobis-
propionimidate, the oxidized form of a reagent differing from that
described here by being one carbon shorter (Peretz, Towbin and Elson,
1976); tartryl diazides, reagents cleavable by treatment with periodate (Lutter, Kurland and Stöffler, 1975); dimethyl suberimidate,
for which methods of cleaving the amidine linkages in high yield have
been reported (Clegg and Hayes, 1974). It is likely that new cleavable reagents with different specificities, perhaps including RNA as
well as protein, will be developed. Mercaptoimidates and diagonal
gel electrophoresis have already yielded important results on the
organization of histones in chromatin (Thomas and Kornberg, 1975).
These and other crosslinking reagents have been employed for studies
on membranes, viruses, mitochondria and enzyme complexes. Methyl 4-
mercaptobutyrimidate has presently been employed in this laboratory
for the investigation of topography of eucaryotic ribosomal proteins
(D. Tolan and C. Horan, unpublished results). Crosslinking with
bifunctional reagents would seem to have a bright future in the
elucidation of a variety of complex cellular structures and interactions.

ACKNOWLEDGMENTS

We would like to thank our many colleagues who have contributed to the work carried out in this laboratory. The generous advice and encouragement of Dr. Alex Glazer during the development of the techniques described here have been invaluable. Dr. John Lambert and Mr. Rodney Jue have made many helpful suggestions during the writing of this manuscript. We thank Ms. Catherine DiBartola for careful typing of this manuscript. This work has been supported by a grant from the United States Public Health Service (grant no. GM17924).

REFERENCES

Acharya, A. S. and Moore, P. B. (1973). Reaction of ribosomal sulfhydryl groups with 5,5'-dithiobis(2-nitrobenzoic acid). J. Mol. Biol., 76, 207-221.

Bickle, T., Hershey, J. W. B. and Traut, R. R. (1972). Spatial arrangement of ribosomal proteins: Reaction of the Escherichia coli 30S subunit with bis-imidoesters. Proc. Nat. Acad. Sci., USA, 69, 1327-1331.

Bollen, A., Heimark, R. L., Cozzone, A., Traut, R. R. and Hershey, J. W. B. (1975). Crosslinking of initiation factor IF-2 to Escherichia coli 30S ribosomal subunit with dimethylsuberimidate. J. Biol. Chem., 250, 4310-4314.

Chang, F. N. and Flaks, J. G. (1972). The specific cross-linking of two proteins from the Escherichia coli 30S ribosomal subunit. J. Mol. Biol., 68, 177-180.

Clegg, C. and Hayes, D. (1972). Introduction de ponts covalents entre proteins voisines du ribosome 50S à E. coli. C. R. Acad. Sci., (Paris), 275, 1819-1822.

Clegg, C. and Hayes, D. (1974). Identification of neighbouring proteins in the ribosomes of Escherichia coli — A topographical study with the cross-linking reagent dimethyl suberimidate. Eur. J. Biochem., 42, 21-28.

Heimark, R. L., Kahan, L., Johnston, K., Herhsey, J. W. B. and Traut, R. R. (1976). Cross-linking of initiation factor IF3 to proteins of the Escherichia coli 30S ribosomal subunit. J. Mol. Biol., 105, 219-230.

Howard, G. A. and Traut, R. R. (1973). Separation and radioautography of micrograms quantities of ribosomal proteins by two-dimensional polyacrylamide gel electrophoresis. FEBS Letters, 29(2), 177-180.

Kenny, J. W., Sommer, A. and Traut, R. R. (1975). Cross-linking studies on the 50S ribosomal subunit of Escherichia coli with methyl 4-mercaptobutyrimidate. J. Biol. Chem., 250(24), 9434-9436.

Knopf, U. C., Sommer, A., Kenny, J. and Traut, R. R. (1975). A new two-dimensional gel electrophoresis system for the analysis of complex protein mixtures: Application to the ribosome of E. coli. Mol. Biol. Rep., 2, 35-40

Lutter, L. C., Bode, V., Kurland, C. G. and Stöffler, G. (1974). Ribosomal protein neighborhoods. III. Cooperativity of ribosome assembly. Mol. Gen. Genet., 129, 167-176.

Lutter, L. C., Kurland, C. G. and Stöffler, G. (1975). Protein neighborhoods in the 30S ribosomal subunit of Escherichia coli. FEBS Letters, 54(2), 144-150.

Lutter, L. C., Zeichhardt, H., Kurland, C. G. and Stöffler, G. (1972). Ribosomal protein neighborhoods. I. S18 and S21 as well as S5 and S8 are neighbors. Mol. Gen. Genet., 119, 357-366.

Peretz, H., Towbin, H. and Elson D. (1976). The use of a cleavable crosslinking reagent to identify neighboring proteins in the 30S ribosomal subunit of Escherichia coli. Eur. J. Biochem., 63, 83-92.

Perham, R. N. and Thomas, J. O. (1971). Reaction of tobacco mosaic virus with a thiol-containing imidoester and a possible application to x-ray diffraction analysis. J. Mol. Biol., 62, 415-418.

Nomura, M., Tissières, A. and Lengyel, P. (eds.). (1974). Ribosomes, Cold Spring Harbor Laboratory, New York.

Shih, C.-Y. T. and Craven, G. R. (1973). Identification of neighbor relationships among proteins in the 30S ribosome: Intermolecular cross-linkage of three proteins induced by tetranitromethane. J. Mol. Biol., 78, 651-663.

Sommer, A. and Traut, R. R. (1974). Diagonal polyacrylamide-dodecyl sulfate gel electrophoresis for the identification of ribosomal proteins crosslinked with methyl-4-mercaptobutyrimidate. Proc. Nat. Acad. Sci., USA, 71, 3946-3950.

Sommer, A. and Traut, R. R. (1975). Identification by diagonal gel electrophoresis of nine neighboring protein pairs in the Escherichia coli 30S ribosome crosslinked with methyl-4-mercaptobutyrimidate. J. Mol. Biol., 97, 471-481.

Sommer, A. and Traut, R. R. (1976). Identification of neighboring protein pairs in the Escherichia coli 30S ribosomal subunit by crosslinking with methyl-4-mercaptobutyrimidate. J. Mol. Biol., 106, 995-1015.

Sun, T.-T., Bollen, A., Kahan, L. and Traut, R. R. (1974). Topography of ribosomal proteins of the Escherichia coli 30S subunit as studied with the reversible crosslinking reagent methyl-4-mercaptobutyrimidate. Biochem., 13, 2334-2340.

Thomas, J. O. and Kornberg, R. D. (1975). Cleavable cross-links in the analysis of histone-histone associations. FEBS Letters, 58, 353-358.

Traut, R. R., Bollen, A., Sun, T.-T., Hershey, J. W. B., Sundberg, J. and Pierce, L. R. (1973). Methyl 4-mercaptobutyrimidate as a cleavable cross-linking reagent and its application to the Escherichia coli 30S ribosome. Biochem., 12, 3266-3273.

Traut, R. R. and Monroe, R. E. (1964). The puromycin reaction and its relation to protein synthesis. J. Mol. Biol., 10, 63-72.

14

ON THE INTRODUCTION OF DISULFIDE CROSSLINKS
INTO FIBROUS PROTEINS AND BOVINE SERUM ALBUMIN

Ch. Ebert, G. Ebert, and H. Knipp
Institute of Polymers, University of Marburg/L
Lahnberge, Geb. H   3550 Marburg/L (West Germany)

ABSTRACT

Hydroxyl groups in serine side chains of collagen, silk fibroin, and bovine serum albumin (BSA) were converted to SH by tosylation. In collagen film, 50% of the serine OH groups could be thiolated at most. In fibroin, only 13% because of its compact β-pleated sheet structure and low susceptibility to swelling. The SH groups introduced are near enough together to form −S−S− bonds by oxidation. The residual SH content after oxidation was 0.1% in collagen and 0.03 to 0.25% in fibroin. Disulfide crosslinking increased the shrinkage temperature of collagen and fibroin and decreased the amount of shrinkage. BSA was crosslinked to dimers (MBSA) according to gel permeation chromatography and sedimentation analysis by the analytical centrifuge. Because these crosslinked proteins can be metabolized by the usual processes, in contrast to those crosslinked by artificial, nonphysiological bridges, they may be used for biological or medical purposes.

INTRODUCTION

Conversion of serine OH groups to SH was investigated by Zervas and Photaki (1), Koshland et al. (2), and Photaki and Bardakos (3). Polgar and Bender (4) transformed the active serine residue in the active center of subtilisin into cysteine, showing that the activity of the enzyme was decreased by this change.

Replacement of OH groups by SH is also interesting for introducing this latter group into proteins free - or almost free - from sulfur for several reasons. One is the possibility of forming covalent crosslinks in proteins like silk fibroin and collagen.

235

This chemical modification should markedly change the physico-chemical behavior of these proteins. Cellulose has been crosslinked in this way by Schwenker (5) and by Sakamoto, Yamada, and Tonami (6).

RESULTS WITH COLLAGEN

Collagen fibers were treated with toluenesulfonyl chloride to esterify the OH group in aqueous phosphate buffer at pH 8 or 9, in dioxane-containing buffer solutions, or in pyridine (7).

$$
\begin{array}{c}
\text{OH} \\
|\\
\text{CH}_2 \\
|\\
\text{-HN-CH-CO-}
\end{array}
\xrightarrow{\text{Tosylchloride}}
\begin{array}{c}
\text{O-Tos} \\
|\\
\text{CH}_2 \\
|\\
\text{-HN-CH-CO-}
\end{array}
$$

$$
\begin{array}{c}
\text{O} \\
\|\\
\text{CH}_3\text{-C-S}^{\ominus} \\
\\
\text{-H}_3\text{C-C}_6\text{H}_4\text{SO}_3^{\ominus}
\end{array}
\longrightarrow
\begin{array}{c}
\text{S-OC-CH}_3 \\
|\\
\text{CH}_2 \\
|\\
\text{-NH-CH-CO-}
\end{array}
\xrightarrow[\text{-CH}_3\text{COO CH}_3]{\text{OH}^-, \text{CH}_3\text{O Na}}
\begin{array}{c}
\text{SH} \\
|\\
\text{CH}_2 \\
|\\
\text{-NH-CH-CO-}
\end{array}
$$

The most tosylation was obtained in a buffer/dioxane medium (2:1); in pyridine no tosylation was observed. This is probably due to failure of collagen to swell in pyridine, so that the reagent cannot penetrate the protein. In purely aqueous medium, on the other hand, tosyl chloride is poorly soluble.

The subsequent transesterification was carried out in 0.5M thio-acetate solution of pH 5.5. Afterward, cleavage of the resulting thioester was performed with 0.5N sodium methylate.

By this procedure 40 to 50% of serine OH groups in collagen could be transformed into cysteine, according to results of amino acid analysis. The limiting factor for the OH to SH exchange seems to be the tosylation step;no more than 55% of the alcoholic serine side chains could be tosylated at most.

For covalent crosslinking, the SH groups introduced were oxidized to -S-S- bonds by air or by iodide (5). In both cases, the residual cysteine content was only 0.1%. This means that almost all SH groups introduced were situated close together enough to be oxidized to disulfide (Table 1).

These covalent -S-S- crosslinks should increase the stability of the collagen structure. For this reason, the temperature of shrinkage, T_s, should be shifted to higher values. As one can see from Figure 1, this is, indeed, the case.

Untreated collagen film (made from bovine achilles tendon by Freudenberg GmbH., Weinheim, West Germany) contracts in water at 54°C; collagen film with 1.43% cystine contracts at 59°C. Further-

Table 1. Serine, cystine and cysteine, and methionine content of treated and untreated collagen films

Sample	Serine	Cysteine + Cystine	Methionine
untreated	3.28	—	0.49
I	2.51	0.58	0.47
II	2.25	1.28	0.30
III	1.82	1.43	0.49

Sample I was tosylated in sodium phosphate buffer at pH 9, sample II in a buffer/dioxane mixture 2 : 1 and sample III was treated twice under the same conditions like sample II.

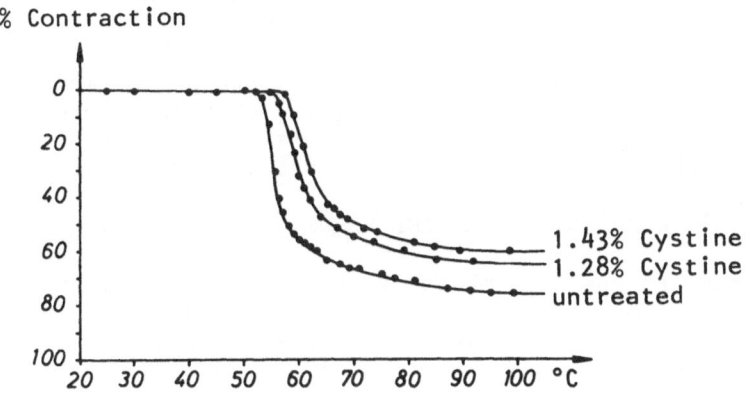

Figure 1. Shrinkage of untreated and cystine-crosslinked collagen film in water.

more the amount of shrinkage is lowered by introducing –S–S– crosslinks because disordering of the macromolecules is restricted in comparison with untreated, disulfide-free collagen.

RESULTS WITH SILK FIBROIN

Similar results were obtained with silk fibroin. Use of buffer/dioxane solutions (2:1) was again the most efficient method for tosylation. By repeating the whole process for the OH to SH exchange,

1.9% of cystine could be introduced. This is only 13% of the
overall serine content of the silk fibroin. This small conversion
is probably limited by the densely packed β-pleated sheet structure,
in which the accessability of the OH groups in the crystalline part
is low.

After oxidation, the residual cysteine content in modified
fibroin was usually somewhat higher than in modified collagen,
namely 0.03 to 0.25%. However, in silk, too, most of the SH groups
introduced are near enough so that they can form -S-S- bonds.
Possibly these SH groups are located in distinct regions near the
less ordered domains between the crystalline parts of the fibroin.

Unlike collagen, silk does not shrink in pure water up to 150°C
because of its highly ordered structure; however, it does contract
in some concentrated electrolyte solutions, such as 6M LiBr.
Figures 2,3, and 4 show the shrinkage behavior of untreated, native
silk fibroin and of disulfide-crosslinked fibroin in 6M, 6.5M and
7M LiBr solution. Again, not only is the temperature, T_s, at which
the ordered structure breaks down increased by the treatment, but
also the amount of contraction is decreased. As with collagen,
increased stability against denaturation by heat and electrolytes
is caused by introducing the naturally occuring cystine crosslink.
Because no racemization occurs during the OH to SH exchange, these
crosslinked proteins can be broken down by the normal metabolic
processes, unlike proteins with artificial crosslinks produced by
tanning agents. For this reason, it seems possible to use -S-S-
stabilized proteins for practical purposes in medicine.

RESULTS WITH BOVINE SERUM ALBUMIN (BSA)

In contrast to the treatments of the insoluble proteins collagen
and fibroin, the modification of BSA could be performed in homogeneous aqueous solutions (8). Tosylation was carried out in an aqueous
phosphate buffer/dioxane - mixture (2:1) as described above. Transesterification (S-acylation) was performed at room temperature in
aqueous 0.4M CH_3COSK at pH 9. To complete the cleavage of the
thioester, the BSA-SAc was treated with alkali at pH 12 for one
hour. The modified BSA (MBSA) was isolated by gel permeation chromatography on a Sephadex G 200 column. Figure 5 shows that the
MBSA obtained eluted approximately 60 ml in front of a very small
amount of unmodified BSA. This means that higher molecular weight
aggregates, probably due to intermolecular covalent crosslinking,
were formed.

Sedimentation analysis made with an analytical ultracentrifuge
(Beckman Model E) showed a slightly asymmetric peak in the
schlierenpattern (Fig. 6). The sedimentation coefficient of MBSA
was determined: S_{20} (1% w NaCl) = 9.9 x 10^{-13} sec, whereas BSA
yielded S_{20} (1% NaCl) = 4.2 - 4.4 x10^{-13} sec. The molecular weights
obtained by sedimentation analysis were M = 6 x10^4 for BSA and
M = 12.2 x10^4 for MBSA. These results show that the main part of
the MBSA is formed by dimerization of BSA. Only a small part

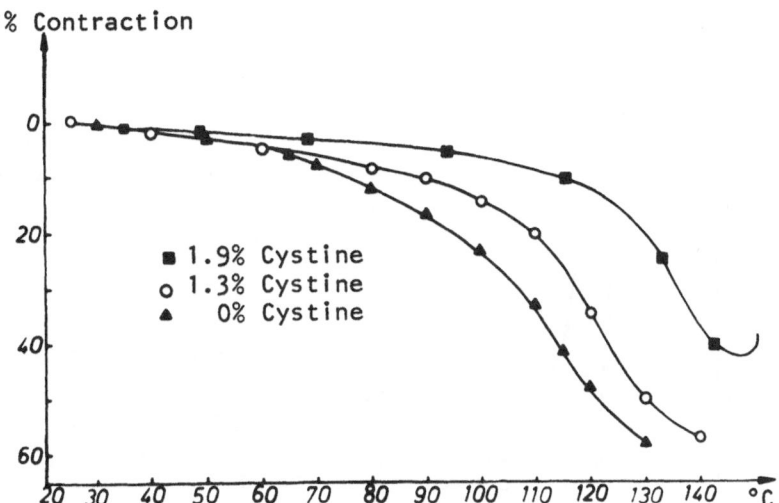

Figure 2.   Temperature-dependent contraction of untreated and cystine-crosslinked silk (bombyx mori) in 6 M LiBr solution.

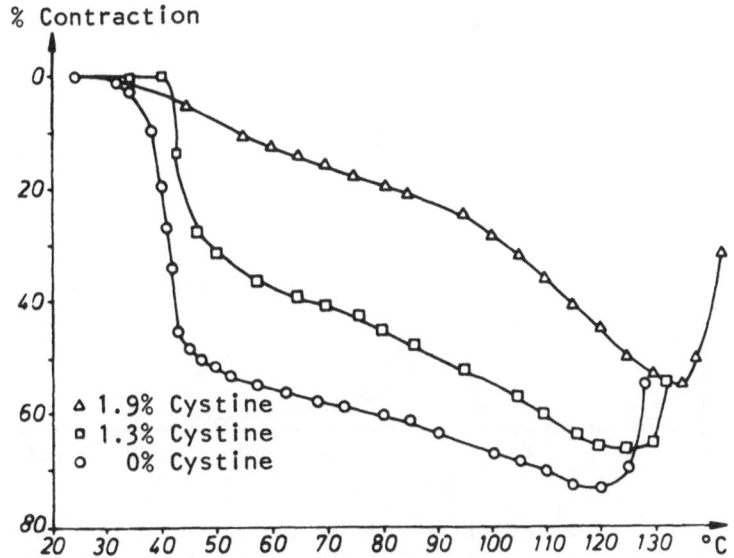

Figure 3.   Temperature-dependent contraction of untreated and cystine-crosslinked silk (bombyx mori) in 6.5 M LiBr solution.

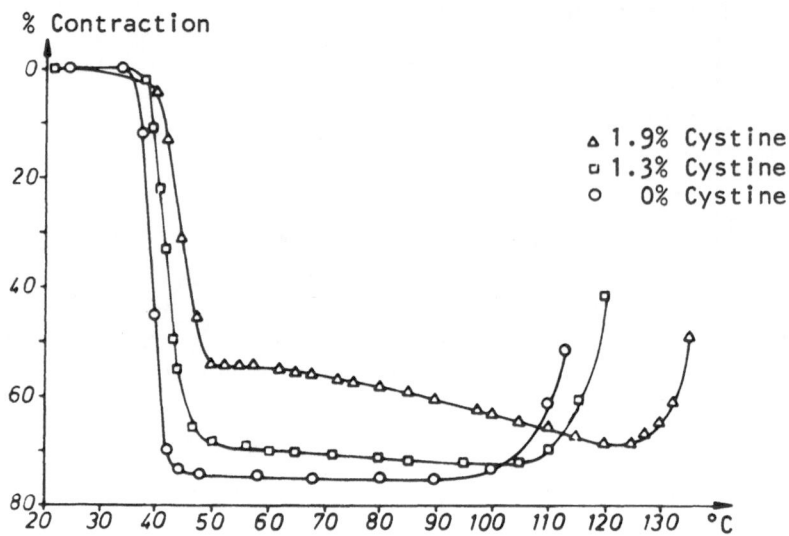

Figure 4.   Temperature-dependent contraction of untreated and cys-
            tine-crosslinked silk (bombyx mori) in 7 M LiBr solution.

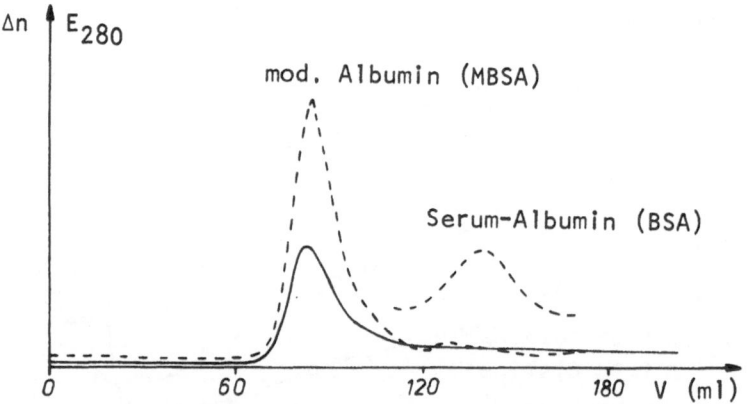

Figure 5.   Gel permeation chromatogram of modified bovine serum al-
            bumin (MBSA). The BSA peak from a test run shows the
            elution volume at which it occurs. Gel: Sephadex G200,
            ——; UV-absorption (E_{280}), ---.

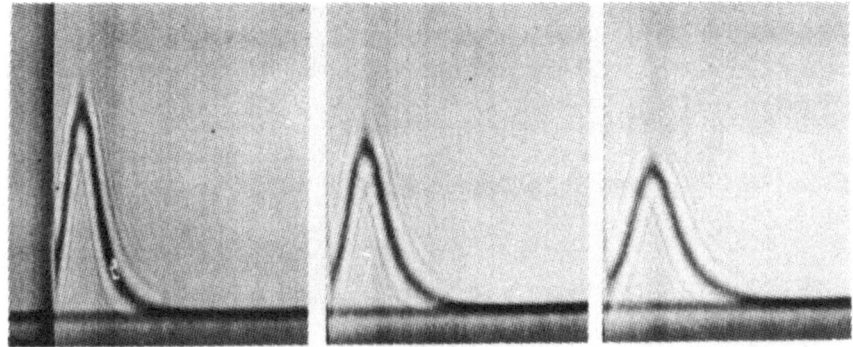

Figure 6.   Schlieren pattern of MBSA in 1% (w) NaCl solution at
            21.5°C, 60,000 r.p.m.; Δt = 2 min and 4 min.

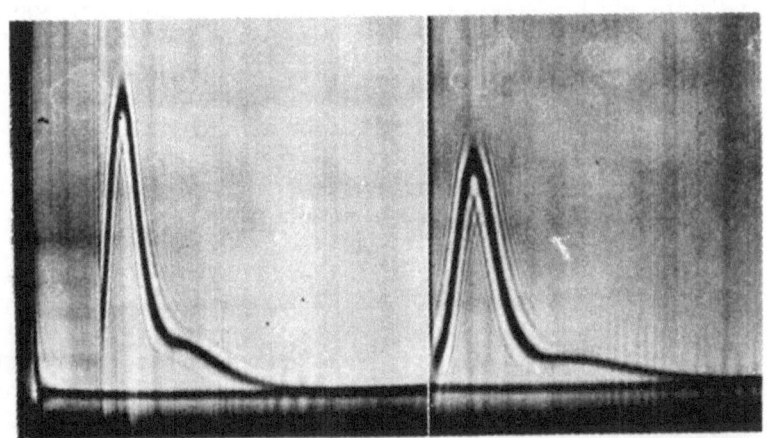

Figure 7.   Schlieren pattern of a mixture of BSA and MBSA in 1% (w)
            NaCl solution at 20°C, 48,000 r.p.m.; Δt = 4 min.

of the BSA has formed larger molecular aggregates.

Figure 7 shows the schlieren pattern of a mixture of BSA and MBSA. The relatively broad schlieren peak of the latter is to be seen at the right of the narrow BSA peak.

Dimerization of unmodified BSA in aqueous solution was investigated by Williams (9). He has found that it takes place at pH values below the isoelectric point, preferably at pH 3.3. Under the conditions used by us – pH ≃ 7 – such dimerization by intermolecular attractive forces should be negligible. On the other hand, native BSA contains only 0.5 – 1 SH group/mole at most (10). By the method of Ellman (11), we found 0.6 Mol-% SH.

The dimerization of MBSA is most probably caused by interchain disulfide bonds formed by oxidation, e.g. by oxygen or by thiol-disulfide interchange (12).

$$2 \text{ MBSA} - (\text{SH})_n \xrightarrow{\text{O}_2} (\text{HS})_{n-1} - \text{MBSA-S-S-MBSA-}(\text{SH})_{n-1}$$
$$\xrightarrow{\text{O}_2} \text{higher aggregates} \qquad \text{MBSA dimer}$$

$$\text{MBSA-SH} + \text{MBSA} {\Large<}^{\text{S}}_{\text{S}} \longrightarrow \text{MBSA-S-S-MBSA-SH}$$
$$\xrightarrow{\text{O}_2} \text{higher aggregates (oligomers)} \qquad \text{MBSA dimer}$$

The SH content obtained by the method of Ellman amounts to 240 µmole SH/g MBSA, which is equivalent to 14 SH/mole MBSA. This means that the SH content of MBSA is approximately 20 times as high as in the original BSA. Evidently oxidation may result in several interchain crosslinks between two MBSA monomers.

In spite of the fact that tosylation was carried out at pH 8-9, below the pK of the ε-amino groups of lysine, the lysine content of the tosylated BSA is lowered to 50% according to amino acid analysis (Fig. 8a and 8b). The tyrosine content has also decreased.

Because of the decrease of the basic amino acid lysine, the proportion between basic/neutral and acidic amino acids is shifted toward the latter. Therefore, the pI becomes smaller. This change in amino acid composition (Table 2) is responsible for the observation that more cystine + cysteine appears to be formed than one obtains by calculation from the decrease of serine + threonine.

In contrast to collagen and fibroin, MBSA retained a relatively high percentage of free SH groups after oxidation: about 20% of the sum of cystine + cysteine. This can be understood if one takes into account not only that the serine content is relatively low but also that many serine residues, because of the completely different structures (conformation etc.) of the fibrous protein and the soluble albumin can not get in contact with one another to form –S-S- bridges.

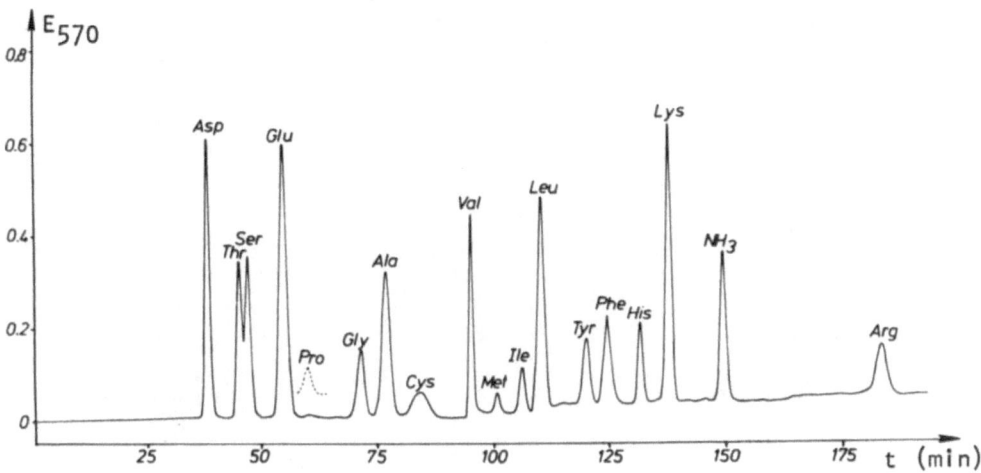

Figure 8a. Amino acid chromatogram of bovine serum albumin (BSA).

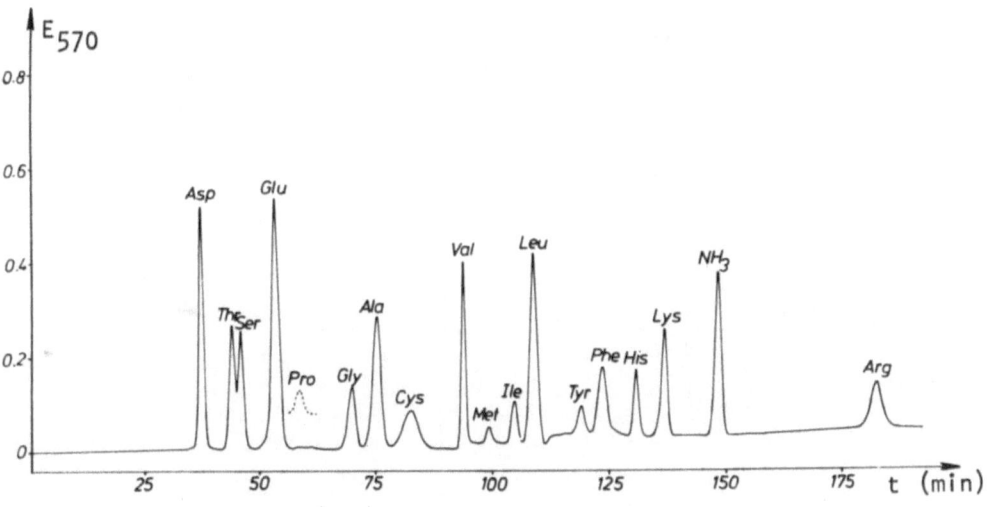

Figure 8b. Amino acid chromatogram of modified bovine serum albumin (MBSA).

Table 2.   Amino acid composition of bovine serum albumin (BSA) and
modified bovine serum albumin (MBSA)

Amino acids	BSA (Mol-%)	MBSA (Mol-%)
Aspartic + Asparagine	8.2	9.6
Threonine	6.3	6.1
Serine	6.2	5.0
Glutamicacid + Glutamine	14.0	15.0
Proline	5.0	5.1
Glycine	3.2	3.6
Alanine	8.0	8.8
1/2 Cystine + Cysteine	4.3	7.1
Valine	5.8	6.0
Methionine	0.6	0.7
Isoleucine	2.1	1.9
Leucine	11.0	10.9
Tyrosine	3.2	1.6
Phenylalanine	4.8	4.8
Histidine	3.0	2.8
Lysine	9.0	4.4
Arginine	4.5	4.3
	Σ 99.7	98.2

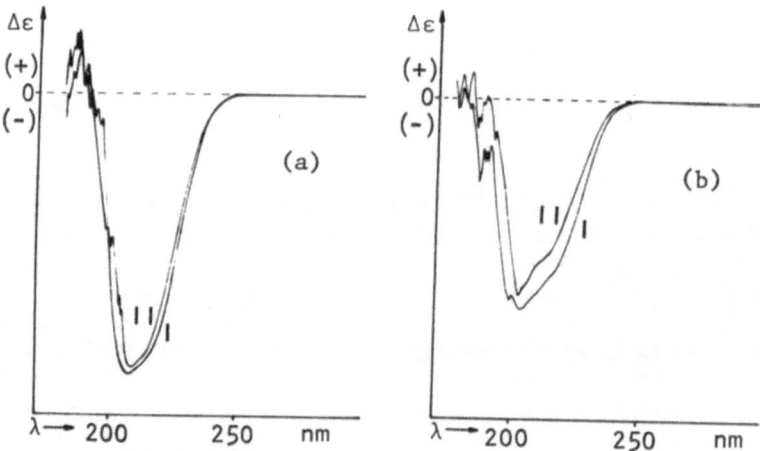

Figure 9.   CD-spectra of BSA (I) and MBSA (II) in 1% NaCl solution
at pH 7 (a) and pH 11 (b).

The α-helical structured regions of the BSA are obviously not influenced by the chemical modification, as one can deduce from circular dichroism (CD) spectra (Fig. 9). The two CD absorption bands at 223 and 206 nm, characteristic of the α-helix, are to be seen in BSA as well as in MBSA. This evidence suggests that only one or at least very few interchain bonds are formed between two BSA molecules, because one would expect the chain-conformation to be distorted to a higher extent, observable in the CD spectra, by multiple bonding.

ACKNOWLEDGMENT

The authors are grateful to the Deutsche Forschungsgemeinschaft for financial assistance.

REFERENCES

1. Zervas, L. and Photaki, I. (1960). Chimia, 13, 375.
2. Struhmeyer, D.H., White, W.N. and Koshland, D.E., Jr. (1963). Proc. Natl. Acad. Sci. U.S. 50, 931.
   Weiner, H., White, W.N. and Koshland, D.E., Jr. (1966). J. Am. Chem. Soc., 88, 3851.
3. Photaki, I. and Bardakos, N. (1965). J. Am. Chem. Soc., 87, 3489.
4. Polgar, L. and Bender, M.L. (1966). J. Am. Chem. Soc., 88, 3153. Biochemistry, 6, 610 (1967).
5. Schwenker, R.F., Jr., Lifland, L. and Pascu, E. (1962). Text. Res. J., 32, 797: ibid. 33, 107 (1963).
6. Sakamoto, M., Yamada, Y. and Tonami, H. (1969). J. Appl. Polym. Sci., 13, 1845.
7. Ebert, Ch. (1971). Thesis, Marburg, West Germany.
8. Knipp, H. (1976). Thesis, Marburg, West Germany.
9. Williams, E.J. and Foster, J.F. (1960). J. Am. Chem. Soc.,82, 3741.
10. Putnam, F. (1965). In "The Proteins", Vol. III, p. 153, Second Edition, Academic Press, New York.
11. Ellman, G.L. (1959). Arch. Biochem. Biophys., 82, 70.
12. Bro, P., Singer, S.T. and Sturtevant, J.M. (1958). J. Am. Chem. Soc., 80, 389.

15

THIOLATION AND DISULPHIDE CROSS-LINKING OF INSULIN TO FORM

MACROMOLECULES OF POTENTIAL THERAPEUTIC VALUE

M. MAHBOUBA AND H.J.SMITH

WELSH SCHOOL OF PHARMACY, UNIVERSITY OF WALES INSTITUTE
OF SCIENCE AND TECHNOLOGY, CARDIFF, S. WALES, U.K.

ABSTRACT

Macromolecules have been prepared containing native insulin carried by a modified insulin skeleton made by partially thiolating the insulin hexamer and forming intermolecular cross-links through disulphide bridges.   Oxidation of partially thiolated insulin (0.5 - 0.7 SH group/mole), formed by reacting insulin with AHTL, with, (a) potassium ferricyanide, (b) Cu^{++}-oxygen gave water soluble macromolecules containing 20-26 and 410-708 monomer units respectively which had rod-random coil shape (light scattering). The larger molecules formed by (b) contained 8g-atom Cu^{++}/hexamer unit and insulin.   The insulin was firmly bound within the macromolecules and was probably bound within an insulin-modified insulin hexamer through coordination to copper.

INTRODUCTION

In concentrated near neutral aqueous solution insulin exists as a hexamer in the presence of certain divalent metallic ions. The two central metallic ions are considered to each be coordinated to three imidazole nuclei supplied by the B-10 histidine residues in the insulin B-chain.   On dilution to concentrations existing in plasma, the hexamer dissociates into the monomer (Pekar & Frank, 1972; Goldman & Carpenter, 1974).

We have recently shown that thiolation of a number of proteins with AHTL followed by oxidation gives disulphide cross-linked aggregates with a high degree of aggregation, e.g. α-chymotrypsin, n = 16-44 (Mahbouba, Pugh and Smith, 1974), globin, n=8-65 (Etemad-Moghadam, Mahbouba, Pugh and Smith), the size of the aggregate being broadly related to the number of thiol groups introduced.   We considered that thiolation of a proportion of the insulin units in the hexamer followed by disulphide cross-linking would give a large water soluble macromolecule where the insulin units were retained within a cross-linked modified insulin skeleton.   This preparation on subcutaneous injection would form a slowly diffusable depot which would slowly be leached into the blood stream with release of active insulin units at the higher dilution in the plasma i.e. a long acting insulin.

METHODS AND RESULTS

Materials

Crystalline bovine insulin containing 24.4 i.u. per mg (B.D.H. Chemical Ltd., England) was used.

Thiolation of Insulin by N-acetyl Homocysteine Thiolactone (AHTL)

A solution of crystalline insulin (60mg) in phosphate buffer pH 8 (0.1M, 20ml) containing AHTL (60mg) was stored at 20° under an atmosphere of carbon dioxide-free nitrogen for a period of 22-36 hours.   The solution was then transferred to a column of Sephadex G-25 (45 x 2.5cm) which had been equilibrated with phosphate buffer.   The column was developed with the same buffer and fractions (4ml) collected after rejection of the forerun (60ml). The fractions were examined spectrophotometrically at λ 276 nm and the fractions (2-8) containing the thiolated protein were combined.

The thiol titre of the modified insulin was determined by Ellman's method (Ellman, 1958).   The initial titre of the thiolated insulin varied with each batch of product and was within the range 0.43-1.0 -SH group/mole (Table 1).

Oxidation of Thiolated Insulin

(a) Using potassium ferricyanide.   A solution of thiolated insulin (40mg, 20ml) was oxidised using potassium ferricyanide reagent (1ml) in the manner previously described for oxidation of thiolated α-chymotrypsin, (Mahbouba, Pugh and Smith, 1974). The protein was isolated from excess reagent by filtration down a column of Sephadex G-25 and collected in the manner previously described.   The thiol titre of the protein fraction was zero confirming that all the thiol groups had been coupled in disulphide linkages.

Table 1

Effect of Time on the Extent of the Thiolation of Insulin

Reaction time (hr)	-SH group/mole insulin monomer
22	0.50
23	0.54
24	0.43
25	0.65
35	0.71
36	1.00

(b) Using cupric chloride and oxygen. A solution of thiolated insulin (40mg, 20ml) in phosphate buffer pH 8 (0.1M) was mixed with a solution (0.6ml) containing cupric chloride (0.6mg Cu^{++}) in hydrochloric acid (N). Oxygen was bubbled through the mixture at a flow rate of 8cc/min for 2 hr. at 20°. The protein was separated from excess reagent by filtration down Sephadex G-25.

Molecular Size of Oxidised Thiolated Insulin Macromolecule

Measurements were made using a Brice-Phoenix instrument at λ 546 nm. Refractive index differences were measured with a differential refractometer (Polymer Consultants Ltd.). Solutions of oxidised thiolated insulin in phosphate buffer pH 8 (0.1M) were clarified by centrifuging at 20,000 r.p.m. with a gravity field of 25,000 g at 4° for 0.5 hr. The protein concentration was determined using Lowry's method (Lowry et al, 1951).

Measurements were made in a cylindrical cell with flat entry and exit faces over angles ranging from 45° to 135°. After applying the appropriate Fresnel corrections (Stacey, 1956) the results were analysed by plotting Kc/R_θ against $\sin^2(\theta/2)$ as described by Zimm (1948). The intercept at zero angle gives the value of 1/M and the average radius of gyration, R_z, was computed from the initial slope by means of the expression,

$$R_z^2 = \frac{3\lambda^2}{16\pi^2 n^2} \cdot \frac{\text{initial slope}}{\text{intercept}} \cdot$$

The average molecular weight of various batches of the oxidised thiolated insulin macromolecule, prepared by either potassium ferricyanide oxidation or Cu^{++} - catalysed oxygenation from thiolated insulin are shown in Table (2). The molecular

size is conveniently expressed in hexamer units, the insulin
hexamer having M = 36,000 (Blundell et al, 1972)

Shape of the Macromolecule

The shape of the macromolecule was deduced from the light
scattering results by plotting the particle scattering factors
(P_θ) against sin ($\theta/2$) and comparing the curve shape with those
of various model shapes i.e. sphere, rod and random coil. The
results for the macromolecules prepared using Cu^{++} - catalysed
oxygenation are summarised in Table (3) and show that the smaller
molecules tended to be rod-shaped whereas coiling became evident in
the larger aggregates.

A plot of M/R_z against M is shown in Fig. (1).

Table 2

Molecular Weight of Oxidised Thiolated Insulin Macromolecules

Method of oxidation	No. of -SH group/mole of insulin monomer	Molecular size of oxidised thiolated insulin(hexamer units).
Potassium ferricyanide	0.50	3.4
	0.65	4.3
Cu^{++} - catalysed oxygenation	0.53	68.6
	0.56	83.7
	0.60	118
	0.78	99.3
	0.67	111

Table 3

Characteristics of Oxidised Thiolated Insulin Macromolecules

No. of -SH groups/mole of thiolated insulin	Molecular Wt. (M)	Molecular Wt. (hexamer units)	Radius of gyration R_z (m)	M/R_z	Shape calculated from Zimm plot
0.54	15.37×10^5	42.7	2.47×10^{-8}	6.24×10^{-13}	Coil
0.60	13.61×10^5	37.8	3.26×10^{-8}	4.17×10^{-13}	Rod
0.70	6.26×10^5	17.4	2.30×10^{-8}	2.72×10^{-13}	Rod
0.78	35.75×10^5	99.3	3.73×10^{-8}	9.58×10^{-13}	Coil or Rod
1.00	33.34×10^5	92.6	2.84×10^{-8}	11.74×10^{-13}	Coil

Effect of Storage of a Solution of Thiolated Insulin
on its Molecular Size

A freshly prepared solution of thiolated insulin (0.58 -SH group/mole) in phosphate buffer pH 8 (0.1M) was clarified by centrifuging and its light scattering properties and thiol titre determined. The solution was then stored at 4° and after periods of 1 and 2 days these determinations were repeated.

An aliquot of the original thiolated insulin solution was oxidised immediately using potassium ferricyanide. The protein was separated and its molecular weight was determined in the usual manner. A solution of this macromolecule formed was stored at 4° for a period of 3 days when its molecular weight was found to be unchanged. The results are summarised in Table (4) and include results from similar experiments using batches of thiolated insulin with a high initial thiol content.

Determination of Zinc and Copper

The zinc content of various protein preparations in phosphate buffer pH 8 (0.1M) was determined on a Varian Techtron Atomic Absorption Spectrophotometer Model 1000 at λ 213.9nm. Copper was determined in a similar manner at λ324.8nm. The protein

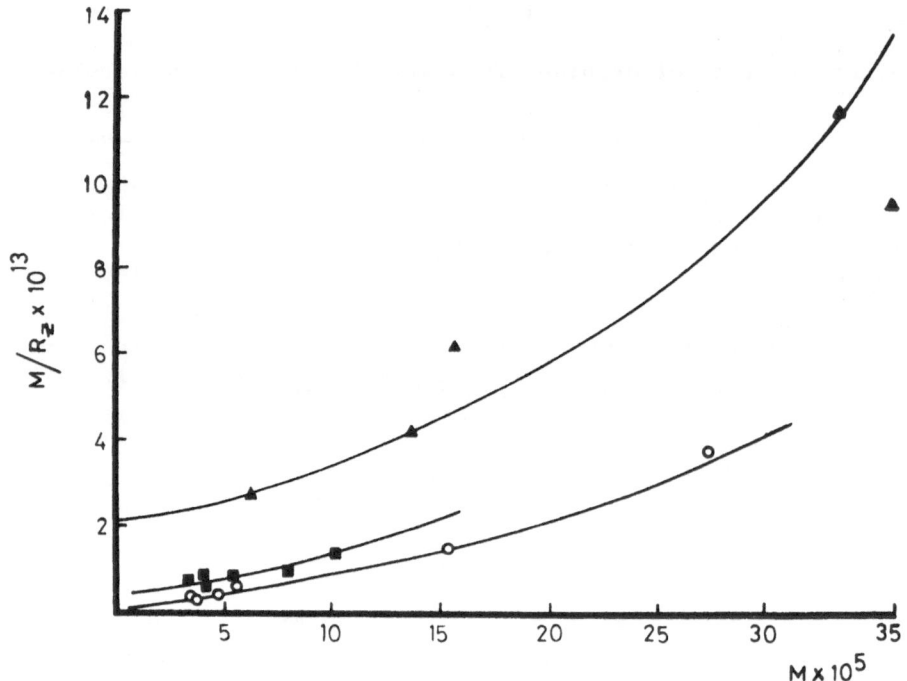

Fig. 1. Change in the ratio M/R_z with increasing M for disulfide crosslinked macromolecule formed from insulin prepared using cupric chloride and oxygen (▲); α-chymotrypsin (■); globin (0).

concentration of the various preparations was determined by Lowry's method (Lowry et al., 1951).

   The thiolated insulin was shown to contain 2g - atom Zn^{++}/ hexamer unit.   Oxidised thiolated insulin prepared using potassium ferricyanide lost the greater part of its zinc content during this process, the extent of the loss being dependent on the initial thiol content of the thiolated insulin preparation (Table 5).

Disc Electrophoresis of Oxidised Thiolated Insulin

   Electrophoresis was performed in a Shandon electrophoresis apparatus by the method of Ornstein and Davis (1964) on 7.5% polyacrylamide gel using a tris discontinuous buffer system at pH 8.5.   The electrophoresis was carried out on a sample containing 40 ug protein for 20 min with 5mA for each running tube. The gels were stained with amido black and excess stain removed by soaking in acetic acid (7%).

Table 4

Effect of Storage on the Molecular Size of Thiolated Insulin

	Initial -SH/mole of thiolated insulin[†]					
	0.58		0.87		1.08	
Storage period	-SH/mole	Molecular size (hexamers)	-SH/mole	Molecular size (hexamers)	-SH/mole	Molecular size (hexamers)
Initial-after centrifugation	0.53	4.2	0.76	8.5	0.95	8.5
1 day	0.44	5.7	0.44	13.9	0.78	14.0
2 days	0.24	11.1	0.18	24.0	0.60	27.8
Initial sample immediately oxidised with ferricyanide	0.0	4.6	0.0	12.0	0.0	14.0

[†] Prepared using Ag^+ - imidazole catalysis (see Mahbouba, Pugh and Smith, 1974) which removes zinc.

Native insulin gave two closely aligned bands midway down the gel.   The lower smaller band was dense and sharp and corresponded to monodesamido-insulin (Schlichtkrull et. al., 1972) while the upper band was more diffuse and less dense and corresponded to insulin.

Thiolated insulin showed the native insulin bands with a pronounced tail.   The surface band was absent.

Oxidised thiolated insulin (Cu^{++} - oxygen oxidation; n=43H) showed the insulin and desamido-insulin bands joined by a faint tail to a prominent band at the surface.

Table 5

Amount of Zinc Retained by Oxidised Thiolated Insulin Prepared Using Ferricyanide

-SH group/mole of thiolated insulin monomer	Zinc retained in oxidised thiolated insulin (%)
0.71	30
0.65	20
0.50	10
0.47	10
0.43	Nil

Dialysis of Oxidised Thiolated Insulin

A freshly prepared solution (40ml, 0.08%) of oxidised thiolated insulin (n=43H, 8.3g-atom Cu^{++}/hexamer) was dialysed in Visking tubing against EDTA (10^{-3}M) at 4° for 40 hr. At the end of this period the macromolecule had decreased in size to 17.4 hexamer units and contained 3.7g-atom Cu^{++}/hexamer unit.

Electrophoresis of the macromolecule before and after dialysis confirmed the presence of insulin and monodesamido-insulin and a decrease in the molecular weight of the macromolecule.

Separation of Mixtures of Oxidised Thiolated Insulin and Zinc-insulin Using Sephadex G-50

A solution (5ml, 0.15%) in phosphate buffer pH 8 (0.1M) of oxidised thiolated insulin (Cu^{++} - oxygen oxidation, n=139 H) was filtered down a column of Sephadex G-50 equilibrated with the same buffer at a flow rate of 50ml/hr. The column was developed with the same buffer and fractions (3ml) collected in the usual manner. The protein appeared in fractions 2-8 as detected spectrophoto-metrically at λ 276nm.

In a similar experiment a solution (5ml) of zinc-insulin (8mg) and blue dextran (4mg) in buffer was filtered down the column. The blue dextran appeared in fractions 3-8 and the insulin in fractions 11-22.

A solution (5ml) of oxidised thiolated insulin (Cu^{++} - oxygen oxidation; 0.094%, n=50.5 H) containing zinc-insulin (5mg) in phosphate buffer was filtered down the column.   The oxidised thiolated insulin appeared in fractions 2-8 and the insulin in fractions 11-24.

A similar result was obtained when the insulin in the previous experiment was increased to 7.5mg.   In both experiments the recovery of insulin was greater than 90%.

DISCUSSION

Crystalline zinc-insulin, which exists as a hexamer in concentrated solution (Cunningham, Fischer and Vestling, 1955), has been partially thiolated using AHTL with the introduction of 3.0-6.0 thiol groups per hexamer (Table I).   Previous work using zinc-free insulin has shown that attempts to prepare a mono-thiolated insulin under controlled conditions with AHTL were unsuccessful, a mixture of dithiolated insulin and native insulin being obtained,(Virupaksha and Tarver, 1964).   Insulin contains three basic groups at which reaction could occur, the terminal α-amino groups of glycine and phenylalanine and the ε-amino group of B-29 lysine, and it would be expected at pH 8 that reaction would preferably occur with the α-amino group (Benesch and Benesch, 1956).   It is well known that oxidative coupling of thiolated proteins occurs with the formation of either intra- or inter-molecular disulphide bonds (Benesch and Benesch, 1958; Stephen, Gallop and Smith, 1966).   Quantitative studies on the average molecular weight of the macromolecule formed from thiolated α-chymotrypsin established that large aggregates (n=16-44) could be formed by this method (Mahbouba, Pugh and Smith, 1974). Subsequent studies with globin (Etemad-Moghadam, Mahbouba, Pugh and Smith) and trypsin (Jivraj) have indicated that the size of the macromolecule is broadly related to the extent of thiolation. However one exception to disulphide coupling in this manner to give large macromolecules has been noted with bovine serum albumin. This protein only forms dimers when quite heavily thiolated molecules are oxidised, e.g. 8 -SH groups/mole, unless the protein solutions have a concentration of >2% when small aggregates are formed (n=6-8)(Alwan, Mahbouba and Smith).

Oxidation of the partially thiolated insulin hexamer using either potassium ferricyanide or Cu^{++} - catalysed oxygenation, gave macromolecules linked through disulphide bonds.   The macromolecules formed using potassium ferricyanide had average molecular weights of 3-4 hexamer aggregates whereas those formed by Cu^{++} - catalysed oxygenation of thiolated insulin with approximately the same thiol content, were very much larger and were in the size range 69-118 hexamers   (Table 2).

Relatively concentrated solutions of the protein preparations
were used throughout this work and these conditions would allow
retention of the hexamer configuration in these products.   Partial
thiolation of the insulin hexamer would provide at least three
thiol groups for cross-linking to other thiolated insulin
molecules with eventual formation of a macromolecule.   Ferricyanide
oxidation of thiolated insulin was accompanied by the loss of most
of the zinc from the oxidised thiolated insulin macromolecule.
Since under these conditions the hexamer would dissociate to dimers
and other small aggregates (Marcker, 1960) only a small
proportion of the molecules present would be capable of forming
intermolecular disulphide bonds associated with chain elongation
and formation of a macromolecule.   Consequently the calculated
average molecular weight of the oxidised thiolated insulin
macromolecule based on the total protein content would be
lowered.

Insulin   is capable of binding considerable amounts of
divalent ions, Zn, Cu, Co, Ni, Cd, and for zinc two distinct
binding sites have been postulated.   The zinc-insulin hexamer
found in the crystalline state and in concentrated solutions
contains two zinc atoms each coordinated to three imidazole nuclei
of the B-10 histidines in a central position inside the hexamer
skeleton (Hodgkin et.al., 1972).   ESR studies on copper-insulin
crystals have given a similar picture (Brill and Venable, 1968).
At least nine atoms of Zn, Cu etc., can be bound to the hexamer in
solution (Cunningham, Fischer and Vestling, 1955) or in the
crystalline state (Hallas - Moller, Petersen and Schlichtkrull, 1952)
the extent of the binding being dependent on pH and the
concentration of metallic ions in the media (Fredericq, 1956).
Excess zinc ions may be removed from the hexamer by dialysis but
the two atoms bound more firmly to imidazole within the hexamer
cannot be removed in this manner (Cunningham, Fischer and Vestling,
1955).

The oxidised thiolated insulin macromolecule prepared using
Cu^{++}-catalysed oxygenation contained very little zinc and 6-8
atom Cu^{++} per hexamer unit.   Since the two atoms of zinc present
in the thiolated insulin precursor are held tightly within the
hexamer it would appear that two atoms of copper had replaced zinc
within the hexamer.   The replacement of zinc by copper within the
hexamer was further indicated when zinc-insulin subjected to the
conditions of Cu^{++}-catalysed oxygenation lost all its zinc and took
up 4.5g-atoms Cu^{++} per hexamer.   Preliminary studies using ESR
showed that the environment of the copper in the oxidised thiolated
insulin was similar to that of the copper present in copper-insulin
(Evans and Mishra).

The synthesis of large macromolecules by Cu^{++} - catalysed

oxygenation of partially thiolated insulin over a reaction period
of two hours is in accord with retention of the hexamer structure
for the protein during oxidation but experiments have shown that
other factors contribute to the formation of such large macromolecules.

Thiol-disulphide interchange is known to occur between thiol
groups introduced into proteins and existing disulphide bonds
(Eldjarn and Pihl, 1957) and it has been shown for thiolated
ribonuclease that this process is rapid (White and Sandoval, 1962).
In our work, storage of a solution of zinc-free thiolated insulin
was associated with an increase in molecular weight with time to
give aggregates with larger molecular weights than the aggregates
formed by immediate ferricyanide oxidation of the fresh thiolated
insulin solution. This aggregation was attributed to thiol-
disulphide interchange between the introduced thiol groups and
disulphide linkages present in either a further molecule of
thiolated insulin, native insulin or the small amount of disulphide
coupled aggregate present. This process would lead to the
gradual gathering up of molecules into aggregates containing an
unchanged number of thiol groups so that subsequent disulphide
intermolecular linking would give larger aggregates than would be
formed by immediate oxidation of thiolated insulin.

Thiol-disulphide interchange is catalysed by copper ion
(Williamson, 1970) and it seems   likely that such a process
occurred to some extent during the two hour reaction period for
Cu^{++} - catalysed oxygenation of thiolated insulin. The covalent
linking of intact insulin units to thiolated insulin units
either by inter- or intramolecular disulphide bonds as a consequence
of such an interchange process was not considered to lead to an
appreciable decrease in the insulin content since disc electrophoresis showed the presence of intact insulin in the oxidised
product.

Aggregation of insulin can occur in the presence of excess
zinc ions to give large aggregates M = 200,000-300,000 (Fredericq,
1956), and the presence of 6-8g-atom Cu^{++}/hexamer in the oxidised
thiolated insulin would assist aggregation into larger aggregates.
Dialysis of the macromolecule against EDTA decreased the molecular
size by about two-thirds and the copper content from 8.3 to 3.7
g-atom/hexamer unit. It seems probable that the excess copper
ions present in the macromolecule contributed to some extent in
forming the large macromolecules observed.

The shape of the large macromolecules prepared by Cu^{++}-
catalysed oxygenation of thiolated insulin were examined using
their light scattering properties. The shape for the smaller
molecules approximated that of a rod whilst for the larger
molecules it resembled that of a random coil. Examination of the

relationship between the radius of gyration (R_z) and the molecular
weight (M) for the macromolecules indicated that the relationship
was not linear.   A plot of M/R_z versus M gave a line of increased
curvature as M increased.   It had been previously noted with
oxidised thiolated α-chymotrypsin and oxidised thiolated globin
that a similar relationship existed and this had been attributed
to the units within the macromolecules becoming more closely packed
as the molecular weight of the aggregate increased (Etemad-Moghadam,
Mahbouba, Pugh and Smith).

     The presence and strength of binding of the intact insulin
units within the oxidised thiolated macromolecule was examined by
disc electrophoresis and by filtration on Sephadex G-50.

     Disc electrophoresis on polyacrylamide gel showed the presence
of the large macromolecule at the surface of the gel and native
insulin and monodesamido-insulin as well distinguished bands
midway down the gel column.   Monodesamido-insulin is known to occur
in crystalline insulin (Mirsky and Kawamura, 1966) and either does
not react in the thiolation reaction or is not completely depleted
under the reaction conditions used.

     Gel filtration of mixtures of oxidised thiolated insulin and
zinc-insulin on a column of Sephadex G-50 showed that the two
components were well separated in the eluate.   Filtration of
oxidised thiolated insulin alone down the column showed the absence
of dissociable insulin in the product.   These results are in
accord with insulin being strongly bound in the macromolecule
within a hexamer structure.

ACKNOWLEDGEMENT

     We wish to thank Dr. W. J. Pugh for help and advice on the
light scattering aspects of this work and Dr. J.C.Evans and
Dr. S.P.Mishra, University College, Cardiff for the ESR results.

REFERENCES

Alwan, S., Mahbouba, M. and Smith, H. J. personal communication.
Benesch, R. and Benesch, R. E. (1956).   Formation of peptide
     bonds by aminolysis of homocysteine thiolactones. J. Amer.
     Chem. Soc., 78, 1597-1599.
Benesch, R. and Benesch, R. E. (1958).   Thiolation of proteins.
     Proc. Nat. Acad. Sci. U.S.A., 44, 848-853.
Blundell, T., Dodson, G., Hodgkin, D. and Mercola, D. (1972).
     Insulin: the structure in the crystal and its reflection
     in chemistry and biology. Advan. Protein Chem., 26, 279-402.
Brill, A. S. and Venable, J. H. (1968).   The binding of

transition metal ions to insulin crystals. J. Mol. Biol., 36, 343-353.

Cunningham, L. W., Fischer, R. L. and Vestling, C. S. (1955). A study of the binding of zinc and cobalt by insulin. J. Amer. Chem. Soc., 77, 5703-5707.

Eldjarn, L. and Pihl, A. (1957). On the mode of action of X-ray protective agents, interaction between biologically important thiols and disulphides. J. Biol. Chem., 225, 499-510.

Ellman, G. L. (1959). Tissue sulfhydryl groups. Archs. Biochem. Biophys., 82, 70-77.

Etemad-Moghadam, F., Mahbouba, M., Pugh, W. J. and Smith, H. J. Macromolecules formed by disulphide cross-linking of thiolated insulin and globin. In press.

Fredericq, E. (1956). The association of insulin molecular units in aqueous solutions. Archs. Biochem. Biophys., 65, 218-228.

Goldman, J. and Carpenter, F. (1974). Zinc binding, circular dichroism and equilibrium sedimentation studies on insulin (bovine) and several of its derivatives. Biochemistry, 13, 4566-4574.

Hallas-Moller, K., Petersen, K. and Schlichtkrull, J. (1952). Crystalline and amorphous insulin-zinc compounds with prolonged action. Science, 116, 394-398.

Hodgkin, D. C., Blundell, T. L., Cutfield, J. F., Cutfield, S. M., Dodson, G. G., Dodson, E. J., Mercola, D. A. and Vijayan, M. (1972). The arrangement in three dimensions of the atoms in insulin molecules and crystals. In "Insulin Action", Fritz, I. ed. p.1-28, Academic Press, New York and London.

Jivraj, S. S. personal communication.

Lowry, O. H., Rosebrough, N. J., Farr, A. L. and Randall, R. J. (1951). Protein measurement with the folin phenol reagent. J. Biol. Chem., 193, 265-275.

Mahbouba, M., Pugh, W. J. and Smith, H. J. (1974). Disulphide cross-linking of thiolated α-chymotrypsin to form macro-molecules. J. Pharm. Pharmac., 26, 952-955.

Mahbouba, M., Mishra, S. P., Evans, J. C. and Smith, H. J. personal communication.

Marocker, K. (1960). Association of Zn-free insulin. Acta Chem. Scand., 14, 194-196.

Mirsky, I. A. and Kawamura, K. (1966). Heterogeneity of crystalline insulin. Endocrinology, 78, 1115-1119.

Ornstein, L. and Davis, B. J. (1964). Disc electrophoresis. Annals N.Y. Acad. Sci., 121, 321-349; 404-427.

Pekar, A. H. and Frank, B. H. (1972). Conformation of proinsulin, a comparison of insulin and proinsulin self-association at neutral pH. Biochemistry, 11, 4013-4016.

Schlichthrull, J., Brange, J., Hein Christiansen, Aa, Hallund, O., Heding, L. G. and Jorgensen, K. H. (1972). Clinical aspects of insulin-antigenicity. Diabetes, 21, 649-656.

Stacey, K. A. (1956). "Light scattering in Physical Chemistry",
    Butterworths, London.
Stephen, J., Gallop, R. G. and Smith, H. (1966).   Separation of
    antigens by immunological specificity, use of disulphide-
    linked antibodies as immunosorbents.   Biochem. J., 101,
    717-720.
Virupaksha, T. K. and Tarver, H. (1964).   The reaction of insulin
    with N-acetyl homocysteine thiolactone: some chemical and
    biological properties of the products.   Biochemistry, 3,
    1507-1511.
White, F. H. and Sandoval, A. (1962).   The thiolation of
    ribonuclease.   Biochemistry, 1, 938-946.
Williamson, M. B. (1970).   Catalysis of formation of mixed
    disulphides between cystine and β-globulins by copper ions.
    Biochem. Biophys. Res. Commun., 39, 379-383.
Zimm, B. H. (1948).   The scattering of light and the radial
    distribution function of high polymer solutions. J. Chem.
    Phys., 16, 1093-1099.

16

CROSSLINKED INSULINS: PREPARATION, PROPERTIES, AND APPLICATIONS*

Dietrich Brandenburg, Hans-Gregor Gattner, Winrich

Schermutzki, Achim Schüttler, Johanna Uschkoreit, Josef

Weimann and Axel Wollmer

Deutsches Wollforschungsinstitut

D-5100 Aachen, Federal Republic of Germany

ABSTRACT

Crosslinked insulins have proved to be valuable for structure-function studies and as proinsulin models. In the first part of the paper, a short review of the literature on analytical investigations, the preparation of A1-B1- and A1-B29-crosslinked derivatives, their biological activities in vivo and in vitro, and CD-spectral properties is given. The results of reduction/reoxidation studies with insulin derivatives containing irreversible and cleavable crosslinks are summarized. In the second part, new A1-B29-crosslinked monomers and 3 symmetrical dimers, linked between A1-A'1, B1-B'1 and B29-B'29, are described, as well as some results of tritium-labelling and of enzymatic degradation experiments with A1-B29-linked insulins.

INTRODUCTION

There are several biological and chemical features which make insulin a particularly attractive molecule for crosslinking studies with bifunctional reagents: It is a vital hormone and essential for the treatment of diabetes. It is a two-chain molecule with labile disulfide bridges, which is rapidly inactivated in the organism.

* This paper is dedicated to Professor Helmut Zahn on the occasion of his 60th birthday (June 13, 1976).

Its biosynthesis proceeds via a single-chain precursor, proinsulin, while the chemical synthesis from the separated chains is very inefficient. Primary structure and three-dimensional crystal structure are known; the monomers have a strong tendency to aggregate. Although a true protein, its size (molecular weight about 6000 daltons) is comparatively small, and its solubility facilitates chemical modification. For reviews on insulin see Blundell et al., 1972, Geiger, 1976).

The purpose of this paper is to give, firstly, a short summary of the work published so far on homo-crosslinking of insulin. Secondly, to report on some recent results obtained in our laboratory and in cooperation with other groups.

PART I

Mohnike et al. (1951, 1953) were the first to react insulin with bifunctional reagents. Hexamethylene diisocyanate, phosgene and dibromophosgene yielded preparations which appeared to contain intermolecular crosslinks and had, in part, the desired property of retarded action. It was assumed, but not proved by analysis, that reactions occurred mainly at the primary amino groups.

In topological investigations of Zahn and Meienhofer (1958) with 1,3-difluoro-2,4-dinitrobenzene, both primary amino groups and tyrosine-OH groups reacted. Under the conditions of indefinite dilution in aqueous solution at pH 8 as well as in dimethylformamide, crosslinking was largely intramolecular. This was shown by the isolation and identification of DNPene-Gly-Phe and DNPene-Gly-Lys from the acid hydrolysates. At higher concentrations, intermolecular reaction occurred to a large extent, as obvious from the formation of DNPene-bis-Gly, DNPene-bis-Phe, and DNPene-bis-Lys. The proximity of A1-glycine and the ε-amino group of B29-lysine in the insulin monomer was demonstrated later by Xray analysis of the two zinc crystals (see Blundell et al., 1972). However, the N-terminals are about 20 Å apart (compare fig. 1 and 3).

Homogeneous Derivatives

The 3 primary amino groups appeared also to be the most promising sites for the preparation of homogeneous crosslinked derivatives, but more specific reagents had to be applied. A reaction scheme is outlined in figure 1. When insulin was reacted with 1 equivalent of m-phenylene diisothiocyanate in 97% pyridine at a relatively high protein concentration of 6.5 μmol/ml, gel filtration revealed that 60% monomers, 23% dimers, 11% trimers and 6% tetramers and higher oligomers had formed. Electrophoretic purification of the monomer peak yielded A1-B1-phenylenedithiocarbamoyl (Pbc)insulin, the first homogeneous crosslinked insulin. Oxidative sulphitolysis gave a

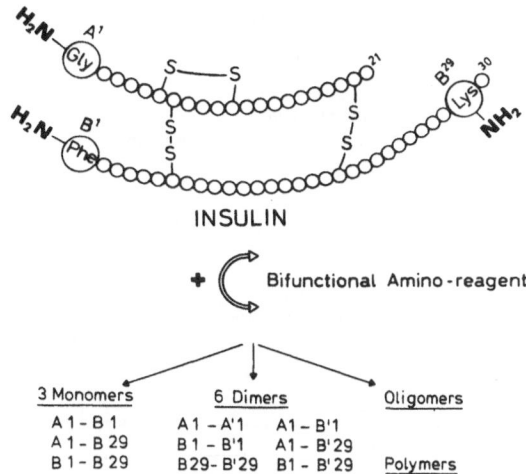

INSULIN

+ Bifunctional Amino-reagent

3 Monomers	6 Dimers	Oligomers
A1 - B1	A1 - A'1 A1 - B'1	
A1 - B29	B1 - B'1 A1 - B'29	
B1 - B29	B29- B'29 B1 - B'29	Polymers

Figure 1.   Schematic representation of insulin and its reaction with a bifunctional amino-specific reagent, leading to the
formation of intra- and intermolecularly crosslinked molecules.
Further derivatives can result from additional substitution of
residual free amino groups.

A-B- hexa-S-sulphonate, which migrated as a single band in paper
electrophoresis. Treatment with trifluoroacetic acid removed the
linked amino acids Gly and Phe as the bis-thiazolinone (Edman
degradation) and gave the shortened A- and B-chain. A1-B1-Pbc-
insulin is biologically almost inactive, the CD-spectrum indicates
severe changes of the three-dimensional structure. Crosslinking
between A1-glycine and B29-lysine could be accomplished with the
same reagent, after previous blocking of B1-phenylalanine with
phenylisothiocyanate, and was less detrimental to the conformation.
A dimer, B1-B'1-Pbc(insulin)$_2$, and a trimer, B1-B'1, A'1-B''1-
(Pbc)$_2$(insulin)$_3$ could also be isolated   (Brandenburg et al., 1971,
1972, 1976).

The finding, that activated esters of dicarboxylic acids react
preferentially with the amino groups of A1-glycine and B29-lysine
(Lindsay 1972, Brandenburg, 1972) was the basis for extended
studies on the crosslinking of insulin. Lindsay obtained A1-B29-
succinyl-insulin by reacting insulin (0.16 μmole/ml) with one
equivalent of the bis-N-hydroxysuccinimide ester in dimethylforma-
mide/triethylamine and subsequent isolation by DEAE-Sephadex
chromatography at pH 7.3. Brandenburg prepared A1-B29-adipoylinsulin

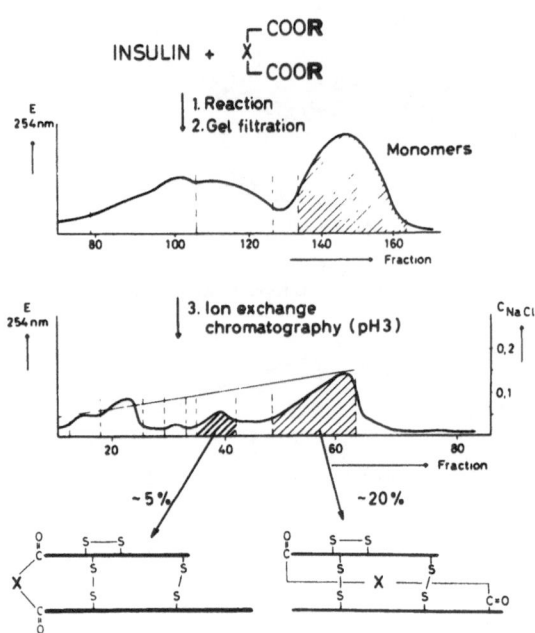

Figure 2.   Crosslinking of insulin with activated esters of
aliphatic dicarboxylic acids (x = 0,2-11) and fractionation of the
reaction products by gel filtration in 10% acetic acid. Subfractio-
nation of the monomers by ion exchange chromatography in 1.5 M
acetic acid/7 M urea at pH 3 yielded the A1-B1- and the A1-B29-
crosslinked derivatives shown (after Brandenburg et al., 1972).

in 13.5% yield from insulin (1.4 μmole/ml) in dimethylsulfoxide/
triethylamine with 1.06 equivalents of the corresponding bis-p-
nitrophenylester. Fractionation was accomplished by gel filtration
and subsequent chromatography of the monomeric material on SP-
Sephadex at pH 3 (see figure 2) and then DEAE-Sephadex at pH 8.4.
Both derivatives had only partially reduced activity in vivo, the
hypoglycaemic activities were 60 and 42%, respectively. However,
the in vitro activity of adipoylinsulin was surprisingly low (5%,
stimulation of lipogenesis in isolated rat fat cells).
In either case, structure and homogeneity were determined by dansyl
end group determination and oxidative sulphitolysis + electrophoresis, in conjunction with isoelectric focussing, gel filtration, and
amino acid analysis after trypsin digestion (Lindsay), or cellulose
acetate electrophoreses, thin layer chromatography, amino acid
analysis of the chain fragments after sulphitolysis, tryptic cleavage and separation, UV- and CD-spectroscopic data, and crystallization (Brandenburg). Additional substitutions at tyrosine and

serine-OH-groups (Zahn and Schade, 1963) were not observed in the purified derivatives.

Nine further derivatives, in which A1 and B29 were linked by bridges from 2 to 13 atoms, were subsequently prepared by reacting insulin with the bis-p-nitrophenylesters of the simple aliphatic dicarboxylic acids (Brandenburg et al., 1972, see also figure 2). The esters of carbonic and malonic acid gave inhomogeneous products.

Crosslinking of insulin with bis-tert-butyl-oxycarbonyl-2,7-diaminosuberic acid bis-2,4,5-trichlorophenylester (Boc_2-Dsu-$(OTcp)_2$ led to the formation of A1,B29-Boc_2-Dsu-insulin in the DL, meso-form (Geiger and Obermeier, 1973) and also the LL-form (Brandenburg et al., 1973b). The former was isolated by partition chromatography on Sephadex LH 20 in n-butanol-acetic acid-water. After deprotection, these derivatives had the same number of positive groups as insulin.

Busse and Carpenter (1974, 1976) designed a cross-bridge on the basis of methionine. The reaction of insulin with carbonyl-bis-methionine-bis-p-nitrophenylester, and fractionation by gel filtration, were similar to the procedure of Brandenburg (1972). Further purification was accomplished by chromatography on DEAE-cellulose at pH 7.2 and, if necessary, on CM-cellulose at pH 3.9 to give the A1-B29-linked derivative in yields of 35-40%. The extensive analyses included amino acid analysis after dinitrophenylation and chymotryptic digestion. Sodium dodecyl sulfate-gel electrophoresis at pH 9.5 confirmed homogeneity of size of the monomer and demonstrated the presence of dimers, trimers and several oligomers in the fractions separated by gel filtration. Recently, Obermeier and Geiger (1975) prepared A1,B29-2,2'-sulfonyl-bis(ethyloxycarbonyl)insulin ($SO_2(Eoc)_2$-insulin) by reacting insulin at high concentration (30 µmoles/ml) in a mixture of 1 N sodium hydrogencarbonate/dimethylformamide (1:2) with 1.05 equivalents of the corresponding N-hydroxysuccinimide ester. After purification by partition chromatography, the yield was 22%.

Several A1-B1-linked derivatives have also been described: Undecanedioyl- and dodecanedioyl-insulin (Brandenburg et al., 1972), suberoyl-insulin (Wollmer et al., 1974), and L.L-2,7-diamino-suberoyl-, α,α'-adipoyl-bis-lysyl- as well as α,α'-bis(benzyl-oxycarbonyl-lysyl) ε,ε'-adipoyl-insulin (Brandenburg et al., 1975). They had formed in low yields of about 5% besides the A1-B29-linked isomers and separated during chromatography on SP-Sephadex (figure 2), although both isomers have equal net charges at this pH. Structures of the adipoyl-lysyl bridges are shown in figure 3.

Structural Considerations. The crosslinks do not allow direct
inferences as to the threedimensional structure of insulin under
native conditions, because the reactions were carried out in organic
solvents. The monofunctional reagents,phenylisothiocyanate and activated esters of carboxylic acids, react with insulin amino groups in
the order Phe⟩ Gly≫ Lys, and Gly⟩ Lys≫ Phe, respectively (Brandenburg, 1969, 1972, Lindsay, 1972). Thus, intramolecular crosslinking
seems to have predominantly been governed by the reactivity of the
amino groups towards the reagents. The length and chemical nature of
the bridging reagents had apparently no significant influence on the
products formed, as obvious from the distribution patterns (see fig.2)
and yields . Dimer formation with the diisothiocyanate is also in
line with these results. Other dimers have so far not been isolated.
Preliminary experiments with suberoyl insulin (Brandenburg, unpubl.)
revealed extensive heterogeneity of the dimeric fraction with no
indication of a preferential species.

Information about structural properties of crosslinked insulins
in aqueous solution has been derived from circular dichroism spectra.
CD spectra of A1-B1-linked insulins have been published by Brandenburg et al., 1971, 1972, 1975, and Wollmer et al., 1974. CD-spectra
of A1-B29-linked insulins are shown by Brandenburg et al., 1972,
Brandenburg and Wollmer, 1973, Busse and Carpenter, 1974, 1976,
Vogt, 1975, and Wollmer et al., 1974. Further data are given by
Brandenburg, 1972, Brandenburg et al., 1973b. In addition to
measurement of spectra in zinc-free solution, the spectral changes
produced upon the addition of zinc ions have been used as a more
dynamic parameter (Wollmer et al., 1974). The results can be
summarized as follows, although there are still various difficulties
in the measurement and interpretation of CD-spectra of insulin itself
and modified insulins:
Crosslinking between A1 and B1 with the Pbc-residue brings about
severe distortions and a loss of ordered structure. This correlates
well with the model (fig. 3). The spectral changes become smaller
with increasing length and flexibility of the bridge, but even a
20-membered crosslink does not allow the molecule to assume the
undisturbed structure. Addition of zinc ions produced changes in the
spectra of the 3 derivatives shown in fig. 3 which indicated that
the ability to form hexamers was still partially retained. The CD
spectra of all A1-B1-crosslinked insulins deviate more from that
of insulin than spectra of derivatives with monofunctional substituents at these sites.

CD spectra of A1-B29-linked insulins show a higher structural
similarity of these molecules to insulin, and particularly to derivatives with 2 monofunctional substituents at A1 and B29, as diacetyl
insulin. When the bridges were systematically elongated, maximal
similarity to the spectrum of insulin, and identity with that of
diacetyl insulin, was found with suberoyl insulin (8 atoms), both in

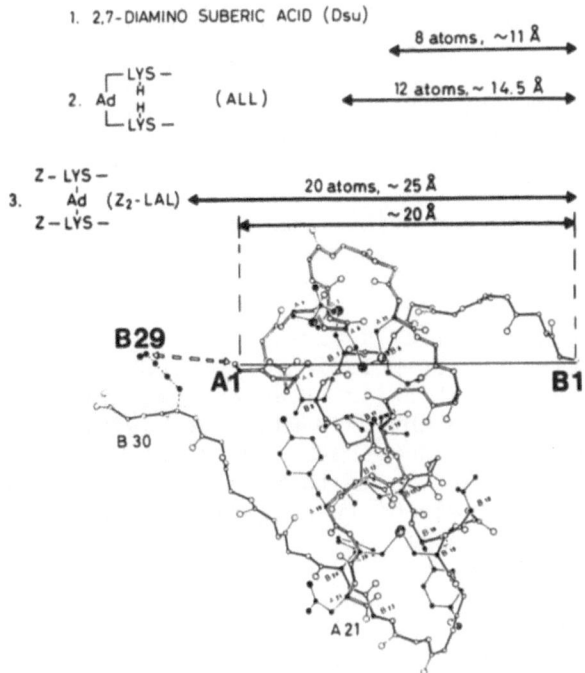

Figure 3. Structure of the insulin molecule viewed perpendicular to the 3-fold axis (Blundell et al., 1972). The distance between the amino group of A1-glycine and B1-phenylalanine is about 20 Å, while A1-glycine and the amino group of B29-lysine are only 5-10 Å apart. Also shown are the maximal distances covered by 3 crosslinking reagents (Brandenburg et al., 1975). Ad = adipoyl.

the far and near uv. These findings (Brandenburg et el., 1972) were in excellent agreement with the crystal structure of insulin (fig. 3). The closest spectral resemblance with insulin was so far found with LL-2,7-diaminosuberoyl-insulin. While a number of A1-B29-crosslinked insulins could already be crystallized, but gave rather poor crystals, this derivative is the only one of all monofunctionally and bifunctionally substituted A1,B29-derivatives which gave excellent 2-zinc rhombohedral crystals. The Xray structural analysis is in progress (S. Cutfield, G.G. Dodson).

The CD spectrum of the B1-B'1-Pbc-insulin dimer indicated high structural similarity to insulin, which again is in good agreement with the arrangement of this part in the molecule (see fig. 3) and its non-involvement in structure-stabilizing interactions (Blundell et al., 1972). These studies confirm our conception that the structure of insulin in the crystal and in solution is essentially equal.

<u>Biological and Immunological Properties</u>. The availability of
linked insulins in preparative amounts and homogeneous state made
structure-activity studies possible. The hypoglycaemic potency of
A1-B29-crosslinked insulins in rabbits or rats is between 30 and 70%,
some assays indicated full insulin activity (Lindsay, 1972, Brandenburg, 1972, Brandenburg et al., 1972, 1973a,b,Robinson et al., 1973).
Lindsay and Loge (1973) found that succinyl-insulin exhibited prolonged action when injected intravenously or into the muscle. A1-B1-
linked derivatives exhibit maximal potencies of 5% (Brandenburg et
al., 1971, 1972).

In contrast, biological activities of A1-B29-derivatives in
vitro, as determined by stimulation of glucose oxidation or lipogenesis in isolated rat fat cells (Brandenburg, 1972, Brandenburg et
al., 1972, 1973a,b,Gliemann and Gammeltoft, 1974, Jones et al., 1976)
or fat tissue (Busse and Carpenter, 1976) are much lower and usually
in the order of 5%. There seems to be a certain increase in activity
when increasing the length of the bridge from oxalyl-insulin (2 atoms,
2%) to undecane dioyl-insulin (11 atoms, 11-14%). Again, A1-B1-
derivatives are less active. Irrespective of the length of the
bridge, most derivatives are less than 1% active.
A1,B29- diacetyl-insulin, as well as the A1,B1-diacetyl-insulin,
both models with monofunctionally blocked amino groups, have activities of 30% (Gliemann and Gammeltoft, 1974).
The activity of the B1-B'1-Pbc-insulin dimer, calculated on a molar
basis, was 34 - 40% (Gliemann and Gammeltoft, 1974).

Receptor binding studies, in which the displacement of labelled
insulin as a function of derivative concentration was studied in
fat cells (Gliemann and Gammeltoft, 1974) or isolated rat liver
membranes (Freychet et al., 1974), were in excellent agreement with
the in vitro potencies. Binding of A1-B29-suberoyl-insulin to isolated intact liver cells was 3-4% (Terris and Steiner, 1975).

Recently, Jones et al.(1976) could show that the apparent dis-
crepancies of bioactivities in vitro and in vivo are due to different
rates at which the derivatives are metabolized. Thus, substitution
and crosslinking of the amino groups at A1 and B29 did not only
change the intrinsic activity of the molecule, but simultaneously
stabilized it against degradation in the organism. A further very
interesting observation is, that crosslinking A1 and B29 by the
dodecane dioyl-residue affects the pattern of glucose turnover
(Tompkins et al., 1974). Similarly to proinsulin, the derivative
has a proportionally greater effect in reducing glucose appearance
from the liver than in stimulating peripheral glucose uptake in
muscle and adipose tissue.

Linking A1-glycine to either B1-phenylalanine or B29-lysine
affects the hormonal activity strongly. Glycine is believed to be

at the periphery of the receptor binding area of insulin (Blundell
et al., 1972, Pullen et al., 1976). The pronounced effect of crosslinking, as compared to monofunctional substitution, may in the
case of the A1-B1-derivatives, be due to conformational changes,
as revealed by CD spectra, and the presence of the bridge. In A1-
B29-linked derivatives, conformational changes are very small. It
appears therefore that the presence of the bridge has a more direct
effect on binding to the receptor and activity.

Only a few immunological data have so far been published. A1-
B1-Pbc-insulin exhibits a relative immuno-reactivity of 3%, A1-B29-
adipoyl-insulin of 5-11% (Brandenburg et al., 1972, Brandenburg,
1972), while Busse and Carpenter (1976) reported values of 90-92%
for A1-B29-carbonyl-bismethionyl-insulin. Thomas et al. (1975)
investigated the binding of several crosslinked insulins to various
selected antibodies raised against insulin. They found that the
apparent immuno-reactivity of a given derivative depends critically
on the antiserum used. These results explain the differences observed above. A binding curve of A1-B29-dodecane dioyl-insulin to a discriminating antiserum has been published by Jones et al. (1976).

Reduction-Reoxidation Studies

Reoxidation of reduced proinsulin by air proceeds with high
yield (Steiner and Clark, 1968), while co-oxidation of the separated
insulin chains gives reproducible yields of insulin in the order
of 10% only. Both from theoretical as well as practical standpoint
it was extremely interesting to examine how crosslinking A- and
B-chain would affect the pairing of SH-groups. Robinson et al.(1973)
found that air oxidation of the reduced hexa-S-sulfonates of A1-B29-
succinyl- and suberoyl-insulin led indeed to a re-formation of the
original derivatives in yields between 24 and 45%, as judged from
the hypoglycaemic activities of the crude reaction products.
In a detailed study, Brandenburg and Wollmer (1973) performed a
series of preparative experiments, each with 10-20 μmoles of adipoyl-
insulin or the corresponding hexa-S-sulfonate. After complete reduction, reoxidations were carried out at a concentration of
0.017 μmole/ml at pH 8.5 - 9.5 for up to 185 hours. Reoxidized adipoyl-insulin was isolated in reproducible yields of 71-76% through
gel filtration. Extensive CD-spectral analyses, in conjunction with
crystallization experiments, determinations of biological activity
in vivo and in vitro demonstrated the identity with the original
adipoyl-insulin within the limits of experimental error. These crosslinked insulins can therefore be regarded as proinsulin models with
reference to their ability to readopt the original conformation.
The role of the connecting peptide with regard to the correct
pairing of cysteine residues can be played by the unbiological
dicarboxylic acid bridges. Variation of the crosslink between 2

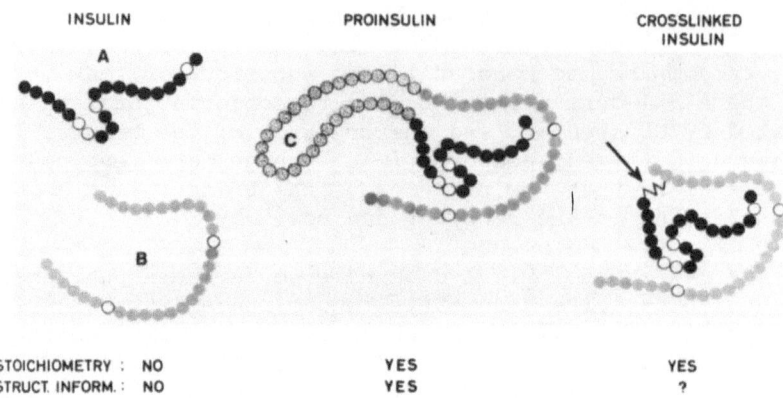

INSULIN   PROINSULIN   CROSSLINKED
                                                            INSULIN

STOICHIOMETRY :	NO	YES	YES
STRUCT. INFORM. :	NO	YES	?

Figure 4.   Schematic representation of insulin A- and B-chains,
proinsulin and a A1-B29-crosslinked derivative in the reduced form
(open circles = cysteine residues). Co-oxidation of separated A-
and B-chains is a statistical, bimolecular reaction, while the
information for folding and intramolecular pairing of SH-groups
of proinsulin as a natural protein is contained within the amino
acid sequence. Artificial crosslinks ensure intramolecular reaction.
The fundamental question was, whether reoxidation would proceed
under re-formation of the insulin structure with correctly formed
disulfide bonds.

and 13 atoms did not significantly affect the yields or the quality
of the reoxidized derivatives (Wollmer et al., 1974). The CD-spectra
of reoxidized oxalyl-, suberoyl- and tridecane dioyl-insulin were
identical with those of the authentic derivatives. Furthermore, the
CD-spectral changes upon addition of zinc could be exactly reproduced.

With respect to the synthesis of insulin, two further problems
were to be solved: Firstly, to find suitable reversible crosslinking
reagents, and secondly, to crosslink the separated A- and B-chain.
Three reagents have been introduced so far (table 1), additional ones
are described in patents (see Geiger, 1976). Reoxidations of reduced
diaminosuberoyl-insulin, the fully protected hexa-S-sulfonate A1-B29-
$(Boc)_2$-Dsu, $B1$-Tfa-AB$(SSO_3^-)_6$, $CO(Met)_2$-insulin and the corresponding
hexa-S-sulfonate, and $SO_2(Eoc)_2$-insulin were carried out under similar conditions as described before at concentrations between
0.0017 and 0.33 µmole/ml in the pH-range 7.8 - 10.6.   The yields
of crosslinked derivatives were between 45 and 85%. According to
Busse and Carpenter (1976), high yields are favored by reoxidation
at high pH.   The diaminosuberoyl bridge could be cleaved by Edman
degradation. The simultaneous removal of B1-phenylalanine was not
considered a draw-back, since des-Phe-insulin is fully active

TABLE 1

Reversible Crosslinks Between the Amino Groups of A1 and B29

Insulin Derivative	Cleavage	% Maximal Yield A	B	Purification	Autors
Dsu (DL, meso)	1. ⬡ NCS			Sephadex G-50	Geiger, Obermeier 1973
Dsu, B1-Tfa	2. Tfa		24		
Dsu (LL)	1. ⬡ NCS 480 eq./NH$_2^-$ 2. Tfa	77 50*	38 23*	Sephadex G-50	Brandenburg et al. 1973
CO(Met)$_2$	CNBr 280 eq./Met	70-80	60	DEAE-cellulose	Busse, Carpenter 1974, 1976
SO$_2$(Eoc)$_2$	0.5 N NaOH 0°C	75	68** 50-60***	Sephadex G-75	Obermeier, Geiger, 1976

A = from authentic, B = from reduced/reoxidized derivative
Tfa = trifluoro acetic acid/acetyl, Eoc = ethyloxycarbonyl
*crystallized, **bioassay of crude product, ***calculated for purified product

(Brandenburg, 1969). It could be prevented by temporary blocking with the trifluoro acetyl group, which was later removed with piperidine. The use of the two other reagents is more simple, since they have no functional groups necessitating temporary protection, and since removal is specific. Minor side reactions and/or incomplete cleavage did occur in all cases and required subsequent purification. The yields obtained have to be taken as preliminary, since so far only a few small-scale experiments have been carried out, and improvements appear possible. Crystalline insulin (or des-Phe-insulin) has been isolated in all cases. These independent findings, in conjunction with biological assays, as well as chemical and CD-spectroscopical analyses of the reoxidized intermediates and final products, demonstrated that correct formation of disulfide bonds had taken place. This could be further proved by partial chymotryptic hydrolysis (Busse and Gattner, 1973) and analysis of the disulfide peptides (Busse and Carpenter, 1976).

The second problem, joining of the separate A- and B-chains,
has so far been investigated in one case. Geiger and Obermeier
(1973) reacted A-chain-S-sulfonate with an excess of Boc-Dsu-tri-
chlorophenylester. The monofunctionally acylated A-chain was then
coupled to the ε-amino group of B1-trifluoroacetylated B-chain-S-
sulfonate to give the hexa-S-sulfonate used for the reduction/oxi-
dation experiment discussed above (see also table 1).

The time-course of reoxidation has been followed by CD-spectroscopy in conjunction with thiol group analysis with diaminosuberoyl-
insulin (Wollmer et al., 1974; Vogt, 1976) and carbonyl-bismethionyl-
insulin (Busse and Carpenter, 1974, 1976). In either case, the spectrum indicated the development of ordered structure from a random
coil state of the fully reduced chain derivative, but synoptical
interpretations of the data are as yet difficult. There were no
indications of ordered structure in the CD-spectra of corresponding
hexa-S-sulfonate or hexa-S-carboxymethyl chain derivatives.

Under denaturing conditions, in the presence of 7 M urea,
reoxidation of reduced suberoyl-insulin gave high yields of monomeric material which, however, had not the structure of the authentic derivative, as shown by CD-spectroscopy (Wollmer et al, 1974).
Similarly, random formation of disulfide bonds occured to a large
extent upon reoxidation of reduced A1-B1-crosslinked insulins.
This was apparent from the formation of polymers and oligomers as
well as from CD-spectra of the monomers.

That these were a mixture of species with correct and incorrect
disulfide structures, could be shown with A1-B1-diaminosuberoyl-
insulin. After Edman degradation of the monomeric material, bioassay and crystallization experiments indicated the presence of
about 30% insulin (Brandenburg et al., 1975). Contrary to expectation, long connecting bridges of 12 and 20 atoms did not favor the
correct pairing of cysteine residues. Joining insulin peptides via
their N-terminals with a polyfunctional amino-protecting group has
already been suggested by Zervas and Photaki (1962) in order to
facilitate the formation of SS-bonds and to effect their stabilization.

In summary, the results show that linking A- and B-chains
can be an efficient way to direct the pairing of cysteine SH-groups
and to effect formation of the correct disulfides bonds. Transformation of the bimolecular into a monomolecular reaction is essential,
but not sufficient. Spatial requirements have to be met which are
consistent with the threedimensional structure of the insulin mole-
kule (figure 3). Length and chemical nature of the bridge between
A1-glycine and B29-lysine, however, are of secondary importance.
The reduced derivatives in which A- and B-chains are linked in an
unnatural way seem to contain structural information for correct

folding, since under denaturing conditions re-formation of the
original derivative is not achieved. The experiments with cleavable
crosslinking reagents have shown that this approach is promising
for the synthesis of insulin.

PART II

Monomers

Derivatives with A1-B29-crosslinks. Five new derivatives were
prepared with the bifunctional reagents listed in table 2, following
the conditions described (Brandenburg, 1972). The yields were comparable with earlier results, i.e. about 20% of oligomers, 20% of
dimers, and 50% of monomers formed, which were further fractionated
by chromatography on SP-Sephadex at pH 3. Boc-protecting groups
were then removed with trifluoroacetic acid. Neither the presence
of hydrophobic blocking groups in α- or ε-position, nor large
differences in the length of the bridging reagent, nor the presence
of sulfur atoms or of 2 carbonamide groups had a significant influence on the formation of the A1-B29-linked derivatives. The
yield of A1-B1-cystinyl-insulin was 6%, the A1-B1-isomers of III,
and IV have been described (Brandenburg et al., 1975). Only the
activated ester of the cyclohexane dicarboxylic acid was much less
reactive. With 2.4 equivalents and after prolonged reaction time,
the yield was only 4%.

The activity of III is the lowest so far found for any A1-B29
linked insulin. This seems to indicate that the cyclic residue
causes additional steric strains on the insulin molecule or contributes to direct interference with receptor binding. As the biological potencies of I and II are compareable to the activity of diaminosuberoyl insulin (Brandenburg et al., 1973), it appears that neither
configuration nor the replacement of two methylene groups by sulfur
affect the receptor interaction. However, the different size and
decreased flexibility of the cystinyl bridge has some conformational consequences as judged from the CD-spectrum and the inability
of the sulfur analogue to crystallize in the 2-zinc form. It is
interesting to note that the ratios of biological activity in vitro/
in vivo of I and II are identical, although the cystinyl bridge is
more labile since it is susceptible to enzymatic cleavage.

Composed bifunctional reagents on the basis of diaminocarboxy-
lic and dicarboxylic acids appeared to be useful, since by varying
both moieties a large number of homologues becomes accessible. The
presence of amino groups does not change the charge of the crosslinked molecule. Removal of the bridge is possible by Edman degradation, if α-amino groups are present, and by trypsin, if ε-amino
groups are free and the bridge is bound in a C^α-N^α-linkage. The

TABLE 2

A1-B29-Intramolecularly Crosslinked Insulins

Derivative	Bifunctional reagent	Bridge atoms	Yield %	Biological activity in vitro[1]	in vivo[2]
I 2,6-DL-meso-Diaminopimeloyl	Boc_2-Dpi$(ONp)_2$	7	17	4.4%	45%
II L,L-Cystinyl	Boc_2-Cys$(OTcp)_2$	8	16	2-3%	32%
III 1,2-cis-Cyclohexane-dioyl	⬡$(COONp)_2$	4	4	0.5%	
IV α,α'-Adipoyl-bis-lysyl	Ad [Lys-ONp | Boc, Boc | Lys-ONp]	12	19	8-10%	
V α,α'-Bis(benzyl-oxycarbonyl-lysyl) ε,ε'-adipoyl	Z-Lys-ONp | Ad | Z-Lys-ONp	20	15	3-4%	

Boc = tert.Butyloxycarbonyl,   Np = p-Nitrophenylester,
Tcp = 2,4,5-Trichlorophenylester

1) Lipogenesis in isolated rat fat cells (J.Gliemann, C.Diaconescu)
2) Hypoglycaemia in rats (W.Puls).

biological potencies of IV and V indicate, in conjunction with
the earlier results (see Part I), optimal activity of A1-B29-cross-
linked insulin when bridges contain 10-12 atoms. According to CD-
spectrum and the spectral changes produced upon the addition of
zinc ions, the structure of IV in solution is very much like that
of insulin. But in the crystalline state the bridge is less well
accommodated, as indicated by the formation of spherical particles
only. The CD-spectrum of V deviates more, the ability of V to
bind zinc is still partially retained.

In further experiments, the amino groups of pimelic acid were protected with the p-biphenylisopropyloxycarbonyl protecting group, which can be removed under very mild acid conditions. Contrary to expectation, this residue turned out to be unstable even at pH 8 and thus did not allow for the isolation of the homogeneous insulin derivatives (Weimann, 1974). When crosslinking of insulin was attempted with dimethylsuberimidate dihydrochloride in aqueous solution, intramolecular reaction occurred to a small extent only, while excess of reagent led largely to intermolecular crosslinking. However, in dimethylsulfoxide/triethylamine, equimolar amounts of reagent gave results similar to avtivated esters of dicarboxylic acids (Schermutzki, 1975).

Labelling Experiments. Labelling of insulin with tritium under reductive conditions (Wilzbach labelling, catalytic dehalogenation) have met with little success due to the instability of the disulfide system. Yip (1972) has therefore incorporated tritium into proinsulin and obtained labelled insulin after enzymatic cleavage. Since pro-insulin is not easily available and since A1-B29 crosslinked insulins can also readopt their structure after disruption of SS-bonds, we have started to examine their use for this purpose. 15 µmoles of A1-B29-diaminosuberoyl-insulin were subjected to Wilzbach labelling with 1 Curie of tritium gas for 4 weeks (H.-J. Machulla). Then, the disulfide bonds were deliberately cleaved by oxidative sulphitoly-sis and the crosslinked chain derivative separated from degradation products and other materials by gel filtration on Sephadex G-50 at pH 8. After reduction and reoxidation (Brandenburg et al., 1973b) gel filtration gave a product which was identical with Dsu-insulin with respect to elution volume, electrophoretic mobility and amino acid composition and had a specific activity of 0.47 Ci/mmole. Although the yield in this first experiment was, at this stage, only 20%, and removal of the crosslink by Edman degradation has not yet been carried out, the approach appears so far promising and is presently being investigated in more detail.

Enzymatic Degradation. In relation to the prolonged half-life which crosslinked insulins posess in vivo (Jones et al., 1976) it was of interest to test whether a stabilizing effect of intra-molecular crosslinking could be demonstrated in in-vitro experiments. First results are summarized in table 3. Carboxypeptidase A and trypsin transform insulin into practically inactive analogues, which lack A21-asparagine + B30-alanine, or the C-terminal peptide B23-30, respectively (Carpenter, 1966). Following a suggestion of G. Dodson and D. Mercola, we prepared the corresponding crosslinked derivatives. The loss of biological avtivity indicated, that the artificial crosslinks were not able to maintain the modified mole-

TABLE 3

Enzymatic Degradation of A1-B29-Crosslinked Insulins

Derivative	Enzyme	Degradation %	Remarks
Oxalyl Suberoyl	Carboxypeptidase A	80-90	des-Asn^{A21}-des-Ala^{B30}-derivatives isolated: biol. activity lost.
Diamino-suberoyl	Trypsin	80	B22/23-split derivative isolated: biol. activity lost.
Suberoyl	Chymotrypsin	much more slowly	conditions for partial cleavage of insulin
Suberoyl	Insulin-Glutathione Transhydrogenase	42	A1,B29-diacetyl-insulin: 100% cleavage (H.THOMAS).

cules in an active conformation and to compensate for the cleavage of the peptide bonds.

A marked difference between insulin and suberoyl-insulin was observed towards hydrolysis by chymotrypsin. Under the conditions described by Busse and Gattner (1973) for partial cleavage of insulin, the crosslinked derivative was practically stable. Cleavage of disulfide bonds was also significantly inhibited. This stabilization appears to be a specific feature of the crosslinked molecule, since the disubstituted, non-linked model, A1,B29-diacetyl-insulin, was cleaved like insulin.

Dimers

Two possibilities exist to obtain insulin dimers: a) To fractionate the products from the reaction of unprotected insulin with bifunctional reagents. b) To effect specific crosslinking with partially protected insulins. The gel chromatographic fraction containing the dimers formed in the reaction of Boc-diaminosuberic acid-bis-p-nitrophenylester and insulin was somewhat better resolved than usually observed (figure 2). After rechromatography on

TABLE 4

Intermolecularly Crosslinked Insulins

Dimer	Yield %	Biological activity[1] in vitro	in vivo	Conformational change[2]
A1-A'1 diamino- suberoyl	1	30%	1%	moderate, zinc binding +
B1-B'1- diamino- suberoyl	37	30-32%	50%	small, zinc binding ++
B29-B'29- suberoyl	40			
sebacoyl	30	10%		

[1] Calculated on molar basis

[2] From CD-spectrum (pH 8) 0, 0.33, 0.66 atoms Zn^{++}/insulin molecule

Sephadex G-75 it appeared homogeneous with respect to molecular weight, but contained at least 5 different components according to electrophoresis at pH 2. By subsequent ion exchange chromatography on SP-Sephadex in 7 M urea at pH 3, followed by rechromatography under similar conditions and then chromatography on DEAE-cellulose at pH 7.3, a homogeneous dimer could be isolated in a yield of 1%. End group determination and oxidative sulphitolysis, followed by analysis of the S-sulphonate chains, showed that the monomers were linked between the A1-amino groups. Deprotection gave the dimer listed in table 4.

In order to link 2 insulin monomers between the B1-amino groups, A1,B29-bis(tert.butyloxycarbonyl)insulin (Geiger et al., 1971) was reacted with the N-protected p-nitrophenylester of diaminosuberic acid (Schermutzki, 1975). The specific joining of 2 molecules via the B29-amino groups was accomplished by A. Schüttler, who reacted A1,B1-bis(trifluoracetyl)insulin (Paselk and Levy, 1974, Friesen, 1976) with suberic as well as sebacic acid nitrophenylesters (figure 5). Even at concentrations above 15 μmoles/ml the reactions in dimethylsulfoxide in the presence of N-methylmorpholine or triethylamine did not go to completion. But after deblocking with trifluoracetic acid or ammonia, respectively, purification was easily accomplished by gel filtration on Sephadex G-50 in 10% acetic acid.

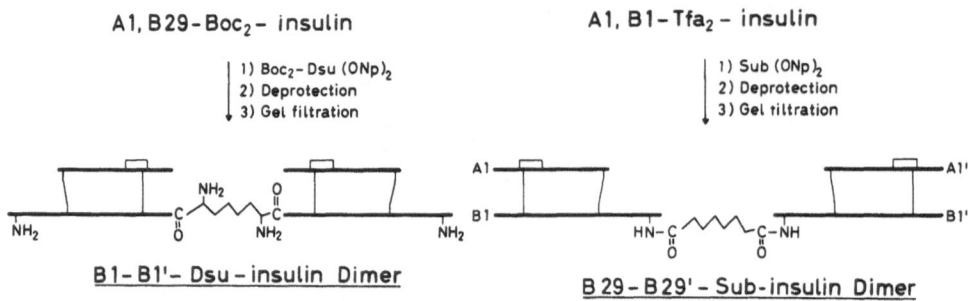

<u>B1-B1'- Dsu - insulin Dimer</u>   <u>B 29 - B 29' - Sub-insulin Dimer</u>

Figure 5.   Scheme of reactions leading to the formation of insulin dimers which are specifically crosslinked with diamino-suberic acid (left) and suberic acid (right).

The CD-spectrum of the B1-B'1-dimer is remarkably similar to that of insulin, the spectral changes upon the addition of zinc ions indicate that the ability to form hexamers is only slightly impaired. Hexamerization of <u>insulin</u> proceeds via the association of 3 dimers (figure 6). In these dimers, the B1-phenylalanines are separated by about 40 Å. Thus, our B1-B'1 dimer cannot represent this type. However, since the B1-phenylalanines of 2 adjacent dimers are in very close contact, this "unnatural" dimer could have been covalently stabilized. Hexamerization would then be feasible via the aggregation of 3 such dimers at the original mono-mer/monomer interfaces (compare figure 6).

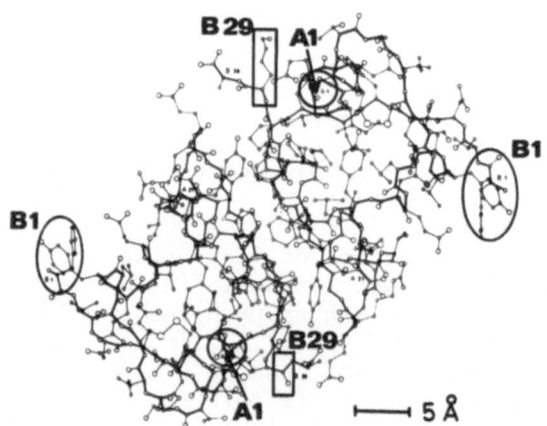

Figure 6.   View of the insulin dimer in the direction of its 2-fold axis (Blundell at al., 1972)

The CD spectra of the A1-A'1-dimer indicate more pronounced conformational changes, but there is still a tendency to aggregate in the presence of zinc. The threedimensional structure (Blundell et al., 1972) seems to rule out that two A1-glycines be linked within one hexamer without major sterical distortions of the monomers. The moderate spectral changes observed would rather imply that hexamerization proceeds in such a way that the 2 monomers of the A1-A'1-dimer belong to 2 different hexamers.

Both dimers retain still moderate biological activity in vitro. This shows that the receptor binding region can only be partially distorted or inaccessible. In contrast, the activity of the B29-B'29-dimer is significantly lower. Figure 6 shows, that in the natural dimer the amino groups involved in the crosslink are about 30 Å apart. However, by simple movement of the lysine side chain they could get as close as about 15 Å. It then would be feasible for sebacic acid, which can maximally span 13 Å, to link the lysines in the "natural" dimer without serious conformational strains. Such a linkage would stabilize the natural dimer. It has been postulated that the receptor binding region of insulin includes the monomer/monomer interface (Blundell et al., 1972, Pullen et al., 1976). The low activity of the B29-B'29-dimer seems to support this view, if sterical integrity can be demonstrated.

CONCLUSIONS

By reacting insulin with bifunctional reagents, a number of intra- and intermolecularly crosslinked homogeneous derivatives has become accessible. They allowed correlations between the structure in solution and in the crystal, have interesting biological properties, and have opened the way for a more efficient synthesis of insulin.

ACKNOWLEDGEMENTS

This work was supported by Bundesministerium für Forschung und Technologie, Bonn (Project 01 VU 093-B13 SN 0015, formerly SN 1054), Minister für Wissenschaft und Forschung des Landes Nordrhein-Westfalen, Düsseldorf, and Deutsche Forschungsgemeinschaft - Sonderforschungsbereich 113 Diabetesforschung Düsseldorf.
The authors wish to express their sincere thanks to Professor H. Zahn for constant help and encouragement, and to S. Cutfield, C. Diaconescu, G. Dodson, J. Föhles, P. Freychet, J. Gliemann, R. Jones, M. Leithäuser, H.-J. Machulla, W. Puls, P. Sönksen and H. Thomas for continuous cooperation and stimulating discussions, and K. Ziegler for having presented this paper.

REFERENCES

Blundell, T., Dodson, G., Hodgkin, D. and Mercola, D. (1972).
    Insulin: The structure in the crystal and its reflection in
    chemistry and biology. Adv. Protein Chemistry 279.

Brandenburg, D. (1969). Des-Phe^{B1}-insulin, ein kristallines Analo-
    gon des Rinderinsulins. Hoppe-Seyler's Z. Physiol. Chem. 350,
    741.

Brandenburg, D., Gattner, H.-G., Weinert, M., Herbertz, L., Zahn, H.
    and Wollmer, A. (1971). Structure-function studies with deriva-
    tives and analogs of insulin and its chains. Diabetes. Proc.
    7th Congr. Intern. Congress Series 231, Excerpta Medica
    Foundation, Amsterdam, 363.

Brandenburg, D. (1972). Preparation of $N^{\alpha A1}, N^{\epsilon B29}$-adipoyl insulin,
    an intramolecularly crosslinked derivative of beef insulin.
    Hoppe-Seyler's Z. Physiol. Chem. 353, 869.

Brandenburg, D., Busse, W.-D., Gattner, H.-G., Zahn, H., Wollmer,A.,
    Gliemann, J. and Puls, W. (1972). Structure-function studies
    with chemically modified insulins.   Peptides. Proc. 12th
    Europ. Peptide Symposium, Reinhardsbrunn Castle, GDR, 270.

Brandenburg, D. and Wollmer, A. (1973). The Effect of a non-peptide
    interchain crosslink on the reoxidation of reduced insulin.
    Hoppe-Seyler's Z. Physiol. Chem. 354, 613.

Brandenburg, D., Gliemann, J., Ooms, H.A., Puls, W. and Wollmer, A.
    (1973a). Structure-function relationships of chemically cross-
    linked, homogenous insulin derivatives. Diabetologia 9, 61.

Brandenburg, D., Schermutzki, W. and Zahn, H. (1973b).
    $N^{\alpha A1}-N^{\epsilon B29}$-crosslinked diaminosuberoylinsulin, a potential
    intermediate for the chemical synthesis of insulin. Hoppe-
    Seyler's Z. Physiol. Chem. 354, 1521.

Brandenburg, D., Schermutzki, W., Wollmer, A., Vogt, H.P. and
    Gliemann, J. (1975). A1-B1-crosslinked insulins for structure-
    activity and reduction-reoxidation studies. Peptides: Chemi-
    stry, Structure and Biology. Ann Arbor Sci. Publ. Inc., 497.

Brandenburg, D., Schermutzki, W. and Weimann, H.-J. (1976). Chemi-
    cal modification of proteins with activated esters and other
    bifunctional reagents. Proc. Int. Wool Res. Conf. Aachen,
    Vol. 3, 182.

Busse, W.-D. and Carpenter, F.H. (1974). Carbonylbis(L-methionine
    p-nitrophenyl ester). A new reagent for the reversible intra-
    molecular cross-linking of insulin. JACS 96, 5947.

Busse, W.-D., Hansen, S.R. and Carpenter, F.H. (1974). Carbonylbis-
    (L-methionyl)insulin. A proinsulin analog which is convertible
    to insulin. JACS 96, 5949.

Busse, W.-D. and Carpenter, F.H. (1976). Synthesis and properties
    of carbonylbis(methionyl)insulin, a proinsulin analogue which
    is convertible to insulin by cyanogen bromide cleavage.
    Biochemistry 15, 1649.

Busse, W.-D. and Gattner, H.-G. (1973). Selective cleavage of one
    disulfide bond in insulin: preparation and properties of insulin

A7-B7-di-S-sulfonate.   Hoppe-Seyler's Z. Physiol. Chem. 354, 147.

Carpenter, F.H. (1966). Relationship of structure to biological activity of insulin as revealed by degradative studies. Amer. J. Med. 40, 750.

Freychet, P., Brandenburg, D. and Wollmer, A. (1974). Receptor-binding assay of chemically modified insulins. Diabetologia 10, 1.

Friesen, H.-J. (1976). Darstellung und Eigenschaften von partiell aminogeschützten und am N-Terminus der A-Kette modifizierten Insulinen aus intaktem Insulin. Thesis TH Aachen.

Geiger, R., Schöne, H.H. and Pfaff, W. (1971). Bis(tert.-butyloxy-carbonyl)-insulin. Hoppe-Seyler's Z. Physiol. Chem. 352, 1487.

Geiger, R. and Obermeier, R. (1973). Insulin synthesis from natural chains by means of reversible bridging compounds. Biochem. Biophys. Res. Comm. 55, 60.

Geiger, R. (1976). Chemie des Insulins. Chemiker Zeitung 100, 111.

Gliemann, J. and Gammeltoft, S. (1974). The biological activity and the binding affinity of modified insulins determined on isolated rat fat cells. Diabetologia 10, 105.

Jones, R.H., Dron, D.I., Ellis, M.J., Sönksen P.H. and Brandenburg, D. (1976). Biological properties of chemically modified insulins. I. Biological activity of proinsulin and insulin modified at A_1-glycine and B_{29}-lysine. Diabetologia 12, 601.

Lindsay, D.G. (1972). Intramolecular cross-linked insulin. FEBS LET. 21, No.1 105.

Lindsay, D.G. and Loge, O. (1973). The biological properties of an intramolecular crosslinked insulin derivative. Diabetologia 9, 78.

Mohnike, G., Schnuchel, G. und Langenbeck, W. (1951). Die Einwirkung von Hexamethylendiisozyanat und Phosgen auf Insulin. Natur-wissenschaften 38, 333.

Mohnike, G., Schnuchel, G., Kupffer, I. und Langenbeck, W. (1953). Darstellung von Derivaten des Insulins durch Einwirkung bifunk-tioneller Verbindungen. Hoppe-Seyler's Z. Physiol. Chem. 294, 12.

Obermeier, R. und Geiger, R. (1975). Ein neues bifunktionelles Rea-gens zur intramolekularen Vernetzung von Insulin. Hoppe-Seyler's Z. Physiol. Chem. 356, 1631.

Paselk, R.A. and Levy, D. (1974). Preparation of several trifluoro-acetyl-insulin derivatives. Biochim. Biophys. Acta 359, 215.

Pullen, R.A., Lindsay, D.G., Wood, S.P., Tickle, I.J., Blundell, T.L., Wollmer, A., Krail, G., Brandenburg, D., Zahn, H., Gliemann, J. and Gammeltoft, S. (1976). Receptor-binding region of insulin. Nature 259, 369.

Robinson, S.M.L., Beetz, I., Loge,O.,Lindsay,D.G.,and Lübke K. (1973). Spaltung und Rückbildung der Disulfidbrücken an Intramokekular vernetzten Insulinen. Tetrahedron Let. 12, 985.

Schermutzki, W. (1975). Darstellung und Eigenschaften reversibel verbrückter Insulinderivate. Thesis, TH Aachen.

Steiner, D.F. and Clark, J.L. (1968). The spontaneous reoxidation of reduced beef and rat proinsulins. Proc. Nat. Acad. Sci. 60, 622.

Terris, S. and Steiner, D.F. (1975). Binding and degradation of ^{125}-I-insulin by rat hepatocytes. J. Biol. Chem. 250, 8389.

Thomas, J.H., Dron, D.I. and Jones, R.H. (1975). Immunospecifity studies with chemically modified insulins. Diabetologia 11, 379.

Tompkins, C.V., Jones, R.H. and Sönksen, P.H. (1974). Effects of structural alterations in the insulin molecule on its effect in regulating glucose production and utilization in vivo. Diabetologia 10, 389.

Vogt, H.-P. (1976). Faltungsstudien am Proinsulin-C-Peptid und L,L-Diaminosuberoylinsulin im Hinblick auf die Struktur des Proinsulins. Thesis TH Aachen.

Weimann, J. (1974). Versuche zur Vernetzung von Insulin mit optisch inaktiver α,ε-Diaminopimelinsäure unter Verwendung der p-Biphenylisopropyloxycarbonylschutzgruppe. Dipl.Arbeit TH Aachen.

Wollmer, A., Brandenburg, D., Vogt, H.-P. and Schermutzki, W. (1974). Reduction/reoxidation studies with crosslinked insulin derivatives. Hoppe-Seyler's Z. Physiol. Chem. 355, 1471.

Wold, F. (1972). Modification reactions. Bifunctional reagents. Methods in Enzymology 25B, 623.

Yip, C.C. (1972). Preparation of ^{3}H-insulin and its binding to liver plasma membrane. Insulin Action. Acad. Press, Now York and London, 115.

Zahn, H. und Meienhofer, J. (1958). Reaktionen von 1,5-Difluor-2,4-dinitrobenzol mit Insulin. 2. Mitt. Versuche mit Insulin. Makromol. Chem. 26, 153.

Zahn, H. und Schade, F. (1963). Notiz über Nitrophenylester. Chem. Ber. 96, 1747.

Zervas L. and Photaki I. (1962). On cysteine and cystine peptides. I. New S-protecting groups for cysteine. J. Amer. Chem. Soc. 84, 3887.

17

THE ENZYMIC DERIVATION OF CITRULLINE RESIDUES FROM ARGININE RESI-
DUES IN SITU DURING THE BIOSYNTHESIS OF HAIR PROTEINS THAT ARE
CROSS-LINKED BY ISOPEPTIDE BONDS

George E. Rogers and Lindsay D. Taylor

Department of Biochemistry, University of Adelaide

South Australia, 5001, Australia

I. ABSTRACT

An enzymic activity present in hair follicles is described
that can convert arginine residues to citrulline residues in
proteins in situ. The Ca^{2+} dependent enzyme activity has been
detected in hair follicle extracts but not in similar extracts of
serum, liver or brain. The enzyme appears to act on proteins
other than hair proteins and the citrulline produced can be
quantitated in acid hydrolysates by a colorimetric procedure.
The formation of citrulline has been confirmed by amino acid
analysis and does not appear to be related to the formation of
isopeptide linkages which is catalysed by the transamidase present
in hair follicles.

II. INTRODUCTION

The proteins of the medulla and inner root sheath cells of
hair follicles (Figure 1a) were shown to be distinct from keratin
(Rogers, 1958, 1962). One of the major distinctions is that they
contain citrulline, not previously described as a constituent of
proteins. In addition they contain relatively large amounts of
glutamic acid and are cross-linked with isopeptide bonds instead
of cystine. Ever since these findings were reported (Rogers,
1962; Steinert, Harding and Rogers, 1969; Harding and Rogers,
1971, 1976) attempts have been made to elucidate the origin of the
citrulline residues to give some explanation for the apparently

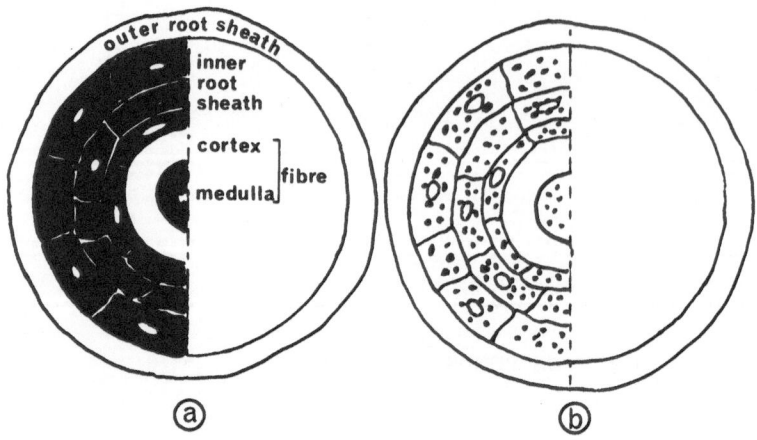

Figure 1. (a) Diagrammatic representation of a hair follicle in
cross-section showing the arrangements of the different cell
layers at a level at which all of the cell contents are hardened
(keratinized in the case of the hair cortex).   It is to be noted
that the cells of the inner root sheath are separated from those
of the medulla by the hair cortex. (b) Similar to (a) but at a
lower level (lower third) of the follicle at which the cells of the
inner root sheath and medulla are not hardened but contain granules
of trichohyalin.

unique occurrence of this amino acid in peptide linkage.

     Early experiments showed that citrulline was derived directly
by modification of arginine at some stage in protein synthesis
(Allen, Lindley and Rogers, 1964).   Recent evidence suggests that
this modification occurs in a protein precursor of the medulla and
inner root sheath (Rogers and Harding, 1976) the material called
trichohyalin (Vörner, 1903) which is abundant in'the cells in the
basal regions of the hair follicle (Figure 1b).   A protein
fraction has been isolated from hair follicles which has
characteristics consistent with it having originated from trichohyalin.   Its amino acid composition is distinct from that of α-
keratin, but it is similar to that of the mature protein of the
medulla and inner root sheath cells except that the arginine content is abnormally high, citrulline is absent and no isopeptide
cross-links are detectable (Table 1).   The protein fraction
appears by polyacrylamide gel electrophoresis to consist of at
least three major molecular weight classes ranging from 50,000 to

TABLE 1

Amino Acid Composition of Hair Follicle Proteins: Comparison of
Medulla and Inner Root Sheath with a Presumptive Trichohyalin (TR)
Fraction and with Hair Keratin

(Residues per 1000 Amino Acid Residues)

Amino acid	Combined[a] medulla and inner root sheath	Trichohyalin protein fraction[b]	Keratin[c]
Cys[d]	23	10	160
Asp	81	62	52
Thr	41	28	59
Ser	75	38	97
Glu	233	291	127
Pro	0	30	52
Cit	67	0	0
Gly	69	41	84
Ala	57	39	50
Val	41	31	54
Met	8	9	0
Ile	30	21	31
Leu	100	108	64
Tyr	24	21	30
Phe	27	30	34
Lys	64	55	24
His	14	14	12
Arg	44	174	71
Isopeptide	Present	Absent	Trace

[a] Total protein solubilised by tryptic digestion of guinea pig hair
follicle residues after extraction with 8 M urea (Rogers and
Harding, 1976).

[b] The protein fraction solubilised from guinea pig hair follicles
by 8 M urea, eluted from DEAE-cellulose at 50 mM KCl and insoluble
in water (Rogers and Harding, 1976).

[c] Total S-carboxymethyl keratin extracted from guinea pig hair by
standard methods (see Gillespie and Inglis, 1965).

[d] Determined as the S-carboxymethyl derivative.

200,000 (Rogers and Harding, 1976 and unpublished observations).

The total protein fraction referred to as TR protein (trichohyalin protein) has been prepared in previous studies in a radioactive form by in vivo labelling of guinea pigs using [^{14}C-
guanidino]arginine. A post-ribosomal supernatant prepared from
hair follicle homogenates was found to convert about 10 per cent
of the arginine residues to citrulline when incubated with TR
protein in vitro for up to 3 hr (Rogers and Harding, 1976 and
unpublished observations). The derived citrulline was detected
by the appearance of radioactivity in the citrulline peak eluted
from an amino acid analyzer. Examination of the radioactive
protein on polyacrylamide gels, after incubation with hair
follicle extract, strongly suggested that isopeptide cross-linking
of the precursor was occurring in vitro (Rogers and Harding,
unpublished observations). Inhibition of the transamidase
activity responsible for cross-linking by addition of anti-factor
XIII (Harding and Rogers, 1972) did not prevent citrulline formation, suggesting two distinct enzymes for cross-linking and for
the formation of citrulline (Rogers and Harding, unpublished
observations).

In the arginine-citrulline conversions produced by hair
follicle extracts using radioactive TR protein as the substrate,
it has not been possible to assay citrulline formation other than
by the amino acid analyzer. The specific radioactivity of the
arginine residues in the TR protein was too low to separate and
detect the small amount of citrulline by high voltage paper
electrophoresis. The necessary loadings of protein hydrolysate
were too high for satisfactory separation of the neutral amino
acids containing the labelled citrulline.

Attempts have been made to use the Archibald colorimetric
procedure for the quantitation of citrulline in protein and the
results are described in the present paper. The method is
considerably less tedious than the [^{14}C-guanidino]arginine procedure and has enabled the presence of the arginine-converting
enzyme in hair follicle extracts to be confirmed and some
characteristics of the enzyme to be defined.

III. DETECTION OF ARGININE-CITRULLINE CONVERSION: THE ARGININE-CONVERTING ENZYME

A. Application of the Colorimetric Procedure for Citrulline
Determination Directly on Protein.

The procedure used the arginine-rich TR protein that was
isolated as a fraction from hair follicle proteins. The TR

protein, acting as the arginine-containing substrate, was incubated with an enzyme-containing hair follicle extract at pH 7.0 and in the presence of Ca^{2+} ions (see Table 2 for details).   After incubation, the protein (substrate plus enzyme protein) was recovered by precipitation with trichloracetic acid (TCA) and any citrulline formed in situ was detected by the Archibald reaction (Table 2) as modified by Guthöhrlein and Knappe (1968).   The color reaction was developed directly from the protein.   Increments of absorbance at 490 nm were observed following incubation of TR protein and the follicle extract (Table 2) indicating the production of citrulline in peptide linkage.   It is to be noted that a trace of citrulline was formed in the absence of TR protein and must be the result of a conversion occurring in the hair follicle extract itself (see also Table 3 and Section IV).   No citrulline was produced when the follicle extract was heated.

Difficulties were experienced with this quantitation of citrulline carried out directly on the protein.   Non-specific color reactions occurred as reported by early workers (Gornall and Hunter, 1941) and trace amounts of suspended material developed. These factors obscured the color development due to citrulline and were the cause of the relatively low citrulline values shown in Table 2.   Hence the procedure was modified as described below.

B.   Determination of Citrulline by Application of the
     Colorimetric Procedure to Protein Hydrolysates

The accuracy of the citrulline determination procedure was greatly improved when the protein, recovered after incubation with follicle extract by TCA precipitation, was hydrolysed completely to amino acids (Table 3).   The hydrolysate was evaporated in vacuo to dryness and the citrulline color then developed as described in Table 2.   The formation of about 50 nmoles citrulline per mg of TR as substrate in 3 hr was demonstrated (Table 3).   The citrulline values are minimal values since they are uncorrected for the hydrolytic destruction to ornithine which is around 20% (Steinert, Harding and Rogers, 1969).   It can be seen from the Table that some citrulline is present initially in the TR substrate and in addition there was some conversion of protein-bound arginine to citrulline in the absence of substrate.   The absolute dependence upon Ca^{2+} ions, not replaceable by Mg^{2+}, is also evident.

TABLE 2

Arginine-citrulline Conversion Catalysed by Hair Follicle Enzyme:
Assay of Citrulline Directly on Protein

Incubation conditions[a]	Citrulline formed[b] nmoles
TR protein (substrate)	
Heated[c] enzyme	0
0 hr	0
1 hr	5.5
3 hr	11
No substrate	
0 hr	0
3 hr	0.5

[a](i) Hair follicle enzyme: prepared by the method described by
        Harding and Rogers (1972) for transamidase, i.e. a
        100,000 x g supernatant from a homogenate of guinea pig
        hair follicles prepared in 5 mM Tris-HCl, 2 mM EDTA,
        pH 7.0.

  (ii) TR protein: the presumptive trichohyalin-derived protein
        was a protein fraction isolated from an S-carboxymethylated
        8 M urea extract of hair follicles and which eluted from
        DEAE-cellulose at 50 mM KCl concentration (Rogers and
        Harding, 1976).

 (iii) Incubation mixture: contained 1 mg substrate protein, 1 mg
        enzyme protein, 100 mM Tris-HCl, pH 7.0, 10 mM cysteine,
        10 mM Ca^{2+} in 1.5 ml.   Incubated at 37° and the reaction
        was stopped by addition of 1.5 ml 10% trichloracetic acid
        (TCA).   The precipitate was sedimented by centrifugation,
        washed twice with 5 ml 5% TCA, once with 5 ml 95% ethanol
        and dried in vacuo.   Protein determinations were carried
        out using the procedure of Lowry, Rosebrough, Farr and
        Randall (1951).   Experimental values for citrulline
        conversion are the average of duplicates.

[b]Citrulline assay: the Archibald reaction as modified by
Guthöhrlein and Knappe (1968) was used and its sensitivity
increased by reducing the volumes of reagents to one-third of
those adopted by those authors.   Thus 0.3 ml of solution to be

Table 2 (Continued)

assayed for citrulline content was mixed with 0.375 ml acid mixture (H_3PO_4 : H_2SO_4 : H_2O in a ratio of 3:1:2), 0.075 ml "redox
buffer" (9% ^{W}/v $NH_4Fe(SO_4)_2$ ·$12H_2O$ and 11% ^{W}/v $(NH_4)_2Fe(SO_4)_2$
·$6H_2O$ in 0.5 M H_2SO_4) and 0.15 ml 0.75% (^{W}/v) diacetylmonoxime
giving a total volume of 0.900 ml.   This was heated in capped
tubes for 20 min at 100°.   The absorbance of the developed color
was measured at 490 nm.

In the results of Table 2 only, the protein from the incubation
mixture was made to a volume of 0.3 ml (H_2O) and dissolved in the
acid mixture by heating the tubes at 60° for 15 min before adding
the remainder of the reagents.

[c]100°C for 5 min.

TABLE 3

Arginine-citrulline Conversion Catalysed by Hair Follicle Enzyme:
Assay of Citrulline after Acid Hydrolysis of Protein

Incubation conditions[a]	Citrulline[b] nmoles		Citrulline increase nmoles
	0 hr	3 hr	
TR protein (substrate)			
in the presence of:			
10 mM Ca^{2+} (normal incubation, see foot-note Table 2)	44	103	59
10 mM Mg^{2+} (replacing 10 mM Ca^{2+})	44	40	0
No enzyme	22	19	0
No substrate	39	51	12

[a]See Table 2 for details.

[b]After incubation, the protein was recovered by TCA precipitation
as described in Table 2 but the protein was hydrolysed in 6 M HCl,
110° for 16 hr, and dried in vacuo.   The determination of citrulline was then carried out as described (Table 2).

IV.   ARGININE-CITRULLINE CONVERSION IN VITRO USING PROTEINS
OTHER THAN THE TRICHOHYALIN PROTEIN

The action of the arginine-converting enzyme might rely on
the recognition of an amino acid sequence containing one or more
arginine residues.   Specific binding to such a sequence might be
necessary before derivation of a citrulline residue can be brought
about.   Alternatively, the major specificity requirement may be
only peptide-bound (i.e., blocked) arginine residues.   Preliminary experiments have not shown direct conversion of free
arginine to citrulline although this amino acid can be produced
by follicle extracts via the urea cycle (unpublished observations).

Two proteins distinct from the natural substrate, trichohyalin protein, were tested as substrates namely, hair keratin,
solubilised as the S-carboxymethylated derivative and a histone
fraction with a high arginine content prepared from chick nuclear
chromatin.   In addition, polyarginine sulphate was tested.
The results are given in Table 4 and it can be seen that not only
was citrulline produced in every one of these substrates but the
amount formed was comparable to that produced from the natural
substrate.

To substantiate the conclusions from colorimetric analysis,
the production of citrulline by the action of hair follicle
extracts was examined in a separate experiment using amino acid
analysis in parallel with the colorimetric procedure (Table 5).
Although the incremental values for the amount of citrulline
formed were not identical from the two procedures, they were
sufficiently close to conclude that approximately 50 nmoles of
citrulline were produced from all of the three proteins when they
were incubated in the presence of follicle extract.   Furthermore,
amino acid analysis confirmed that the proteins in the follicle
extract itself (no substrate present) can undergo conversion of
some of their arginine residues to citrulline residues.

V.   PROPERTIES AND OCCURRENCE OF THE ARGININE-CONVERTING ENZYME

To the present time, the arginine-converting enzyme has only
been detected in hair follicle tissue.   No enzyme has been
detected in serum (guinea pig) or in high-speed supernatants
prepared from guinea pig liver and brain (data not presented).
Serum and liver were chosen because they are known sources of
transglutaminase activity (Loewy, Matačić and Darnell, 1966;
Clarke, Mycek, Neidle and Waelsch, 1959) and brain (white matter)
was tested in view of a recent report that citrulline is present
in myelin proteins (Finch, Wood and Moscarello, 1971).

TABLE 4

The Derivation of Citrulline from Arginine in Several Proteins and
Polyarginine by the Action of Hair Follicle Enzyme

Substrate[a] (arginine content, moles per cent, in parenthesis)	Citrulline nmoles		Citrulline increase nmoles
	0 hr	3 hr	
No substrate	20	29	9
TR protein (13.4)	23	63	40
S-carboxymethylated hair keratin[b] (6.7)	28	71	43
Histone[c] (10.2)	29	63	34
Polyarginine[d]	15	53	38

[a]TR protein, hair follicle enzyme and incubation conditions were
as described in Table 2. Citrulline was assayed by the procedure
described in Tables 2 and 3, i.e., after acid hydrolysis of the
product recovered from incubation.

[b]S-carboxymethylated keratin was prepared from washed albino
guinea pig hair essentially according to the method of Gillespie
and Inglis (1965).

[c]Histone was a fraction precipitated by 5% perchloric acid from a
total histone preparation isolated from chick cell nuclei according to the method of Nelson and Yunis (1969).

[d]Polyarginine sulphate Sigma product M.W. 15,000 - 50,000.

The optimum pH of the follicle enzyme is close to 7. It has
an absolute dependence on Ca^{2+} with an optimum of 15 mM. No
other co-factor dependency has yet been detected. It may not be
an -SH enzyme since cysteine did not activate and some 20%
inhibition was observed at 20 mM concentration in the present
experiments. Ammonia, presumably the other product of the conversion, does appear to produce some inhibition at 100 mM concentration (up to 50%). Glutamine, which by analogy with the
isopeptide cross-linking reaction might be suspected as participating,

TABLE 5

Comparison of Colorimetric and Amino Acid Analysis for the Assay
of Citrulline Produced by the Action of Hair Follicle Enzyme on
Several Proteins

Substrate[a]	Citrulline (nmoles)					
	Colorimetric[b]			Amino acid analysis[c]		
	0 hr	3 hr	Increase	0 hr	3 hr	Increase
No substrate	13	35	22	10	30	20
TR protein	26	75	49	20	68	48
S-carboxy- methylated keratin	24	73	49	25	90	65
Histone	19	76	55	<11	86	>75

[a]TR protein, S-carboxymethylated keratin and histone were prepared
as described in Tables 2 and 4.   Hair follicle enzyme and incubation conditions were as given in Table 2.

[b]Colorimetric assay is described in Tables 2 and 3.

[c]Technicon $21^1/_2$ hr analysis system adapted to a Beckman 120c
analyzer.

had no effect when tested at concentrations up to 25 mM.

VI. DISCUSSION

Sufficient preliminary work has now been carried out to confirm the existence of an enzymic activity in hair follicles
which catalyses the derivation of citrulline residues from arginine
residues in situ.   This reaction and the formation of isopeptide
links apparently are coincident processes occurring in vivo
specifically in the medulla and inner root sheath cells.   However, there is no evidence that the processes are interlinked and
they appear to be brought about by distinct enzymes.   Presumably
the arginine-converting enzyme is confined to the cells of the
medulla and inner root sheath of the hair follicle, since it has
been shown to act on solubilised keratin, but no citrulline

residues are to be found in keratin formed in vivo.

Much remains to be done to clarify the understanding of the properties and function of this enzymic activity.   For example -

(a)   can the enzyme be purified to high specific activity from hair follicle extracts?

(b)   what is the reaction mechanism for the conversion?

Even though some progress has been made by utilising the colorimetric procedure for determining citrulline, the analysis is still inadequate in sensitivity.   A simpler arginine-containing substrate is needed since the TR protein itself is a complex fraction which also demands separate study.   If a simpler substrate can be found, then it should be possible to devise a highly-sensitive radiochemical procedure for measuring rates of citrulline formation.

The later coalescence of further enzyme studies and protein chemical studies of the TR proteins should help in elucidating the significance of citrulline formation and isopeptide cross-linking in the process of hair and wool growth.   It is possible too, that the arginine-converting enzyme might have some application in protein chemistry.

VII.   ACKNOWLEDGMENTS

This work was supported by the Australian Wool Corporation and the Australian Research Grants Committee whose financial aid is gratefully acknowledged.

REFERENCES

Allen, A.K., Lindley, H. and Rogers, G.E. (1964). Metabolic relationships of protein-bound arginine and citrulline in hair follicles. VIth Int. Cong. Biochem. N.Y. Abstract Volume V-B-20.

Clarke, D.D., Mycek, M.J., Neidle, A. and Waelsch, H. (1959). The incorporation of amines into protein. Arch. Biochem. Biophys. 79, 338-354.

Finch, P.R., Wood, D.D. and Moscarello, M.A. (1971). The presence of citrulline in a myelin protein fraction. FEBS Letters 15, 145-148.

Gillespie, J.M. and Inglis, A.S. (1965). A comparative study of high-sulphur proteins from α-keratins. Comp. Biochem. Physiol. 15, 175-185.

Gornall, A.G. and Hunter, A. (1941). A colorimetric method for
    the determination of citrulline. Biochem. J. 35, 650-658

Guthöhrlein, G. and Knappe, J. (1968). Modified determination of
    citrulline. Anal. Biochem. 26, 188-191.

Harding, H.W.J. and Rogers, G.E. (1971). ε-(γ-glutamyl)lysine
    cross-linkage in citrulline-containing protein fractions from
    hair. Biochemistry 10, 624-630.

Harding, H.W.J. and Rogers, G.E. (1972). Formation of the ε-(γ-
    glutamyl)lysine cross-link in hair proteins. Investigation
    of transamidases in hair follicles. Biochemistry 11,
    2858-2863.

Harding, H.W.J. and Rogers, G.E. (1976). Isolation of peptides
    containing citrulline and the cross-link ε-(γ-glutamyl)lysine
    from hair medulla protein. Biochim. Biophys. Acta 427,
    315-324.

Loewy, A.G., Matačič, S. and Darnell, J.H. (1966). Transamidase
    activity of the enzyme responsible for insoluble fibrin
    formation. Arch. Biochem. Biophys. 113, 435-438.

Lowry, O., Rosebrough, N., Farr, A. and Randall, R. (1951).
    Protein measurement with the Folin-Phenol reagent. J. Biol.
    Chem. 193, 265-275.

Nelson, R.D. and Yunis, J.J. (1969). Species and tissue specifi-
    city of very lysine-rich and serine-rich histones. Exptl.
    Cell Res. 57, 311-318.

Rogers, G.E. (1958). Some observations on the proteins of the
    inner root sheath of hair follicles. Biochim. Biophys. Acta
    29, 33-43.

Rogers, G.E. (1962). Occurrence of citrulline in proteins.
    Nature 194, 1149-1151.

Rogers, G.E. and Harding, H.W.J. (1976). The isolation of a
    unique protein fraction from hair follicle tissue and its
    significance in the structure of hair. Proc. Int. Wool
    Textile Res. Conf. Aachen 1975 in Schriftenreihe Deutsches
    Wollforschungsinstitut an der Technischen Hochschule Aachen.
    in press.

Steinert, P.M., Harding, H.W.J. and Rogers, G.E. (1969). The
    characterisation of protein-bound citrulline. Biochim.
    Biophys. Acta 175, 1-9.

Vörner, H. (1903). Ueber Trichohyalin Ein Beitrag zur Anatomie
    des Haares und der Wurzelscheiden. Dermatol. Z. (Berlin)
    10, 357-376.

18

THERMODYNAMICS OF CROSS LINKS

John A. Rupley, Robert E. Johnson, and Patricia H. Adams

Department of Chemistry, University of Arizona

Tucson, Arizona   85721

ABSTRACT

Comparison of the melting temperatures of native lysozyme and
a cross-linked ester derivative of lysozyme (Imoto and Rupley, 1973)
yielded a value of 5.5 kcal/mole for the free energy of stabilization
developed through forming the cross link and a value of zero for the
corresponding enthalpy. There is close agreement between experiment
and calculation from the statistical theory of polymers.

INTRODUCTION

Cross links in proteins can be either intramolecular or intermolecular. In either case, a principal function of cross links is
presumed to be stabilization of an organized structure. Estimates
of the stabilization due to cross-linking have been based on the
statistical theory of chain polymers (Schellman, 1955; Flory, 1956).
Because of the assumptions made in calculations of this sort, it
would be desirable to have an experimental value for the free energy
of stabilization developed through cross-linking.

Experiments that measure the thermodynamics of cross link formation are difficult to design. Figure 1 states the problem for an
intramolecular cross link, for which the disorganization process is
denaturation. One seeks to determine the change in the free energy
of unfolding associated with introduction of the cross link. There
is in principal no barrier to measuring the free energies of
reactions 2 and 3 of Figure 1. The difficulty lies in interpreting
the difference in free energy that presumably would be observed.
Specifically, we must have information about reactions 1 and 4 of

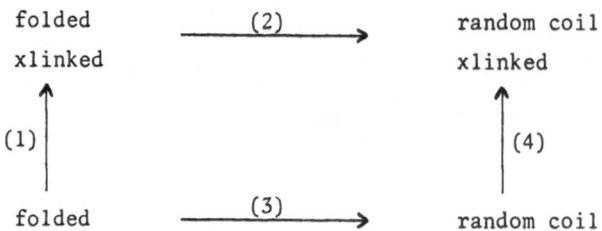

Figure 1. The problem can be simplified by ignoring the chemistry
of the cross-linking reaction. Thus in the present discussion we will
consider only the contributions to the thermodynamics from changes
in conformation that follow cross-linking. With regard to
reaction 1, one must have sufficient structural information for both
the cross-linked and non-cross-linked folded molecules to allow
an estimate of the cost of establishing the cross link in the
folded state. With regard to reaction 4, one must know which
particular chain elements are cross-linked, as a basis for interpretation and so that experiment can be compared to theory. Calcu-

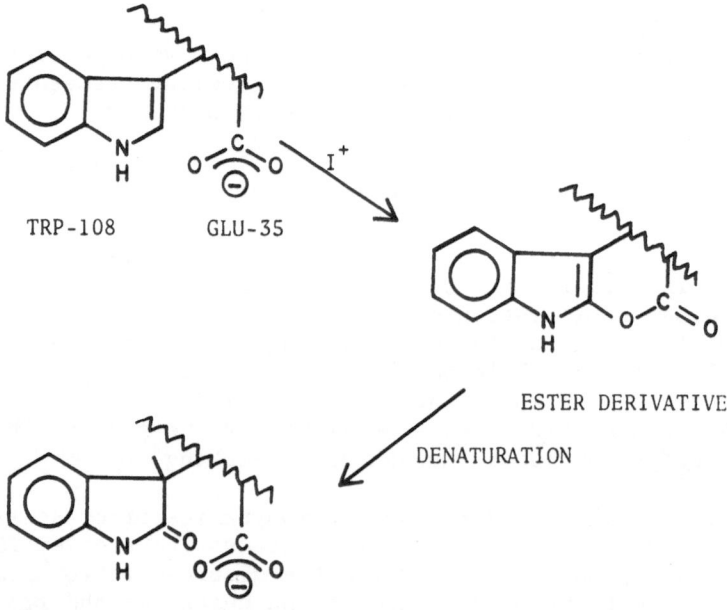

Figure 2. Oxidation of Lysozyme by Iodine

lations based on polymer statistics give an estimate of the entropy change associated with reaction 4, i.e. introduction of the cross link into a random coil polypeptide.

The experiments to be described were carried out with one particular cross-linked molecule for which we have the structural and chemical information needed to evaluate the effect of cross-linking on the thermodynamics of unfolding. More than twenty years ago it was observed that lysoyzme was inactivated by iodine (Fraenkel-Conrat, 1950). Subsequent study of this reaction using both chemical (Hartdegen and Rupley, 1973; Imoto and Rupley, 1973; Imoto, Hartdegen, and Rupley, 1973) and x-ray diffraction approaches (Beddell, Blake, and Oatley, 1975) showed that the product of iodine oxidation contains a curious cross link, an ester of the side-chain carboxylate of Glu-35 with the enol form of oxindoleala-nine 108. Figure 2 indicates the chemistry of this reaction. The ester derivative is labile. Denaturation of the cross-linked molecule results in hydrolysis of the ester to produce the free acid and oxindolealanine as products. Refolding of the molecule does not regenerate the ester. It is important for the present discussion that the crystallographic analysis showed that the establishment of the cross link resulted in no significant changes in structure of the molecule except for rotations about side-chain bonds of the two cross-linked residues. These rotations, which serve to bring a carboxylate oxygen of Glu-35 from van der Waals contact with the ring of Trp-108 to within covalent bonding distance of the δ_1 carbon of the ring, are the following: for Glu-35, 25° (α-β), 108° (β-γ), 45° (γ-δ); for Trp-108, 12° (α-β), 15° (β-γ). The rotations do not produce a high energy conformation nor do they reduce the number of configurations accessible to the folded molecule. Thus we can justify the assumption that for reaction 1 of Figure 1 there is no conformational contribution to the free energy, and for purposes of our discussion ΔG_1 is zero.

THERMODYNAMICS OF UNFOLDING OF THE ESTER DERIVATIVE OF LYSOZYME

The following approach was taken in the experiments described below that were designed with the aim of evaluating the effect of cross-linking on the stability of lysozyme. First, the melting temperature for the cross-linked molecule was determined. At this temperature, the free energy of unfolding (ΔG_2 of Figure 1) is zero. The free energy of unfolding of the native molecule was calculated for the same temperature and the same solvent (ΔG_3 of Figure 1). This calculation could be carried out owing to the detailed studies from Tanford's laboratory (Aune and Tanford, 1969a & b; Tanford and Aune, 1970) on the unfolding of lysozyme as a function of pH, temperature, and guanidine hydrochloride (GuHCl) concentra-

tion. Finally, the difference between the free energies of unfolding of native lysozyme and the cross-linked derivative were related
to the expectation from theory for the introduction of this
particular cross link into the lysozyme random coil.

The unfolding of the cross-linked ester derivative can be
monitored using one of the optical methods suitable for study of
protein denaturation. The change in uv absorbance spectrum associated with unfolding of the ester is a typical tryptophan difference
spectrum and is like that found for denaturation of native lysozyme,
except for the absence of the change above 300 nm associated with
perturbation of Trp-108. Because of the lability of the ester
cross link in the unfolded molecule, the absorbance spectrum of
the denatured molecule changes rapidly and irreversibly with time.
This is shown in Figure 3, which is a tracing of a Cary spectrophotometer record of absorbance changes following addition of
protein to 2M GuHCl at pH 2. The absorbance changes at 62.0° C show
a rapid jump followed by a change which is substantial even during
the first minute after mixing. At the lower temperature for which
data is given in Figure 3 and at which the ester intermediate is
folded, there is no time dependence of the absorbance and of course
little deviation from the reference absorbance (regions A of the
record of Figure 3). In order to extract the contribution of the
unfolding process to the absorbance changes and to eliminate the
contribution of subsequent covalent changes (conversion of the
ester to oxindole and carboxylate products), the data were extrapolated to the time of mixing, as indicated in Figure 3. In
justification of this extrapolation, the progress curve for the
change is first order, and the product of the reaction is by
spectrum the free oxindole.

Figure 4 gives the melting curve for the cross-linked derivative. Results from two separate experiments are described in
Figure 4, and there is close agreement between them. The scatter
in the data for higher temperatures relates to the extrapolation
described in the preceding paragraph.

Reversibility of the unfolding reaction was tested as follows.
A sample of the protein was held for a known period of time at
approximately 70° C, at which temperature unfolding of the ester
derivative is complete (Figure 4). The sample was then diluted
into solvent at low temperature. The absorbance spectrum of this
reversed sample was measured against a reference sample which had
not been exposed to high temperature. The data in Figure 5 show
that as expected there is an irreversible change that is presumably
associated with ester hydrolysis and that increases with the
period of time the protein was held at 70°. The important point
is that extrapolation to zero time at 70° gives the absorbance
of the control protein sample which had not been denatured. Thus

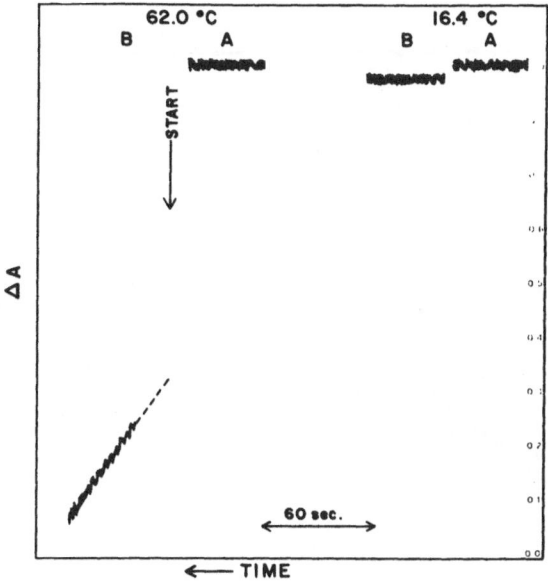

Figure 3. Temperature and time dependence of the absorbance at 292 nm of the cross-linked ester derivative.

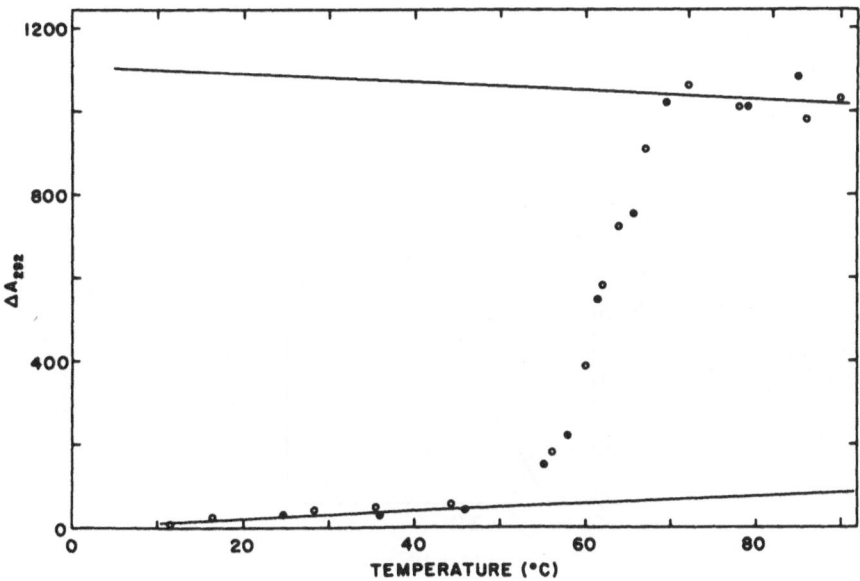

Figure 4. Melting curve for the ester derivative in 2M GuHCl at pH 2.

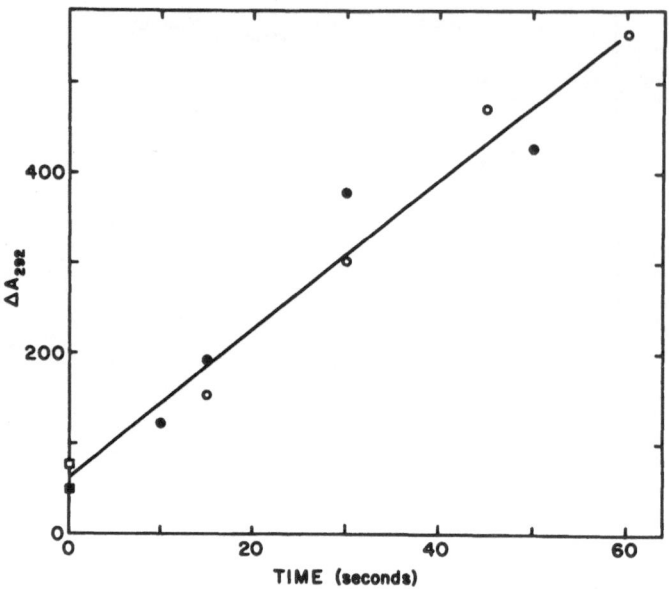

Figure 5.   Reversibility of the absorbance changes associated with
unfolding of the cross-linked ester derivative.

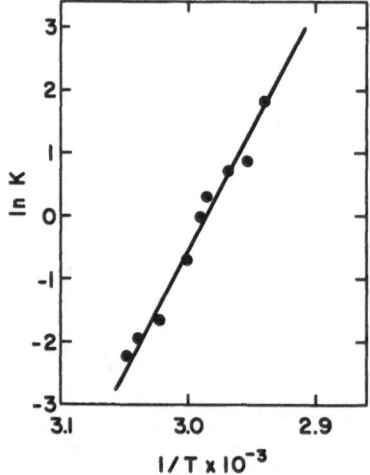

Figure 6.   Van't Hoff plot of the data of Figure 4.

we can conclude that the unfolding process is reversible.

Figure 6 is a van't Hoff plot of the data of Figure 4. The enthalpy of unfolding of the cross-linked molecule is 76.4 kcal/mole at the melting temperature, 61.8° C.

Table I summarizes the thermodynamic values for unfolding of native lysozyme and the ester derivative in 2M GuHCl at pH 2. The values calculated for native lysozyme are based on the results of Tanford and Aune (Aune and Tanford, 1969a & b; Tanford and Aune, 1970). In this work it is shown that the thermodynamics of unfolding of lysozyme can be separated into contributions that reflect the thermodynamics of unfolding in a reference solvent (water of low pH) and the effects on the thermodynamics associated with change in pH or change in guanidine hydrochloride concentration. These relationships, which are summarized in Table II, are needed to specify the thermodynamics of unfolding of the native molecule at a temperature much removed from its melting temperature. Without the work of Tanford and Aune, which required analysis of an extensive body of experimental data, it would not be possible to interpret the results described here on the thermodynamics of unfolding of the cross-linked molecule.

In order to test the agreement between the equations of Tanford and Aune and results obtained in this laboratory, measurements were made on the unfolding of native lysozyme in the same solvent used for the measurements with the cross-linked molecule. The top two lines of data of Table I show that the agreement between experiment and calculation is good (0.5 kcal/mole for the free energy and 1.5 kcal/mole for the enthalpy).

The melting temperature for the cross-linked molecule (61.8° C; Table I) is almost 30° greater than the melting temperature of native lysozyme. This corresponds to a change in free energy of unfolding of 5.5 kcal/mole (Table I). It is important that cross-linking produces no significant change in the enthalpy of unfolding (76 kcal/mole measured for the ester derivative and calculated for native lysozyme).

We believe that the free energy difference of 5.5 kcal/mole, from the comparison of the unfolding reactions of the cross-linked derivative and native lysozyme, can be equated to the free energy change that follows introduction of this particular cross link into the random coil molecule. We have already discussed the justification for assuming that the free energy is zero for reaction 1 of Figure 1 (the configurational contribution to the introduction of the cross link into the folded molecule). With regard to reaction 4 of Figure 1, Tanford's laboratory has examined the conformation of lysozyme and other proteins in guanidine hydrochloride solution

TABLE I

Thermodynamics of Unfolding of Native and
Cross-Linked Lysozyme in 2M GuHCl, pH 2

		T (°C)	ΔG^O (kcal/mole)	ΔH^O (kcal/mole)	ΔS^O (e.u.)
Native	Experimental	32.4	0	+46.7	+153
	Calculated*	(32.4)	+0.4	+48.2	+157
Cross-linked	Experimental	61.8	0	+76.4	+228
Native	Calculated*	(61.8)	-5.5	+76.2	+244

*Aune and Tanford, 1969a & b; Tanford and Aune, 1970

TABLE II

Relationships Used to Calculate the Equilibrium Constant
(Free Energy) and Enthalpy of Unfolding for Native
Lysozyme (Aune and Tanford, 1969a & b; Tanford and Aune, 1970)

$$\underline{K_{ND} = K^O_{ND} \; F(a_H)(1 + 1.2 \; a_{\pm})^{14}}$$

$$\log K^O_{ND} = -1365.2 + 478.060 \log T + \frac{52.8881 \times 10^3}{T}$$

$$F(a_H) = \frac{(1 + 10^{-3.9}/a_H)(1 + 10^{-4.4}/a_H)}{(1 + 10^{-1.9}/a_H)(1 + 10^{-1.9}/a_H)}$$

$$\log a_{\pm}^2 = -0.5191 + 1.4829 \log C - 0.2562(\log C)^2 + 0.5884(\log C)^3$$

$$\underline{\Delta H^O_t = \Delta H^O_{25°} + \int_{25°}^{t} \Delta C_p \, dt}$$

$$\Delta H^O_t = +41.2 \text{ kcal/mole}$$

$$\Delta C_p = +950 \text{ cal/deg-mole}$$

(Tanford, 1968). These studies show that guanidine hydrochloride is a "good" solvent in which polypeptide chains behave as random coils. It is also important that the enthalpies of unfolding of the cross-linked and non-cross-linked molecules are identical (Table I). Thus one can assign the free energy difference of 5.5 kcal/mole to an entropy effect and to reaction 4 of Figure 1. This is the entropy contribution which can be directly estimated from the statistical theory of polymers, which will be considered in the following section. As additional justification for equating the free energy difference between the unfolding reactions of the cross-linked and non-cross-linked molecules to an entropy effect, we note calorimetric measurements made on the enthalpy of disulfide bond reduction for lysozyme (Lapanje and Rupley, 1973), which showed that there is no significant enthalpy contribution associated with loop formation.

THEORY

The intent of the discussion of this section is to compare the experimental value for the change in free energy associated with introduction of a cross link into lysozyme, described in the preceding section, with a theoretical estimate for the configurational contribution to the free energy change of reaction 4 of Figure 1, i.e. the introduction of a cross link into the random coil lysozyme molecule. The approach most suitable for the present case is that of Flory (1956), who considered the introduction of a cross link into a polymer already containing cross links. In this regard, native lysozyme has four disulfide bridges, with the cysteine residues closest to Glu-35 being residues 30 and 64, and those closest to Trp-108 being 94 and 115.

Flory's method is described in Figure 7 and Table III. The new cross link joining chain elements A and A' is to be introduced between other elements of the chain already part of cross links, i.e. elements C_1 and C_2, and C_3 and C_4, respectively. Flory (1956) estimated the reduction in configurational probability associated with formation of the (A,A') cross link as follows: (a) Both chains are cut, at A or A'. (b) The chain segments between C_j and the cut (at A or A') are treated as Gaussian chains with end C_j fixed at its mean position. (c) The probability of formation of the cross link is the product of the probabilities that all four end-to end vectors (C_j to A or A') meet in the same volume element $\underline{d\tau}$, with integration over all space. Equation (1) of Table III expresses this relationship, divided by the probabilities that the several ends pair up to give the original unsevered chains. The function $\underline{P}$ of Table III is therefore the change in configurational probability associated with formation of the cross link. (d) Integration of equation (1) and appropriate averaging of parameters

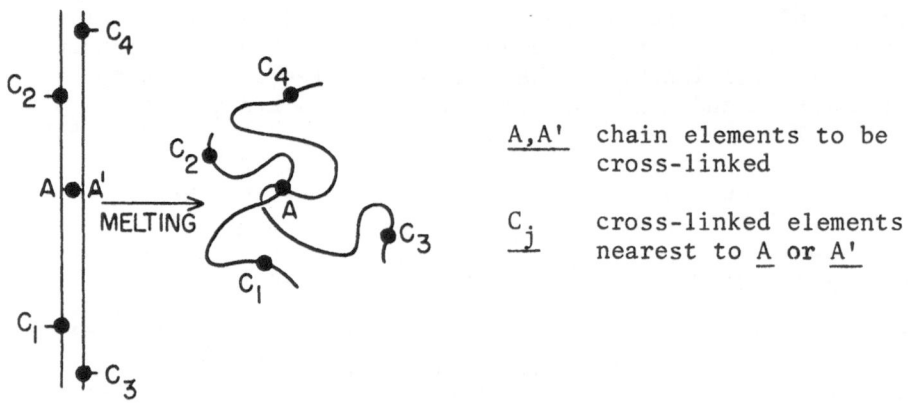

Figure 7. From Flory (1956)

TABLE III

Relationships Used to Estimate the Thermodynamics of
Reaction 4 of Figure 1. From Flory (1956)

$$P = \Delta\tau \; \frac{\int W_1(r_1)W_2(r_2)W_3(r_3)W_4(r_4)d\tau}{\int W_1(r_1)W_2(r_2)d\tau \; \int W_3(r_3)W_4(r_4)d\tau} \qquad (1)$$

$$= \Delta\tau \; \left(\frac{3}{2\pi e^2 1^2}\right)^{3/2} \left(\frac{(1/n_1 + 1/n_2)(1/n_3 + 1/n_4)}{(1/n_1 + 1/n_2 + 1/n_3 + 1/n_4)}\right)^{3/2} \qquad (2)$$

$\Delta\tau$ volume containing A and A' in ordered state

r_j vector from C_j to $d\tau$

W_j Gaussian probability function for end-to-end vector

1 length of statistical element of chain

n_j number of statistical elements between C_j and A,A'

(described by Flory, 1956) gives equation (2), into which values
appropriate for a particular cross link can be substituted.

Flory's (1956) discussion was based on the need to understand
the properties of polymer networks, i.e. cross-linked fibers like
collagen. Thus he generalized equation (2) to accomodate such
systems, and he related statistical and phenomenological parameters.

Application of equation (2) to estimating the entropy change
associated with reaction 4 of Figure 1 is straightforward. We take
the volume $\Delta\tau$ to be that of a sphere of diameter 7.5 Å, the distance
between the C_α carbons of residues 35 and 108; 1 is set at 3.8 Å,
the $(C_\alpha-C_{\alpha+1})$ distance along the polypeptide chain; the values
of n_i are 5, 29, 14, and 7, respectively. Substitution into
equation (2) gives ΔS_4 =   R ln P as -11.9 e.u. and ΔG_4 as 4.0 kcal/
mole.

The above estimate of 4 kcal/mole for ΔG_4 of Figure 1 is in
reasonable agreement with the experimental estimate of 5.5 kcal/mole.
The following two facts bear on this comparison: (a) a decrease
of $\Delta\tau$ would raise the theoretical estimate of ΔG_4; (b) the most
obvious source of error in the experimental estimate of ΔG_4 is that
ΔG_1 may be greater than zero, which would result in the 5.5 kcal/mole
value being lower than the true value.

CONCLUSIONS

Three points brought forward in the above discussion deserve
emphasis. (a) It is possible to estimate through experiment the
contribution a cross link makes to stabilization of an organized
structure. Arriving at such an estimate requires, however, a
substantial background of structural and chemical information.
(b) The stabilizing effect of a typical cross link is large,
about 5 kcal/mole. (c) There is remarkably good agreement
between experiment and simple statistical theory for the
thermodynamics of cross link formation.

REFERENCES

Aune, K.C., and Tanford, C. (1969). Thermodynamics of the
     Denaturation of Lysozyme by Guanidine Hydrochloride. I.
     Dependence on pH at 25°. Biochemistry, 8, 4579.

Aune, K.C., and Tanford, C. (1969). Thermodynamics of the Denatura-
     tion of Lysozyme by Guanidine Hydrochloride. II. Dependence
     on Denaturant Concentration at 25°. Biochemistry, 8, 4586.

Beddell, C.R., Blake, C.C.F., and Oatley, S.J. (1975). An X-ray Study of the Structure and Binding Properties of Iodine-Inactivated Lysozyme. J. Mol. Biol., 97, 643.

Flory, P.J. (1956). Theory of Elastic Mechanisms in Fibrous Proteins. J. Am. Chem. Soc., 78, 5222.

Fraenkel-Conrat, H. (1950). The Essential Groups of Lysozyme, With Particular Reference to Its Reaction with Iodine. Arch. Biochem., 27, 109.

Hartdegen, F.J., and Rupley, J.A. (1973). Oxidation of Lysozyme by Iodine: Identification of Oxindolealanine 108. J. Mol. Biol., 80, 649.

Imoto, T., and Rupley, J.A. (1973). Oxidation of Lysozyme by Iodine: Identification and Properties of an Oxindolyl Ester Intermediate; Evidence for Participation of Glutamic Acid 35 in Catalysis. J. Mol. Biol., 80, 657.

Imoto, T., Hardegen, F.J., and Rupley, J.A. (1973). Oxidation of Lysozyme by Iodine: Isolation of an Inactive Product and Its Conversion to an Oxindolealanine-Lysozyme. J. Mol. Biol., 80, 637.

Lapanje, S., and Rupley, J.A. (1973). Enthalpy of Reduction of Disulfide Cross Links in Denatured Lysozyme. Biochemistry, 12, 2370.

Schellman, J.A. (1955). The Stability of Hydrogen-Bonded Peptide Structures in Aqueous Solution. Compt.-rend. Lab. Carlsberg. Ser. Chim., 29, 230.

Tanford, C. (1968). Protein Denaturation. In "Advances in Protein Chemistry." Volume 23, p. 122, Academic Press, New York.

Tanford, C., and Aune, K.C. (1970). Thermodynamics of the Denaturation of Lysozyme by Guanidine Hydrochloride. III. Dependence on Temperature. Biochemistry, 9, 206.

19

PHYSICAL AND CHEMICAL CONSEQUENCES OF KERATIN CROSSLINKING, WITH APPLICATION TO THE DETERMINATION OF CROSSLINK DENSITY

Emory Menefee

Western Regional Research Laboratory, Agricultural Research Service, U.S. Department of Agriculture, Berkeley, California 94710

ABSTRACT

The high levels of covalent disulfide crosslinking in keratins strongly affect (1) structural stability, (2) viscoelasticity, and (3) chemical reactivity. This paper briefly reviews recent work on these subjects, with critical emphasis on methods by which chemical and physical properties can be related to inter- and intra-molecular crosslink density in heterogeneous systems like keratins. Detailed attention is drawn to effects of crosslinking on the hydrolysis of keratin by acids or enzymes. Within the limits of reasonable assumptions, it is possible to account quantitatively for crosslink dependent variations in the hydrolysis rate of different keratins, and also to derive a formula for calculating the absolute intermolecular crosslink density from the amount of keratin dissolved after partial hydrolysis and the number of chain ends appearing in the soluble fraction.

INTRODUCTION

Keratins appear in one or another of several forms in all vertebrate phyla: mammals, birds, reptiles, amphibia, and fish. Their most obvious role is as a mechanically tough protective coat that is insoluble and resists degradation. Fraser et al. (1972) have reviewed the evolutionary development of keratins, and cite evidence that they have existed for at least 450 million years, an estimate based on the presence of a keratin-like protein in a genus of lamprey eels (Rudall, 1947) extending from that time. For several reasons, including genetic modification, there is considerable variation in the composition of keratins from different sources (Bradbury et al., 1970;

307

Bradbury, 1973; Baden et al., 1973). On the average, however,
in order of decreasing abundance, the preponderant amino
acids are cystine, glutamic acid, serine, glycine, and proline.
One main characteristic of the keratins that distinguishes them
from other structural proteins is the presence of extensive
intermolecular cystine crosslinking. Even though other natural
crosslinks have been reported (Zahn et al., 1969; Hinton, 1974),
it is the disulfide link that essentially defines a keratin. The
density, spatial distribution, and kind (whether intra- or intermolecular) of disulfide crosslinks are important factors in determining whether the keratin is of the "soft" type such as that
in stratum corneum, or "hard" as in wool or human hair, though
crosslinks are by no means the only factors differentiating these
types.

Understanding the nature of crosslinking in wool and human
hair and how it can be manipulated has great technical importance. Furthermore, the connection between the structure of a
keratin and the physiological condition of the animal producing
it is of increasing interest (Sims, 1970).

Physical and chemical properties of wool have been more
thoroughly studied that those of other keratins, for obvious commercial reasons. General reviews of keratin structure and
composition include those of Bradbury (1973) and Fraser et al.
(1972). Reviews dealing more specifically with relations between structure and mechanical behavior are those of Bendit and
Feughelman (1968) and Chapman (1969). However, a brief review of certain aspects of keratin structure at this point may
help clarify the role of crosslinking.

Figure 1, taken from Menefee (1968), shows a concept of
keratin fiber microstructure schematically diagrammed in a
longitudinal section about 25 nm across. The heavy vertical
lines represent protofibrillar strands of 2 or 3 alpha helical
protein chains, which are embedded in a less well-ordered,
possibly globular, matrix (G). These two portions comprise a
mutually interdependent composite in which some helical chains
(H_2) are organized axially into clusters of lightly crosslinked
regions (microfibrils), while others (H_1) are though to be contained in the matrix in a more highly stabilized and crosslinked
form. The matrix resists dimensional changes in extension,
but not radially or in compression, whereas the microfibrils
resist dimensional change in compression and radially, but are
readily extensible. This model can be made to account for various observed mechanical properties, as well as their change
after treatments that affect crosslinking, helix content, or
secondary interactions (Menefee, 1968, 1971). A strong implication of the model is the near impossibility of affecting the
properties of one component of a fibrous keratin without in some

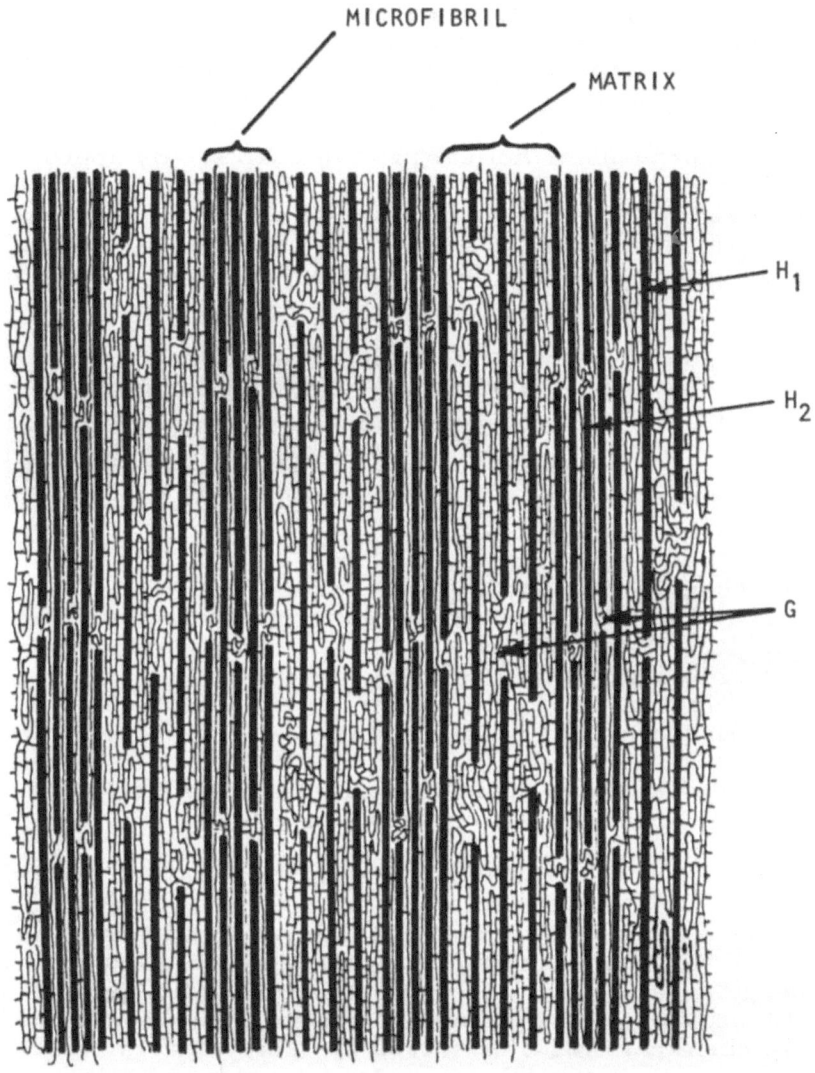

Figure 1. Representation of fibrous keratin structure in a section about 25 mm wide. Vertical bars are protofibrils containing 2 or 3 alpha helical protein chains. H_1 is high melting helical protein in matrix. H_2 is low melting helical protein in microfibrils. G represents crosslinked partially oriented amorphous (possibly globular) matrix protein.

way affecting the others. Thus, when some disulfide crosslinks
in the matrix region are reduced, this change will decrease the
stability of the helical protein chains within the microfibrils
(Menefee and Yee, 1965a), lowering the helix disordering temperature. Still, structural interdependence should not be con-
strued to mean that it is impossible to ascribe specific chemical and physical effects to such well defined entities as crosslinks, and amorphous or helical protein chains. By appropriate
mathematical analysis of the composite structures of different
kinds of keratin fiber, and by selectively subjecting portions of
various fibers to chemical change, it is possible to obtain a
fairly convincing picture of the component behavior in wool and
other fibrous keratins.

Examination of experimental studies shows that most functions of crosslinking in keratins can be grouped into the following categories:

1. Structural stability
   a. Memory effects
   b. Helix stability
   c. Swelling limits
2. Mechanical properties
   a. Deformation in tension, compression and torsion
   b. Relaxation effects
3. Chemical stability
   a. Dissolution of components by crosslink scission
   b. Hydrolysis of peptide bonds

In the following sections these categories will be discussed in
relation to the specificity of crosslinking effects.

STRUCTURAL STABILITY

Memory effects. A primary function of keratin crosslinking
seems to be that of keeping the structure intact. Keratin fibers
can undergo great elongation, up to 100% in water, and still exhibit full recovery, if the extension is done under conditions that
minimize chemical crosslink scission. This behavior contributes to the excellent (though often slow) wrinkle recovery properties of wool fabrics. On the other hand, if the disulfide
crosslinks are chemically treated to allow breakage under strain
and then reformation, the fiber will be set into the new configuration. The kinetics of setting and long-term relaxation in keratins seem to be governed largely by sulfhydryl-disulfide interchange (Caldwell et al., 1965; Feughelman, 1966; Weigmann et
al., 1966; Weigmann and Dansizer, 1969, 1971a; Asquith and
Puri, 1970). In this process, when strain is applied to a fiber
in an appropriate aqueous environment, sulfhydryl groups react
with adjacent disulfide bonds, yielding new disulfide bonds and

sulfhydryls, but under less strain. Although the net number of crosslinks may be constant during an interchange, the reaction can be made to occur more rapidly by increasing the pH, or producing more free sulfhydryl groups initially by partial reduction.

The interchange view of setting is widely accepted. However, evidence is accumulating that the formation of additional crosslinks such as lanthionine and lysinoalanine is also important (Orwell et al., 1966; Robson et al., 1969), since these and similar crosslinks have been shown to form under conditions conductive to setting (Ziegler, 1964; Miró and García-Domínguez, 1966, 1967a, b, 1968; Sweetman, 1967; Crewther et al., 1967; Corfield et al., 1967; Asquith and García-Domínguez, 1968a, b; García-Domínguez et al., 1971; Otterburn, 1975).

Structural stabilization also may be achieved by purposefully creating additional crosslinks, either by heating (Menefee and Yee, 1965a; Asquith and Otterburn, 1970; Watt, 1975), by radiation (Beevers and McLaren, 1974), or by treatment with formaldehyde or other crosslinking reagents (Griffith and Mason, 1966; Weigmann and Dansizer, 1971b; Watt, 1971; Zahn et al., 1971; Reddie and Nicholls, 1971; Caldwell and Milligan, 1972).

Helix stabilization. The "melting" of the alpha helical protein of wool occurs at about 215°-235°C when dry (Menefee and Yee, 1965a; Crighton et al., 1971), and at about 130°C in water (Zahn, 1943; Feughelman and Mitchell, 1966; Cook and Delmenico, 1968). These temperatures are lowered by decreasing the cystine content of the fiber (Menefee and Yee, 1965a; Ebert and Müller, 1966). Figure 2, redrawn from data by Ebert and Müller (1966), shows the effects of cystine reduction on the helix disordering temperatures of Lincoln wool, kid mohair, and Austral-A wool in water. Because of the high initial crosslink density, a large decrease in cystine content is necessary before the melting temperature is appreciably lowered. The figure also illustrates an apparent anomaly, namely that the melting temperatures of the native keratins increase with decreasing initial cystine content. This indicates that the disulfide crosslinking of these keratins is by no means the only factor determining their composite properties. To support this observation, unpublished torsion pendulum studies in our laboratory indicated that the melting transition temperature of dry human hair is only slightly higher than that of wool, even though the cystine content of hair is about 50% more.

It is difficult to increase the melting temperature of the alpha helical component of particular keratins by additional crosslinking, since the microfibrillar portion is not normally involved in crosslinking to a significant extent (Watt and Morris,

1968). For example, treatments with formaldehyde solutions or benzoquinone (Watt, 1965), or difluorodinitrobenzene (Zahn, 1955), have no effect, although treatment with an aqueous mixture of $CaCl_2$, HCl, and formaldehyde, or with gaseous formaldehyde, raises the melting temperature of wool by 12°C (Watt, 1965).

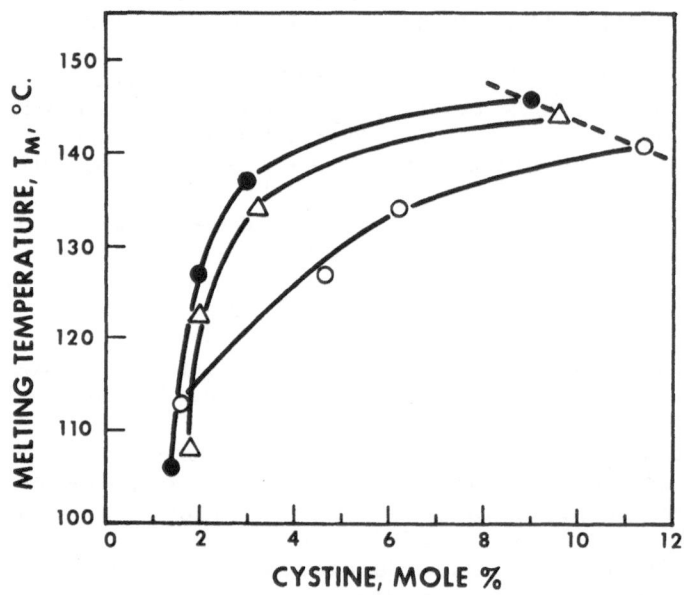

Fig. 2 Helix melting temperature of keratins in water as a function of cystine content in moles cystine per total moles of residues: ● Lincoln wool, △ kid mohair, ○ Austral-A-wool. After Ebert and Müller (1966).

The melting transition temperature of extensively reduced wool fibers has been used by Frazer et al. (1968) as a measure of the effect of various crosslinking reactants and monofunctional substituents. The transition region of the reduced wool is around 55-65°C in water. It is lowered by bulky monofunctional substituents but raised by difunctional crosslinking reagents. Similar effects may occur in native fibers after such treatments but they are likely to be masked by the high intrinsic crosslink density.

There is some evidence that the helix melting behavior is determined largely by crosslinks to the cysteine residues, and not by crosslinks introduced into other chain positions (Watt and Morris, 1970).   Thus, if a reduced fiber is crosslinked with formaldehyde, its thermal stability is that of the native fiber; however, if the fiber is reduced and alkylated before the formaldehyde treatment, the melting temperature drops sharply with the extent of reduction.

The complexity of factors influencing the melting temperature of keratin proteins suggests that its measurement is of little value by itself in determining an unknown crosslink density of keratin, though it has undoubted application in correlating changes in relative crosslinking after chemical treatments.

Swelling limitation.   Sorption of liquids by semicrystalline polymers is thought to occur almost entirely in the amorphous regions, which in the case of keratins is mostly the non-helical matrix.   Fraser et al. (1971) report that X-ray study of dry and hydrated porcupine quill shows that the volume swelling of the matrix in water is 53% and that of the microfibrils is 11%.

Since the matrix is the repository of most of the crosslinks, the maximum swelling in appropriate solvents, such as formic acid, would be expected to correlate well with the crosslink density.   For fibers in which the disulfide content has been chemically reduced, this is indeed true.   Feughelman and Chapman (1966) and Caldwell and Milligan (1970) reported such correlations and proposed the method for use in determining unknown crosslink densities.   Holt and Milligan (1970) extended the method to enable estimation of other kinds of crosslinks introduced into wool by chemical treatment.   In all these applications some caution is called for, since alteration of swelling can sometimes occur merely with the introduction of bulky constituents that are not at all involved in crosslinking.

By similar reasoning, Gillespie (1970) postulated that the equilibrium swelling of different keratins should depend largely on the proportion of high sulfur protein present, and its sulfur content.   Results for the swelling of several different keratins in formic acid (Gillespie, 1970; Feughelman and Chapman, 1966; Whiteley et al., 1970) are shown on Figure 3.   No simple correlation is apparent.   Gillespie proposed that the lack of correlation may be due to differing proportions of inter-and intra-chain crosslinking, with only the former acting to restrict swelling.   Taking this suggestion, and assuming that the envelope of lowest points on Figure 3 represents the curve for maximum interchain crosslinking (100%), estimates can be made of the amounts of intramolecular crosslinking necessary to bring the aberrant points back to the curve.   The largest estimates of intramolecular crosslinking as percentages of total crosslinking are shown on the figure near the corresponding points.

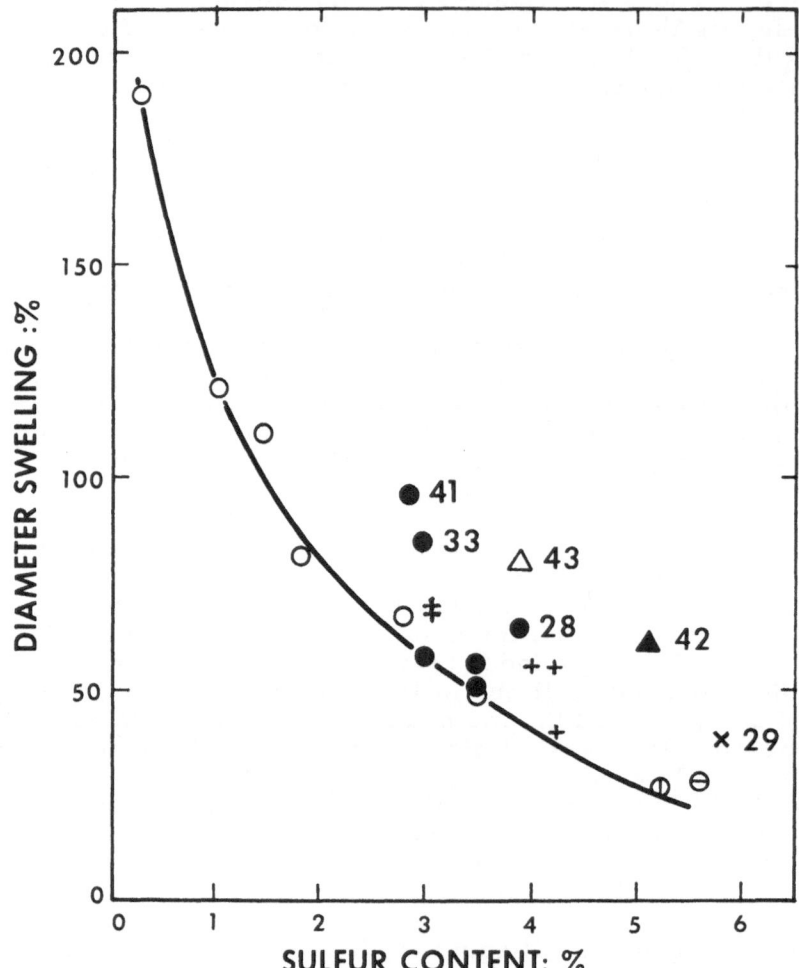

Fig. 3. Diameter swelling in 98-100% formic acid as a function
of weight percent sulfur, for various keratins. Swelling is calculated relative to the dry fiber. O Corriedale wool with vari-
ously decreased disulfide content (Feughelman and Chapman,
1966). Sulfur content of the native fiber was taken as 3.5 %,
with lower percentages calculated as though disulfide reduction
was equivalent to sulfur loss. + various native wools (Whiteley
et al., 1970). ● various native wools, △ calf hair, ▲ seal fur,
Φ rhesus monkey fur, θ raccoon hair, x poodle hair (Gillespie,
1970).

Previous studies of the amount and composition of the soluble fraction of wool after partial hydrolysis (Menefee and Yee, 1965b) indicated that a fraction amounting to 10% of the weight of a Lincoln wool was intramolecularly crosslinked. Assuming this fraction to be in the high-sulfur region, which accounts for about 25% of the weight of this wool, it follows that about 40% of the crosslinking could be intramolecular. This estimate is consistent with those shown in Figure 3. Nevertheless, these estimates are largely speculative; considerably more work is needed before either method can be applied reliably to an unknown keratin.

MECHANICAL PROPERTIES

If the disulfide crosslinking of keratins is progressively decreased or if additional crosslinks are introduced, regular changes appear in most mechanical properties such as Young's modulus, torsional modulus, yield points, break strength, and elongation to break (Feughelman, 1963; Feughelman and Mitchell, 1964; Crewther, 1965; Menefee and Yee, 1965a; Chapman, 1969; Friedman and Tillin, 1974). Those properties involving high strain are usually most affected by cystine levels, a relationship to be expected from consideration of models in which the ultimate extension depends on internal restraints such as crosslinking. However, properties such as Young's modulus in the Hookean region are also affected, supporting the hypothesis (Menefee, 1968, 1971) that the matrix bears a considerable portion of the stress at all extensions as well as when under torsional strain. On this basis keratins from different sources, showing large variations in sulfur content and the amount of high sulfur protein present, would be expected to show related differences in mechanical properties. According to Crewther (1965) there is a common curve relating the strain at the transition between yield and post yield regions to disulfide content, for Lincoln wool, mohair, horsehair, mutant Merino wool, and human hair. The curve was extended to very low disulfide levels by using reduced, carboxymethylated fibers. Not surprisingly, there is a large increase in the yield strain at low levels of disulfide crosslinking.

Rather different conclusions were reached in studies of torsional and extensional properties of a number of wools from sheep fed per abomasum to increase the cystine content (Feughelman and Reis, 1964; Armstrong and Feughelman, 1969a, b; Whiteley et al., 1970; Campbell et al., 1972). In most cases there were no significant changes in static mechanical properties. Although Armstrong and Feughelman (1969a) noted a decrease in the extent of relaxation of torque in water with increasing sulfur enrichment, this effect was attributed to differences in aspartic acid content rather than to sulfur.

Similarly, measurements of the bending and Young's modulus of human hair, Cotswold wool, and Lincoln wool (Khayatt and Chamberlain, 1947) showed differences, but these did not match qualitatively the estimated sulfur contents of these keratins. A similar property-composition inconsistency is shown for the mechanical behavior of wool, mohair, and human hair at the break point (Susich and Zagieboylo, 1953).

On the other hand, shear modulus measurements of Lincoln wool and monkey fur (Mason, 1965) suggest that for these fibers there is some correlation between modulus and the content of sulfur and high sulfur protein (Gillespie, 1967). Firm conclusions from only two samples are, however, not possible. Unfortunately, mechanical data are lacking for a larger series of keratins in which the variation in sulfur content appears to be due almost entirely to variation in the composition and amount of the high sulfur fraction (Gillespie, 1967).

The tentative conclusion was reached (Broad et al., 1970) that in wool, any sulfur in excess of about 3.0% is incorporated in a form that does not contribute any desirable qualities to wool; however, 2.7% sulfur appears to be a lower limit for satisfactory mechanical behavior. Whether the excess above 3% occurs as intramolecular crosslinking remains to be demonstrated.

CHEMICAL STABILITY

As a necessary preliminary to chemical characterization of keratin proteins, nearly all the interchain disulfide crosslinks must be broken before the proteins can be dissolved in suitable disaggregating solvents that disrupt the remaining hydrogen bonding and hydrophobic interactions. The most commonly used methods for solubilization are those of Gillespie and his colleagues, summarized in recent reviews (Bradbury, 1973; Fraser et al., 1972). The disulfide bonds are usually reduced by sodium thioglycollate or mercaptoethanol at moderate to high pH in the presence of urea as a disaggregating solute, after which the free sulfhydryl groups are alkylated with sodium iodoacetate to yield S-carboxymethyl derivatives that are unable to reoxidize to the disulfide form. Ordinarily not more than 85% of wool and even less of hair is solubilized by this method. When the soluble fraction is acidified to pH 4, the precipitate contains most of the alpha helical protein and has a much smaller sulfur concentration than the original keratin. This low-sulfur fraction contains a large proportion of helix-favoring amino acids such as aspartic acid, glutamic acid, and leucine, as well as an excess of anionic or acidic side chains over basic side chains. Furthermore, it tends to associate readily, a cause for uncertainty in determining its molecular weight. Recovery of the soluble protein in the supernatant by dialysis or addition of ethanol yields a sulfur-rich fraction, which is low in helix content and high in those amino acids that hinder helix formation, such as cystine, proline, threonine, and serine. The side chains are predominantly basic, and the sulfur-rich fraction does not tend to associate readily.

Aside from the use of solubilization in keratin protein analysis, it has been made the basis for tests to determine the presence of crosslinks in wool that do not undergo the scission reactions of the disulfide bonds. The urea-bisulfite test has been commonly used, but it has been found to be unreliable in distinguishing introduced crosslinks from monofunctional substituents (Caldwell et al., 1966); in fact, urea-bisulfite solubility can be appreciably changed merely by treating wool with hot water or hot ethanol (Sotiriou-Provata and Vassiliadis, 1966). On the other hand, reduction by tributyl phosphine (Kilpatrick and Maclaren, 1970; Caldwell et al., 1966), or oxidation by performic acid or reduction by thioglycollate in urea (Caldwell et al., 1966) appear to permit solubility distinctions between true crosslinks and bulky substituents. These various uncertainties in interpretation appear to limit severely the usefulness of solubilization methods for quantitative estimation of crosslink densities.

Hydrolysis. Aside from dissolution by crosslink scission, it is possible to hydrolyze peptide linkages partially or completely and effect solution of keratin oligopeptides or, if hydrolysis proceeds long enough, simple amino acids. Partial acid hydrolysis, usually with HCl, has long been used for keratin study (Gordon et al., 1941; Leach et al., 1964; Menefee and Yee, 1965b). Hydrolysis can also be effected enzymatically, often with more specific peptide cleavage (Cole et al., 1971; Milligan et al., 1971).

In this section a simple model for keratin is used to analyze effects of crosslink density on the rate of peptide bond hydrolysis. The model allows the absolute density of intermolecular crosslinking to be calculated from results of partial hydrolysis of keratins or other crosslinked proteins. As we have seen, there are indications that a large proportion of crosslinking in keratins is intramolecular. Since the total disulfide concentration and the free sulfhydryl content can be found by chemical methods, a separate determination of total intermolecular crosslinking will provide more complete structural information for the keratins. For example, intramolecular crosslinking generally indicates a globular protein. Although such a structure for the keratin matrix has been postulated (Crewther, 1965), more definite evidence for the kind of crosslinking present in the matrix would test this notion.

During hydrolysis, composition changes in the soluble fraction and residue seem roughly consistent with the presence of low- and high-sulfur components as already described. However, hydrolysis of a material like wool, which is a three dimensional gel of high crosslink density, presumably entails some cleavage of peptide bonds in all chains, with liberation to the sol fraction of fragments that are no longer attached to the gel. The crosslink density of the sol fraction is always less than half that of the gel fraction (Charlesby, 1960). This view of general hydrolysis suggests that the compositions of the sol and gel fractions are almost certainly not linear functions of the high- and low-sulfur components, or of any other simple compositional groupings. Detailed analysis of such a complex hydrolyzing system would seem to be impossible, but by making reasonable assumptions one can approach the dissolution problem in several tractable ways. The simplest is to consider keratin to be a randomly crosslinked network of monodisperse protein chains, with peptide hydrolysis occurring randomly. From this model the amount and composition of the soluble fraction and residue can be related to the extent of hydrolysis. Progressively more complex procedures are similar, except that more than one component is considered. These components can be independent or interconnected by crosslinks, each liberating its own set of oligopeptides during hydrolysis. Although these more complex models accord better with what is known of the actual keratin structure, the computational details become much more complicated. Therefore only the simplest model will be considered in this section.

In the simplified analysis, three assumptions are needed. First, the keratin is taken to be composed of monodisperse protein chains, each containing U amino acid residues. Even though keratin is known to be heterodisperse, a posteriori analysis shows that polypeptide molecular weight is not important for the results of this paper. Second, a fraction Q of the total number of residues is randomly crosslinked. Examination of available sequences of keratin proteins shows that although some spatial pattern of crosslink distribution may exist, it is not obvious. For present purposes the assumption of randomness appears strong, but possibly superfluous, since it is uncertain whether this assumption has any effect on the subsequent derivation. The third assumption is that peptide bond cleavage occurs randomly. For all possible pairwise combinations of 20 amino acids there are 400 hydrolysis rate constants, some of which have been determined experimentally, using dipeptides (Synge, 1945) or polypeptides (Heyns et al., 1958). Even though these rates differ widely, the pseudo-random occurrence of residue pairs should insure the validity of the random attack assumption. Relaxation of this assumption immediately precludes analytical solutions, and greatly increases computational difficulty.

The simple model has already been solved exactly to give a closed-form solution for the fraction S of keratin that becomes soluble after cleavage of a fraction P of the peptide bonds (Menefee and Bartulovich, 1965). The results (Eqs. 28 and 33 of that paper) have been rearranged into the following rather more simple form:

$$S = S_o (1-Q + QS_o) \tag{1}$$

$$S_o = a^{U-1} + \frac{P}{U}(1-a)^{-2} \left[2(1-\frac{P}{1-a})(1-a^U) + 2U(1+a)^U - U(2-P)(1+a^{U-1}) \right] \tag{2}$$

$$a = (1-P)(1-Q + QS_o) \tag{3}$$

The sterile coefficient S_o (Charlesby, 1960) is the probability that a given crosslink on a particular molecule is sterile; that is, its link to another molecule does not tie it via other links on that molecule to the gel. For present purposes it should be regarded merely as a computational parameter. For keratins, U is known to be in the range of 100 or greater, so that still further reduction of Eq. 2 may be achieved by ignoring terms like a^U. This is a valid approximation as long as Q is fairly large, and permits reduction of Eq. 2 to

$$S_o = \frac{P(2+PU)(1-a) - 2P^2}{U(1-a)^3} \tag{4}$$

Eq. 4 (or 2) can be expanded as a power series in P, to give the amount of keratin dissolved after a small amount of hydrolysis:

$$S \cong \frac{2PU}{(QU)^2} + \cdots \tag{5}$$

To bring Eq. 5 into a form suitable for experimental testing, peptide bond cleavage is assumed to follow first order kinetics:

$$P = 1 - \exp(-kt) \tag{6}$$

If Eq. 6 is substituted into Eq. 5, and S differentiated with respect to the time t, then the initial rate of hydrolysis is found to be

$$\left(\frac{dS}{dt}\right)_{t=o} = \frac{2k}{Q^2 U} \tag{7}$$

According to Eη. 7 the initial rate of hydrolysis is inversely proportional to the square of the crosslink density or amount of sulfur (since Q is nearly proportional to the cystine or sulfur content). This expression was found to be valid for the hydrolysis of mohair, wool, and human hair by a mixture of formic and hydrochloric acid (Menefee and Yee, 1965b). Insofar as chemical resistance implies difficulty of dissolution, it might be expected that other kinds of chemical attack would follow similar principles, and show a strong dependence on the amount of crosslinking.

<u>Determination of crosslink density from hydrolysis rates.</u> From the same assumptions the crosslink density of keratins can be estimated by appropriate inversion of the theoretical expressions for solubility after partial hydrolysis. The theory already published (Menefee and Bartulovich, 1965), which is too long to include here, can be extended to calculate the number of protein chain ends appearing in the soluble keratin extract after breaking a fraction P of the peptide bonds. The complete expression is

$$\mu = \frac{1}{U(1-P)}\left(\frac{aP}{(1-a)^2}\left[(2-2a-P)(1-a^{U-1})+P(U-1)(1-a)\right]+a^U\right) \qquad (8)$$

where μ is the number of moles of chain ends (2 per molecule) in the soluble fraction per mole of amino acid residues in the original sample. The other terms have the same meaning as before.

For given values of P, U, and Q, the theoretical soluble fraction S was calculated according to Eqs. 1-3, and the chain end density μ calculated by Eq. 8. Figure 4 is a plot of log μ versus log S for several levels of crosslinking Q. From these calculations the surprising result emerged that there is essentially no dependence on U of the S-μ plots, except at very low crosslink densities and solubilities. Therefore, for this particular correlation it is unnecessary to use the U-dependent expressions in Eqs. 1-3 and 8. Instead, expressions were derived for infinite U, as follows:

$$S_o = \left(\frac{P}{1-a}\right)^2 \qquad (9)$$

$$\mu = \frac{aP^2}{(1-P)(1-a)} \qquad (10)$$

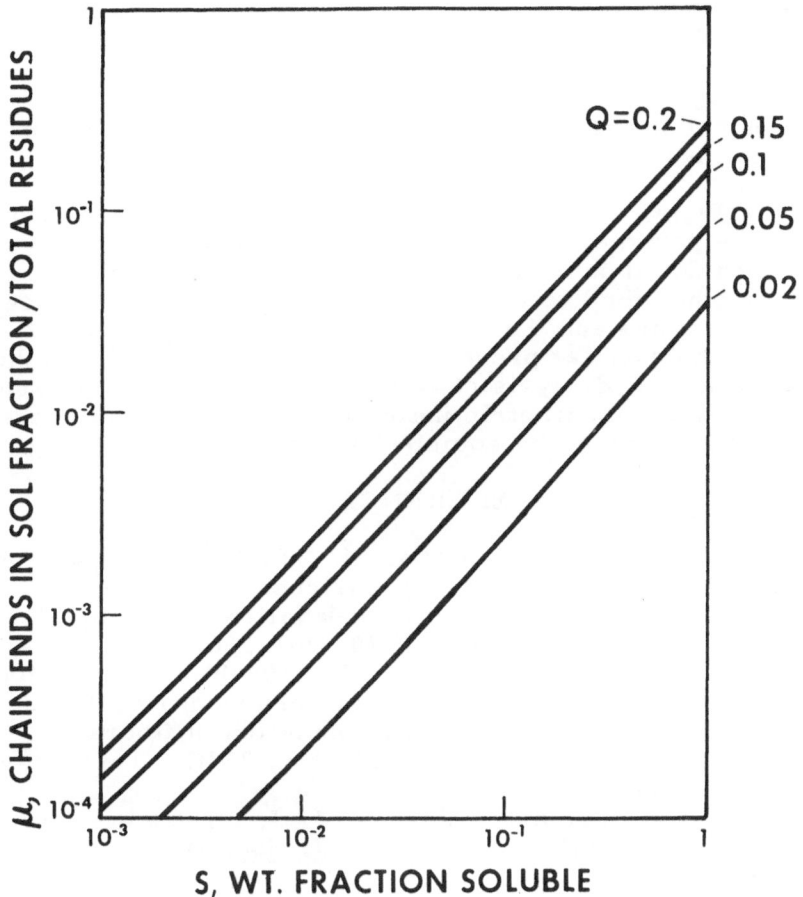

Fig. 4. Number of double chain ends in soluble fraction per total number of original residues, as a function of the weight fraction soluble after partial cleavage of main chain bonds. Curves calculated using Eqs. 11 and 12 of the text, with various values of initial crosslink density Q (fraction of residues originally carrying a crosslink).

Final forms suitable for simple calculation are obtained by combining Eqs. 1 and 3 with Eqs. 9 and 10, yielding

$$S_o = \left(\frac{\sqrt{1 + \frac{4}{\mu}(\frac{S}{\mu} - 1)} - 1}{2(\frac{1}{\mu} - \frac{1}{S})} \right)^2 \tag{11}$$

$$Q = \frac{S_o - S}{S_o(1-S_o)} \qquad (12)$$

These equations imply that the crosslink density of an un-known crosslinked protein can be determined by a single measurement each of solubility and chain end concentration after partial hydrolysis. These results would be used in Eq. 11 to determine the parameter S_o, which in turn would be used in Eq. 12 to find Q. It is necessary only to be sure that the hydrolysis (or other method used to cleave peptide linkages) does not affect crosslinks and that the cleavage is not carried too far. Experiments are at present underway to apply this method to the partial acid hydrolysis of keratins, using trinitrobenzene-sulfonic acid as a colorimetric indicator to measure the concentration of amine ends in the soluble fraction.

REFERENCES

Armstrong, L. D. and Feughelman, M. (1969a). The torsional properties of normal and sulfur-enriched wool fibers. I. Modulus of rigidity and torsional relaxation of wet and dry fibers at 20°C. Text. Res. J., 39, 261-266.

Armstrong, L. D. and Feughelman, M. (1969b). The torsional properties of normal and sulfur-enriched wool fibers. II. Modulus of rigidity and torsional relaxation of fibers in water at temperatures between 20°C and 85°C. Text. Res. J., 39, 267-272.

Asquith, R. S. and García-Domínguez, J. J. (1968a). New amino acids in alkali treated wool. J. Soc. Dyers Col., 84, 155-158.

Asquith, R. S. and García-Domínguez, J. J. (1968b). Cross-linking reactions occurring in keratin under alkaline conditions. J. Soc. Dyers Col., 84, 211-216.

Asquith, R. S. and Otterburn, M. S. (1970). Self crosslinking in heated keratin. J. Text. Inst., 61, 569-577.

Asquith, R. S. and Otterburn, M. S. (1971). Self crosslinking in keratin under the influence of dry heat. Appl. Polymer Symp., 18, 277-281.

Asquith, R. S. and Puri, A. K. (1970). The formation of mixed disulfides by the action of thioglycollic acid on wool cystine and its relationship to wool setting. Text. Res. J., 40, 273-280.

Baden, H. P., Goldsmith, L. A. and Fleming, B. (1973). A comparative study of the physicochemical properties of human keratinized tissues. Biochim. Biophys. Acta, 322, 269-278.

Beevers, R. B. and McLaren, K. G. (1974). Effect of low doses of cobalt-60 gamma radiation on some physical properties and the structure of wool fibers. Text. Res. J., 44, 986-994.

Bendit, F. G. and Feughelman, M. (1968). Keratin. Encycl. of Polymer Sci. and Tech., 8, 1-44.

Bradbury, J. H. (1973). Structure and chemistry of keratin fibers. Adv. Protein Chem., 27, 111-211.

Bradbury, J. H., Chapman, G. V., King, N. L. R. and O'Shea, J. M. (1970). Keratin fibers: amino acid analyses of histological components. Aust. J. Biol. Sci., 23, 637-643.

Broad, A., Gillespie, J. M. and Reis, P. J. (1970). The influence of sulfur-containing amino acids on the biosynthesis of high-sulfur wool proteins. Aust. J. Biol. Sci., 23, 149-164.

Caldwell, J. B., Leach, S. J. and Milligan, B. (1965). The mechanism of setting and the release of set in water. Text. Res. J., 35, 245-251.

Caldwell, J. B., Leach, S. J. and Milligan, B. (1966). Solubility as a criterion of crosslinking in wool. Text. Res. J., 36, 1091-1095.

Caldwell, J. B. and Milligan, B. (1970). The estimation of crosslinks in wool from the extent of swelling in formic acid. J. Text. Inst., 61, 588-596.

Caldwell, J. B. and Milligan, B. (1972). The sites of reaction of wool with formaldehyde. Text. Res. J., 42, 122-124.

Campbell, M. E., Whiteley, K. J. and Gillespie, J. M. (1972). Compositional studies of high- and low-crimp wools. Aust. J. Biol. Sci., 25, 977-987.

Chapman, B. M. (1969). A review of the mechanical properties of keratin fibers. J. Text. Inst., 60, 181-207.

Charlesby, A. (1960). "Atomic Radiation and Polymers," Pergamon Press, New York.

Cole, M., Fletcher, J. C., Gardner, K. L., and Corfield, M. C. (1971). A study of enzymatic hydrolysis applicable to the examination of processed wools. Appl. Polymer Symp., 18, 147-161.

Cook, J. R. and Delmenico, J. (1968). Measuring the "melting point" of wool. J. Text. Inst., 59, 157-160.

Corfield, M. C., Wood, C., Robson, A., Williams, M. J. and Woodhouse, J. M. (1967). The formatition of lysinoalanine during the treatment of wool with alkali. Biochem. J., 103, 15C-16C.

Crewther, W. G. (1965). The stress-strain characteristics of animal fibers after reduction and alkylation. Text. Res. J., 35, 867-877.

Crewther, W. G., Dowling, L. M., Inglis A. S. and McLaren, J. A. (1967). The formation of various crosslinkages in wool and their effect on the supercontraction properties of fibers. Text. Res. J., 37, 736-745.

Crighton, J. S., Findon, W. M. and Happey, F. (1971).
    Application of thermoanalytical methods in the study of ker-
    atins and related proteins. Appl. Polymer Symp., 18, 847-
    856.

Ebert, G. and Müller, F. H. (1966). Differential calorimetri-
    metrische Untersuchungen über den Einfluss von Cystin-
    und Thioätherbindungen auf Umlagerungsvorgänge in
    Keratinfasern. Koll.-Z. u. Z. für Polymere, 214, 38-45.

Feughelman, M. (1963). The mechanical properties of per-
    manently set and cystine reduced wool fibers at vaious rela-
    tive humidities and the structure of wool. Text. Res. J.,
    33, 1013-1022.

Feughelman, M. (1966). Sulfhydryl-disulfide interchange and
    the stability of keratin structure. Nature, 211, 1259-1260.

Feughelman, M. and Chapman, B. M. (1966). The swelling of
    wool fibers with reduced disulfide content in 98% formic
    acid. Text. Res. J., 36, 1110-1111.

Feughelman, M. and Mitchell, T. (1964). The torsional prop-
    erties of single wool fibers. III. Disulfide reduced and
    permanently set wool fibers. Text. Res. J., 34, 593-597.

Feughelman, M. and Mitchell, T. W. (1966). The melting of
    alpha keratin in water. Text. Res. J., 36, 578-579.

Feughelman, M. and Reis, P. J. (1964). The longitudinal
    mechanical properties of wool fibers and their relationship
    to the low-sulfur keratin fraction. Text. Res. J., 34, 334-
    336.

Fraser, R. D. B., MacRae, T. P., Millward, G. R., Parry,
    D. A. D., Suzuki, E. and Tulloch, P. A. (1971). The
    molecular structure of keratins. Appl. Polymer Symp., 18,
    65-83.

Fraser, R. D. B., MacRae, T. P. and Rogers, G. E. (1972).
    "Keratins," Charles C. Thomas, Springfield, Illinois.

Frazer, L. A., Leach, S. J. and Milligan, B. (1968).
    Thermal transitions in reduced wool fibers. J. Appl.
    Polymer Sci., 12, 1992-1996.

Friedman, M. and Tillin, S. (1974). Partly reduced alkylated
    wool. Text. Res. J., 44, 578-580.

García-Domínguez, J. J., Miró, P., Reig, F. and Anguera, S.
    (1971). The formation of beta-aminoalanino-alanine in alka-
    line treated wool. Appl. Polymer Symp., 18, 269-275.

Gillespie, J. M. (1967). The high-sulfur proteins of alpha-
    keratins: their relation to fiber structure and properties.
    J. Polymer Sci., C20, 201-214.

Gillespie, J. M. (1970). The swelling of keratins in formic
    acid. Text. Res. J., 40, 853-855.

Gordon, A. H., Martin, A. J. P. and Synge, R. L. M. (1941).
    A study of the partial acid hydrolysis of some proteins, with
    special reference to the mode of linkage of the basic amino
    acids. Biochem., 35, 1369-1387.

Griffith, J. C. and Mason, P. (1966). The effect of formaldehyde on wool before, during, and after contraction in lithium bromide solution. Text. Res. J., 36, 1021-1022.

Heyns, K., Walter, W. and Grützmacher, H. F. (1958). Zur Frage der Unterscheidung sterischer und induktiver Einflüsse beim Abbau synthetischer Polypeptide. J. Polymer Sci., 30, 573-579.

Hinton, E. H., Jr. (1974). A survey and critique of the literature on crosslinking agents and mechanisms as related to wool keratin. Text. Res. J., 44, 233-292.

Holt, L. A. and Milligan, B. (1970). The introduction of amide and ester crosslinks into wool. J. Text. Inst., 61, 597-603.

Khayatt, R. M. and Chamberlain, N. H. (1948). The bending modulus of animal fibers. J. Text. Inst., 38, T185-197.

Kilpatrick, D. J. and Maclaren, J. A. (1970). A solubility test for wool and its application in studies on chemically modified wools. Text. Res. J., 40, 25-28.

Leach, S. J., Rogers, G. E. and Filshie, B. K. (1964). The selective extraction of wool keratin with dilute acid. I. Chemical and morphological changes. Arch. Biochem. Biophys., 105, 270-287.

Mason, P. (1965). Thermal transitions in keratin. IV. Experiments in lateral compression. Text. Res. J., 35, 736-742.

Menefee, E. (1968). A mechanical model for wool. Text. Res. J., 38, 1149-1163.

Menefee, E. (1971). Relation of keratin structure to its mechanical behavior. Appl. Polymer Symp., 18, 809-821.

Menefee, E. and Bartulovich, J. J. (1965). Crosslinking in keratins. I. Theory for solubility on simultaneous chain scission and crosslink cleavage. J. Appl. Polymer Sci., 9, 2819-2827.

Menefee, E. and Yee, G. (1965a). Thermally induced structural changes in wool. Text. Res. J., 35, 801-812.

Menefee, E. and Yee, G. (1965b). Crosslinking in keratins. III. Acid hydrolysis of keratins. J. Appl. Polymer Sci., 9, 2835-2846.

Milligan, B., Holt, L. A. and Caldwell, J. B. (1971). The enzymic hydrolysis of wool for amino acid analysis. Appl. Polymer Symp., 18, 113-125.

Miró, P. and García-Domínguez, J. J. (1966). Bestimmung von Lysinoalanin in Hydrolysaten von Wolle nach Hitze- und Alkalibehandlung. Melliand Textilber., 47, 676-680.

Miró, P. and García-Domínguez, J. J. (1967a). Methode zur gleichzeitigen Bestimmung von Lanthionin und Lysinoalanin in Hydrolysaten von Wolle. Melliand Textilber., 48, 558-560.

Miró, P. and García-Domínguez, J. J. (1967b). Action of nucleophilic reagents on wool. J. Soc. Dyers Col., 83, 91-95.

Miró, P. and García-Domínguez, J. J. (1968). Action of nucleophilic reagents on wool. II. Action of sodium sulfite at pH 8.6. J. Soc. Dyers Col., 84, 310-313.

Orwell, R. L., Datyner, A. and Nicholls, C. H. (1966). Disulfide bond breakdown in wool during high temperature steam treatments. J. Soc. Dyers Col., 82, 441-446.

Otterburn, M. S. (1975). Crosslinking by lysinoalanine in set and alkali-treated wool. Text. Res. J., 45, 88-89.

Reddie, R. N. and Nicholls, C. H. (1971). Some reactions between wool and formaldehyde. Text. Res. J., 41, 841-852.

Robson, A., Williams, M. J. and Woodhouse, J. M. (1969). The formation of lysino alanine and lanthionine in wool fibers stretched in boiling water, and their relation to permanent set. J. Text. Inst., 60, 140-151.

Rudall, K. M. (1947). X-ray studies of the distribution of protein chain types in the vertebrate epidermis. Biochim. Biophys. Acta, 1, 549-562.

Sims, R. T. (1970). Hair as an indicator of incipient and developed malnutrition and response to therapy - principles and practice. In "An Introduction to the Biology of the Skin," R. H. Champion, T. Gillman, A. J. Poole and R. T. Sims (Editors), Blackwell Scientific Publ., Oxford.

Sotiriou-Provata, M. aand Vassiliadis, A. (1966). Treatments of wool in organic solvents and their effect on the urea-bisulfite test. Text. Res. J., 36, 1031-1037.

Susich, G. and Zagieboylo, W. (1953). The tensile behavior of some protein fibers. Text. Res. J., 23, 405-417.

Sweetman, B. J. (1967). The hydrothermal degradation of wool keratin. II. Chemical changes associated with the treatment of wool with water or steam at temperatures above 100°. Text. Res. J., 37, 844-851.

Synge, R. L. M. (1945). The kinetics of low temperature acid hydrolysis of gramicidin and of some related dipeptides. Biochem. J., 39, 351-355.

Watt, I. C. (1965). The modification of wool fibers by cross-linking reactions. Proc. Int. Wool Textile Res. Conf., Paris, II, 259-270.

Watt, I. C. (1971). Modification of physical properties of wool fibers. Appl. Polymer Symp., 18, 905-914.

Watt, I. C. (1975). Properties of wool fibers heated to temperatures above 100°C. Text. Res. J., 45, 728-735.

Watt, I. C. and Morris, R. (1968). Selective crosslinking in the microfibrillar component of keratin. Text. Res. J., 38, 674-675.

Watt, I. C. and Morris, R. (1970). Evidence for methylene-dithion crosslinks in formaldehyde-treated keratin. Text. Res. J., 40, 952-953.

Weigmann, H.-D. and Dansizer, C. (1969). The stabilization of irreversibly deformed keratin fibers. Text. Res. J., 39, 692-699.

Weigmann, H.-D. and Dansizer, C. (1971a). The stabilization of irreversibly deformed keratin fibers. II. Mechanism of stabilization. Text. Res. J., 41, 576-586.

Weigmann, H.-D. and Dansizer, C. (1971b). Effects of cross-links on the mechanical properties of keratin fibers. Appl. Polymer Symp., 18, 795-807.

Weigmann, H.-D., Rebenfeld, L. and Dansizer, C. (1966). Kinetics and temperature dependence of the chemical stress relaxation of wool fibers. Text. Res. J., 36, 535-542.

Whiteley, K. J., Balasubramaniam, E. and Armstrong, L. D. (1970). The swelling and supercontraction of sulfur-enriched wool fibers. Text. Res. J., 40, 1047-1048.

Zahn, H. (1943). Über thermisch verkurzte Keratinfasern. Naturwissenschaften, 31, 137-139.

Zahn, H. (1955). Crosslinking reactions with amino acids and fibrous proteins. Proc. Int. Wool Textile Res. Conf., Australia, C, 425-451.

Zahn, H., Beyer, H., Hammoudeh, M. M. and Schallah, A. (1969). Vernetzungs- und Selbstvernetzungsreaktionen bei Wolle. Melliand Textilber., 11, 1319-1324.

Zahn, H., Schallah, A., Scharff, D. and Meichelbeck, H. (1971). Crosslinking of wool during treatment with carboxylic acid chlorides. Appl. Polymer Symp., 18, 163-174.

Ziegler, Kl. (1964). New crosslinks in alkali treated wool. J. Biol. Chem., 239, 2713-2714.

20

AN X-RAY DIFFRACTION STUDY OF THERMALLY-INDUCED STRUCTURAL CHANGES IN α-KERATIN[*]

Kay Sue (Lee) Gregorski

Western Regional Research Laboratory, Agricultural Research Service, U. S. Department of Agriculture, Berkeley, California 94710

ABSTRACT

A series of events takes place as wool is heated under vacuum from room temperature to 250°C: Loosely and strongly bound water molecules are removed at temperature at and below 150°C; a glass transition of the amorphous keratin occurs at 160-175°C; helices melt at 215 and 235°C. The structural changes take place between the glass transition and the helix melting temperature are observed as reflected in the low-angle X-ray diffraction patterns: The 39 A meridional reflection is intensified; a 4-point diagram at azimuthal angle of 45 degree with spacing around 46 A appears; the intensity of the 33 A meridional reflection decreases, and the 66 A meridional reflection is the most heat-resistant.

INTRODUCTION

The effect of dry heat on wool and other proteins has been studied extensively. Studies prior to 1965 have been adequately reviewed (Howitt, 1964; Menefee and Yee, 1965). More recent studies are briefly summarized below.

Janowski and Speakman (1965) noted that the yellowing of wool from heating at 150°C in dry air and nitrogen depended on reactions of the hydroxylic side chains of serine and/or threonine. Pande (1965) investigated wool and other protein fibers in a nitrogen atmosphere by differential thermal analysis (DTA). He associated the shoulder on the DTA curve at 160°C with loss of tightly bound water and the doublet at 210-300°C with the para-ortho structure in α-keratins and possible depolymerization of the polypeptide chain. Bendit (1966) studied the melting of α-keratin in vacuo and found that the helices in α-keratin melt in the range of 200-225°C. Haly and Snaith (1967) used DTA and high-angle X-ray diffraction to observe phase transition endotherms. They found the first major melting endotherm to begin at 220°C and peak at 240°C. When Corriedale wool was heated in vacuo at 210°C for 8 hr, the 5.1 A and 9.8 A reflections were partially oriented, and on heating at 220°C for 2 hr, the 9.8 A reflection became disoriented. Crighton and Happey (1968) examined keratin and related protein fibers by DTA under continuous evacution. They discovered a glass transition at 160-165°C and two endotherm peaks at 220-230°C and 230-250°C in both Merino and Lincoln wool. They associated these endotherms with the para-ortho structure in wool and partial decomposition of disulfide bonds. Alter and Kivimagi (1969) found the melting region of human hair keratin to be around 213-218°C when hair was heated in an inert fluid. Rusznak et al. (1971) found that a 30-sec thermal treatment in the range of 150-190°C in the presence of air was sufficient to induce irreversible chemical changes in wool. Kassenbeck and Stay (1975) noted that when wool fibers of bilateral structure were heated, the ortho cortex, which is thought to be lower in sulfur, is less stable than the para cortex, which is higher in sulfur.

Amino acid analyses of heated wool were reported by several authors. Weclawowicz et al. (1965) observed that when wool was heated dry at 200°C, the cystine, tyrosine, serine, valine, threonine and basic amino acids contents decreased with time and the lanthionine content increased with time. Horio et al. (1965) studied the pyrolysis of wool in sealed tubes and found losses in cystine, aspartic acid, serine, threonine, and basic amino acids. Asquith and Otterburn (1969, 1971) heated wool in sealed tubes at 100-180°C for 48 hr and found losses in arginine, lysine, and histidine when the temperature was 140°C or higher. At the same time, new amino acids such as lysinoalanine, β-aminoalanine, ornithine, lanthionine, ε-(γ-glutamyl)lysine and ε-(β-aspartyl)lysine were formed. Milligan et al. (1971) isolated both ε-(γ-glutamyl)lysine and ε-(β-aspartyl)lysine from enzyme digests of wool that had been heated at 160-170°C under 2 mm Hg pressure for 2 hr. Kulkarni (1975) analyzed the amino acid composition of wool which had been heated in a sealed tube at 170°C for 6 hr. He found losses in cystine, tyrosine, lysine, histidine, and serine, and formation of lanthionine.

The formation of amide crosslinks in thermally-treated proteins at temperature around 165°C was first postulated by Mecham and Olcott (1947). Further verification of amide bond formation at this temperature was obtained by Fox et al. (1962) through syntheses of proteinoids from mixtures of dry amino acids. For wool heated in vacuo, Menefee and Yee (1965) observed a glass transition at around 165 °C, with possible concomitant formation amide crosslinks. In addition, two melting points at 215°C and 235°C were established. Studies by authors cited earlier support the occurrence of thermal transitions in wool reported by Menefee and Yee (1965). The chemical changes induced by dry heat include degradation of amino acid residues and the production of new amino acids and/or new crosslinks. The corresponding amide-crosslinked amino acids in wool heated near the glass transition were isolated from enzyme digests of the sample by Milligan et al. (1971). The present paper is concerned with the question of whether these new crosslinks in heated wool occur regularly in wool keratin, as is suggested by low-angle X-ray diffraction studies.

EXPERIMENTAL

Sample preparation for heat treatment and subsequent X-ray examination was as follows. A small bundle of degreased Lincoln wool fibers was pulled through a 1 cm long glass capillary with a diameter of about 0.7 mm. A cut was made on the glass capillary at the midpoint. The glass capillary was then pulled apart to expose 1.5 to 2 mm of the wool. The two halves of the capillary were bound onto a small aluminum holder by a thin wire. The aluminum holder was notched in the middle so the wool was not directly in contact with the metal. The sample was then heated under vacuum at constant temperature for various times, after which the assembly was placed in the X-ray camera for testing.

X-ray diffraction patterns were taken on films in an evacuated Debye-Scherrer type camera. Nickel-filtered copper K_α radiation was used. The specimen to film distance was calibrated to be 10.4 cm.

RESULTS AND DISCUSSION

The low-angle X-ray diffraction method was first applied to the study of structural regularities in α-keratins by Corey and Wyckoff (1936). Evidence of chemical effects to the lateral periodicity of α-keratin was first obtained by Zahn and Kohler (1950) and to the longitudinal periodicity by Kratky et al. (1955). Both groups noted the intensification of certain low-angle reflections in horse and/or human hair after treatment with nitric acid. Since then, the periodicities of various amino acid residues in α-keratins have been postulated from the low-angle X-ray diffraction patterns of chemically treated α-keratins. Isomorphous replacement of tyrosine residues by 3,5-diiodo-tyrosine residues intensified the 33 A reflection (Fraser and MacRae, 1957; Fraser et al., 1960; Haly and Feughelman, 1960; Richards and Speakman, 1955). Specific staining of the less reactive fraction of the cystine in the tips of porcupine quills with methyl mercuric iodide intensified the 66 A reflection (Dobb et al., 1965). Acetylation of the ε-amino groups of lysine residues (Heidemann and Halboth, 1967; Spei et al., 1968) and esterification of the carboxyl groups of acidic amino acid residues (Spei, 1970) of α-keratins intensified the 39 A reflection. The presence of a reflection near 200 A has been observed in chemically treated mohair fibers (Spei et al., 1968, Spei 1971). Under specific chemical treatment, this reflection split into a 4-point diagram (Bonart and Spei, 1972; Spei, 1972). Addition of various heavy metal salts to human hair (Simpson and Wood, 1960) and staining of the most reactive (matrix) cystine with silver nitrate (Wilson, 1972) intensified the 25 A reflection, which has been interpreted as a matrix reflection (Spei, 1972, 1973). Staining of the least reactive (microfibril) cystine with silver nitrate intensified the 66 A, 39 A, and 25 A reflections (Wilson, 1972).

With this background, it is possible to interpret the low-angle X-ray diffraction data for thermally-treated Lincoln wool. Spacings and relative intensities are summarized in Tables 1-4. The data are categorized in essentially four temperature regions, namely, heating below the first melting point, at the first melting point, above the first melting point, and at the second melting point.

As shown in Tables 1-4, the most important changes found in the low-angle X-ray diffraction patterns of samples heated under vacuum at temperatures ranging from 170°C to 235°C were the following: (1) the intensity of the 39 A reflection first increased and then decreased; (2) a 4-point diagram appeared at azimuthal angle of 45 degrees at spacing of about 45-48 A; and (3) the intensity of the 33 A reflection decreased and soon disappeared. The rate at which these changes occurred increased with higher temperatures. These changes also can be seen by comparing the photograph of the low-angle X-ray diffraction pattern of the unheated sample (Figure 1) with that of a heated sample (Figure 2). In these examples, the X-ray pattern of the unheated sample was obtained after 30 hr of exposure and that of the heated sample after only 8 hr. Yet, in the unheated sample the 39 A reflection is barely visible, indicating that the increase in intensity of the 39 A reflection in the heated sample is significant. Since the 39 A reflection has been associated with the basic (Heidemann and Halboth, 1967; Spei et al., 1968) as well as the acidic (Spei, 1970) amino acid residues in wool, this intensification is consistent with formation of a significant amount of amide crosslinks in this sample, which was heated at 226.5°C for 6 min.

Amide crosslinks formed between the ε-amino groups of lysine residues and the carboxyl groups of glutamic and aspartic acids residues have been demonstrated in heated wool by Asquith

and Otterburn (1971) and Milligan et al. (1971). Recently Wilson
(1972) showed that the 39 A reflection could also be associated
with the least reactive (microfibril) cystine. So, the presence
of lysinoalanine (Asquith and Otterburn, 1971; Milligan et al.,
1971) and lanthionine (Asquith and Otterburn, 1969, 1971;
Kulkarni, 1975; Milligan et al., 1971; Weclawowicz et al., 1965)
in the heated wool could also contribute to the intensification of
the 39 A reflection. However, for wool heated at 2 mm Hg pressure for 2 hr at temperature 160-170°C, the concentration of
amide crosslinks, reported by Milligan et al. (1971), was almost
twice the combined concentration of lanthionine and lysinoalanine.
It is likely that amide crosslinks contribute a substantial part of
the intensity of the 39 A reflection. The presence of the 4-point
diagram, shown distinctly in <u>Figure 2</u>, and the shape of the spots
are consistent with these new crosslinks forming a highly
ordered structure in wool.

   Previous authors postulated the 33 A reflection to be the period of the tyrosine residues in α-keratins (Fraser and MacRae,
1957; Fraser et al., 1960; Haly and Feughelman, 1960; Richards
and Speakman, 1955). The disappearance of the 33 A reflection
in the heated samples showed that the tyrosine residues are unstable to heat. Amino acid analyses of wool heated under various dry conditions showed loss in tyrosine (Horio et al., 1965;
Kulkarni, 1975; Weclawowicz, 1965) and the loss increased with
the time of treatment (Horio et al., 1965; Weclawowicz, 1965).

   Other changes were observed in each of the four temperature
regions. Heating below the first melting point at temperatures
173.6; 194.7; and 205°C for a long time (<u>Table 1</u>) did not destroy
the general low-angle X-ray diffraction pattern of wool. The
lipid ring at 47 A decreased in intensity and lost its sharpness.
The spacings agree well among the samples and those reported
by Heidemann (1966) for Lincoln wool. The decomposition of the
disulfide linkages is small. Since wool containing lanthionine
and lysinoalanine was found to be more resistant to heat and
urea-bisulfite (Ruznak et al., 1971) than the control, the increase in relative torsional modulus of the heated wool as shown
by curves (1) and (2) of <u>Figure 3</u> (Menefee and Yee, 1965) could
be caused by both the amide crosslinks and these non-amide
crosslinks.

   Heating at the first melting point (215.7°C) for a long time
(<u>Table 2</u>) destroyed most of the low-angle reflections of wool.
After a 20-min and also the 30-min heating the most prominent
meridional reflections remaining are the 66 A, 39 A, and 25 A.
After a 60-min heating the 66 A reflection, which is associated
with the least reactive (microfibril) cystine, became barely
visible, which could be due to considerable disulfide decomposition. The 5.1 A reflection, representing the helical turns

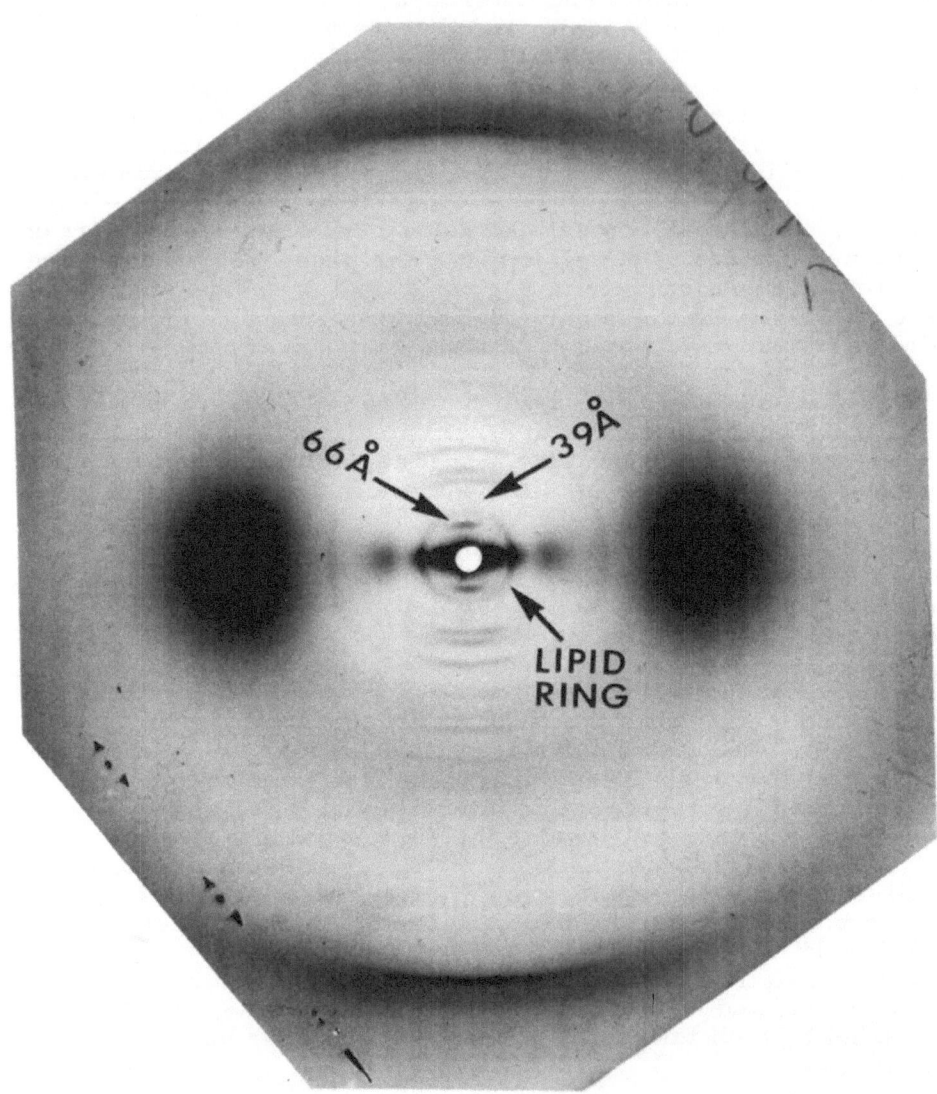

Fig. 1. Low-angle X-ray diffraction pattern of unheated Lincoln wool. X-ray exposure time: 30 hr.

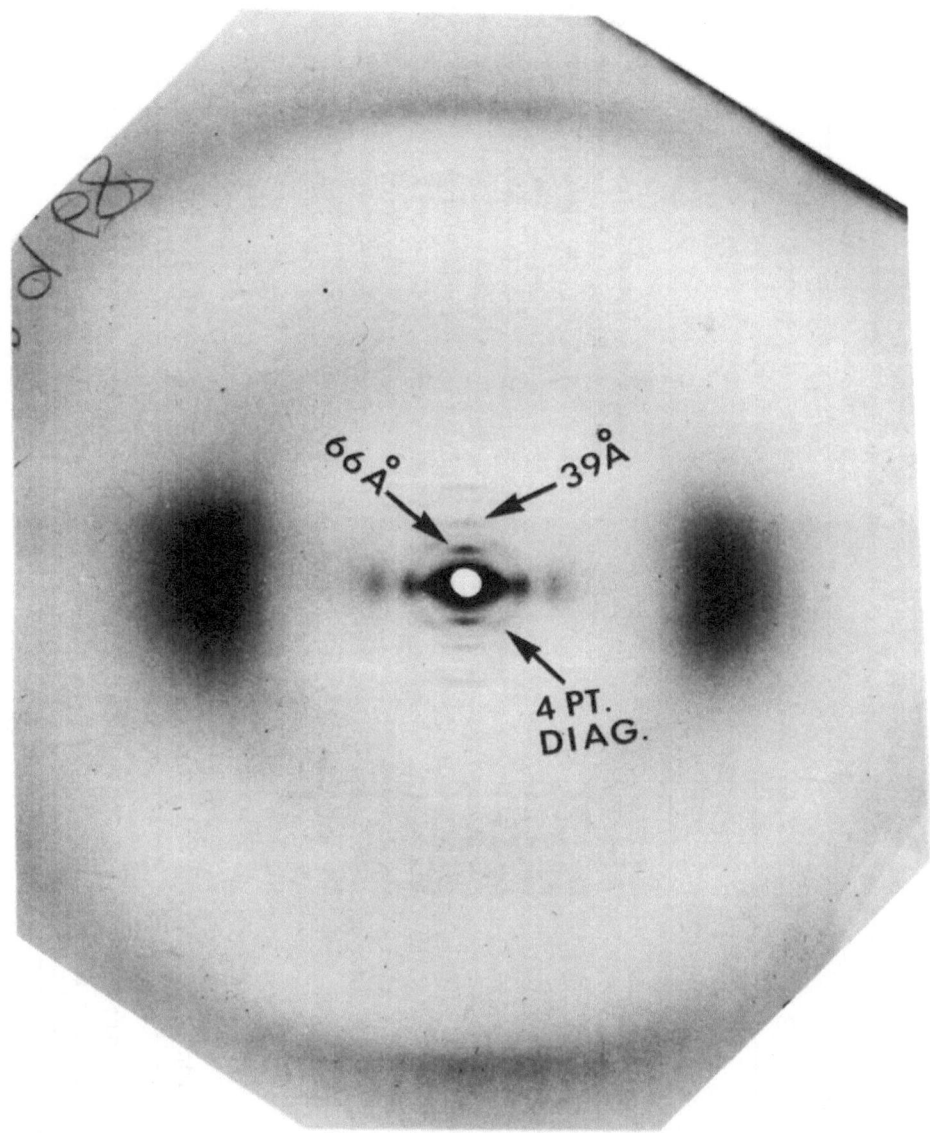

Fig. 2. Low-angle X-ray diffraction pattern of Lincoln wool heated at 226°C for 6 min. X-ray exposure time: 8 hr.

Table 1

Effect of Heating Below the First Melting Point

Temp. (°C)	20	173.6	194.7	194.7	205.0	205.0
Time (Min.)		90	40	80	10	60
	Spacing (Å)	Spacing (Å)	Spacing (Å)	Spacing (Å)	Spacing (Å)	Spacing (Å)
Meridional	66.7 (vs)	66.4 (vs)	66.5 (vs)	66.2 (s)	66.2 (vs)	65.7 (vs)
	48.2 (w)	47.2 (vw)	-	-	-	-
	39.6 (vw)	38.9 (m)	38.9 (s)	38.9 (m)	39.3 (s)	38.5 (s)
	33.2 (vw)	32.7 (w)	-	-	-	-
	28.2 (m)	28.5 (m)	28.8 (m)	27.9 (w)	28.4 (m)	27.9 (w)
	24.7 (m)	24.9 (s)	24.6 (s)	24.4 (m)	24.7 (s)	24.3 (s)
	19.8 (m)	20.0 (s)	19.7 (s)	19.7 (m)	19.8 (m)	19.7 (m)
	5.16 (s)	5.18 (vs)	5.17 (vs)	5.15 (ms)	5.16 (vs)	5.14 (vs)
Equatorial	77.0	79.2	79.2	79.4	78.7	79.2
	47.3	43.7	42.6	44.5	43.3	42.1
	26.4	26.5	26.5	26.6	26.9	26.8
	9.5	9.5	9.5	9.5	9.6	9.5
4-Point Diagram	none	46.8 (m)	45.5 (m)	47.9 (m)	46.1 (m)	46.1 (m)

All patterns had same X-ray exposure time. Relative intensities are enclosed in parentheses. vs = very strong, s = strong, ms = medium strong, m = medium, w = weak, vw = very weak.

Table 2

Effect of Heating at First Melting Point: 215.7°C

Time (Min)	1 Spacing (Å)	6 Spacing (Å)	20 Spacing (Å)	30 Spacing (Å)	45 Spacing (Å)	60 Spacing (Å)
Meridional	64.8 (vs)	64.4 (vs)	64.8 (vs)	63.2 (m)	62.9 (w)	64.8 (vvw)
	47.9 (w)	48.4 (vw)	-	-	-	-
	38.4 (w)	38.2 (s)	37.3 (s)	36.8 (w)	36.4 (vw)	-
	-	-	-	-	-	-
	28.0 (w)	28.2 (w)	-	-	-	-
	24.5 (s)	24.3 (s)	24.1 (s)	23.8 (w)	23.4 (vw)	-
	19.7 (s)	19.6 (s)	-	-	-	-
	5.18 (vs)	5.16 (vs)	5.12 (vs,d)	5.12 (s,d)	5.04 (m,d)	5.01 (w,d)
Equatorial	80.2	79.7	83.0	80.9	81.6	84.1
	43.5	43.2	43.3	42.9	44.2	-
	26.4	26.6	26.7	26.0	-	-
	9.5	9.6	9.6	9.4	9.4	9.3 (a)
4-Point Diagram	46.8 (w)	46.6 (s)	45.0 (s)	45.8 (w)	47.9 (vw)	none

Same X-ray exposure time. Relative intensities are enclosed in parentheses.
vs = very strong, s = strong, m = medium, w = weak, vw = very weak, vvw = very, very
weak, d = diffused, a = arcing.

along the fiber axis, became weak and diffuse. The 9.3 A equatorial reflection, representing the lateral helical spacing, became partially disoriented, as indicated by the arcing of the reflection. Similar changes in the orientation of these two reflections were observed by Haly and Snaith (1967). The 39 A reflection and the 4-point diagram became invisible, indicating disappearance of the ordered crosslinks formed earlier. However, the equatorial microfibrillar reflection at 84 A is still present. This bigger spacing could be due to isotropic expansion in the radial direction of the fiber. At this temperature, the relative torsional modulus decreased steadily as shown by curves (3) and (4) of Figure 3 (Menefee and Yee, 1965).

Heating above the first but below the second melting point (226.5°C) for 20 min gave no observable low-angle meridional reflection (Table 3). The 80 A equatorial microfibrillar spacing was still present; the microfibrils having not yet completely melted. The 5.1 A and the 9.4 A reflections showed a high amount of disorientation. The relative torsional modulus decreased rapidly as shown by curves (5) and (6) of Figure 3 (Menefee and Yee, 1965).

Heating at the second melting point (232.2°C) for 8 min gave no observable low-angle X-ray diffraction pattern (Table 4). The 80 A microfibrillar spacing disappeared; the microfibrils having undergone melting. The 5.1 A and the 9.4 A reflections became completely disoriented; the helices now completely disordered. The torsional modulus could not be measured because the fiber became too weak to support the weight of the bob (Menefee and Yee, 1965).

Heating from the first melting point to higher temperature caused a slight shrinking of the meridional spacings (averaging 3%) and at the same time a slight increase in the microfibrillar spacing (averaging 2.9%), probably caused by partial melting of helices and cleavage of crosslinks.

CONCLUSIONS

Formation of crosslinks in wool heated at temperatures ranging from 170-235°C was inferred from the low-angle X-ray diffraction patterns of Lincoln wool. These thermally-induced new crosslinks form an ordered structure in wool, as shown by the appearance of the 4-point diagram and the shape of the spots on the 4-point diagram. The intensification of the 39 A meridional reflection shows that these crosslinks may be amide crosslinks formed between basic and acidic amino acid residues. However, the presence of non-amide crosslinks such as lysinoalanine and lanthionine cannot be excluded.

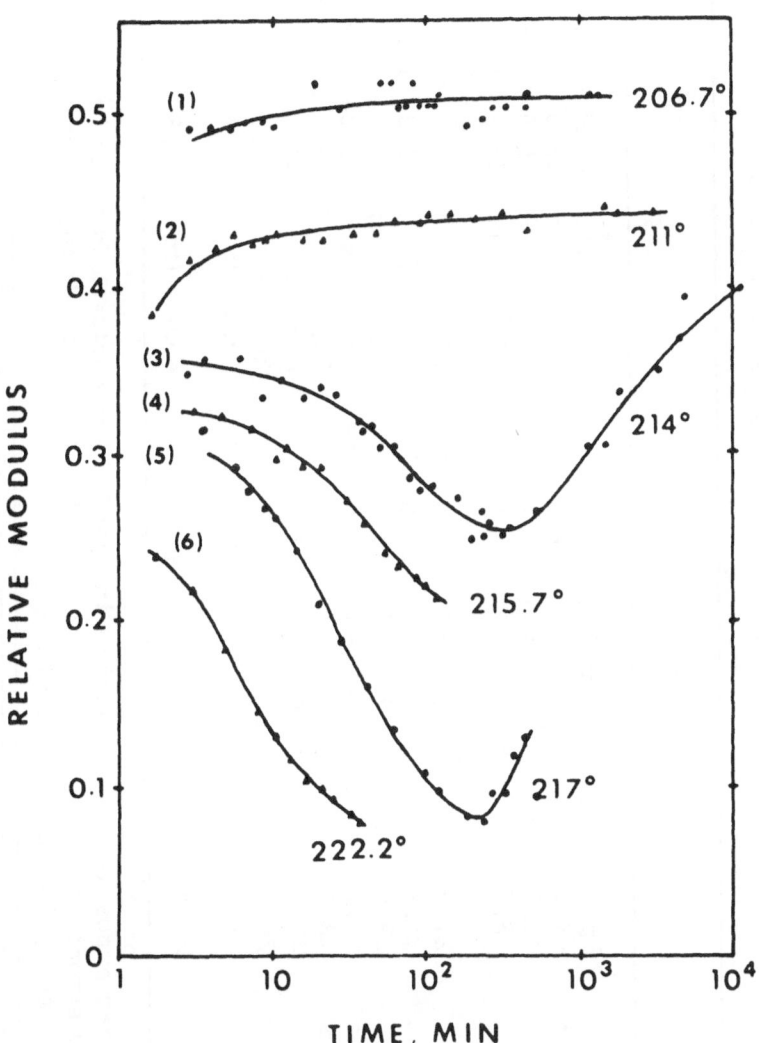

Fig. 3. Relative torsional modulus vs. time at various temperatures.

Table 3

Effect of Heating Above First Melting Point: 226.5°C

Time (Min)	2		4		6		8		10	
	Spacing (Å)		Spacing (Å)		Spacing (Å)		Spacing (Å)		Spacing (Å)	
Meridional	64.8	(vs)	65.3	(vs)	62.9	(vs)	63.5	(ms)	62.0	(s)
	46.9	(vw)	47.8	(vw)	47.2	(vw)	46.9	(vw)	45.5	(vw)
	38.5	(s)	38.3	(s)	37.5	(s)	37.2	(s)	36.4	(m)
	–		–		–		–		–	
	27.8	(m)	27.5	(m)	27.6	(m)	26.9	(vw)	–	
	24.1	(s)	24.3	(s)	24.0	(s)	23.7	(s)	23.3	(m)
	19.6	(m)	19.7	(m)	19.3	(m)	19.4	(vw)	–	
	5.08	(vs)	5.10	(vs)	5.06	(vs)	5.06	(s)	5.04	(s, d)
Equatorial	78.0		78.5		76.7		80.5		82.1	
	43.2		43.0		42.9		42.7		43.3	
	26.6		27.1		26.8		26.5		27.4	
	9.4		9.5		9.3	(a)	9.4	(a)	9.4	(a)
4-Point Diagram	44.7	(m)	45.5	(s)	45.0	(s)	45.0	(m)	45.3	(m)

Same X-ray exposure time. Relative intensities are enclosed in parentheses.
vs = very strong, ms = medium strong, s = strong, m = medium, w = weak,
vw = very weak, d = diffused, a = arcing.

Table 4

Effect of Heating at the Second Melting Point:
232. 2°C

Time (Min)	1		3		4	
	Spacing (Å)		Spacing (Å)		Spacing (Å)	
Meridional	66. 5	(vs)	65. 4	(ms)	64. 8	(s)
	47. 7	(w)	47. 9	(vw)	47. 3	(w)
	38. 4	(w)	38. 4	(m)	37. 7	(m)
	-		-		-	
	28. 2	(m)	27. 3	(vw)	-	
	24. 9	(s)	24. 0	(m)	23. 9	(w)
	20. 0	(m)	19. 2	(vw)	-	
	5. 19	(vs)	5. 11	(ms)	5. 12	(s)
Equatorial	80. 2		81. 4		82. 2	
	43. 8		44. 1		44. 1	
	27. 1		27. 2		27. 2	
	9. 4		9. 5		9. 3	(a)
4-Point Diagram	47. 6	(m)	45. 1	(m)	45. 6	(m)

Same X ·ray exposure time. Relative intensities
are in parentheses.
vs = very strong, s = strong, m = medium, w =
weak, vw = very weak, a = arcing.

The fact that the 66 A meridional reflection persists longer
than the 25 A meridional reflection when wool is heated may indicate that the 66 A reflection is related to the less reactive
cystine, in the microfibril, and the 25 A reflection to the more
reactive cystine, in the matrix.

The disappearance of the 33 A meridional reflection is additional confirmation that the tyrosine residues in wool are unstable to heating at and above 170°C under vacuum.

ACKNOWLEDGEMENT

I thank Dr. Kenneth J. Palmer for teaching me the X-ray
diffraction technique and Dr. Emory Menefee for the heated
wool samples.

REFERENCES

Alter, H. and Kivimagi, K. (1969). The melting of keratin.
  Text. Res. J., 39, 608-610.
Asquith, R.S. and Otterburn, M.S. (1969). Basic amino acids
  in heated keratin. J. Text. Inst., 60, 208-210.
Asquith, R.S. and Otterburn, M.S. (1971). Self-crosslinking
  in keratin under the influence of dry heat. Appl. Polym.
  Symp. No. 18, 277-287.
Bendit, E.G. (1966). Melting of α-keratin in vacuo. Text. Res.
  J., 36, 580-581.
Bonart, R. and Spei, M. (1972). Grundlagen fur die Interpreta-
  tion der Rontgenkleinwinkeldiagramme von α-Keratin.
  Kolloid-Z. Z. Polym., 250, 385-393.
Corey. R.B. and Wyckoff, R.W.G. (1936). Long spacings in
  macromolecular solids. J. Biol. Chem., 114, 407-416.
Crighton, J.S. and Happey, F. (1968). Differential thermal
  analysis of keratin and related protein fibres, Symposium on
  Fibrous Proteins, Australia 1967, Ed. by W. G. Crewther,
  New York, Plenum Press, p. 409-420.
Dobb, M.G., Fraser, R.D.B. and MacRae, T.P. (1965). The
  structure of the keratin filament, 3e Congres International de
  la Recherche Textile Lainiere, Paris (CIRTEL) I, 95-102.
Fox, S.W., Harada, K. and Rohlfing, D.L. (1962). The ther-
  mal copolymerization of α-amino acids, In "Polyamino Acids,
  Polypeptides, and Proteins," Ed. by M.A. Stahmann, Univ.
  Wisconsin Press, Madison, Wisconsin, p. 47-54.
Fraser, R.D.B. and MacRae, T.P. (1957). Evidence of regu-
  larities in the chemical structure of α-keratins, Nature, 179,
  732-733.
Fraser, R.D.B., MacRae, T.P. and Rogers, G.E. (1960).
  Recent observations on the structure of α-keratin. J. Text.
  Inst. Trans., 51, 497-505.
Haly, A.R. and Feughelman, M. (1960). The X-ray diffraction
  pattern of iodinated wools. Text. Res. J., 30, 622-623.
Haly, A.R. and Snaith, J.W. (1967). Differential thermal anal-
  ysis of wool - the phase-transition endotherm under various
  conditions. Text. Res. J., 37, 898-907.
Heidemann, G. (1966). Project No. UR-E10-(20)-8, Grant No.
  FG-Ge-103, Report No. 3, 1966, German Wool Research
  Institute, Aachen, W. Germany.
Heidemann, G. and Halboth, H. (1967). X-ray evidence of
  regularly distributed lysine in α-keratin. Nature, 213, 71-
  72.
Horio, M., Kondo, T., Sekimoto, K. and Funatsu, M. (1965).
  Pyrolysis of wool in sealed tubes, 3e Congres International
  de la Recherche Textile Lainiere, Paris (CIRTEL) II, 189-
  200.
Howitt, F.O. (1964). The yellowing of wool: a survey of the
  literature. J. Text. Inst. Trans., 55, 136-145.

Janowski, Z. and Speakman, J.B. (1965). The action of heat on wool, 3e Congres International de la Recherche Textile Lainiere, Paris (CIRTEL) II, 157-165.

Kassenbeck, P. and Stay, A. (1975). Denaturierung keratinischer Proteine unter Einwirkung von Hitze, Progr. Colloid & Polymer Sci., 57, 123-132.

Kratky, O., Sekora, A., Zahn, H. and Fritze, E.R. (1955). Kleinwinkel-Rontgen-interferenzen bei nitrierten Faserkeratinen. Z. Naturforsch., 10b, 68-72.

Kulkarni, V.G. (1975). The separation of cortical cells and the pyrolysis of wool keratin. Text. Res. J., 45, 89-90.

Mecham, D.K. and Olcott, H.S. (1947). Effect of dry heat on proteins. Ind. Eng. Chem., 39, 1023-1027.

Menefee, E. and Yee, G. (1965). Thermally-induced structural changes in wool. Text. Res. J., 35, 801-812.

Milligan, B., Holt, L.A. and Caldwell, J.B. (1971). The enzymic hydrolysis of wool for amino acid analysis. Appl. Polym. Symp. No. 18, 113-125.

Pande, A. (1965). Differential thermal analysis and its application. Lab. Practice, 14, 1048-1051.

Richards, H.R. and Speakman, J.B. (1955). The iodination of wool. J. Soc. Dyers Col., 71, 537-544.

Rusznak, I., Trezl, L., Bereck, A. and Bidlo, G. (1971). Influence of short thermal treatments on wool. Appl. Polym. Symp. No. 18, 175-183.

Simpson, W.S. and Woods, H.J. (1960). Enhancement of the high-spacing meridional reflextions in the X-ray photograph of keratin impregnated with heavy-metal salts. Nature, 185, 157.

Spei, M., Heidemann, G. and Halboth, H. (1968). Further x-ray evidence of regularly distributed lysine in α-keratin. Nature, 217, 247.

Spei, M., Heidemann, G. and Zahn, H. (1968). X-ray evidence of the "198 A Period" in α-keratin. Naturwiss., 55, 346.

Spei, M. (1970). Rontgenkleinwinkeluntersuchungen an Carboxylgrupppenmarkiertem α-Keratin. Kolloid-Z. Z. Polym., 238, 436-438.

Spei, M. (1971). The influence of detergents on low-angle x-ray diffraction patterns of α-keratin. Appl. Polym. Symp. No. 18, 659-662.

Spei, M. (1972). Rontgenographische Nachweise der fundamentalen 198 A Periodizitat in α-Keratin. Kolloid-Z. Z. Polym., 250, 207-213.

Spei, M. (1972). Rontgenkleinwinkeluntersuchungen zur Frage einer geordneten Matrix in α-Keratin. Kolloid-Z. Z. Polym., 250, 214-221.

Spei, M. (1973). Structure of α-keratin. Text. Res. J. 43, 692-693.

Wilson, G.A. (1972). Low-angle x-ray diffraction studies of reduced and silver-stained α-keratin. Polymer, 13, 63-68.

Weclawowicz, M., Reitzer, M.-Th. and Schutz, R.A. (1965).
    Evolution des caracteristiques et des proprietes chimiques
    de fibres proteiniques naturelles de structures differenciees
    sous l'effet de traitements a haute temperature, 3e Congres
    International de la Recherche Textile Lainiere, Paris
    (CIRTEL) II, 179-185.
Zahn, H. and Kohler, K. (1950). The action of nitric acid on
    albuminoids. I. New rontgen interference action of nitrated
    fibroin and keratin. Z. Naturforsch., 5b, 137-138.

*Parts of this manuscript have been accepted for publication by
Textile Research Journal.

21

INTRODUCTION OF NEW CROSSLINKS INTO PROTEINS

Klaus Ziegler, Irene Schmitz, and Helmut Zahn

Deutsches Wollforschungsinstitut, Aachen, Federal
Republic of Germany

ABSTRACT

Analysis of the crosslinks ε-(γ-glutamyl) lysine and ε- (β-aspartyl) lysine present in treated wool has been improved by modifying the enzymic digestion.

Treatment of wool with either monocarboxylic acid chlorides in dimethylsulfoxide or with 1-fluoro-2,4-dinitrobenzene in the presence of acetate considerably decreased ε-amino groups and solubility. Since no information of interchain amide crosslinks was observed, the hypothesis of so-called self-crosslinking postulated by ZAHN has to be withdrawn. The effects of both treatments are explained in the light of new results.

The reaction of wool with glutaraldehyde leads to a stabilization of the fiber. Experiments with glutaraldehyde and primary alkyl amines as model compounds revealed that the cyclic form of the aldehyde gave the unstable N-alkyl-2,6-dihydroxypiperidine, which either looses water to give N-alkyldihydropyridine or condenses with 2,6-dihydroxytetrahydropyran to yield a copolyether which was isolated. According to recent publications, crosslinking of proteins by glutaraldehyde is due to the formation of quaternary pyridinium compounds.

ENZYMIC DIGESTION OF CROSSLINKED WOOL

The number of the various types of crosslinks in keratins has been enlarged by two isopeptides, ε-(γ-glutamyl) lysine and ε-(β-aspartyl) lysine, which were detected in enzymic digests of untreated or treated wool (Asquith et al., 1970; Milligan et al., 1971; Cole et al., 1971). However, with increasing crosslinks in treated wool, the material becomes less digestible by enzymes, requiring modification of the hydrolytic procedure.

In a study of effects of heat treatment, dinitrophenylation, and acylation of wool (Schmitz, 1975; Schmitz et al., 1976), a considerable improvement of the procedure was obtained when the enzymic digestion was started by predigesting with Pronase E at 39°C for at least 24 hr. After this pretreatment, the modified wools were much more susceptible to the attack of dithiothreitol as reducing agent for the disulfides. Since the addition of urea as swelling agent was not necessary, dialysis, in which low molecular protein fragments usually get lost, was avoidable. After reduction and subsequent acylation of the sulfhydryl groups by iodoacetic acid, the fiber material is once more digested with Pronase E followed by treatment with aminopeptidase M and prolidase.

Quantitative amino acid analysis of the enzymic digests was carried out with a Biocal analyzer (BC 201) using lithium citrate buffer for the determination of both isopeptides as well as for serine, threonine, asparagine, and glutamine.

In heated wool, more than 20 μmole/g of ε-(γ-glutamyl) lysine was found, whereas formation of traces of ε-(β-aspartyl) lysine seems to begin at 120°C (see Table 1). There was no further appreciable increase in the amount of either isopeptide, even after the wool was predried, when it was subsequently heated to 180°C.

Table 1

Contents of Isopeptides in Wool Heated for 48 hr at various temperatures (in μmole/g)

Temp. $^\circ$C	ε-(γ-Glutamyl) lysine		ε-(β-Aspartyl) lysine	
untreated	12	(15)[a]	0	(10)
60	12	(17)	0	(17)
100	12	(20)	0	(20)
120	20	(25)	2	(20)
140	23	(40)	3	(35)

[a]Values in brackets are those determined by Asquith et al. (1971).

The application of these proteolytic enzymes to detect acid labile crosslinks is of further importance when the content of glutamine and asparagine in modified wool has to be determined.

However, it is generally known that the content of glutamine decreases in solution because of cyclization to pyrrolidone-4-
carboxylic acid. Experiments with insulin, glucagon, and synthetic glutamine-containing peptides resulted in a correction factor
of 1.33 for glutamine, whereas asparagine values agreed with
theory. The amino acid values obtained for aspartic acid, glutamic acid, and their amides are compared with those published in
Table 2.

Table 2

Contents of Glutamic Acid, Glutamine, Aspartic Acid and Asparagine
in Untreated Wool[a]

	Glu	Gln	Asp	Asn	Total
Enzymic hydrolysis:					
Schmitz, 1975	500	466	238	230	1434
Milligan, 1971	530	509	210	364	1603
Cole, 1971	547	440	209	308	1504

[a]Expressed in μmole/g dry protein; corrected data.

The data given in Table 2 are comparable to some extent. The
differences in the amino acid contents can be ascribed the different enzymic digestion procedures as well as the mode of calculation, i.e. by subtraction.

DINITROPHENYLATION IN SODIUM ACETATE

The reaction of proteins with 1-fluoro-2,4-dinitrobenzene
(FDNB) yielding, after acid hydrolysis, bright yellow dinitrophenyl derivatives of amino acids, has received much attention in
wool research, particularly to study the accessibility of side
chain groups such as amino-, hydroxyl-, imidazole-, and
sulfhydryl groups in modified wool. Despite the great variety of
conditions applied, complete dinitrophenylation of the ε-amino
groups of the lysyl residues has not been accomplished. This effect was ascribed to reaction of FDNB with the carboxylate anions
of glutamic and aspartic acid residues to form DNP-esters (Zahn et
al., 1969). Being very active acylating agents for nucleophilic
groups, these DNP-esters were assumed to react with neighboring
ε-amino groups of lysyl residues to yield crosslinks of the structure of ε-(γ-glutamyl)-lysine and/or ε-(β-aspartyl) lysine. Such
a hypothetical scheme is given in Figure 1.

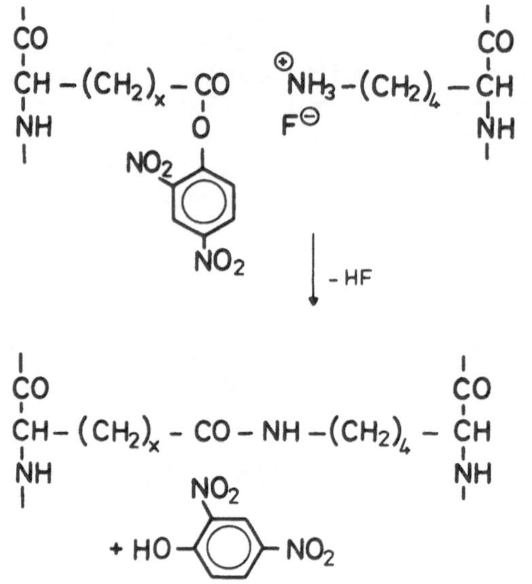

Fig. 1. Hypothetical scheme of formation of ε-(γ-glutamyl)-lysine or ε-(β-aspartyl) lysine after reaction of FDNB with glutamyl or aspartyl residues (Zahn et al., 1969).

When dinitrophenylation was carried out in acetate buffer solution, pH 6, the carboxylate anion reacted more rapidly than the protonated ε-amino group, resulting in a yield of only 20% dinitrophenylated lysyl residues. From these data it was tentatively concluded that the rest of lysyl groups might have been involved in formation of the above-mentioned isopeptides. Besides, it seemed that some data from solubility tests, swelling, and super-contraction supported the self-crosslinking theory indirectly.

However, a series of investigations (Reinert, 1972; Schmitz, 1975) conclusively demonstrated that the long maintained hypothesis, shown in Fig. 1, had to be abandoned. FDNB does in fact react with carboxylate anions which stem from the acetate anions of the buffer solution instead of the glutamic and aspartic acid residues of the wool. Therefore dinitrophenylacetate is formed, which, in turn, acetylates many ε-amino groups in wool; these then become unavailable for further dinitrophenylation. These results are based on experiments using labeled acetate in the buffer solution, transferring radioactivity to wool and Nε -acetyl-lysine has been isolated in exactly the amount corresponding to the decreased yield of ε-DNP-lysine. Further proof of the invalidity of the self-crosslinked hypothesis brought about by FDNB came from amino acid analysis of the enzymic digests, which showed no difference in isopeptide content of wool dinitrophenylated in acetate buffer solution at pH 6 to that of untreated wool.

ACYLATION OF WOOL IN THE PRESENCE OF DIMETHYLSUFLOXIDE

When wool is treated with chlorides of monocarboxylic acids in organic solvents, acylation of the various amino-, hydroxyl-, and sulfhydryl groups is likely to occur. In the case of dimethyl-sulfoxide (DMSO) as solvent, almost none of the acyl groups were incorporated. On the other hand, the number of available ε-amino groups, analyzed by estimating the content of Nε- DNP lysine was considerably decreased (Zahn et al., 1971). Further investigations of the treated wool showed a drastic decrease of solubility in performic acid/ammonia as well as in thioglycolate/urea. From these results it was assumed that the acyl chlorides might have reacted with the side-chain carboxyl groups of aspartic- and glutamic acid residues, producing mixed anhydrides. Subsequent reaction of these mixed anhydrides, which would act as acylating agents for the ε-amino groups of combined lysine, would give rise to the formation of interchain amide crosslinks, already known as isopeptides (Zahn et al., 1969; Zahn et al., 1971). This scheme of reaction set out in Figure 2 is hypothetical; and the postulated so-called self-crosslinking theory has to be withdrawn since additional results have become available (Scharff, 1972). Reaction of the solvent DMSO with carboxylic acid chlorides was found to give rise to formaldehyde, methylmercaptan, and the carboxylic acid, according to a modified PUMMERER rearrangement (see Figure 3). The reaction of the solvent with the acylating reagent takes place at a higher speed than the originally expected acylation of nucleophilic groups in wool. The fact that no differences have been observed between the chemical properties of formaldehyde-treated wool and acylated wool, both treated in DMSO as solvent (Scharff, 1972), supports the proposed mechanism of rearrangement. The effects of crosslinking wool are easily explained by reaction of formaldehyde with primary amino groups forming N-methylene derivatives. No additional self-crosslinking reaction is involved: the wool sample that had been acylated in DMSO showed no increase in the isopeptide content of the enzymic digest.

CROSSLINKING OF PROTEINS BY GLUTARALDEHYDE

Reaction of the bifunctional glutaraldehyde with proteins has been applied in various fields such as leather tanning, enzymology, and in the processing of sheep skin. Glutaraldehyde is also known for wool processing because the treated fibers showed a remarkable increase in strength and enhanced settability. Furthermore, increased stability in alkaline solutions is observed (Happich et al., 1965; Kusch et al., 1972).

The yellowish brown discoloration can be avoided if the aldehyde is used as its sodium bisulfite adduct (Kusch et al., 1972).

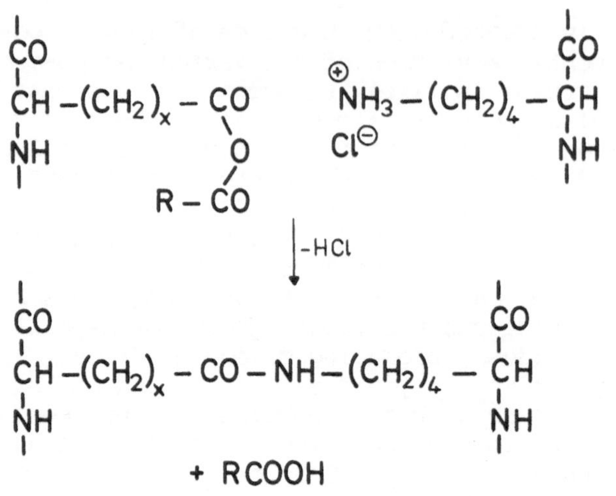

Fig. 2. Hypothetical scheme of reaction of the formation of an isopeptide crosslink in wool. Mixed anhydride formed by carboxylic acid chloride reacts in DMSO and pyridine with ε-amino groups.

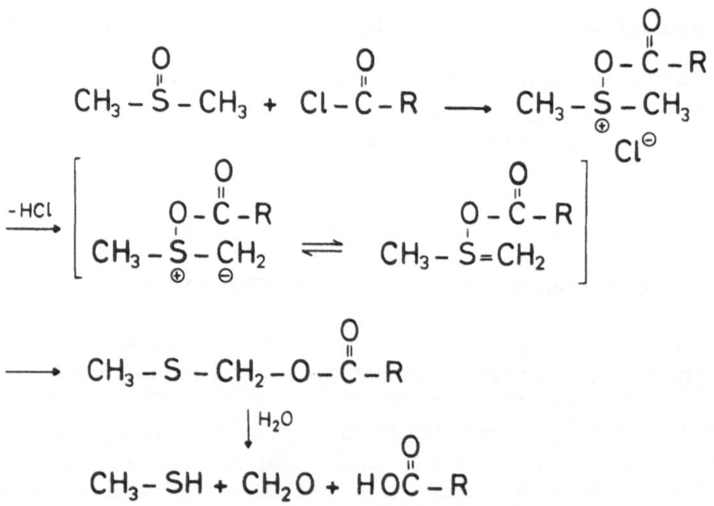

Fig. 3. Scheme of reaction of dimethylsulfoxide with carboxylic acid chloride.

$$
\begin{array}{ccc}
\text{H} & & \text{H} \\
\text{C} - (\text{CH}_2)_3 - & \text{C} \\
\text{O} & & \text{O}
\end{array}
$$

$$\Updownarrow \quad + \ 2 \ \text{NaHSO}_3$$

$$
\begin{array}{ccc}
& \text{H} & & \text{H} \\
\text{NaO}_3\text{S} - & \text{C} - (\text{CH}_2)_3 - & \text{C} - \text{SO}_3\text{Na} \\
& \text{OH} & & \text{OH}
\end{array}
$$

In solution, the two components of this adduct react with wool quite differently. The glutaraldehyde moiety causes crosslinking, whereas sodium bisulfite initially cleaves cystine disulfide linkages into sulfhydryl- and S-sulfocysteine groups; this reaction is followed, under oxidizing conditions, by rebuilding cystine bonds.

Amino acid analyses published before 1972 by various workers (Di Modica and Marzona, 1971; Happich, 1971) indicate that the side chains of lysine and probably histidine take part in the reaction of glutaraldehyde to give strong crosslinks. However, the exact structure of the crosslinks was not understood. Another series of investigations was planned to analyze the amino acid contents of glutaraldehyde-treated wool after enzymic digestion. A decrease of the lysine content by about 10% was observed (Ziegler and Liesenfeld, 1976).

Parallel to the treatment of wool with glutaraldehyde, we decided to investigate the reaction of glutaraldehyde with simple alkyl amines as model compounds (Lubig, 1972). In the reaction mixture, a polyether consisting of 2,6-dihydroxy-tetrahydropyran and N-alkyl-2,6-dihydroxypiperidine was isolated. The formation of such an ether could be explained by reaction of the cyclic form of glutaraldehyde with the amine, resulting in an unstable intermediate, N-alkyl-2,6-dihydroxypiperidine which, upon losing water would give N-alkyldihydropyridine. Condensation of this with 2,6-dihydroxy-tetrahydropyran would yield the coproduct shown in Figure 4.

Recently, a very interesting contribution to the much debated nature of protein crosslinking by glutaraldehyde has been published (Hardy et al., 1976a). On the basis of model experiments with glutaraldehyde and ε-aminocaproic acid, α-acetyl-lysine, and later with ovalbumin (Hardy et al., 1976b), the crosslinking action was inferred to be compounds with a chromophore with λmax at 265 nm (see Figure 5).

When the derivative present in glutaraldehyde-treated ovalbumin was subjected to acid hydrolyses, 1-(5-amino-5-carboxypentyl)-pyridinium chloride hydrochloride was isolated (see Hardy et al., 1976b).

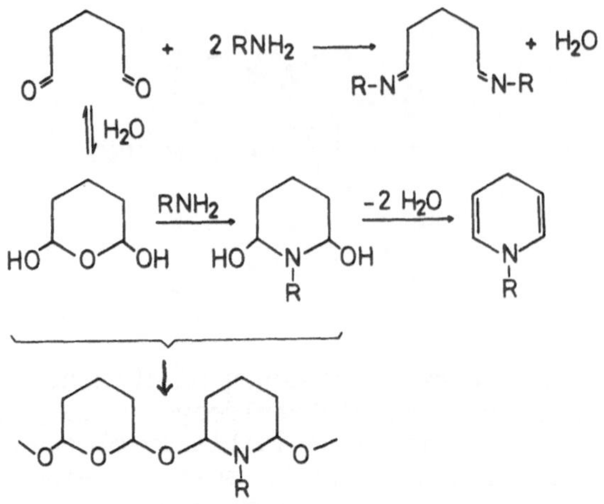

Fig. 4. The reaction of glutaraldehyde with primary amines.

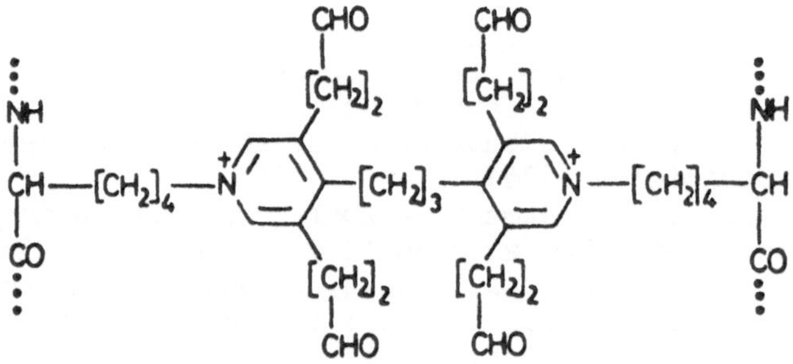

Fig. 5. Quaternary pyridinium compound in glutaraldehyde—treated
ovalbumin (Hardy et al., 1976b).

ACKNOWLEDGMENTS

We thank the Forschungskuratorium Gesamttextil for the
financial support of this research project (AIF No. 2872). The
support was granted by the Bundeswirtschaftsministerium through
the Arbeitsgemeinschaft Industrieller Forschungsvereinigungen.
This work was also supported by the Minister fur Wissenschaft und
Bildung des Landes Nordrhein-Westfalen and the International Wool
Secretariat, London.

REFERENCES

Asquith, R. S. and Otterburn, M. S. (1971). Self-crosslinking in keratin under the influence of dry heat. Appl. Polymer Symp., 18, 277.

Asquith, R. S., Otterburn, M. S., Buchana, J. H., Cole, M., Fletcher, J. C. and Gardner, K. L. (1970). The identification of ε-(γ-L-glutamyl)-L-lysine cross-links in native wool keratins. Biochim. Biophys. Acta, 221, 342.

Cole, M., Fletcher, J. C., Gardner, K. L. and Corfield, M. C. (1971). A study of enzymatic hydrolysis applicable to the examination of processed wools. Appl. Polymer Symp., 18, 147.

Di Modica, G. and Marzona, M. (1971). Cross-linking of wool keratin by bifunctional aldehydes. Text. Res. J., 41, 701.

Happich, W. F., Windus, W. and Naghski, J. (1965). Stabilization of wool by glutaraldehyde. Text. Res. J., 35, 850.

Happich, W. F. (1971). New process expands uses for wool-skins. Appl. Polymer Symp., 18, 1483.

Hardy, P. M., Nicholls, A. C. and Rydon, H. N. (1976a). The nature of cross-linking of proteins by glutaraldehyde. Part I. Interaction of glutaraldehyde with the amino-groups of 6-aminohexanoic acid and of α-N-acetyl-lysine. J. Chem. Soc. Perkin T.I., 9, 958.

Hardy, P. M., Graham, J. H. and Rydon, H. N. (1976b). Formation of quaternary pyridinium compounds by the action of glutaraldehyde on proteins. J. Chem. Soc. Chem. Comm., 5, 157.

Kusch, P., Lubig, R. and Topert, G. (1972). Ein neues Verfahren zur chemischen Fixierung von Krimmergarn. Melliand Textilber., 53, 1393.

Lubig, R. (1972). Ein Beitrag zur Umsetzung von Wolle mit Glutaraldehyd. Dissertation, Technische Hochschule Aachen.

Milligan, B., Holt, L. A. and Caldwell, J. B. (1971). The enzymic hydrolysis of wool for amino acid analysis. Appl. Polymer Symp., 18, 113.

Reinert, F. (1972). Reaktionen des 1-Fluor-2,4-dinitrobenzols mit Faserproteinen, seine Nebenreaktionen und sich daraus ergebende Eigenschaftsanderungen. Dissertation, Technische Hochschule Aachen.

Scharff, H. D. (1972). Reaktionen von Faserproteinen mit Carbonsäurechloriden in dipolaren aprotischen Lösungsmitteln. Dissertation, Technische Hochschule Aachen.

Schmitz, I., Baumann, H. and Zahn, H. (1976). Ein Beitrag zur enzymatischen Totalhydrolyse von Wollkeratin. Proc. Int. Wool Text. Res. Conf., Aachen 1975 II, 313.

Schmitz, I. (1975). Analyse von saurelabilen Aminosaurederivaten nach enzymatischer Totalhydrolyse von Proteinen und Keratinen. Dissertation, Technische Hochschule Aachen.

Zahn, H., Beyer, H., Hammoudeh, M. M. and Schallah, A. (1969).
    Vernetzungs- und Selbstvernetzungsreaktionen bei Wolle.
    Melliand Textilber., 50, 1319.
Zahn, H., Schallah, A., Scharff, H. D. and Meichelbeck, H.
    (1971). Crosslinking of wool during treatment with carbox-
    ylic acid chlorides. Appl. Polymer Symp., 18, 163.
Ziegler, K. and Liesenfeld, I. (1976). Studies on glutaraldehyde
    treated wool. Proc. Int. Wool Text. Res. Conf., Aachen 1975
    III, 88.

22

COMPARISON OF WOOL REACTIONS WITH SELECTED MONO-
AND BIFUNCTIONAL REAGENTS

N. H. Koenig and Mendel Friedman

Western Regional Research Laboratory, Agricultural
Research Service, U.S. Department of Agriculture,
Berkeley, California 94710

ABSTRACT

The molecular structure of wool is discussed in relation to
chemical reactivity and the role of disulfide crosslinks. Ideal
characteristics of an effective medium (e.g. dimethylformamide)
for modifying wool include the ability to penetrate and swell wool
without interfering with reagents used. The extent of reaction of
wool or reduced wool is compared for mono- and bifunctional activated vinyl compounds, isocyanates, acid chlorides, acid anhydrides, sulfonyl chlorides, and alkyl halides. The degree of
crosslinking is assessed by solubility, supercontraction, and
tensile tests. Optical and electron scanning microscopy can give
evidence of external polymer deposition in contrast to internal
chemical modification. Effects of crosslinking by bifunctional
reagents are related to changes in mechanical, chemical, and
biological (moth-resisting) properties of the modified wool.

INTRODUCTION

Wool is a composite protein material highly crosslinked by
the natural amino acid cystine, which makes up about 12% of the
wool weight. Chemical modification of wool often aims to add
additional crosslinks that are more stable to some environments
than the original disulfide crosslinks. New crosslinks can be
introduced by reaction with bi- or polyfunctional reagents,
thereby connecting two or more suitably located reactive sites in
the protein. Effects of crosslinking wool include decreased
swelling in solvents that can swell wool, increased wet strength,
and greater resistance to chemical and biological attack. However, elongation is decreased and the fibers are often more

brittle. In this paper we describe some structural features of wool, compare wool reactions mainly those studied by the authors, describe applicable methods to measure crosslinking, and show how crosslinking can change wool's properties.

WOOL STRUCTURE AND APROTIC SWELLING MEDIA

The complex morphology of the wool fiber (Lundgren and Ward, 1962) is illustrated in Fig. 1. Fine wool fibers contain two types of cells: flattened cuticle cells on the surface and under them long, polyhedral cortical cells aligned in the fiber direction. Coarse wool fibers contain a third structure, the medulla, which forms a spongy central core. A cell membrane complex separates the cells. The cuticle cells are made up of epicuticle, exocuticle, and endocuticle. The cuticle is important in controlling diffusion of chemicals into the interior. The cortex, which controls bulk properties, has orthocortex and paracortex regions differing in swelling properties and reactivity. Microfibrillar rods embedded in a disordered protein matrix comprise the substructure of the cortex. The microfibrils are believed to consist of a group of protofibrils. Bradbury (1973, 1975) thinks that the protofibrils probably consist of two-strand α-helical polypeptide units.

Wool, like other keratin proteins, consists of amino acid residues linked by peptide bonds. The extended peptide chains are crosslinked by disulfide bonds, resulting in a strong, resilient, insoluble material stable to most environmental attack. Some special characteristics (Friedman et al., 1973; Friedman and Masri, 1974, Friedman, 1977) that may give wool special utility are (a) low solubility; (b) steric accessibility to water and aqueous solutes; (c) physical form -- as crimped and resilient fibers; (d) relatively high content of reactive groups that may be chemically modified to provide desirable properties.

Wool can be modified in two general ways: surface deposition or internal chemical modification. The present discussion is restricted to internal modification, but to highlight the difference we shall first mention one treatment that forms a surface polymer, probably poly(4-vinylpyridine) with coordinately bound zinc (Koenig and Friedman, 1974). Optical microscopic examination showed extensive deposits on fiber surfaces (Fig. 2).

In contrast, we examined wool chemically modified by a mixture of tetrabromophthalic anhydride and tolylene diisocyanate (Koenig and Friedman, 1977). Simultaneous reaction occurred in hot cresol, which swells wool. The treated wool was both flame resistant and machine washable. Native wool (Fig. 3) and treated wool (Fig. 4) were examined by scanning electron microscopy. In this case, comparison shows no deposit of non-wool material on the surface. Therefore, we infer that treatment results primarily in internal chemical modification.

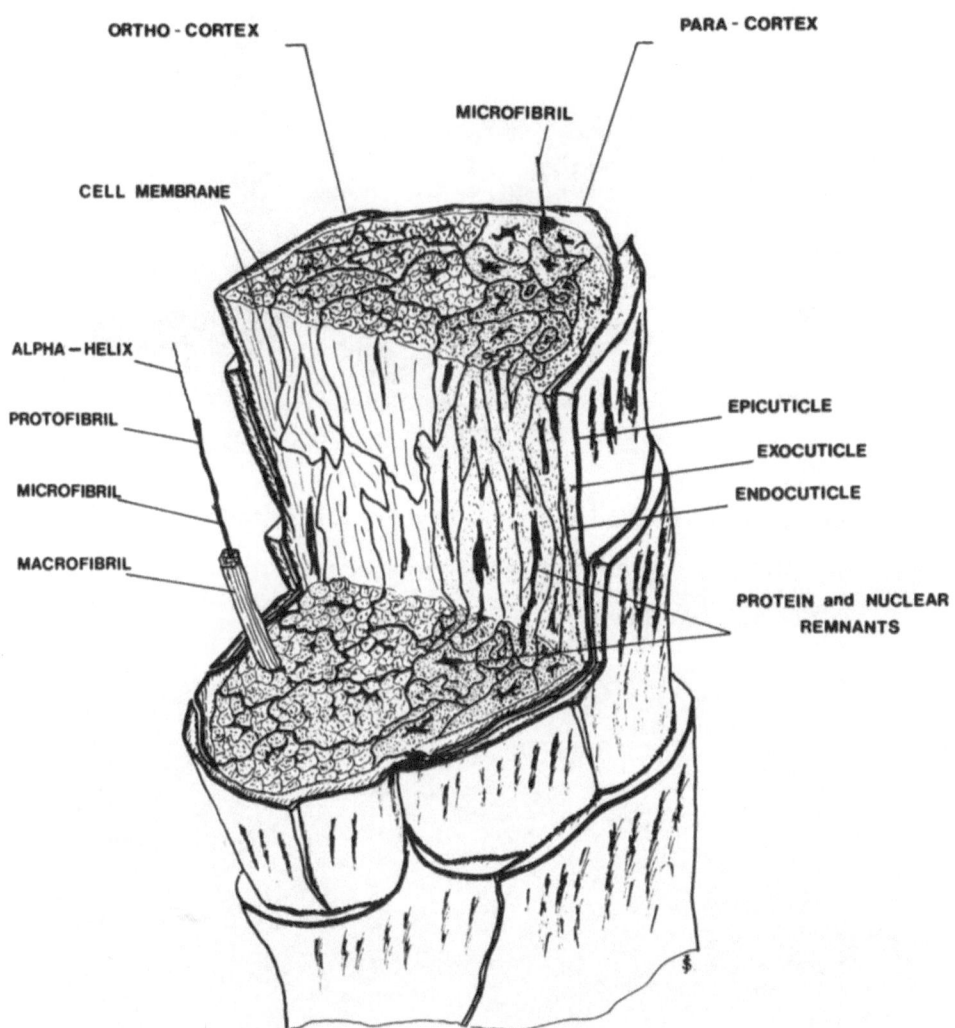

Fig. 1. Schematic representation of a wool fiber. (We thank Professor J. Sikorski, University of Leeds, for this figure; cf also Dobb et al,. 1961).

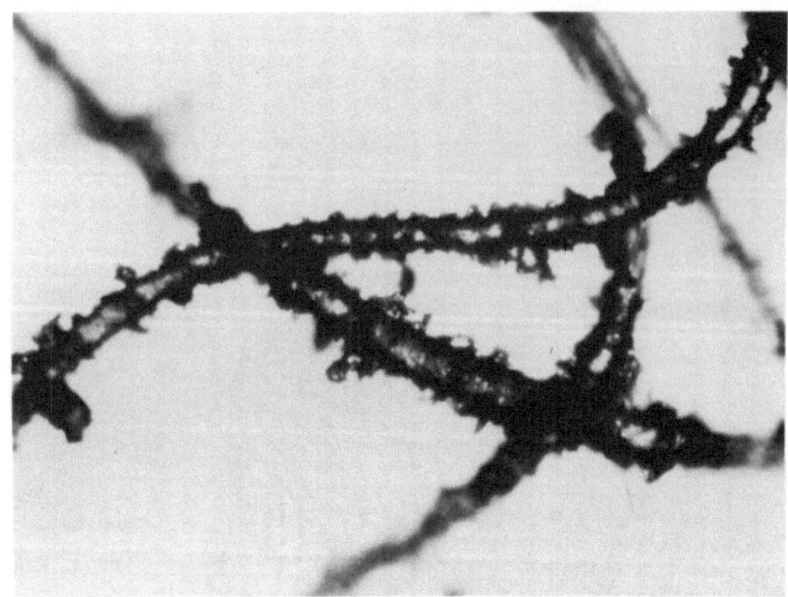

Fig. 2. Wool fibers treated with 4-vinylpyridine-zinc chloride (x170).

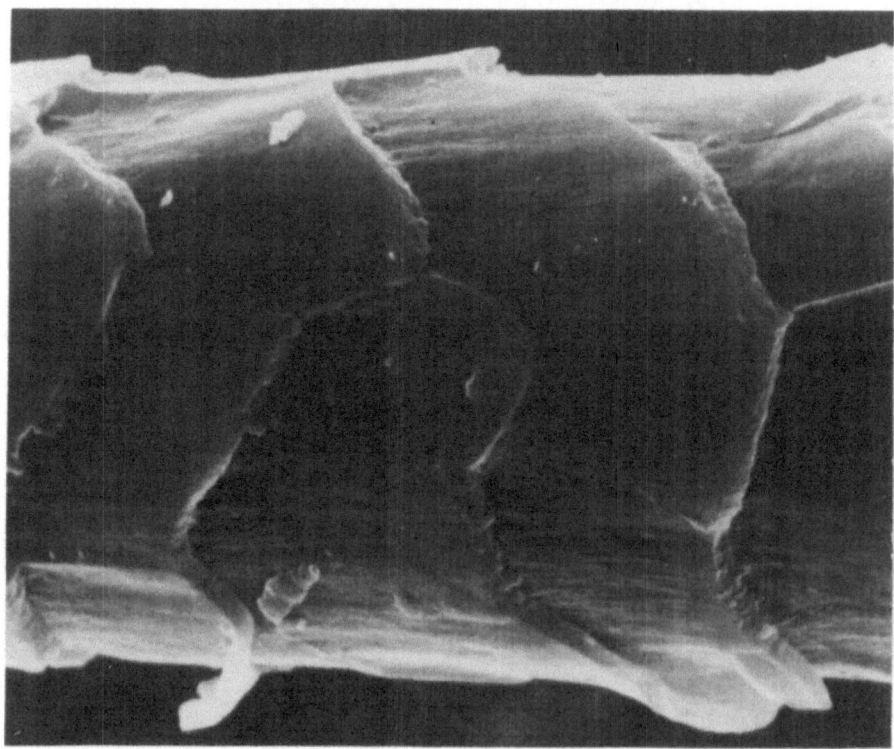

Fig. 3. Electron scanning microscopy (X3000) of native wool.

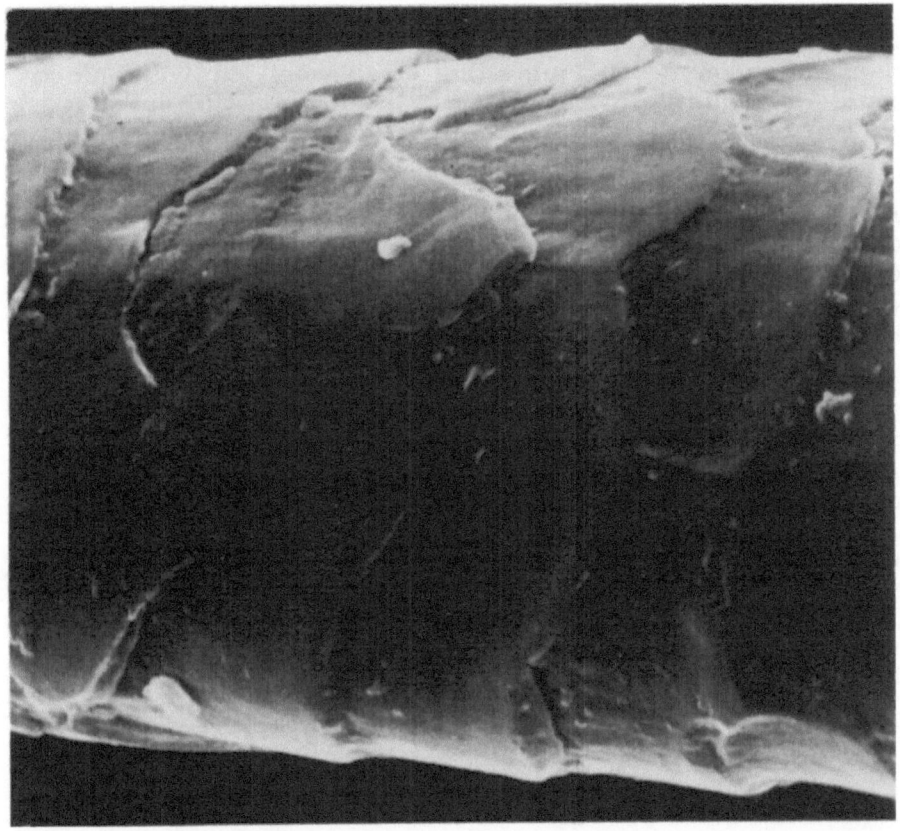

Fig. 4. Electron scanning microscopy (X3000) of wool treated with tetrabromophthalic anhydride and tolylene diisocyanate (19% uptake).

Chemical modification of wool requires penetration into the internal keratin structure. Wool keratin consists of polypeptide chains with attached reactive groups such as amino, hydroxyl, amide, sulfhydryl, phenolic, guanidino, imidazole, and carboxyl. The amino acid chains of the wool molecule are held together tightly by electrostatic bonds, hydrogen bonds, hydrophobic interactions, dipole attractions, dispersion forces, and disulfide bonds. Therefore, some molecules that react with soluble proteins will not modify wool because they are too large or of unsuitable polarity to penetrate the wool structure. However, in the presence of liquids that swell wool, medium-sized molecules can enter. In recent years, a few special nonaqueous liquids (media), particularly dimethylformamide (DMF) (Koenig, 1962) and dimethyl sulfoxide (DMSO) (Koenig, 1961), have been found useful for internal chemical modification of wool.

The course of many chemical reactions is profoundly affected by such dipolar, aprotic solvents (Parker, 1969). Protein-solvent interactions as illustrated with DMSO are as follows (Friedman, 1967, 1968; Friedman and Koenig, 1971): (1) Preferential solvation of positive charges by DMSO leaves negative charges and other nucleophilic centers free from destabilizing influences of positive charges. (2) The strong dipole of the sulfoxide group in DMSO acts as a hydrogen-bonding acceptor. (3) Proteins are polyelectrolytes so that their ionization is governed by the medium. (4) Hydrogen-bonding and hydrophobic interactions influence protein solution conformations. (5) Degree of ionization of the protein functional groups determines the concentration of each nucleophilic species at any given pH.

Aprotic swelling media suitable for chemical modification of wool include DMSO, DMF, 1-methyl-2-pyrrolidinone, and 4-butyrolactone. An effective medium for acylating-type internal modification of wool should have low molecular weight, lack hydrogen atoms, be highly polar, swell wool fibers, have a high boiling point, and exert a catalytic effect (Koenig, 1976).

CROSSLINKING TESTS

The wool was modified as previously described (Koenig et al., 1973; Koenig, 1976), and uptake was calculated as percent increase in dry weight. Crosslinking in wool has been extensively studied. Various methods have been developed to measure the extent of crosslinking (McPhee, 1958; Alexander et al., 1963; Whitfield and Wasley, 1964; Zahn et al., 1969; Caldwell and Milligan, 1970; Wold, 1972; Hinton, 1974). Evidence for crosslinking is often indirect. The results can be ambiguous because changes in wool properties caused by substituents introduced by monofunctional reagents can be similar to changes by crosslinks formed by bifunctional reagents. Moreover, bifunctional reagents can react at one wool site only (that is, react without crosslinking) and they may polymerize. It is therefore desirable to compare bifunctional reagents with monofunctional ones which are similar in chemical structure. It is also desirable to obtain evidence by two or more kinds of test. Effects of mono- and bifunctional reagents were shown to be strongly differentiated by several crosslinking tests of reduced wool treated with closely related maleimides (Moore and Ward, 1956). Some of the procedures used to evaluate selected (crosslinking) reactions include:

Alkali Solubility. This test measures the percent weight loss of the wool after 1 hr in 0.1 M NaOH at 65°C (Harris and Smith, 1936). Since alkali cleaves disulfide bonds, solubility may be reduced by new alkali stable crosslinks. The effect of these crosslinks may be offset if the treatment also cleaves some disulfide bonds, attacks peptide bonds, or increases the acid character of the wool (Koenig, 1965).

Sodium Sulfide Solubility. Percent weight loss is determined for a wool sample exposed to 3% $Na_2S \cdot 9H_2O$ for 1 hr at 65°C. Nine-tenths of normal wool is dissolved by the combined attack of sulfide ion and hydroxide ion on the disulfide bonds. A drastic reduction in solubility was reported for wool treated with glutaraldehyde, which presumably crosslinks the wool (Happich et al., 1965).

Acid Solubility. Percent weight loss is measured after exposing the wool sample to 5 M HCl for 1 hr at 65°C (Zahn and Würz, 1954). Solubility may be reduced if the treatment introduces acid-stable crosslinks. On the other hand, an opposing effect results if the treatment lowers the molecular weight by bond cleavage or increases the relative number of basic groups.

Peracetic Acid-Ammonia Solubility. Percent weight loss is determined after 24 hr at room temperature in 2% peracetic acid followed by 24 hr in 0.3% NH_3 (Alexander et al., 1952). Peracetic or performic acid oxidizes disulfide bonds to sulfonic acid groups. Up to 90% of the wool then dissolves in dilute alkali. The test is very sensitive to a small number of stable crosslinks. Bifunctional reagents reduce the solubility to a greater extent than monofunctional reagents of a similar size (Caldwell et al., 1966).

Supercontraction. Single fibers are removed from treated fabrics and their length is measured before and after immersion in 5% $NaHSO_3$ for 1 hr at 97°C. Because the properties of single fibers are quite variable, several tests must be made to get a statistically valid value. The fibers shorten (supercontract) to about 3/4th of their original length. Addition of new stable crosslinks often prevents supercontraction by stabilizing the original molecular structure. Monofunctional reagents, especially if they introduce bulky groups, may also decrease supercontraction.

<u>Fiber Tensile Tests.</u> Single fibers are extended under water at 21°C at a constant rate. Most test values are the average of 17 tests. An applied stress causes elongation (strain) of the fibers. Hydrogen bonds are broken, so that it is easier to detect new crosslinks because they usually increase molecular constraints. However, deformation also produces changes in entropy and rearrangement of intermolecular forces; interpretation at a molecular level is therefore difficult. Moreover, different types of crosslink may affect fiber properties differently.

<u>Insect Tests.</u> Sometimes resistance to insect enzymes is related to crosslinking. Some fabrics were evaluated for resistance to moths and other insects. Black carpet beetle larvae, which attack wool more severely than moth larvae, were allowed to feed on the samples at the Stored-Product Insects Research and Development Laboratory, Agricultural Research Service, Savannah, Georgia. Treated and untreated samples (0.5 g) were exposed to the larvae and excrement weight was measured. Wool is considered satisfactorily resistant if the average quantity of excrement per larva is not over 0.5 mg, provided no single value is over 0.6 mg and the untreated check is over 1.5 mg (the values ranged from 1.9-3.2). The data are reported relative to untreated wool as 100%; the lower the value, the greater the insect resistance of the treated sample.

REACTIONS WITH MONO- AND BIFUNCTIONAL REAGENTS

In this paper we compare effects of modifying wool with selected monofunctional and structurally related bifunctional reagents. Chemical, physical, and biological properties of the modified wool keratin are described. Our main objective has been to use differences in properties resulting from these two types of treatment to evaluate effects of crosslinking.

ACTIVATED VINYL COMPOUNDS

Wool was modified with activated mono- and bifunctional vinyl compounds in aqueous DMSO and mixed media (Koenig et al., 1973). Amino acid and other chemical analyses showed that lysine and histidine side chains in wool reacted in proportions dependent on the particular vinyl compound. The mechanism of reactions of activated vinyl compounds with soluble proteins was studied by Friedman (1967). The reactivity of these compounds with water is less than that of acylating reagents so that reactions can be carried out in aqueous solution. One way to get a high degree of reaction with wool is to reduce the disulfide bonds first, forming reactive sulfhydryl groups (Moore,

1960; Schöberl, 1960). Native wool has been treated with acrylonitrile (Bikales et al., 1957) and vinyl sulfone (Schoene, 1951), but in the presence of NaOH, which risks wool damage. Our present study is with native wool in slightly alkaline aqueous media (pH 8.4), as well as in DMSO, with or without triethylamine (Et_3N).

Wool has side chains with basic amino, imidazole, and guanidino groups. These groups can in principle react with activated vinyl groups by Michael-type nucleophilic addition (Suminov and Kost, 1969). The vinyl double bond is activated by an adjacent strong electron-withdrawing group such as cyano or sulfone. From studies on amino acids and peptides (Friedman and Wall, 1966), reactivity with wool is expected to increase with the electron-withdrawing power (electronegativity) of the group next to the double bond.

The carbon-nitrogen bonds formed in reactions with lysine or histidine are stable to hydrolysis with strong HCl. Consequently, reaction of these amino acids can be determined by amino acid analysis. In general, other amino acids that might react, e.g., serine and tyrosine, did not. Reaction of lysine and histidine wool side chains with monofunctional acrylonitrile forms dicyanoethyllysine and cyanoethylhistidine.

Bifunctional compounds can react in a more varied manner than monofunctional compounds. For example, Fig. 5 shows one possible reaction of vinyl sulfone. In this case, reaction with lysine and histidine has crosslinked two chains of the wool structure.

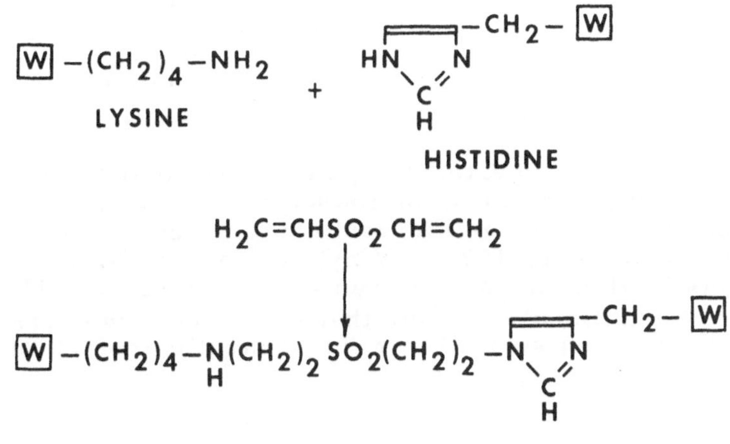

Fig. 5. Amino acids of two wool (W) chains crosslinked by vinyl sulfone.

Compounds with other possible activating groups were compared with vinyl sulfone in DMSO (Table 1). As anticipated from analogous reactions (Friedman and Wall, 1966; Ring et al., 1967), the highly electronegative sulfone group has a greater activating effect than the ester group. Accordingly, reaction with vinyl sulfone (83% lysine, 74% histidine) is much faster than with the ester, tetramethylene diacrylate (30% lysine, 8% histidine). The ester data also show that lysine reacts faster than histidine. The low reactivity of divinylbenzene suggests that the activating group needs a positive field, not merely ability to accept charge by resonance. Bis(β-chloroethyl) vinyl phosphonate was also unreactive with native wool, although this compound reacts readily with reduced wool (Friedman and Tillin, 1970).

Water-soluble vinyl compounds were compared in aqueous borate (Table 2). Vinyl sulfone reacted with 86% of the lysine in wool and 90% of the histidine in 1 hr. N,N'-methylenebisacrylamide reacted with less than 1/2 as much lysine as did vinyl sulfone because the amide group is not so strongly electron-withdrawing as the sulfone group (Friedman and Wall, 1966). Methyl vinyl ketone reacts with less lysine than does vinyl sulfone. On the other hand, it is very interesting that histidine reacts completely with this reagent. In no other case was reaction complete with either histidine or lysine. Acrylonitrile reacted less with lysine (55%) and histidine (67%) than did vinyl sulfone. This parallels the 28°C data (Friedman and Koenig, 1971), and confirms that the cyano group is less activating than the sulfone group. The most noticeable reaction of arginine was with vinyl sulfone (18% in 60 min).

Peracetic acid-ammonia solubility studies (Table 3), have control and treated fabrics grouped according to the medium, DMSO or aqueous.

For samples in DMSO, the medium alone had a very slight effect, reducing solubility from 87% (untreated wool) to 76%. As expected, vinyl sulfone further decreased solubility to 60%. Tetramethylene diacrylate, which could form crosslinks unstable to the test conditions (Holt and Milligan, 1970), did not reduce solubility.

Fabrics treated in borate buffer were also tested. The aqueous borate medium alone decreased solubility since mildly alkaline conditions cause self-crosslinking reactions of wool (Weideman and Wevers, 1971). Vinyl sulfone and N,N'-methylenebisacrylamide caused even lower solubility. They are examples of bifunctional reagents that introduce new crosslinks detectable by reduced solubility in the peracetic acid-ammonia test.

ACYLATING REAGENTS

Acid Chlorides. When wool is treated with mono- and bifunctional acid chlorides in hot DMF, the primary reaction is

Table 1. Comparative Treatments of Wool with Vinyl Compounds in DMSO, at 60°C for 2 hr

Compound	Structure	Moles reacted, %		
		Lysine	Histidine	Arginine
None (wool in Et$_3$N/DMSO)		8	12	3
Vinyl sulfone	CH$_2$=CHSO$_2$CH=CH$_2$	83	74	7
Tetramethylene diacrylate	CH$_2$=CHCO$_2$(CH$_2$)$_4$O$_2$CCH=CH$_2$	30	8	
Divinylbenzene	CH$_2$=CHC$_6$H$_4$CH=CH$_2$	6	14	
Bis(β-chloroethyl) vinyl phosphonate	CH$_2$=CHPO(OCH$_2$CH$_2$Cl)$_2$	10	5	0

Table 2. Comparative Treatments of Wool with Vinyl Compounds in pH 8.4 borate, 80°C, 1 hr

Compound	Structure	Moles reacted, %		
		Lysine	Histidine	Arginine
None (wool in borate)		10	90	3
Vinyl sulfone	CH$_2$=CHSO$_2$CH=CH$_2$	86	90	18
N,N'-Methylenebis-acrylamide	CH$_2$=CH(CONH)CH$_2$(NHCO)CH=CH$_2$	39	19	3
Methyl vinyl ketone	CH$_2$=CHCOCH$_3$	52	100	
Acrylonitrile	CH$_2$=CHCN	55	67	1

Table 3

Peracetic Acid-Ammonia Solubility Test for Crosslinking of
Controls and Treated Wools

Compound	Moles reacted, %			Solubility %
	Lysine	Histidine	Arginine	
None (untreated wool)				87
In DMSO (0.13 M Et$_3$N)				
None (wool in Et$_3$N/ DMSO	8	12	3	76
Vinyl sulfone	83	74	7	60
Tetramethylene diacrylate	60	34	7	87
Divinylbenzene	6	14		75
In aqueous borate, pH 8.4				
None (wool in borate)	10	15	3	63
Vinyl sulfone	83	89	14	35
N,N'-methylenebis- acrylamide	39	19	3	49
Methyl vinyl ketone	59	100	5	62

probably acylation; the HCl formed may be volatilized, extracted, or bound to basic sites in wool (Koenig and Crass, 1975).
The maximum uptake of a monofunctional reagent, myristoyl
chloride, is about 35% by weight (Koenig, 1976). The maximum
is the same in four swelling media, although the rates vary.
The uptake is about 1.67 mmoles/g wool, which may represent
reaction of primary amino groups, 0.21 mmoles/g wool, and
aliphatic hydroxyl groups, 1.46 mmoles/g wool. Other potential reactive groups in wool (in mmoles/g wool) are: arginine,
0.55; histidine, 0.07; phenolic, 0.29; and free carboxyl, 0.84.

The effect of concentration on maximum uptake was studied (Koenig and Crass, 1975) with a bifunctional reagent, dodecanedioyl chloride (Table 4). With an acid chloride-DMF
ratio of 1:5 (v/v), maxium uptake was 22% in 30 min (no further
change in 120 min). As the proportion of acid chloride increased, however, uptakes up to 29% were obtained. This acid
chloride is bifunctional; crosslinking is favored by low concentration. At higher concentrations, all available wool sites react
before the primary acid chloride adduct can form crosslinks
through the second acid chloride. Hence a larger number of
acid chloride molecules can react with the fixed number of wool
sites, giving a higher maximum uptake than that realizable by
crosslinking.

Modification alters the resistance of wool to chemical attack, including oxidation by peracetic acid (Table 5). Treatments with short- and medium-chain length monoacid chlorides (propionyl and octanoyl) increase the peracetic acid-ammonia solubility, possibly as a result of a wedging process whereby the small substituent facilitates penetration (Koenig, 1965). On the other hand, long-chain monoacid chlorides decrease solubility. This increased resistance to peracetic acid oxidation may reflect steric shielding of the disulfide bonds by long hydrocarbon chains.

Table 4

Effect of Dodecanedioyl Chloride Concentration on Maximum Uptake, 1.2 g wool, 105°C

Acid chloride, ml	DMF, ml	Reaction time, min	Uptake of acid chloride, %
1	5	10	16
1	5	30	22
1	5	120	22
2	4	60	25
2	4	90	26
3	3	90	28
3	3	120	29

Table 5

Peracetic Acid-Ammonia Solubilities of Treated Wool

Acid chloride	No. of carbons	Uptake, %	Solubility, %
Untreated wool			84
Monoacid chlorides			
Propionyl	3	10	100
Octanoyl	8	17	92
Lauroyl	12	16	60
Myristoyl	14	21	31
Myristoyl	14	25	17
Diacid chlorides			
Adipoyl	6	14	47
Sebacoyl	10	18	15
Dodecanedioyl	12	16	27
Dodecanedioyl	12	20	11

Long-chain diacid chlorides also decrease the peracetic acid-ammonia solubility. Diacid chlorides are much more effective in this respect than monoacid chlorides of corresponding chain length. This increased effectiveness is evidence for new crosslinks formed by diacid chlorides. The new crosslinks hold the wool chains together and resist solubilization by compensating for the cleaved disulfide crosslinks. Even a few of new crosslinks can greatly reduce solubility.

Mechanical properties and supercontraction of wet single wool fibers are shown in Table 6. Overall, treatment decreases tensile strength, as shown by decreased stress needed to break the fibers and by smaller extension (elongation) at break. Strength loss decreases as chain length increases.

Table 6

Wet Single Wool Fiber Strength, Elasticity, and Supercontraction

Acid chloride	Uptake, %	Stress at break, g/tex	Extension at break, %	Super-contraction, %
Untreated wool		8.1	47	24
Propionyl chloride	11.2	3.8	36	18
Lauroyl chloride	24.0			16
Myristoyl chloride	21.3	5.6	46	
Succinyl chloride	10.3	2.5	20	
Adipoyl chloride	16.0			4
Sebacoyl chloride	18.3	3.9	22	3

A very significant effect is the large difference in extensibility between the monoacid and diacid chloride treated fibers. Values for extension at break of the fibers treated with monoacid chlorides, 36% and 46%, are roughly twice as great as for fibers treated with the diacid chlorides, 20% and 22%. The lower extensibility of the latter can be attributed to the restrictive influence of the added crosslinks.

Supercontraction was measured on single fibers exposed to hot $NaHSO_3$. The monoacid chlorides cause a moderate lowering of the supercontraction value. In contrast, diacid chloride treament almost elminates supercontraction. Since the decrease in supercontraction corresponds to increased acid chloride functionality, the data suggest that the diacid chlorides introduce new crosslinks. Such crosslinks may stabilize the structure and in this way prevent supercontraction that normally occurs when $NaHSO_3$ cleaves the disulfide crosslinks of native wool. In summary, supercontraction, extension-at-break, and peracetic acid-ammonia solubility all suggest crosslinking by diacid chlorides.

Isocyanates. Wool has been chemically modified with mono- and diisocyanates in DMSO (Koenig, 1961) and DMF (Koenig, 1962). With phenyl isocyanate, the maximum uptake is about 38% by weight, or 3.18 mmoles/g wool. Rates of modification in various effective media decrease in the order: DMSO > DMF > N-methylpyrrolidinone > 4-butyrolactone. Although reaction rate varies in these media, the final wool reaction products are similar and do not include the medium. These and earlier studies (Fraenkel-Conrat et al., 1945; Farnworth, 1955) indicate that the isocyanate adds to wool side chains (amino, guanidino, hydroxyl, phenolic, amide, carboxyl, etc.) with reactive hydrogen atoms. Some sites have been identified by analysis of enzyme hydrolysates (Caldwell et al., 1973).

Effects of crosslinking have been examined in Table 7 by comparing two solubility tests of wool treated with structurally similar mono- and bifunctional reagents, $CH_3(CH_2)_7NCO$ (octyl isocyanate) and $OCN(CH_2)_6NCO$ (hexamethylene diisocyanate).

Table 7

Peracetic Acid (AcOOH) and Sodium Sulfide (Na_2S) Solubility

Treatment	Uptake, %	AcOOH solubility, %	Na_2S solubility, %
Untreated wool	0	82	88
Octyl isocyanate	8	48	54
Hexamethylene diisocyanate	8	9	12
Hexamethylene diisocyanate	9	-	5
Hexamethylene diisocyanate	19	2	-
Octyl isocyanate	20	31	23

To alter solubility, the bond with the isocyanate adduct must be stable, under test conditions, to oxidation by peracetic acid or reduction by Na_2S. All solubility values are lower than those of wool suggesting that both treatments are stable to both tests. The effect of a given reagent in reducing solubility increases with uptake. For a given uptake, i.e. 8%, the solubilities are much lower for the diisocyanate than for the monoisocyanate, which indicates that hexamethylene diisocyanate crosslinks wool.

The same two isocyanate treatments were further tested by supercontraction and tensile tests shown in Table 8 (Koenig, 1976).

Table 8

Supercontraction and Tensile Properties of Wool Fibers after
Mono- and Disocyanate Treatments

Reagent	Uptake, %	Supercon- traction, %	Extension at break, %	Break stress, g/tex
Control (untreated wool)	0	24	40	7.0
$CH_3(CH_2)_7NCO$	20	22	39	6.4
$OCN(CH_2)_6NCO$	19	0.4	18	3.1

The monofunctional octyl isocyanate has only a small effect on
these tests. On the other hand, hexamethylene diisocyanate prevents supercontraction almost entirely and drastically lowers
breaking stress and elongation. These results can be attributed
to molecular constraints of stable crosslinks introduced by the
diisocyanate. In a related study, Watt (1965) provided further
evidence for crosslinking by hexamethylene diisocyanate, in that
it reduces swelling of treated wool in formic acid to a greater
degree than the monofunctional phenyl isocyanate. In summary,
experiments by Watt and us with four different tests designed to
measure crosslinking all support the view that bifunctional isocyanates crosslink wool.

Acid Anhydrides. Many organic acid anhydrides react with
wool in the presence of DMF (Koenig, 1965). Modification with
cyclic anhydrides, e.g. succinic anhydride, increases the number of carboxyl groups. The reaction attaches the substituent to
wool by an amide link and forms a carboxyl group from the
cleaved anhydride ring.
Noncyclic anhydrides, e.g., butyric anhydride, also acylate
the amino group, but do not introduce new carboxyl groups. The
modified wool will also be more acid in character, since some of
the basic groups have been acylated to nonbasic amide side
chains. The original acid groups will be less neutralized by
basic groups than in the native wool. However, this effect will
be increased, in the case of the cyclic dianhydride, by the additional carboxyl group introduced.
Acid or alkali "solubility" measures the extent of chemical
attack by HCl or NaOH. Modification with acid anhydrides, by
decreasing the proportion of basic groups to acid groups, might
be expected to decrease acid solubility. Conversely, the increase in acidic groups and accompanying decrease of basic
groups tend to increase alkali solubility. Steric effects of modi-

fication also influence chemical solubility. A small substituent may facilitate chemical attack by acting as a wedge in the wool molecule. On the other hand, a large substituent may decrease solubility by shielding vulnerable areas.

Another factor influencing solubilities is the polarity of the substituent. In particular, nonpolar substituents such as long-chain hydrocarbons can provide an environment less favorable to acid or alkali attack. Accordingly, the net effect of modification on acid and alkali solubility reflects various substituent influences including change of acid-base balance, steric factors, and polarity effects. Alkali and acid solubilities of anhydride-modified fabrics are shown in Table 9.

Table 9

Alkali and Acid Solubility of Treated Wool

Acid anhydride	Uptake %	Number adduct carbons	Alkali solubility, %	Acid solubility, %
Untreated wool	0	0	13	15
Propionic	9	4	53	24
Butyric	6	5	37	20
Heptanoic	15	8	30	12
Succinic	15	4	42	21
n-Octenylsuccinic	10	12	24	10
	18	12	28	7
Dodecenylsuccinic	11	16	20	6
	15	16	22	3
n-Octadecenylsuccinic	19	22	10	3
11-Tricosenylsuccinic	15	27	10	3

The data show that only very large nonpolar reagents reduce alkali solubility. Correspondingly, very small reagents increase acid solubility. The increase, which may be ascribed to a wedge effect, is small because it is opposed by a decrease in basic character.

Our approach based on studying the influence of systematic variations in the modifying compounds may permit analysis of relative contributions of polar, steric, and structural factors as they relate to observed changes in wool properties.

Some cyclic anhydride-treated wools were evaluated by the peracetic acid-ammonia solubility test (Table 10). The smaller two monofunctional anhydrides, succinic and glutaric, give products that are about as soluble as untreated wool. The third and

bulkier monofunctional compound, hexahydrophthalic anhydride,
has a moderate effect in lowering solubility, to 46%. However,
pyromellitic and benzophenonetetracarboxylic dianhydrides give
a much greater effect for a lower molar uptake — evidence for
crosslinking by these two bifunctional compounds.

Table 10

Peracetic Acid-Ammonia Solubility Test for Crosslinking

| Treatment (in DMF, 110°C) | Anhydride uptake | | Peracetic acid-NH$_3$ solubility, % |
	% (g per 100 g wool)	mmoles per g wool	
None (untreated wool)	0	0	84
Succinic anhydride	17	1.70	85
Glutaric anhydride	23	2.02	81
Hexahydrophthalic anhydride	22	1.43	46
Pyromellitic dianhydride	18	0.83	26
Benzophenonetetra- carboxylic anhydride	27	0.84	25

One biological consequence of modifying wool with acid anhydrides was evaluated by carpet beetle larvae feeding studies
shown in Table 11. Change in acid-base character, protective
groups, or crosslinks could affect the ability of insect enzymes
to digest the wool. The greatest resistance and at the lowest
molar uptakes are given by the two dianhydrides. However, the
lesser effect of succinic anhydride could be because it is a
smaller molecule and not necessarily because it is monofunctional and unable to form new crosslinks.

In summary, the lowered solubility in the peracetic acid-
ammonia test indicates dianhydride treatments crosslink wool,
while the increased resistance to insects can be ascribed partially to crosslinking.

SULFONYL CHLORIDES

Wool has also been treated with mono- and bifunctional sulfonyl halides in aprotic swelling media (Koenig, 1967; Koenig and
Friedman, 1976). Typical experiments are summarized in
Table 12. The most favorable solvent volume ratio is about 1.5
ml methyl pyrrolidinone to 4.5 ml 4-butyrolactone, with a 19%
uptake in 1.5 hr. The limited weight uptake despite large excess

of reagent or longer reaction time is evidence that the sulfonyl chloride is reacting with a limited number of reactive sites in wool.

Table 11

Insect Resistance of Acid Anhydride Treated Wool

Treatment	Anhydride uptake		Relative larval excrement, %
	% (g per 100 g wool)	mmoles per g wool	
None (untreated wool)	0	0	100
Succinic anhydride	16	1.60	63
Pyromellitic dianhydride	26	1.18	23
Benzophenonetetra- carboxylic dianhydride	27	0.84	26

Table 12

Benzenesulfonyl Chloride Treatments:
2 ml reagent, 1 g wool, 105°C

Methyl pyrrolidinone, ml	Butyrolactone, ml	Time, hr	Uptake, %
0	6.0	1.0	9
6.0	0	1.0	12
2.0	4.0	1.0	15
1.5	4.0	1.5	19
1.8	4.0	1.5	18
1.5	4.0	2.0	20

Steric accessibility of the reagent appears to influence the extent of modification, since the molar uptake of the larger bromo derivative is only one-third that of unsubstituted benzenesulfonyl chloride.

Amino acid analyses (Table 13) show that, in contrast to acylated wool, the sulfonylated wool contains lysine, histidine, and tyrosine derivatives that survive hydrolysis in 6 N HCl at 110°C for 24 hr. Further evidence that the sulfonyl chlorides react covalently with wool is that samples treated with benzene-

or methanesulfonyl chloride have increased resistance to sodium hypochlorite, an oxidizing agent that damages natural wool.

Table 13

Amino Acid Analyses of Treated Wool

Sulfonyl chloride	Uptake, %	Moles reacted, %			
		Lysine	Histidine	Arginine	Tyrosine
Benzene	18	42	41	25	12
Dansyl	7	43	37	28	90

Reactions with diphenyl ether disulfonyl chloride are shown in Table 14. The limiting uptake is about 13%. Maximum uptake of this bifunctional reagent may be less than that of benzenesulfonyl chloride (20%) because of crosslinking, since each reagent molecule can combine with two reacting sites. However, it is also possible that fewer sites are accessible to the larger diphenyl ether disulfonyl chloride.

Table 14

Diphenyl Ether Disulfonyl Chloride Treatments, 1 g wool, 110°C

Disulfonyl chloride, g	Methyl pyrrolidinone, ml	Butyrolactone, ml	Time, hr	Uptake, %
0.5	0	5.0	.75	6
0.5	5.0	0	.75	12
1.0	1.5	4.0	.75	13
1.0	1.5	4.0	1.5	13

A biological effect of the treatment is that of resistance to moths and carpet beetle larvae. Table 15 shows that treated samples are more resistant to damage, since the larval excrement value is less. The greater resistance of the two disulfonyl chloride treatments compared to benzenesulfonyl chloride may be the result of new crosslinks introduced by the bifunctional disulfonyl chlorides. The difference is even more striking when the data are considered on a molar uptake basis.

Table 15

Insect Resistance of Sulfonyl Chloride Treated Wool

Treament	Reagent uptake		Relative larval excrement, %
	% (g per 100 g of wool)	mmoles per g wool	
Untreated wool	0	0	100
Dansyl chloride	7	0.30	77
Trichlorobenzene sulf.	16	0.65	49
Benzenesulfonyl	19	1.35	43
Benzenedisulfonyl	14	0.69	22
Diphenyl ether disulf.	15	0.51	24

KERATIN DISULFIDE REACTIONS

Sulfhydryl-Disulfide Interchange. The thiolate anion is a strong nucleophile able to react with disulfide bonds by nucleophilic displacement. Since these displacements are usually reversible, they are often described as sulfhydryl-disulfide equilibria or interchange reactions, reviewed by Friedman (1973). When native wool keratin is stressed, e. g. in distilled water at 100°C, scission of stressed disulfide bonds can occur by sulfhydryl-disulfide interchange (Caldwell et al., 1965; Feughelman, 1966; Asquith and Puri, 1970; Weigmann and Dansizer, 1971). Beyond a certain level, the process becomes irreversible. The force of retraction due to stress on the original disulfide bonds provides part of the "memory" directing the structure back to its native state. However, disulfide interchange can interpose new crosslinks that interfere sterically with reformation of the native α-keratin helices. Formation of disulfide bonds conformed to the extended state may stabilize the extended β-keratin crystallites.

DMSO has special properties that make it useful for modifying wool with water-sensitive compounds. At room temperature, effects of DMSO on wool are subtle: primarily swelling the wool to enhance reactivities of basic functional groups and facilitate sulfhydryl-disulfide interchange. For example, to elucidate the mechanism by which DMSO permeates and swells keratin proteins such as hair, Price and Menefee (1967) examined stress relaxation of human hair in DMSO. The decay of stress applied to keratin fibers immersed in DMSO proved to be a sensitive index of induced structural changes. At 24°C, stress relaxation was more rapid and more complete in 80-100% DMSO than in an aqueous medium. The difference is attributed mainly to DMSO-initiated sulfhydryl-disulfide interchange. Price and Menefee discussed possible mechanisms.

Friedman and Koenig (1971) believe the effect of DMSO can be usefully interpreted as follows: First, the rate and extent of sulfhydryl-disulfide interchange reactions of proteins are direct- ly related to the degree of ionization of the sulfhydryl group; i. e. it is the S⁻ (thiolate) anion that initiates interchange rather than the SH (sulfhydryl) group. They suggest that DMSO enhances in- terchange by hydrogen-bonding to the sulfhydryl proton (as in A, below), thus lowering its pK value and increasing its degree of ionization at any pH. An analogous effect may operate in mixed aqueous DMSO media, either as in A, directly, or through water as illustrated in B. Second, the strong hydrogen-bonding ability of water will result in effective solvation of the thiolate anion, as in C. The electron density on the sulfur anion is thereby de- creased and its nucleophilicity lowered. In DMSO, hydrogen bonding to the anion as in C is not possible because DMSO is only a strong proton accepter in hydrogen-bonding. In summary, DMSO can facilitate sulfhydryl-disulfide interchange by increas- ing both the concentration and the nucleophilic reactivity of the conjugate base of a protein sulfhydryl group.

$$(CH_3)_2 \text{--}S^+\text{--}O^-\text{----}H\text{--}S\text{--}W \qquad\qquad (A)$$

$$(CH_3)_2 \text{--}S^+\text{--}O^-\text{----}H\text{--}\underset{\underset{H}{|}}{O}\text{:----}H\text{--}S\text{--}W \qquad (B)$$

$$H\text{--}O\text{--}H\text{---}S^-\text{--}W \qquad\qquad (C)$$

Crosslinking Reduced Wool. Disulfide bonds are key struc- tural elements of wool. They stabilize keratin fiber structure and determine its characteristic tensile behavior. Reduction of disulfide bonds generates sulfhydryl groups, which are potential sites for further modification. Reduction of protein disulfide bonds followed by alkylating the generated sulfhydryl groups has been studied extensively (Alexander et al., 1963; Friedman and Noma, 1970; Maclaren, 1971; Friedman, 1973; Hinton, 1974). In such a study, Friedman and Tillin (1974) showed that partial reduction of wool disulfide bonds followed by alkylation of the generated sulfhydryl groups with 2-vinylpyridine could be controlled so that the treated wool fibers were not importantly weakened (cf. also Crewther, 1965).

The reduction of wool (W) with mercaptans and rebuilding of bonds with alkyl dihalides has been studied by Harris and colleagues (Patterson et al., 1941; Geiger et al., 1942; Harris and Brown, 1947).

$$W-S-S-W+2\ HS-CH_2COOH \rightleftarrows 2-W-SH+HOOCCH_2SSCH_2COOH \quad (1)$$

$$2-W-SH+Br(CH_2)_nBr \longrightarrow W-S(CH_2)_nS-W+2\ HBr \quad (2)$$

After reducing wool with thioglycollate and alkylating with ethylene dibromide, Zahn (1954) and Crewther (1967) identified the crosslinking amino acid derivative dimethylene-S, S'-dicysteine.

In the work of Harris and colleagues, about one-half of the available disulfide bonds were cleaved and recrosslinked. Table 16 compares two properties of reduced wool treated with a mono- and bifunctional reagent, respectively.

Table 16

Tensile Properties and Alkali Solubility of Reduced and
Alkylated Wool

Alkyl bromide	30% index	Alkali solubility, %
Control (untreated)	0.99	10.5
C_2H_5Br	0.70	18.3
$BrCH_2CH_2Br$	0.92	6.6

When a wool fiber is stretched in water up to 30% elongation (the yield limit) and then relaxed, it requires the same work to stretch it a second time. If a calibrated fiber is chemically treated so as to break chemical bonds, less work is required to stretch it again. The ratio to the original work is the "30% index" (Speakman, 1947). Treatment with ethyl bromide, which alkylates sulfhydryl groups so they cannot be reoxidized to disulfide bonds, reduced the 30% index from 0.99 to 0.70. A much smaller reduction of 30% index results when ethylene dibromide is used because original disulfide bonds are replaced by thioether crosslinks.

The difference between the two treatments is even more strikingly shown by alkali solubility. The ethyl bromide treatment raises alkali solubility from 10.5 to 18.3% because reduction plus alkyation without crosslinking decreases the molecular weight and allows reagent to penetrate more easily because swelling is less restricted. On the other hand, ethylene dibromide lowers the solubility to 6.6%. Reaction with the dibromide maintains molecular weight and replaces the disulfide bonds, which are unstable toward alkali, with stable thioether crosslinks. Other studies, with $Br(CH_2)_3Br$, showed that the wool was also more resistant to acids, oxidation, and reduction.

Although native wool resists attack by proteolytic enzymes, it becomes more susceptible if the fiber is mechanically damaged. If wool is reduced, then crosslinked with an alkyl dihalide, it is not attacked by pepsin, even after mechanical injury. On the other hand, enzyme digestion does occur if the reduced wool is methylated so that crosslinks are not reformed.

Two enzymes in moths appear to be involved in wool digestion, a cystine reductase and a protease (Powning and Irzykiewicz, 1960, 1962). By this system, cysteine is formed, which reduces disulfide bonds to sulfhydryl groups. The reduced protein is then hydrolyzed by the protease. In support of this hypothesis, wool in which disulfide bonds were first reduced and then crosslinked with an alkyl dibromide is moth resistant (Geiger, 1942). The crosslinking treatment not only removes sites of attack (disulfide bonds) by moth enzymes, but also restores the wool structure to nearly its native state, but with moth-resistant crosslinks. Effects of reduction followed by treatment with trimethylene dibromide on insect attack are shown in Table 17. As indicated by visual examination and loss in wool weight, the modified wools are attacked less by moth larvae and the resistance increases as the content of unchanged cystine decreases. Carpet beetle larvae attack wool more severely; nevertheless, visual appraisal and decrease in excrement weight show that the modified wool is decidedly more resistant to attack at 6. 5% cystine content. Since the new crosslinks are built into the molecular structure, this insectproofing effect is not impaired by drycleaning or laundering.

Table 17

Effect of Reduction Followed by Alkylation with $Br(CH_2)_3Br$ on Larval Feeding of Moths and Carpet Beetles (Geiger, 1942).

Cystine content, %	Weight loss due to moth larvae, %	Beetle larval excrement, mg
12. 2 (control)	15. 2	52
10. 0	8. 0	45
6. 5	0. 3	21

CONCLUSIONS

In summary, our results show that (a) wools modified with monofunctional compounds have properties that differ appreciably from those obtained with structurally related bifunctional ones and (b) structures and chemical properties of the modifying reagents can be chosen to achieve desired properties. These results suggest that one can develop general structure-reactivity relationships between structural and functional characteristics of the modifying reagents and physicochemical and biological properties of the modified proteins.

ACKNOWLEDGMENTS

We thank R. E. Bry and R. Davis for carpet beetle feeding tests and W. H. Ward and T. L. Hayes for scanning electron microscopy.

REFERENCES

Alexander, P., Fox, M., Stacey, K. A. and Smith, F. L. (1952). The reactivity of radiomimetic compounds. I. Crosslinking of proteins. Biochem. J., 52, 177-184.

Alexander, P., Hudson, R. F. and Earland, C. (1963). "Wool: Its Chemistry and Physics," Chapman and Hall, London.

Asquith, R. S. and Puri, A. K. (1970). The formation of mixed disulfides by the action of thioglycollic acid on wool cystine and its relationship to wool setting. Text. Res. J., 40, 273-280.

Bikales, N. M., Black, J. J. and Rapoport, L. (1957). The cyanoethylation of wool. Text. Res. J., 27, 80-81.

Bradbury, J. H. (1973). Structure and chemistry of keratin fibers. Adv. Prot. Chem., 27, 111-211.

Bradbury, J. H. (1975). The morphology and chemical structure of wool, IUPAC International Symposiun on Macromolecules, Jerusalem, Israel, (July 13-18) Abstracts, p. 49-50.

Caldwell, J. B., Leach, S. J. and Milligan, B. (1965). The mechanism of setting and the release of set in water. Text. Res. J., 35, 245-251.

Caldwell, J. B., Leach, S. J. and Milligan, B. (1966). Solubility as a criterion of crosslinking in wool. Text. Res. J., 36, 1091-1095.

Caldwell, J. B. and Milligan, B. (1970). The estimation of crosslinking in wool from the extent of swelling in formic acid. J. Text. Inst., 61, 588-596.

Caldwell, J. B., Milligan, B. and Roxburgh, C. M. (1973). The sites of reaction of phenyl isocyanate with wool. J. Text. Inst., 64, 461-467.

Crewther, W. G. (1965). Stress-strain characteristics of animal fibers after reduction and alkylation. Text. Res. J., 35, 867-877.

Crewther, W. G., Dowling, L. M., Inglis, A. S. and Maclaren, J. A. (1967). The formation of various cross linkages in wool and their effect on the supercontraction properties of the fibers. Text. Res. J., 37, 736-745.

Dobb, M. G., Johnston, F. R., Nott, J. A., Oster, L.,
   Sikorski, J. and Simpson, W. S. (1961). Morphology of the
   cuticle layer in wool fibers and other animal hairs. J. Text.
   Inst., 51, T153-T170.
Farnworth, A. J. (1955). The reaction between wool and
   phenyl isocyanate. Biochem. J., 59, 529-533.
Feughelman, M. (1966). Sulfhydryl-disulfide interchange and
   the stability of keratin structure. Nature, 211, 1259-1260.
Fraenkel-Conrat, H., Cooper, M. and Olcott, H. S. (1945).
   Action of aromatic isocyanates on proteins. J. Am. Chem.
   Soc., 67, 314-319.
Friedman, M. (1967). Solvent effects in reaction of amino
   groups in amino acids, peptides, and proteins with α,β-
   unsaturated compounds. J. Amer. Chem. Soc., 89, 4709-
   4713.
Friedman, M. (1968). Solvent effects in reactions of protein
   functional groups. Quart. Rept. Sulfur Chem., 2, 124-144.
Friedman, M. (1973). "Chemistry and Biochemistry of the
   Sulfhydryl Group in Amino Acids, Peptides, and Proteins,"
   Pergamon Press, Oxford, England and Elsmford, New York.
Friedman, M. (1977). Flame-resistant wool and wool blends.
   In "Flame-Retardant Polymeric Materials," Vol. 2, M.
   Lewin, S. M. Atlas and E. M. Pearce (Editors), Plenum
   Press, New York.
Friedman, M., Harrison, C. S., Ward, W. H. and Lundgren,
   H. P. (1973). Sorption behaviour of mercuric and methyl-
   mercuric salts on wool. J. Appl. Polym. Sci., 17, 377-390.
Friedman, M. and Koenig, N. H. (1971). Effect of dimethyl
   sulfoxide on chemical and physical properties of wool. Text.
   Res. J., 40, 605-609.
Friedman, M. and Masri, M. S. (1974). Interaction of mer-
   cury compounds with wool and related biopolymers. In
   "Protein-Metal Interactions," M. Friedman (Editor), Plenum
   Press, New York, Vol. 48, pp. 505-550.
Friedman, M. and Noma, A. T. (1970). Cystine content of
   wool. Text. Res. J., 40, 1073-1078.
Friedman, M. and Tillin, S. (1970). Flame-resistant wool.
   Text. Res. J., 40, 1045-1047.
Friedman, M. and Tillin, S. (1974). Partly reduced alkylated
   wool. Text. Res. J., 44, 578-580.
Friedman, M. and Wall, J. S. (1966). Additive linear free-
   energy relationships in reaction kinetics of amino groups
   with α,β-unsaturated compounds. J. Org. Chem., 31, 2888-
   2894.
Geiger, W. B., Kobayashi, F. F. and Harris, M. (1942).
   Chemically modified wools of enhanced stability. J. Res.
   Natl. Bur. Standards, 29, 381-389.
Happich, W. F., Windus, W. and Naghski, J. (1965). Stabilization of wool by glutaraldehyde. Text. Res. J., 35, 850-
   852.

Harris, M. and Brown, A. E. (1947). New developments in the chemical modification of wool. Amer. Dyestuff Reporter, 36, 316-319

Harris, M. and Smith, A. L. (1936). Oxidation of wool: alkali solubility test for determining the extent of oxidation. J. Res. Natl. Bur. Standards, 17, 577-583.

Hinton, E. H., Jr. (1974). A survey of critique of the literature on crosslinking agents and mechanisms as related to wool keratin. Text. Res. J., 44, 233-292.

Holt, L. A. and Milligan, B. (1970). The introduction of amide and ester crosslinks into wool. J. Text. Inst., 61, 597-603.

Koenig, N. H. (1961). Isocyanate modification of wool in dimethyl sulfoxide. Text. Res. J., 31, 592-596.

Koenig, N. H. (1962). Modification of wool in dimethylformamide with mono- and diisocyanates. Text. Res. J., 32, 117-122.

Koenig, N. H. (1965). Wool modification with acid anhydrides in dimethylformamide. Text. Res. J., 35, 708-715.

Koenig, N. H. (1976). Chemical modification of wool in aprotic swelling media. J. App. Polymer Sci., 120, in press.

Koenig, N. H. and Crass, R. A. (1975). Acid chloride modification of wool. Text. Res. J., 45, 178-182.

Koenig, N. H. and Friedman, M. (1972). Surface modification of wool and other fibrous materials by 4-vinylpyridine and zinc chloride. Text. Res. J., 42, 319-320.

Koenig, N. H. and Friedman, M. (1976). Properties of wool treated with sulfonyl chlorides. 172nd Meeting of the American Chemical Society, San Francisco, California, August 31-September 3, Abstracts, p. CELL 77.

Koenig, N. H. and Friedman, M. (1977). Combined application of reactive compounds in nonaqueous swelling solvents for flame- and shrink-resistant wool. Text. Res. J., 47, 139-141.

Koenig, N. H., Muir, M. W. and Friedman, M. (1973). Wool modification by activated vinyl compounds. Text. Res. J., 43, 682-688.

Lundgren, H. P. and Ward, W. H. (1962). Levels of molecular organization in α-keratins. Arch. Biochem. Biophys., Suppl. 1, 78-111.

Maclaren, J. A. (1971). Quantitative reduction and alkylation of wool. Text. Res. J., 41, 713.

McPhee, J. R. (1958). The reaction of formaldehyde with wool and its effect on digestion by insects. Text. Res. J., 28, 303-314.

Moore, J. E. (1960). Modification of keratins with sulfones and related compounds. U. S. Patent 2,955,016.

Moore, J. E. and Ward, W. H. (1956). Cross-linking of bovine plasma albumin and wool keratin. J. Amer. Chem., 78, 2414-2418.

Parker, A. J. (1969). Protic-dipolar aprotic solvent effects on rates of biomolecular reactions. Chem. Rev., 69, 1-32.

Patterson, W. I., Geiger, W. B., Mizell, L. R. and Harris, M. (1941). Role of cystine in the structure of the fibrous protein, wool. J. Res. Natl. Bur. Standards, 27, 89-103.

Powning, R. F. and Irzykiewicz, H. (1960). Cystine and glutathione reductases in clothes moth Tineola bisselliella Aust. J. Biol. Sci., 13, 59-68.

Powning, R. F. and Irzykiewicz, H. (1962). The digestive proteinase of clothes moth larvae. I. Partial purification of the proteinase. J. Insect Physiol., 8, 267-274.

Price, V. H. and Menefee, E. (1967). On the effect of dimethyl sulfoxide on hair keratin. J. Invest. Dermatol., 49, 297-301.

Ring, R. N., Tesoro, G. C. and Moore, D. R. (1967). Kinetics of the addition of alcohols to activated vinyl compounds. J. Org. Chem., 32, 1091-1094.

Schöberl, A. (1960). New reactions in reduced wool fibers. J. Text. Inst., 51, T613-T629.

Schoene, D. L. (1951). Divinyl sulfone tanned proteins. U. S. Patent 2,579,871.

Speakman, J. B. (1947). Mechano-chemical methods for use with animal fibres. J. Text. Inst., 38, T102-T126.

Suminov, S. I. and Kost, A. N. (1969). Nucleophilic addition of amino groups to an activated carbon-carbon double bond. Russ. Chem. Rev., 38, 884-899.

Watt, I. C. (1965). The modification of wool fibers by crosslinking reactions. Proc. Int. Wool Textile Res. Conf., Paris, II, 259-270.

Weidemann, E. and Wevers, H. W. (1971). The influence of mild alkali treatment of wool for short periods and at different temperatures. SAWTRI Bull., 5 (March), 21-29.

Weigmann, H.-D. and Dansizer, C. (1971). The stabilization of irreversibly deformed keratin fibers. II. Mechanism of stabilization. Text. Res. J., 41, 576-586.

Whitfield, R. E. and Wasley, W. L. (1964). Reactions of proteins. In "Chemical Reactions of Polymers," E. M. Fettes (Editor), Interscience, New York.

Wold, F. (1972). Bifunctional reagents. In "Methods in Enzymology," Vol. 25, C. H. W. Hirs and S. N. Timasheff (Editors), Academic Press, New York.

Zahn, H. (1954). Preparation of microbiologically resistant wool by means of chemical modification. Part II. Paper chromatographic investigation of the samples. Text. Res. J., 24, 26-31.

Zahn, H., Beyer, H., Hammoudeh, M. M. and Schallah, A. (1969). Vernetzungs- und Selbstvernetzungsreaktionen bei Wolle. Melliand Textilber., 50, 1319-1324.

Zahn, H. and Würz, A. (1954). Solubility in acid as a means of determining changes in wool. J. Text. Inst., 45, 88-92.

23

THE EFFECTS OF ETHYLENE GLYCOL ON WOOL FIBERS

Sandra J. Tillin, Richard A. O'Connell, Allen G.
Pittman and Wilfred H. Ward
Western Regional Research Laboratory, Agricultural
Research Service, U.S. Department of Agriculture,
Berkeley, California 94710

INTRODUCTION

We have previously shown that ethylene glycol can be a useful solvent for certain continuous processes in wool textile
finishing. A brief treatment in hot ethylene glycol can (1) impart considerable stretch properties to woven wool fabric when
treated in a slack condition (Pittman and Wasley, 1974; Pittman
and Wasley, 1975), (2) produce permanent set, i.e. creases or
crimp, to wool held in a constrained configuration during treatment (Pittman et al., 1975) and (3) produce rapid and thorough
dyeing when used as a dyebath medium (Pittman et al., 1976).

For treatment, wool is immersed for 10-60 sec in ethylene
glycol heated to 140-160°C. An acid, such as p-toluenesulfonic
acid, can be added to the glycol to act as an esterification catalyst for reaction between ethylene glycol and the carboxyl groups
present in wool protein. Other workers have reported increases
in both setting (Holt et al., 1969) and dyeing rate (Alexander et
al., 1951) for esterified wool.

Our working hypothesis was that the changes in wool properties were mainly due to the formation of new crosslinks. The
presence of an acid catalyst should significantly increase the
rate of esterification, possibly leading to cross-esterification by
glycol between carboxyl groups on adjacent protein chains. New
crosslinks, especially in the α-helix region, would stabilize the
helix and in effect raise the crystalline melting temperature.
Examination of various treated fibers by x-ray diffraction has
shown that fibers treated with glycol containing 0.2% p-toluenesulfonic acid at 150°C for 60 sec retain the characteristic α-

helix pattern at 5.14 A, while fibers treated under the same
conditions without the acid lose the 5.14 A reflection (Pittman
and Wasley, 1974).

To investigate the possibility that new crosslinks may be
formed during the glycol treatment, especially when an acid catalyst is present, we used several well known methods which
have been found useful in detecting additional crosslinks in wool.
Comparisons were made between glycol treated wool and
both wool esterified with methanol and wool crosslinked with
formaldehyde.

EXPERIMENTAL

Ethylene Glycol Treatment. Wool top (64's quality), which
had been dried under vacuum at 100°C for 2 hr and then conditioned at 21°C and 65% RH, was treated at 140°, 145°, 150° and
155°C in both ethylene glycol and ethylene glycol containing 0.2%
p-toluenesulfonic acid. Treatment times at each temperature
were 10, 20, 30, 45 and 60 sec. Immediately after treatment
the fibers were quenched in cold water and then thoroughly
rinsed in running water. After air drying, the wool was further
dried under vacuum and conditioned at 21°C and 65% RH.

Supercontraction in Ethylene Glycol. Single wool fibers
were mounted between plastic tabs and their length measured
before and after glycol treatment. At least 10 fibers were used
to obtain each average length.

Crosslink Estimation by Formic Acid Swelling. The procedure developed by Caldwell and Milligan (1970) was used as
described except that all samples were immersed for 10 min in
97+% formic acid.

Stress-Strain Curves. Stress-strain curves of treated and
untreated fibers were obtained for fibers tested in both pH 6.5
phosphate buffer and in 97+% formic acid. An Instron Tensile
Tester was used at an extension rate of approximately 50% per
min.

Supercontraction in $NaHSO_3$ and LiBr. Single fibers
mounted between plastic tabs were first measured, then exposed
for 1 hr in 5% $NaHSO_3$ at 97°C, washed one half hour in water,
dried and remeasured to determine change in length. Alternatively, single fibers were exposed for 1 hr at 97°C to a saturated solution of LiBr in water, dip rinsed in 95% ethanol, dried,
and remeasured.

Percent Solubility. The procedure using performic acid

followed by ammonia digestion was used as described by
Caldwell et al. (1966).

Esterification with methanol. Wool top was esterified with
ethanol by the method of Holt et al. (1969).

Formaldehyde treatment. Wool top was treated with formaldehyde according to Caldwell and Milligan (1970).

RESULTS AND DISCUSSION

When wool fabric is treated with hot ethylene glycol the
amount of stretch potential obtained depends on whether or not
an acid catalyst is used. More stretch is produced without the
acid than with the acid. This may be due in part to the inhibition
of supercontraction when an acid is present. Supercontraction
in wool fibers is associated with the disruption and melting of
the crystalline α-helix portion of the fiber. It is also associated
with hydrogen bond rearrangement and disulfide rupture with or
without disulfide interchange. As can be seen in Table 1, all
fibers treated in ethylene glycol above 140°C without acid supercontract; supercontraction increases with an increase in temperature and also with increasing time at each temperature. In
comparison, fibers treated in the presence of 0.2% p-toluene-

Table 1

Supercontraction of Wool Fibers in Ethylene Glycol

Treatment temperature (°C)	Treatment time (sec)	% Supercontraction	
		No acid catalyst	With acid[a] catalyst
140	30	0	0
145	30	3	-
145	45	4	-
145	60	8	-
150	15	1	-
150	30	15	<1
150	60	24	<1
160	20	32	-
160	30	39	3
160	40	42	6

[a] 0.2% p-toluenesulfonic acid.

sulfonic acid show no supercontraction at the treatment times indicated except at 160°C and then only a small amount.

Kassenbeck (1975) has reported observations on wool fibers heated in ethylene glycol. He finds that a wool fiber heated in glycol without acid will begin to loose its natural crimp at 70-80°C. From 100-135°C the fiber will straighten out and remain in a linear extended form. Above 135°C reverse curling occurs, with maximum curling obtained at 150°C. This sequence of events is attributed to the differences in heat stability of the low and high sulfur components of the bilateral wool structure. He also finds that when 0.2% p-toluenesulfonic acid is added to the glycol the straightening and crimp reversal phenomena are inhibited.

Following the procedure of Caldwell and Milligan (1970), we measured the swelling behavior in formic acid of glycol treated fibers. Using the Flory-Rehner equation to estimate the cross-link density of a swollen polymer, they developed a technique for the semiquantitative determination of crosslinks in wool. The crosslinks are expressed in terms of disulfide equivalents per gram of wool. As Table 2 indicates, wool treated with formaldehyde, a compound known to form crosslinks in wool shows a definite increase in apparent crosslink density by this method. On the other hand, glycol treatment and esterification with methanol both show an apparent decrease in crosslinking. For glycol treated fibers, the apparent crosslink density decreases with increasing time of treatment at each temperature and with increasing temperature. However, the decrease in crosslink density is less when acid is present. The increased swelling behavior of the glycol-treated fibers may indicate changes in the interactions of the fiber protein chains, possibly as a result of esterification, rather than a true decrease in total crosslink density. Esterification would tend to cause an increase in swelling because of the lessened electrostatic interaction between carboxyl and basic groups. The similar swelling behavior of methylated wool and wool treated with acidified glycol could be interpreted as an indication of esterification.

When wool top is dyed in ethylene glycol, for example at 145°C for 10 sec, much deeper dye penetration occurs when an acid catalyst is present in the dyebath. The effect of acid in glycol may be the same as the effect of acid in conventional aqueous wool dyeing. However, wool treated with ethylene glycol and then dyed by conventional methods achieves the same dye penetration in less time than untreated wool. The increased dyeability may be due in part to changes in the wool structure making it more accessible to the dye molecules. It may also be due to esterificaation. Esterification would change the acid-base balance in a way favoring uptake of an acid dye.

Table 2

Formic Acid Swelling of Glycol Treated Fibers[a]

Treatment temperature-time	Equivalent disulfide content μg/mole	
Untreated control	465	
Methanol[b]	425	
Formaldehyde[c]	760	
	Acid[d]	No acid
145°C, 30 sec	419	402
145°C, 60 sec	381	238
150°C, 20 sec	-	360
150°C, 30 sec	435	301
150°C, 45 sec	432	-
155°C, 20 sec	-	270
155°C, 30 sec	409	217
155°C, 45 sec	382	-

[a] Calculated according to Caldwell, J. B. and
  Milligan B. (1970).
[b] Methanol/wool ratio 40/1, containing 0.1 M
  HCl, 60°C, 24 hr.
[c] Solution containing 2 M formaldehyde, 2 M HCl
  and 2 M $CaCl_2$, 20°C, 2 hr.
[d] 0.2% p-toluenesulfonic acid.

The stress-strain properties of treated and untreated fibers
show similar differences. If new crosslinks have been introduced, the stress-strain curve should show an increase in the
moduli of the Hookian and yield regions and a change in the turnover point between the yield and post yield regions. Figure 1
illustrates the stress-strain curves in water for untreated and
glycol treated fibers. The curves for the treated fibers lie below
that for the untreated fibers. This means that the treated fibers
are more easily extended, indicating less molecular interaction.
Table 3 shows clearly that glycol treatment without acid produces a greater change than when an acid is present. Although
elongation at break does not change significantly with a change in
treatment conditions, the yield stress, stress at 20% and stress
to break decrease both when the temperature is raised and when
acid is omitted. The difference between treatment with and
without acid is particularly evident at 150°C. A 20-sec treatment without acid produces a greater decrease in stress properties than treatment for 45 sec with acid. Stress-strain curves

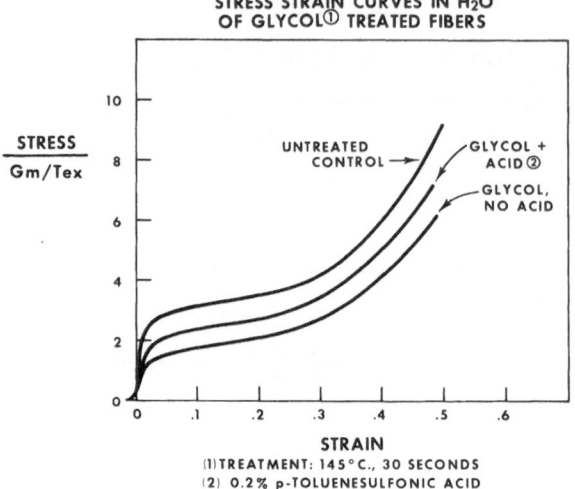

STRESS STRAIN CURVES IN H$_2$O
OF GLYCOL[1] TREATED FIBERS

(1)TREATMENT: 145°C., 30 SECONDS
(2) 0.2% p-TOLUENESULFONIC ACID

Fig. 1

Table 3

Effect of Acid[a] in Ethylene Glycol on Wet Single Fiber Stress-
Strain Properties

Treatment temperature-time	Yield stress	Stress at 20%	Stress to break	% Elonga-tion
Untreated	2.60 ± 0.2[b]	3.66 ± 0.1	9.30 ± 1.0	48.8 ± 3
Acid				
145°C, 30 sec	1.93 ± 0.2	2.95 ± 0.2	7.97 ± 1.0	47.5 ± 3
150°C, 45 sec	1.74 ± 0.1	2.59 ± 0.1	7.68 ± 0.6	49.6 ± 3
No acid				
145°C, 30 sec	1.44 ± 0.1	2.22 ± 0.2	6.64 ± 0.7	50.2 ± 3
150°C, 20 sec	1.38 ± 0.1	2.09 ± 0.2	6.19 ± 0.8	47.7 ± 3

[a]0. 2% p-toluenesulfonic acid
[b]Stress in grams/tex; 95% confidence limits.

of treated fibers measured in formic acid instead of water
showed the same pattern of decrease in stress for the treated
fibers over the entire stress-strain curve. Formic acid is
more effective than water in breaking hydrogen bonds, therefore

any changes in the stress-strain curve resulting from a difference in number of covalent crosslinks would be more evident.
No such changes were found.

Further evidence that treatment with glycol does not produce
additional crosslinking was obtained by measuring the supercontraction of the fibers in 5% sodium bisulfite and in concentrated lithium bromide. As a reducing agent, sodium bisulfite
breaks existing disulfide bonds, while lithium bromide acts to
break hydrogen bonds. Table 4 shows that in both media

Table 4

Percent Supercontraction

	$NaHSO_3$	LiBr
Untreated control	28%	38%
Acid[a]	28	37
No acid[b]	25	38
Formaldehyde[c]	3	18

[a] Treated in ethylene glycol containing 0.2% p-toluenesulfonic acid at
145°C for 30 sec.
[b] Treated in ethylene glycol at 145°C
for 30 sec.
[c] Solution containing 2 M formaldehyde, 2 M HCl and 2 M $CaCl_2$,
20°C, 2 hr.

supercontraction of both acid and nonacid treated fibers was not
significantly different from the control fibers. In marked contrast, stabilization of the fiber by additional crosslinks, such as
those obtained by treatment with formaldehyde, greatly
decreases supercontraction. Percent solubility by treatment
with performic acid followed by ammonia digestion also showed
no significant differences between treated and untreated fibers.
This treatment has been used to estimate the crosslink density
of wool. A decrease in solubility indicates an increase in the
number of crosslinks.

CONCLUSION

In conclusion, none of the data supports the hypothesis that
new crosslinks are introduced by treatment of wool with hot

ethylene glycol. While ethylene glycol can, in principal, give a crosslinked ester, its large excess when used as a hot treating media would tend to favor reaction of only one hydroxyl group per glycol molecule. The evidence does suggest, however, some disordering of the fiber structure at the molecular level, both in the amorphous and crystalline regions. There is also some indication of possible esterification, particularly when acid is present, as evidenced by the increased swelling and dyeability of the treated wool. An alternative explanation for the relative stability of the fibers treated in the presence of acid, i. e. retention of the α-helix pattern at 5. 14 A and little or no supercontraction during treatment, may be due to inhibition of disulfide interchange. Disulfide interchange is an important factor in supercontraction and is known to be favored in basic media by the presence of the sulfur anion. Acidification would suppress anion formation, thus decreasing the likelihood of the exchange.

ACKNOWLEDGEMENT

We wish to thank Mrs. B. Molyneux for both the stress-strain and the supercontraction data.

REFERENCES

Alexander, P. , Carter, D. , Earland, C. and Ford, O. E. (1951). Esterification of carboxyl groups in wool. Biochem. J. , 48, 629-637.

Caldwell, J. B. , Leach, S. J. and Milligan, B. (1966). Solubility as a criterion of cross-linking in wool. Text. Res. J. , 36, 1091-1095.

Caldwell, J. B. and Milligan, B. (1970). The estimation of cross-links in wool from the extent of swelling in formic acid. J. Text. Inst. , 61, 588-596.

Holt, L. A. , Leach, S. J. and Milligan, B. (1969). Wool setting: the effect of esterification. Text. Res. J. 39, 290-293.

Kassenbeck, P. and Stay, A. (1975). Denaturierung Keratinischer Proteine unter Einwirkung von Hitze. Prog. Coll. & Polym. Sci. , 57, 123-132.

Pittman, A. G. and Wasley, W. L. (1974). Two-way stretch wool and wool blend fabrics. Book of Papers, 1974 National Technical Conference, AATCC, 68-75.

Pittman, A. G. and Wasley, W. L. (1975) Two-way stretch in woven wool fabrics. Amer. Dyest. Reptr. 64(5), 48-49.

Pittman, A. , Wasley, W. and O'Connell, R. (1976). A new treatment for producing stretch or set in wool. Proc. Fifth Int. Wool Text. Res. Conf. , Aachen, Germany, III, 450-460.

Pittman, A. G. , Ward, W. H. and Wasley, W. L. (1976). Rapid dyeing and finishing of wool in hot ethylene glycol. Text. Res. J. , 46, 921-924.

24

PROTEIN: POLYANION INTERACTIONS. STUDIES ON THE TREHALOSE-P

SYNTHETASE AS A MODEL SYSTEM

Alan D. Elbein and Y. T. Pan

Department of Biochemistry, University of Texas Health

Science Center, San Antonio, Texas 78284

INTRODUCTION

The interaction of proteins with various types of polyelectrolytes occurs commonly in nature. In the case of enzymes, the interaction may be quite specific with regard to the nature of the polyelectrolyte. The addition of the polyelectrolyte to an enzyme may result in either an increase in enzyme activity or an inhibition of activity depending on the enzyme in question. Thus, polyanions such as polynucleotides, or sulfated glycoaminoglycans like heparin and chondroitin sulfate, may be activators of some enzymes but are inhibitors of others. A similar situation apparently applies to polycations such as polylysine or polyornithine. There are apparently two important requirements for these polyelectrolytes to be enzyme effectors: 1) the macromolecular nature of the polyelectrolyte and 2) the presence of strong positive or negative charges (1).

A considerable number of papers on this subject have appeared in the literature. Several excellent reviews have been published on the activation (1) or inhibition (2) of enzymes by polyelectrolytes. A comprehensive review containing information on the interactions of polyelectrolytes with proteins has also appeared recently (3).

TREHALOSE PHOSPHATE SYNTHETASE

For a number of years, our laboratory has been concerned with the interaction of polyanions with the trehalose phosphate synthetase of Mycobacterium smegmatis. The trehalose phosphate

synthetase represents an interesting example of a protein-poly-
electrolyte interaction in which a polyanion apparently binds to
the enzyme and somehow alters its conformation to enable the
enzyme to utilize a given substrate. This system thus appears to
be a good model system for studying the phenomena of protein:
polyanion interactions. It should be mentioned at the onset that
the role of this interaction in the physiology of M. smegmatis is
not understood at this time.

The biosynthesis of trehalose phosphate can occur by either
one of two reactions:

$$\text{UDP-Glc} + \text{glucose-6-P} \rightarrow \text{trehalose-P} + \text{UDP} \qquad (1)$$

$$\text{GDP-Glc} + \text{glucose-6-P} \rightarrow \text{trehalose-P} + \text{GDP} \qquad (2)$$

Reaction 1 was found in yeast (4), insects (5), Mycobacterium
tuberculosis (6), and Dictyostelium discoideum (7). Reaction 2
was first shown in Streptomyces hygroscopicus (8). In all species
of Streptomyces examined (9), UDP-Glc would not serve as a glucosyl
donor. However, in other actinomycetes, principally Mycobacteria,
Liu et al. (10) found that crude cell free extracts of a number of
these organisms were able to utilize both uridine diphosphate-D-
glucose and guanosine diphosphate-D-glucose as the glucosyl donors
for the synthesis of trehalose phosphate. In order to determine
whether these organisms contained two different enzymes for
trehalose phosphate synthesis, a crude extract of M. smegmatis was
fractioned on DEAE-cellulose and two fractions (Fraction I and II)
were separated (Fig. 1). Fraction I could catalyze trehalose
phosphate synthesis from GDP-Glc but was relatively inactive with
UDP-Glc. Fraction II, which eluted from the column at a higher
KCl concentration, had no detectable enzyme activity with either
substrate. However, when Fraction II was added back to Fraction I,
the enzymatic activity for UDP-Glc was restored. Under this condition (i.e. in the presence of Fraction II) GDP-Glc was still an
active glucosyl donor. The active component in Fraction II was
later identified as an RNA species with a molecular weight of
between 50,000 and 100,000, since it gave a typical RNA ultraviolet
absorption spectrum and was inactivated by incubation with RNase
but not by incubation with trypsin or DNase.

With respect to substrate specificity, Lapp et al. (11) found
that the enzyme isolated from DEAE-cellulose (Fraction I) was able
to use all five glucose nucleotides as the glucosyl donors in the
reaction, although TDP-Glc and CDP-Glc were not nearly as effective as the other sugar nucleotides (Table I). However, there was
a significant difference between those nucleotides with pyrimidine
bases and those with purine bases. It can be seen that when CDP-
Glc, TDP-Glc or UDP-Glc were used as substrates, there was an

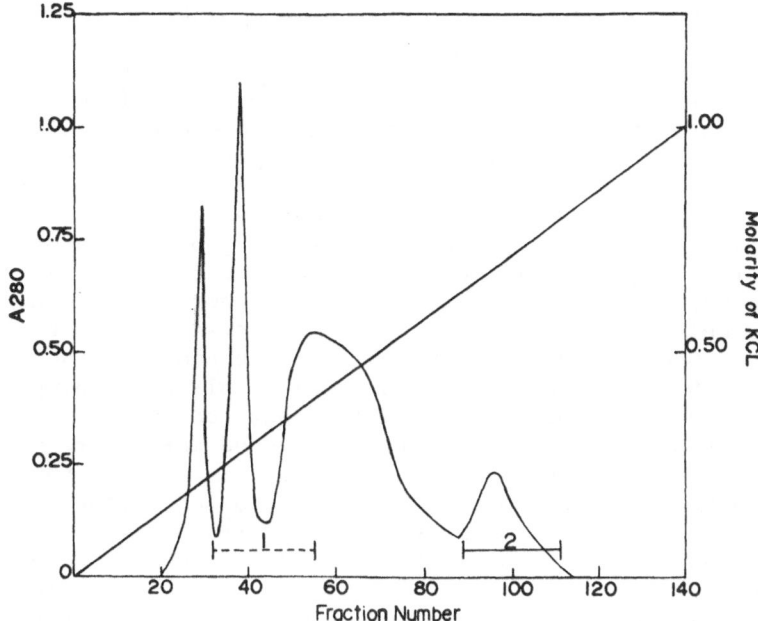

Fig. 1. Separation of an extract of M. smegmatis on DEAE-cellulose.
Crude enzyme preparation was placed on a column of DEAE-cellulose
and eluted with a gradient of KCl from 0 to 1 M. Fractions were
combined as shown by ------(Fraction 1) or _____(Fraction 2).

Sugar nucleotide donor	Trehalose-P (mμmoles) formed in the presence of		
	No CS	0.025 μg CS	0.1 μg CS
UDP-Glc	4.2	171.7	177.3
TDP-Glc	3.6	70.1	60.1
CDP-Glc	0	39.8	49.7
ADP-Glc	61.7	191.0	209.7
GDP-Glc	68.3	112.7	136.2

Table I. Sugar nucleotide specificity of trehalose phosphate synthetase. Trehalose phosphate formation was assayed in the presence
of various glucose nucleotides. CS represents chondroitin sulfate.

almost complete dependence on the presence of polyanion, whereas
with the purine nucleotides (ADP-Glc and GDP-Glc) significant activity was observed in the absence of polyanion. In the latter two
cases, activity was increased about 3 fold by polyanion whereas
with the pyrimidine nucleotide, polyanion increased the activity
20 fold or more.

EFFECT OF POLYELECTROLYTES ON SYNTHETASE

In addition to the RNA from M. smegmatis (Fraction II), a number of different high molecular-weight polyanionic compounds were also able to activate the synthetase to use UDP-Glc as the glycosyl donor. As shown in Fig. 2 and Table 2, these various polyanions had considerably different abilities to stimulate the enzyme. It appeared that polyanions containing both uronic acid residues and sulfate groups were all fairly good activators of the enzyme. Thus, heparin, which has both N- and O-sulfate groups as well as uronic acid residues and has the highest sulfate content (charge density) of the various glycosaminoglycans, was the best activator. Other glycosaminoglycans such as chondroitin 4-sulfate, chondroitin 6-sulfate and dermatan sulfate, all containing uronic acid residues and O-sulfate groups and about one half of the sulfate content of heparin, were all fairly good activators of the enzyme, although less effective than heparin. On the other hand, those polyanions which contained either O-sulfate groups or uronic acid residues alone, such as keratan sulfate, hyaluronic acid and D-galacturonan, were not activators of the synthetase. Some polynucleotides were also effective activators. High-molecular-weight poly(cytidylic acid) and poly(uridylic acid) were fairly good activators of the synthetase, although they appeared to be somewhat less effective than chondroitin sulfate. On the other hand, high-molecular-weight poly(adenylic acid) and poly(guanylic acid) were poor activators of the enzyme.

The stimulation of enzymatic activity by polyanion depended not only upon the nature of the anionic charge but also upon the size of the polyanion. Degradation of chondroitin sulfate with hualuronidase led to the rapid disappearance of activity ability. As shown in Fig. 3, after a hydrolysis of 4 hours, the chondroitin sulfate still had good activating ability but this decreased with longer times of enzyme treatment. When the 4 hour and 22 hour hydrolyzates were fractionated on a column of Sephadex G-100, it was found that only the high-molecular-weight components, which emerged in the same location as intact chondroitin sulfate, had good activating ability. On the other hand, the low-molecular-weight materials obtained from the Sephadex G-100 column, as well as a sulfated tetrasaccharide and a sulfated octasaccharide, were ineffective in activating the enzyme. It was estimated that the molecular weight of the polyanion had to be in the order of 10,000-12,000 daltons or higher (12).

As might be expected, synthetase was inhibited by a number of polycations. Thus, the polycation, poly-D,L-ornithine, was found to inhibit the activity of synthetase when UDP-Glc was used as the substrate (12). Fig. 4 shows the effect of adding increasing amounts of poly-D,L-ornithine to incubation mixtures in which

Polyanion	Trehalose phosphate formed after addition of activator(μg)[a]				
	0.02	0.05	0.1	0.2	0.4
Chondroitin 4-sulfate	0.29	0.32	0.38	0.44	0.50
Chondroitin 6-sulfate	0.30	0.32	0.35	0.45	0.58
Oversulfated chondroitin sulfate	0.31	0.39	0.45	0.63	0.83
Dermatan sulfate (sample A)	0.40	0.46	0.61	0.81	0.95
Dermatan sulfate (sample B)	0.36	0.51	0.72	0.86	1.02
Heparin	0.37	0.54	0.79	0.89	1.09
Heparan sulfate	0.29	0.38	0.58	0.67	0.82
Keratan sulfate (sample A)	0.28	0.23	0.28	0.36	0.53
Keratan sulfate (sample B)	0.24	0.23	0.2	0.2	0.18
Dextran sulfate	0.19	0.21	0.23	0.22	0.23
Hyaluronic acid	0.21	0.22	0.21	0.21	0.15
γ-Carrageenan	0.28	0.37	0.55	0.78	0.97
Ribonucleic acid (Peak-2)	0.39	0.48	0.65	0.74	0.94
None	0.21				

[a]Absorbance at 620 nm.

Table 2. Activation of trehalose phosphate synthetase by various polyanions.

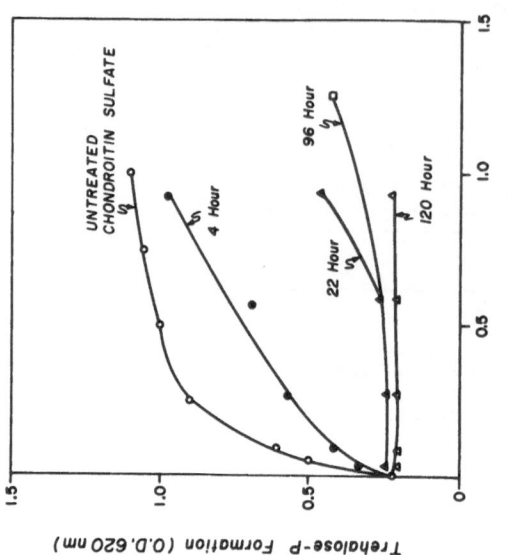 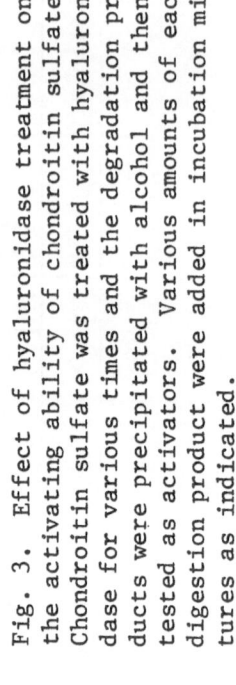

Fig. 3. Effect of hyaluronidase treatment on the activating ability of chondroitin sulfate. Chondroitin sulfate was treated with hyaluronidase for various times and the degradation products were precipitated with alcohol and then tested as activators. Various amounts of each digestion product were added in incubation mixtures as indicated.

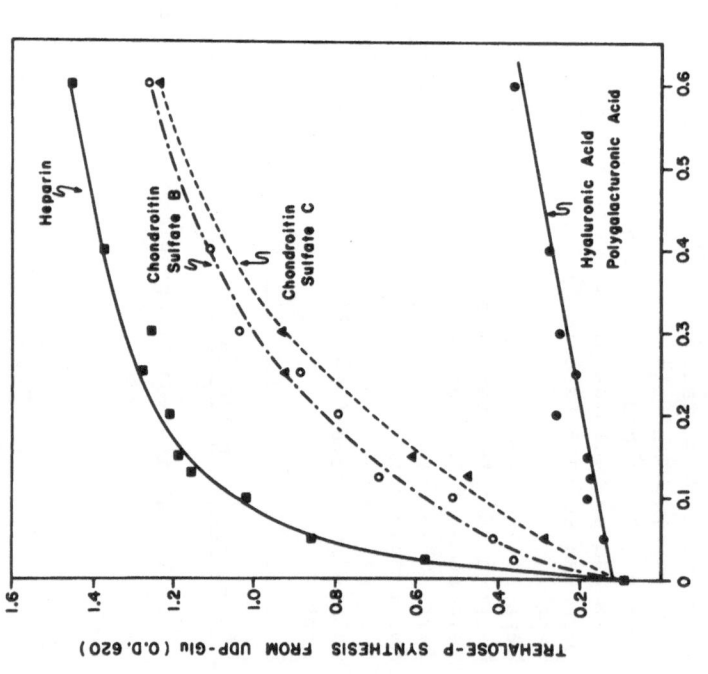

Fig. 2. Effect of various polyanions on the activity of trehalose phosphate synthetase with UDP-Glc as substrate. Various polyanions were added to incubation mixture in the concentration as shown.

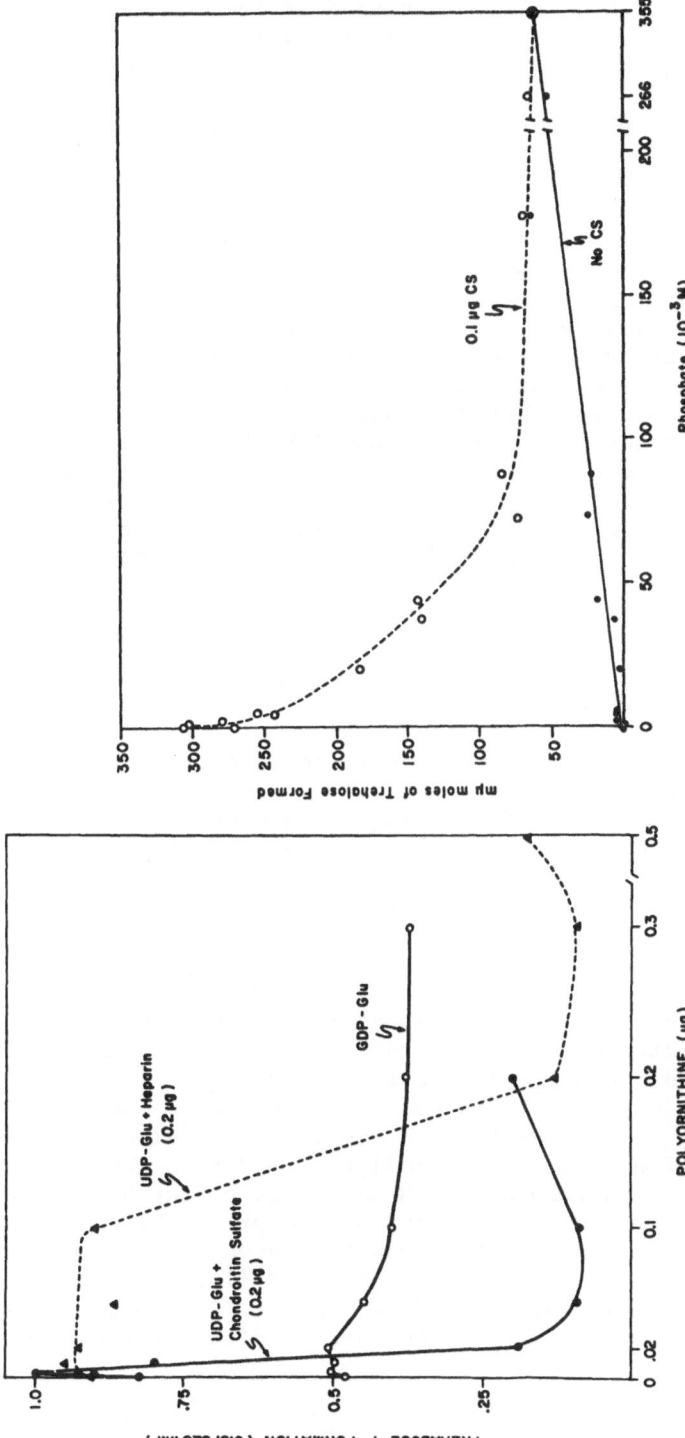

Fig. 5. Effect of increasing concentrations of phosphate on the synthesis of trehalose phosphate from UDP-Glc and glucose-G-P. The concentration of potassium phosphate buffer (pH 7.0) in the incubation mixture was varied as indicated. Experiments were run in the presence and absence of chondroitin sulfate (CS).

Fig. 4. Effect of poly-D,L-ornithine on the activity of trehalose phosphate synthetase. Incubation mixtures contained 0.2 μg of heparin or chondroitin sulfate and various amounts of poly-D,L-ornithine as shown. All components of the incubation mixture including polyanion and polycation were mixed before the addition of enzyme.

either heparin or chondroitin sulfate was used as the activator of
the enzyme. It should be noted that considerably larger amounts
of poly-D,L-ornithine were necessary to inhibit the enzyme when
heparin was used as the activator than when chondroitin sulfate
was used. These results are consistent with the abilities of
these polyanions to activate the enzyme, since heparin is a better
activator than chondroitin sulfate (Fig. 2) and presumably binds
more tightly to the enzyme. Moreover, poly-D,L-ornithine did not
significantly inhibit the activity when GDP-Glc was used as the
substrate. The inhibition by other polyamino acids such as polylysine, poly-L-histidine and poly-L-leucine, was dependent upon
their charges and their molecular weight. A molecular weight of
4,000-16,000 appeared to be optimum for inhibition. In addition
to an optimum molecular size, the polycations must have a positive
charge at pH 7.0. It appeared that the inhibition of activity by
various polycations was due to a competition with the enzyme for
the polyanion which consequently decreased the concentration of
polyanion available for binding to the enzyme. Thus, less inhibition was observed when enzyme and polyanion were mixed before
the addition of polycation than when polyanion and polycation were
mixed prior to enzyme addition.

POSSIBLE MECHANISM OF POLYELECTROLYTE-
ENZYME INTERACTION

With regard to the mechanism involved in polyelectrolyte-
protein interaction, it seems likely that the two molecules are
attracted to each other because of opposing surface charges. For
example, activation of the enzyme by polyanion was abolished by
high salt concentration. As shown in Fig. 5, when the enzyme was
assayed at various buffer concentrations with UDP-Glc and chondroitin sulfate, enzyme activity decreased as the buffer concentration was increased. This observation suggested that the
interaction between enzyme and polyanion was likely to be an electrostatic bonding which might occur as a result of basic amino
acids on the protein interacting with acidic groups (presumably
sulfate ester group) on the polyanion to form a complex. The
results of preliminary experiments in our laboratory have suggested
that the basic amino acids of the protein could be histidine residues on the surface. Thus, binding of radioactive RNA to the
synthetase is optimum at pH values of 6-8 and falls off at higher
pH's. In addition, activation of the enzyme, and presumably
binding of the polyanion, is inhibited when the enzyme is treated
with ethoxyformic anhydride, a known histidine modifier. The
inhibition of enzyme activity by ethoxyformic anhydride could be
partially overcome by preincubation of the enzyme with heparin
(Fig. 6).

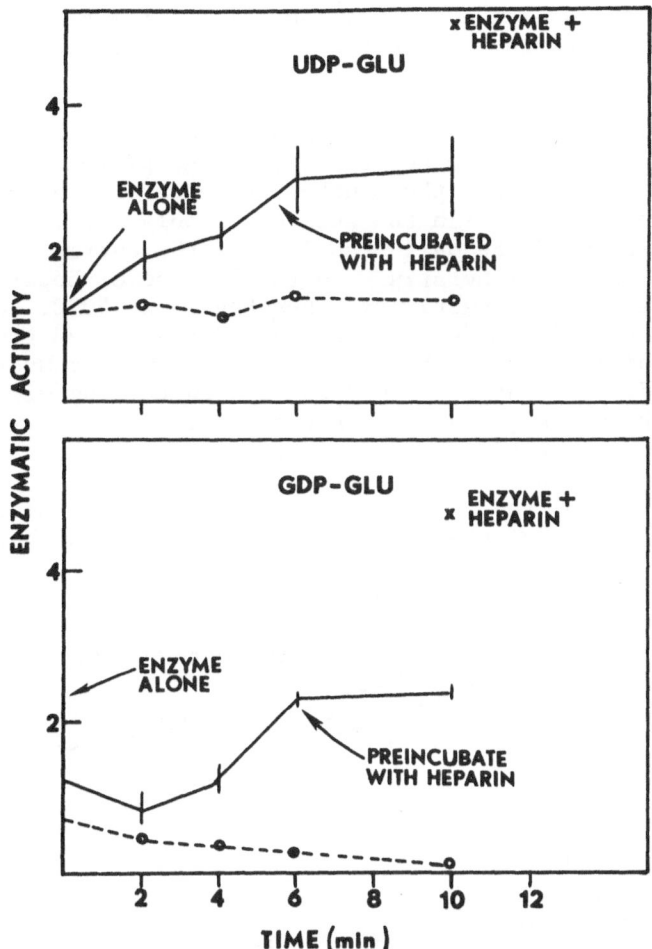

Fig. 6. Effect of preincubation of heparin with trehalose phosphate synthetase on the prevention of inhibition by ethoxyformic anhydride. After preincubation with either heparin (3 μg) (–·–·–) or ethoxyformic anhydride (0.2 mM) (··o···o··o) for various times as indicated, trehalose phosphate synthetase was incubated with either ethoxyformic anhydride or heparin and then stood at room temperature for another 10 min. prior to enzyme activity assay.

Thus, these preliminary studies suggest that histidine residues may be involved in this interaction. Moreover, the synthetase
could be immobilized on a heparin-sepharose gel and still retained
good enzymatic activity. No exogenous polyanion was required for
maximum activity under these conditions (13). The fact that the
enzyme bound to Heparin-Sepharose columns was utilized in pufi-
cation of the synthetase (See below).

How the interaction of polyanion and synthetase can lead to
activation of the enzyme is not clear. In the presence of a
polyanion such as heparin, the synthetase appears to be much more
stable than is the enzyme in the absence of polyanion. Fig. 7
shows a comparison of the heat stability of the enzyme both in the
presence and absence of heparin. In the absence of heparin, the
synthetase was rapidly inactivated when heated at 50°C. However,
in the presence of 0.5 mg/ml heparin the enzyme showed greatly
increased stability with either substrate, UDP-Glc or GDP-Glc. On
the other hand, low concentrations of heparin (0.5 µg/ml) did not
protect the enzyme to any significant extent. This fact indicated
that the polyanion stabilized the protein to heat denaturation.
Nevertheless, protecting the enzyme from denaturation may not be
the only function of the polyanion. In an experiment designed to
determine the nature of the activation, enzyme was incubated with

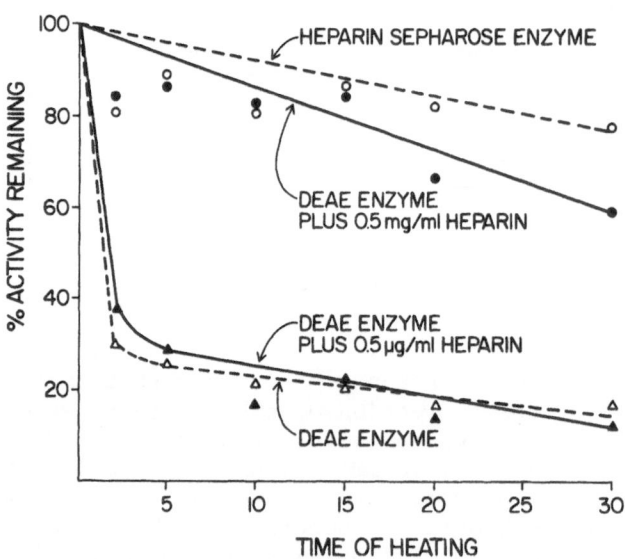

Fig. 7. Effect of heparin on the stability of trehalose phosphate
synthetase. Enzymes were heated at 50°C. for varying times in the
absence or presence of the indicated amount of heparin added.

UDP-Glc in the presence or absence of polyanion (Fig. 8). In the presence of polyanion, formation of trehalose phosphate began immediately and proceeded in a linear fashion for about 20 min., whereas in the absence of polyanion no significant amount of trehalose phosphate was formed. However, when a polyanion such as chondroitin sulfate was added to these incubation mixtures at later times, trehalose phosphate formation began immediately and proceeded at the same rate as in those assays which contained polyanion from the start. The results of this experiment thus suggested that the polyanion might bind to the enzyme and consequently alter its conformation to a more active form.

The trehalose phosphate synthetase of M. smegmatis has recently been purified to homogeneity (up to 160 fold) by ammonium sulfate fractionation. DEAE-cellulose chromatography, Sephadex G-200 chromatography and Heparin-Sepharose 4B affinity chromatography (Table 3). It is interesting to note that purification of enzyme resulted in an increased requirement for polyanion for enzymatic activity. Although in some cases, activation of proteins by polyanion does seem to involve the removal of an inhibitor (14,15), this does not seem to be the case for the synthetase. If an inhibitor were present and were removed by the polyanion, one might expect a decreased requirement upon purification. The results of purification also clearly indicated that both UDP-Glc and GDP-Glc activities were associated with the same protein.

Table 3. Purification of Trehalose Phosphate Synthetase

Fraction	Total Protein	Specific Activity				Total Activity (specific activity x total protein)			
		UDP-Glc	UDP-Glc plus Heparin (2.0 μg)	GDP-Glc	GDP-Glc plus Heparin (2.0 μg)	UDP-Glc	UDP-Glc plus Heparin (2.0 μg)	GDP-Glc	GDP-Glc plus Heparin (2.0 μg)
Crude sonicate, 200 ml	7700.0	2.2	4.9	1.9	3.9	16,940	37,730	14,630	30,000
(NH$_4$)$_2$SO$_4$ (0-50%), 43 ml	2429.5	10.3	12.4	9.7	10.7	25,024	30,126	23,566	25,996
DEAE-cellulose, 7.8 ml	180.9	36.2	74.1	34.5	46.6	6,549	13,405	6,241	8,430
Sephadex G-200, 23 ml	69.0	18.5	206.0	171.7	220.0	12,765	14,214	11,847	15,180
Heparin-Sepharose 4B, 28 ml	13.4	157.3	796.8	317.7	583.3	2,108	10,677	4,257	7,816
Purification factor		72	162	166	149				
Recovery (%)						12.4	28.3	29.1	26.0

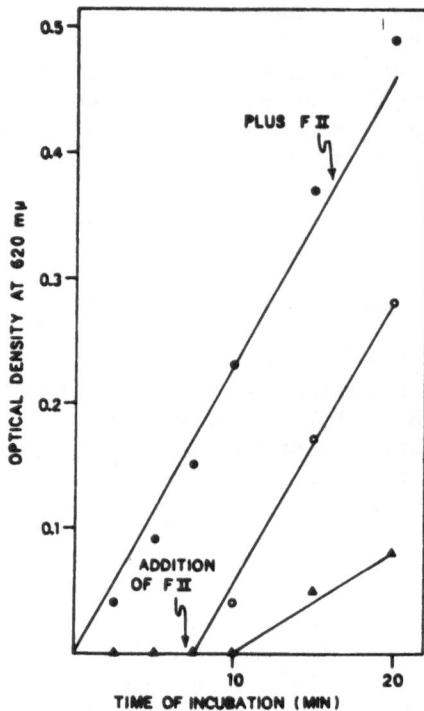

Fig. 8. Time course of trehalose phosphate formation with UDP-Glc
as substrate. Two parallel experiments were run: One in the
presence of RNA (0.75 μg) and one in its absence. Both incubation
mixtures were incubated at 30° and an aliquot was withdrawn and
assayed for trehalose phosphate content. After 7.5 min. of incubation, the incubation mixture lacking RNA was withdrawn and added
to a third tube containing 0.3 μg of FII (RNA); ●-●, Enzyme plus
FII (RNA);▲-▲, Enzyme; o-o, FII (RNA)added to enzyme after 7.5
min.

     The purified enzyme appeared to be homogenous in polyacryla-
mide gels (Fig. 9) but it had a heterogenous subunit composition
as determined by sodium dodecyl sulfate-polyacrylamide gel electrophoresis (Fig. 9). The molecular weights of the two major bands
on SDS-gel was estimated at approximately 90,000 and 45,000,
respectively. It was found that the apparent molecular weight of
the active synthetase when determined by gel filtration on Sephadex
G-200 column in the presence of 1 M KCl was approximately 40,000-
50,000. Apparently, aggreation of the protein occurs. The function of polyanion on the synthetase could be its ability to bind
the enzyme and alter its conformation or it could be to cause
aggregation of enzyme molecules or both. The participation of
polyelectrolyte in controlling association-dissociation of the
enzyme has been suggested in several enzymes such as β-glucuroni-

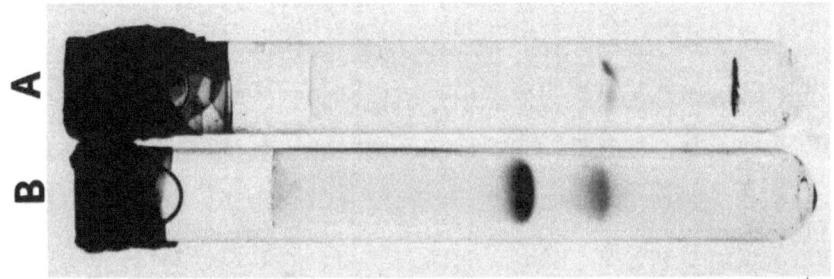

Fig. 9. Analysis of purified trehalose phosphate synthetase by electrophoresis in (A) polyacrylamide gel and (B) polyacrylamide gel containing sodium dodecyl sulfate.

dase (16) testicular hyaluronidase (17) and rabbit muscle aldolase, lactate dehydrogenase, pancreatic α-amylase and sweet potato β-amylase (18). Further studies on this interesting phenomena of protein:polyanion interaction are in progress.

The interaction of proteins with polyanions may have important implications in various disease processes such as atherosclerosis. For example, Tracy et al. (19) found evidence for complex formation between low density lipoproteins and glycosaminoglycans in the intimal layer of human aorta. They showed a correlation between the amount of lipoprotein sequestered by the intima and the severity of the atherosclerosis. The complex demonstrated by these workers appeared to be similar to that observed by Bihari-Varga et al. (20) by adding glycosaminoglycans from human aorta to whole serum. Srinivasan et al. (21,22) extracted complexes of lipoprotein-glycosaminoglycan from fatty streaks of human aorta with sodium chloride. Partial characterization of these complexes indicated that they were composed of LDL and VLDL and chondroitin sulfate and/or heparitin sulfate. Complexes of lipoproteins and glycosaminoglycans have also been shown to form in vitro by mixing lipoproteins with various glycosaminoglycans (23-27). Thus, a thorough study of the mechanism of these interactions may have significant implications in the etiology of various diseases.

REFERENCES

1. Bernfeld, P., in The Amino Sugars, E. Balazs and R. Jeanloz, Eds., Vol. II, Academic Press, New York, 1966, p. 213.
2. Bernfeld, P., in Metabolic Inhibitors, R. M. Hochster and J. H. Ouastil, Eds., Vol. II, Academic Press, New York, 1963, p. 437.

3. Elbein, A. D., in Advances in Enzymology, A. Meister, Ed.,
    Vol. 40, John Wiley & Sonds, Inc., New York, 1974, p. 29.
4. Cabib, E., and Leloir, L. F., J. Biol. Chem., 231, 259 (1958)
5. Murphy, T. A., and Wyatt, G. R., J. Biol. Chem., 240, 1500
    (1965)
6. Goldman, D. S., and Lornitzo, F. A., J. Biol. Chem. 237, 3332
    (1962)
7. Roth, R., and Sussman, M., Biochem. Biophys. Acta., 122, 225
    (1966)
8. Elbein, A. D., J. Biol. Chem., 242, 403 (1967)
9. Elbein, A. D., J. Bacteriol., 96, 1623 (1968)
10. Liu, C., Patterson, B. W., Lapp, D., and Elbein, A. D., J.
    Biol. Chem., 224, 3728 (1969)
11. Lapp, D., Patterson, B. W., and Elbein, A. D., J. Biol. Chem.,
    246, 4567 (1971)
12. Elbein, A. D. and Mitchell, M. Carbohydrate Res., 37, 223
    (1974).
13. Elbein, A. D., and Mitchell, M. Arch. Biochem. Biophys., 168,
    369 (1975)
14. Horowitz, M. I., Pamer, T., and Glass, G. B. J., Proc. Soc.
    Exp. Biol. Med., 133, 853 (1970)
15. Lornitzo, F. A., and Goldman, D. S., J. Biol. Chem., 239,
    2730 (1964)
16. Bernfeld, P., Bernfeld, H. C., Nisselbaum, J. S., and Fishman,
    W. H., J. Amer. Chem. Soc., 76, 4872 (1964)
17. Bernfeld, P., Tuttle, L. P., and Hubbard, R. W. Arch. Biochem.
    Biophys., 92, 232 (1961)
18. Bernfeld, P., Berkeley, B. J., and Bieker, R. E., Arch.
    Biochem. Biophys., 111, 31 (1965).
19. Tracy, R. E., Dzoga, K. R., and Wissler, R. W. Proc. Soc.
    Exptl. Biol. Med., 118, 1095 (1965)
20. Bihari-Varga, M., Gergely, J. and Giro, S. J. Ather. Res., 4,
    106 (1964)
21. Srinivasan, S. P., Dolan, P., Nadharkrishnamurthy, B. and
    Berenson, G. S. Atherosclerosis, 16, 95 (1972)
22. Srinivasan, S. P., Dolan, P., Nadharkrishnamurthy, B. and
    Berenson, G. S. Prep. Biochem., 2, 83 (1972)
23. Bernfeld, P., Nisselbaum, J. S., Berheley, B. J. and Hanson,
    R. W. J. Biol. Chem., 235, 2852 (1960)
24. Levy, R. and Day, C. E., in Atherosclerosis (R. Jones, ed.)
    Springer-Verlag, New York, 1970, p. 186.
25. Iverius, P. J. Biol. Chem., 247, 2607 (1972)
26. Bihari-Varga, M. and Vegh, M. Biochem. Biophys. Acta, 144,
    202 (1967)
27. Nakashima, Y., DiFerrante, Jackson, R. L., and Pownall, W. J.
    J. Biol. Chem. 250, 5386 (1976).

Kinetic Studies of Immobilized α-Chymotrypsin

in Apolar Solvents

Myron L. Bender, A. B. Cottingham, Lee K. Sun,
and K. Tanizawa

Division of Biochemistry, Dept. of Chemistry
Northwestern University, Evanston, Ill. 60201

ABSTRACT

The mechanism of α-chymotrypsin action has been probed by extending studies of native chymotrypsin to immobilized chymotrypsin, where the organic content of the solution can be raised to much higher levels and thus one can explicitly look at the role of water. When one does this, one finds that water only appears in the deacylation reaction. The premise that one can go from native chymotrypsin (soluble) to immobilized chymotrypsin (insoluble) has been tested by several criteria. It has been found in many instances that the two are identical: in absolute rate, in pK_a. They are, however, not identical to one another in binding, due to differences in diffusion, which is to be expected. Thus, mechanistically immobilized and native chymotrypsin are identical to one another and the use of immobilized chymotrypsin can be used to specify the mechanism even more: it must proceed through two tetrahedral intermediates and two acyl-enzyme intermediates.

INTRODUCTION

The chymotrypsin mechanism has been worked on for a very long time (Bender and Kezdy, 1965). It is known that the reaction occurs by equilibrium binding, followed by two kinetic steps with an acyl-enzyme as intermediate. It is known that a hydroxyl group of a serine moiety is the prime nucleophile, that an imida-

zole group of a histidine residue is the prime catalyst, and that a
carboxylate group of an aspartate moiety is also involved. This
triad is called the "charge-relay system" and we have shown its
existence in simple systems (Komiyama, Breaux and Bender,
1976). Evidence for the formation of a tetrahedral intermediate
in both acylation and deacylation comes from kinetics. A hydrolysis has to involve a water molecule, but its mode of action was
not known. Therefore, we attempted to find this out by reducing
the water in the medium. In other words, organic solvent-water
mixtures were used. But this met with failure, for above 30% organic solvent, the enzyme would precipitate and no further experiments could be carried out. Therefore, we resorted to immobilized chymotrypsin. This is an insoluble system to begin with,
and thus no precipitation of the enzyme can occur. The question
may be reasonably asked if immobilized chymotrypsin is related
to native chymotrypsin. Our tests, to be described later, indicate
that indeed it is, and thus one can use the results found with immobilized chymotrypsin for the general mechanism of chymotrypsin
action. With immobilized chymotrypsin, we were able to go up to
95% dioxane-water and were able to show that the main effect of
increasing the dioxane concentration (and thus decreasing the water concentration) was to reduce the rate of deacylation. Thus,
water must be involved in the deacylation step only, and there
must be two acyl-enzymes, one which is bound to the leaving
group, and one which is bound to water.

EXPERIMENTAL PROCEDURE

Materials. α-Chymotrypsin (α-CHT), lots CDI-1IA and CDI-34M
608, were obtained from the Worthington Biochemical Corp. Chymotrypsin bound to Sepharose 4B (S-CHT) was also a Worthington
product (lot IEACDI 94J012). Chymotrypsin bound to glass was
synthesized from a porous glass support containing primary amine
groups (Biomaterial Supports GAO-3940, meq per g). This material was obtained from the Corning Biological Products Group. The
synthesis was achieved by coupling α-CHT with diazotized porous
glass, which was converted from the commercial porous glass by
p-nitrobenzoylation, reduction, and diazotization, according to usual procedures (Weetall, 1970). N-acetyl-L-tyrosine ethyl ester
(ATEE), lot 22C-3180, was obtained from Sigma Chemical Co.
N-acetyl-L-tryptophan p-nitrophenyl ester (lot H-4285) was obtained from the Cyclo Chemical Co. and used without further purification. N-benzoyl-L-tyrosine p-nitroanilide (lot 129B-5340)

was obtained from Sigma Chemical Co. Dioxane (spectral grade) was obtained from International Chemical and Nuclear Corp. and used without further distillation.

Methods. Stock solutions of α-CHT were prepared and titrated as described previously (Schonbaum, Zerner and Bender, 1960).

The concentration of S-CHT was determined as follows. 1.4 ml of S-CHT was added to 20 ml of pH 8.3, 0.1 M Tris buffer. 0.2 ml of 0.01 M p-nitrophenyl-p'-guanidinobenzoate hydrochloride (NPGB) in dimethyl formamide was added to the above solution, and the reaction mixture was kept well-stirred. 2 ml aliquots were withdrawn at different time intervals and were filtered through millipore filters. The clear filtrate was then measured at 401.5 nm. The absorbances, after correcting for spontaneous hydrolysis of NPGB, were extrapolated back to time zero. The burst of p-nitrophenolate was taken as a measure for the active site concentration of S-CHT. The active site concentration thus obtained for the commercially available S-CHT was 1.0×10^{-4} M. The protein concentration determined by amino acid analysis was 1.1×10^{-4} M (aspartic acid content was used as the standard for calculation). For kinetic experiments, commercial S-CHT was washed with de-ionized double-distilled water, resuspended in water containing 0.2% NaN_3, and was used within two weeks. Active site concentration of the washed S-CHT was determined using NPGB as described above.

Steady state kinetics of ATEE was studied on a Radiometer pH stat; 4 ml of buffer and 0.4 ml of ATEE in CH_3CN were equilibrated in the thermostatted titration vessel, a suitable amount of α-CHT stock solution or well-stirred gel suspension of S-CHT (50-200 μl) was added. The contents in the vessel were stirred at a speed where the reaction rate was independent of the stirring speed (vide infra). The concentration of base used for titration was varied from 0.03 N to 1.6 N, depending on the substrate concentration in each run.

The hydrolytic rates of N-acetyl-L-tryptophan p-nitrophenyl ester and N-benzoyl-L-tyrosine p-nitroanilide in aqueous organic solvent were determined spectrophotometrically by using a Gilford spectrophotometer at 340 nm and 405 nm, respectively (Kezdy, Clement and Bender, 1964; Bundy, 1962). The reactions were carried out in a thermostatted reaction vessel at 25° with stirring. The optical density change of this reaction medium during a certain period was recorded as a function of substrate or dioxane concentration. It was confirmed that the stirring speed used in these reactions was sufficient enough, that is, the velocities of these

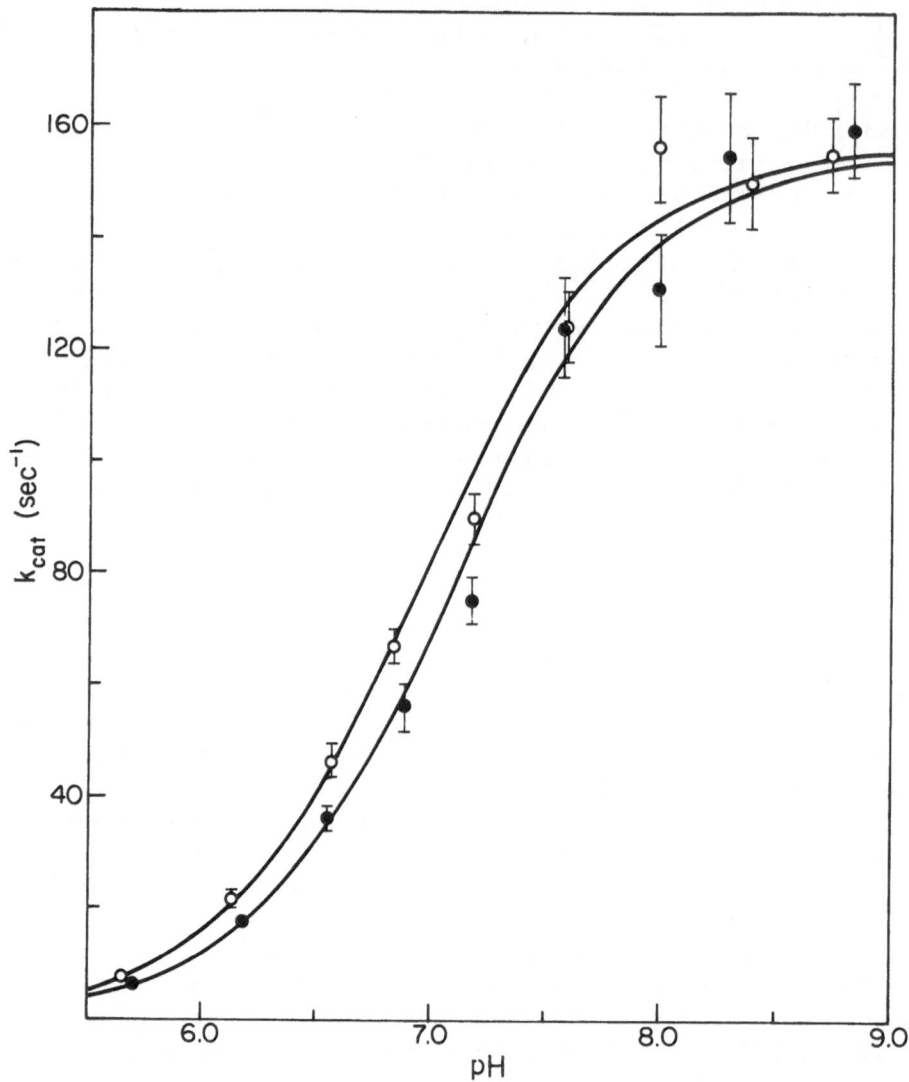

Fig. 1.   The α-chymotrypsin-catalyzed hydrolysis of N-acetyl-
           L-tyrosine ethyl ester at 25° .(k_{cat} / pH profile).

   O   Native chymotrypsin (soluble) in 9.1% acetonitrile-
           water. k_{cat} (lim) = 157.03 ± 4.10 sec⁻¹; pK_a = 6.96
           ± 0.02.

   ●   Chymotrypsin bound to Sepharose (insoluble) in
           9.1% acetonitrile-water. k_{cat} (lim) = 156.17 ± 5.68
           sec⁻¹; pK_a = 7.09 ± 0.02.

heterogeneous reactions were independent of the stirring speed used. The reaction rate was corrected for spontaneous hydrolysis and also corrected by using the corresponding $\Delta\varepsilon$ in each aqueous dioxane system.

RESULTS

The kinetics of native chymotrypsin (soluble) and chymotrypsin attached to Sepharose (insoluble) were compared using ATEE as a substrate in the pH region 5.5 - 9, to ascertain whether these two forms of chymotrypsin were indeed the same. The hydrolysis of ATEE was used as the reaction for both soluble and insoluble chymotrypsin. The kinetic constants, k_{cat} and $K_{m(app)}$ were obtained from the total progress curve of the reaction by use of a computer-assisted analysis based on the chord method. Experimental conditions and results for the reactions of ATEE with S-CHT and α-CHT are summarized in Figs. 1 and 2.

In Fig. 1, limiting k_{cat} values at high pH and pK_a values are calculated by use of the following equation:

$$k_{cat} = k_{cat(lim)} - k_{cat} \frac{[H^+]}{K_a}$$

As shown in Fig. 1, for α-CHT, k_{cat} follows a sigmoid curve of $pK_a = 6.96 \pm 0.02$ and $k_{cat(lim)} = 157.03 \pm 4.10$ sec^{-1}. For Sepharose-bound α-CHT, $pK_a = 7.09 \pm 0.02$ and $k_{cat(lim)} = 156.17 \pm 5.68$ sec^{-1}. These numbers are probably within experimental error.

However, the $K_{m(app)}$ vs. pH profiles for soluble and insoluble α-CHT are very different. (See Fig. 2) This is probably due to the fact that when the reaction is fast (as at high pH), the rate-determining step may be the diffusion of the product out of the Sepharose beads. At lower pH, the reaction rate is slower, and depletion of products as they are formed is more efficient, therefore $K_{m(app)}$ approaches the value for α-CHT.

The results given above indicate that the properties of insolubilized α-chymotrypsin are kinetically identical with those of the free enzyme (with the exception of the binding, which is not important mechanistically). Therefore, the effect of increasing organic solvent (dioxane) concentration on the various kinetic parameters was probed. There is essentially no effect of dioxane on the binding step (except at very low concentrations of dioxane where the dielectric constant of the solvent is changing rapidly). Furthermore, there is only a small effect on the rate in going from 0% to 95% dioxane. This small effect is of the order of a factor of two.

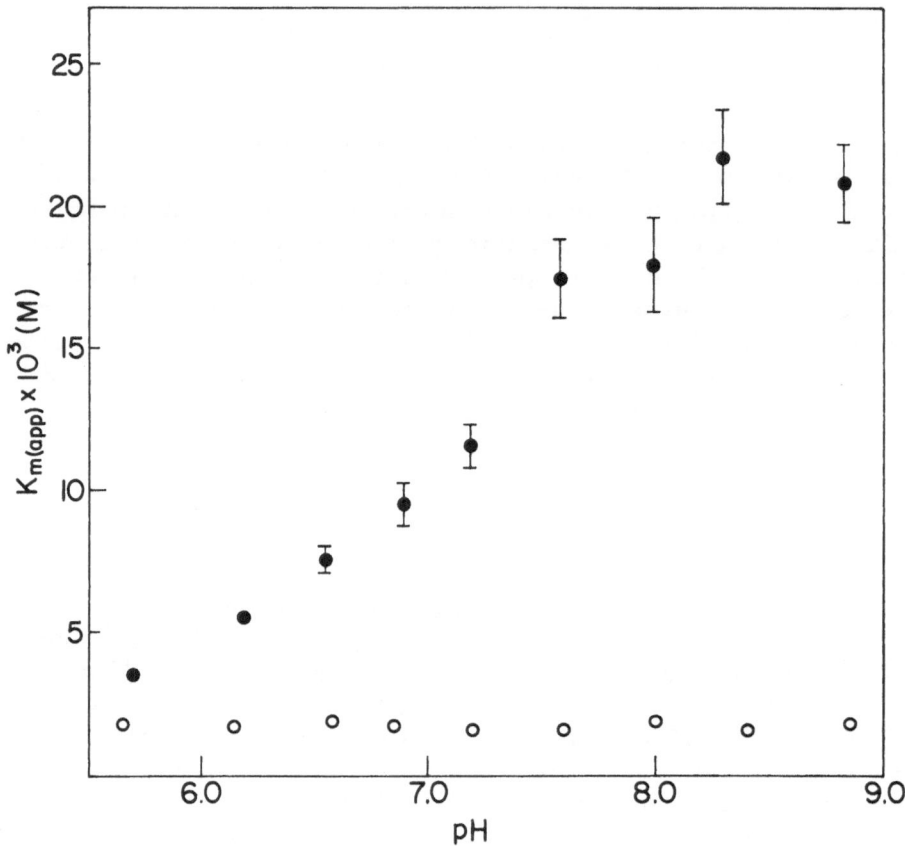

Fig. 2.   The α-chymotrypsin-catalyzed hydrolysis of N-acetyl-
L-tyrosine ethyl ester at 25° ($K_{m(app)}$/ pH profile).

O   Native chymotrypsin (soluble) in 9.1% acetonitrile-
       water. $K_{m(app)}$ = 1.47 - 1.81 x 10^{-3} M.

●   Chymotrypsin bound to Sepharose (insoluble) in 9.1%
       acetonitrile -water. $K_{m(app)}$ = 3.46 - 21.72 x 10^{-3} M.

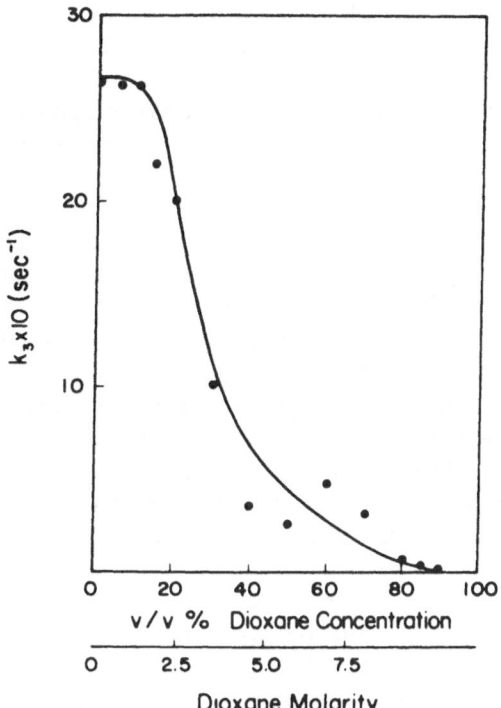

Fig. 3.   The effect of dioxane concentration on the deacylation
            rate of porous glass α-chymotrypsin-catalyzed hydroly-
            sis of N-acetyl-L-tryptophan p-nitrophenyl ester at 25°,
            pH 5.8, 0.005 M acetate-dioxane [S] = 1.04 x 10⁻⁴ , and
            6.72 x 10⁻⁴ M; [E] = 2.54 x 10⁻⁷ M.

One may legitimately ask if there is any effect in increasing the
dioxane concentration. The answer is a decided yes. The effect
occurs in the deacylation step. Increasing the dioxane concentration from 0-95% results in a large diminution of the deacylation
rate constant. The diminution amounts to over a factor of thirty
or if you believe the bottom point in Fig. 3, it amounts to an infinite effect. A more reasonable statement is that the effect is
somewhere between thirty and infinity.

DISCUSSION

Chymotrypsin bound to Sepharose and to glass has been investigated. One part of the investigation indicates that chymotrypsin

bound to a solid surface behaves in no way differently from native
(soluble) chymotrypsin. Thus, the results found with insoluble
chymotrypsin can be used with those of the soluble enzyme. In
going to high dioxane concentrations, only chymotrypsin attached
to porous glass was used, for we wanted a support which would
neither swell nor shrink.

The effect of dioxane concentration appears almost exclusively in deacylation. The effect of dioxane concentration on the deacylation rate constant can be explained by the Scheme. The deacylation rate should be dependent on the water concentration. The components of the transition state of deacylation must include the acyl-
serine ester, imidazole, and a molecule of water. A water molecule and imidazole are involved in the rate-determining proton
transfer reaction. The first order dependence on the nucleophile
water is expected by analogy to the kinetics of methanolysis reported previously (Bender, Clement, Gunter and Kezdy, 1964). As
shown in Fig. 3, it is hard to show that the reaction is linear in
water concentration, since the reaction is also dependent on secondary effects of the aqueous organic solvent such as that on the
structure of the solvent. Taking into account microscopic reversibility in respect to acylation and deacylation, the assumption that
the solvent polarity affects only the deacylation step can be ruled
out.

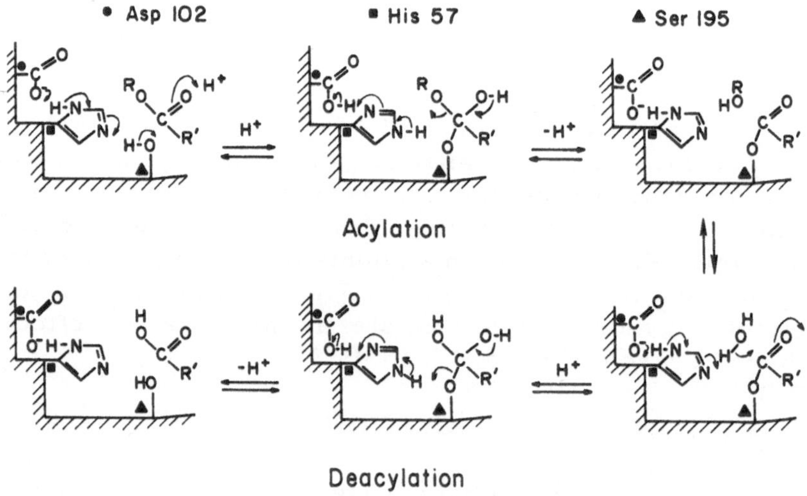

Acylation

Deacylation

Dioxane as employed does not cause irreversible denaturation of the porous glass-bound enzyme, but the question can arise whether this organic solvent causes reversible denaturation, or a conformational change of the enzyme, or both. This also appears unlikely for, again in contrast to the experimental findings, such effects should be reflected by changes in both acylation and deacylation rates. The actual decrease in the availability of water with increasing dioxane appears to best account for both the small effect on acylation and the large effect on deacylation.

ACKNOWLEDGEMENT

This research was supported by Grant GM-20853 of the National Institutes of Health.

REFERENCES

Bender, M. L. , Clement, G. E. , Gunter, C. R. , and Kezdy, F. J. (1964). The Kinetics of α-Chymotrypsin Reactions in the presence of added Nucleophiles. J. Amer. Chem. Soc. ,86, 3697.

Bender, M. L. and Kezdy, F. J. (1965). Mechanism of Action of Proteolytic Enzymes. Ann. Rev. Biochem. , 34,49.

Bundy, H. F. (1962). A New Spectrophotometric Method for the Determination of Chymotrypsin Activity. Anal. Biochem. , 3, 431.

Kezdy, F. J. , Clement, G. E. , and Bender, M. L. (1964). The Observation of Acyl-Enzyme Intermediates in the Chymotrypsin-catalyzed Reactions of N-acetyl-L-tryptophan Derivatives at low pH. J. Amer. Chem. Soc. , 86, 3690.

Komiyama, M. , Breaux, E. J. , and Bender, M. L. (1976). Cyclo-amylose-catalyzed Hydrolysis as a Model for the "Charge-Relay" System. Submitted to Bioorganic Chemistry.

Schonbaum, G. R. , Zerner, B, and Bender, M. L. (1960). The Spectrophotometric Determination of the Operational Normality of an α-Chymotrypsin Solution. J. Biol. Chem. , 235, 2930.

26

FACTORS AFFECTING CYANOBOROHYDRIDE REDUCTION OF AROMATIC SCHIFF'S BASES IN PROTEINS

Leroy Chauffe[*] and Mendel Friedman

Western Regional Research Laboratory, Agricultural Research Service, U. S. Department of Agriculture, Berkeley, California 49710

ABSTRACT

Reductive alkylation of bovine serum albumin (BSA) and wool by aromatic aldehydes and sodium cyanoborohydride has been investigated. The aldehydes used were chosen to allow convenient quantitative measurement of binding by ultraviolet spectroscopy. Alkylation of BSA occurred primarily at two highly reactive sites. Variation of time, pH, reactant concentration, and addition of urea had little effect on the extent of alkylation of BSA. However, more extensive alkylation was achieved in buffered aqueous dimethyl sulfoxide. The unusual reactivity of two ϵ-amino groups on the BSA molecule is attributed to closely placed lysine residues in the primary sequence rather than to favorable placement of unrelated, distant reactive centers. Similarly, only a few of the potentially available ϵ-amino groups of wool were observed to react.

INTRODUCTION

This paper describes part of a research program for preparing protected proteins for improved ruminant nutrition (Broderick, 1975). In this particular study, we have sought to develop useful reaction conditions for reductive alkylation of lysine ϵ-amino groups. In addition, we wished to establish whether reductive alkylation with aromatic aldehydes can be useful for measuring the available lysine content of proteins. Because simple assay techniques that allow determination of the extent of protein modification are of paramount importance, the aldehydes were selected for convenient measurement of uptake by means of their characteristic ultraviolet (UV) absorption spectra.

Sodium cyanoborohydride has been introduced as a reducing agent in selective alkylation of ϵ-amino groups in proteins (Friedman et al., 1974). This reducing agent is particularly appropriate because of its acid stability (Kreevoy et al., 1969) and its selective reduction of imines (Schiff's bases) in neutral and acidic media (Borch et al., 1971). Reductions at acid pH are not accompanied by deleterious side reactions, such as desulfurization, which often occur when reductions are carried out in alkaline media. Reaction conditions that may affect the reducing properties of the cyanoborohydride ion have been investigated. In some cases, results of parallel reductions with the borohydride ion are included for comparison.

The aromatic aldehyde most used in this study was pyridoxal. Its choice was due to its biological importance, water solubility and characteristic UV absorptivity. Demspey and Christensen (1962) have reported specific binding of pyridoxal-5-phosphate to bovine serum albumin (BSA). They interpreted their results as showing two high-affinity binding sites on BSA. Site I ($K > 10^6$) is occupied when the molar ratio of aldehyde to protein is 1:1 and site II ($K \approx 10^5$) when the molar ratio is 2:1. Binding to both sites is thought to occur by formation of Schiff's bases. This high binding affinity is also thought to depend on specific interaction between the anionic phosphate ester moiety of pyridoxal-5-phosphate and certain cationic groups of the protein (Means and Feeney, 1971). Above the ratio 2:1, binding of pyridoxal-5-phosphate is considered nonspecific. Our choice of pyridoxal was intended to exclude specific binding through the phosphate group and to allow more carbonyl group interaction with the 50 moles of ϵ-amino groups potentially available per mole of BSA. Consequently, we sought to identify conditions that would lead to maximum modification of the available ϵ-amino groups. We report effects of concentration, pH, time, solvent, and aldehyde structure on the reductive alkylation of a soluble protein (BSA) and an insoluble protein (wool). Reductive alkylation by several isomeric sets of aromatic aldehydes are included to show the stereospecificity of the reactions.

MATERIALS AND METHODS

The BSA used was obtained from the Sigma Chemical Co. (Lot 35C-8050). The wool was a fine top, 1961 clip, from Dubois, Idaho. Other substances used were of reagent grade obtained from usual commercial sources and were used without further purification.

The reductive alkylation protocol which follows differs from that previously published (Friedman et al., 1974). A typical run follows: The protein (BSA, 100 mg) and aromatic aldehyde (pyridoxal, 70.6 mg) were placed, dry, into a flask. A solution of sodium cyanoborohydride (21.7 mg) in 20 ml of the reaction solvent (equilibrated to 30°) was added to a single portion. The

resultant mixture was held at 30° with shaking for one hour.
BSA dissolved completely during reaction (wool reactions
remained heterogeneous). At the end of that time, the reaction
was quenched by adding 3 ml cold, concentrated (37%) hydrochloric acid. The protein was dialyzed against 0.01N acetic
acid for 24 hours, then freeze-dried. The extent of reaction was
estimated as previously reported (Friedman et al., 1974) from
the ultraviolet spectrum of the modified protein and sometimes
by amino acid analysis.

RESULTS AND DISCUSSION

Effect of Reactant Concentration. Alkylation of BSA in pH
7.2 borax buffer with pyridoxal (5:1 molar ratio of pyridoxal to
ε-amino groups in lysine) yielded approximately 2 moles of ε-
amino groups modified per mole of protein. This extent of modification corresponds to reaction at the two strong binding sites
for pyridoxal-5-phosphate reported by Dempsey and Christensen.
Since additional but weaker binding sites for pyridoxal-5-
phosphate exist in the presence of high concentration of the phosphate, we increased the concentration of pyridoxal (Figure 1) and

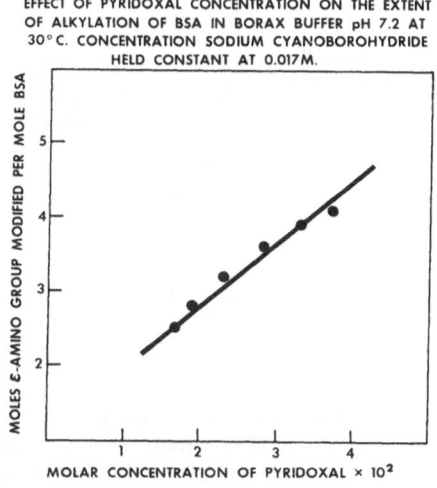

EFFECT OF PYRIDOXAL CONCENTRATION ON THE EXTENT
OF ALKYLATION OF BSA IN BORAX BUFFER pH 7.2 AT
30°C. CONCENTRATION SODIUM CYANOBOROHYDRIDE
HELD CONSTANT AT 0.017M.

MOLES ε-AMINO GROUP MODIFIED PER MOLE BSA

MOLAR CONCENTRATION OF PYRIDOXAL × 10²

Figure 1

sodium cyanoborohydride (Figure 2) in an effort to increase
alkylation. Pyridoxal concentration was varied from a fifteen-
fold to a seventy-five-fold excess relative to the potentially
available ε-amino groups. The increase in alkylation was less
than two-fold. A similar variation in the sodium cyanoborohydride concentration did not increase alkylation.

These results are best interpreted in terms of the mechanism

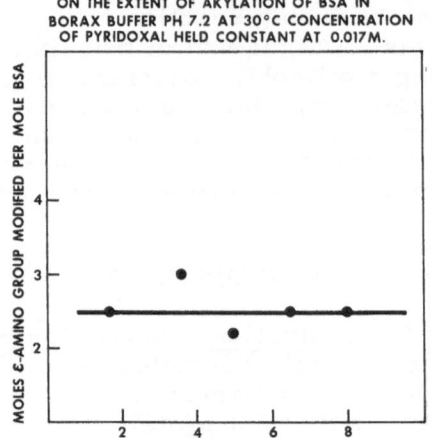

EFFECT OF SODIUM CYANO BOROHYDRIDE CONCENTRATION
ON THE EXTENT OF AKYLATION OF BSA IN
BORAX BUFFER PH 7.2 AT 30°C CONCENTRATION
OF PYRIDOXAL HELD CONSTANT AT 0.017M.

Figure 2

proposed by Borch et al. (1971):

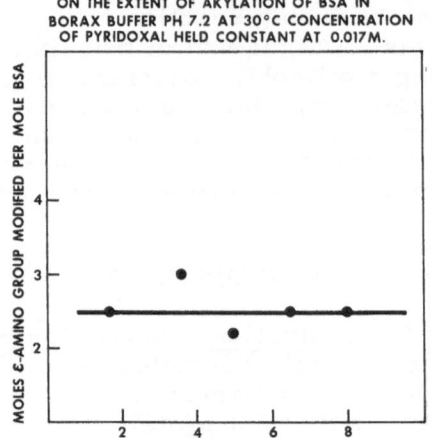

Reduction of aldehydes and ketones is negligible in the pH range
6-7. Since reduction of the Schiff's base is the fast step in the
reaction, the extent of alkylation is expected to be unchanged by
increasing the amount of cyanoborohydride anion above the stoi-
chiometric quantity. This conjecture is supported by the results.
However, increasing the pyridoxal concentration should give
more alkylation. The increase observed is much smaller than
anticipated. It appears that only two lysine amino groups are
sufficiently nucleophilic at pH 7.2 to react rapidly. Imine
formation by protonated amino groups might also occur if
longer reaction times were allowed.

Effect of Reaction Time. To test this possibility, reaction
times as long as seventy-two hours were used (Table 1). Even

Table 1

Reaction parameter	Moles ϵ-amino group modified per mole BSA
Time (hours)	
1	2.3
2	2.9
4	2.6
18	3.1
72	3.1
Urea (molar conc.)	
1	2.7
2	2.1
4	2.5
8	2.5

after such long times, cyanoborohydride anion was still available for reduction. However, the extent of alkylation after seventy-two hours was hardly greater than after two hours.

Effect of pH and Buffer Concentration. Alkylations were carried out at several acid concentrations to test the possibility producing more accessible ϵ-amino groups by varying the protein's hydrogen bonding and configuration. Variation of pH from 2.7 to 7.0 (citrate buffer) caused a linear increase in the number of the amino groups alkylated (Figure 3). The increase was from zero to approximately three moles of amino group modified per mole of BSA. Three-fold change in the buffer concentration had no effect on the number of amino groups modified. The anticipated increase in the extent of alkylation near pH 4 because of the configurational change in the protein (Taylor et al., 1975a) did not occur. This observation is not unreasonable because decreased imine formation even at the two strongly nucleophilic sites might result from the decreased concentration of unprotonated amino group.

Effect of Urea. Denaturation in neutral solution with urea was tested to avoid interference by protonation. The amount of alkylated BSA produced in one hour in the presence of urea concentrations as high as 8M (Table 1) was no greater than the amount in the absence of urea. Since it must be assumed that extensive unfolding of the protein occurs in the urea solution, the unreactive amino groups are not prevented from reacting by the specific configuration of the native protein. The two specific protein-pyridoxal complexes must be energetically favored to an

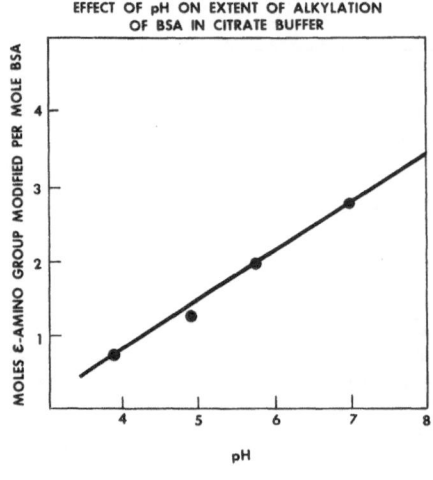

Figure 3

extent which allows their formation and subsequent Schiff's base reaction even in 8M urea.

Effect of Solvent. Reactions of amino groups of peptides as nucleophiles have been shown to be accelerated by dimethyl sulfoxide, (DMSO) (Friedman, 1967). The rates of these reactions typically pass through a maximum and then decrease with increasing DMSO concentration. Alkylation of BSA by pyridoxal behaves in this way. The extent of reaction increased with increasing DMSO content in a semi-aqueous solution to a maximum near 50% DMSO-50% buffer (Figure 4). The net increase was approximately four-fold. This observed increase apparently results from increased ionization due to the added DMSO of the normally protonated ϵ-amino groups. Also, nucleophilicities in DMSO solvent medium are inherently greater than in aqueous media because of selective solvation of positive species (electrophilic agents) by the DMSO molecules which frees negative (nucleophilic) reagents from the stabilizing influence of positive species.

Effect of Aldehyde Structure. Several isomeric sets of aromatic aldehydes (Figures 5 & 6) were reacted with BSA and with wool. The extents of alkylation by these aldehydes are listed in Table 2. Electromeric as well as steric factors appear important. With both proteins, those substituted aldehydes with electron-releasing groups (p-N, N-dimethyl- and methoxy-benzaldehydes) react to lesser extents than those substituted aldehydes with electron-withdrawing groups (nitrobenzaldehydes and pyridine carboxaldehydes). The greater reaction by those aldehydes with electron-withdrawing groups is consistent with

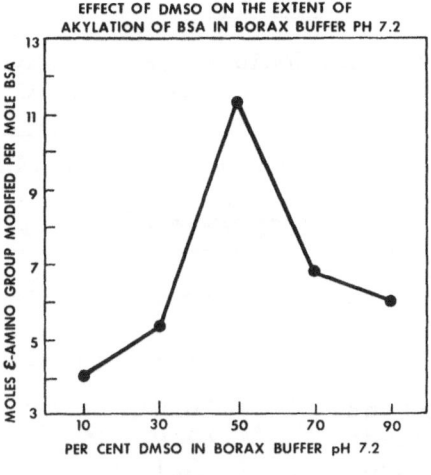

EFFECT OF DMSO ON THE EXTENT OF
AKYLATION OF BSA IN BORAX BUFFER PH 7.2
MOLES Ɛ-AMINO GROUP MODIFIED PER MOLE BSA
13
11
9
7
5
3
10
30
50
70
90
PER CENT DMSO IN BORAX BUFFER pH 7.2

Figure 4

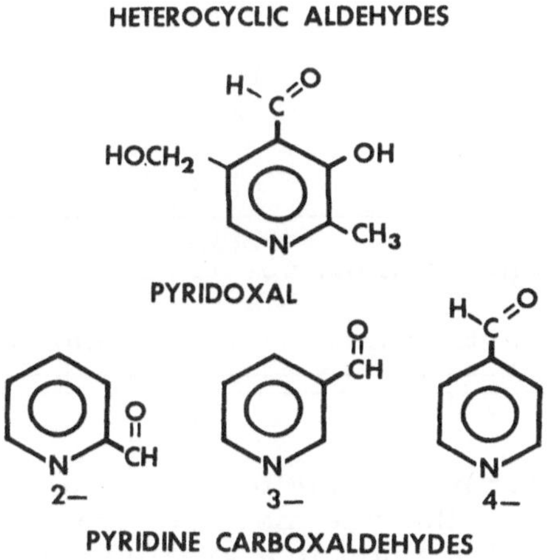

HETEROCYCLIC ALDEHYDES
HOCH2
OH
CH3
PYRIDOXAL
2–
3–
4–
PYRIDINE CARBOXALDEHYDES

Figure 5

SUBSTITUTED BENZALDEHYDES
X = o–O2N–
m–O2N–
p–O2N–
2–CH3O–
4–CH3O–
p–(CH3)2N–

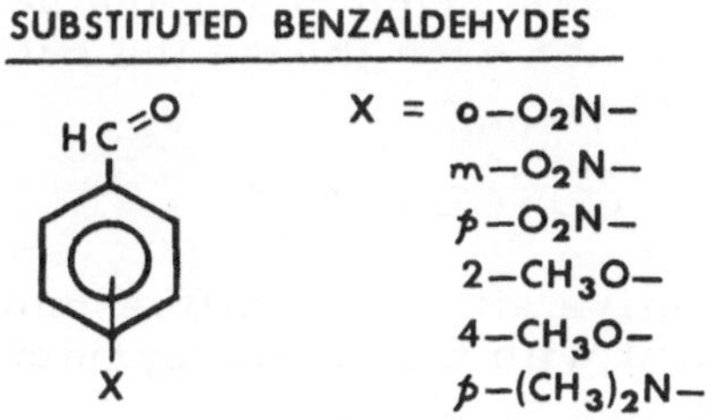

Figure 6

Table 2

Reductive Alkylation with Sodium Cyanoborohydride in 50%
DMSO-50% Borax Buffer at 30°C

Alkylating agent	Moles ε-amino group modified per mole BSA[a]	Mmoles ε-amino group modified per 100 G wool[b]
o-Nitrobenzaldehyde	14.7	--
m-Nitrobenzaldehyde	13.1	3.5
p-Nitrobenzaldehyde	1.2	1.3
2-Pyridine carboxaldehyde	16.9	9.7
3-Pyridine carboxaldehyde	17.9	7.1
4-Pyridine carboxaldehyde	3.1	4.9
p-N,N-dimethylamino benzaldehyde	0	0.4
2-Methoxybenzaldehyde	10.5	8.4
4-Methoxybenzaldehyde	2.0	0.9
Pyridoxal	11.3	2.7

[a] Extent modification determined spectrophotometrically.

[b] Extent modification determined by amino acid analysis.

easier Schiff's base formation due to more reactive carbonyl
functions. It is striking that the more sterically hindered ortho
substituted benzaldehydes react to a greater extent than the less
hindered para isomers. This favorable ortho group interaction
facilitating Schiff's base formation could be hydrophobic or electrostatic. Electrostatic interaction seems the more likely
because 2-pyridine carboxaldehyde shows the same increased
reactivity over its 4-isomer even though the bulk factors for the
two isomers are identical.

CONCLUSIONS

The two highly reactive sites on the BSA molecule are
assumed to be coincident with the unique pyridoxal-5-phosphate
(PP) binding site identified by Dempsey and Christensen (1962).
These sites have been reported by Taylor and co-workers (1975b)
as catalytic sites for the enzyme-like catalysis of the decompo-

Table 3

Comparison of the Extent Reductive
Alkylation of BSA by Pyridoxal using
Cyanoborohydride and Borohydride as
Reducing Agents

	pH	Buffer	Moles ϵ-amino group modified per mole BSA
NaBH$_3$CN	5	Citrate	1.3
	7	Citrate	2.8
	7.2	Borax	2.8
NaBH$_4$	5	Citrate	0.9
	7	Citrate	1.3
	7.2	Borax	2.6

sition of the Meisenheimer complex, 1,1-dihydro-2,4,6-trini-trocyclohexadienate. The catalysis requires an unprotonated base on the protein (pKa = 8.4) and is completely inhibited by PP. The following structure, A, has been suggested for the

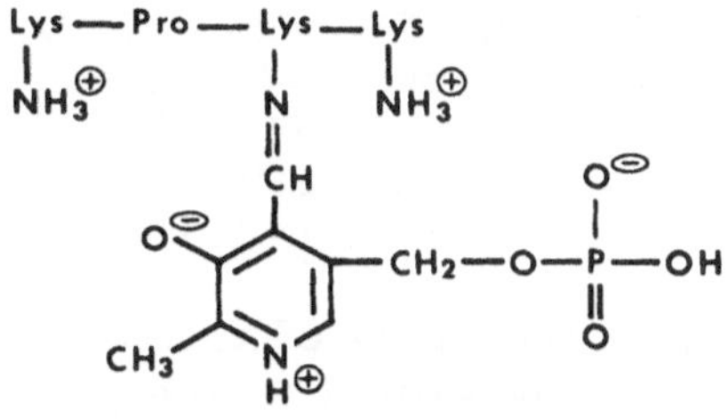

A 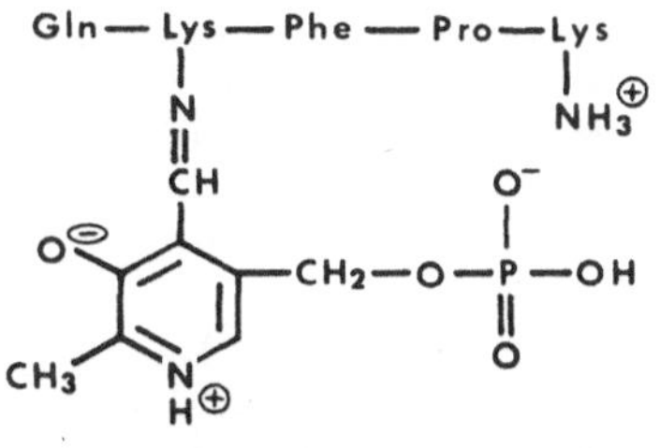   B

PP-BSA complex (Anderson et al., 1971). Another sequencing study gave the additional structure, B (Anderson et al., 1971). The close proximity of lysine residues is deduced to give the microenvironment necessary for the aldehydes to react rapidly to form Schiff's bases. Other microenvironments are apparently less favorable with or without the presence of denaturing agents. The unusually low pKa values for the reactive amino groups probably results from the close proximity of similarly charged groups. It has been shown that the influence of one ammonium group on another, three carbons away, is to lower the pKa approximately 2 units (Edsall and Wyman, 1958).

In summary, the reductive alkylation of BSA using pyridoxal is unaffected by most changes in reaction conditions. Reaction of BSA is characterized by two highly reactive ε-amino groups. Similarly, only a few of the potentially available ε-amino groups of wool were observed to react.

REFERENCES

*Present address is the Department of Chemistry, California State University, Hayward, California 94542.

Anderson, J. A., Chang, H. W. and Grandjean, C. J. (1971). Nature of the binding site of pyridoxal-5-phosphate to bovine serum albumin. Biochemistry, 10, 2408-2415.

Anderson, L. O., Rehnstrom, A. and Eaker, D. C. (1971). Studies on nonspecific binding: The nature of the binding of fluorescein to bovine serum albumin. Eur. J. Biochem., 20, 371-380.

Borch, R. F., Berstein, M. D. and Durst, H. D. (1971). The cyanohydridoborate anion as a selective reductive agent. J. Amer. Chem. Soc., 93, 2897-2902.

Broderick, C. A. (1975). Factors affecting ruminant responses to protected amino acids and proteins, In "Protein Nutritional Quality of Foods and Feeds," M. Friedman (Editor), Dekker, New York, Part 2, pp. 211-259. cf. also papers by Broderick and by Friedman and Broderick, this volume.

Dempsey, W. B. and Christensen, H. N. (1962). The specific binding of pyridoxal-5-phosphate to bovine serum albumin. J. Biol. Chem., 237, 1113-1120.

Edsall, J. T. and Wyman, J. (1958). "Biophysical Chemistry," Vol. 1, Academic Press, Inc., New York, pp. 457-463.

Friedman, M., Williams, L. C. and Masri, M. S. (1974). Reductive alkylation of proteins with aromatic aldehydes and sodium cyanoborohydride. Int. J. Peptide Protein Res., 6, 183-185.

Friedman, M. (1967). Solvent effects in reactions of amino acids, peptides and proteins with α,β-unsaturated compounds. J. Amer. Chem. Soc., 89, 4709-4713.

Kreevoy, M. M. and Hutchins, J. E. C. (1969). Acid-catalyzed hydrolysis and isotope exchange in $LiBH_3CN$. J. Amer. Chem. Soc., 91, 4329-4330.

Means, G. E. and Feeney, R. E. (1971). "Chemical Modification of Proteins," Holden-Day, Inc., San Francisco, pp. 132-134.

Taylor, R. P., Berga, S., Chau, V. and Bryner, C. (1975a). Bovine serum albumin as a catalyst. III. Conformational studies. J. Amer. Chem. Soc., 97, 1943-1948.

Taylor, R. P., Chau, V., Bryner, C. and Berga, S. (1975b). Bovine serum albumin as a catalyst. II. Characterization of the kinetics. J. Amer. Chem. Soc., 97, 1934-1943.

27

CHEMISTRY OF THE CROSSLINKING OF COLLAGEN DURING TANNING

J. W. Harlan and S. H. Feairheller

Eastern Regional Research Center, Agricultural Research

Service, U.S. Department of Agriculture, Philadelphia,

Pennsylvania 19118

ABSTRACT

The materials commonly used for crosslinking collagen as part of the process of converting animal hides into leather fall into three main groups: mineral tannages, aldehyde tannages, and "vegetable" tannages. The most important mineral crosslinking agents are hydrated basic chromium III sulfate complexes. These compounds form extended polynuclear coordination complexes containing hydroxol, oxo, and sulfato bridges into which ionized carboxyl groups on collagen enter readily as coordinating ligands accomplishing crosslinking. On pH adjustment and partial drying, highly stable complexes are formed with oxo bridges predominating and protein amide groups entering the coordination complex. The aldehyde tannages proceed through aldehyde condensation reactions with collagen amino groups to give alpha-hydroxyamines which can condense with other collagen amine groups to effect crosslinking. The vegetable type tanning agents, whether natural plant extracts or synthesized, are complex, high molecular weight polyhydroxy compounds that do not rely on crosslinking as such to be effective. Their effectiveness appears to depend on other properties. This and additional information concerning these commercial tannages are reviewed.

INTRODUCTION

The hide matrix from which leather is made is composed primarily of the protein collagen ordered in microcrystalline

helical units. The crystalline structure is stabilized by a large
number of hydrogen bonds between peptide nitrogens and opposite
carbonyl groups in the helix. These microcrystallites are then
connected by a secondary amorphous structure in which polar and
nonpolar side chains participate and covalent crosslinking is
established. This has been discussed thoroughly in earlier papers
in this symposium.

Native hide collagen is chemically and biologically inert
when dry but is stiff, brittle, and relatively useless. However,
when wet, it is subject to biological attack and is hydrothermally
unstable. The crystallites melt at 58°C in pure water and as low
as 37°C in salt solutions.

The purpose of tannage is primarily to increase hydrothermal
stability of the crystallites, secondarily to increase biological
inertness, and finally, to improve the utility of the hide's
physical properties.

A simple empirical shrink temperature test has been used in
most tanning research as the criterion of hydrothermal stability.
To interpret the literature one must understand precisely what
this test measures. When native or tanned hide collagen absorbs
water, it swells and extends in length. As the swollen collagen
is heated in water it retains its dimensions until a critical
temperature is recorded as the shrink temperature. It had been
suggested (Garrett and Flory, 1956) that the shrink temperature
corresponded to the melting point (Tm) of the water solvated collagen crystallites. This concept was explored at the Eastern
Regional Research Center (ERRC) (Witnauer and Fee, 1957), where
the temperatures at which this shrinkage of native and tanned
collagen occurred were measured over a range of moisture and other
solvent contents. This data (Figure 1) supports four postulates
critical to understanding and analyzing the effect of tannages on
the hydrothermal stability of collagen and the meaning of shrink
temperature.

(1) Native collagen crystallite melting points are depressed
by nonreactive solvents as predicted by thermodynamics of a crystalline polymer-solute system.

(2) The melting point Tm decreases with increasing amounts of
dissolved solute according to an equation of the form

$$\Delta Tm = Tm° - Tm = A\ Tm\ v_s (1 - \chi v_s)$$

where $Tm°$ is the melting point of the polymer at zero solute content, v_s is the volume fraction of the solute, and A and χ are
constants for a given polymer-solute pair. The magnitude of A

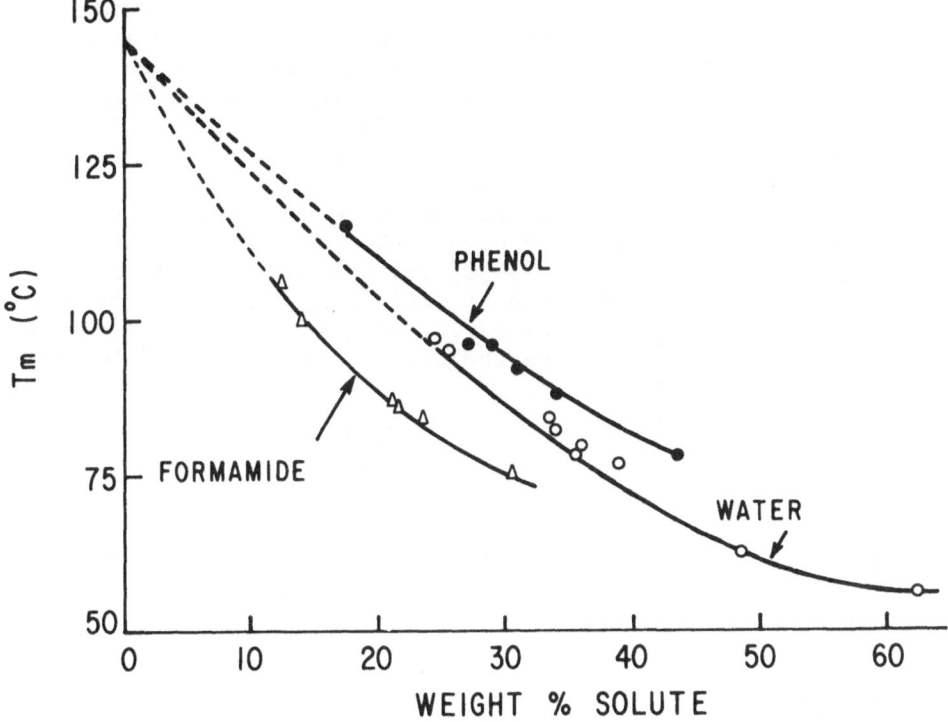

Figure 1. Melting Point (Tm) Depression of Native Collagen.

depends on the energy change in the crystalline polymer transition being observed and the sign and magnitude of χ depend on the energy of interaction between the polymer and solute (Flory, 1953).

(3) The maximum melting point depression with a given solute is limited by the extent of solubility (the last points for each of the three solutes in Figure 1). Further additions of (or immersion in) the solute will not further decrease the melting point.

(4) Native collagen has an intrinsic (extrapolated) melting point (Tm°) of 145°C, when completely "dry" (no solute present). Thus if a treatment or "tannage" of collagen does not alter the native collagen crystallites, the observed Tm° should be 145°C. Conversely, observed values of Tm° other than 145°C indicate that the structure of the collagen crystallites has been changed by the treatment.

Interpretation of observed increases in Tm° as indicating increased crosslinking and decreases in Tm° as indicating a decrease in crosslinking in the collagen crystallites, along with the four postulates discussed above, leads to the following interpretation of the data of Kawamura et al. (1975) in Figure 2. Figure 2 shows the melting point depression curves (with water as the solute) for untanned, chrome-tanned, and vegetable (mimosa)-tanned calfskin. As in the previous data, Tm for untanned hide extrapolates to give a Tm° of 145–150°C at 0% H_2O. However, the Tm° for chrome-tanned skin is distinctly higher than 145°C, indicating cross linking. Tm° for the vegetable-tanned skin is definitely lower than 145°C. This fact, plus the slope of the curve, suggests that the vegetable tannin is actually destabilizing the helix at low moisture levels

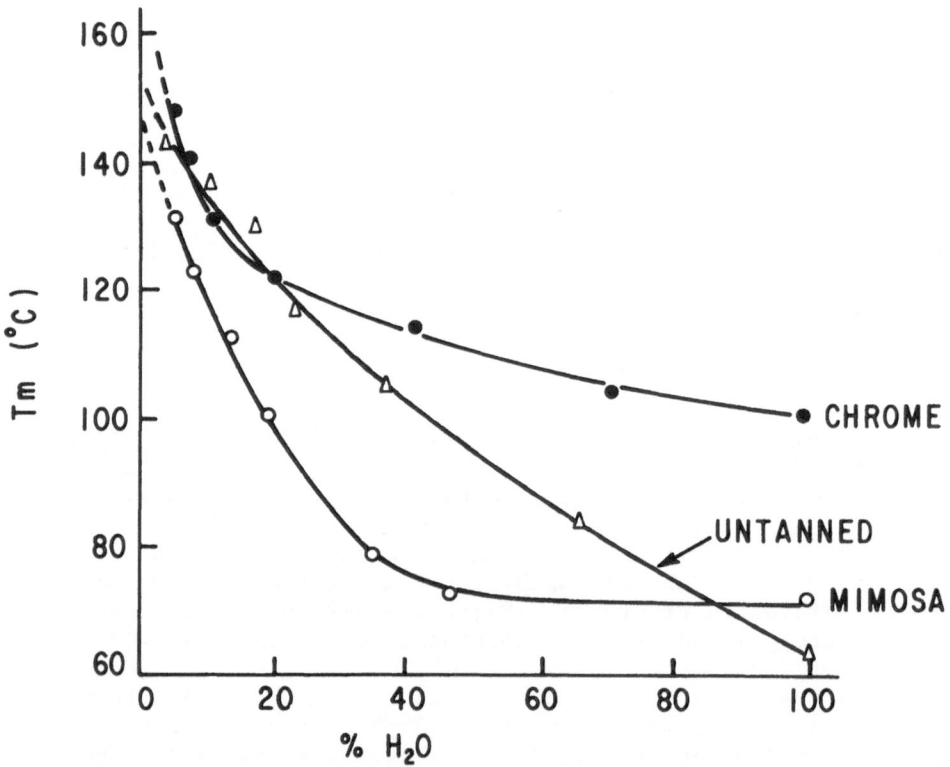

Figure 2. Melting Point (Tm) Depression of Tanned and Untanned Calfskin.

(probably by competing for hydrogen bonding sites normally involved in crosslinking). The fact that vegetable tannins are practical commercial tannages appears due to their ability to limit the effective solubility of water. In this example, the vegetable-tanned hide is hydrothermally stable up to 75°C, even in aqueous solutions, while the shrinkage temperature of the untanned hide falls to 60°C when immersed in water.

The evidence in the literature on the mechanism of hydrothermal stabilization of the major tannages, using this approach, is summarized in (Table I). The changes in the intrinsic crystallite melting points ($\Delta Tm°$) are referred to native collagen. The observed polymer-water interaction energy constants χ of equation 1, are expressed here as B values. $B = \chi RT/V_s$, where V_s is the molar volume of water. The B values reflect the degree of curvature in the curves of Figure 1 and 2. Positive values of B indicate that the polymer is a poor solvent for water. Negative values indicate high solvency power. Relative torsional modulus measurements (Witnauer and Fee, 1957) of untanned, vegetable-tanned, and formal-dehyde-tanned hides led to the postulation that an increase in modulus indicated increased crosslinking. Estimations of the crosslinks remaining (by tensile modulus measurements after destroy-ing the hydrogen bonding in the crystallites) gave the data in the last column for aldehyde tannages (Cater, 1963). Unfortunately, none of these methods has been applied to compare directly all of these important commercial tannages. However, we can conclude that aldehyde and chrome tannages involve crosslink formation while vegetable tannages do not.

The depression of Tm° and a large positive interaction coefficient for vegetable tannages suggest that they act as solutes, competing with and blocking the effectiveness of water as a solute and thereby providing hydrothermal stability. Formaldehyde, on the other hand, results in an increased Tm° and a negative B, suggesting a crosslinking mechanism. The only relevant study including glu-taraldehyde (Cater, 1963) suggests substantial crosslinking.

The surprisingly limited data on chrome tanning shows both a large increase in crystallite melting point, indicating crosslinking and a large positive B suggesting that the high degree of hydro-thermal stability imparted by this tannage is due to operation of both stabilizing mechanisms: reducing the solute effect of water and providing additional crosslinking.

CHEMISTRY OF CROSSLINKING REACTIONS

The types of functional groups available for reaction and their relative molar amounts are shown on the top row of Table II. The

TABLE I

Effect of Tannage on Cowhide and Solute Activity of Water

System	$\Delta Tm°$[a] (°C)	B[a] (cal/cc)	Modulus[a] (Relative)	Crosslinks[b] ($\#/10^5$g)
Untanned	0 (ref.)	+2	1.0 (ref.)	0.2
Vegetable Tanned	-25	+35	1.5	
Formaldehyde Tanned	+6	-9	5.0	5
Chrome Tanned	+ (<20)[c]	+(large)[c]	-	-
Glutaraldehyde Tanned	-	-	-	10

[a]Witnauer and Fee, 1957. $\Delta Tm° = Tm°$ observed $- Tm°$ of untanned collagen. B is the energy of interaction of the polymer-water system $= \chi RT/V_s$ (see text).

[b]Cater, 1963.

[c]Kawamura, 1975.

peptide and hydroxyl groups are present in greatest number. They are polar, but not ionized, and form hydrogen bonds with natural vegetable and synthetic (syntan) type tannins.

Mineral tannages crosslink through coordination with the relatively abundant carboxyl ions on collagen and provide the most hydrothermally stable and commercially important tannages. While chrome, aluminum, zirconium and combinations of these are used, chrome tanning is by far the most important.

Aldehyde tannages form stable covalent crosslinked compounds through the relatively limited number of amine groups available.

Chromium Tanning

Chrome tanning is generally carried out by adding the hide collagen to an aqueous solution of 33% basic trivalent chromium

TABLE II

Functional Groups Important In Tanning

	$P\vert -\overset{\overset{\textstyle O}{\Vert}}{C}-OH$	$P\vert -NH_2$	$P\vert -\overset{\overset{\textstyle O}{\Vert}}{C}-NH-R$	$P\vert -OH$
Relative Number	3	1	19	3
Type of Tannage	"Mineral" Chrome Zirconium Alum	Aldehyde Formaldehyde . Glutaraldehyde	Vegetable Syntans	
Type of Bonds Formed	Coordinate	Covalent	Hydrogen	

sulfate. By using a combination of the latest gel permeation, gel electrophoresis, ion exchange chromatography, and spectroscopic techniques, Slabbert (1975) established that there are at least 10 ionic and neutral complexes in a 33% basic chrome sulfate solution. The structures of eight of these were determined along with the relative amounts of each. The six compounds present in highest concentration are shown in Figure 3.

The simplest ion, Cr^{+3}, with six coordinated waters, was present as 9% of the total chrome. The routes of formation of the complexes resulting from replacement of water with sulfate ion (left side of the figure) or by hydroxide ion (right side of the figure) are indicated.

The salient features of the route of formation of the most abundant species, the $^{+2}$ charged binuclear complex with a bidentate sulfate bridge, are the replacement of H_2O by $SO_4^=$ in $Cr(H_2O)_6^{+3}$ to give the +1 monosulfate ion, the formation of the third complex in a series of steps in which OH^- replaces H_2O in a pair of ions, and condensation to form the olate bridged binuclear structure with monodentate sulfate groups. One sulfate group then rotates into the plane of the other, displacing it from the coordination complex, to form the bidentate sulfate bridge.

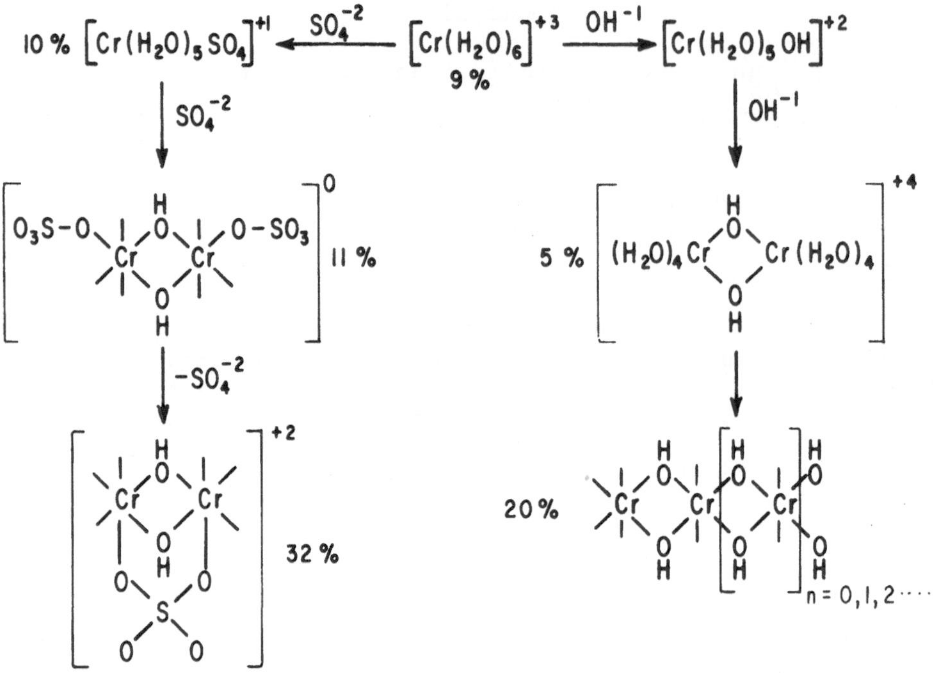

Figure 3. Composition of Typical Chromium Sulfate Tanning Solution (33% basic; 0.4 molar in Cr-III).

This is, of course, a purely descriptive and oversimplified mechanism, but it gives an idea of the type of reactions possible consistent with experimental evidence and the thermodynamics and stereochemistry of chrome complexes as recently reviewed (Irving, 1974).

In Figure 4, we show some of the same types of reactions that can take place as the carboxylate ion attached to the collagen enters these complexes. The carboxylate group can displace water from the $[Cr(H_2O)_6]^{+3}$ ion (9% abundance) to form monodentate bonds, as shown in the upper equation, or bidentate coordinate bonds with binuclear complexes like the one shown in the lower equation. Two mechanisms by which crosslinking can occur (Figure 5) are:

(1) Straightforward entry of two carboxylate ions into the same chrome complex.

(2) Olation involving elimination of water and formation of a linkage between two complexes.

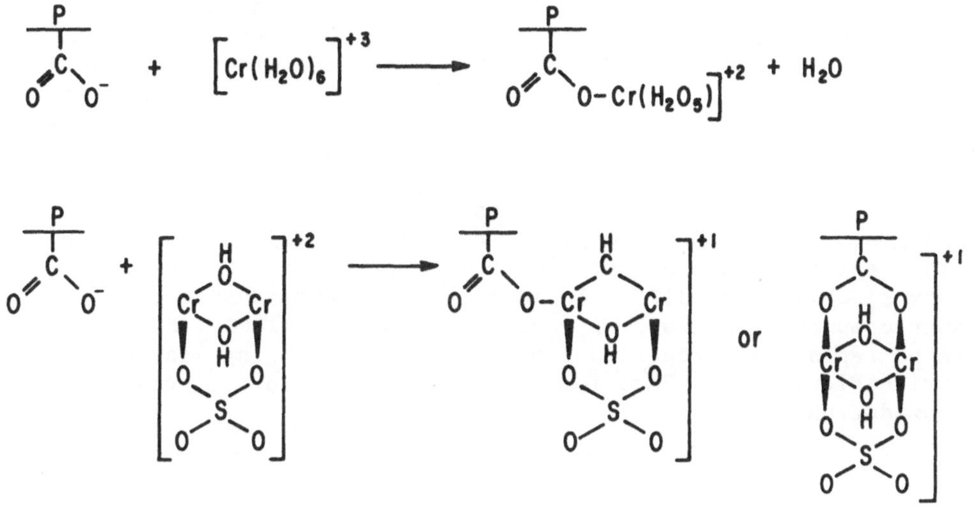

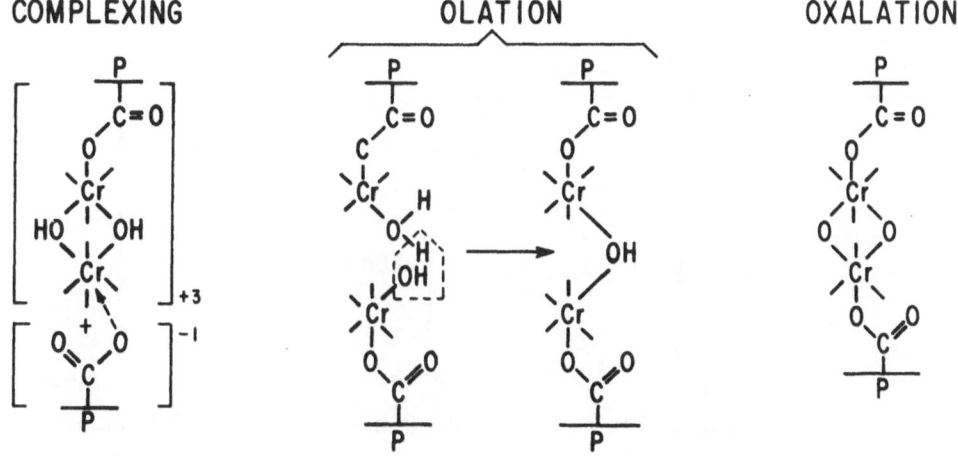

Figure 4.   Complex Formation with Protein Carboxyl Groups.

Figure 5.   Crosslink Formation from Chromium Complexes.

The olation reaction is favored by increasing the alkalinity of the reaction mixture. As the reactions proceed and multinuclear complexes form with multiple olate bridges, hydronium ions are released and highly stable oxalate bridges are formed, as shown on the right hand side of Figure 5. While not shown here for the sake of simplicity, there is evidence that bidentate sulfate groups remain in the final complex after curing and drying. Apparently they play a role in improving stability of the complexes.

Similar coordination complexes are involved in other mineral tannages. All of these complexes can be reversed or modified by acids, salts, strong bases, and chelating agents. Chromium compllexes, while more difficult to form, have the advantage over other complexing cations of reacting much more slowly in these ligand replacement reactions and therefore producing leather more serviceable in use. Chromium^{+3} is also unique in its resistance to oxidation.

Aldehyde Tanning

The most important collagen functional group involved in aldehyde crosslinking is the amino group. Some potential reactions which need to be considered are shown in Figure 6. Formation of a methylol or substituted methylol derivative is certainly the first

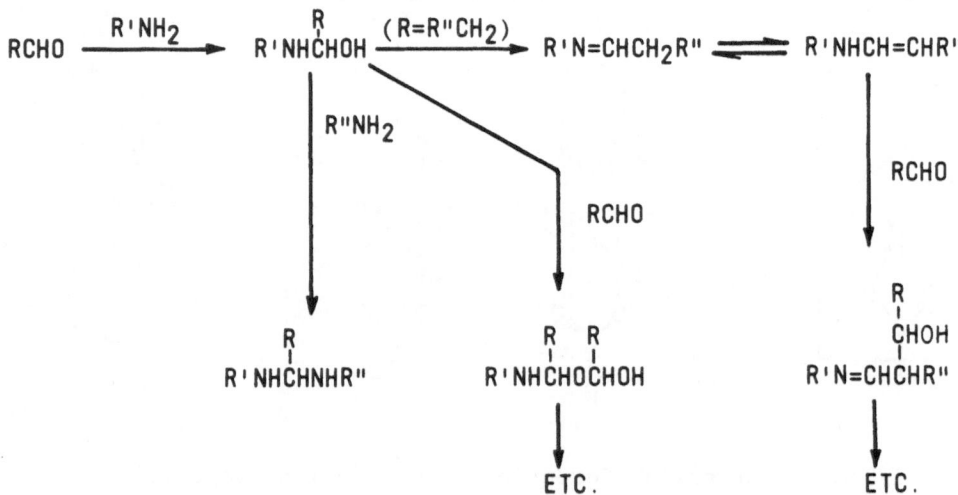

Figure 6.   Aldehyde-Amine Reactions.

step. Several alternatives are then possible. Apparently direct
substitution of an amino group for the hydroxyl is a possibility.
This leads directly to a crosslink in a protein, but a hydrolyzable
one. Addition of more aldehyde can take place and, except for
formaldehyde, Schiff's base formation can take place. In this
latter case, further steps lead to condensation reactions of the
type shown here, which have recently been proposed for glutaraldehyde crosslinking. While other aldehydes have been investigated
for their tanning ability, only formaldehyde and glutaraldehyde
are used industrially. Since these two aldehydes crosslink by
different mechanisms, they are discussed separately.

The results of early studies on the reactions of protein with
formaldehyde have been reviewed (Mellon, 1958; Walker, 1964). Most
of these studies were based on the use of model compounds, and none
resulted in the isolation of characterized products that were
directly related to potential crosslinks in proteins. The most
extensive of these studies were those of Fraenkel-Conrat (Fraenkel-
Conrat et al., 1945; Fraenkel-Conrat and Olcott, 1948), who demonstrated which functional groups of proteins could react, either
alone or in combination, with formaldehyde. These groups included
primary amino and amido groups, secondary amide groups of the peptide bond, guanidino groups, imidazoyl groups, indolyl groups, thiol
groups, and phenolic groups. Of these, all but the indolyl and
thiol groups are found in collagen and are thus potenially capable
of taking part in crosslinking. However, since the industrial process is carried out under relatively mild conditions of pH and
temperature and in relatively dilute solution, definite limitations
are placed on the types of reactions that can be involved. The
reactions of interest must also be reversible since simple washing
of formaldehyde-tanned collagen with water slowly extracts formaldehyde, while acid hydrolysis accompanied by steam distillation
results in a quantitative recovery. This is not true for all
protein, but it is true for collagen. Therefore, only those types
of products which are formed under mild conditions and which are
hydrolyzable should be considered.

A study of the reactions of formaldehyde with wool keratin
probably indicates most clearly the types of compounds formed with
collagen, even though this protein has a different amino acid composition (Caldwell and Milligan, 1972). The formaldehyde treatment
was carried out under extremely mild conditions using ^{14}C-formalde-
hyde, and the product was enzymatically hydrolyzed. It is interesting that histidine and tyrosine were recovered quantitatively and
thus were probably not involved in the reaction. Several ^{14}C-
labelled products were detected, but only two were sufficiently
separated to permit some judgment to be made concerning their iden-
tities. They were reported to be δ N-hydroxymethylglutamine (I) and
ε-N,ε-N'-methylenedilysine (II). The latter represents a crosslink.

$$\underset{I}{HOCH_2NHC\overset{\overset{\displaystyle O}{\|}}{C}H_2CH_2\overset{\overset{\displaystyle NH_2}{|}}{C}HCO_2H}$$

$$\underset{II}{CH_2(NHCH_2CH_2CH_2CH_2\overset{\overset{\displaystyle NH_2}{|}}{C}HCO_2H)_2}$$

The formation of this latter product involves simple addition of a lysinyl residue side chain amino group to formaldehyde, followed by condensation of the resulting N-methylol derivative with another lysinyl residue side chain. Acid hydrolysis of the product leads to recovery of the formaldehyde and lysine.

The crosslinks formed in collagen as a result of formaldehyde tannage are probably of this simple and hydrolytically unstable type, indicated in this study on wool (Caldwell and Milligan, 1972). Perhaps further effort should be given to an extension of this approach to elucidate the structures formed in formaldehyde tannage of collagen.

Glutaraldehyde, unlike formaldehyde, irreversibly crosslinks collagen and other proteins. The crosslinks that are formed have not been fully characterized but progress is being made toward achieving this. Before these crosslinks are discussed, however, some discussion of glutaraldehyde itself is felt to be needed since uncertainty still exists concerning the nature of this crosslinking agent. Proposals have been made in the past (Richards and Knowles, 1968) and continue to be made (Mansan, Puzo, and Mazarquil, 1975) that the reactive species is a condensation polymer of glutaraldehyde. That this is most certainly not the case has been demonstrated by a number of studies (Hardy, Nicholls, and Rydon, 1969; Korn, Feairheller, and Filachione, 1972; Whipple and Ruta, 1974; Blass, Verriest, Lean, and Weiss, 1976). A variety of derivatives of glutaraldehyde, which have been proposed as being present in aqueous solutions of the reagent, are shown in Figure 7, along with glutaraldehyde itself. Perhaps all of these are present, although various studies have proposed specific combinations. The important fact is that all are in equilibrium with glutaraldehyde and glutaraldehyde itself is probably the reactive species.

Of all the crosslink characterization studies made thus far, the most recent (Hardy, Nicholls, and Rydon, 1976) is perhaps the most significant. In this study 6-amino-hexanoic acid was utilized as a model compound for lysinyl residues, the residues in the protein which are involved in the crosslinking reaction. A structure

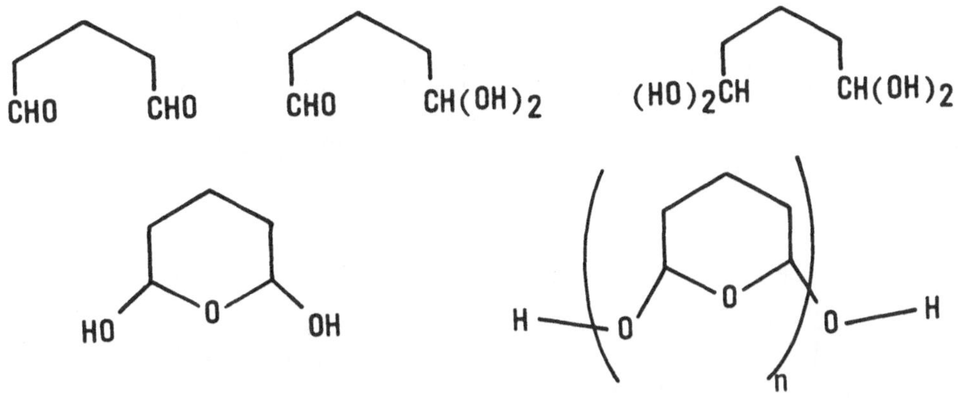

Figure 7. Glutaraldehyde Derivatives Present in Aqueous
Solution.

proposed for a product Hardy et al. isolated and characterized had
all of the spectroscopic and chromatographic characteristics of
protein crosslinks. By analogy with the proposed structure for
this product, it seems probable that the inital product is formed
in the protein from a lysinyl residue and three molecules of
glutaraldehyde which condense to form the heterocyclic ring compound
as shown in Figure 8. The first step is the amine to aldehyde
addition reaction which must be followed by aldehyde condensation
reactions. At some point in the sequence, an oxidation step must
be included to arrive at the pyridinium ring. Further reactions of
the same types can then yield crosslinked products of various
types, two of which are shown at the bottom of this figure.

Our studies (Korn, Feairheller, and Filachione, 1972) have
shown that three products are isolatable from the reaction of
glutaraldehyde with collagen, casein, polylysine, or α-N-carboxy-
benzoyl-lysine. The properties of two of these products are such
that they could correspond to products like those shown here.
However, it is not entirely clear why glutaraldehyde should be
unique among the dialdehydes in forming stable crosslinks of this
type in proteins. Succinaldehyde forms crosslinks, but they are
not stable to hydrolysis, while adipaldehyde forms crosslinks only
to a very limited extent (Cater, 1965). If these are the structures
of the crosslinks, it is not obvious why one less or one more
methylene group between the aldehyde groups is important to stability

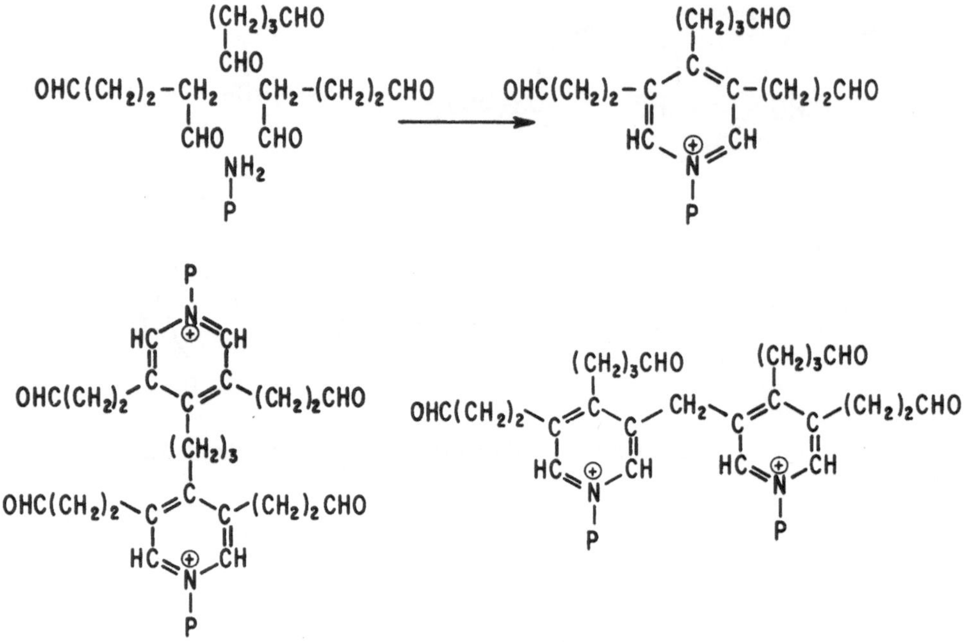

Figure 8.   Reaction Products of Glutaraldehyde with Proteins.

or even crosslink formation.   There appears to be continued interest
in this problem and these studies will hopefully continue.

SUMMARY

From the limited information available, it appears that
chemical crosslinking does occur as a result of tanning collagen
with commercial chromium and aldehyde tanning agents.   This,
however, does not appear to be the case with vegetable tanning
agents.   The leather resulting from vegetable tanning has adequate
stability for many uses without requiring chemical crosslinking.
The other properties contributed by vegetable tanning materials are
more important.   Chromium tanning agents form relatively stable
coordinate covalent bonds with collagen carboxyl groups, resulting
in a product that has a good hydrothermal stability but is somewhat
unstable to the action of acids and bases.   Formaldehyde forms
covalent bonds with collagen amino groups but they are unstable to
hydrolytic conditions and the tannage is not stable.   Glutaraldehyde, on the other hand, forms covalent crosslinks with collagen
amino groups, and these crosslinks have a good hydrothermal

stability as well as being stable to the action of dilute acids and bases. For this reason, glutaraldehyde tannages are often used in conjunction with chrome tannages to impart perspiration resistance and washability to leathers. In spite of all of the extensive literature on tanning and tanning agents, serious gaps remain concerning the nature and extent of crosslinking and stabilization caused by all of these tanning agents. Additional research is needed to develop a better understanding of the reactions involved and the structures of the crosslinks formed. This understanding would lead to better utilization of existing tanning agents and development of new tanning agents.

REFERENCES

Blass, J., Verriest, C., Leau, A., and Weiss, M. (1976). Monomeric glutaraldehyde as an effective crosslinking reagent for proteins. J. Am. Leather Chem. Assoc., 71, 121.

Caldwell, J. B., and Milligan, B. (1972). Sites of reaction of wool with formaldehyde. Text. Res. J., 42, 122.

Cater, C. W. (1963). The evaluation of aldehydes and other difunctional compounds as cross-linking agents for collagen. J. Soc. Leather Trades Chemists, 47, 259.

Cater, C. W. (1965). Investigations into the efficiency of dialdehydes and other compounds as cross-linking agents for collagen. J. Soc. Leather Trades Chemists, 49, 455.

Flory, P. J. (1953). "Principles of Polymer Chemistry." Cornell University Press, Ithaca, N.Y.

Fraenkel-Conrat, H., Cooper, M., and Olcott, H. S. (1945). The reaction of formaldehyde with proteins. J. Amer. Chem. Soc. 67, 950.

Fraenkel-Conrat, H., and Olcott, H. S. (1948). Reaction of formaldehyde with proteins. J. Biol. Chem., 174, 827.

Garrett, R. R., and Flory, P. J. (1956). Evidence for a reversible first-order phase transition in collagen-diluent mixtures. Nature, 177, 176.

Hardy, P. M., Nicholls, A. C., and Rydon, H. N. (1969). The nature of glutaraldehyde in aqueous solution. Chem. Commun., 565.

Hardy, P. M., Nicholls, A. C., and Rydon, H. N. (1976). The nature of the cross-linking of proteins by glutaraldehyde. Part I. Interaction of glutaraldehyde with the amino groups of 6-aminohexanoic acid and α-N-acetyl-lysine. J. Chem. Soc. Perkin I, 958.

Irving, H. M. N. H. (1974). Fact or fiction? How much do we really know about the chemistry of chromium today? J. Soc. Leather Technologists and Chemists, 58, 51.

Kawamura, A., Vada, K., Wehara, K., and Takata, E. (1975). Studies on chrome tanning liquor by gel chromatography and electrophoresis. Proceeding of the XIV Congress of the

International Union of Leather Chemists and Technologists
        Societies, Vol. I, 222.
Korn, A. H., Feairheller, S. H., and Filachione, E. M. (1972).
        Glutaraldehyde: nature of the reagent. J. Mol. Biol., 65,
        525.
Mellon, E. F. (1958). Aldehyde Tannage. "Chemistry and Techno-
        logy of Leather." Vol. II, Chapter 17 (O'Flaherty, F.,
        Roddy, W. T., and Lollar, R. M., eds.), Reinhold Publishing
        Corp., New York, N.Y.
Mansan, P., Puzo, G., and Mazarquil, H. (1975). Study of the
        mechanism of glutaraldehyde-protein bond formation. Biochimie,
        57, 1281.
Richards, F. M., and Knowles, J. R. (1968). Glutaraldehyde as a
        protein crosslinking reagent. J. Mol. Biol., 37, 231.
Slabbert, N. P. (1975). Chromium-III complexes which constitute
        chrome sulfate tanning solution. Proceedings of the XIV
        Congress of the International Union of Leather Chemists and
        Technologists Societies, Vol. I, 240.
Walker, J. F., (1964) "Formaldehyde." Reinhold Publishing Corp.,
        New York, N. Y.
Whipple, E. B., and Ruta, M. (1974). Structure of aqueous
        glutaraldehyde. J. Agr. Chem., 39, 1666.
Witnauer, L. P., and Fee, J. G., (1957). Effect of diluents on
        fusion temperature of the crystalline region in plain and
        tanned cowhide. J. Polymer Science, 26, 141.

28

CHEMICAL MODIFICATION OF COLLAGEN AND THE EFFECTS ON ENZYME-BINDING: MECHANISTIC CONSIDERATIONS

JACK R. GIACIN and SEYMOUR G. GILBERT

Food Science Department, Cook College, Rutgers

University, New Brunswick, New Jersey, 08903

ABSTRACT

The effect of structural modification on the enzyme-binding capacity of collagen has been studied using β-galactosidase ($\underline{E}$. $\underline{coli}$ K_{12}) immobilized to collagen membrane by the impregnation procedure. The apparent steady-state activities of the resultant collagen-enzyme complexes were determined as a means of evaluating the enzyme-binding capacity of the modified collagen. In addition, the amount of enzymic protein bound to the collagen support was determined by the tryptophan content of the complex.

The tertiary structure of the collagen matrix was modified by cross-linking with the difunctional reagent, glutaraldehyde, and by aging in the dry state. Such structural modifications were found to markedly reduce the enzyme (β-galactosidase) binding capacity of collagen films. The enzyme-binding capacity of the cross-linked collagen membrane was completely restored by proteolytic enzyme treatment of the aged film but only partly so for the glutaraldehyde treated films. Proteolytic enzymes used to treat a dispersion of collagen microfibrils prior to casting into a membrane also resulted in an increase in enzyme-binding. The effect of structural modification of collagen on enzyme-binding and the locus of enzyme attachment are discussed.

INTRODUCTION

It has been long recognized that a number of enzymes in a living cell are bound to the various membraneous structures of the cell. Green et al. (1965) have demonstrated a pH-dependent binding of several glycolytic enzymes to the erythrocyte membranes and concluded that the entirety of the glycolytic system is associated with the membrane and is not free in solution.

A similar conclusion was drawn by Arnold and Pette (1968) from studies carried out on the in vitro binding of aldolase, glyceraldehyde phosphate dehydrogenase, fructose-6-phosphate kinase, phosphoglycerate kinase, pyruvate kinase and lactate dehydrogenase to the structural proteins: F-actin, myosin, acto-myosin and stroma-protein. The binding of the glycolytic enzymes to F-actin is in agreement with in vivo histochemical findings which revealed a localization of the main activity of aldolase and a series of other glycolytic enzymes within the isotropic zones of cross-striated muscle (Brandau and Pette, 1966).

A number of enzymes appear therefore to be localized in a specific micro-environment, which can influence their biocatalytic activity. Because of the complexity of biological membranes, our understanding of the influence of micro-environmental effects on membrane-bound enzymes is minimal. An important contribution to better understanding the mode of action of a membrane-bound enzyme has been the development of the concept of heterogeneous catalysis by enzymes artifically immobilized to water-insoluble supports, where these immobilized enzymes were viewed as models for the cellular system (Goldman, 1973; Goldman and Katchalski, 1971)

In 1972, Vieth, Gilbert and Wang reported the use of the structural protein collagen as a carrier for the immobilization of enzymes. Subsequent studies by these and other workers have established the general utility of reconstituted collagen as a carrier for enzyme and whole microbial cell binding. (The following references and the references cited therein provide a useful review of the literature on enzyme and whole microbial cell attachment to collagen: Wang and Vieth, 1973; Vieth and Venkatasubramanian, 1973; Vieth, Wang and Saini, 1973; Barndt et al., 1975; Venkatasubramanian et al.1974; Lin, et al. 1976)

Apart from our studies, gels and sponges of collagen

were used for binding papain via bis-diazobenzidine-2, 2'-disulfonic acid (Silman et al., 1966). The covalent coupling of enzymes to chemically activated collagen film has also been described in the literature (Coulet et al.,1974; Coulet et al., 1975; Lee et al., 1976)

The mechanism of non-covalent attachment of enzymes and whole microbial cells to collagen has been discussed by Venkatasubramanian et al., (1974) in terms of the formation of a network of multiple physico-chemical interactions.

Some approximations can be made as to the number of collagen-enzyme interactions required to provide a cumulative bond energy equal to or greater than a covalent bond (approximately 100 Kcal/mole). Assuming a bond energy of 20-25 Kcal/mole for an ionic bond, a minimum of 4-ionic interactions per tropocollagen molecule are required to provide a cumulative bond energy of 100 Kcal/mole, while a total of 15 to 50 cooperative hydrogen bonds (assumed 2-7 Kcal/mole for hydrogen bonding; (Cartmell & Fowles, 1956) are required to provide an analogous cumulative bond energy.

If one considers that the tropocollagen molecule contains approximately 1000 potential binding sites (ionic and hydrogen bonding sites) per molecule (Table 1) only about 1.0 to 2% of the potential sites are required to interact with the enzymic protein molecule to provide cumulative bond energies equal to or greater than a covalent bond. Arnold and Pette (1968) reported that the binding of aldolase to F-actin involved two binding sites per actin molecule, with the binding sites having different dissociation constants.

While this discussion provides supportive evidence that the mechanism of enzyme protein-carrier protein binding involves the formation of a stable network of physico-chemical bonds, the nature of the functional group and/or groups within the collagen molecule involved in the formation of such non-covalent linkages with the enzyme and the locus of enzyme attachment on collagen have not been elucidated.

Native collagen is built up of a highly ordered array of linear and lateral aggragations of essentially rod-like, linear macromolecules called tropocollagen. The tropocollagen sub-unit is comprised of three hydrogen-bonded polypeptide chains intertwined in a helical configuration. Protruding from the ends of the triple-

TABLE 1
ESTIMATED NUMBER OF POTENTIAL BINDING SITES PER COLLAGEN MOLECULE

AMINO ACID RESIDUES CONSIDERED	NO. OF RESIDUES CON-DUES PER 1000 RESIDUES [a]	NO. MOLES PER MOLE COLLAGEN [b] [c]	NO. OF MOLES PER GRAM COLLAGEN $(\times 10^{-4})$	TOTAL MOLE EQUIVALENT ACTIVE BINDING SITES PER GM. COLLAGEN $(\times 10^{-3})$	NO. OF POTENTIAL BINDING SITES PER TROPOCOLLAGEN MOLECULE [d]
ACIDIC AMINO ACIDS					
1. Aspartic Acid	45.0	135.0	4.0	1.1	330
2. Glutamic Acid	72.0	214	6.5		
BASIC AMINO ACIDS					
1. Lysine	26	78.0	2.5		
2. Histidine	4.5	13.0	0.4		
3. Arginine	50.0	150.0	4.5	.7	210
HYDROXYLIC AMINO ACIDS					
1. Hydroxyproline	98.0	294.0	9.0		
2. Threonine	17.0	51.0	1.5		
3. Serine	33.6	101.0	3.0		
4. Tyrosine	3.5	10.5	0.30	1.4	400

a. Rubin et al., 1965.　b. Assumed 1000 residues/αstrand.　c. Assumed Mol. Wt. of collagen = 300,000 or 1 gm - 3×10^{-6} moles.　d. An example of the calculation is shown in Appendix I.

helical body or crystalline region are non-helical or amorphous regions termed telopeptides (Ramachandran, 1967). Piez (1967) has proposed that the telopeptide region (involving ten to fifteen amino acid residues/α chain) is continuous with the principle axis of the helical portion. The amorphous or telopeptide region of the tropocollagen molecule can be cleaved by proteolytic enzyme treatment of collagen without loss of helical structure (Hodge et al., 1966; and Rubin and Stangel, 1969).

We have proposed to take advantage of the well defined structure of the tropocollagen molecule (crystalline and amorphous regions) and the susceptibility of the telopeptide region of the molecule to protease attack, to determine the structural features of the collagen molecule which provides the loci of binding sites for enzyme attachment.

In this paper, we present data which indicates the loci of binding sites within the tropocollagen molecule are primarily within the crystalline or helical portion of the molecule. Related studies to establish the specific functional groups within the collagen molecule involved in the formation of non-covalent linkages with enzymic proteins are currently under investigation in our laboratory. Preliminary results of this study are described.

MATERIALS AND METHODS

Enzymes - β-galactosidase (E.coli K_{12}), swine stomach pepsin (2883 units/mg) and bovine pancreas trypsin (221 units/mg) were obtained from Worthington Corporation, Freehold, New Jersey.Pronase (Grade 13) a non-specific protease (Streptomyces griseus) was purchased from Calbiochem Co., San Diego, California.

Collagen -Cattle-hide collagen was obtained from Devro, Inc., Somerville, New Jersey. The fibrous collagen was washed with 10% sodium chloride solution, followed by washing with distilled water. The collagen was freeze-dried and stored at ambient temperature.

Substrate - 0.15 M Lactose in 0.02 M sodium phosphate buffer at pH 7.0 was used as a substrate for β-galactosidase.

   Preparation of collagen membranes - Collagen dispersion was prepared from hide collagen by dispersing 3.5g
of freeze-dried collagen in 500ml of aqueous lactic acid
solution at pH 2.8. The collagen dispersion was cast to
form a membrane, using a Gardner knife (Lieberman et al.,
1972). The membrane was left to dry at room temperature
for 48 hours. The thickness of the dried membrane was
between 0.028 and 0.038 mm.

   Preparation of cross-linked collagen membranes -
Collagen membrane was prepared by the above method.
Cross-linking of the collagen membrane was carried out
by two procedures; (a) by a chemical cross-linking agent
and (b) by a natural process associated with drying.

   The collagen membrane was cross-linked chemically
by immersing the film in an alkaline glutaraldehyde solution (pH 7.6, 0.1M phosphate buffer). The tanning
time and glutaraldehyde concentrations employed are tabulated in Table 2. The tanned membrane was washed thoroughly with distilled water (10 liters) and the resulting
membranes were swollen in 0.1M tris-HCl buffer, pH 8.6
for sixty minutes. The swelled membrane was washed in
running tap water and immersed into the enzyme impregnation bath.

   Cross-linking by procedure (b) is accomplished by
cutting a large piece of freshly cast collagen membrane
into equal sized sections and drying the samples for different periods of time at ambient temperature. After
drying the membrane for the desired period of time, the
membrane was swollen in 0.1M tris-HCl buffer, pH 8.6 for
sixty minutes. The swollen membrane was then washed in
running tap water and immersed into the enzyme impregnation bath.

   Proteolytic enzyme treatment of cross-linked collagen membrane. The naturally cross-linked and glutaraldehyde tanned membranes were further structurally modified by treatment with the proteolytic enzymes, pepsin,
pronase and trypsin.

   The proteases, pepsin and pronase were dissolved in
0.05% acetic acid solution, pH 3.5 and 0.1M calcium acetate buffer solution, pH 7.2, respectively. For the
trypsin studies, a 0.1M tris-HCl buffer solution, pH 8.1
was employed. The proteolytic enzymes were dissolved in
the respective solutions at an enzyme concentration of
0.4% (w/v). The collagen films were immersed in the di-

gestion bath in the ratio of 1:100, enzyme:collagen
(wt/wt).   The films were treated for three hours,
with mechanical shaking. The protease treated films
were removed and washed with 15% NaCl solution in 0.02M
disodium phosphate buffer, followed by distilled water
washing. The resulting protease treated films were
then immersed into the enzyme impregnation bath.

Proteolytic enzyme treatment of collagen microfibrils. Two different proteolytic enzymes, pronase
and pepsin, were used to modify the structure of collagen microfibrils prior to the preparation of collagen
membrane:

Impurities present in the pronase preparation were
removed by ultra-filtration through an Amicon PM 10
membrane (cut-off range >10,000) followed by several
washings with 0.1M calcium acetate buffer, pH 7.2.
The filtered pronase solution was sterilized by filtration through a millipore HA 0.45µ filter. Pepsin was
used directly without further purification. The buffer
solutions used for pronase and pepsin digestion of the
collagen microfibril dispersion were similar to the
method described above.

Two gm of shredded cowhide collagen was dispersed
in 200 ml of protease solution containing 40 mg of
enzyme to give an enzyme to collagen ratio of 1:50
(wt/wt). The digestion was carried out in a mechanical
shaker at 20°C for varying periods of time. Following
the desired digestion time, the protease-treated collagen microfibrils were recovered from solution by washing twice with 15% NaCl in 0.02M disodium phosphate
buffer of pH 9.0, followed by washing with distilled
water and centrifugation. The resultant collagen fibrils were dispersed by the method previously described
and cast to form a membrane. The resulting membranes
were then immersed into the enzyme impregnation bath.

The supernatants and washings from the recovery of
the collagen fibrils were combined and passed through
an Amicon PM10 ultrafiltration membrane to remove proteolytic enzymes and dissolved collagen. The ultra-
filtrate was assayed by an ultraviolet spectrophotometric procedure for determination of released aromatic
amino acid residues, particularly tyrosine. The extent
of telopeptide cleavage was determined from the tyrosine
content of the supernatant.

Preparation of collagen-enzyme complexes. Collagen-enzyme complexes were prepared by the membrane impregnation method (Vieth et al., 1972, Giacin et al.,
1974). In this procedure, collagen membrane was first
formed and then impregnated with the enzyme. A solution
containing 20 units of enzymic activity (O-nitrophenyl,
β-D-galactopyranoside, 20°C) per mg of collagen film in
0.02M phosphate buffer at pH 7.0 was employed as the impregnation bath. All enzyme impregnation reactions were
carried out at 4°C for a twenty-four hour period. At the
end of the impregnation period, the membrane was removed
from the bath and dried for eighteen hours. The resultant complex was washed with distilled water followed by
washing with 0.02M phosphate buffer of pH7.0. The activity of the collagen-enzyme complex was determined immediately after washing the membrane. The preparation of
collagen-enzyme complexes from structurally modified collagen was carried out in a similar manner.

Enzyme Assay Procedure. The catalytic potency of
the immobilized β-galactosidase was determined in a plug
flow reactor (Vieth et al., 1972). Glucose liberated
by the catalytic activity of β-galactosidase on lactose
was determined by the glucose oxidase-chromogen method
(Worthington Biochem. Co., 1972) with some modifications.

Chemical analysis for bound enzyme was based on
the tryptophan content of the complex. A modification
of the method of Gaitonde and Dovey (1970) was employed
(Eskamani, et al., 1974).

Measurement of molecular weight between elastically active linkages (i.e., network property of aged collagen membrane. The molecular weight between the network
linkages was determined from the stress-strain behavior
of the membrane according to the procedure developed by
Wiederhorn et al., (1951) and modified by Lieberman (1971).
The stress-strain measurements were performed at the
Ethicon Corporation, Somerville, New Jersey, courtesy of
Mr. J. Olivo.

RESULTS AND DISCUSSION

Enzyme Binding Capacity Determination

The enzyme binding capacity of collagen membranes
was evaluated by means of two procedures, namely by (a)
determination of the catalytic activity of the collagen
enzyme complex, and (b) direct determination of the en-

zymic protein immobilized on collagen by the tryptophan
content of the complex. Good agreement between the two
procedures was obtained. That the effect of structural
modification (e. g. chemical cross-linking) on the enzyme
binding capacity of collagen membrane could be evaluated
by determining the activity of the resultant collagen
enzyme complexes is implied by the equivalency of the
catalytic activity and direct determination methods.

Unless otherwise stated, the enzyme binding capacity of membraneous collagen was measured by catalytic
potency of the resultant collagen-enzyme complexes.

Cross-Linking of Collagen Membrane - The Effect
of Structural Modification on the Enzyme-Binding
Capacity of Membraneous Collagen

In using collagen as a matrix for enzyme immobilization, good mechanical strength and dimensional stability of the collagen membrane is necessary to withstand
the high shear flow rates encountered in continuous operation (Vieth and Venkatsubramanian, 1973). To impart
the required mechanical strength to collagen membranes
prepared from lactic acid and/or cyanoacetic acid disper-
sions, it is necessary to modify the collagen structure
by glutaraldehyde tanning (Wang and Vieth, 1973) or by
annealing (Vieth et al., 1972, a,b; Eskamani, 1972).
Wang et al.,(1974) have recently reported that collagen
membranes prepared from a hydrochloric acid dispersion
possessed good mechanical strength and no exogenous crosslinking was required.

To effectuate the desired mechanical properties
to collagen, the collagen matrix structure has been modified by chemical means with the difunctional reagent,
glutaraldehyde. The glutaraldehyde cross-linking step
promotes the formation of interhelical cross-linkages
between tropocollagen molecules through the formation
of a Schiff's base (Veis, 1968) or a Michael-type addition product (Richards and Knowles, 1968). However,
while cross-linking imparts the desired mechanical properties to collagen membrane, it can effectively reduce
the number of potential sites for enzyme-binding and
modify the physical properties of the collagen membrane
(Lieberman et al., 1972;Lieberman and Gilbert, 1973).

Table 2 presents the stable limits of activity of
several collagen-lactase complexes prepared by impreg-
nating collagen membranes cross-linked to varying degrees

TABLE 2

FACTORIAL STUDY OF THE EFFECTS OF GLUTARALDEHYDE CON-
CENTRATION AND TREATMENT TIME ON THE ENZYME-BINDING
CAPACITY OF COLLAGEN MEMBRANE

RUN NO.	GLUTARALDEHYDE CONCENTRATION %	TREATMENT TIME (MIN)	APPARENT SPECIFIC ENZYME ACTIVITY[a,b] (UNITS/g COMPLEX)		
			Expt. I	II	AVERAGE
1.	0	0	300.39	301.36	300.87
2.	0.05	2	100.22	104.52	102.37
3.	0.05	5	96.40	91.00	93.70
4.	0.125	3.5	53.05	58.70	55.87
5.	0.20	2	48.61	50.61	49.61
6.	0.20	5	33.90	31.51	32.70

a. Enzyme impregnated collagen membranes were not dried before washing with 0.02 M phosphate buffer solution.

b. A unit is defined as one u mole glucose produced per min.

by varying the glutaraldehyde concentration and contact time. Data of Table 2 clearly demonstrate the enzyme-binding capacity of membraneous collagen is markedly reduced by glutaraldehyde tanning prior to enzyme immobilization. Complex #6, which was prepared from a membrane contacted with 0.2% glutaraldehyde solution for five minutes had an enzymic activity 90% less than complex #1, prepared from a non-cross-linked control.

Statistical analysis of the data (analysis of variance of the apparent specific enzymic activities) revealed that both glutaraldehyde concentration and contact time are highly significant ($P<0.05$) in their influence on enzymic binding. These variables were found to act in concert (interaction effect) to affect enzyme-binding as shown by a computer-generated response surface. Thus, the cross-link network formed by the tanning process appears to modify the matrix structure of the collagen membrane and restrict diffusion of the enzyme into the preformed carrier matrix during impregnation (Venkatasubramanian et al., 1974).

This argument is substantiated by the results shown in Table 3, where the complexation process was carried out by immobilizing β-galactosidase on collagen in the presence of sucrose and sorbitol (Eskamani, 1972). Such compounds have a high propensity to form hydrogen bonds and were expected to compete with the enzyme for active hydrogen bonding sites within the collagen matrix. The results of these studies showed a 12% and 81% reduction in the amount of enzyme bound when the immobilization is carried out in the presence of 20% sucrose and 20% sorbitol, respectively. The greater impedance to enzyme-binding in the presence of the polyol, sorbitol, is thought to result from the "zippering" action of sorbitol, which effectively cross-links the collagen microstructure through the formation of a network of hydrogen bonds (Lieberman et al., 1972), and restricts enzyme diffusion into the membrane bulk phase during impregnation.

A further test to establish that the complexation mechanism involves interactions within the membrane bulk and not just surface interactions is described below, where cross-linking was effected by a natural cross-linking process associated with aging in the dry state. Collagen in the native state exists as an aggregate of triple-helical molecules. With advancing age, solubility, extensibility, and swelling of the protein decrease and the tensile strength and shrinkage temperature increase. These changes in the physical properties of collagen are

TABLE 3

THE EFFECT OF HYDROGEN BONDING SUB-
STRATES ON THE ENZYME BINDING CAPACITY
OF COLLAGEN MEMBRANE

RUN NO.	FILM PRESWELLED IN A SOLUTION OF:	APPARENT SPECIFIC ACTIVITY (Units/g Complex)
1	Control (buffer)	866
2	20% sucrose	760
3	20% Subitol	163

consistent with the gradual accumulation of inter and intra-molecular cross-linkages. LaBella (1971) has summarized the potential cross-links in aged collagen as involving lysine, tyrosine and glutamic acid residues and carbohydrates. Tanzer (1973) in a recent review article reiterated the importance of lysyl ε-amino groups in both inter and intra-molecular cross-linking in aged collagen. The increase in the number of hydrogen bonds, as a function of collagen fiber age, has also been described (Verzar, 1964).

When collagen membrane was subjected to varying storage times, prior to the enzyme impregnation step, the stable limit activities of the resultant collagen-enzyme complexes were found to decrease as a function of the storage time (film age). Table 4 shows the apparent specific activities of the resultant collagen-lactase complexes obtained from three independent experiments. The level of bound enzyme for representative complexes, based on the tryptophan content of the complex, is also presented in this table. As shown, the amount of bound enzyme decreased as a function of storage time. These results are in agreement with the corresponding specific activity determinations.

Our results with lactase immobilized to aged collagen membrane showed a complex prepared from a membrane stored for twenty-eight days, prior to impregnation, had a catalytic potency approximately 35% of the activity of the non-aged control. After storage for a hundred and fifty-three days, the activity of the resultant complex was only 1.6% that of the non-aged control.

These findings imply that membrane aging results in the formation of a cross-link network which acts as a barrier to diffusion of the enzyme into the membrane bulk phase, in a manner analogous to the action of sorbitol.

THE NATURE AND LOCI OF ENZYME BINDING SITES

A. The Effect of Telopeptide Cleavage on the Enzyme
    Binding Capacity of Membraneous Collagen

For the membrane impregnation method, it has been shown that the accessibility of the enzyme to the active binding sites in collagen is restricted by inter- and

TABLE 4

THE EFFECT OF STORAGE IN THE DRY STATE ON
THE ENZYME-BINDING CAPACITY OF COLLAGEN
MEMBRANE

STORAGE TIME (DAYS)	APPARENT SPECIFIC ENZYME ACTIVITY (UNITS/g COMPLEX)			AMOUNT OF ENZYME[d,e] BOUND (mg LACTASE PER g COMPLEX)
	Expt. I[a]	II[b]	III[c]	III
0	431.26	589.90	568.16	139
7	325.38	-	-	-
14	236.08	-	-	-
21	247.33	406.12	474.34	87
28	148.62	-	-	-
42	-	-	422.94	66
49	-	329.50	-	-
58	-	210.82	-	-
113	5.55	-	-	-
153	7.05	-	-	-

a. Prepared from freeze-dried collagen. Film thickness: 0.038 mm. The values are the average of two reactors.

b Prepared from wet collagen. Film thickness: 0.028 mm

c Prepared from freeze-dried collagen. Film thickness: 0.028 mm.

d These values represent the loading factor at steady state activity

e Determination of protein immobilized by tryptophan content

intramolecular cross-linkages which act as a barrier to
enzyme diffusion. Furthermore, it was shown that a high
proportion of the cross-links found in aged collagen occur within the telopeptide or non-helical region of the
tropocollagen molecule (Drake et al., 1966; Veis et al.
1967; La Bella, 1971). One might, therefore, expect that
the removal or disruption of cross-links within the non-
helical region of collagen would facilitate the diffusion process and a greater amount of enzyme would be
bound. This was found to be the case. Figure 1 shows
the activities of collagen-lactase complexes prepared
from aged collagen membranes which were pepsin treated
to remove telopeptide appendages, prior to the enzyme
impregnation step. Such attack on the amorphous region
of the tropocollagen molecule by proteolytic enzymes is
well documented in the literature Rubin et al., 1965;
Drake et al., 1966; Steven, 1966. The lower curve in
Figure 1 is a plot of the activities of the corresponding non-protease treated control samples. As shown by
the upper curve in Figure 1, a lactase-collagen complex
prepared from a membrane which was stored for twenty-
one days prior to pepsin treatment, had an enzymatic activity 1.7 fold greater than the complex prepared from
the corresponding non-pepsin treated membrane. A comparison of the activities of the collagen-lactase complexes prepared from a membrane which had been stored
for 153 days, showed pepsin treatment of the membrane
(prior to enzyme impregnation) resulted in a 67 fold increase in catalytic potency over the complex prepared
from the corresponding non-pepsin treated sample.

Examination of the network properties of aged collagen membranes provides additional evidence that the
inter- and intramolecular cross-links formed during membrane storage occur primarily in the amorphous or telopeptide region of the tropocollagen molecule. In Figure
2, the variation in the average molecular weight between
network linkages is shown as a function of membrane storage time for a series of collagen membranes and the corresponding samples following protease treatment. The reported values for the molecular weight between network
linkages are the average of twelve determinations, involving six samples per data point.

The initial drop in average molecular weight between
network linkages of both the curves is due to an increase
in the number of measurable cross-links per tropocollagen
molecule. As shown by the lower curve of Figure 2, no
measurable increase in the number of cross-linkages per

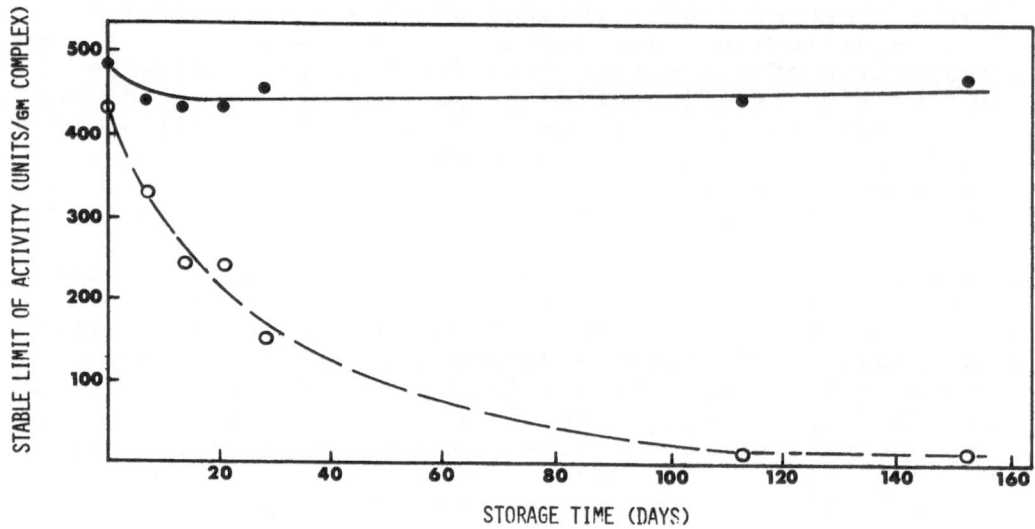

Fig. 1.   Effect of pepsin treatment on the enzyme binding
capacity of aged collagen membrane; -o- lactase immobilized on aged collagen membrane; -●- lactase immobilized
on pepsin treated collagen membrane.

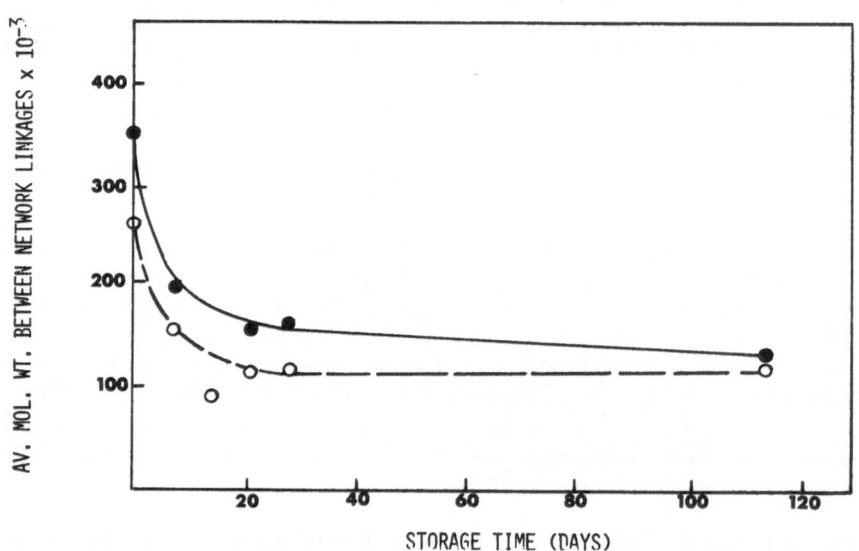

Fig. 2.   Variation of average molecular weight between
elastically active cross-links with membrane storage for
pepsin and non-pepsin treated collagen membranes; -o-
aged collagen membrane; -●- pepsin treated membrane.

tropocollagen molecule was observed after 21 days of
storage. Analysis of variance revealed that the difference in average molecular weight between network linkages, following pepsin treatment, was statistically significant (P<0.01) for freshly cast membrane. For the
long-term storage samples (113 days), the difference in
molecular weight between elastically-active network linkages following pepsin treatment was not statistically
significant. This is attributed to the formation of a
limited number of cross-linkages between the helical regions of the tropocollagen molecule.

As previously discussed, the cross-linkages formed
during collagen aging occur primarily within the telopeptide region of the tropocollagen molecule. Furthermore, the stress-strain properties of collagen membranes
measure primarily cross-linkages between rigid chains
(e.g. inter-helical cross-linking) and not cross-linking
between long flexible chains (e. g. telopeptide appendages) (Bartenov and Zuyev, 1968; Chien and Chang 1973).
Thus cross-linking between the helical regions contributes a greater influence on the network properties of
collagen than telopeptide associated cross-linkages.
It is the inter-helical cross-links which are primarily
measured by the Wiederhorn and Reardon procedure(1951).
While telopeptide associated inter- and intramolecular
cross-linking between tropocollagen molecules and fibrils
will not be readily reflected by the number of elastically active network linkages measured, such a process
can result in the loss of potential binding sites for
the enzyme and/or to the modification of the matrix structure of the collagen membrane. The latter may affect
the swelling characteristics of the collagen membrane
bulk phase during impregnation. The lower catalytic activity observed for the aged membranes and the resultant
increase in enzyme binding capacity following proteolytic enzyme treatment are consistent with this hypothesis.

In Table 5, enzyme binding data for aged collagen
membranes treated with pepsin, pronase and trypsin are
presented. From Table 5, it is seen that the enzyme
binding capacity of aged collagen membrane is increased
significantly by pepsin or pronase digestion of the membrane prior to enzyme immobilization. This is consistent with the above discussion. The inability of the
protease, trypsin, to increase the enzyme binding capacity of the aged collagen may be attributed to the specificity of trypsin and to it's pH optimum (Worthington
Biochemical Corp., 1972). Furthermore, the effective-

TABLE 5

THE EFFECT OF PROTEASE DIGESTION ON THE ENZYME-BINDING CAPACITY OF AGED COLLAGEN MEMBRANES

PROTEASE SOURCE	MEMBRANE STOR-AGE TIME (days)	APPARENT SPECIFIC ENZYME ACTIVITY (units/g COMPLEX)	AMOUNT OF EN-ZYME BOUND (mg enzyme/g complex) [a]	RATIO [b] (units/mg enzyme)	EFFECTIVE-NESS FACTOR [c]
Untreat-ed con-trol	0	568.16	139	4.1	0.63
	21	474.34	87	5.4	0.83
	42	422.94	66	6.4	0.98
Pepsin	0	672.85	193	3.5	0.54
	21	714.62	204	3.5	0.54
	42	593.34	174	3.4	0.54
Pronase	0	634.29	181	3.5	0.54
	21	638.78	184	3.5	0.54
	42	582.71	172	3.4	0.52
Trypsin	0	541.84	123	4.4	0.68
	21	481.22	94	5.0	0.77
	42	430.32	76	5.6	0.86

[a] Determination of protein immobilized by tryptophan content

[b] Apparent specific enzyme activity/amount of enzyme bound

[c] Ratio/free enzyme activity (6.5 units per mg enzyme)

ness factor for the enzyme bound to protease treated membrane (pronase or pepsin) is much lower than for the corresponding collagen-lactase complex prepared for non-protease treated membrane (storage time forty-two days).
This implies that for the protease treated samples, more
enzyme has been bound within the membrane bulk phase than
was expressed by the apparent enzyme activity. For the
aged membrane (non-pepsin treated), such mass transfer
limitations are reduced.

Hsieh et al., (1976) observed a similar effect of
protease treatment in the preparation of collagen-im-
mobilized urease. Pepsin treatment of the collagen membrane, prior to enzyme immobilization, resulted in a 220%
increase in enzymic activity over the complex prepared
from the corresponding non-pepsin treated membrane.

In contrast, the enzyme loading capacity of glutaraldehyde tanned membranes was not increased significantly
by protease treatment. Collagen-lactase complexes prepared from glutaraldehyde tanned membranes, which were
pepsin treated prior to enzyme impregnation, showed a
13 to 32% increase in enzymic activity, as compared to
the complexes prepared from tanned membrane with no pepsin treatment (Lin et al., 1976)

Although the enzyme-binding capacity of the tanned
membrane was increased, the activities of the resultant
complexes were only 30 to 45% of the activity of the
non-tanned control. Glutaraldehyde crosslinking involves
the side chains (e.g. lysine, arginine, etc.) of the collagen helical region, to a much greater extent than
cross-linking effectuated by storage of the film in the
dry state (Lieberman, 1971; Barndt et al., 1975). Further, these cross-linkages are not readily cleaved by
proteases. The interhelical cross-linkages reduce the
swelling characteristics of the collagen membrane and
therefore are considered to decrease the accessibility
of the enzyme to the active binding sites.

B.   Loci of Enzyme-Binding Sites

As outlined in Figure 1, the removal of telopeptide
appendages from aged collagen membranes by protease digestion, resulted in a marked increase in enzyme-binding
capacity of the modified carrier matrix. These findings
implied that the complexation process involves interactions within the membrane bulk. They further implied
that the binding mechanism involves localization or

regio-specific binding of the enzyme within the crystalline or helical region of the collagen microstructure
(Gilbert, 1973).

A further test to establish that the complexation
mechanism involves localization of the enzyme within
the helical domain of the collagen microstructure is
described below. When a dispersion of collagen-micro-
fibrils was pepsin treated to remove telopeptide appendages, prior to casting the membrane, and the enzymic
activities of the resultant collagen-lactase complexes
determined, the activity increased in proportion to the
level of telopeptide cleavage. Figure 3 shows the
variation in stable limits of activity of several collagen-lactase complexes prepared from pepsin treated dispersion, as a function of the percent telopeptide tyrosine released. The level of enzymic protein bound has
been superimposed in Figure 3, to illustrate the agreement between evaluating the enzyme binding capacity of
membraneous collagen by direct determination of bound
enzyme and by determination of the catalytic activity
of the collagen-enzyme complex.The data of Figure 3 demonstrate that enzyme binding increases with pepsin digestion of a collagen dispersion and is directly proportional to the extent of telopeptide released. Similar
results were obtained by pronase treatment of dispersed
collagen microfibrils (Lin, 1975).

Examination of the enzyme-binding data for collagen-
lactase complexes prepared from protease treated membranes and/or dispersion establishes the equivalency of
the binding mechanism involved in both processes. These
complexes have essentially the same catalytic potency
(see Figures 1 and 3).

Our results from protease treatment of collagen and
the affect on enzyme binding are inconsistent with the
possible mechanism of enzyme-collagen complexation previously suggested (Bernath and Vieth, 1974). The primary sites of enzyme binding were proposed to be localized within the telopeptide region of the tropocollagen
molecule. Our studies, however, show that the enzyme
binding capacity of membraneous collagen increases following protease treatment (as shown by Figures 1 and 3)
and implies that the collagen-enzyme complexation mechanism involves primarily side chains localized within
the helical region or domain of the collagen microstructure and that amino acid residues within the telopeptide
region do not play a significant role in enzyme-binding.

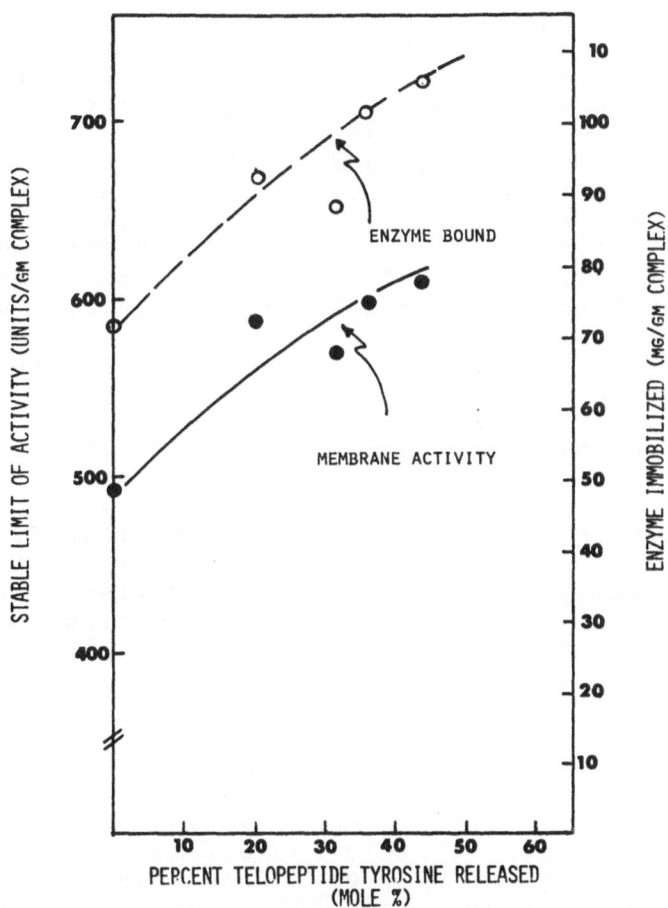

Fig. 3. Effect of pepsin treatment of dispersed collagen microfibrils on the enzyme binding capacity of membranous collagen.

C.   Nature of Active Binding Sites of Collagen

In elucidating the mechanism of collagen-enzyme complexation, it is necessary to establish both the loci of
binding sites and the nature or type(s) of interactions
involved. Accordingly, the contribution of ionic interactions involving the ε-amino group of lysyl side chains
of collagen has been determined by a chemical modification
procedure (Luo et al., 1976). In Figure 4, the variation
in relative apparent specific activities are shown as a
function of percent ε-amino groups modified for three
samples. These are: a collagen membrane modified by a
carbamylation reaction (Stark, 1965), a membrane modified by a diazotization reaction, (Bowes and Kenton,
1949), and a sample modified by a succinylation reaction
(Klapper and Klotz, 1972). As shown, chemical modification of the lysyl residues in collagen has a marked effect on the binding of lactase to the modified membrane.
A collagen-lactase complex prepared from a film in which
15% (mole percent) of the available ε-amino groups had
been converted to the carbamate derivative had an enzyme
activity 33% less than the complex prepared from the unmodified control film. A 60% decrease in relative apparent activity was observed for a complex prepared from
a membrane in which 30 mole percent of lysyl ε-amino
groups had been converted to the hydroxyl derivative
via the diazotization reaction. The decrease in enzyme-
binding capacity of the chemically modified membrane implies that the lysyl ε-amino groups function as receptor
sites for enzyme binding and that the complexation mechanism involves ionic interactions of lysyl ε-amino groups
of collagen with enzyme amino acid side chains (e. g.
carboxyl groups) as a principle step in the formation of
a stable network of physico-chemical bonds. As discussed previously, a minimum of four ionic interactions
per tropocollagen molecule would provide a cumulative
bond energy of approximately 100Kcal/mole. The decrease
in enzyme binding activity of the modified collagen membrane can be attributed to either a reduction in the
average binding constant for the ε-amino groups, or to
the modification of a small number of sites having a
high binding constant (Grossberg and Pressman, 1963).
Since the level of bound enzyme decreases linearly as
a function of the percentage of ε-amino groups modified,
it appears that the average binding constant is reduced.

These findings provide additional supportive evidence
that collagen-enzyme complexation involves regio-specific
binding of the enzyme within the crystalline domain of
the collagen microstructure, since the lysyl residues of

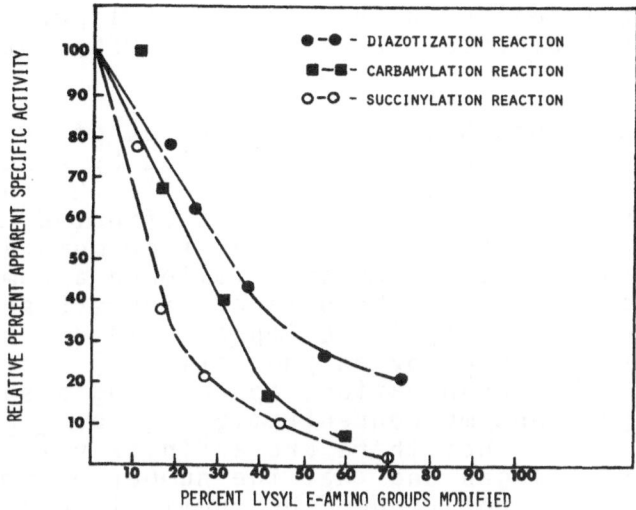

Fig. 4. Effect of chemical modification of lysyl ε-amino groups on the binding of β-galactosidase to collagen membrane.

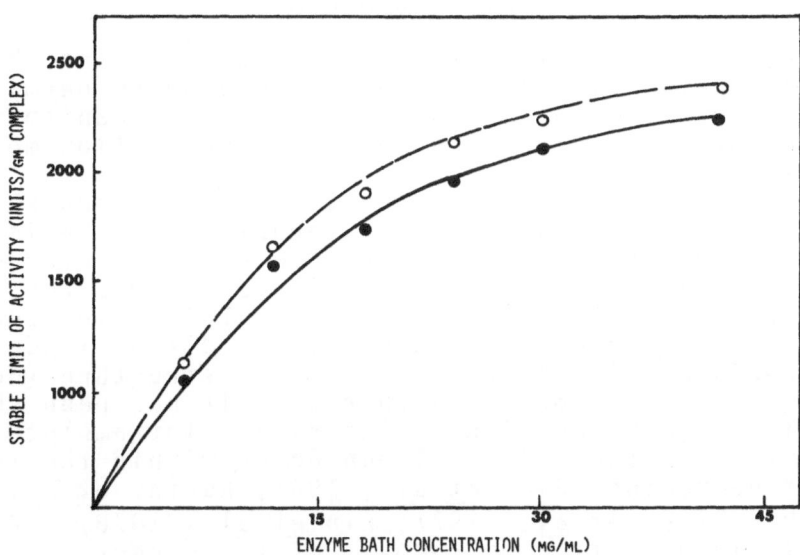

Fig. 5. Effect of enzyme bath concentration on the enzyme loading capacity of collagen membrane; -o- lactase immobilized on modified collagen membrane; -●- lactase immobilized on unmodified membranous collagen.

collagen are localized within the helical portion of the
tropocollagen molecule (Drake et al., 1966; Rubin et al.,
1965).

When the side chain carboxyl groups of collagen (aspartic and glutamic acid) were modified by reaction with
carbodiimide (Hoare and Koshland, 1966), the enzyme binding capacity of the membrane was not affected. Sorption
data for β-galactosidase on the modified membrane and the
corresponding unmodified control followed a common Lang-
muirian-type isotherm. The data of Figure 5 shows that
the amount of β-galactosidase impregnated on the collagen
membrane (as evaluated by enzyme activity) increased with
increasing bath concentration, approaching a saturation
value at higher enzyme concentrations (Luo et al., 1976).
This demonstrates that there are a finite number of binding sites on collagen and that the number of potential
binding sites was not reduced by chemical modification
of the carboxyl residues. These studies were carried
out with a collagen membrane in which approximately 12
mole percent of the available carboxyl residues had been
modified.

ENZYME BINDING MECHANISM

Our findings on the effect of structural and chemical modification of collagen on enzyme binding permit
certain general conclusions regarding the mechanism of
collagen-enzyme complexation, by the impregnation method.

1. An analysis of the microstructure of collagen shows
    that the amorphous region, external to the helical
    portion of the tropocollagen molecule, is located
    within the hole zone of the collagen microstructure
    (Piez, 1967; Stark et al., 1971). This hole zone
    may constitute either continuous or discontinuous
    channels within the collagen fibril structure (Hodge
    and Petraska, 1963). Furthermore, it has been shown
    that a high proportion of inter- and intramolecular
    cross-links found in collagen occur within the telo-
    peptide region (Piez et al., 1961; Rubin, et al.,
    1965; Stark et al., 1971; Lin et al., 1976). Pro-
    teolytic digestion or removal of these peptides
    disrupt telopeptide associated cross-linkages with-
    in and between collagen molecules. This leads to a
    more open structure, which facilitates the swelling
    of collagen, and results in an increase in enzyme
    diffusion into the membrane bulk phase during enzyme
    impregnation. In other words, the formation of

inter- and intramolecular cross-links within the telopeptide region of the collagen molecule act as a barrier or "gate" to diffusion of the enzyme into the internal surfaces of the collagen membrane during enzyme impregnation. Proteolytic enzyme cleavage of the telopeptide appendages effectively removes this restriction or barrier to diffusion and accounts for the increase in enzymic activity of collagen-lactase complexes prepared from protease treated collagen.

2.   The binding mechanism involves localization or regio-specific binding of the enzyme within the crystal-line region of the collagen microstructure. The lysyl ε-amino groups, localized within the helical portion of the tropocollagen molecule, function as principle receptor sites for enzyme binding. The complexation mechanism involves ionic interactions of the lysyl ε-amino groups with enzyme amino acid side chains (e. g. carboxyl groups) as a principle step in the formation of a stable collagen-enzyme supra-network.

Elucidation of the binding mechanism involved in collagen-enzyme complex formation has both practical and theoretical importance. Collagen bound enzymes can pro-vide a reasonable model for studying interactions be-tween enzymes and cellular membrane and for determining the effect of such a microenvironment on the activity of the enzyme. Enzymes covalently bound to chemically activated collagen films had previously been proposed as simplified models for mitochondrial membrane bound en-zymes (Coulet et al., 1974; Julliard et al., 1971). Re-actions involving covalent bond formation are not charac-teristic of a biological system. On the other hand, col-lagen bound enzymes obtained by complexation (formation of a network of non-covalent linkages) may provide a more reasonable model. In addition, knowledge of the binding mechanism should be useful in the design of more efficient immobilized enzyme reactors for practical application of immobilized enzyme catalysis.

CONCLUSIONS

The results of this study have delineated the mechan-ism of collagen-enzyme binding in terms of the loci of ac-tive binding sites and the nature of the functional group and/or groups within the helical region of the collagen molecule involved in binding (by the process of membrane

impregnation). Further studies are in progress in our
laboratory to determine the contribution of other amino
acid residues of collagen to enzyme-binding, and to determine the equivalency of the binding mechanism involved
in immobilization by the process of impregnation and macromolecular complexation.

APPENDIX

Total No. of Active Binding Sites/gm Collagen

$$=(6.02 \times 10^{23})(\text{Moles of Active Sites/gm})$$

$$= (6.02 \times 10^{23})(3.14 \times 10^{-3})$$

$$= 1.9 \times 10^{21}$$

(where 6.02×10^{23} equals Avagadro's Number of bind-

ing sites per mole of binding sites).

Total No. of Ionic Binding Sites/gm Collagen

$$= (6.02 \times 10^{23})(\text{Moles of Ionic Binding Sites})$$

$$= (6.02 \times 10^{23})(1.8 \times 10^{-3})$$

$$= 1.1 \times 10^{21}$$

Total No. of Hydrogen Bonding Sites/gm Collagen

$$= (6.02 \times 10^{23})(\text{Moles of Hydrogen Bonding Sites/gm})$$

$$= (6.02 \times 10^{23})(1.40 \times 10^{-3})$$

$$= 0.8 \times 10^{21}$$

Total Potential Binding Sites/molecule of Collagen

$$\frac{(1.9 \times 10^{21})(300,000)}{6.02 \times 10^{23}} = 946 \text{ Potential Binding Sites/ Molecule}$$

When 1.9×10^{21} = Total No. of Active Binding Sites/gm
Collagen, 300,000 = Collagen Molecular Weight; and 6.02
$\times 10^{23}$ = Avagadro's No. of Molecules per mole of Collagen.

REFERENCES

Arnold, H., and Pette, D. (1968). Binding of Glycolytic Enzymes to Structure Proteins of the Muscle. European J. Biochem. 6:163.

Barndt, R. L., Leeder, S. G., Giacin, J. R. and Kleyn, D. H. (1975). Sanitation of a biocatalytic reactor used for the hydrolysis of acid whey. J. Food Sci. 40:291.

Bartenov, G. M. and Zuyev, Y. S. (1968). Strength and failure of viscoelastic materials. Pergamon Press, New York.

Bernath, F. R. and Vieth, W. R. (1974). Collagen as a carrier for enzymes. Materials Science and Process Engineering Aspects of Enzyme Engineering.

Bowes, J. H. and Kenton, R. H. (1949). The effect of deamination and esterification on the reactivity of collagen. J. Biochem. 44:142.

Brandau, H. and Pette, D. (1966). Enzyme localization and enzyme activity. II. Localization of enzymes of energy metabolism in cross-striated muscle. Enzyml. Biol. Clin. 6:123.

Cartmell, E. and Fowles, G. W. Z. (1956), Valency and Molecular Structure. Butterworths Scientific Publications. London, England.

Chien, J. C. W. and Chang, E. P. (1973). Influence of telopeptide on the morphology and physicomechanical properties of reconstituted collagen. Biopolymers, 12:2045.

Coulet, P. R., Julliard, J. H. and Gautheron, D. C. (1974). A mild method of general use for covalent coupling of enzymes to chemically activated collagen films. Biotech. and Bioengr. 16:1055.

Coulet, P. R. Godinot, C. and Gautheron, D. C. (1975). Surface-Bound Aspartate Aminotransferase on Collagen films. Compared Properties with Native Enzyme. Biochemica et Biophsica Acta. 391:272.

Drake, M. P., Davison, P. E., Bumps, S. and Schmitt, F. O. (1966). Action of proteolytic enzymes on tropocollagen and insoluble collagen.   Biochemistry 5(1), 301

Eskamani, A. (1972).   Characterization of lactase immobilized on collagen. Ph. D. thesis, Rutgers University, New Brunswick, NJ.

Eskamani, A., Chase, T., Jr., Frendenberger, J. and Gilbert, S. G. (1974).   Determination of protein immobilized on solid support by tryptophan content. Anal. Biochem. 57:421

Gaitondi, M. K., and Dovey, T. (1970).   A rapid and direct method for the quantitative determination of tryptophan in the intact protein.   Biochem. J. 117:907.

Giacin, J. R., Jakubowski, J.,Leeder, J. G., Gilbert, S. G. and Kleyn, D. H. (1974).   Characterization of lactase immobilized on collagen.   Conversion of whey lactose by soluble and immobilized lactase. J. Food Sci. 39:751.

Gilbert, S. G. (1973).   Enzyme Engineering User's Conference.   December 7, 1973.   Rutgers Inter-disciplinary Enzyme Technology Group.   Rutgers University, New Brunswick, N. J.

Goldman, R., Goldstein, L., and Katchalski, E. (1971). Biochemical Aspects of Reactions on Solid Supports.   G. R. Stark, Ed. Academic Press, New York.

Goldman, R., (1973).   Enzyme membrane model systems and their implication in biological research.   Biochimie, 55:953.

Green, D. E., Murer, E., Hultin, H. O., Richardson, S. H., Salmon, B., Brierley, G. P. and Baum, H. (1965). Association of integrated metabolic pathways with membranes.   I.   Glycolytic enzymes of the red corpuscle and yeast.   Arch. Biochem, Biophys. 112:635.

Grossberg, A. L. and Pressman, D. (1963).   Effect of Acetylation on the Active Site of Several Antihapter Antibodies: Further Evidence for the Presence of Tyrosine in Each Site.   Biochemistry, 2:90.

Hoare, D. G. and Koshland, D. E. (1966).   Procedure for the Selective Modification of Carboxyl Groups in Proteins.   J. Am. Chem. Soc. 88(9):2057

Hodge, A. J. and Petraska, J. A. (1963). Aspects of Pro-
        tein Structure. G. N. Ramachandran, Ed. Academic
        Press, New York.

Hsieh, F., Davidson, B., and Vieth, W. R. (1976). Model-
        ing of continual mass transfer level effects in
        an enzyme-membrane reactor system; with a direct
        approach to the identification of intrinsic rate
        parameters. J. Applied Chemistry and Biotech-
        nology. In Press.

Julliard, J. H. Godinot, C. and Gautheron, D. A. (1971).
        Some modifications of the kinetic properties of
        bovine liver glutamate dehydrogenase (NAD(P)) Co-
        valently bound to a solid matrix of collagen.
        (FEBS Letters, 14:185.

Klapper, M. H. and Klotz, I. M. (1972). Acylation with
        dicarbonylic Acid Anydrides. Methods in Enzym-
        ology, C. H. W. Hirs, Ed. Volume XXV, Academic
        Press, New York.

La Bella, F. S. (1971). Cross-links in elastin and colla-
        gen. Biophysical properties of the skin. H. R.
        Eldin, Ed. Wiley Interscience, New York.

Lee, K. H., Coulet, P. R., Gautheron, D. C. (1976).
        Grafting of enzymes on collagen films using Wood-
        ward's reagent "K" and a water soluble carbodil-
        mide derivative. Biochimie, 58:489.

Lieberman, E. R. (1971). Studies on the permeation of
        gases through collagen films. Ph. D. Thesis,
        Rutgers University, New Brunswick, N. J.

Lieberman, E. R., Gilbert, S. G. and Shrinivasa, V., (1972)
        The use of gas permeability as a molecular probe
        for the study of cross-linked collagen structures.
        Trans. N. Y. Acad. Sci. 34:694.

Lieberman, E. R. and Gilbert, S. G. (1973). Gas permea-
        tion of collagen films as affected by cross-link-
        ages, moisture and plasticizer content. J.
        Polymer Sci.: Symposium No. 41:33.

Lin, P. M. (1975). Modification of collagen for enzyme
        Immobilization. Ph. D. Thesis, Rutgers University,
        New Brunswick, NJ.

Lin, P. M., Giacin, J. R., Leeder, J. G. and Gilbert, S.
    G. (1976). Chemical and enzymatic modification
    of collagen: the effect of cross-linking on the
    enzyme-binding capacity of collagen. J. Food
    Sci., 41: 1056.

Luo, K. M., Giacin, J. R. and Gilbert, S. G. (1976).
    Chemical modification of collagen for enzyme im-
    mobilization. Presented 36th Annual IFT Meeting,
    Anaheim, California, June 7-9.

Piez, K. A., Lewis, M. S. Martin, G. R. and Gross, J.
    (1961). Subunits of the collagen molecule.
    Biochem. Biophys. Acta. 53:596.

Piez, K. A. (1967). Treatise on Collagen, Vol. I, Chemistry
    of Collagen, G. N. Ramachandran, Ed., Academic
    Press, New York.

Ramachandran, G. N. (1967). Treatise on Collagen, Vol. I,
    Structure of collagen at the molecular level, G.
    N. Ramachandran, Ed., Academic Press, New York.

Richards, F. M. and Knowles, J. P. (1968). Glutaraldehyde
    as a protein cross-linking reagent. J. Mol. Biol.
    37:231.

Rubin, A. L., Drake, M. P., Davison, P. E., Pfahl, D.,
    Speakman, P. T. and Schmitt, F. O. (1965). Ef-
    fects of pepsin treatment on the interaction
    properties of tropocollagen macromolecules.
    Biochem. 4:181.

Rubin, A. L. and Stangel, K. H. (1969). Biomaterials,
    L. Stark and G. Agarual, Ed., Plenum Press, New
    York.

Silman, I. H., Alba-Weissenberg, M. and Kachalski, E.
    (1966). Some water-insoluble papain derivatives.
    Biopolymers, 4:441.

Stark, G. R. (1965). Reaction of cyanate with functional
    groups of proteins. III. Reactions with amino
    and carboxyl groups. Biochemistry, 4:1030.

Stark, M., Rauterberg, J. and Kuhn, K. (1971) Evidence
    for a non-helical region at the carboxyl terminus
    of the collagen molecule. FEBS letters, 13:101.

Stevens, F. S. (1966). The depolymerizing action of pepsin on collagen. Molecular weights of the component polypeptide chains. Biochim. Biophys. Acta. 130:190.

Tanzer, M. L. (1973). Cross linking of collagen. Science 180:54.

Venkatasubramanian, K., Saini, R. and Vieth, W. R. (1974). On the mechanisms of enzyme and whole microbial cell attachment to collagen. J. Ferm. Technol. 52:268.

Verzar, F. (1964). Aging of the collagen fiber. In "International Review of Connective Tissue Research". Vol. 2: Academic Press, New York.

Veis, A., Anesey, J. and Massell, S. (1967). A limited microfibril model for the three-dimensional arrangement within collagen fibrils. Nature, London, 215:931.

Vieth, W. R., Gilbert S. G. and Wang, S. S. (1972a). Urea hydrolysis on collagen urease complex membrane.

Vieth, W. R., Gilbert, S. G. and Wang, S. S. (1972b). Performance of collagen-invertase complex membrane in a biocatalytic module. Trans. New York Academy of Sci. 34:454.

29

STRATEGIES IN THE RACEMIZATION-FREE SYNTHESIS

OF POLYTRIPEPTIDE MODELS OF COLLAGEN

Rao S. Rapaka, D.E. Nitecki and Rajendra S. Bhatnagar

University of California, San Francisco

San Francisco, California 94143

ABSTRACT

Synthesis of racemization-free sequential polypeptides is a major challenge to the peptide chemist. The loss of optical integrity at a single residue can drastically alter the properties of the polymer. We have investigated racemization in the synthesis of polytripeptide models of collagen. The conformational features of collagen are determined by a large imino content and the presence of Gly in every third position. Thus polymers of tripeptides containing Pro and Gly with an asymmetric residue A permit evaluation of the contribution of A to collagen conformation. In the synthesis of such polymers, the position of Pro in the tripeptide is of importance because of its variable reactivity at the N or C terminal residue. N-terminal Pro is more reactive but also sterically hinders polycondensation. Terminal Pro also participates in undesirable side reactions such as formation of urea or diketopiperazine derivatives and cyclization. Cyclization can be avoided by using concentrated solutions of monomer. Racemization occurs during the synthesis of OCl_5Ph or ONp esters but not in the synthesis of ONSu esters. This, however, does not guarantee formation of optically pure products. Best results are obtained with ONSu esters and limited use of base during polymerization. Polymerization appears to be stereoselective since the largest molecular weight polymers are purest, smallest molecular weight products highly racemized.

INTRODUCTION

The synthesis of high molecular weight racemization free sequential polypeptides is a major challenge to the peptide chemist because of a large number of factors that can influence the yield and the properties of the synthetic polymers. The protected starting monomeric peptide has to be easily obtainable as an optically pure compound. The choice of the starting mono- mer should be based on the steric influence of the N- and C- termini as well as the propensity of the C-terminal amino acid to racemize during the activation step. The conditions used for polymerization, mainly the extent of neutralization of the amine salt and the choice of solvent should be carefully considered. All these factors affect the degree of polymerization as well as the optical integrity of the polymer product.

The following discussion is concerned largely with the strategic approaches that must be considered in order to obtain large molecular weight, optically pure, sequential polypeptide models of collagen, but it is also relevant to the synthesis of other repeating polymeric peptides.

CONFORMATIONAL MODELS OF COLLAGEN

We have synthesized a number of sequential polypeptides as conformational models of collagen (Rapaka and Bhatnagar, 1975a,b, 1976c; Rapaka, Bhatnagar and Nitecki, 1976a,b). The conforma- tional properties of collagen are derived at least in part, by stereochemical contribution of the imino residues proline and hydroxyproline present in large concentration in that protein, as well as by the presence of glycine in every third position in the sequence. Many collagen models are based on sequences containing proline, glycine and a third residue which is usually an asymmetric amino acid. A polymeric sequence such as -Gly-X-Pro- or -Gly-Pro- X-,* approximating the collagen sequence, may be synthesized in principle from monomers containing any permutative sequence of the three residues. Thus the polymerization of either Gly-X-Pro, X- Pro-Gly or Pro-Gly-X would yield the polymeric sequence (Gly-X- Pro)$_n$. In practice careful consideration has to be given to steric problems arising from the position of the proline in the trimeric sequence and racemization problems encountered in the activation of the optically active acid when placed in the C- terminus of a peptide.

*Because collagen may be considered to be a polymer of glycine-led tripeptides, polymeric sequences modeling collagen are usually denoted as polymers of glycine-led monomeric subunits.

STERIC INFLUENCES

The course of polymerization is expected to be strongly influenced by steric hindrance of bulky residues in the N-terminus. The secondary amino acids, such as proline or sarcosine in the N-terminus of the polymerizing monomer could be expected to yield lower molecular weight polymers. On the other hand, the more pronounced basicity (and therefore nucleophilicity) of a secondary amino group may exert a compensating effect. In practice, it is difficult to arrive at any clear-cut conclusions. Brown et al (1972) utilized monomeric sequences containing N-terminal proline, Pro-X-Gly, to obtain high molecular weight (Gly-Pro-Ala)$_n$ and (Gly-Pro-Ser)$_n$. DeTar, Alberts and Gilmore (1972) also synthesized high molecular weight (Gly-Pro-Hyp)$_n$ from the monomer Pro-Hyp-Gly. Other workers have found the monomers with C-terminal proline more satisfactory. Bell, Jones and Webb (1975b) synthesized sequences of the type (Gly-Pro-X)$_n$ utilizing monomers X-Gly-Pro because sequences with N-terminal proline and C-terminal optically active amino acid were found to be racemized. These observations corroborated the earlier observations by Blout and colleagues (Bloom et al, 1966) that (Gly-Pro-Ala)$_n$ obtained from monomer Gly-Pro-Ala was racemized during the activation steps. Bonora and Toniolo (1974) initially utilized the monomeric sequence Pro-Nva-Gly in the synthesis of oligomeric peptides, since glycine at the C-terminus eliminates the possibility of racemization. However, Nva-Gly-Pro- monomeric sequence was found to yield better results in the stepwise synthesis of oligomers as well as the high molecular weight polymer (Nva-Gly-Pro)$_n$. Extending these observations from collagen-like polypeptides to the polypeptide models of elastin (Bell et al, 1975a) synthesized (Gly-Val-Gly-Val-Pro)$_n$ by polymerizing the monomer containing N-terminal glycine and C-terminal proline in preference to monomers containing valine at either terminus, thereby avoiding some of the steric hindrance problems. Yet the same polypeptide was synthesized via the monomer Val-Pro-Gly-Val-Gly in very good yield (Urry et al, 1975).

We have successfully synthesized several collagen-like polymers from tripeptides containing either N-or C- terminal proline. Some of these syntheses are summarized in Figures I-VI. The (Gly-Val-Pro)$_n$ polymer was obtained by polymerization of either Gly-Val-Pro or Pro-Gly-Val monomers (Rapaka and Bhatnagar, 1975). The monomer with C-terminal proline and N-terminal glycine yielded the higher molecular weight polymer. This difference may be due to the fact that in the Gly-Val-Pro monomer the N-terminal glycine offers no steric restraints whereas in Pro-Gly-Val monomer both termini are sterically bulky. Polymers (Gly-Leu-Pro)$_n$ and (Gly-Phe-Pro)$_n$ were obtained from monomers Pro-Leu-Gly and Pro-Gly-Phe correspondingly (Rapaka and Bhatnagar, 1975a; Rapaka, Bhatnagar and Nitecki, 1976a,b). Sterically hindered N-terminal

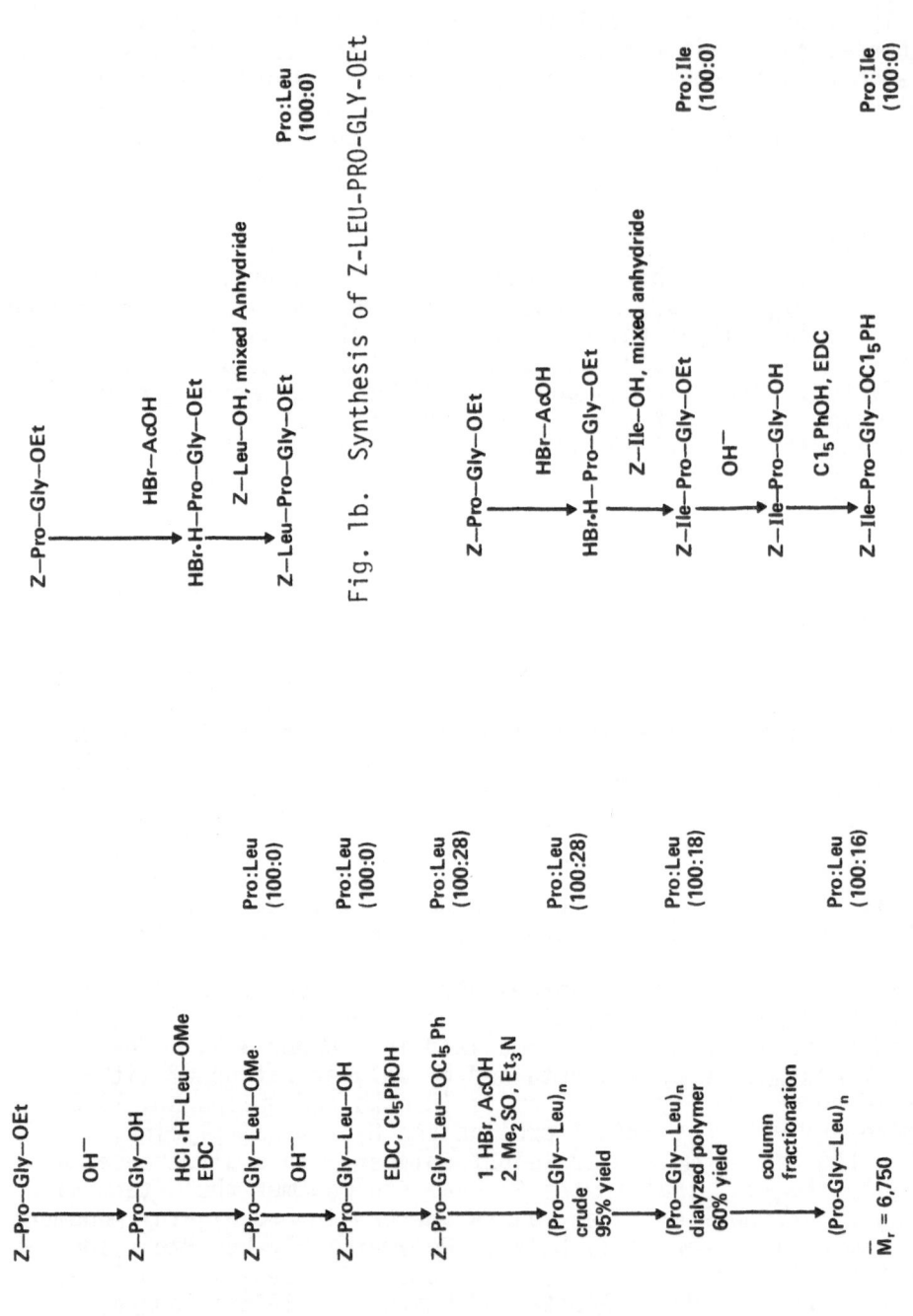

Fig. 1b.   Synthesis of Z-LEU-PRO-GLY-OEt

Fig. 1c.   Synthesis of Z-ILE-PRO-GLY-OC1₅Ph

Fig. 1a.   Synthesis of (PRO-GLY-LEU)ₙ

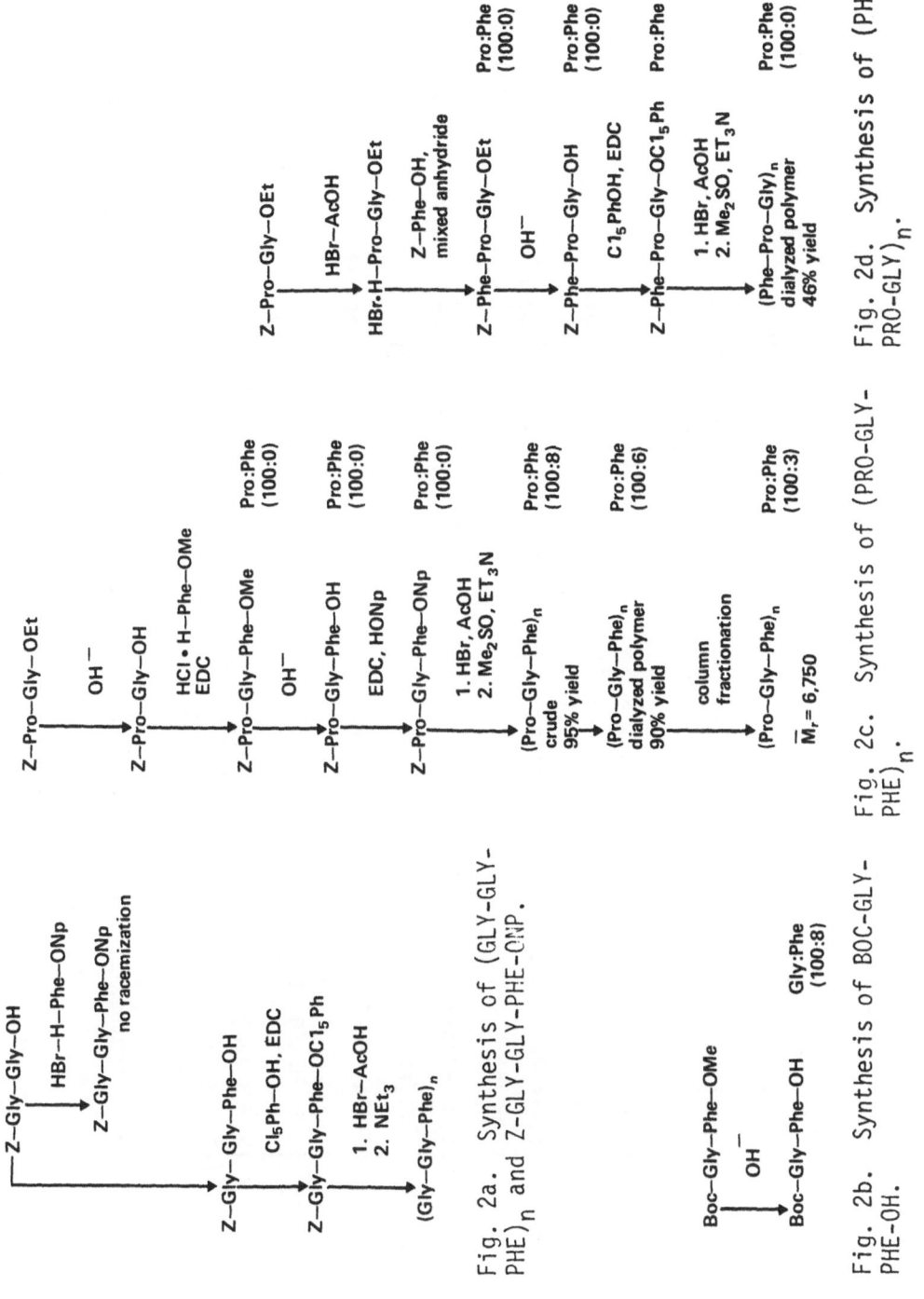

Fig. 2a. Synthesis of (GLY-GLY-PHE)$_n$ and Z-GLY-GLY-PHE-ONP.

Fig. 2b. Synthesis of BOC-GLY-PHE-OH.

Fig. 2c. Synthesis of (PRO-GLY-PHE)$_n$.

Fig. 2d. Synthesis of (PHE-PRO-GLY)$_n$.

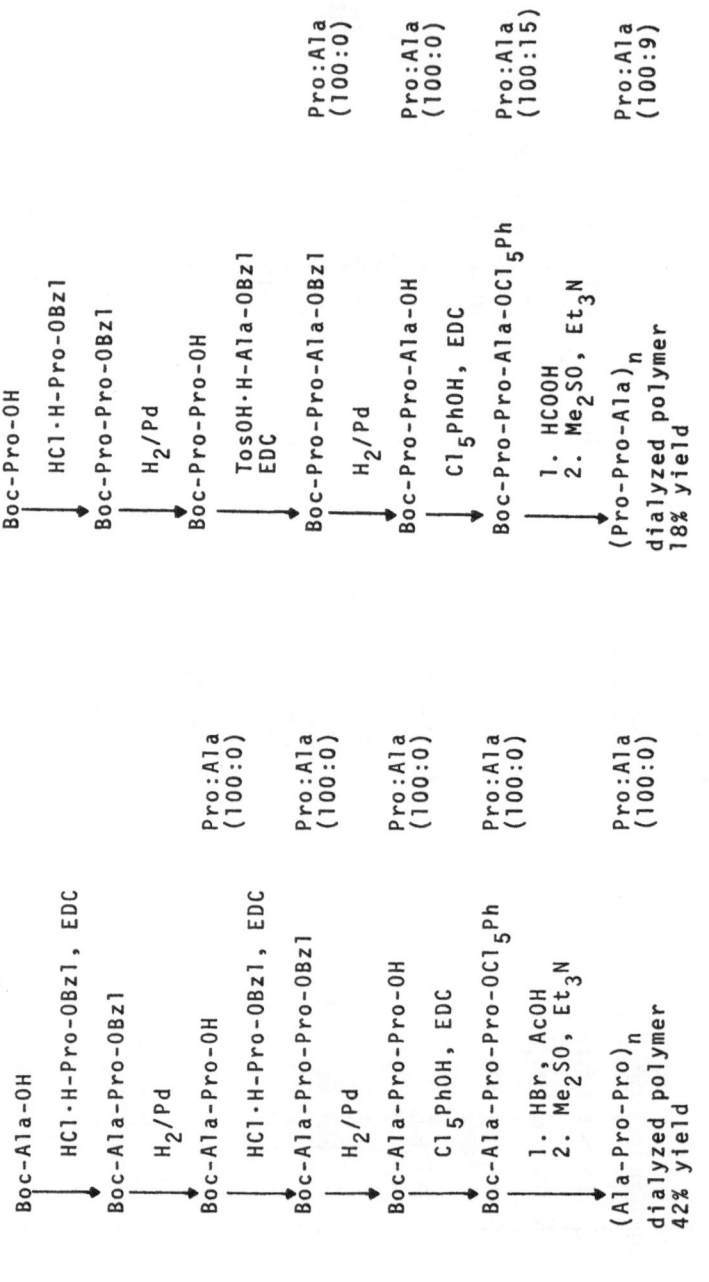

Boc-Ala-OH
    HCl·H-Pro-OBzl, EDC
Boc-Ala-Pro-OBzl
    H_2/Pd
Boc-Ala-Pro-OH   Pro:Ala (100:0)
    HCl·H-Pro-OBzl, EDC
Boc-Ala-Pro-Pro-OBzl   Pro:Ala (100:0)
    H_2/Pd
Boc-Ala-Pro-Pro-OH   Pro:Ala (100:0)
    Cl_5PhOH, EDC
Boc-Ala-Pro-Pro-OCl_5Ph   Pro:Ala (100:0)
    1. HBr, AcOH
    2. Me_2SO, Et_3N
(Ala-Pro-Pro)$_n$ Pro:Ala (100:0)
dialyzed polymer
42% yield

Fig. 3a.   Synthesis of (ALA-PRO-PRO)$_n$.

Boc-Pro-OH
    HCl·H-Pro-OBzl
Boc-Pro-Pro-OBzl
    H_2/Pd
Boc-Pro-Pro-OH
    TosOH·H-Ala-OBzl
    EDC
Boc-Pro-Pro-Ala-OBzl   Pro:Ala (100:0)
    H_2/Pd
Boc-Pro-Pro-Ala-OH   Pro:Ala (100:0)
    Cl_5PhOH, EDC
Boc-Pro-Pro-Ala-OCl_5Ph   Pro:Ala (100:15)
    1. HCOOH
    2. Me_2SO, Et_3N
(Pro-Pro-Ala)$_n$ Pro:Ala (100:9)
dialyzed polymer
18% yield

Fig. 3b.   Synthesis of (PRO-PRO-ALA)$_n$.

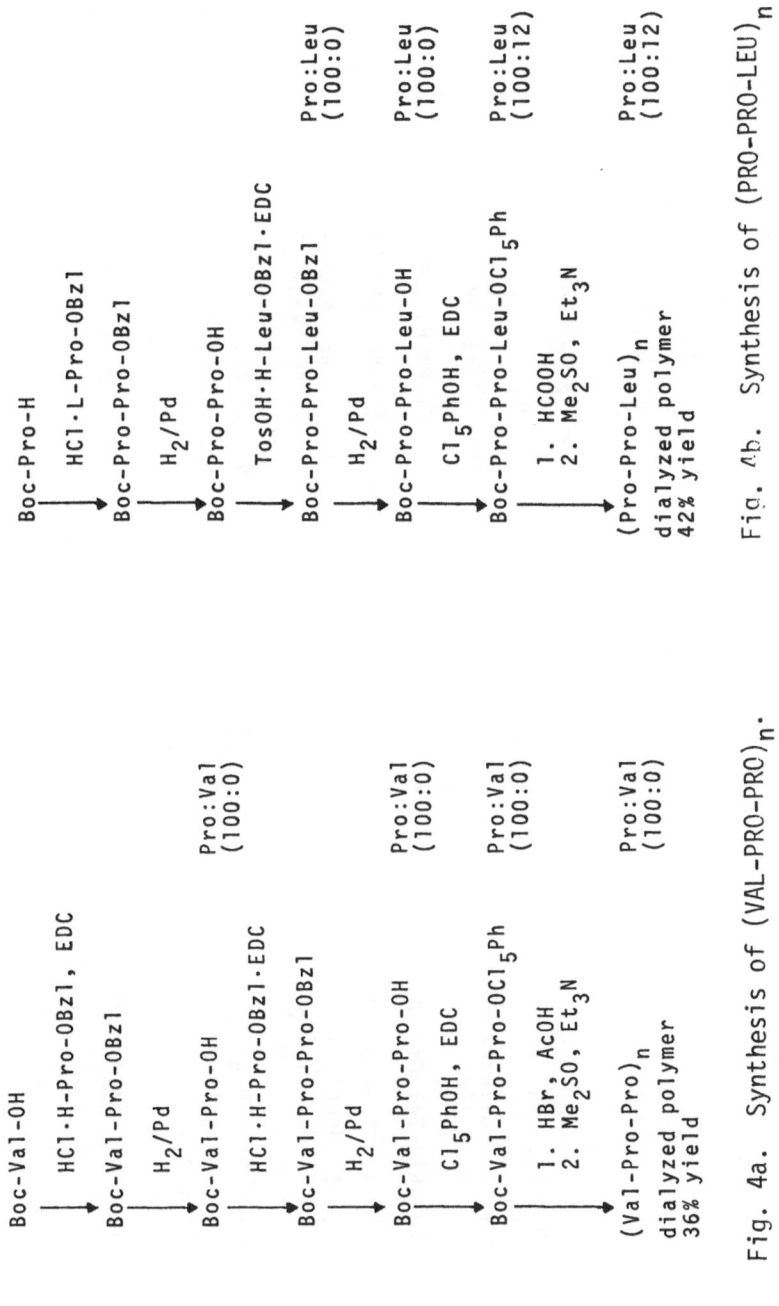

Fig. 4a. Synthesis of (VAL-PRO-PRO)$_n$.

Fig. 4b. Synthesis of (PRO-PRO-LEU)$_n$

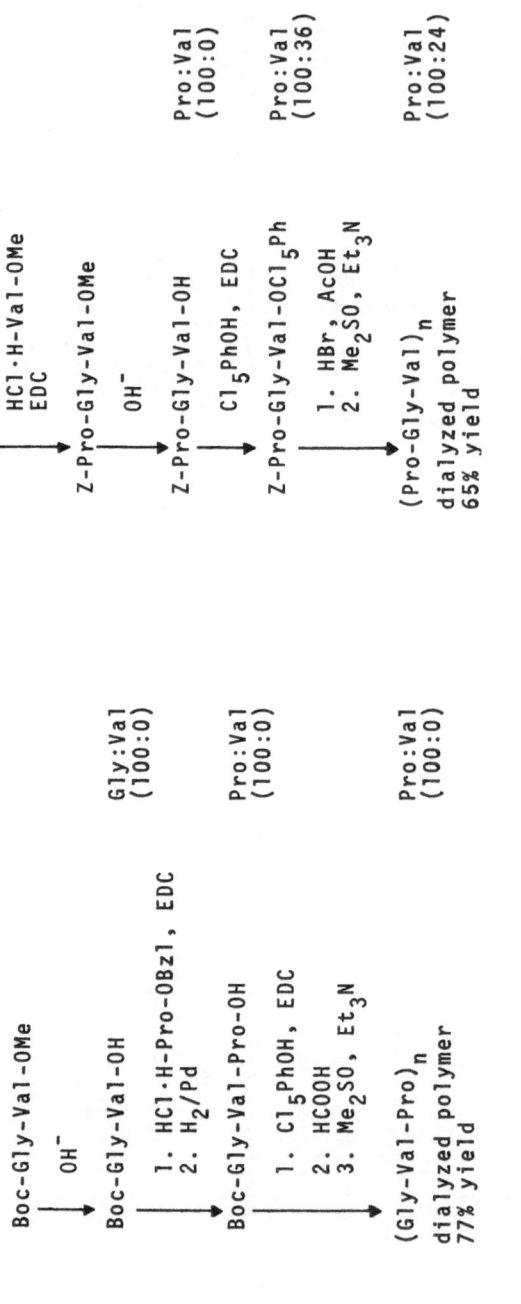

Fig. 5a. Synthesis of (GLY-VAL-PRO)$_n$.

Fig. 5b. Synthesis of (PRO-GLY-VAL)$_n$.

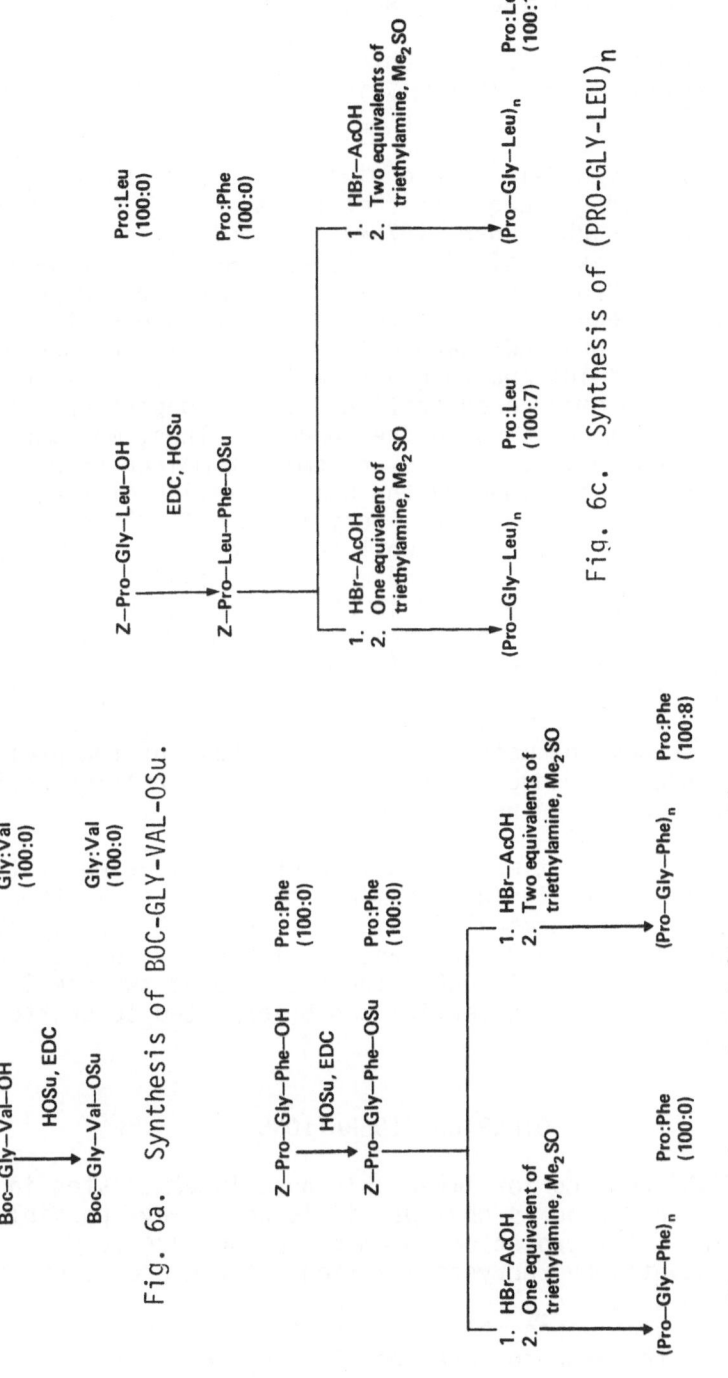

Fig. 6a. Synthesis of BOC-GLY-VAL-OSu.

Fig. 6b. Synthesis of (PRO-GLY-PHE)$_n$.

Fig. 6c. Synthesis of (PRO-GLY-LEU)$_n$

isoleucine did not prevent polymerization of Ile-Pro-Gly to yield (Gly-Ile-Pro)$_n$ polymer (Rapaka, Bhatnagar and Nitecki, 1976a).

Satisfactory results were obtained in the synthesis of poly-tripeptide polymers that lack glycine residues and are thereby more sterically hindered at both termini (Rapaka and Bhatnagar, 1976a,b,c; Rapaka, Bhatnagar and Nitecki, 1976a,b). Monomers Ala-Pro-Pro and even Val-Pro-Pro yielded the corresponding polymers in good yield although their molecular weight was not established (Rapaka and Bhatnagar, 1976c). A polymer of the sequence (β-Ala-Pro-Pro)$_n$ was obtained by polymerizing either Pro-Pro-β-Ala or Pro-β-Ala-Pro monomers. The latter monomer unexpectedly yielded lower molecular weight product (Rapaka and Bhatnagar, 1975b). Katakai, Oya and Iwakura (1975) presented evidence indicating that yields and molecular weights of the polymer are independent of monomer concentrations, but do depend on the sequence of the amino acids in the monomer. Thus, monomer Val-Ala-Gly resulted in lower yields and smaller molecular weight polymers than were obtained by the polymerization of the monomer Ala-Val-Gly, In our laboratory, very large differences were observed in polymerization of two tripeptides containing two secondary amino acids (Rapaka and Bhatnagar, 1975a). Polymerization of monomer Pro-Sar-Gly resulted in 20,000 molecular weight polymer (Gly-Pro-Sar)$_n$, while polymerization of monomer Sar-Pro-Gly yielded polymer (Gly-Sar-Pro)$_n$ of 4000 molecular weight. Both of these monomer sequences have C-terminal glycine and a secondary amino acid in the N-terminus. It is not unexpected that a more restricted freedom of motion in the side chain of the proline residue may present reduced steric hindrance to polymerization when compared to sarcosine residue.

It seems at this time that neither polymerization of tripeptide monomers containing C-terminal proline (Fairweather and Jones, 1972), nor N-terminal proline (Brown et al, 1972; DeTar, Alberts and Gilmore, 1972) precludes formation of polymers of molecular weights above 10,000. The presence of two sterically hindered residues at both termini can be expected to create difficulties.

OTHER CONSIDERATIONS

Although the need for impeccably pure intermediates in the synthesis of this kind is obvious, it is not always possible to achieve. The fully protected monomer peptides are easy to purify, frequently by recrystallization. The N-protected peptide active esters are already in the realm of active reagents, but it is frequently possible to recrystallize these. The next intermediate, the N-deprotected peptide active ester ammonium salt is indeed programmed to react and it is virtually impossible to ascertain its degree of purity or to do anything about it, although attempts have been made (Brown et al, 1972). These salts, especially the products of HBr deprotection are usually highly hygroscopic (Fairweather and Jones, 1972).

In the most commonly used variants of design, the N-terminus is protected by either benzyloxycarbonyl (Z) or t-butyloxycarbonyl (Boc) groups. The Boc group is more acid labile and can be removed under considerably milder conditions than the Z group. Moreover, the resulting trifluoroacetate or hydrochloride salts tend to be somewhat less hygroscopic, and therefore easier to handle. The Boc group can also be removed by formic acid (Halpern and Nitecki, 1967) under very mild conditions; however, in our hands, the resulting molecular weights of the polymers were lower than by the use of their reagents (Rapaka and Bhatnagar, 1976a).

The protection of the C-terminus for the synthesis of the desired monomer presents another series of choices. If glycine is next to the N-terminal amino acid, the amino group of which is protected by the benzyloxycarbonyl group, alkaline saponification of the C-terminal esters may result in considerable formation of urea derivatives (Bodansky and Ondetti, 1966; Bodansky et al, 1963) Alkaline saponification of Z-Pro-Gly-Leu-OMe and Z-Pro-Gly-Phe-OMe yielded the corresponding acids satisfactorily, but the same reaction with Z-Pro-Gly-Val-OMe resulted in very poor yield, presumably due to urea formation (Rapaka and Bhatnagar, 1975a). In our experience, whenever possible, the Boc group would be the choice for amino protection and the benzyl ester for carobxyl protection.

Cyclization of the peptides can occur in the synthesis of the monomers as well as during the polymerization with drastic reduction in the yield. The formation of diketopiperazines is particularly easy with dipeptides containing secondary amino acids such as sarcosine and proline. Fairweather and Jones (1973) experienced difficulties in the synthesis of sequences containing Gly-Sar-Ala. Alternate synthetic intermediates Z-Ala-Gly-Sar-OCl$_5$-Ph and Z-Sar-Ala-Gly-OCl$_5$Ph were easily prepared. However, upon polymerization these yielded only low molecular weight dialyzable oligomers and pentachlorophenol, presumably because of competition between polymerization and cyclization reaction. The syntheses of (Sar-Pro-Gly)$_n$ and (Pro-Sar-Gly)$_n$ proceeded smoothly and in good yields (Rapaka and Bhatnagar, 1975a). Generally, formation of cyclic tripeptides is unlikely and is favorable only if the tripeptides are composed entirely of secondary amino acids such as proline, hydroxyproline or sarcosine (Rothe, 1975; Dale and Titlestad, 1969; Deber, Torchia and Blout, 1971; Kopple, 1972). Condensation of two tripeptide units-cyclodimerization is another competing reaction and is facilitated by the β-turn in the peptide (Kopple, 1972). The occurence of this side reaction is not well documented because of predisposition of many peptide chemists to seek only the higher molecular weight polymers.

The reaction conditions during the polymerization, such as solvent, temperature, the amount of base used for neutralization and concentration of the monomer play an important role in the yield and molecular weight of the product. Kovacs, Schmit and Ghatak (1968) systematically investigated the effect of solvents and polymerization procedures on the degree of polymerization. High molecular weight polymers were obtained utilizing penta-chlorophenyl.esters in dimethyl formamide in concentrated solu-tions; dilute solutions yielded cyclic peptides. In our hands, concentrated solutions of pentachlorophenyl esters in dimethyl sulfoxide yielded most satisfactory results. Brack and Spach (1970) reported an alternate method of polymerization of the active esters as a concentrated suspension (1,000 m mole/l) in benzene. Zeiger, Lange and Maurer (1973) studied the polymeriza-tion of the pentachlorophenyl esters at a concentration of 50 m mole/l, both as suspension in benzene and as a solution in di-methyl formamide. Higher molecular weights were obtained by the suspension method.

RACEMIZATION

Racemization is an important but little investigated para-meter in the synthesis of sequential peptides. It can occur at various stages of synthesis: attachment and removal of protecting groups, peptide coupling, formation of the active monomer and polymerization. We have synthesized a number of polytripeptide models of collagen and investigated the optical purity of the intermediates at various steps of the synthesis of the monomers and the fractionation of the polymers (Figures I-VI), (Rapaka and Bhatnagar, 1975a,b, 1976c; Rapaka, Bhatnagar and Nitecki, 1976a,b). The strategic choice for a racemization free synthesis is utilization of glycine or proline as the C-terminus, since the activation of a peptide with C-terminal amino acid frequently results in racemization. Since this is not always possible, it was improtant to investigate the synthesis of polymers via activation of peptides with a C-terminal optically active amino acid.

DeTar, Silverstein and Rogers (1966) investigated the problem of racemization in the synthesis of Z-A^1-A^2-ONp from Z-A^1-A^2-OH and p-nitrophenol. Their results revealed extensive racemiztion in the stage of active ester formation.

In the preparation of Z-Pro-Val-ONp from the corresponding dipeptide and p-nitrophenol was racemized up to 50%, while only 0.6% racemization occurred if this active ester was prepared from Z-Pro-OH and H-Val-ONp (Tomeda, Kayahara and Iriye, 1973a). Active esters of peptides containing C-terminal alanine (Bloom et al, 1966; Stewart, 1965) and serine (Chakravarty, Mathur and Dhar, 1973) were synthesized and employed for polymer synthesis, however the optical purity of the polymers was not reported. Many syntheses of optically pure polymers have been achieved by a variety of methods; among others, the sequences were (Glu-Ala-Glu)$_n$, (Kovacs, Giannotti and Kapoor, 1966); (Tyr-Ala-Glu)$_n$, (Ramachandran, Berger and Katchalski, 1971); (Gly-Gly-Phe)$_n$ and (Gly-Pro-Ala)$_n$ (Cowell and Jones, 1972) and (Gly-Leu)$_n$, (Hardy, Rydon and Thompson, 1972).

An interesting "backing-off" procedure, proposed by Goodman and Steuben (1959) circumvent the racemizing monomer-activation step. It involves coupling of a hydrohalide salt of an amino acid active ester to the carboxyl component of an N-protected amino acid or peptide, i.e., Z-A^1-A^2-OH + HBr·H-A^3-ONp (or other ester) ----- Z-A^1-A^2-A^3-ONp. DeTar and Estrin (1966) utilized this procedure to synthesize optically pure (Gly-Gly-Phe)$_n$. This "backing-off" procedure eliminates the problem of racemization during peptide activation, but the yields obtained are unsatisfactory (Doyle et al, 1970) and this method has seen little use.

We have investigated the problem of racemization in the following peptide sequences: -Pro-Gly-Leu-, -Leu-Pro-Gly-, -Ile-Pro-Gly-, -Pro-Gly-Phe-, -Phe-Pro-Gly-, -Pro-Gly-Val-, -Gly-Val-Pro-, -Pro-Pro-Ala-, -Ala-Pro-Pro-, -Val-Pro-Pro-, -Pro-Pro-Leu-, -Gly-Val- and -Gly-Phe-. Pentachlorophenyl, p-nitropehnyl and N-hydroxysuccinimide esters were used for polymerization. The optical purity of the intermediates and polymers was determined using Crotalus adamamteus L-amino acid oxidase on the peptide hydrolyzates. The remaining D-amino acids were quantitatively determined by amino acid analysis. All monomeric peptides contained one asymmetric amino acid and the other two amino acids were either both proline or one proline or one proline and one glycine. Proline is not oxidized by L-amino acid oxidase; it served as an internal standard. In sequences, where only glycine and an asymmetric amino acid were present (and proline was absent), glycine was used as the standard, since it has been shown that L-amino acid oxidase does not react with glycine (Holme and Goldberg, 1975). These procedures were described elsewhere in detail (Rapaka, Bhatnagar Nitecki, 1976a,b).

No detectable racemization occurred in peptide coupling steps, utilizing either mixed anhydride or carbodiimide procedure. This is not always the case. Coupling between Z-Pro-Val-OH and

H-Pro-OMe was known to produce extensive racemization (Tomida, Kayahara and Iriye, 1973a,b). This problem was averted in the synthesis of (Pro-Val-Pro)$_n$ by the stepwise elongation starting from Boc-Val-OH (Figure 4). Saponification of methyl esters proceeded without racemization, except in the saponification of Boc-Gly-Phe-OMe (Figure II). Hydrogenation of benzyl esters proceeded without racemization, as expected.

Preparation of N-protected peptide active esters, containing C-terminal optically active amino acid (other than proline) is prone to racemization. Preparation of such pentachlorophenyl esters from Z-Pro-Gly-Leu-OH, Z-Pro-Gly-Val-OH, Boc-Pro-Ala-OH and Boc-Pro-Pro-Leu-OH resulted in racemization and the different amounts of D-amino acids formed indicate that racemization may depend on the nature of the C-terminal amino acid and also on the sequence. Contrary to our expectation, no racemization could be detected in the formation of Z-Pro-Gly-Phe-ONp. Formation of N-hydroxysuccinimide esters from Z-Pro-Gly-Leu-OH, Z-Pro-Gly-Phe-OH and Boc-Gly-Val-OH proceeded without racemization and hence N-hydroxysuccinimide ester method of activation seems to be superior to pentachlorophenyl ester procedure.

Polymerization of active esters produced unexpected results. Z-Pro-Gly-Phe-ONp, which was optically pure, yielded a polymer with 8% D-phenylalanine. Our results agreed with the results reported by Kovacs et al (1972) for (Gly-Gly-Phe)$_n$. Polymerization of HBr·H-Pro-Gly-Phe-OSu, using one equivalent of triethylamine, yielded an optically pure polymer; utilization of two equivalents of triethylamine produced a polymer containing 8% of D-phenylalanine. Polymerization of HBr·H-Pro-Gly-Leu-OSu, with one equivalent of triethylamine produced a polymer with 7% D-leucine and 14% D-leucine with two equivalents of triethylamine. These results show that racemization is possible even after the activation step, i.e. during polymerization; the polymerization step is base sensitive and even if an optically pure active ester is obtained either by N-hydroxysuccinimide ester procedure or "backing off" procedure, it carries no guarantee for an optically pure polymer. Racemization is also sequence dependent. No racemization could be detected in the polymerization of HBr·H-Pro-Gly-Phe-OSu with one equivalent of triethylamine, whereas 7% of D-leucine was detected in the polymerization of HBr·H-Pro-Gly-Leu-OSu with one equivalent of triethylamine.

In all our experiments, the monomers containing either glycine or proline as the C-terminal amino acid polymerized without racemization as shown in the polymers (Ala-Pro-Pro)$_n$, (Val-Pro-Pro)$_n$ and (Phe-Pro-Gly)$_n$.

In many cases, polymerization was found to be somewhat stereo-selective. The polymers fractionated by dialysis or by gel filtration were found to contain less racemized product in the higher molecular weight fraction, than the crude product or the starting monomeric tripeptide. This implies that polymerization reaction between the epimeric peptides is slower than between peptides of identical configuration. Polymerization results in higher molecular weight polymers if the monomer is entirely composed of either L or D amino acids as compared to a monomer containing both L and D residues. The presence of a L-D or D-L residue might introduce an elbow-like bend, and thereby enhance cyclization and result in a lower molecular weight polymer. This is in agreement with the results obtained by Fairweather and Jones (1972).

CONCLUSIONS

In conclusion, the results from our limited investigations reveal that:

1. Active ester formation of a C-terminal optically active amino acid is prone to racemization and in the sequences studied active ester formation proceeded without racemization if N-hydroxysuccinimide ester method of activation is employed.

2. Polymerization is sensitive to the amount of base used and controlled use of base is warrented.

3. Racemization can occur during polymerization and is probably sequence dependent.

ACKNOWLEDGEMENTS

This study was supported in parts by NIH Grants DE-03861 (RSB) and AI-05664 and AI-11938 (DEN). Some of the flow charts based on the authors' work were redrawn with permission from Biopolymers.

The authors are grateful to Miss Inge M. Stoltenberg and Mr. K.R. Sorensen for expert technical assistance; the authors also wish to acknowledge Ms. Sandra Hodess for her assistance in the preparation of this manuscript.

Dr. Rapaka's present address is:
                    Laboratory of Molecular Biophysics
                    University of Alabama Medical Center
                    Birmingham, Alabama 35294

REFERENCES

Bell, J., Boohan, R., Jones, H. and Moore, R. (1975a) Sequential
    polypeptides. Part IX. The synthesis of two sequential poly-
    peptide elastin models. Int. J. Peptide Protein Res. 7 227.
Bell, J., Jones, J. and Webb, C., (1975b) Sequential polypeptides
    Part X. The synthesis of some sequential polypeptide collagen
    models with functional side chains. Int. J. Peptide Protein
    Res. 7 235.
Bhatnagar, R. and Rapaka, R. (1976) Synthetic polypeptide models
    of collagen synthesis and applications in The Biochemistry
    of Collagen, G. N. Ramachandran, Ed., Chicago, Plenum Press,
    p. 479.
Bloom, S., Dasgupta, S., Patel, R. and Blout, E., (1966) The
    synthesis of glycyl-L-prolylglycyl and glycyl-L-prolyl-L-
    alanyl oligopeptides and sequential polypeptides. J. Amer.
    Chem. Soc. 88 2035.
Bodanszky, M., Sheehan, J., Ondetti, M. and Lande, S. (1963) Glycine
    analogs of bradykinin. J. Amer. Chem. Soc. 85 991.
Bodanszky, M. and Ondetti, M. (1966) in Peptide Synthesis, New
    York, Interscience Publishers, p. 26.
Bonora, G. and Toniolo, C. (1974) Sequential oligopeptides:
    Synthesis and characterization of the oligopeptides and
    polypeptides with the repeating sequence L-norvalyl-glycyl-L-
    proline. Biopolymers 13 1055.
Brack, A. and Spach, G. (1970) Nouvelles conformations de co-
    polypeptides ordonnes a base de L-alanine et de glycine,
    Comptes Rendus Acad. Sci. Ser. C. 271 916.
Brown, F., DiCorato, A., Lorenzi, G. and Blout, E.,(1972) Synthesis &
    structural studies of two collagen analogues: Poly (L-Prolyl-
    L-Seryl-Glycyl) and Poly (L-Prolyl-L-Alanyl-Glycyl). J. Mol.
    Biol. 63 85.
Chakravarty, P., Mathur, K. and Dhar, M. (1973) Synthesis of poly
    (Glu-Ala-Ala-Ala-Ser) and Poly (Ala-Ala-Glu-Ala-Ser) as model
    receptors for acetylcholine. Indian J. Biochem. Biophysics
    10 233.
Cowell, R. and Jones, J. (1971), Sequential polypeptides, Part I.
    Use of mono-esters of catechol in the synthesis of sequential
    polypeptides. J. Chem. Soc (C) 1082.
Dale, J. and Titlestad, K. (1969) Cyclic oligopeptides of sarcosine
    (N-methylglycine), J. Chem. Soc. Chem. Commun. 656.
Deber, C., Torchia, D. and Blout, E. (1971) Cyclic peptides I.
    Cyclic (tri-L-prolyl) and derivatives: Synthesis and molecular
    conformation from nuclear magnetic resonance, J. Amer. Chem.
    Soc 93 4893.
DeTar, D. and Estrin, N., (1966) An optically pure sequence peptide
    Poly·Gly·Gly·Phe. Tetrahedron Lett. 48 5985.

DeTar, D., Silverstein, R. and Rogers, F. Jr. (1966) Reactions
    of carbodiimides. III. The reactions of carbodiimides with
    peptide acids, J. Amer. Chem. Soc. 88 1024.
DeTar, D., Albers, R. and Gilmore, F. (1972) Synthesis of
    sequence peptide polymers related to collagen, J. Org.Chem.
    37 4377.
Doyle, B., Traub, W., Lorenzi, G., Brown, F. III, Blout, E. (1970)
    Synthesis and structural investigation of poly (L-alanyl-L-
    alanyl-glycine) J. Mol. Biol. 51 47.
Fairweather, R. and Jones, J., (1972) Sequential polypeptides.
    Part IV. The synthesis of poly- (L-alanyl-glycyl-L-proline)
    and its stereoisomers. J. Chem.Soc.Perkin Trans. I 1908.
Fairweather, R. and Jones, H., (1973) The antigenicity of sequential
    polypeptides I. The synthesis of some sequential collagen models.
    Immunology 25 241.
Goodman, M. and Steuben, K., (1959) Peptide synthesis via amino acid
    active esters, J. Amer. Chem. Soc. 81 3980.
Halpern, B. and Nitecki, D. (1967) The deblocking of t-butyloxy-
    carbonyl peptides with formic acid. Tetrahedron Lett. 31 3031.
Hardy, P., Rydon, H. and Thompson, R. (1972) Polypeptides part XVII.
    The synthesis of some sequential polypeptides of γ-benzyl D-
    glutamate and L-leucine. J. Chem. Soc.Perkins Trans. I 5.
Holme, D. and Goldberg, D., (1975) Coupled optical rate determina-
    tions of amino acid oxidase activity, Biochim. Biophys. Acta
    377 61.
Johnson, B. (1974) Synthesis, structure and biological properties
    of sequential polypeptides J. Pharm. Sci 63 313.
Katakai, R., Oya, M. and Iwakura, Y. (1975) Synthesis and conforma-
    tional study of sequential polypeptides, (L-Ala-L-Val-Gly)$_n$
    and (L-Val-L-Ala-Gly)$_n$ Biopolymers 14 1315.
Kopple, K., (1972) Synthesis of cyclic peptides J. Pharm. Sci. 61
    1345.
Kovacs, J., Giannotti, R. and Kapoor, A. (1966) Peptides with
    known repeating sequence of amino acids. Synthesis of poly-
    L-glutamyl-L-alanyl-L-phenylalanine pentachlorophenyl
    active ester. J. Amer. Chem. Soc. 88 2282.
Kovacs, J., Schmit, G. and Ghatak, U. (1968) Polypeptides with
    known repeating sequences of amino acids. Comparison of
    several methods used for the synthesis of poly-γ-D and L-
    glutamylglycine and investigation of its serological reaction
    with antianthrax immune serum. Biopolymers 6 817.
Kovacs, J., Meyers, G., Johnson, R., Giannotti, R., Cortegiano, H.
    and Roberts, J. (1972) in Progress in Peptide Research Vol.
    2, p. 185-193, Saul Lande, Ed., Gordon and Breach, New York.
    "On the problem of racemization during the synthesis of
    sequential polypeptides".

Ramachandran, J., Berger, A. and Katchalski, E. (1971) Synthesis and physicochemical properties in aqueous solution of the sequential polypeptide poly (Tyr-Ala-Glu)$_n$. Biopolymers 10 1829.

Rapaka, R. and Bhatnagar, R. (1975a) Synthesis of polypeptide models of collagen. Int. J. Peptide Protein Res. 7 119.

Rapaka, R. and Bhatnagar, R. (1975 b) Polypeptide models of collagen. Synthesis of (Pro-Pro-β-Ala)$_n$. Int. J. Peptide Protein Res 7 475.

Rapaka, R., Bhatnagar, R. and Nitecki, D. (1976a) Racemization in the synthesis of polytripeptide models of collagen. Biopolymers 15 317.

Rapaka, R., Bhatnagar, R. and Nitecki, D. (1976b) Racemization in the synthesis of sequential polypeptides using N-hydroxysuccinimide esters. Biopolymers 15 1585.

Rapaka, R. and Bhatnagar, R. (1976c) Polypeptide models of collagen: synthesis of (Pro-Pro-Ala)$_n$ and (Pro-Pro-Val)$_n$. Int. J. Peptide Protein Res. 8 371.

Rothe (1965) Synthesis of cyclotri-L-prolyl, a cyclotripeptide having a nine membered ring, Angew. Chem. Int. Ed. 4 356.

Stewart, F. (1965) The synthesis and polymerization of peptide p-nitrophenyl esters, Aust. J. Chem. 18 887.

Tomida, I., Kayahara, H. and Iriye, R., (1973a) Racemization in the coupling reaction, Pro-Val + Pro with the activated ester methods, Agr. Biol. Chem 37 2549.

Tomida, I., Kayahara, H. and Iriye, R. (1973b) Racemization in the coupling reaction with the several methods beside the activated ester. Methods Agr. Biol. Chem. 37 2557.

Urry, D., Mitchell, L., Ohnishi, T. and Long, M., Proton and carbomagnetic resonance studies of the synthetic poly-pentapeptide of elastin. J. Mol. Biol. 96 101.

Zeiger, A., Lange, A. and Maurer, P. (1973) Synthesis of two sequential polypeptides by dispersion in benzene and their circular dichroism spectra in aqueous solution: poly (L-Glu-L-Lys-L-Ala-Gly) and Poly (L-Ala-D-Glu- L-Lys-D-Ala-Gly). Biopolymers 12 2135.

30

CONFORMATIONAL PROPERTIES OF POLYPEPTIDE MODELS OF COLLAGEN

Rajendra S. Bhatnagar and Rao S. Rapaka
University of California, San Francisco
San Francisco, California 94143   and
V.S. Ananthanarayanan
Indian Institute of Science
Bangalore, India

ABSTRACT

Individual collagen chains exist as Polypro II helices because of their large imino content and their super-coiling into the collagen triple helix is facilitated by Gly in every third position. Because of this, collagen may be considered as being made up of Gly-led triplets. One fourth of such triplets in collagen have the sequence Gly-Pro-X and another one fifth, Gly-X-Hyp, where X is an α-amino residue. The stereochemical properties of the imino peptide bond, the position of the imino residue in the sequence and its interactions with neighboring residues, determine the conformation. Synthetic polytripeptides of sequence $(Gly\text{-}Pro\text{-}X)_n$ generate collagen-like conformations in aqueous solutions whereas $(Gly\text{-}X\text{-}Pro)_n$ sequences usually do not. This difference has been attributed to different H-bonding properties of X-NH in the two sequences. Stabilizing interactions are said to occur between the side chain of X and Pro ring atoms in X-Pro but not in Pro-X. We investigated this question using $(Gly\text{-}Pro\text{-}Sar)_n$ (I) and $(Gly\text{-}Sar\text{-}Pro)_n$ (II). (I) generated collagen-like conformations in aqueous solutions but (II) did not although some order was elicited in helix promoting solvents. Since NH-H bonds are not possible at Sar residues, they may not play a final role in stabilizing the collagen helix. It appears likely that non-bonding interactions of the imino residue with the residue on its C-terminal may play a significant role in stabilizing collagen-like conformations.

INTRODUCTION

Collagen, the principal skeletal protein in the body is eminently well-equipped for its various structural and mechanical functions by virtue of a unique, highly ordered triple helical conformation, aggregation properties, and an unusual amino acid composition expressed in a polymer-like sequence. While co-valent cross-links play a significant role in defining the ultimate properties of large polymeric aggregates of collagen, the individual triple-helices derive their stability from the unusual features of the amino acid composition and sequence. Interactions between neighboring residues in the sequence play a major role in defining the conformational features of collagen. Our discussion concerns some of these interactions.

SIGNIFICANT FEATURES OF THE COLLAGEN SEQUENCE

Collagen has a very high content of glycine, which accounts for one-third of the total amino acid residues and appears in every third position in the helical region extending over 10 per-cent of the length of the molecule. Each chain in the triple helix can thus be regarded as being made up of glycine-led triplets and may therefore be considered to be a polymer $(Gly-2-3)_n$. Another major difference between the composition of collagen and most other proteins is the unusually high content of the imino acid residues proline and hydroxyproline. These account for one-fourth of the total residues in collagen and appear in over half of the glycine-led triplets. Proline occurs in the second position, with an α-amino acid (A) as the third position, as -(Gly-Pro-A)- in over one fourth of the triplets. Another one fifth of the triplets contain an α-amino acid residue in the second position followed by an imino residue, usually hydroxy-proline, as -(Gly-A-Hyp)-. Nearly one-tenth of the triplets con-tain two imino residues, -(Gly-Pro-Hyp)-. The placement of the imino residue in relation to the α-amino residue in a triplet, profoundly affects the conformational features of the sequence.

STABILIZATION OF THE COLLAGEN HELIX AND ITS RELATION TO THE SEQUENCE

Homopolymers of glycine, $(Gly)_n$ and proline, $(Pro)_n$ form helical structures which share many similarities. Because of the preponderance of these residues in collagen, collagen helices may be expected to be similar to $(Gly)_n$ and $(Pro)_n$ helices. In fact the individual helices of collagen bear strong conformational resemblance to these helices but differ in the fact that unlike the homopolymers, they exist in a triple helical "coiled-coil" conformation. Super-coiling of three peptide chains into the triple-helix is facilitated by the presence of glycine in every third position (Ramachandran and Ramakrishnan, 1976). The glycine residues stack at the axis of the triple-helix. The triple helix is stabilized by interchain hydrogen bonds, involving the -NH groups of glycine and other α-amino acids in the sequence.

Factors other than hydrogen bonding play an important role in the stability of the collagen helix since imino residues which constitute nearly one fourth of all residues, lack free -NH groups and consequently cannot participate in hydrogen bonds. Unlike other helical forms which are stabilized to a large extent by hydrogen bonds, the collagen helix derives its conformational stability from the stereochemical properties of the imino peptide bonds and the interactions of the imino ring with neighboring residues.

CONFORMATION DIRECTING ROLE OF THE IMINO RESIDUES

In a peptide bond involving an imino residue, the N-C bond is part of a rigid five-membered ring and therefore does not have any rotational freedom. A major consequence of this is the increased rigidity in the immediate vicinity of an imino residue. This restriction stabilizes the conformation of collagen and related helices (Ramachandran and Ramakrishnan, 1976) Interactions of proline ring atoms with neighboring residue side-chains may impose further restrictions on the conformation of proline-containing peptide sequences (Schimmel and Flory, 1968). Thus the placement of an imino residue in relation to an optically active α-amino acid residue in the triplet sequence may have a profound effect on conformational stability.

STUDIES ON COLLAGEN-LIKE POLYPEPTIDES

In order to determine the relative contributions of different features of its composition and sequence to the stability of the collagen helix, it is necessary to examine the conformational properties of polymeric polypeptides which mimic in a reiterative manner the collagen sequence under examination. The repeating nature of polymeric polypeptides serves to amplify the conformational features of the monomeric sequence.

Conformational studies have shown that polymers of glycine-led triplets which do not contain an imino acid $(Gly-A_1-A_2)_n$, do not generate collagen-like conformations (Anderson et al, 1970; Andries et al, 1971; Andries and Walton, 1971; Doyle et al, 1970). Many such triplets are interspersed in the collagen sequence and in the native protein; They conform to the overall collagen helix. Structural studies on the polyhexapeptide (Gly-Ala-Ala-Gly-Pro-Pro)$_n$ showed that the -Gly-Ala-Ala- sequence is

capable of existing in the collagen-like helical state since the
overall conformation of the polyhexapeptide is essentially similar
to that of collagen (Segal et al, 1969).   (Gly-Ala-Ala)$_n$ by
itself generates predominantly the β-conformation (Oriel and
Blout, 1966).   Since (Gly-Pro-Pro)$_n$ generates very stable collagen-like helices (discussed below), the conformational features
of the -Gly-Pro-Pro- sequence appear to override the conformational
propensities of -Gly-Ala-Ala- in the above hexapeptide sequence
and in general this may be considered to be the case wherever
sequences which generate collagen-like conformations, exist in
the vicinity of sequences which by themselves may not acquire
such conformations.

The conformational roles of glycine and imino residues are
best demonstrated by the polytripeptide (Pro-Pro-Gly)$_n$ which
generates collagen-like triple helical aggregates in the solid
state (Yonath and Traub, 1969; Okuyama *et al*, 1972) as well as
in aqueous solutions (Engel *et al*, 1966; Kobayashi *et al*, 1970;
Brown *et al*, 1972b).   The overriding conformational influences
of proline were also observed in the polymer (Pro-Pro-β-Ala)$_n$
which generates collagen-like conformations in aqueous solutions
(Bhatnagar and Rapaka, 1975).

The stereochemical features of imino residues suggest that
their presence in a peptide sequence may impart at least a
tendency to generate the polyproline-collagen helix in their
immediate vicinity.   Examination of the available data suggests
that the presence of an imino residue in the glycine-led triplet
sequence permits the formation of collagen-like structures in
the solid state in many polymers, however, the stability of the
triple-helical conformation in the aqueous milieu is dependent
on the placement of the α-amino residue in relation to the
imino residue in the sequence (for detailed reviews, see
Traub and Piez, 1971; Walton and Blackwell, 1973; Fraser and
McRae, 1973; Bhatnagar and Rapaka, 1976; Ramachandran and
Ramakrishnan, 1976).   An important fact that emerges from studies
on polytripeptide models of collagen is that interactions of the
imino ring with the side chains of neighboring residues may
play a significant role in stabilizing the collagen-like triple
helix.

Polytripeptides of the sequence (Gly-Gly-Pro)$_n$ (which may
also be expressed as (Gly-Pro-Gly)$_n$), lack side chains on the
α-amino residues.   (Gly-Gly-Pro)$_n$ form ordered aggregates in
the solid state, but these bear very little resemblance to the
collagen triple-helix (Traub, 1969).   The chains form hydrogen-
bonded sheets of polyproline II type helices rather than triple
helices.   In aqueous solutions also the polymer exhibits only

limited resemblance to the collagen conformation, although the chains do form aggregates (Oriel and Blout, 1966).

In contrast to $(Gly-Gly-Pro)_n$ when the tripeptide sequence includes an α-amino residue with as side chain, structures with greater resemblance to collagen are generated. Both $(Gly-Ala-Pro)_n$ and $(Gly-Pro-Ala)_n$ exist in the solid state in several forms including the collagen triple helix (Segal and Traub, 1969; Schwartz et al, 1970; Doyle et al, 1971; Brown et al, 1972 a,b). Similar observations were reported by Scatturin et al (1975) for $(Gly-Leu-Pro)_n$ and $(Gly-Pro-Leu)_n$. Both polymers exist in the solid state as collagen-like triple helices. Traub and Piez (1971) reported that $(Gly-Ser-Pro)_n$ may exist in the solid state as polyproline II helices whereas $(Gly-Pro-Ser)_n$ forms triple helical aggregates. The solid state structures of several poly-tripeptide models of collagen are enumerated in Table I.

SOLUTION CONFORMATIONS OF POLYTRIPEPTIDE MODELS OF COLLAGEN

While the presence of a glycine and a proline residue in the monomeric tripeptide ensures the potential for generating the triple-helix in the polymer analogue of collagen, stability of the triple helix is governed by factors which are poorly understood. Differences in the stability of the collagen-like triple helical conformation of different polytripeptide analogues of collagen become obvious when the polymers are exposed to various solvents (Table II). These differences are manifested in the optical rotatory dispersion and circular dichroism spectra of the polymers. The circular dichroism spectra of collagen (Carver and Blout, 1967; Tiffany and Krimm, 1972) and several poly-tripeptide analogues including $(Gly-Pro-Pro)_n$ (Engel et al, 1966; Kobayashi et al, 1970; Brown et al, 1972b); $(Gly-Ala-Pro)_n$, (Doyle et al, 1971); $(Gly-Pro-Ala)_n$ and $(Gly-Pro-Ser)_n$ (Brown et al, 1972b) indicate that the major features of the circular dichroism of the collagen-like triple-helix include a large negative ellipticity in the 200 nm region and a positive peak centered around 220 nm. Characteristically, polymers which retain triple-helical order in solution exhibit these features, whereas polytripeptides which do not exist as similarly ordered aggregates in solution, do not show these features. The 200 nm trough and 220 nm peak are observed in $(Gly-Pro-Ala)_n$ in solution (Brown et al, 1972b) whereas $(Gly-Ala-Pro)_n$ which exists in the solid state in a triple helical form, loses this structure in aqueous solution as evidenced by the lack of these features. In the case of $(Gly-Ala-Pro)_n$, even the exposure of triple-helical polymer in the solid state, to water vapor results in the loss of helical order. Similar differences in helical stability in solutions have been shown for

Table I

SOLID STATE CONFORMATIONS OF POLYTRIPEPTIDE MODELS OF COLLAGEN

Tripeptide Sequence[a]	Conformation	References
Gly-Gly-Pro	Polyproline II	Traub, (1969)
Gly-Ala-Pro[b]	Polyproline II	Segal and Traub, (1969)
Gly-Ala-Pro[c]	Polyproline II	Doyle *et al*, (1971)
Gly-Ala-Pro[d]	Triple Helix	Doyle *et al*, (1971)
Gly-Pro-Ala	Triple Helix	Traub and Yonath, (1967)
Gly-Leu-Pro	Triple Helix	Scatturin *et al*, (1975)
Gly-Pro-Leu	Triple Helix	Scatturin *et al*, (1975)
Gly-Pro-Lys	Triple Helix	Traub *et al*, (1969)
Gly-Ser-Pro	Polyproline II	Traub and Piez, (1971)
Gly-Pro-Ser	Triple Helix	Brown *et al*, (1972b)
Gly-Pro-Phe	Triple Helix	Tamburro *et al*, (1967)
Gly-Pro-Pro	Triple Helix	Yonath and Traub, (1969)
Gly-Pro-Hyp	Triple Helix	Rogulenkova *et al* (1964)
Gly-Hyp-Pro	Triple Helix	Andreeva *et al* (1967)
Gly-Hyp-Hyp	Triple Helix	Andreeva *et al* (1970)

[a]In order to maintain uniformity, all collagen-related sequences
are listed as polymers of glycine-led triplets, regardless of
the starting monomers which may have been used to synthesize
the polymer.

[b](Gly-Ala-Pro)$_n$ forms different types of aggregate depending on
the last solvent to which the polymer was exposed. Solvent in
b = water

[c]solvent = organic

[d]solvent = trifluoracetic acid

Table II

SOLUTION CONFORMATIONS OF POLYTRIPEPTIDE MODELS OF COLLAGEN

Tripeptide Sequence	H$_2$O	Glycol[1]	Solvent HFIP[2]	TFE[3]	References
Gly-Gly-Pro	-			*	Oriel & Blout, 1966
Gly-Ala-Pro	-		±	±	Doyle *et al*, 1971
Gly-Pro-Ala	+		±	±	Brown *et al*, 1972a,b
Gly-Leu-Pro	-	±	±		Scatturin *et al*, 1975
Gly-Pro-Leu	±	+	±		Scatturin *et al*, 1975
Gly-Ser-Pro	-	±			Brown *et al*, 1972a,b
Gly-Pro-Ser	±		-	-	Brown *et al*, 1972a,b
Gly-Phe-Pro				±	Brahmachari *et al*, 1977
Gly-Pro-Pro	+				Engel *et al*, 1966
Gly-Sar-Pro	-	±	±	±	Ananthanarayanan *et al*, 1976
Gly-Pro-Sar	±	+	+	+	Ananthanarayanan *et al*, 1976
β-Ala-Pro-Pro	+				Bhatnagar & Rapaka, 1975

[1]glycol = ethylene glycol

[2]HFIP = hexafluoroisopropanol

[3]TFE = trifluoroethanol

+ = Triple helical

- = not triple helical

± = partially ordered conformations

* (Gly-Gly-Pro)$_n$ showed evidence for increased order in 1.4 M acetic acid.

the isomeric polytripeptide pairs (Gly-Ser-Pro)$_n$ and (Gly-Pro-Ser)$_n$ (Brown *et al*, 1972b) and (Gly-Leu-Pro)$_n$ and (Gly-Pro Leu)$_n$ (Scatturin *et al*, 1975).

We synthesized (Gly-Sar-Pro)$_n$ and (Gly-Pro-Sar)$_n$ as models for collagen conformation (Rapaka and Bhatnagar, 1975). Examination of the solution conformations of these polymers showed that while (Gly-Sar-Pro)$_n$ lacks helical order in aqueous solutions, (Gly-Pro-Sar)$_n$ exists in solution as triple helical aggregates (Ananthanarayanan *et al*, 1976a). The circular dichroism spectrum of (Gly-Sar-Pro)$_n$ (Fig. 1) lacked the positive peak at 220 nm and the trough located at 200 nm ($\theta \simeq -13,500$). No differences in the magnitudes of either the trough or the peak were observed on heating the solution to 90° indicating that no loss of order occurred. In contrast, the circular dichroism spectrum of (Gly-Pro-Sar)$_n$ (Fig. 2) show a peak at 220 nm with a magnitude of θ = +3000 and a trough at 198 nm with $\theta \simeq -15,500$. As seen in Fig. 2, the magnitudes of both the trough at 198 nm and peak at 220 nm were decreased on heating to 90°. Brown *et al* (1972a) have used the magnitude of the 220 nm peak as an index of triple-helical order in following the helix $\rightleftarrows$ coil transition in solution. According to this criterion, (Gly-Pro-Sar)$_n$ exists in ordered state at the lower temperature, and on heating its triple-helix is destroyed. Additional evidence that this polymer is collagen-like in aqueous solution is the position of the negative circular dichroism at 198 nm which nearly coincides with that for collagen (Tiffany and Krimm, 1972; Mandel and Holzwarth, 1973).

Our studies suggested that the sarcosine residue resembles the α-amino residues alanine, serine and leucine in imparting different conformational stability to the polytripeptide depending on its position in the triplet, in relation to the proline residue. Sarcosine on the N-terminal side of the proline residue acts like the α-amino residues in generating a triple helix of low stability. In contrast, as seen with the α-amino acids, sarcosine imparts a high degree of conformational stability to the triple helix when it is present on the C-terminus of the proline residue.

As discussed above, the collagen triple helix is stabilized by imino peptide bonds, the presence of glycine in every third position and by hydrogen bonds. The first two features are shared by the (Gly-A-Pro)$_n$ and (Gly-Pro-A)$_n$ polymers and in explaining the differences in the stabilities of the collagen-like triple helices of (Gly-Ala-Pro)$_n$ and (Gly-Pro-Ala)$_n$, Doyle *et al* (1971) considered the configuration of the alanine -NH group in the two polytripeptides. In the triple helix of (Gly-Ala-Pro)n, the alanine -NH group would be pointed inward

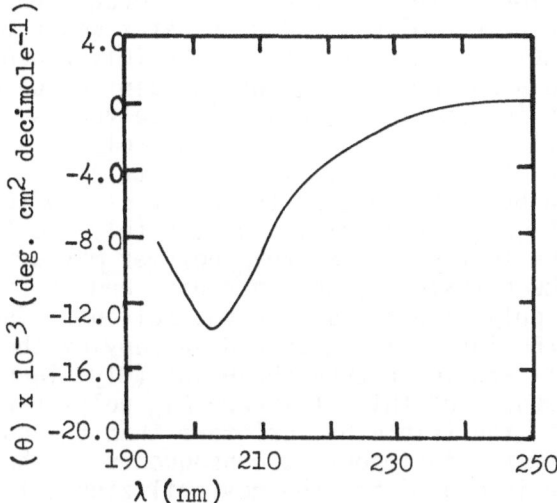

Fig. 1: CIRCULAR DICHROISM OF (GLY-SAR-PRO)$_n$ IN WATER
     The polymer ($\underline{M}$ = 23,250) was dissolved in water to a con-
centration of 1.0 $\overline{m}$g/ml. The spectrum shown was recorded at 25°.
The CD remained unaltered on heating the solution to 90°.

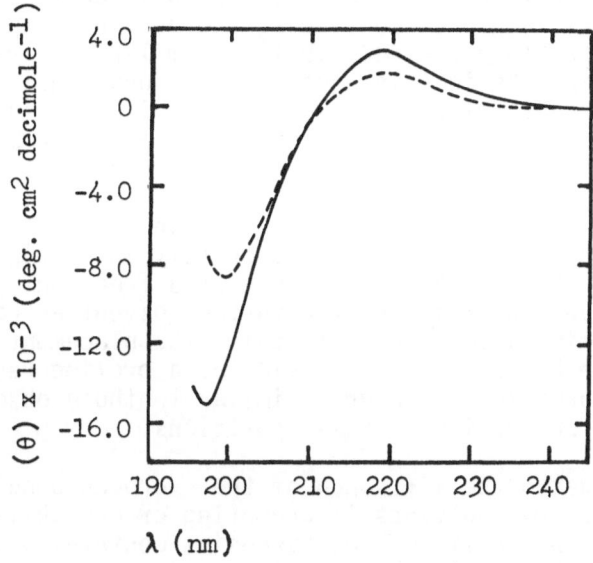

Fig. 2: CIRCULAR DICHROISM OF (GLY-PRO-SAR)$_n$ IN WATER
     The concentration of the polymer ($\underline{M}$ = 22,000) was 1.0 mg/ml.
CD at 25° ——————— ;   CD at 90° --------

towards the interior of the helix and its hydrogen-bonding interactions with polar solvents such as water would tend to disrupt the triple-helix. In contrast, in the (Gly-Pro-Ala)$_n$ triple helix, the alanine -NH group would point outwards from the helix and its interaction with a polar solvent would not have the same helix disrupting effect. The role of solvent-polypeptide interactions in stabilizing the triple helix was further discussed by these workers (Brown et al, 1972b). They suggested that the greater stability of the (Gly-Pro-A)$_n$ polymer in comparision to the (Gly-A-Pro)$_n$ polymer may arise from differences in the enthalpies, and entropies and the interactions of these polytripeptides with the solvent. They determined that the internal energy and entropy of the (Gly-Pro-Ala)$_n$ triple helix are lower than those of (Gly-Ala-Pro)$_n$ triple helix. Because of this, (Gly-Pro-A)$_n$ polymer would be more restricted to the triple helical form than the corresponding (Gly-A-Pro)$_n$ polymer. An important consequence of these differences would be in minimizing the destabilizing effect of solvent-polymer interactions. Brown and coworkers (1972a) examined the role of solvent in stabilizing the collagen-like helices of (Gly-Pro-Ala)$_n$ and (Gly-Pro-Ser)$_n$, using solvents of different hydrogen bonding strengths. The helix was disordered in strong hydrogen bond forming solvents such as hexafluoro-isopropanol and trifluoroacetic acid. Weak hydrogen bond forming solvents such as polyhydric alcohols promoted helix formation. The polymer (Gly-Ala-Pro)$_n$ which is unstructured in aqueous solutions, acquired considerable order in the presence of ethylene glycol (Doyle et al, 1971). Similarly (Gly-Leu-Pro)$_n$ which is not triple helical in aqueous solutions acquires greater order in the presence of polyhydric alcohols (Scatturin et al, 1975). The effects of various solvents on the conformation of several collagen-like polypeptides are listed in Table II.

The role of solvent -NH interactions in stabilizing the helix is not supported by our studies on (Gly-Sar-Pro)$_n$ and (Gly-Pro-Sar)$_n$. Because of the absence of a free -NH group, the sarcosine residue cannot interact with the solvent as has been suggested for amino acids. Despite this, the placement of sarcosine on the N- or C- terminal side of a proline residue results in conformational features similar to those observed with alanine, serine or leucine in these positions.

We investigated the influence of the hydrogen-bonding properties of several solvents in promoting or destabilizing the collagen-like triple-helix of the sarcosine-containing collagen analogues. (Gly-Sar-Pro)$_n$ showed evidence for much greater order in ethylene glycol than in water and this was confirmed by the loss of helical order as heating as seen by a decrease in the magnitude of the 200 nm trough (Fig. 3). These observations

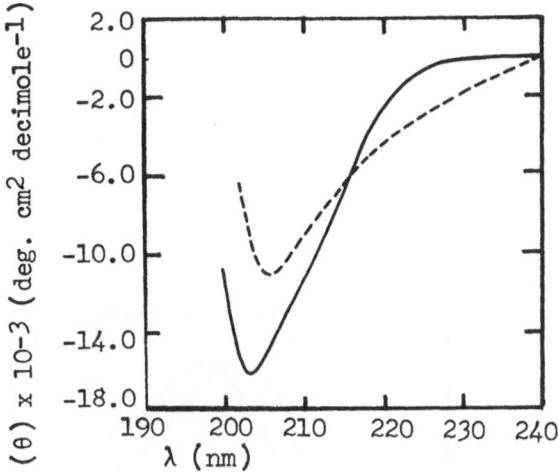

Fig. 3: INCREASED ORDER IN (GLY-SAR-PRO)$_n$ IN ETHYLENE GLYCOL
    The polymer was dissolved to a final concentration of
1 mg/ml in ethylene glycol and spectra were recorded at 25°
——————— and at 95°-------.

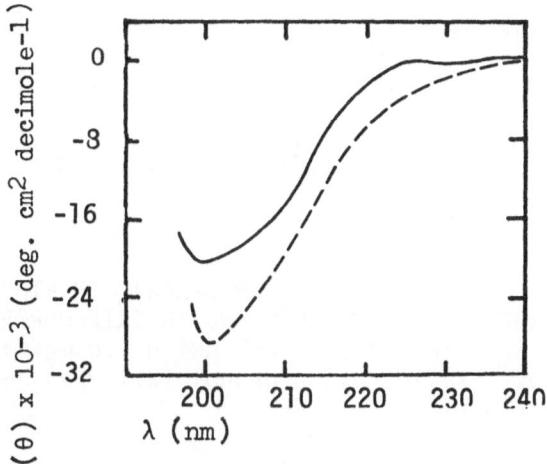

Fig. 4: CIRCULAR DICHROISM OF (GLY-PRO-SAR)$_n$ IN ETHYLENE GLYCOL-
        HEXAFLUOROISOPROPANOL AND IN TRIFLUOROETHANOL
    The spectra were recorded at 25° using 1.o mg/ml polypeptide
in ethylene glycol-hexafluoroisopropanol (2:1. v/v) ———————
and in trifluoroethanol -------.

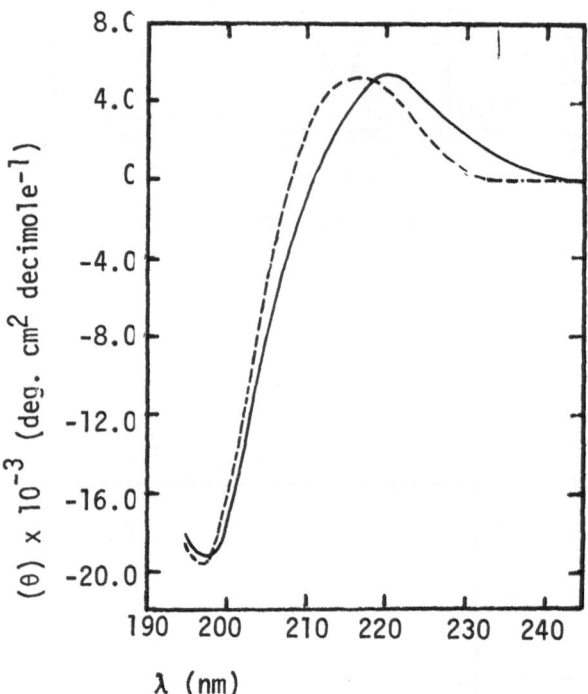

Fig. 5: CIRCULAR DICHROISM OF (GLY-PRO-SAR)$_n$ IN ETHYLENE GLYCOL-
              HEXAFLUOROISOPROPANOL AND IN TRIFLUOROETHANOL
        The spectra were recorded at 25° using 1.o mg/ml polymer
in ethylene glycol-hexafluoroisopropanol (2:1 v/v) ─────────
and in trifluoroethanol ----------.

parallel the observations on (Gly-Ala-Pro)$_n$ (Doyle *et al*, 1971) and (Gly-Leu-Pro)$_n$ (Scatturin *et al*, 1975). In contrast to the observations on the alanine and leucine polymers, however, we observed greater order in (Gly-Sar-Pro)$_n$, in trifluoroethanol than in a mixture of ethylene glycol with hexafluoroisopropanol, as seen by the greater magnitude of the trough at 200 nm. The ellipticity in trifluoroethanol was θ = -28,500 and in ethylene glyco-hexafluoropropanol, θ = -20,250 (Fig. 4). The isomeric ordered polymer (Gly-Pro-Sar)$_n$ was not soluble to an appreciable extent in ethylene glycol, however, it also showed greater order in trifluoroethanol and ethylene glycol-hexafluoroisopropanol than in water (Fig. 5). The spectra in the two solvents differed very little. Our data in organic solvents show an opposite trend to that observed in the case of the alanine, serine and leucine. The correlation between the hydrogen-bonding strength of the solvent and the triple helix in solution does not appear to hold in the case of the sarcosine polymers. While our observations suggest a role for solvent polymer interaction in stabilizing the triple-helix, firm conclusions concerning the nature of such interactions cannot be made at present.

ROLE OF THE INTRAHELICAL INTERACTIONS IN THE STABILITY OF THE COLLAGEN TRIPLE HELIX

As discussed above, the collagen triple helix derives its conformational features from the stereochemical restrictions imposed by the iminopeptide bonds and the presence of glycine in every third position. The three chains are stabilized together by a system of intra- and interchain -NH---O=C hydrogen bonds. In the various models proposed for collagen (Ramachandran and Ramakrishnan, 1976) either one or two such hydrogen bonds may be formed for every three residues. In the two bonded structure one of the interchain hydrogen bonds is formed directly between the backbone atoms of neighboring chains. The other hydrogen bond is formed between the -NH of the first peptide unit and a water molecule which is also hydrogen bonded to a C=O group on the neighboring chain. In polytripeptides where there is only one -NH group available for hydrogen bonding, as in the case of (Gly-Pro-Pro)$_n$ the hydrogen-bonded water may play a stabilizing role (Traub and Yonath, 1966).

In the case of (Gly-Sar-Pro)$_n$ and (Gly-Pro-Sar)$_n$, the same number of hydrogen bonding sites are available as in (Gly-Pro-Pro)$_n$, however, only (Gly-Pro-Sar)$_n$ exhibits conformational resemblance ot (Gly-Pro-Pro)$_n$. It appears thus that inter- and intrachain hydrogen-bond formation may not fully explain the differences in the stabilities of (Gly-Pro-Sar)$_n$ and (Gly-Sar-Pro)$_n$.

The conformational role of interactions of the proline peptide unit with neighboring residue has been considered by several workers. Schimmel and Flory (1968) calculated conformational energies of proline peptides and concluded that proline ring may interact with the side chain of the α-amino acid residue preceding it in the sequence, resulting in restriction of the possible conformations to a range which includes collagen-like structures. No such interactions occur with the residue succeeding the proline residue. These workers also concluded that there are no significant interactions between the proline ring and residues not immediately adjacent. It is difficult to reconcile these conclusions with the conformational features of (Gly-A-Pro)$_n$ and (Gly-Pro-A)$_n$ sequences which appear to imply stabilizing interactions between the proline residue and the side chain of the succeeding residue. This is seen in the generation of more stable collagen-like conformations when there is an α-amino residue other than glycine present on the C-terminus of the proline, but not on its N-terminal side. Such interactions may presumably be more prominent in peptides where the proline residue is followed by another proline residue, or a sarcosine residue, each of which has an N-substituent.

Further evidence for the role of non-bonded interactions of the proline peptide bond in generating and stabilizing specific helical forms was provided by Leach *et al*, (1966). These workers determined that interaction may occur between the bulky C^δ atom of the proline ring with the amide O atom of the third previous residue. Leach and coworkers suggested that in peptides containing several proline residues, such interactions may stabilize collagen-like conformations. In this connection, the observations of Young and Deber (1975) are also pertinent. These workers observed that amino acid substituents may influence molecular rotations several bonds away and they suggested that these interactions may be of significance in generating and stabilizing protein conformations.

CONCLUSIONS

The unique triple-helical conformation of collagen is stabilized by a number of factors inherent in its sequence. Studies on polypeptide models of collagen show that the presence of imino residues and the occurrence of glycine in every third position in the sequence permits the generation of triple-helices but the stability of the helix appears to depend on the position of the α-amino residue in relation to the proline residue, in the glycine-led triplet. It has been suggested that the triple-helical aggregates are stabilized by -NH--O=C hydrogen bonds.

Our studies using sarcosine containing polytripeptide models of collagen indicate that previous conclusions emphasizing the role of solvent-Polypeptide interactions in stabilizing collagen-like triple helices, need to be modified. Our studies with the sarcosine-containing collagen analogues suggest that non-bonded interactions of the imino peptide bond may play a significant role in the stability of the triple helix. This conclusion is in accord with previous suggestions that certain interactions of the proline ring may restrict the conformational range of a peptide sequence to the collagen conformation. The nature of such non-bonded interactions remains to be elucidated.

ACKNOWLEDGEMENTS

The authors wish to express their sincere thanks to Prof. G.N. Ramachandran for his counsel and continuous interest in these studies. One of the authors (RSB) is especially grateful to Prof. Ramachandran for introducing him to the complexity of the collagen structure during a fruitful sabbatical year. This research was supported by NIH Grants DE-03861 (in San Francisco) and AM-15964 (in Bangalore) and a grant from the Council of Scientific and Industrial Research, India. Dr. Rapaka's present address is Laboratory of Molecular Biophysics, University of Alabama Medical Center, Birmingham, Alabama 35294. The authors would also like to thank Ms. Sandra Hodess for her kind and patient assistance in the preparation of this manuscript. Figures 1-5 have been redrawn by permission from Biopolymers.

REFERENCES

Ananthanarayanan, V.S., Brahmachari, S.K., Rapaka, R.S. and Bhatnagar, R.S. (1976). Polypeptide models of collagen: Solution properties of $(Gly-Pro-Sar)_n$ and $(Gly-Sar-Pro)_n$. Biopolymers 15, 707.

Anderson, J.M., Rippon, W.B. and Walton, A.G. (1970). Model tripeptides for collagen. Biochem. Biophys. Res. Comm. 39,802.

Andreeva, N.S., Esipova, N.G., Millionova, M.I., Rogulenkova, V.N. and Shibnev, V.A. (1967). Polypeptides with regular sequences of amino acids as the models of collagen structure. in Conformation of Biopolymers, G.N. Ramachandran, ed., Academic Press, N.Y., V. 2, p. 469.

Andreeva, N.S., Esipova, N.G., Millionova, M.I., Rogulenkova, V.N., Tumanyan, V.G. and Shibnev, V.A. (1970). Synthetic regular polytripeptides and proteins of collagen class. Biofizika 15, 198.

Andries, J.C. and Walton, A.G. (1971). The morphology of poly (Gly-Ala-Glu(OEt)). J. Mol. Biol. 56, 515.

Andries, J.C., Anderson, J.M. and Walton, A.G. (1971). Morphological and structural studies of poly (Gly-Gly-Ala). Biopolymers 10, 1049.

Bhatnagar, R.S. and Rapaka, R.S. (1975). Polypeptide models of collagen: Properties of (Pro-Pro-β-Ala)$_n$. Biopolymers 14, 597.

Bhatnagar, R.S. and Rapaka, R.S. (1976). Synthetic polypeptide models of collagen: Synthesis and applications in Biochemistry of Collagen, G.N. Ramachandran and A.H. Reddi, eds., Plenum Press, N.Y., p. 479.

Brahmachari, S.K., Ananthanarayanan, V.S., Rapaka, R.S. and Bhatnagar, R.S. (1977). The role of phenylalanine in collagen: Solution studies of (Gly-Phe-Pro)$_n$. in press.

Brown, F.R. III, DiCorato, A., Lorenzi, G.P. and Blout, E.R. (1972a). Synthesis and structural studies of two collagen analogues: Poly (L-prolyl-L-seryl-glycyl) and Poly (L-prolyl-L-alanyl-glycyl). J. Mol. Biol. 63, 85.

Brown, F.R. III, Hopfinger, A.J. and Blout, E.R. (1972b). The collagen-like triple helix to random chain transition: Experiment and theory. J. Mol. Biol. 63, 101.

Doyle, B.B., Traub, W., Lorenzi, G.P. and Blout, E.R. (1971). Conformational investigations on the polypeptide and oligopeptides with the repeating sequence L-alanyl-L-prolyl glycine. Biochemistry 10, 3052.

Doyle, B.B., Traub, W., Lorenzi, G.P. and Blout, E.R. (1971). Synthesis and structural investigations of poly (L-alanyl-L-alanyl-glycine). J. Mol. Biol. 51, 47.

Engel, J., Kurtz, J., Katchalski, E. and Berger, A. (1966). Polymers of tripeptides as collagen models II: Conformational changes of poly(L-prolyl-glycyl-L-prolyl) in solution, J. Mol. Biol. 17, 255.

Fraser, R.D.B. and MacRae, T.P. (1973). Conformation in fibrous proteins and related synthetic polypeptides. Academic Press, N.Y.

Kobayashi, Y., Sakai, R., Kakiuchi, K. and Isemura, T. (1970). Physicochemical analysis of (Pro-Pro-Gly)$_n$ with defined molecular weight-temperature dependence of molecular weight in aqueous solution. Biopolymers 9, 415.

Leach, S.J., Nemethy, G. and Scheraga, H.A. (1966). Computation of the sterically allowed conformations of peptides. Biopolymers 4, 369.

Mandel, R. and Holzwarth, E. (1973). Ultraviolet circular dichroism of polyproline and oriented collagen. Biopolymers 12, 655.

Okuyama, K., Tanaka, N., Ashida, T., Kakudo, M., Sakakibara, S. and Kishida, Y. (1972). An X-ray study of the synthetic polypeptides (Pro-Pro-Gly)$_{10}$. J. Mol. Biol. 72, 571.

Oriel, P.J. and Blout, E.R. (1966). On the structure of Gly-Pro-Gly and Gly-Pro-Ala oligopeptides and sequential polypeptides. J. Am. Chem. Soc. 88, 2041.

Ramachandran, G.N. and Ramakrishnan, C. (1976). Molecular Structure, in Biochemistry of Collagen, G.N. Ramachandran and A.H. Reddi, eds., Plenum Press, N.Y. p. 45.

Rogulenkova, V.N., Millionova, M.I. and Andreeva, N.J. (1964). On the close-structural similarity between poly-Gly-L-Pro-L-Hypro and Collagen. J. Mol. Biol. 3 483.

Scatturin, A., Tamburro, A.M., Del Pra, A. and Bordignon, E. (1975). Conformational studies on sequential polypeptides. Part IV: Structural investigations on (Pro-Leu-Gly)10, (Pro-Leu-Gly)n and (Leu-Pro-Gly)n. Int. J. Peptide Protein Res. 7, 425.

Schimmel, P.R. and Flory, P.J. (1968). Conformational energies and configurational statistics of copolypeptides containing L-proline. J. Mol. Biol. 34, 104.

Schwartz, A., Andries, J.C. and Walton, A.G. (1970). Structural and morphological investigations of poly (Gly-Ala-Pro). Nature 226, 161.

Segal, D.M. and Traub, W. (1969). Polymers of tripeptides as collagen models VI: Synthesis and structural investigation of poly (L-alanyl-L-prolyl-glycine). J. Mol. Biol. 43, 487.

Segal, D.M., Traub, W. and Yonath, A. (1969). Polymers of tripeptides as collagen models VIII: X-rays studies of four polyhexapeptides. J. Mol. Biol. 43, 519.

Tamburro, A.M., Scatturin, A., Marchiori, F. (1968). Conformational studies of sequential polypeptides III. Synthesis of poly (L-prolyl-L-phenylalanylglycine). Gazz. Chim. Ital. 98, 638.

Tiffany, M.L. and Krimm, S. (1972). Effect of temperatures on the circular dichroism spectra of polypeptides on the extended state. Biopolymers 11, 2309.

Traub, W. (1969). Polymers of tripeptides as collagen models V. An X-ray study of poly (L-prolyl-glycyl-glycine). J. Mol. Biol. 43, 479.

Traub, W. and Piez, K.A. (1971). The chemistry and structure of collagen. Adv. Protein Chem. 25, 243.

Traub, W. and Yonath, A. (1967). Polymers of tripeptides as collagen models III. Structural relationship between two forms of poly (L-prolyl-L-alanyl-glycine). J. Mol. Biol. 25, 351.

Traub, W., Yonath, A. and Segal, D.M. (1969). On the molecular structure of collagen. Nature 221, 914.

Walton, A.G. and Blackwell, J. (1973) Biopolymers, Academic Press, N.Y.

Yonath, A. and Traub, W. (1969). Polymers of tripeptides as collagen models. IV. Structure analysis of poly (L-Pro-Gly-L-Pro). J. Mol. Biol. 43, 461.

Young, P.E. and Deber, C.M. (1975). Long-range steric effects on rotational barriers in peptide chains. Biopolymers 14, 1547.

31

IONIZING RADIATION-INDUCED CROSSLINKING IN PROTEINS

Osamu Yamamoto

Research Institute for Nuclear Medicine and Biology

Hiroshima University, Kasumi 1-2-3, Hiroshima, Japan

INTRODUCTION

A number of reports describe effects of ionizing radiation on proteins both in solid state and in solution. Such studies have been done mainly in relation to inactivation of enzymes (Augenstein, 1962, 1964; Luse, 1964; Dale, 1966; Garrison, 1968). Recently Grossweiner (1976) reviewed photochemical and radiochemical reactions for enzyme inactivation. These reports discuss aggregation, chain cleavage, opening and interchange of disulfide linkages, production and disappearance of sulfhydryl groups, decomposition of amino acid residues, and their negative responses. The diversity of the results presented makes it difficult to generalize about mechanism of radiation-induced transformation. Various factors are involved in the radiation-induced crosslink-formation in proteins.

ESR techniques have been useful in determining the radical species produced in proteins by radiation, especially in the solid state. Pulse radiolysis is used to study the nature of species derived from water that are important radiation chemistry of proteins. Ultracentrifugal sedimentation analysis and gel filtration have also provided much information on protein crosslinking.

Large doses of radiation, too high to study biological effects, have often been used in studies of radiation-induced changes in proteins. The results obtained have proven useful in radiation food chemistry. At the present time, radiation is used to sterilize foods. Problems in food hygiene caused by radiation may develop in the future.   The author hopes that this report will contribute to an understanding of radiation food chemistry.

EVIDENCE OF RADIATION-INDUCED AGGREGATION IN PROTEINS

Many workers have reported on protein crosslinking or aggregation formed by exposure to ionizing radiation (TABLE 1).   As early as 1915 Fernau and Pauli presented a report on coagulation of proteins by exposure to radiation from radium.   During the early stage, the detection of radiation-induced aggregates was confined to coagulation of proteins (Wels, 1923; Fernau and Spiegel, 1929; Arnow, 1935).   From 1915 it was thought that, due to aggregation, the stability of proteins with respect to denaturation was lowered by ionizing radiation.   However, studies showed that radiation-induced denaturation was due not only to aggregation but also to other factors such as peptide-chain cleavage, changes in amino acid composition, and destruction of secondary or tertiary structures.

Table 1.   Findings of Radiation-Induced Protein Aggregation

Protein	State	Radiation	Worker	
Albumin	solid	γ-ray	Alexander *et al.*	1956
		X-ray α-ray	Rosen *et al.*	1958
		electron (2 MeV)	Alexander *et al.*	1960
	solution	α-ray	Fernau and Pauli	1915
		α-ray	Wels	1923
		α-ray	Fernau and Spiegel	1929
		α-ray	Arnow	1935
		X-ray	Barron and Finkelstein	1952
		X-ray	Carroll *et al.*	1952
		X-ray α-ray	Alexander *et al.*	1956
		γ-ray	Rosen and Boman	1957
		γ-ray	Rosen *et al.*	1957
		X-ray	Rosen *et al.*	1958

		X-ray (140 kV)	Rosen	1959
		γ-ray	Hay and Zakrzewski	1968
		γ-ray	Radola	1968
Fibrinogen	solid	γ-ray	Koenig *et al.*	1960
	solution	X-ray (soft)	Koenig and Perrings	1952
		electron (1 MeV)	Sowinski *et al.*	1958 1959
		γ-ray	Koenig *et al.*	1960

Gamma-
  globulin
	solution	X-ray	Barron and
			  Finkelstein
				1952

		X-ray (220 kV)	Morton	1960
		X-ray (250 kV)	Luzzio	1963
Hemocyanin	solution	X-ray	Pickels and Anderson	1947
Lactate de- hydrogenase	solution	γ-ray	Winstead and Reece	1970
		X-ray (200 kV)	Schüessler and Denkl	1972
		X-ray (200 kV)	Schüessler *et al.*	1975
Lysozyme	solid	γ-ray	Stevens *et al.*	1967 1969
		γ-ray	Friedberg	1969
		γ-ray	Marciani and Tolbert	1972
	solution	X-ray	Alexander *et al.*	1956
		γ-ray	Aldrich and Cundall	1969
Myoglobin	solution	γ-ray	Radola	1968
Oxytocin	solution	γ-ray	Purdie and Lynn	1973

Peroxidase	solution	γ-ray	Delincée *et al.*	1971
		γ-ray	Delincée and Radola	1974a 1974b
Polypeptidase	solid	γ-ray	Hayden *et al.*	1966
	solution	γ-ray	Korgaonkar and Joshi	1968
Ribonuclease	solid	γ-ray	Shapira	1963
		γ-ray	Haskill and Hunt	1965 1967a 1967b
		γ-ray	Ray and Hutchinson	1967c 1967b
		γ-ray	Friedberg	1969
		γ-ray	Stevens *et al.*	1969
		γ-ray	Delincée and Radola	1975
	solution	γ-ray	Bridge	1963
		γ-ray	Shapira	1963
		γ-ray	Hayden and Friedberg	1964
		γ-ray	Jung and Schüessler	1966
		γ-ray	Schüessler and Jung	1967
		γ-ray	Mee *et al.*	1972
		X-ray (200 kV)	Schüessler	1973
		γ-ray	Delincée and Radola	1975
Trypsin	solid	γ-ray	Stevens *et al.*	1969

The ultracentrifuge was first used by Pickels and Anderson (1947) to study protein denaturation. They irradiated hemocyanin with X-rays. Thereafter, ultracentrifugal sedimentation analysis has become the main technique used for this purpose, although viscometry and light scattering measurements are also used. In 1963 Shapira employed column chromatographic gel filtration to

analyse protein radiolytic products.   This method is now commonly
employed as a technique of choice.

In the solid state, aggregation or crosslinking should be
mainly due to the direct action of radiation, while in solution,
it should be due to both the direct action of radiation and the
indirect effects of H·, OH·, O_2H, and e_{aq}^- derived from irradiated
water.   Radiation effects on protein aggregation is more pronounced
in solution than in the solid state (Fig. 1) (Alexander *et al.*,
1956).   Shapira (1963) also observed aggregation in solutions of
ribonuclease but not in the dry state.

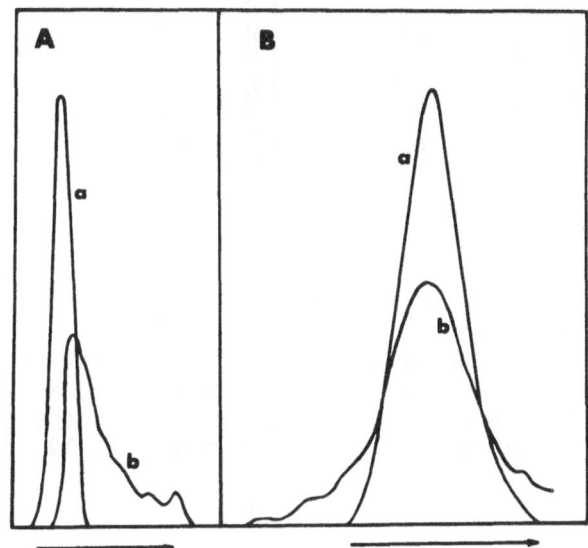

Fig. 1.   The Lamm scale ultracentrifuge diagrams of irradiated bovine serum albumin.   (A) Indirect effect: The control solution (a),
and the irradiated solution (b) which has received 100,000 R, are
1 per cent.   (B) Direct effect: The solid has received 8 x 10^5 R
and is measured as a 1 per cent solution (b) compared with a control solution (a).   (Alexander *et al.*, 1956)

Figure 2 shows an evident increase in aggregation yield
with increasing radiation dose (Hay and Zakrzewski, 1968).   On
the other hand, inhibitory effects on aggregate formation have
been reported by Shapira (1963) and Purdie and Lynn (1973) who
observed suppression of aggregate formation rates in the presence of O_2.   Shapira (1963) also observed a protective effect
of 3,3'-guanidinopropyl disulfide against aggregate formation.

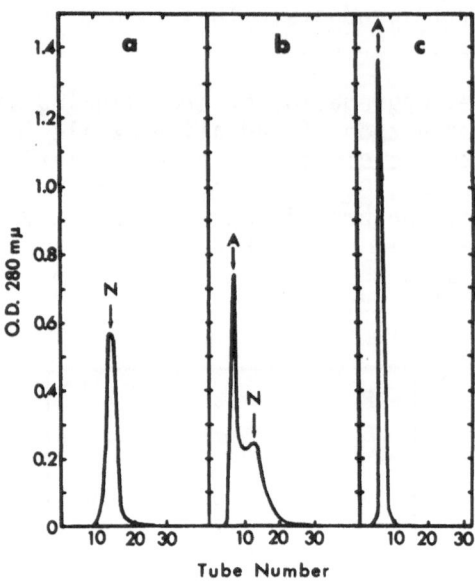

Fig. 2. Sephadex G-200 filtration patterns of horse serum albumin irradiated with variable doses. Column 1.5 x 60.0 cm; 0.85% NaCl, pH 4.5 to 5.5. (a) Native; (b) irradiated with 750 kR; (c) irradiated with 2,000 kR. A denotes the unretarded fraction, the fraction which emerged at the void volume of the column; N denotes the retarded fraction, which emerged from the column at the same volume as the native albumin. (Hay and Zakrzewski, 1968)

EDTA is also a suppressor (Fig. 3) (Schüessler, 1973). High salt concentrations also impedes aggregation (Schüessler and Denkl, 1972).

COVALENT CROSSLINKING MECHANISMS

A. In the Solid State

As an example of radiation-induced covalent crosslinkage in a protein, guanidine hydrochloride does not disassemble aggregates produced in ribonuclease and lysozyme (Friedberg, 1969). According to Augenstein (1958) ionizing energy can migrate along the covalently bonded structure and eventually becomes localized

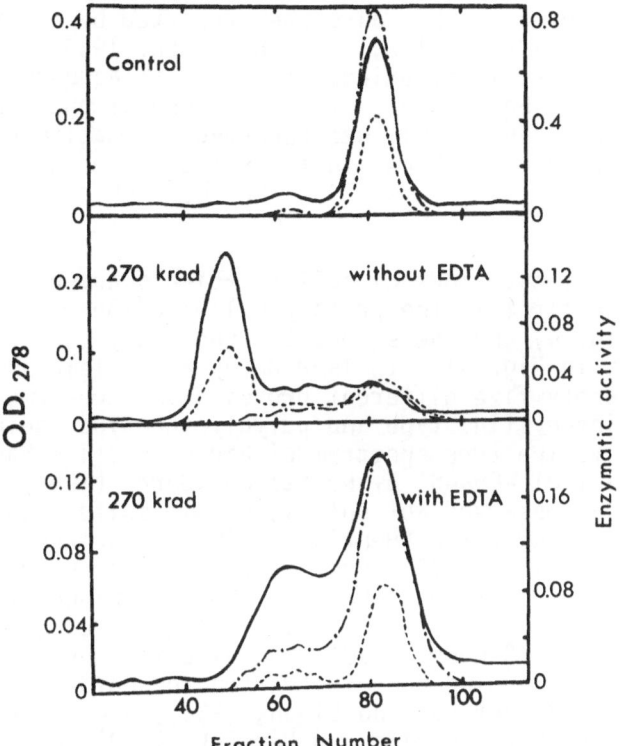

Fig. 3. Separation of native and irradiated ribonuclease. The enzyme was irradiated in a concentration of 0.5 mg/ml 0.01 M phosphate buffer. The concentration of Na_2H_2 EDTA was 2×10^{-3} M. ————: Protein content measured by $O.D._{278}$; -----: hydrolase activity; -·-·-: depolymerase activity. In the case of the control the values for the hydrolase activity must be multiplied by 2. (Schüessler, 1973)

at a restricted number of weaker bonds (e.g. disulfide bonds) of the protein molecule. Ray *et al.* (1960) concluded that radiation opened the -S-S- linkage. The broken disulfide bond may interchange and reform. Alexander *et al.* (1960) have attributed intermolecular bonds formed by disulfide links to the aggregation of serum albumin. According to Stevens *et al.* (1967) "incorrect" intermolecular disulfide bonds are formed after rupture of the disulfide because there is no SH production nor significant amino acid destruction in irradiated lysozyme.

Disulfide bond breakage has also been reported by other workers (Pechèr *et al.*, 1959; Yalow, 1959; Blombäck, 1969). Though
Sulfhydryl disappearance in irradiated proteins in aqueous systems
(Barron and Johnson, 1954; Pihl *et al.*, 1958; Romani and Tappel,
1959; Lange and Pihl, 1960, 1961) was reported, formation of SH
groups from the broken -S-S- bonds in the solid state (Williams
and Hunt, 1963; Hunt and Williams, 1964) has been observed.

The formation of two kinds of sulfur radicals has been confirmed by ESR. The first is the primary sulfur anion radical
-$\dot{S}$-S^-- (left in Fig. 4) and the second the secondary sulfur radical -CH_2-$S^\cdot$ (right in Fig. 4). In 1958 Gordy and Shields obtained
ESR spectra for twenty five different proteins and confirmed the
presence of cysteine-cystine type and polyglycine type radicals.
A broad cysteine-cystine type spectrum of higher g value has been
ascribed to the radical -CH_2-$S^\cdot$ by Kurita and Gordy (1961). This
type of spectrum was observed not only in simple sulfur-containing compounds such as cysteine (Henriksen, 1962; Henriksen *et al.*,
1963; Shields and Hamrick, 1970), cystine (Kurita and Gordy, 1959),
reduced glutathione (Henriksen, 1962; Ormerod and Singh, 1966),
oxidized glutathione (Ormerod and Singh, 1966), and reduced and
oxidized penicillamine (Sanner, 1970), but also in many sulfur-
containing proteins including albumin (Patten and Gordy, 1960;
Henriksen *et al.*, 1963; Ormerod and Singh, 1966), α-chymotrypsin
(Stratton, 1968), feather (Shields and Hamrick, 1970), hair (Shields and Hamrick, 1970), horn (Patten and Gordy, 1960; Shields and
Hamrick, 1970), insulin (Henriksen *et al.*, 1963),   lysozyme
(Henriksen, 1966; Stratton, 1967, 1968), nail (Patten and Gordy,
1960), pepsin (Stratton, 1968), ribonuclease (Henriksen *et al.*,
1963; Hunt and Williams, 1964; Stratton, 1967, 1968; Copeland,
1975), and trypsin (Henriksen, 1966; Ormerod and Singh, 1966;
Stratton, 1968).

Henriksen *et al.* (1963) irradiated sulfur-containing proteins
at $77^\circ K$, and observed an uncharacteristic spectrum not observed
before. Similar results were reported by Patten and Gordy (1960)
and Singh and Ormerod (1965a, b). This spectrum is asymmetric,
originating from a line on the low field side with a g value of
about 2.01. On warming, the characteristic secondary sulfur type
spectrum was formed. Ormerod and Singh (1966) referred to it as
a primary sulfur radical and proposed the following reactions;
For disulfide compounds

$$e + \text{-S-S-} \longrightarrow \text{-}\dot{S}\text{-}S^-\text{-}$$

For sulfhydryl compounds

$$e + \text{-SH} \longrightarrow \text{-}S^-H$$
$$\text{-}S^-H + HS\text{-} \longrightarrow \text{-}\dot{S}\text{-}S^-\text{-} + H$$

Lysozyme

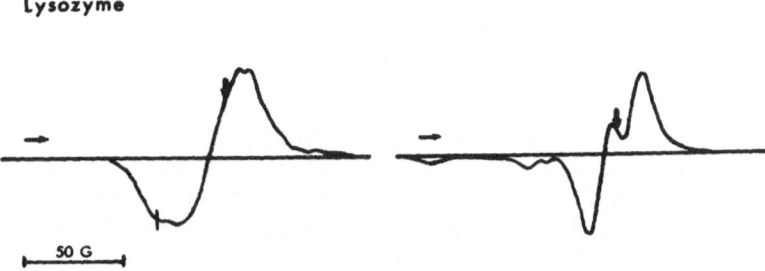

Fig. 4. ESR spectra of lysozyme *in vacuo*, observed 30 min after
irradiation with 2 Mrad of cobalt-60 gamma-rays. In Figs. 6 and 7,
the spectrum on the left-hand side is at 100°K after irradiation at
77°K, and that on the right-hand side is at 295°K after irradiation
at 295°K. The arrow marks the position of g = 2.003 and the short
vertical line the position of g = 2.02. The left-hand side of this
figure demonstrates that some fine structure is visible in the low-
temperature spectrum, as well as the peak near 2.02, which is characteristic of proteins with disulfide bonds. (Stratton, 1968)

Histone

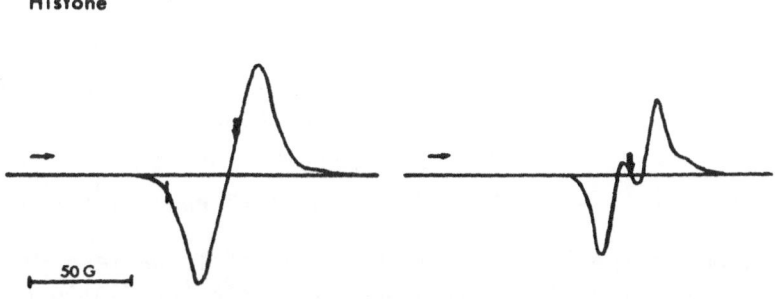

Fig. 5. ESR spectra of histone *in vacuo*, observed 30 min after
irradiation with 0.5 Mrad of cobalt-60 gamma-rays. Markers as in
Fig. 4. These symmetrical spectra are characteristic of proteins
without disulfide bonds. (Stratton, 1968)

On warming

$$-\dot{S}-S^{-}- \longrightarrow S_{II} \xrightarrow{+R^{+}} 2 -CH_2-S\cdot + R\cdot$$

The primary type spectrum has been observed in irradiated simple compounds such as cysteine (Henriksen, 1962), reduced glutathione (Henriksen, 1962), oxidized glutathione (Ormerod and Singh, 1966), and proteins including albumin (Henriksen *et al.*, 1963; Singh and Ormerod, 1965a, b; Ormerod and Singh, 1966), alcohol dehydrogenase (Stratton, 1968), α-chymotrypsin (Stratton, 1968), lysozyme (Stratton, 1968), pepsin (Stratton, 1968), ribonuclease (Henriksen *et al.*, 1963; Stratton, 1968; Copeland, 1975), trypsin (Stratton, 1968), and urease (Stratton, 1968).

Polymerization initiated by -CH$_2$-S$\cdot$ radicals has been discussed by Marciani and Tolbert (1972). The disulfide bond through formation of thiyl radicals must be the origin of one type of crosslinking in proteins. However, not all kinds of proteins contain -SH or -S-S- groups. Furthermore, sulfur bridges have not been observed in some proteins containing such groups (Ray and Hutchinson, 1967a; Hayden and Friedberg, 1964).

Drew and Gordy (1963) have observed ESR spectra of polyglycine and polyalanine which were doublets and tetraplets, respectively, and were ascribed to -NH-$\dot{C}$R-CO-. Polyglycine type spectra (right in Fig. 5) have been observed in various proteins (Gordy and Shields, 1958, 1960; Patten and Gordy, 1960; Singh and Ormerod, 1965b; Ormerod and Singh, 1966; Stratton, 1968). The stable doublet spectrum is also included in sulfur-containing proteins (see Fig. 4). Singh and Ormerod (1965b) have proposed the following reaction scheme between the sulfur and the doublet type radicals.

RH $\xrightarrow{}$ R$\cdot$	Initial radical formation	(1)
R$\cdot$ $\longrightarrow$ R$\cdot$	Migration to a glycine to give "doublet" radical	(2)
R$\cdot$ + -SH $\longrightarrow$ RH + -S$\cdot$	-CH$_2$-S$\cdot$ formation	(3)
RH + -S$\cdot$ $\longrightarrow$ R$\cdot$ + -SH	"Doublet" radical formed	(4)

As shown on the left in Fig. 5, a symmetric pattern was obtained when nonsulfur-containing proteins were irradiated at 77°K (Singh and Ormerod, 1965a, b; Ormerod and Singh, 1966; Stratton, 1968) as well as polyglycine type (Patten and Gordy, 1960; Drew and Gordy, 1963). This specific radical at 77°K has not been identified with certainty, but this low-temperature pattern might be attributable to an ionized or charged segment of a polymer such as -HN^{+}CHR-CO- or -HN-RH$\dot{C}$C^{+}O-, as suggested by Drew and Gordy (1963).

Denaturation of a protein involves peptide chain breaks which produce molecules smaller than the native protein molecule. Ray and Hutchinson (1967b) report that 15% of the peptide backbone was broken when RNase lost 80% activity. Garrison *et al.*, (1967) have described a scission mechanism for acetylalanine:

$$-NH-CHR-CO- \quad \longrightarrow\!\!\!\!\!\sim\!\!\!\!\!\longrightarrow \quad -\dot{N}H + \dot{C}HR-CO- \qquad\qquad (1)$$

$$-\dot{N}H + -NH-CHR-CO- \quad \longrightarrow \quad -NH_2 + -NH-\dot{C}R-CO- \qquad (2)$$

$$-NH-CHR-CO- \quad \longrightarrow\!\!\!\!\!\sim\!\!\!\!\!\longrightarrow \quad -NH-\dot{C}R-CO- + H \qquad\qquad (3)$$

$$H + -NH-CHR-CO- \quad \longrightarrow \quad -NH_2 + \dot{C}HR-CO- \qquad\qquad (4)$$

$$\dot{C}HR-CO- + -NH-CHR-CO- \quad \longrightarrow \quad CH_2R-CO- + -NH-\dot{C}R-CO- \quad (5)$$

ESR studies also suggest formation of amino acid residue radicals in polymers and peroxide radicals in the presence of O_2 (Drew and Gordy, 1963):

Tyrosine radicals   Lysine radical

Tryptophan radicals   Peroxide radical

The above radicals may interact with each other giving both intra- and intermolecular crosslinkings. According to Garrison (1968), recombination of radicals to form new intramolecular bonds may be restrained by secondary and tertiary structures. Because molecules have lower mobilities in the solid state, the crosslinking yield should be lower in the solid state than in solution.

B.   In Solution

Alexander *et al.* (1956) studied X-ray-induced aggregates of bovine serum albumin. He reports that smaller aggregates are covalently crosslinked because the proportion of protein in the main ultracentrifuge peak was unaltered by addition of urea. Rosen (1959) reported an irreversible intramolecular crosslink. He explains that it yields following the reaction between the protein molecule and a free radical formed in the radiolysis of the solvent. Crosslinking mechanisms in aqueous solution could be different from those in the solid state because the indirect action of radiation is mainly effective in solution.

The following radicals have been reported based on ESR studies in aqueous solutions.

Glycine $\quad$ $H_2N\text{-}\dot{C}H\text{-}COOH$ $\quad$ (Smith *et al.*, 1970; Paul and Fischer, 1969, 1971; Paupko *et al.*, 1971)

Alanine $\quad$ $H_2N\text{-}\dot{C}\text{-}COOH$ $\quad$ (Smith *et al.*, 1970; Paupko *et al.*, 1971)
$\quad\quad\quad\quad$ CH_3

$\quad\quad\quad\quad$ $H_2N\text{-}CH\text{-}COOH$ $\quad$ (Paul and Fischer, 1969; Smith *et al.*, 1970)
$\quad\quad\quad\quad$ $\dot{C}H_2$

Serine $\quad$ $H_2N\text{-}CH\text{-}COOH$ $\quad$ (Taniguchi *et al.*, 1968; Paul and Fischer, 1969)
$\quad\quad\quad\quad$ $\dot{C}HOH$

Threonine $\quad$ $H_2N\text{-}CH\text{-}COOH$ $\quad$ (Taniguchi *et al.*, 1968)
$\quad\quad\quad\quad$ $\dot{C}(CH_3)OH$

$\quad\quad\quad\quad$ $H_2N\text{-}CH\text{-}COOH$ $\quad$ (Taniguchi *et al.*, 1968)
$\quad\quad\quad\quad$ $C(\dot{C}H_2)OH$

Valine $\quad$ $H_2N\text{-}CH\text{-}COOH$ $\quad$ (Taniguchi *et al.*, 1968; Paul and Fischer, 1969)
$\quad\quad\quad\quad$ $\dot{C}H(CH_3)_2$

$\quad\quad\quad\quad$ $H_2N\text{-}CH\text{-}COOH$ $\quad$ (Taniguchi *et al.*, 1968; Paul and Fischer, 1969)
$\quad\quad\quad\quad$ $CH(\dot{C}H_2)CH_3$

Leucine $\quad$ $H_2N\text{-}CH\text{-}COOH$ $\quad$ (Taniguchi *et al.*, 1968)
$\quad\quad\quad\quad$ $CH_2CH(\dot{C}H_2)CH_3$

Isoleucine $\quad$ $H_2N\text{-}CH\text{-}COOH$ $\quad$ (Taniguchi *et al.*, 1968)
$\quad\quad\quad\quad$ $CH(CH_3)\dot{C}HCH_3$

Aspartic acid $\quad$ $H_2N\text{-}\dot{C}\text{-}COOH$ $\quad$ (Paul and Fischer, 1969)
$\quad\quad\quad\quad$ CH_2COOH

$$H_2N-CH-COOH \\ \quad\ \ | \\ \quad\ \ CHCOOH$$

(Armstrong and Humphreys, 1967; Paul and Fischer, 1969)

Glutamic acid

$$H_2N-C-COOH \\ \quad\ \ | \\ \quad\ \ CH_2CH_2COOH$$

(Paul and Fischer, 1969)

$$H_2N-CH-COOH \\ \quad\ \ | \\ \quad\ \ CHCH_2COOH$$

(Paul and Fischer, 1969)

$$H_2N-CH-COOH \\ \quad\ \ | \\ \quad\ \ CH_2CHCOOH$$

(Paul and Fischer, 1969)

These aliphatic radicals contribute little to crosslinking or molecular degradation in aqueous systems. Indeed, the destructive amino acid residues in proteins are mostly confined to sulfur-containing and aromatic amino acid residues, as listed in TABLE 2.

Table 2. Destructive Amino Acid Residues in Protein in Aqueous Solution by Ionizing Radiation

Protein	Destructive amino acid residue	Worker	
Alcohol dehy-drogenase	cySH	Pihl *et al.*	1958
	cySH	Romani and Tappel	1959
Aldolase	cySH	Romani and Tappel	1959
Alkali phos-phatase	try, tyr, cySS, his, met, arg	Lynn and Skinner	1974
Amylase	his, arg, lys, phe, tyr	Tanabe	1973
Apocatalase	cySS, his, tyr	Lynn and Raoult	1973
ATP creatine phospho-transferase	his, met, phe, arg, cySH	Friedberg and Hayden	1961
Catalase	cySS, met, his, phe, tyr	Shimazu and Tappel	1964
	cySS, his, tyr	Lynn and Raoult	1973

Chymotrypsin	tyr	Baverstock *et al.*	1974
Collagen	acidic, basic, ring-having amino acids	Bowes and Moss	1962
	met, phe, thr	Cassel	1959
Cytochrome c	met, his, cySS, phe	Kumuta and Tappel	1962
Deoxyribo-nuclease	try	Armstrong and Charlsby	1967
Elastase	try, tyr	Lynn and Skinner	1975
Fibrinogen	try, tyr	Chanderkar *et al.*	1976
Glyceralde-hyde-3-phosphate dehydro-genase	cySH	Pihl *et al.*	1958
	cySH	Lange and Pihl	1960 1961
Haematin	cySS, his, tyr	Lynn and Raoult	1973
Hemoglobin	met, his, phe	Kumuta and Tappel	1962
Insulin	cySS, tyr, phe, his	Drake *et al.*	1957
Lysine vaso-pressin	tyr, phe, cySS	Klassen *et al.*	1974
Ovalbumin	his, cySS, met, phe	Kumuta and Tappel	1962
	his, cySS, met, phe	Shimazu and Tappel	1964
Oxytocine	cySS, tyr	Purdie and Lynn	1973
	tyr, phe, cySS	Klassen *et al.*	1974
Proteinase	tyr, phe, arg	Hatano *et al.*	1963
Ribonuclease	cySS, tyr	Williams and Hunt	1963
	cySS, met, tyr, phe	Hayden and Friedberg	1964

	met, tyr, phe, lys, his, cySS	Jung and Schüessler	1966
	cySS, met, tyr	Mee *et al.*	1972
Subtilisin BPN'	tyr, try, met, his	Ohtsuki *et al.*	1970
Superoxide dismutase	his, tyr, lys	Barra *et al.*	1975

According to results from pulse radiolysis studies, sulfur-containing and aromatic amino acids have a high reactivity towards OH˙ radicals and hydrated electrons (e^-_{aq}) (TABLE 3 and TABLE 4).

Radiolysis of these specific amino acids has been studied in detail. The main radiolytic degradations are:

Cysteine ⟶ Cystine ⟶ Cystine disulfoxide

⟶ Cysteic acid

> (Whitcher, 1953; Markakis and Tappel, 1960; Grant *et al.*, 1961; Brdička *et al.*, 1963; Packer, 1963; Rokushika *et al.*, 1966; Purdie, 1967; Wilkening *et al.*, 1967, 1968; Owen *et al.*, 1968; Packer and Winchester, 1968; Yamamoto, 1972a; Lynn and Louis, 1973; Al-Thannon *et al.*, 1974)

Methionine ⟶ Methionine sulfoxide ⟶ α-Aminobutylic acid

Methionine sulfone   Methylsulfonic acid

> (Koloušek *et al.*, 1956, 1957; Kumuta *et al.*, 1957; Kopoldová *et al.*, 1958; Shimazu *et al.*, 1964; Ohara, 1966; Wacker *et al.*, 1966; Kopoldová *et al.*, 1967; Yamamoto, 1972b)

Phenylalanine ⟶ Tyrosine ⟶ Dopa

> (Rowbottom, 1955; Nosworthy and Allsopp, 1956; Vermeil and Lefort, 1957; Fletcher and Okada, 1961; Ibraginov and Brodskaya, 1962; Brodskaya and Sharpatyi, 1967a, b; Wheeler and Montalvo, 1969; Yamamoto and Okuda, 1975)

Tryptophan ⟶ Hydroxytryptophan ⟶ Formylkynurenine

⟶ Kynurenine ⟶ Hydroxykynurenine ⟶ Hydroxyanthra-

Table 3.   Reactivity of Amino Acids Towards OH· Radicals

Amino acid	Rate constant with OH· $(M^{-1}sec^{-1})$	
	Scholes *et al.* (1965)	
	Thiocyanate method (pH 5-8)	Thymine method (pH 2-3)
Glycine	1.0×10^7	4.5×10^6
Alanine	4.6×10^7	2.8×10^7
Valine	—	4.2×10^8
Leucine	9.8×10^8	1.2×10^9
Isoleucine	—	1.1×10^9
Serine	1.9×10^8	1.7×10^8
Threonine	—	2.3×10^8
Proline	—	1.8×10^8
Hydroxyproline	—	2.1×10^8
Aspertic acid	4.5×10^7	1.9×10^7
Glutamic acid	—	7.9×10^7
Arginine	2.0×10^9	4.5×10^8
Lysine	—	3.7×10^8
Histidine	3.0×10^9	—
Phenylalanine	3.5×10^9	4.4×10^9
Tyrosine	—	5.8×10^9
Tryptophan	8.5×10^9	4.4×10^9
Methionine	4.9×10^9	3.7×10^9
Cystine	—	3.2×10^9

nilic acid

(Jayson *et al.*, 1954; Nofre *et al.*, 1960; Peter and Rajewsky, 1963; Armstrong and Swallow, 1969)

Table 4. Reactivity of Amino Acids Towards Hydrated Electron (e_{aq}^-)

Amino acid	Rate constant with e_{aq}^- $(M^{-1}sec^{-1})$ e_{aq}^- decay method (pH 5-8)	
	Davies *et al.* (1965)	Braams (1965)
Glycine	8.5×10^6	—
Alanine	5.9×10^6	—
Valine	5.2×10^6	—
Leucine	3.3×10^6	—
Serine	1.5×10^7	—
Threonine	2.0×10^7	—
Asparagin	—	1.5×10^8
Arginine	—	1.5×10^8
Lysine	—	2.0×10^7
Histidine	—	6.0×10^7
Phenylalanine	1.5×10^8	1.1×10^8
Tyrosine	4.0×10^8	1.6×10^8
Tryptophan	2.6×10^8	4.0×10^8
Cysteine	—	8.7×10^9
Cystine	—	1.3×10^{10}

Histidine $\longrightarrow$ Imidazole derivatives and ring scission products

(Liebster and Kopoldová, 1964)

It has been shown that these specific amino acids can bind to proteins (Yamamoto, 1967, 1975). TABLE 5 compares the binding yields of various amino acids to serum albumin (Yamamoto, 1973a). Friedberg (1972) has shown that the binding activity of cysteine is much higher than that of aspartic acid. Gamma-irradiation of

Table 5. Comparison of the Radiation-Induced Binding of Amino
Acids with Serum Albumin (SA) and with RNA in Air
(Yamamoto, 1973a)

Amino acid (10^{-4} M)	Binding yield (%)	
	SA (1 mg/ml)	RNA (1 mg/ml)
Alanine	3.9 ± 0.2	< 1
Leucine	2.9 ± 0.1	< 1
Mehionine	32.2 ± 0.2	4.5 ± 0.2
Cysteine	43.1 ± 1.0	4.5 ± 0.2
Cystine	42.3 ± 2.4	4.4 ± 0.5
Histidine	40.3 ± 3.0	4.3 ± 0.7
Phenylalanine	54.6 ± 1.8	4.6 ± 1.0
Tryptophan	59.0 ± 2.3	5.4 ± 1.0

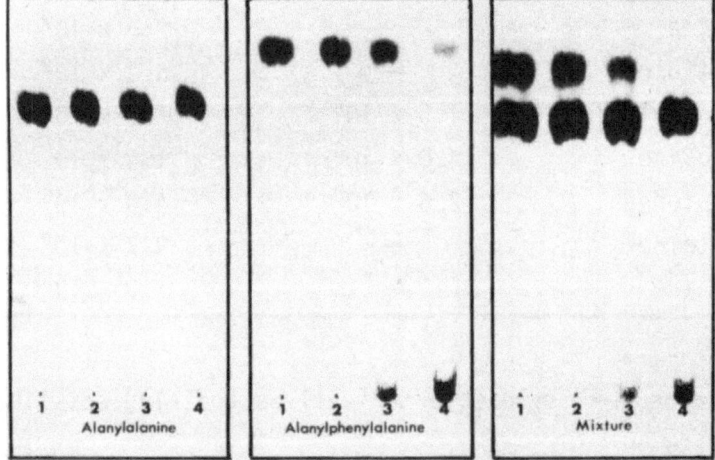

Fig. 6. Paper chromatograms of alanylalanine (10M), alanylphenylalanine (10^{-3}M) and the mixture (both 10^{-3}M) irradiated in aqueous
solution under air. 1) 0 rad; 2) 2.41 x 10^5 rad; 3) 4.83 x 10^5
rad; 4) 9.65 x 10^5 rad. (Yamamoto, unpublished)

alanylalanine and alanylphenylalanine in aqueous solutions, markedly changed only the latter (Fig. 6) (Yamamoto, unpublished).
Specific amino acids do not bind to polyglutamic acid which does
not contain any specific amino acid residues (Yamamoto, 1973b).
These results indicate that aliphatic side chains or peptide bonds
do not participate in binding.

Yamamoto (1967) has proposed a crosslinking model in which
radical-radical binding of specific amino acid residues takes
place (Fig. 7).

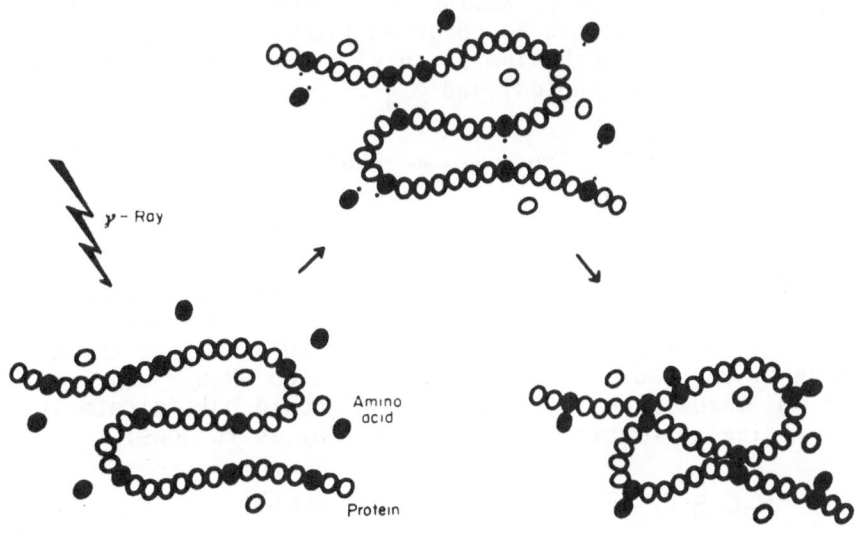

Fig. 7. Model of the radiation-induced binding of the specific
amino acids with protein. ●: Specific amino acid or specific
amino acid residue in protein; o: nonspecific amino acid or non-
specific amino acid residue in protein for the radiation-induced
binding. (Yamamoto, 1967)

If the process is a radical-radical binding, cysteine or cystine
should bind to methionine but they did not (Yamamoto, 1972b),
whereas cysteine, cystine and methionine did bind to aromatic
amino acids (Yamamoto, 1972a, b). Since the formation of biphe-
nyl type dimers from phenylalanine has been reported (Yamamoto,
1973b), aromatic amino acids should bind mutually. It is therefore concluded that radicals formed in specific amino acids attack the ring of aromatic amino acids and bind by substitution
for H atoms.

Active radicals which can attack the ring of aromatic amino

acids can be classified into three groups.
1. Thiyl Radicals:
     Various mechanisms for formation of thiyl radicals from sulf-
hydryl or disulfide compounds have been presented by many workers
(Littman *et al.*, 1957; Markakis and Tappel, 1960; Grant *et al.*,
1961; Packer, 1963; Brdička *et al.*, 1963; Armstrong and Wilkening,
1964; El Samahy *et al.*, 1964; Purdie, 1967; Adams *et al.*, 1967;
Wilkening *et al.*, 1967, 1968; Al-Thannon *et al.*, 1968, 1974; Owen
et al., 1968; Packer and Winchester, 1968; Jayson *et al.*,
1971; Wu and Kunz, 1975). In binding of cysteine and cystine to
albumin, scavenging effects of nitrous oxide, which is a e_{aq}^- scavenger, and of methanol and potassium thiocyanate, which are OH·
radical scavengers, could not be observed (Yamamoto, 1972a).
These results suggest a mechanism of thiyl radical formation without participation of OH· radical and e_{aq}^- .

$$\begin{matrix} R\text{-}S\text{-}H \\ \\ R\text{-}S\text{-}S\text{-}R \end{matrix} \Bigg\rangle\!\!\!\longrightarrow R\text{-}S\cdot$$

The binding yield of methionine, however, was affected by scavengers. The yield decreased in the presence OH· radical scavengers
and increased in the presence of e_{aq}^- scavengers. This increase in
the presence of the e_{aq}^- scavenger, nitrous oxide, must be due to
the increase of OH· radical because the reaction $N_2O + e_{aq}^- \longrightarrow$
$N_2 + OH\cdot + OH^-$ takes place. Therefore, OH· could participate in
methionine radical formation (Yamamoto, 1972b) as follows:

$$H_3C\text{-}\ddot{\underset{..}{S}}\text{-}R \xrightarrow{\;+OH\cdot\;} H_2\dot{C}\text{-}\ddot{\underset{..}{S}}\text{-}R \quad + \quad (H_2O)$$

$$\Updownarrow$$

$$H_2C{=}\dot{\underset{..}{S}}\text{-}R$$

$$\Updownarrow$$

$$H_2C\overset{..}{\underset{\underset{R}{|}}{\cdots}}\ddot{S}: \;\rightleftharpoons\; H_2C\text{-}\underset{\underset{R}{|}}{S}\cdot$$

2. Conjugated Ring Radicals:
     Binding yields of aromatic amino acids with serum albumin
were also compared under different conditions. For phenylalanine,
tryptophan, and histidine, the yield decreased in the presence of
OH· radical scavengers (Yamamoto, 1973b). These results suggest
participation of OH· radicals in radical formation of these amino
acids. Similar result was obtained for nicotinic acid (Yamamoto,
unpublished). Two types of radicals, for example the phenyl- and
cyclohexadienyl radicals are possible from the action of OH·
towards benzene. Partially saturated compounds as final products

from conjugated ring compounds were not found and ring-ring bound
dimers were obtained as main products in aqueous systems (Stein
and Weiss, 1949; Phung and Burton, 1957; Baxendale and Smithies,
1959; Dorfman *et al.*, 1962; Yamamoto, 1973b). Yamamoto (1973b),
therefore, concluded that only conjugated ring radicals such as
phenyl radicals are active in crosslinking reactions. In the formation of conjugated ring radicals, an abstraction of one hydrogen
atom by an OH· radical takes place.

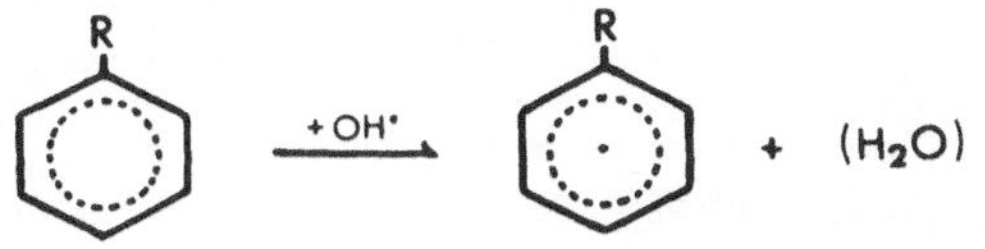

3. Conjugated Ring-Oxyl Radicals:

For OH-substituted aromatic amino acids, OH· radicals and e_{aq}^-
scavenger effects decreased with increasing number of OH-groups
(Fig. 8). Therefore, conjugated ring radicals formed by OH· and
conjugated ring-oxyl radicals such as phenoxyl radicals formed by
unknown mechanisms may participate in crosslinking reactions.

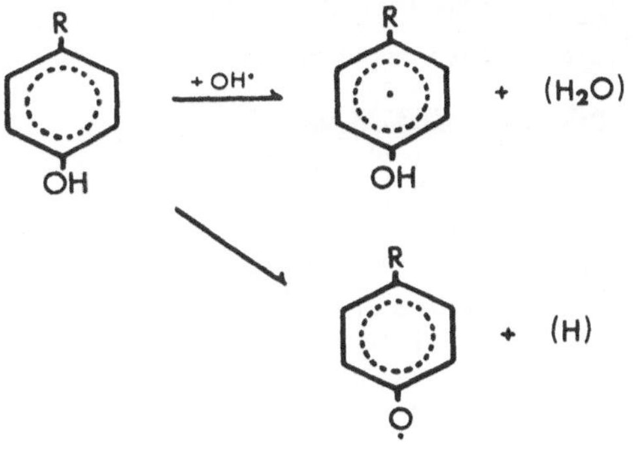

The lack of an OH· radical and e_{aq}^- scavenging effects in binding of OH-substituted aromatic amino acids, as observed in binding of cysteine and cystine, suggests that the mechanism of radical
formation for the O atom is similar to that for the S atom. As shown
in TABLE 4, the reaction rate of e_{aq}^- with cysteine and cystine is
quite high. The e_{aq}^- scavenger effect in binding of cysteine and
cystine can not be explained if their reaction rates with e_{aq}^- are
much higher than that of the e_{aq}^- scavenger. However, if e_{aq}^-
participates also in oxyl radical formation, the rate constant of
e_{aq}^- with tyrosine should be much higher than that with phenylalanine. Since there is no great difference in these rate constants
(TABLE 4), a mechanism of thiyl or oxyl radical formation exists
without the participation of OH· and e_{aq}^- .

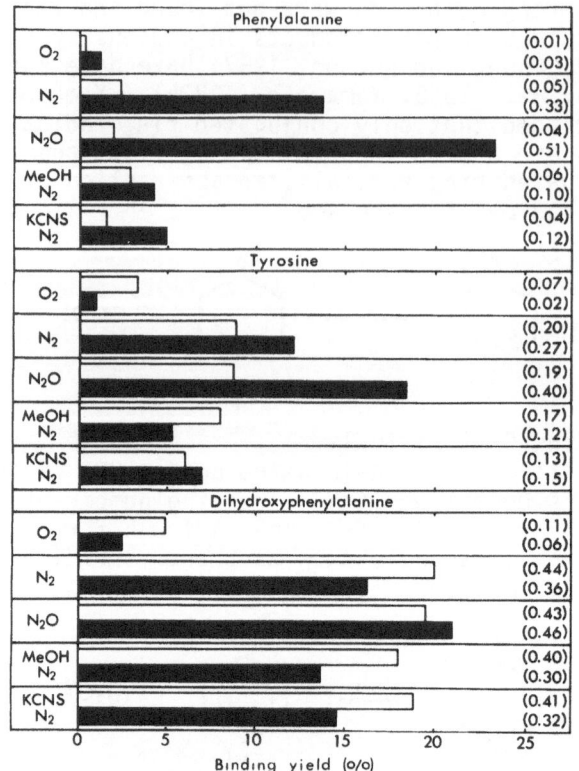

Fig. 8. Radiation-induced binding yields of phenylalanine, tyrosine and dopa (10^{-4}M, ^{14}C 0.5 μCi/ml) with serum albumin (1 mg/ml)
in aqueous solution under various atmospheres. MeOH: 10^{-2}M;
KCNS: 10^{-3}M; O_2, N_2, and N_2O: saturated by bubbling for 15 min.
Radiation dose: 4.83 x 10^4 rads. Standard deviation < ± 2%.
■: In neutral solution; □: in 0.1 N HCl solution. Numbers in
parentheses = G values. (Yamamoto and Okuda, 1975)

Oxidized sulfur-containing amino acids, such as cysteic acid,
cystine disulfoxide, methionine sulfoxide, and methylsulfonic
acid, do not bind to proteins (Yamamoto, 1972a, b). Therefore,
oxidized thiyl radicals such as R-OS· or ·SOH, R-O_2S· or ·SO_2H,
and ·SO_3H may have no binding activity toward proteins even if
they are formed. This result could be due to a decrease in the
electronegativity of the S atom by O atom(s) and suggests that the
specific amino acid radical may act as an electrophilic agent, as

illustrated (Yamamoto, 1972b, 1973b; Yamamoto and Okuda, 1975):

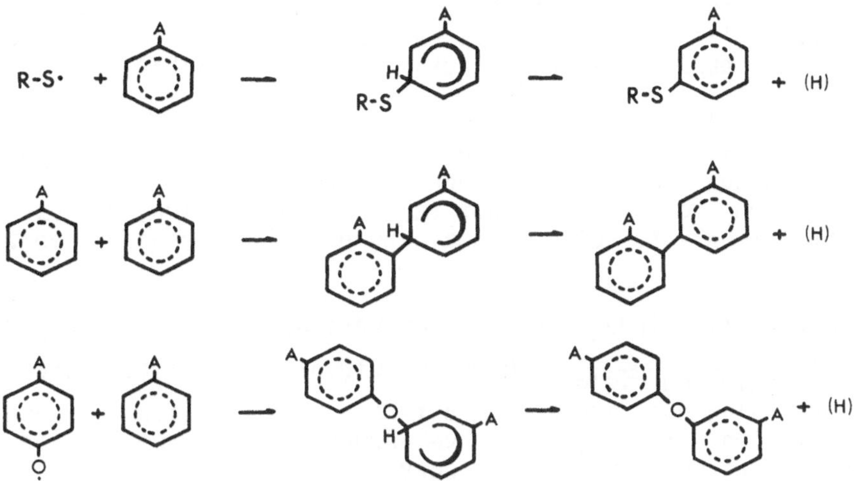

Specific electrophilic amino acid radical reagents could attack nucleophilic sites on the ring of aromatic amino acids. The bound form can become stabilized because of resonance of unshared electrons of oxygen, sulfur, or π electrons of the aromatic ring with those of the bound ring.

In radiolysis of phenylalanine, tyrosine, and dihydroxyphenylalanine (DOPA) (Yamamoto and Okuda, 1975), the decomposition rate decreased with increasing numbers of OH-groups on the aromatic ring. The decomposition rate of amino acid molecule as a whole was almost equal to the mutual binding yield and the decomposition rate of the alanyl side chain was very low in the N_2-saturated solution. This indicates that the modification of aromatic amino acids occurrs mainly by ring dimerization. The decrease of the mutual binding rate with increasing numbers of OH-groups on the ring appears inconsistent with the equivalent of the three amino acids with albumin (Fig. 8). However, this inconsistency may be resolved if one takes into account the ratio of binding of the aromatic amino acids to one another to the binding with albumin. This ratio was found to be greater for those aromatic amino acids with fewer OH-groups.

Aliphatic amino acids show very little reactivity, as mentioned earlier. The longer the chain length of the aliphatic amino acid, the greater the binding yield (Yamamoto, unpublished). Their binding yields with albumin was almost equal to that with nucleic acids. An OH' radical scavenging effect was also observed. The binding of aliphatic amino acids to proteins and nucleic acids might be due to a radical-radical reaction.

Binding reaction, $-S^\cdot + {}^\cdot S- \rightarrow -S-S-$, could be the main contribution to the protein crosslinking in the solid state as discussed earlier. This may not be true in aqueous solution because of the numerous sites attackable by the flexible $-S^\cdot$ radicals. Horváth *et al*, (1972), Horváth and Cságly (1974a, b), and Horváth and Holland (1976) reported on the binding of sulfur-containing proteins with MEG, an SH compound. They concluded that disulfide bonds are formed between the protein and MEG. However, disulfide bond formation may not be the main reaction, because MEG can bind not only to sulfur-containing but also nonsulfur-containing proteins.

In radiation-induced binding of various amino acids to proteins, $1 - \dfrac{1}{\text{binding yield}}$ shows an evident exponential increase with increasing dose suggesting a first order reaction in N_2-saturated solution. In air, the curve abruptly rises from 10^5 rad. With O_2-saturated solution, the curve rose at about 2.5×10^5 rad. This indicates a high suppressive effect of O_2 on the covalent binding reaction and a high O_2 consumption in the solution system.

It may be concluded that the thiyl radicals are produced from cysteine, cystine and methionine residues, the conjugated ring radicals from phenylalanine, tyrosine, tryptophan, and histidine residues, and the conjugated-oxyl radicals from tyrosine (and DOPA and hydroxytryptophan produced from tyrosine and tryptophan, respectively, by oxidation) residues, and that these then bind to the ring of aromatic amino acid residues. These radicals then interact to form a variety of crosslinks.

NONCOVALENT CROSSLINKING MECHANISM

Alexander *et al*. (1956) reported that irradiation of serum albumin in the dry state produced a insoluble and a soluble parts. The water-soluble part contained large aggregates which could be dispersed by concentrated urea to some extent. This observation indicates noncovalent bindings exist in some parts of the aggregates. These appear to behave in a manner similar to those produced by indirect action in aqueous solution. This type of interaction has also been observed after dissolving irradiated dry proteins in water. It is not apparent whether this interaction was produced before or after solution of the irradiated proteins in water. A direct detection of such interactions in the dry state has not yet been reported.

Irradiation of aqueous solution of serum albumin gives rise to aggregation (Rosen *et al*., 1958; Rosen, 1959). Secondary forces appear to contribute to this aggregation. The molecular

weight of radiation-induced aggregates from serum albumin is dependent on ionic strength and is reduced by urea (Alexander *et al*.,
1956).

Delincée and Radola (1974a) observed a spontaneous dissociation, on repeated gel chromatography, of radiation-induced aggregates from peroxidase and myoglobin, strongly suggesting the
involvement of noncovalent bonds. Schüessler *et al*. (1975) found
that radiation-induced aggregates of lactate dehydrogenase are
split into fragments by addition of concentrated sodium dodecyl
sulfate (Fig. 9). They attribute aggregate formation to hydrophobic and electrostatic interactions. Foss (1961) also discusses
formation of hydrophobic binding in irradiated lysozyme, ribonuclease, and chymotrypsin.

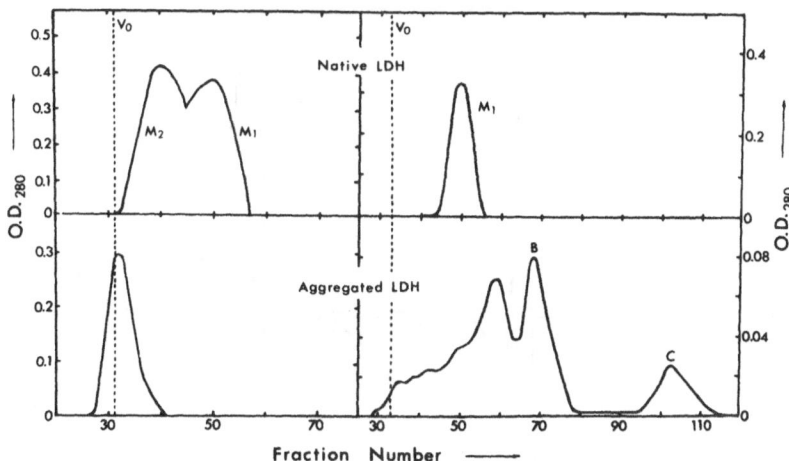

Fig. 9. Gel-filtration of native and aggregated lactate dehydrogenase (LDH) on Sephadex G-200. SDS/LDH ratio: 0.33 g/g on the
left-hand side and 1.4 g/g on the right-hand side. V_0 = void
volume. M_1 = monomer of the sub-unit; M_2 = dimer of the sub-unit.
A, B and C = fragments of sub-unit of different size. (Schüessler
et al., 1975)

Fricke *et al*. (1957), Leone (1960a, 1960b, 1960c) concluded
from studies on ovalbumin that structural breakdown of the
protein is probably caused by disruption of hydrogen bonds.
According to Platzman and Franck (1958), a number of hydrogen

bonds will temporarily develop near the ionization sites because the sudden induction of a charge disrupts the electrical dipoles and induces a rearrangement of polar groups in the vicinity of the charge. Since this rearrangement can rupture several hydrogen bonds, secondary structures of the molecule can be destroyed. Such hydrogen-bond breakages probably bring about reformation of hydrogen bonds which gives rise to aggregates. Moreover, some modified portions, for example, formation of C=O (Jayko and Garrison, 1958) might produce new noncovalent bonds.

Additional possible noncovalent bond-formations in irradiated proteins will be presented here. These formed by interactions of N-oxides and sulfoxides with SH groups. Sulfoxides could be produced from sulfur-containing amino acid residues and N-oxides may

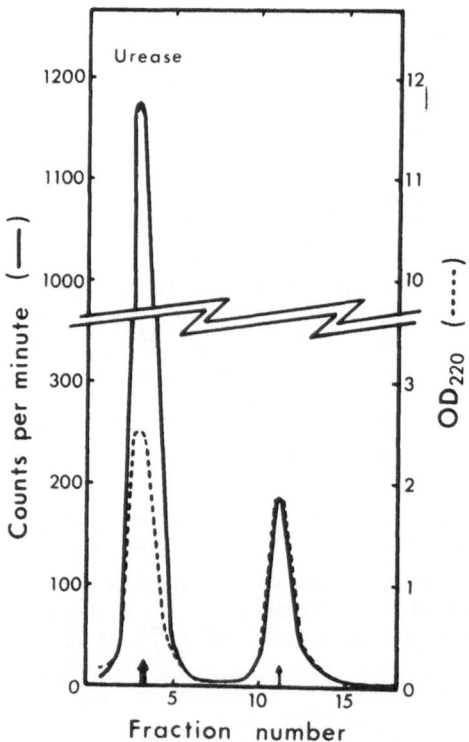

Fig. 10. Gel-filtration of urease (2 mg/ml) mixed with 4NQO (5 x 10^{-4}M, 0.25 μCi/ml) in 0.4 mM NaHCO$_3$ solution. ----: O.D.$_{220}$; ———: radioactivity of 4NQO in TCA precipitate. ⬆: Urease peak; ↑: 4NQO peak. (Yamamoto, unpublished)

be formed from histidine in irradiated protein. There is no direct evidence for such noncovalent bonding but suggestive evidence exists.

Yamamoto (1976) proposed that N-oxides may bind to carboxyl group of acidic proteins by hydrogen bonding. However, later Yamamoto (unpublished) confirmed that one of the N-oxides, 4-nitroquinoline-N-oxide, binds to SH-containing proteins but not to acidic proteins (Fig. 10). The binding of 4-nitroquinoline-N-oxide to the SH-containing proteins markedly decreased in the presence of sodium dodecyl sulfate. These results suggest the presence of a kind of hydrogen bond between NO group and SH groups.

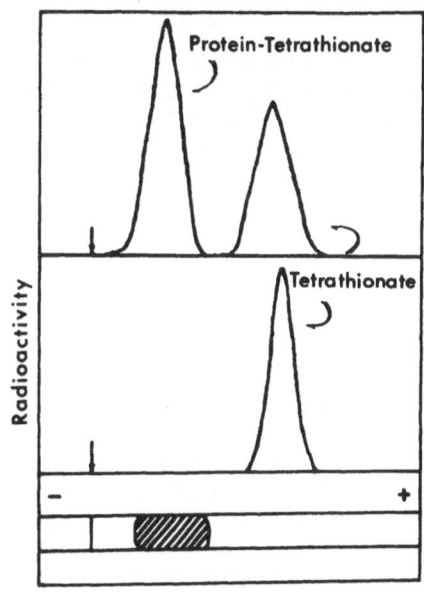

Fig. 11. The paper electrophoretic separation of the labeled reaction products after incubation of glyceraldehyde-3-P dehydrogenase with ^{35}S-tetrathionate. The enzyme (1×10^{-4} M) was incubated with ^{35}S-tetrathionate (3×10^{-4} M) for 20 min at $0°$ in 0.66 M phosphate buffer, pH 7.4. The electrophoresis was carried out in a 0.067 M phosphate buffer, pH 7.4. The current was 1 ma per paper strip (width, 1.7 cm), and the time was 30 min. The protein was located by staining of the paper strip with Amidoblack. (Pihl and Lange, 1962).

Pihl and Lange (1962) found that tetrathionate binds to one of SH-containing proteins, glyceraldehyde-3-phosphate dehydrogenase (Fig. 11):

$$NaO_3\text{-S-S-S-S-}O_3Na + Prot.\text{-SH} \longrightarrow Prot.\text{-S-S-S-}O_3Na$$

If such reaction takes place, cystamine, cystine, and oxidized glutathione (GSSG) should also bind to this enzyme. However, the SH groups of the protein completely failed to interact with these disulfides. When they used GSSG synthesized from GSH with O_2, they observed enzyme inhibition and H_2O_2 production. This implies formation of further oxidized products:

$$G\text{-SH} \longrightarrow G\text{-S-S-}G \longrightarrow \underset{\underset{O\ \ O}{\parallel\ \parallel}}{G\text{-S-S-}G} \longrightarrow 2\ \underset{\underset{O}{\parallel}}{\overset{\overset{O}{\parallel}}{G\text{-S-OH}}}$$

These might bind to the SH-containing protein. It is possible that S=O groups can participate in the noncovalent binding to SH-containing proteins as well as NO groups in the N-oxides. In contrast to covalent crosslinking, noncovalent crosslinking yields may be increased in the presence of O_2, because the formation rates of N-oxides and oxidized sulfur products could be higher than in its absence.

CONCLUSION

1. Since the report of Fernau and Pauli in 1915 on the coagulation of albumin following irradiation in aqueous solution, aggregate formation has been reported for many proteins irradiated in solution and in the solid state. Ultracentrifugal sedimentation and gel-filtration have been mainly used as useful techniques for aggregate analysis. Aggregation yields are much higher following irradiation of aqueous solutions compared to dry proteins.

2. In solid state irradiation stable thiyl (-S') and peptidyl (-NH-ĊR-CO-) radicals are formed. These radicals participate in covalent crosslinking by a radical-radical interchange and recombination.

3. In irradiated aqueous systems, sulfur-containing and aromatic amino acid residues of proteins are particularly reactive. Aliphatic amino acid residues show little activity. Covalent crosslinking is produced by thiyl, conjugated ring, and conjugated ring-oxyl radicals.

4. The existence of noncovalent crosslinking in solid state irradiation of proteins has not been completely confirmed. However, in solution irradiation, some definite evidence exists which

indicates for the presence of hydrophobic and electrostatic interactions. It has been suggested that hydrogen bonding between NO
and SO groups produced by radiation and SH groups can be formed.

REFERENCES

Adams, G. E., McNaughton, G. S., and Michael, B. D. (1967). The
pulse radiolysis of sulphur compounds Part 1. Cysteamine and
cystamine. "The Chemistry of Ionization and Excitation"
(G. R. A. Johnson and G. Scholes, eds.), pp. 281-293,
Taylor and Francis LTD, London.

Aldrich, J. E. and Cundall, R. B. (1969). The radiation-induced
inactivation of lysozyme. *Int. J. Radiat. Biol.*, *16*, 343-358.

Alexander, P., Fox, M., Stacey, K. A., and Rosen, D. (1956).
Effects of some direct and indirect effects of ionizing radiations in protein. *Nature*, *178*, 846-849.

Alexander, P., Hamilton, L. D. G., and Stacey, K. A. (1960). Irradiation of proteins in the solid state. I. Aggregation and
disorganization of secondary structure in bovine serum albumin. *Radiat. Res.*, *12*, 510-525.

Al-Thannon, A. A., Barton, J. P., Packer, J. E., Sims, R. J., Trumbore, C. N., and Winchester, R. V. (1974). The radiolysis of
aqueous solutions of cysteine in the presence of oxygen. *Int.
J. Radiat. Phys. Chem.*, *6*, 233-248.

Al-Thannon, A. A., Peterson, R. M., and Trumbore, C. N. (1968).
Studies in the aqueous radiation chemistry of cysteine. I.
Deaerated acidic solutions. *J. Phys. Chem.*, *72*, 2395-2399.

Armstrong, D. A. and Wilkening, V. G. (1964). Effects of pH in the
γ-radiolysis of aqueous solutions of cysteine and methyl mercaptan. *Can. J. Chem.*, *42*, 2631-2635.

Armstrong, R. C. and Charlesby, A. (1967). The role of tryptophan in the inactivation of deoxyribonuclease. *Int. J.
Radiat. Biol.*, *12*, 523-534.

Armstrong, R. C. and Swallow, A. J. (1969). Pulse- and gamma-
radiolysis of aqueous solutions of tryptophan. *Radiat. Res.*,
40, 563-579.

Armstrong, W. A. and Humphreys, W. G. (1967). Amino acid radicals
produced chemically in aqueous solutions. Electron spin resonance spectra and relation to radiolysis products. *Can. J.
Chem.*, *45*, 2589-2597.

Arnow, L. E. (1935). Physicochemical effects produced by the irradiation of crystalling egg albumin solution with α particles.
J. Biol. Chem., *110*, 43-59.

Augenstein, L. G. (1958). A proposed mechanism of protein inactivation. "Symposium on Information Theory in Biology" (H.
Yockey, ed.), pp. 287-292, Pergamon Press, New York.

Augenstein, L. G. (1962). The effects of ionizing radiation of enzymes. *Advan. Enzymol.*, *24*, 359-413.

Augenstein, L. G., Brustad, T., and Mason, R. (1964). The relative roles of ionization and excitation processes in the radiation inactivation of enzymes. "Advances in Radiation Biology" (L. G. Augenstein, R. Mason, and H. Quastler, eds.), Vol.
I, pp. 227-266, Academic Press, New York.

Barra, D., Bossa, F., Calabrese, L., Rotilio, G., Roberts, P. B.,
and Fielden, E. M. (1975). Selective destruction of amino
acid residues in irradiated solutions of superoxide dismutase.
Biochem. Biophys. Res. Commun., *64*, 1303-1309.

Barron, E. S. G. and Finkelstein, P. (1952). Studies on the mechanism of action of ionizing radiations. X. Effect of X-rays on
some physicochemical properties of proteins. *Arch. Biochem.
Biophys.*, *41*, 212-232.

Barron, E. S. G. and Johnson, P. (1954). Studies on the mechanism
of action of ionizing radiations. XI. Inactivation of yeast
alcohol dehydrogenase by X-irradiation. *Arch. Biochem. Biophys.*, *48*, 149-153.

Baverstock, K., Cundall, R. B., Adams, G. E., and Redpath, J. L.
(1974). Selective free radical reactions with proteins and
enzymes: The inactivation of α-chymotrypsin. *Int. J. Radiat.
Biol.*, *26*, 39-46.

Baxendale, J. H. and Smithies, D. (1959). X-Irradiation of aqueous benzene solutions. *J. Chem. Soc.*, 779-783.

Blombäck, B. (1969). The N-terminal disulphide knot of human fibrinogen. *Brit. J. Haemat.*, *17*, 145-157.

Bowes, J. H. and Moss, J. A. (1962). The effect of gamma radiation on collagen. *Radiat. Res.*, *16*, 211-223.(1962)

Braams, R. (1966). Rate constants of hydrated electron reactions
with amino acids. *Radiat. Res.*, *27*, 319-329.

Brdička, R., Spurný, Z., and Fojtík, A. (1963). Effect of the
dose intensity on the rate of radio-oxidation of cystine in
aqueous solutions. *Colln Czech. Chem. Commun.*, *28*, 1491-1498.

Brodskaya, G. A. and Sharpatyi, V. A. (1967a). Radiolysis of phenylalanine aqueous solutions. *Zhur. Fiz. Khim.*, *41*, 1108-
1113.

Brodskaya, G. A. and Sharpatyi, V. A. (1967b). Radiolysis of aqueous solutions of tyrosine, concentration, and oxygen effect.
Zhur. Fiz. Khim., *41*, 2850-2856.

Carroll, W. R., Mitchell, E. R., and Callanan, M. J. (1952). Polymerization of serum albumin by X-rays. *Arch. Biochem. Biophys. 39*, 232-233.

Cassel, J. H. (1959). Effects of gamma radiation of collagen. *J.
Am. Leather Chemists' Assoc.*, *54*, 432-449.

Chanderkar, L. P., Gurnani, S., and Nadkarni, G. B. (1976). The
involvement of aromatic amino acids in biological activity of
bovine fibrinogen as assessed by gamma-irradiation. *Radiat.
Res.*, *65*, 283-291.

Copeland, E. S. (1975). Characterization of secondary radical reactions in irradiated ribonuclease. *Radiat. Res.*, *61*, 63-75.

Dale, W. M. (1966). Irradiation effects of enzyme (*in vitro*). "Radiation Biology" (A. Zuppinger, ed.), Vol. 1, pp. 214-235, Springer-Verlay, Heiderberg.

Davies, J. V., Ebert, M., and Swallow, A. J. (1965). Reactions of the hydrated electron with glycine and other amino acids and peptides. "Pulse Radiolysis" (M. Ebert, J. P. Keene, A. J. Swallow, and J. H. Baxendale, eds.), pp. 165-170, Academic Press, London and New York.

Delincée, H. and Radola, B. J. (1974a). The effect of γ-irradiation on the change and size properties of horseradish peroxidase. *Radiat. Res., 58*, 9-24.

Delincée, H. and Radola, B. J. (1974b). Effect of γ-irradiation on the change and size properties of horseradish peroxidase: Individual isoenzymes. *Radiat. Res., 59*, 572-584.

Delincée, H. and Radola, B. J. (1975). Structural damage of gamma-irradiated ribonuclease revealed by thin-layer isoelectric focussing. *Int. J. Radiat. Biol., 28*, 565-579.

Delincée, H., Radola, B. J., and Drawart, F. (1971). Isoelectric properties of gamma-irradiated horseradish peroxidase. *Int. J. Radiat. Biol., 19*, 93-97.

Dorfman, L. F., Taub, I. A., and Bühler, R. E. (1962). Pulse radiolysis studies. I. Transient spectra and reaction-rate constants in irradiated aqueous solutions of benzene. *J. Chem. Phys., 36*, 3051-3061.

Drake, M. P., Giffee, J. W., Johnson, D. A., and Koenig, V. L. (1957). Chemical effects of ionizing radiation on protein. I. Effect of γ-radiation on the amino acid content of insulin. *J. Am. Chem. Soc., 79*, 1395-1401.

Drew, R. C. and Gordy, W. (1963). Electron spin resonance studies of radiation effects on polyamino acids. *Radiat. Res., 18*, 552-579.

El Samahy, A., White, H. L., and Trumbore, C. N. (1964). Scavenging action in γ-irradiated aqueous cysteine solutions. *J. Am. Chem. Soc., 86*, 3177-3178.

Fernau, A. and Pauli, W. (1915). Über die Einwirkung der durchdringenden Radiumstrahlung auf anorganische und Biokolloide. I. *Biochem. Z., 70*, 420-441.

Fernau, A. and Spiegel, A. M. (1929). Physikalisch-chemische Untersuchungen bestrahlter Proteine. *Biochem. Z., 204*, 14-27.

Fletcher, G. L. and Okada, S. (1961). Radiation-induced formation of dihydroxyphenylalanine from tyrosine and tyrosine-containing peptides in aqueous solution. *Radiat. Res., 15*, 349-354.

Foss, J. G. (1961). Hydrophobic bonding and conformational transitions in lysozyme, ribonuclease and chymotrypsin. *Biochim. Biophys. Acta, 47*, 569-579.

Fricke, H., Leone, C. A., and Landmann, W. (1957). Role of structural degradation in the loss of serological activity of ovalbumin irradiated with gamma-rays. *Nature, 180*, 1423-1425.

Friedberg, F. (1969). Effect of irradiation on some lyophilized proteins. *Radiat. Res., 38*, 34-42.

Friedberg, F. (1972). Covalent binding of amino acid to proteins due to gamma irradiation. *Z. Naturforsch., 27b*, 85.

Friedberg, F. and Hayden, G. A. (1962). The effect of radiation on the amino acid content and the enzymatic activity of ATP-creatine phosphotransferase. *Arch. Biochem. Biophys., 98*, 485-491.

Garrison, W. M. (1968). Radiation chemistry of organo-nitrogen compounds. "Current Topics in Radiation Research" (M. Ebert and A. Howard, eds.), Vol. IV, pp. 43-94, North-Holland Publishing Company, Amsterdam.

Garrison, W. M., Jayko, M. E., Weeks, B. M., and Sokol, H. A. (1967). Chemical evidence for main-chain scission as a major decomposition mode in the radiolysis of solid peptides. *J. Phys. Chem., 71*, 1546-1547.

Gordy, W. and Shields, H. (1958). Electron spin resonance studies of radiation damage to proteins. *Radiat. Res., 9*, 611-625.

Gordy, W. and Shields, H. (1960). Structure and orientation of free radicals formed by ionizing radiations in certain native proteins. *Proc. Nat. Acad. Sci. USA, 46*, 1124-1136.

Grant, D. W., Mason, S. N., and Link, M. A. (1961). Products of the γ-radiolysis of aqueous cystine solutions. *Nature, 192*, 352-353.

Grossweiner, L. I., Kaluskar, A. G., and Baugher, J. F. (1976). Flash photolysis of enzymes. *Int. J. Radiat. Biol., 29*, 1-16.

Hatano, H., Ganno, S., and Ohara, A. (1963). Radiation sensitivity of amino acid in solution and in protein to gamma rays. *Bull. Inst. Chem. Res. Kyoto Univ., 41*, 61-70.

Haskill, J. S. and Hunt, J. W. (1965). Radiation damage to crystalline ribonuclease; An assay method for damage involved in the refolding of the reduced protein. *Biochim. Biophys. Acta, 105*, 333-340.

Haskill, J. S. and Hunt, J. W. (1967a). Radiation damage to crystalline ribonuclease; Identification of the physical alterations by gel filtration on Sephadex. *Radiat. Res., 31*, 327-342.

Haskill, J. S. and Hunt, J. W. (1967b). Radiation damage to crystalline ribonuclease; Importance of free radicals in the formation of denatured and aggregated products. *Radiat. Res., 32*, 606-624.

Haskill, J. S. and Hunt, J. W. (1967c). Radiation damage to crystalline ribonuclease; Identification of polypeptide chain breakage in the denatured and aggregated products. *Radiat. Res., 32*, 827-848.

Hay, M. and Zakrzewski, K. (1968). Molecular aggregation and serological specificity of human serum albumin irradiation in solution. *Radiat. Res., 34*, 396-410.

Hayden, G. A. and Friedberg, F. (1964). Effects of gamma radiation on ribonuclease. *Radiat. Res., 22*, 130-135.

Hayden, G. A., Rogers, S. C., and Friedberg, F. (1966). Radiation degradation of polyamino acid in the solid state. *Arch. Biochem. Biophys.*, *113*, 247-250.

Henriksen, T. (1962). ESR studies on the formation of sulfur radicals in irradiated cysteine, glutathione, and djenkolic acid. *J. Chem. Phys.*, *36*, 1258-1262.

Henriksen, T. (1966). Effect of the irradiation temperature on the production of free radicals in solid biological compounds exposed to various ionizing radiations. *Radiat. Res.*, *27*, 694-709.

Henriksen, T., Sanner, T., and Pihl, A. (1963). Secondary processes in proteins irradiated in the dry state. *Radiat. Res.*, *18*, 147-162.

Horváth, M. and Cságoly, E. (1974a). Disulphide proteins in the binding reaction with radioprotector MEG. *Int. J. Radiat. Biol.*, *25*, 87-94.

Horváth, M. and Cságoly, E. (1974b). Haemoglobin, a sulphhydryl-protein in the binding reaction with radioprotective MEG. *Int. J. Radiat. Biol.*, *25*, 351-359.

Horváth, M. and Holland, J. (1976). Effect of ^{60}Co γ-irradiation on the reaction of mixed disulphides of mercaptoethylguanidine with enzymes of rat-liver cytoplasm. *Int. J. Radiat. Biol.*, *29*, 137-144.

Horváth, M., Fóris, G., Cságoly, E., Sztanyik, L., and Dalos, B. (1972). The binding of radioprotective AET to proteins. *Int. J. Radiat. Biol.*, *21*, 263-278.

Ibraginov, A. P. and Brodskaya, G. A. (1962). Action of X-rays on aqueous solutions of tyrosine and phenylalanine. "Proc. Second All-Union Conf. Radiat. Chem.", pp. 268-276, Acad. Science, USSR.

Jayko, M. E. and Garrison, W. M. (1958). Formation of $>$C=O bonds in the radiation-induced oxidation of protein in aqueous systems. *Nature*, *181*, 413-414.

Jayson, G. G., Scholes, G., and Weiss, J. (1954). Formation of formylkynurenine by the action of X-rays on tryptophan in aqueous solution. *Biochem. J.*, *57*, 386-390.

Jayson, G. G., Stirling, D. A., and Swallow, A. J. (1971). Pulse- and X-radiolysis of 2-mercaptoethanol in aqueous solution. *Int. J. Radiat. Biol.*, *19*, 143-156.

Jung, H. and Schüessler, H. (1966). Zur Strahleninaktivierung von ribonuclease I. Auftrennung der Bestrahlungsprodukte. *Z. Naturforsch.*, *B 21*, 224-231.

Klassen, N. V., Purdie, J. W., Lynn, K. R., and D'Iorio, M. (1974). Pulse radiolysis of oxytocin and lysine vasopressin. *Int. J. Radiat. Biol.*, *26*, 127-132.

Koenig, V. L. and Perrings, J. D. (1952). Physicochemical effects of radiation. I. Effect of X-rays on fibrinogen as revealed by the ultracentrifuge and viscosity. *Arch. Biochem. Biophys.*, *38*, 105-119.

Koenig, V. L., Sowinski, R., and Oharenko, L. (1960). Physicochemical effects of radiation. V. Effects of gamma radiation on bovine fibrinogen. *Radiat. Res., 13*, 432-444.

Koloušek, J., Liebster, J., and Babický, A. (1956). Radiochemische Zersetzung von DL-Methionin. *Colln Czech. Chem. Commun., 22*, 874-878.

Koloušek, J., Liebster, J., and Babický, A. (1957). Radiochemical degradation of DL-methionine. *Nature, 179*, 521-523.

Kopoldová, J., Koloušek, J., Babický, A., and Liebster, J. (1958). Degradation of DL-methionine by radiation. *Nature, 182*, 1074-1076.

Kopoldová, J., Liebster, J., and Gross, E. (1967). Radiation chemical reactions in aqueous solutions of methionine and its peptides. *Radiat. Res., 30*, 261-274.

Korgaonkar, K. S. and Joshi, S. V. (1968). Gamma irradiation studies with synthetic polyamino acids: Studies with poly-L-tyrosine and poly-DL-alanine using monolayer. *Radiat. Res., 35*, 213-226.

Kumuta, U. S., Gurnani, S. U., and Sahasrabudhe, M. B. (1957). *In vitro* lability of methionine to ionizing radiations. *J. Sci. Ind. Res., 16C*, 25-29.

Kumuta, U. S. and Tappel, L. (1962). Decrease of radiation damage to proteins by sulfhydryl protectors. *Radiat. Res., 16*, 679-685.

Kurita, Y. and Gordy, W. (1961). Electron spin resonance in a gamma-irradiated single crystal of L-cystine dihydrochloride. *J. Chem. Phys., 34*, 282-288.

Lange, R. and Pihl, A. (1960). The mechanism of X-ray inactivation of phosphoglyceraldehyde dehydrogenase. *Int. J. Radiat. Biol., 2*, 301-308.

Lange, R. and Pihl, A. (1961). The radiosensitizing effect of thioglycolic acid, dithioglycolic acid, and homocystine on muscle glyceraldehyde-3-phosphate dehydrogenase. *Int. J. Radiat. Biol., 3*, 249-258.

Leone, C. A. (1960a). Effects of γ-rays on the serologic properties of ovalbumin I. Irradiated lyophilized protein. *J. Immunol., 85*, 107-111.

Leone, C. A. (1960b). Effects of γ-rays on the serologic properties of ovalbumin II. Fractions from irradiated lyophilized protein. *J. Immunol., 85*, 112-119.

Leone, C. A. (1960c). Effects of γ-rays on the serologic properties of ovalbumin III. Irradiated solutions. *J. Immunol., 85*, 268-274.

Liebster, J. and Kopoldová, J. (1964). The radiation chemistry of amino acids. "Advances in Radiation Biology" (L. G. Augenstein, R. Mason, and H. Quastler, eds.), Vol. 1, pp.157-226., Academic Press, New York and London.

Littman, F. E., Carr, E. M., and Brady, A. P. (1957). The action of atomic hydrogen on aqueous solutions. *Radiat. Res., 15*, 159-173.

Luzzio, A. J. (1963). The serologic specificity of radiation al-tered-human-serum γ-globulin. *J. Immunol.*, *90*, 224-227.

Lynn, K. R. and Louis, D. (1973). The effects of γ-radiolysis on solutions of papain. *Int. J. Radiat. Biol.*, *23*, 477-485.

Lynn, K. R. and Raoult, A. P. D. (1973). The effects of γ-irra-diation on solutions of catalase, apocatalase and haematin. *Int. J. Radiat. Biol.*, *24*, 25-31.

Lynn, K. R. and Skinner, W. J. (1974). Radiolysis of an alkaline phosphatase. *Radiat. Res.*, *57*, 358-363.

Lynn, K. R. and Skinner, W. J. (1975). The γ-radiolysis of elas-tase. *Radiat. Res.*, *63*, 245-252.

Marciani, D. J. and Tolbert, B. H. (1972). Analytical studies of fractions from irradiated lysozyme. *Biochim. Biophys. Acta*, *271*, 262-273.

Markakis, P. and Tappel, A. L. (1959). Products of γ-irradiation of cysteine and cystine. *J. Am. Chem. Soc.*, *82*, 1613-1617.

Mee, L. K., Adelstein, S. J., and Stein, G. (1972). Inactivation of ribonuclease by the primary aqueous radicals. *Radiat. Res.*, *52*, 588-602.

Morton, J. I. (1960). The effects of X-irradiation on the anti-genic-combining properties of some purified human-serum gamma globulins. *Int. J. Radiat. Biol.*, *2*, 45-53.

Nofre, C., Cier, A., Michou-Saucet, C, and Parnet, J. (1960). Action des radicaux libres hydroxyles sur les acides aminés *Compt. Rendu*, *251*, 811-813.

Nosworthy, J. and Allsopp, C. B. (1956). Effects of X-rays on dilute aqueous solutions of amino acids. *J. Colloid Sci.*, *11*, 565-574.

Ohara, A. (1966). On the radiolysis of methionine in aqueous so-lution by gamma irradiation. *J. Radiat. Res. (Japan)*, *7*, 18-28.

Ohtsuki, K., Fukuhara, M., and Sumizu, K. (1970). Chemical and enzymatic properties of Co γ-ray irradiated subtilisin BPN'. *J. Radiat. Res. (Japan)*, *11*, 113-119.

Ormerod, M. G. and Singh, B. B. (1966). The formation of un-paired electrons on sulphur in irradiated dry proteins as studied by electron spin resonance. *Biochim. Biophys. Acta*, *120*, 413-426.

Owen, T. C., Rodriguez, M., Johnson, B. G., and Rorch, J. A. G. (1968). The radiation chemistry of biochemical disulfides. I. The low-dose X-radiolysis of cystine. *J. Am. Chem. Soc.*, *90*, 196-200.

Packer, J. E. (1963). The action of ^{60}Co-gamma-rays on aqueous so-lutions of hydrogen sulphide and of cysteine hydrochloride. *J. Chem. Soc.*, 2320-2325.

Packer, J. E. and Winchester, R. V. (1968). Radiolysis of neutral aqueous solutions of cysteine in the presence of oxygen. *Chem. Commun.*, 826-827.

Patten, F. and Gordy, W. (1960). Temperature effects on free radical formation and electron migration in irradiated proteins. *Proc. Nat. Acad. Sci. USA*, *46*, 1137-1144.

Paul, H. and Fischer, H. (1969). Electronenspinresonanz kurzlebiger Radikale aus einigen Aminosäuren und Amiden. *Ber. Bunsenges.*, *73*, 972-980.

Paul, H. and Fischer, H. (1971). ESR.-Untersuchung zur Reaktion von Hydroxylrakikalen mit Glycin. *Helv. Chim. Acta*, *54*, 485-491.

Paupko, R., Loewenstein, A., and Silver, B. L. (1971). Electron spin resonance study of radicals derived from simple amines and amino acids. *J. Am. Chem. Soc.*, *93*, 580-586.

Pechère, J. F., Dixon, G. H., Maybury, R. H., and Neurath, H. (1958). Cleavage of disulfide bonds in trypsinogen and α-chymotrypsinogen. *J. Biol. Chem.*, *233*, 1364-1372.

Peter, G. and Rajewsky, B. (1963). Die indirekte Wirkung von Röntgenstrahlen auf Aminosäuren. II. Bestrahlung von Tryptophan. *Z. Naturforsch.*, *B 18*, 110-114.

Phung, P. V. and Burton, M. (1957). Radiolysis of aqueous solutions of hydrocarbons benzene, benzene-d_6, cyclohexane. *Radiat. Res.*, *7*, 199-216.

Pickels, E. G. and Anderson, R. S. (1947). Molecular association of hemocyanin produced by X-rays as observed in the ultracentrifuge. *J. Gen. Physiol.*, *30*, 83-99.

Pihl, A. and Lange, R. (1962). The interaction of oxidized glutathione, cystamine, monosulfoxide, and tetrathionate with -SH groups of rabbit muscle D-glyceraldehyde 3-phosphate dehydrogenase. *J. Biol. Chem.*, *237*, 1356-1362.

Pihl, A., Lange, R., and Eldjarn, L. (1958). Alleged susceptibility of sulphydryl enzymes to ionizing radiation. *Nature*, *182*, 1732-1733.

Platzman, R. and Franck, J. (1958). A physical mechanism for the inactivation of proteins by ionizing radiation. "Symposium on Information Theory in Biology", pp. 262-275, Pergamon Press, New York.

Radola, B. J. (1968). Radiation-induced hybridization of proteins. *Biochim. Biophys. Acta*, *160*, 469-472.

Ray, D. K. and Hutchinson, F. (1967a). Inactivation of dry ribonuclease by ionizing radiation. I. Search for the breakage of sulfur bridges. *Biochim. Biophys. Acta*, *147*, 347-356.

Ray, D. K. and Hutchinson, F. (1967b). Inactivation of dry ribonuclease by ionizing radiation. II. A suggested mechanism. *Biochim. Biophys. Acta*, *147*, 357-368.

Ray, D. K., Hutchinson, F., and Morowitz, H. J. (1960). A connection between S-S bond breakage and inactivation by radiation of a dry enzyme. *Nature*, *186*, 312-313.

Rokushika, S. (1974). Effects of gamma irradiation on the function and conformation of ribonuclease A in dilute solution. *Radiat. Res.*, *57*, 349-357.

Romani, R. J. and Tappel, A. L. (1959). An aerobic irradiation of alcohol dehydrogenase, aldolase and ribonuclease. *Arch. Biochem. Biophys.*, *79*, 323-329.

Rosen, D. (1959). Intermolecular and intramolecular reactions of human serum albumin after its X-irradiation in aqueous solution. *Biochem. J.*, *72*, 597-602.

Rosen, D., Alexander, P., Goldberg, R., and Hamilton, L. D. G. (1958). A comparison of the effects of X- and α-rays on serum albumin. *Radiat. Res.*, *9*, 172-173.

Rosen, D. and Boman, H. G. (1957). Effects of gamma rays on solutions of human serum albumin. II. Chromatographic studies. *Arch. Biochem. Biophys.*, *70*, 277-282.

Rosen, D., Brohult, S., and Alexander, P. (1957). Effects of gamma rays on solutons of human serum albumin. I. Sedimentation studies. *Arch. Biochem. Biophys.*, *70*, 266-276.

Rowbottom, J. (1955). The radiolysis of aqueous solution of tyrosine. *J. Biol. Chem.*, *212*, 877-885.

Sanner, T. (1970). Intermolecular transfer of radiation energy at 77°K. An ESR study of irradiated mixtures of macromolecules and oxidized penicillamine. *Radiat. Res.*, *44*, 13-23.

Scholes, G., Shaw, P., Willson, , R. L., and Ebert, M. (1965). Pulse radiolysis studies of aqueous solutions of nucleic acid and related substances. "Pulse Radiolysis" (M. Ebert, J. P. Keene, A. J. Swallow and J. H. Baxendale eds.), pp. 151-164, Academic Press, New York and London.

Schüessler, H. (1973). X-ray inactivation of ribonuclease in the presence of EDTA. *Int. J. Radiat. Biol.*, *23*, 175-182.

Schüessler, H. and Denkl, P. (1972). X-ray inactivation of lactate dehydrogenase in dilute solution. *Int. J. Radiat. Biol.*, *21*, 435-443.

Schüessler, H. and Jung, H. (1967). Zur Strahleninaktivierung von Ribonuclease. II. Aminosäure-Zusammensetzung der Bestrahlungsprodukte. *Z. Naturforsch.*, *B 22*, 614-621.

Schüessler, H., Niemczyk, P., Eichhorn, M., and Pauly, H. (1975). On the radiation-induced aggregates of lactate dehydrogenase. *Int. J. Radiat. Biol.*, *28*, 401-408.

Shapira, R. (1963). Radiation-induced aggregation of bovine pancreatic ribonuclease. *Int. J. Radiat. Biol.*, *7*, 537-548.

Shields, H. and Hamrick, Jr. P. J. (1970). Relative stability of the characteristic sulfur and doublet resonances in X-irradiated native proteins as measured with ESR. *Radiat. Res.*, *41*, 259-267.

Shimazu, F., Kumuta, U. S., and Tappel, A. L. (1964). Radiation damage to methionine and its derivatives. *Radiat. Res.*, *22*, 276-287.

Shimazu, F. and Tappel, A. L. (1964). Comparative radiolability of amino acids of proteins and free amino acids. *Radiat. Res.*, *23*, 203-209.

Singh, B. B. and Ormerod, M. G. (1965a). Primary radical formation in irradiated proteins. *Nature, 206*, 1314-1315.

Singh, B. B. and Ormerod, M. G. (1965b). The effect of sulphur compounds on free radical fractions and formation in irradiated dry proteins. *Biochim. Biophys. Acta, 109*, 204-213.

Smith, P., Fox, W. M., McGinty, D. J., and Stevens, R. D. (1970). Electron paramagnetic resonance spectroscopic study of radicals derived from glycine, *dl*-α-alanine, and β-alanine in aqueous solution. *Can. J. Chem., 48*, 480-491.

Sowinski, R., Oharenko, L., and Koenig, V. L. (1958). Physicochemical effects of radiation. III. Effects of high-speed electrons on bovine fibrinogen as revealed by the ultracentrifuge and viscosity. *Radiat. Res., 9*, 229-239.

Sowinski, R., Oharenko, L., and Koenig, V. L. (1959). Physicochemical effects of radiation. IV. Effect of high-speed electrons on human fibrinogen. *Radiat. Res., 11*, 90-100.

Stein, G. and Weiss, J. (1949). Chemical actions of ionizing radiations on aqueous solutions. Part II. The formation of free radicals. The action of X-rays on benzene and benzoic acid. *J. Chem. Soc.*, 3245-3254.

Stevens, C. O., Long, J. L., and Upjohn, D. (1969). Radiation produced aggregation in crystalline preparations of ribonuclease, lysozyme, and trypsin. *Proc. Soc. Exptl. Biol. Med., 132*, 951-956.

Stevens, C. O., Sauberlich, H. E., and Bergstrom, G. R. (1967). Radiation-produced aggregation and inactivation in egg white lysozyme. *J. Biol. Chem., 242*, 1821-1826.

Stratton, K. (1967). Electron spin resonance studies on proton-irradiated ribonuclease and lysozyme. *Radiat. Res., supp. 7*, 102-115.

Stratton, K. (1968). Temperature effects on the formation and reactions of free radicals in gamma-irradiated dry enzymes. *Radiat. Res., 35*, 182-201.

Tanabe, R. (1973). Radiation effect of gamma-rays on α-amylase in aqueous solution. *Bull. Inst. Chem. Commun. Kyoto Univ., 51, No. 1*, 37-43.

Taniguchi, H., Fukui, K., Ohnishi, S., Hatano, H., Hasegawa, H., and Maruyama, T. (1968). Free-radical intermediates in the reaction of the hydroxyl radical with amino acids. *J. Phys. Chem., 72*, 1926-1931.

Vermeil, C. and Lefort, M. (1957). Production de tyrosine par action des rayons γ sur les solutions aqueuses de phénylalanine. *Compt. Rendu, 244*, 889-891.

Wacker, A., Moustafa, Z. H., and Lochmann, E.-R. (1966). Über die Wirkung von Röntgenstrahlen auf Methionin. *Biophysik, 3*, 207-212.

Wels, P. (1923). Der Einfluß der Röntgenstrahlen auf Eiweißkorper. *Pflügers Arch., 199*, 226-236.

Wheeler, O. H. and Montalvo, R. (1969). Radiolysis of phenylalanine and tyrosine in aqueous solution. *Radiat. Res., 40*, 1-10.

Whitcher, J. R. (1963). Determination of molecular weights of proteins by gel filtration on sephadex. *Anal. Chem., 35*, 1950-1953.

Wilkening, V. G., Lal, M., Arends, M., and Armstrong, D. C. (1967). The γ-radiolysis of cysteine in deaerated 1N $HClO_4$ solutions. *Can. J. Chem., 45*, 1209-1214.

Wilkening, V. G., Lal, M., Arends, M., and Armstrong, D. C. (1968). The cobalt-60 γ radiolysis of cysteine in deaerated aqueous solutions at pH values between 5 and 6. *J. Phys. Chem., 72*, 185-190.

Williams, J. F. and Hunt, J. W. (1963). Molecular lesions produced in ribonuclease by gamma-rays. *Nature, 200*, 779-781.

Winstead, J. A. and Reece, T. C. (1970). Effects of gamma radiation on the chemical and physical properties of lactate dehydrogenase. *Radiat. Res., 41*, 125-134.

Wu, J.-T. and Kuntz, R. R. (1975). The reactions of hydrogen atoms in aqueous solutions: Effect of pH on reactions with cysteine and penicillamine. *Radiat. Res., 64*, 662-666.

Yallow, R. S. (1959). Production of sulfhydryl groups as a result of the indirect or direct effect of ionizing radiation. "1st Proc. Natl. Biophys. Conf.", p. 169, Yale University Press, New Haven.

Yamamoto, O. (1967). Biochemical studies of radiation damage. I. Inactivation of the pH 5 fraction in amino acyl sRNA synthesis *in vitro* and the binding of amino acids with protein and nucleic acid by gamma-ray irradiation. *Int. J. Radiat. Biol., 12*, 467-476.

Yamamoto, O. (1972a). Radiation-induced binding of cysteine and cystine with aromatic amino acids or serum albumin in aqueous solution. *Int. J. Radiat. Phys. Chem., 4*, 227-236.

Yamamoto, O. (1972b). Radiation-induced binding of methionine with serum albumin, tryptophan or phenylalanine in aqueous solution. *Int. J. Radiat. Phys. Chem., 4*, 335-345.

Yamamoto, O. (1973a). Radiation-induced binding of nucleic acid constitutents with protein constitutents and with each other. *Int. J. Radiat. Phys. Chem., 5*, 213-229.

Yamamoto, O. (1973b). Radiation-induced binding of phenylalanine, tryptophan and histidine mutually and with albumin. *Radiat. Res., 54*, 398-410.

Yamamoto, O. (1975). Radiation-induced binding of some protein and nucleic acid constitutents with macromolecular components in cell systems. *Radiat. Res., 61*, 261-273.

Yamamoto, O. (1976). Ionizing radiation-induced DNA-protein crosslinking. "Aging, Carcinogenesis, and Radiation Biology" (K. C. Smith ed.), pp. 165-192, Prenum Publishing Corporation, New York.

Yamamoto, O. and Okuda, A. (1975). Radiation-induced binding of OH-substituted aromatic amino acids, tyrosine and dopa, mutually and with albumin in aqueous solution. *Radiat. Res., 61*, 251-260.

32

PEROXYDISULFATE ANION-INDUCED CROSSLINKING OF PROTEINS

Howard L. Needles

Division of Textiles and Clothing

University of California, Davis, California 95616

ABSTRACT

Peroxydisulfate anion in aqueous solutions is a strong oxidizing agent decomposing to sulfate free radicals and subsequently to hydroxyl radicals. In the presence of proteins, aqueous solutions of peroxydisulfate undergo more rapid induced decomposition to these free radical species, which in turn leads to oxidative attack and crosslinking and/or degradation of the protein. Under mild conditions and at low peroxydisulfate concentrations, protein crosslinking predominates over degradation. Peroxydisulfate-induced crosslinking of gelatin, fibroin, and other proteins is reviewed. The mechanism of crosslinking is interpreted in light of physical and chemical data, amino acid analyses, peroxydisulfate's known mode of decomposition, the effect of protein modification on oxidative crosslinking, and related model compound studies.

INTRODUCTION

Peroxydisulfate (persulfate) anion is one of the strongest oxidizing agents known in aqueous solution (House, 1962). Peroxydisulfate undergoes first order uncatalyzed decomposition by the following scheme in the absence of an oxidizable substrate:

$$S_2O_8^= \rightarrow 2SO_4^{\cdot -} \quad \text{Rate determining (1)}$$
$$SO_4^{\cdot -} + H_2O \rightarrow HSO_4^- + OH\cdot \quad (2)$$
$$S_2O_8^= + OH\cdot \rightarrow HSO_4^- + SO_4^{\cdot -} + 1/2\ O_2 \quad (3)$$
$$SO_4^{\cdot -} + OH\cdot \rightarrow HSO_4^- + 1/2\ O_2 \quad (4)$$

The rate determining initial step (1) is characteristic of all peroxydisulfate oxidations and is very slow at room temperature. The resulting sulfate radical rapidly reacts with water to form a hydroxyl radical and hydrogen sulfate anion (step 2). Hydroxyl radical can induce further decomposition of peroxydisulfate anion (step 3), and chain termination occurs through recombination of sulfate and hydroxyl free radicals (step 4). In the presence of an oxidizable substrate (X-H), the rate of peroxydisulfate decomposition increases due to an induced free radical decomposition process as follows:

$$X-H \xrightarrow[\text{or OH·}]{SO_4^-} X· + HSO_4^- \quad \text{or} \quad H_2O \xrightarrow{S_2O_8^=} X-OSO_3^- + SO_4^- \qquad \text{etc.}$$

MODEL COMPOUND STUDIES

Aqueous peroxydisulfate solutions have been used to oxidize several classes of organic compounds related to proteins.

N-substituted amides are attacked by sulfate and/or hydroxyl free radicals at the carbon _alpha_ to the amino group to form an aldehyde or ketone and the corresponding dealkylated amide (Needles and Whitfield, 1964). Kinetic studies (Remy, Whitfield and Needles, 1967) show that in the absence of an _alpha_ carbon adjacent to the amide nitrogen as in acetamide, the reaction remains first order, indicating that this amide has little effect on the overall decomposition of peroxydisulfate; however, with N-substituted amides a marked acceleration in peroxydisulfate decomposition is noted, and the order of the reaction becomes 3/2 order showing the amide-induced decomposition of peroxydisulfate occurs.

Only limited information is available on decomposition of amino acids by peroxydisulfate (Lang, 1936). Simple amino acids are oxidatively attacked by aqueous potassium peroxydisulfate to give aldehydes containing one less carbon atom than the starting amino acid and quantitative yields of ammonia and carbon dioxide. The action of alkaline peroxydisulfate on the more complex amino acid tryptophan leads to a mixture of products including anthranilic acid, 3-hydroxyanthranilic acid and o-aminophenol (Boyland, Sims and Williams, 1956). The findings of these studies are consistent with free radical attack by sulfate and/or hydroxyl radical on the amino acids and extensive oxidation of the amino acids due to the high reactivity of these radicals. N-acetylation of the amino acid would be expected to protect the amino group of the amino acid from attack and to change the overall course of reaction. Aqueous solutions of peroxydisulfate cause oxidative sission of N-acetylamino acids to acetamide, aldehyde or ketone, and carbon dioxide (Needles and Whitfield, 1966). When reactive side chains are present in the N-acetylamino

acid, oxidative attack of the amino acid side chain is favored
and formation of acetamide, aldehyde, and carbon dioxide decreases. N-acetyltryptophan and N-acetyltyrosine are oxidized
and crosslinked to melanine-type structures. N-acetylamino acid
attack by peroxydisulfate is 3/2 order; however, the rate of decomposition is slower than found for the corresponding amides,
possibly due to the overall deactivation of the α-carbon adjacent
to the amide nitrogen by the free carboxyl group.

REACTION OF PROTEINS WITH PEROXYDISULFATE ANION

Wool

The formation of free radicals on wool substrate by aqueous
peroxydisulfate have been demonstrated by electron spin resonance
(ESR) (Burke, Kenny and Nicholls, 1962). The ESR spectra from
treatment of wool with aqueous peroxydisulfate is similar to spectra
from treatment of wool with high energy radiation. Peroxydisulfate-
induced attack on wool keratin has been attributed to attack by
sulfate and hydroxyl radicals on selected amino acid side chains
and on the backbone of the wool (Needles, 1965a). Significant
attack of cystine, methionine, arginine, histidine, proline, tyrosine, and possibly phenylalanine side chains was found, accompanied
by non-hydrolytic cleavage of the protein chains, and attack of
terminal amino acids. Loss of cystine was accompanied by an increase in the cysteic acid content of the wool. Although radical-
induced degradation of the wool predominated, it is probable that
attack of tyrosine and phenylalanine led to oxidative crosslinking of the wool, thereby moderating degradation. Acetylation
or esterification do not markedly alter the sites of attack by
peroxydisulfate on wool, although the rate of attack is reduced
by acetylation (Needles, 1965b), presumably due to the lower
concentration of peroxydisulfate anion present within the fiber.

Gelatin

A study of the action of peroxydisulfate anion on gelatin
indicates that both crosslinking and degradation of the gelatin
occurs simultaneously with crosslinking predominating during the
initial reaction period and at moderate peroxydisulfate concentrations (Needles, 1967a). At higher peroxydisulfate concentrations, the aqueous gelatin crosslinks immediately to a water
insoluble gel followed by resolublization of the protein due to
degradative attack by peroxydisulfate anion. The concentration
of gelatin must exceed 3% (dry weight) in the aqueous solution
for interchain crosslinking and insolublization of the gelatin
to occur. At concentrations of 0.04-0.02 $\underline{M}$ peroxydisulfate,
crosslinking occurs within less than 1 hr at 70°C, but as the
concentration of peroxydisulfate is increased to 0.4 $\underline{M}$ immediate

gelatin occurs followed by degradation of the crosslinked gel, and solution of the gelatin with an overall drop in viscosity compared to untreated gelatin. Addition of phosphate as a buffer or bisulfate to the gelatin solution do not contribute to crosslinking, but do slow the rate of crosslinking. Silver nitrate, a known catalyst of peroxydisulfate decomposition, increases the rate of crosslinking, whereas hydroquinone, an oxygen savenger, prevents crosslinking. Acetylation or esterification of the gelatin does not prevent crosslinking, but does alter the rate of crosslinking, whereas deamination of gelatin greatly reduces gelatin's capacity to crosslink. These data suggest that modification of free amino or carboxyl groups in the gelatin does not prevent crosslinking, but destruction of free amino and side chain amino groups in gelatin greatly reduces the number of sites available for crosslinking.

The rate of induced free radical attack on the crosslinking of gelatin by peroxydisulfate anion is only slightly affected by the presence of specific cations or by the pH of the solution. The rate of decomposition of peroxydisulfate increases by a factor of three when gelatin is introduced and is consistent with peroxydisulfate's known mode of induced homolytic decomposition in the presence of an oxidizable substrate.

Analysis of crosslinked gelatin shows that crosslinking occurs by a number of complex reactions. Small losses in Van Slyke, and amide nitrogen are attributed to some loss of terminal amino groups and hydrolytic cleavage of amide groups respectively. Increase in the sulfur content in gelatin is attributed to grafting of sulfur as sulfate into the gelatin.

Amino acid analysis show that significant loss of histidine, methionine, tyrosine, and phenylalanine and increases in cystine and cysteic acid in the crosslinked gelatin. Histidine, tyrosine, and phenylalanine are known to undergo oxidation to quinone intermediates (Thompson, 1962), and quinones are known to bind to active sites in proteins (Mason, 1955). Therefore, the observed crosslinking in gelatin is believed to be due to peroxydisulfate-induced oxidation of these amino acids to quinone intermediates followed by condensation of these intermediates with other reactive sites (amino, hydroxyl, guanidino, histidine imino, etc.) to give interchain crosslinks. At higher peroxydisulfate concentrations, initial crosslinking by the above mechanism occurs until the sites available are exhausted, then extensive hydrolytic and free radical-induced main chain cleavage of the protein occurs thereby reversing the crosslinking of the gelatin protein chains and causing further degradation of susceptible amino acid side chains.

Two other reports have confirmed that other water-soluble proteins derived from collagen are crosslinked by peroxydisulfate anion (Needles and Whitfield, 1969; Slocum, 1966).

Fibroin

Silk fibroin, a protein rich in tyrosine residues, is cross-linked by treatment with 0.1-1.0 $\underline{M}$ ammonium peroxydisulfate at 70°C or in the presence of silver ion catalyst at 40°C (Needles, 1967b). As in the case of gelatin, the fibroin causes induced decomposition of the peroxydisulfate anion. The fibroin is not totally insolubilized by the crosslinking treatment, but the soluble portion of the fibroin is crosslinked to some extent, since its viscosity is greater than dissolved untreated fibroin at the same concentration. At higher peroxydisulfate concentrations degradation predominates over crosslinking. Methylation of tyrosine greatly reduces the capacity of fibroin to crosslink in the presence of peroxydisulfate, whereas modification of trypto-phan in fibroin has little effect on crosslinking. The sulfur content of the fibroin increases providing further evidence of induced peroxydisulfate decomposition. On peroxydisulfate treat-ment fibroin shows losses in tyrosine, histidine, and tryptophan, amino acids capable of crosslinking through quinoid intermediates, as well as lysine, arginine, proline, and methionine. Attack of amino acid residues in fibroin by peroxydisulfate is less discriminating in the presence of silver ion and is probably ex-plained by the competition of silver-ion $\underline{vs}$ fibroin-induced de-composition peroxydisulfate in the reaction.

Other Proteins

The peroxydisulfate-induced crosslinking of other proteins including deamidated gliadin, soybean globulin, alkali hydrolyzed soybean protein, phosvitin, and pepsin has been reported (Slocum, 1966). Aqueous solutions of >3% protein and 0.01-0.5% peroxydi-sulfate were used at temperatures of 60°C to 150°C (under pressure) for periods of 3 to 30 minutes. These crosslinked pro-teins were proposed substitutes of egg albumin (white) in food applications.

CONCLUSIONS AND FUTURE WORK INDICATED

In the presence of protein, peroxydisulfate undergoes induced decomposition and in turn leads to competitive crosslinking and degradation of the protein. Initially crosslinking predominates over degradation, but as the sites of crosslinking are diminished, degradation predominates. The peroxydisulfate anion decomposes to sulfate and hydroxyl free radicals through auto decomposition and protein-induced decomposition processes. Free radical attack

Crosslinking and Degradative Reaction Scheme for Peroxydisulfate Attack on Proteins

$$1/2 \ S_2O_8^= \longrightarrow SO_4^{-} \xrightarrow{H_2O} OH \cdot + HSO_4^{-}$$

Peroxydisulfate Decomposition

$$P\text{-}H \xrightarrow[\text{or } OH\cdot]{SO_4^{-}} P\cdot + HSO_4^{-} \text{ or } H_2O$$

$$P\cdot \xrightarrow{S_2O_8^=} P\text{-}OSO_3^{-} + SO_4^{-}$$

$$POSO_3^{-} \xrightarrow[H^+]{H_2O} P\text{-}OH + HSO_4^{-}$$

Protein-induced Decomposition

$$P\text{-}Ar\Big\langle\begin{smallmatrix}H\\H\end{smallmatrix} \xrightarrow[\text{or } OH\cdot]{SO_4^{-}} P\text{-}Ar\Big\langle\begin{smallmatrix}\cdot\\H\end{smallmatrix} + SO_4^{-}$$

$$P\text{-}Ar\Big\langle\begin{smallmatrix}\cdot\\H\end{smallmatrix} \xrightarrow{S_2O_8^=} P\text{-}Ar\Big\langle\begin{smallmatrix}OSO_3^{-}\\H\end{smallmatrix} + SO_4^{-}$$

$$P\text{-}Ar\Big\langle\begin{smallmatrix}OSO_3^{-}\\H\end{smallmatrix} \xrightarrow[H^+]{H_2O} P\text{-}Ar\Big\langle\begin{smallmatrix}OH\\H\end{smallmatrix} \xrightarrow{etc.} P\text{-}Ar\Big\langle\begin{smallmatrix}OH\\OH\end{smallmatrix}$$

$$P\text{-}Ar\Big\langle\begin{smallmatrix}OH\\OH\end{smallmatrix} \xrightarrow{S_2O_8^=} P\text{-}Ar\Big\langle\begin{smallmatrix}O\\O\end{smallmatrix} \xrightarrow{H\text{-}X\text{-}P} P\text{-}Ar\Big\langle\begin{smallmatrix}OH\\X\text{-}P\\OH\end{smallmatrix}$$

Crosslinking via Quinoid Intermediates

$$\text{\textasciitilde}NH\text{-}\underset{R}{CH}\text{-}\overset{O}{\overset{\|}{C}}\text{\textasciitilde} \xrightarrow[\text{or } OH\cdot]{SO_4^{-}} \text{\textasciitilde}NH\underset{R}{\overset{\cdot}{C}}\overset{O}{\overset{\|}{C}}\text{\textasciitilde} + HSO_4^{-} \text{ or } H_2O$$

$$\text{\textasciitilde}NH\underset{R}{\overset{\cdot}{C}}\overset{O}{\overset{\|}{C}}\text{\textasciitilde} \xrightarrow{S_2O_8^=} \text{\textasciitilde}NH\underset{R}{\overset{OSO_3^{-}}{C}}\overset{O}{\overset{\|}{C}}\text{\textasciitilde} + SO_4^{-}$$

$$\text{\textasciitilde}NH\text{-}\underset{R}{\overset{OSO_3^{-}}{C}}\text{-}\overset{O}{\overset{\|}{C}}\text{\textasciitilde} \xrightarrow[H^+]{H_2O} \text{\textasciitilde}NH\underset{R}{\overset{OH}{C}}\text{-}\overset{O}{\overset{\|}{C}}\text{\textasciitilde} \to \text{\textasciitilde}NH_2 + R\overset{OO}{\overset{\|\|}{CC}}\text{\textasciitilde}$$

Degradation

P-H   - Oxidizable Protein

$P\text{-}ArH_2$ - Protein with Aromatic Side Chain

P-x-H   - Protein with Reactive Side Chain

on the protein substrate is of only limited selectivity, with oxidative attack of reactive amino acid side chains including histidine, proline, cystine, methionine, tyrosine, tryptophan occurring. Oxidative coupling (crosslinking) of histidine, tyrosine, and tryptophan via quinoid intermediates is postulated as the major crosslinking reaction. On exhaustion of accessible amino acid sites in the presence of excess peroxydisulfate, free radical attack at α-carbon atoms along the main chain and subsequent cleavage of the chain into amide and α-ketoacid groups occurs.

Peroxydisulfate reactions on proteins provide a method to examine the relative reactivity of various sites in a protein to hydroxyl free radicals (the predominant reactive species under most reaction conditions), and the rate and degree of protein crosslinking possible. Further studies on the peroxydisulfate attack of proteins of known structure will provide information concerning the accessibility and reactivity of selected active sites with hydroxyl free radicals. Peroxydisulfate-induced crosslinking of proteins provides a method to insolubilize and immobilize proteins and may find use in such areas as encapsulation of materials with water permeable, but water insoluble biodegradable protein substrates and as heat set natural adhesives.

REFERENCES

Boyland, E., Sims, P. and Williams, D. C. (1956). Oxidation of Tryptophan and Some Related Compounds with Persulfate. Biochem. J., 62, 546-550.

Burke, M., Kenny, P. and Nicholls, C. H. (1962). Formation by Oxidizing Agents of Free Radicals in Wool and Silk. Nature, 196, 667-668.

House, D. A. (1962). Kinetics and Mechanism of Oxidations by Peroxydisulfate. Chem. Revs., 62, 185-203.

Lang, K. (1936). The Action of Persulfate in Alkaline Solution on Amino Acids. Z. Physiol. Chem., 241, 68-70.

Mason, H. S. (1955). Comparative Biochemistry of Phenolase Complex. Advan. Enzymol., F. F. Nord, ed., Vol. 16, 105-184.

Needles, H. L. (1965a). Persulfate Degradation of Wool. Textile Res., 35, 298-303 (1965).

Needles, H. L. (1965b). Degradation of Acetylated and Esterified Wools by Aqueous Persulfate. Textile Res. J., 35, 953-955.

Needles, H. L. (1967a). Crosslinking of Gelatin by Aqueous Peroxydisulfate, J. Polym. Sci., A-1, 5, 1-13.

Needles, H. L. (1967b). Crosslinking of Silk Fibroin by Aqueous Peroxydisulfate. J. Appl. Polym. Sci., 11, 719-726.

Needles, H. L. and Whitfield, R. E. (1964). Free Radical Chemistry of Peptide Bonds. I. Dealkylation of Substituted Amides. J. Org. Chem., 29, 3632-3634.

Needles, H. L. and Whitfield, R. E.   (1966).   Decarboxylation of
    N-Acetylamino-acids by Aqueous Peroxydisulfate.   Chem. and
    Ind., 287-288.
Needles, H. L. and Whitfield, R. E.   (1969).   Crosslinking of
    Collagens Employing A Redox System Comprising Persulfate and
    a Reducing Agent, U. S. Patent, 3,427,301 (Feb. 11, 1969).
Remy, D. E., Whitfield, R. E., and Needles, H. L.   (1967).   Study
    of the Decomposition Rates of Aqueous Peroxydisulfate in the
    Presence of Selected Amides and N-Acetylamino-acids.   Chem.
    Commun., 681-682.
Slocum, D. H.   (1966). Treatment of Proteinaceous Materials with
    Peroxydisulfate Salts, U. S. Patent 3,272,639 (Sept. 13, 1966).

33

CROSS LINKING IN THE RADIOLYSIS OF SOME ENZYMES AND RELATED PROTEINS

K.R. Lynn

Division of Biological Sciences, National Research

Council of Canada, Ottawa, Ontario, Canada K1A 0R6

Abstract

In non-covalently bound complexes of serveral serine proteases and of ribonuclease with DNA the enzymes were protected against the effects of ionizing radiation. No scavenging by the nucleic acids was observed. Similarly, complexing trypsin with silica protected the enzyme from radiolytic destruction.

Irradiation of solutions of serine proteases required about twice the D_{37} dose to produce about 10% polymerization: significantly lower relative doses were effective in causing polymerization in both lima bean protease inhibitor and in the octapeptidal hormone oxytocin.

Several sulfhydryl enzymes which have been examined were very efficiently inactivated by ionizing radiation. There was, at the same time, apparent formation of novel intra-molecular -S-S- bonds.

Introduction

During studies of the effects of ionizing radiation on some enzymes and related proteins, non-covalent cross-linking of the target molecules to both inorganic and organic compounds considerably modified the effects of the radiolysis. Results obtained with complexes formed between trypsin and SiO_2 (Lynn, 1972) were extended, in studies with several serine proteases and with ribonuclease when those enzymes were associated (Hofstee, 1962) in reversible complexes with DNA (Lynn, 1974, 1976). It was shown that the modifications in

557

radiolytic effects observed were not caused by simple scavenging by
the complexing partner (DNA). Those modifications may be the
result of conformational changes undergone by the proteins on formation of the complexes, or of the DNA diverting attack from susceptible residues in the enzymes (Lichtin et al., 1972).

Incidental to some studies of the effects of ionizing radiation
on serine enzymes (Lynn 1972a, 1970; Lynn and Orpen, 1969, 1969a;
Lynn 1970, 1971) a serine-protease inhibitor from lima beans (Haynes
and Feeney, 1967; Lynn and Raoult, 1976) and a peptidal hormone,
oxytocin (Purdie and Lynn, 1973) observations of relevance to a
consideration of cross linking in proteins were made. Under comparable conditions the larger molecule showed greater ability to
maintain its biological (enzymatic) activity than the peptide (m.w.
1000). Polymerization, which occurred significantly at a dose of
$2 \times D_{37}$ in the enzyme became a factor of measureable size at doses
to oxytocin approximating those of the D_{37}. It is also worth noting
that no evidence of an effect on the radiation sensitivity of
chymotrypsin was observed on changing the state of the enzyme from
that of a monomer (5×10^{-6} M) to a dimer (10^{-3} M).

On irradiation of several sulfhydryl enzymes such as papain
(Lynn and Louis, 1973) ficin, bromelain and chymopapain (Lynn,
unpublished work) no losses of amino acid residues could be
detected at doses where there was ca. 70% loss of activity. On
examining the products of such irradiation after peptidal digestions
it was found that changes in the enzymes had occurred following
radiolysis. These data suggest that there may be re-shuffling
between -SH and -S-S bonds during irradiation, losses of activity
then occurring.

Effects of Gamma Radiolysis on Aqueous Solutions/ Suspensions of Some Complexed Enzymes

Trypsin can be attached as an esterolytically active enveloping
monolayer to SiO_2 (Haynes and Walsh, 1969). The attachment is made
using glutaraldehyde and the final compound, which is stable in
borate buffer, pH 8.5, contains only the enzyme, silica and methylene groups. The enzyme lost esterolytic activity on irradiation as
a free solution (5×10^{-6} M) or as a SiO_2-trypsin suspension
(equivalent to a trypsin concentration of 10^{-6} M). Changing the
buffer concentration from 7×10^{-2} M to 7×10^{-4} M was without
effect. D_{37} values (Fig. 1) are collected in Table 1 where it is
seen that the complexing of the enzyme afforded it about fifty fold
protection. Silica is not expected to act as a scavenger for the
OH. or e_{aq}^-, which are the major products from irradiation of water
and so the species interacting with the enzyme. Similarly, the
methylene groups which hold the trypsin molecules in the envelope
surrounding the silica would not be notably reactive (Anbar and

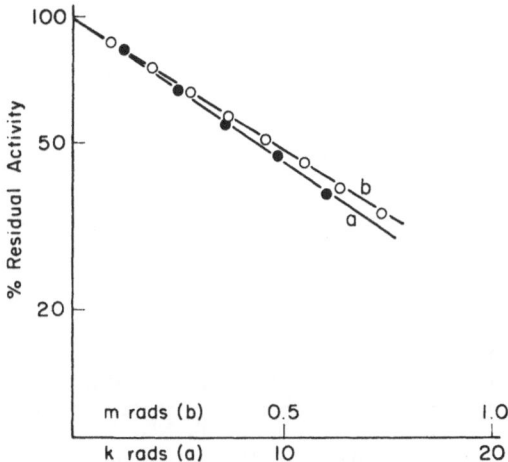

Fig. 1. Percentage residual tryptic activity, measured with TAME, plotted semilogarithmically as a function of dose for: (a) 5×10^{-6} M trypsin ($\bullet$); (b) trypsin-SiO$_2$ (O). Irradiations in 7×10^{-2} M borate buffer (pH 8.5).

Neta, 1967). It is thus reasonable to attribute the lower radio-sensitivity of the SiO$_2$-bound trypsin to modifications of its conformation - which are known to occur from, for example, inhibition studies (Haynes and Walsh, 1969).

More biologically relevant examples of the effects of cross-linking in protecting enzyme activity from radiolysis are afforded by complexes formed between trypsin, chymotrypsin, chymotrypsinogen, and ribonuclease with DNA.

These complexes, which are simply formed (Fig. 2) are soluble and readily dissociable (by changing ionic strength). It was, thus,

Table I. D_{37}'s from: (A) irradiated 5×10^{-6} M solutions of trypsin in 7×10^{-2} M, 7×10^{-4} M (brackets) borate buffer (pH 8.5); and (B) irradiated SiO$_2$-trypsin compound in the same buffer. All experiments at room temperature and in air.

Substrate	D_{37} (A)[a] in krads	D_{37} (B)[a] in mrads Batches	
		GC-19259	GC-19175
N-Z-L-TyrNp	10.9 ± 1.3 (10.7 ± 1.2)	0.60 ± 0.06	0.53 ± 0.10
BAEE	12.1 ± 0.6	—	—
TAME	12.9 ± 1.0	0.62 ± 0.07	0.67 ± 0.03
BANA	7.4 ± 0.3 (7.5 ± 0.4)	0.71 ± 0.07	0.64 ± 0.09

[a] ±Average deviations.

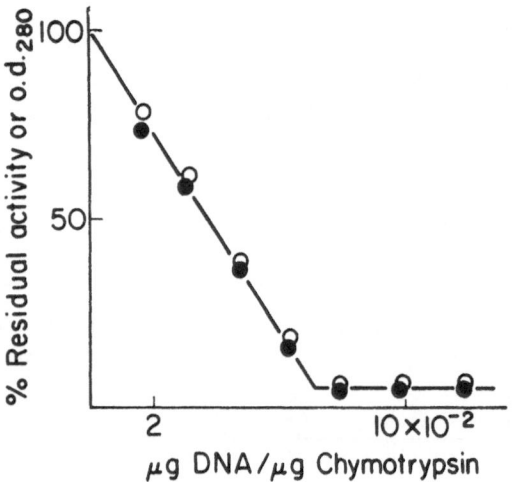

Fig. 2.   Residual activity (O) or optical density (280 nm; ●)
plotted as a percentage versus the weight ratio of DNA to chymotrypsin.

possible to examine the effects of radiolysis on the components in
non-associated as well as in associated states. Each complex with
each serine protease reacted in a unique manner on radiolysis, as
is shown in Fig. 3, and responded to assays uniquely also - as the
data of Fig. 4 show. Similarly, the damage sustained by the enzyme
varied from one preparation to another of the complexes, though the
data of Fig. 5 are typical. Comparison of the results from radiolysis of dissociated complexes of the enzymes and DNA, which were
reproducible (Lynn, 1974), with those measured earlier for the
proteases alone (e.g. Lynn, 1970) showed that the nucleic acid was
not acting as a radical scavenger. The D_{37}'s of about 16 k rads
(the precise figure depended on the substrate employed in the
assays) which were measured with the dissociated complexes were
comparable with those obtained with free solutions of the same
enzymes (Lynn, 1970, 1971). The variable nature of the results
obtained with the complexes thus suggested that the precise way in
which the protein and DNA interacted is critical in affecting the
nature of the damage sustained by the complexed enzyme. This conclusion has some relevance in considering the nature of the
radiolytic damage to individual components of multicomponent
systems such as living cells.

The results of amino acid analysis of the enzymes after
irradiation as complexes with DNA (Fig. 5 summarizes one set of

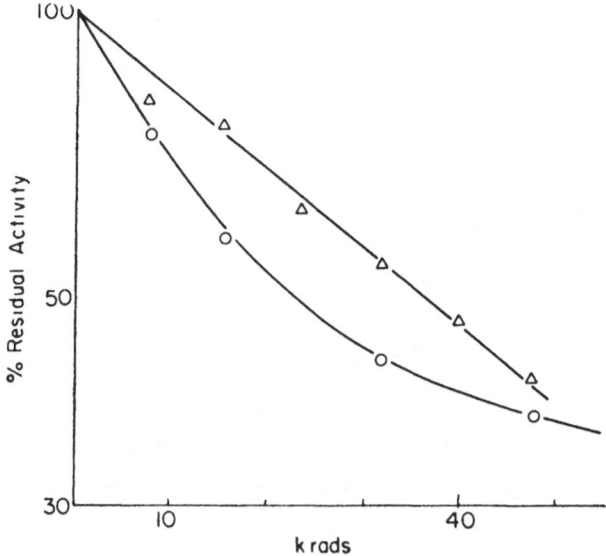

Fig. 3. Residual activity of chymotrypsin for BTEE plotted as a function of the dose delivered to two different preparations of DNA/chymotrypsin complex. Irradiation under oxygen.

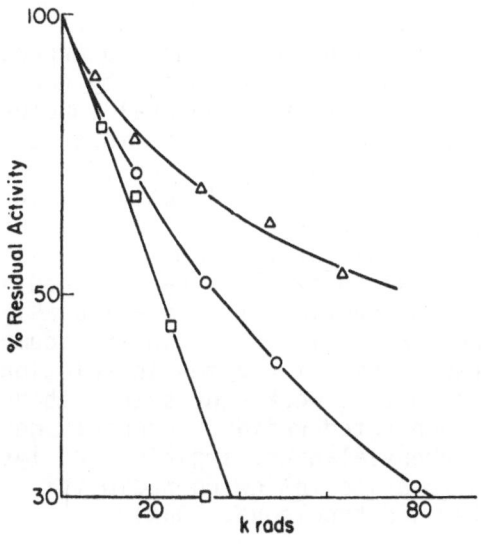

Fig. 4. Residual chymotryptic activity of a DNA/chymotrypsin complex irradiated under oxygen. Assays with Δ-N-Z-L-TyrNp; O-GPANA; □-BTEE.

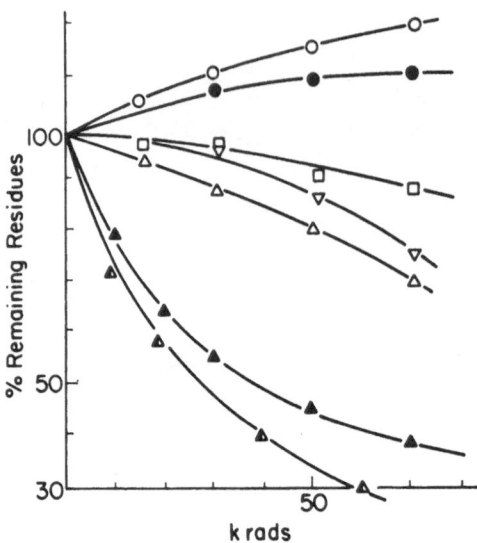

Fig. 5. Percentage of residues remaining after irradiation
of a DNA/trypsin complex under N₂. O-ala; ●-gly, asp, glu; △-tyr,
phe; □-ile, his; ▲-trp; (▲-trp, after irradiation under oxygen);
∇-cys.

data) could not be readily compared with those from irradiations of
the enzymes alone in solution (e.g. Lynn and Orpen, 1968; Lynn, 1971)
because of the uniqueness of the responses by each preparation of
the enzyme/DNA complex. However, a general conclusion was that in
the complex, attack by OH. and e_{aq}^- may be more random in the targets
found: a greater variety of amino acid residues were affected, at
doses comparable in terms of residual activity, in the complexed
than in the free enzyme. It is also noteworthy that the quality
of the damage, on radiolysis of the complexes, was different from
that observed on irradiation of the free components. There was,
for example, apparent production of some amino acid residues such
as gly, asp, glu and ala in Fig. 5. Such an accumulation is not
commonly found on radiolysis of enzymes in solution at the concentrations employed in this work - or even a thousand-fold greater
(Lynn, 1972a). The generated residues probably derive from
cysteine, tyrosine, phenylalanine, arginine and lysine from which
they have been seen to arise following radiolysis of the free
amino acids (Liebster and Kopoldova, 1964).

Complexes between DNA and ribonucleases have the same properties
as those discussed above (Lynn, 1976) with the added advantage that
radiolysis of them produces reproducible results. This may be

accounted for by the similarities between DNA and RNA, the natural
substrate of this enzyme. With this system it was shown (Fig. 6)
that there was clearly significant protection afforded the enzyme
on complexing with the nucleic acid: D_{37}'s were raised about six
fold. Demonstration that the nucleic acid was not acting simply as
a radical scavenger was also possible with this system, using
mixtures, under dissociating conditions, of the components of the
complex (Fig. 6).

     The manner in which interactions between enzymes and other
molecules protects them from radiation damage is still not clear.
In part it is probably by a changed conformation of the proteins
in the cross-linked compounds that susceptible centres relevant to
the enzymic activity are no longer exposed to reactions of OH. or
e_{aq}^{-}. Though there is extensive damage to the enzymes there is
also retention of considerable catalytic activity, as a comparison
of Figures 6 and 7 show.

     Further relevant to an examination of the role of cross-linking
in the radiolysis of proteins is some data obtained from complexes
of chymotrypsin with cardiolipin, and of concanavalin-A with glycogen-
that is, of protein with lipid and protein with protein. The former
complex, which was insoluble but formed a cloudy suspension in water,
was not formed in the presence of 0.2 M sodium perchlorate; the
latter complex, in which all sites on the con-A were not occupied
(Cifionelli et al., 1958) so that a soluble product was formed, was
not produced except in the presence of salt. So it was possible to
irradiate the components of the two complexes in the presence of
each other when complexed and not. Although measurements of
biological activities could not be made in these experiments, comparisons of the effects of gamma irradiation on the enzyme and
con-A were possible using amino acid analyses. These show (Fig. 8)
that several residues were more rapidly destroyed in the complexed
than in the non-complexed state.

        Cross-linking During Radiolysis of Proteins in Solution

     Measurement of the polymerization of proteins on irradiation
in the dry state at high doses have been reported from a number of
laboratories (e.g. Haskill and Hunt, 1967; Marciani and Tolbert,
1972; Delincée and Radola, 1975). Observations made during a study
of the effects of radiolysis on comparatively dilute solutions of
biologically active polypeptides - such as enzymes (Lynn and Orpen,
1969; Lynn, 1971, 1972a), a tryptic/chymotryptic inhibitor isolated
from lima beans (Lynn and Raoult, 1976), and a peptidal hormone,
oxytocin (Purdie and Lynn, 1973) have produced results which, while
semi-quantitative, are of interest when collected together.

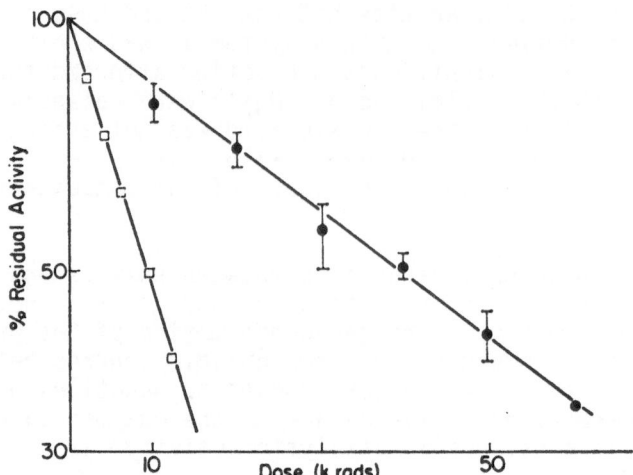

Fig. 6.   Radiolyses of DNA/ribonuclease complex in oxygenated water: Averages of four experiments assayed with cytidine 2':3' cyclic phosphate (●).   Estimation of D_{37} for ribonuclease irradiated in the presence of 0.10 M $NaClO_4$.   Sufficient DNA was present to completely complex the enzyme in the absence of the perchlorate (see text) (□).

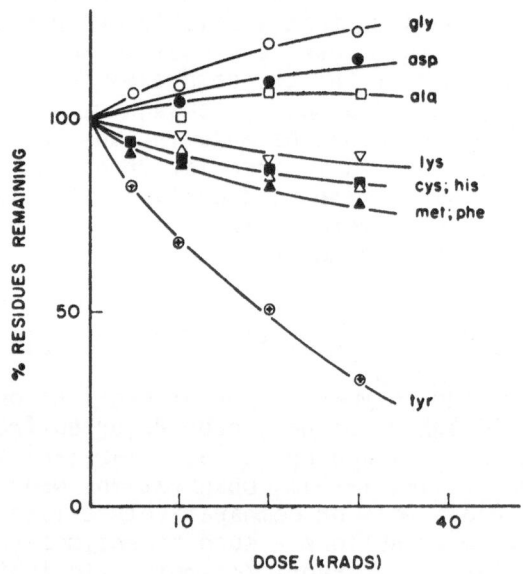

Fig. 7.   Analysis of ribonuclease after irradiation, under O_2, as a dissolved complex with DNA.   (○) gly; (●) asp; (□) ala; (▽) lys; (■) cys; (△) his; (▲) met, phe; (⊠) tyr.   Other residues were unaffected.

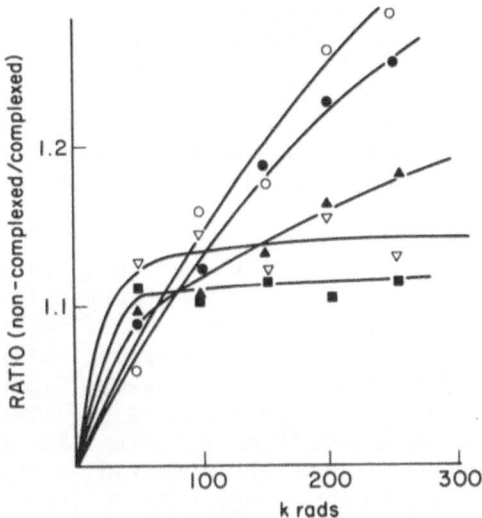

Fig. 8. The ratio of residual amino acids in "non-complexed" to those in "complexed" Con-A or chymotrypsin, plotted <u>versus</u> dose. Con-A/glycogen-his (o); tyr (∇): chymotrypsin/cardiolipin-phe (●); trp (▲); tyr (■).

Comparison of results from irradiations of chymotrypsin as a 5×10^{-6} M and a 1×10^{-3} M solution (Lynn, 1970) showed that no effects from simply increasing the concentration of the enzyme were apparent: there was a proportionality between the D_{37}'s measured and the concentrations. Similar results were observed with the oxytocin.

It has been found, in work cited above, that 10^{-3} M chymotrypsin, with a D_{37} of about 2 Mrads yielded 10-15% polymer on irradiation with about 4 Mrads. The lima bean protease inhibitor, when irradiated under nitrous oxide to a dose of 50-100 krads, displayed clear signs of polymerization though the D_{37} found under these conditions of radiolysis was about 160 krads. Further, oxytocin with a D_{37} of 230 krads, showed significant formation of higher polymers at 100 krads dose - when there was also apparent formation of (soluble) dimers in high yields.

So it is clear that, in relation to the D_{37}, for losses of biological (enzymatic, inhibitory or hormonal) abilities there was a marked trend in susceptibility to polymer formation with a lowering of the molecular weight of the irradiated proteinous substance. This result is one that might reasonably be anticipated. The tertiary structure of an enzyme such as chymotrypsin would require considerable modification before polymerization would be a significant possibility. By contrast, an octapeptide such as oxytocin (I) may

$$\overline{\text{Cys.Tyr.Ileu.Gln.Asn.Cys}}\text{.Pro.Leu.Gly.NH}_2$$

(I)

readily be envisaged as dimerizing following ring opening. This occurs at the -S-S- bond which is notoriously susceptible to radiolytic attack.

On irradiation of papain in dilute acetate solution remarkably efficient inactivation of the enzyme was observed (Lynn and Louis, 1973) with negligible destruction of amino acid residues (Table II). To confirm that sulfhydryl enzymes of the same family behave similarly we have lately irradiated the enzymes ficin, bromelain and chymopapain under the same conditions as employed for the papain. The results for ficin are summarized in Table III. Again remarkably efficient inactivation of the enzymes occurred, despite the fact that there was again only minor damage evident from amino acid analyses. One explanation of these results is that novel -S-S- bonds are formed on radiolysis, possibly involving, also, in the case of ficin, a limited chain reaction. However, no evidence of intermolecular binding was found on gel chromatography of irradiated ficin. The hypothesized new sulfur-sulfur linkages must, then, be intramolecular and incorporate the active site sulfhydryl groups. If this is so, one might anticipate the formation of new peptides in peptide maps prepared from the irradiated enzymes.

Table II

Residue dose‡ (krads)	0	2	6	10
	Residues/molecule			
Cys. acid	0	0·06	0·16	0·23
Asp.	19·0	19·3	18·7	18·9
Thr.	7·9	8·0	7·9	8·0
Ser.	12·8	12·8	12·7	12·9
Glu.	19·8	19·8	20·1	19·8
Pro.	10·0	10·2	10·1	10·0
Gly.	27·9	27·9	27·9	28·1
Ala.	14·0	14·0	14·0	14·2
1/2 Cys.	7·1	6·9	7·0	6·6
Val.	18·2	18·1	17·9	18·1
Met.	0	0	0	0
Ile.	11·8	12·0	11·9	12·0
Leu.	11·0	10·9	10·9	10·9
Tyr.	19·1	19·0	19·0	19·1
Phe.	4·0	4·0	4·0	4·0
His.	2·0	2·0	2·0	2·0
Lys.	10·1	10·0	10·0	9·9
Arg.	12·1	12·0	11·9	12·0
Trp.†	5·0	4·9	5·1	5·0

† Determined by method of Spies and Chambers (1949).
‡ Data were also obtained for 4 and 8 krad doses and conformed completely with trends shown in the results presented.

Table III. Summary of D_{37} values for irradiation of 5 x 10^{-5} M ficin in 10^{-2} M NaOAe.

Irradn. under

Substrate	O_2				N_2			N_2O	
	R*	U	D	R	U**	D	R	U	D***
CGN	5.6	4.2	7.5	12.4	4.0	7.7	12.5	2.5	6.0
BAME				12.0	4.6	7.7			
BAPA				14.9	5.5	7.6			
Casein				18.6	9.5	7.7			

*   - Repaired with dithiothreitol before assay.

**   - Unrepaired with reducing agent before assay.

*** - Sulfhydryl content assay with DTNB.

    Following radiolysis of ficin to its D_{37} dose, we reacted the product with iodoacetate (to block the sulfhydryl and prevent either self digestion or repair) digested the substituted enzyme with pepsin and prepared maps using thin layer electrophoresis and chromatography. In Fig. 9 a comparison of samples is possible, and some changes in the distribution of the peptidal products of the digestions are evident. This result, in conjunction with those obtained from the amino acid analyses supports the hypothesis of intramolecular rearrangements occurring. That these involve -S-S- bonds was confirmed by comparing "diagonal" maps prepared using the technique of Brown and Hartley (1966). Again differences are

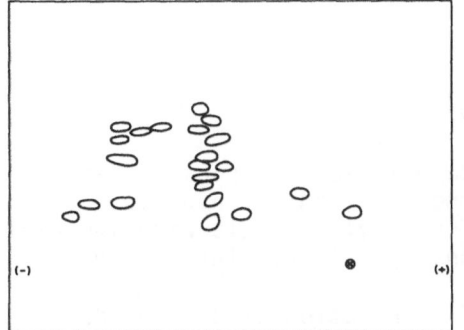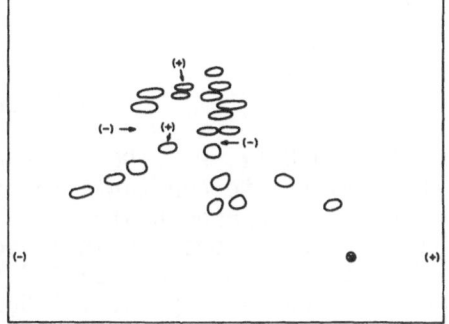

    Fig. 9. Thin layer peptide maps from pepsin digests of (left) non-irradiated and (right) irradiated ficin, -S-carboxy-methylated. Movement in horizontal direction by electrophoresis at pH 4.4, in vertical direction by chromatography.

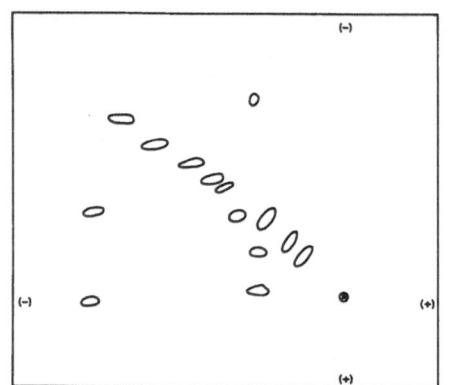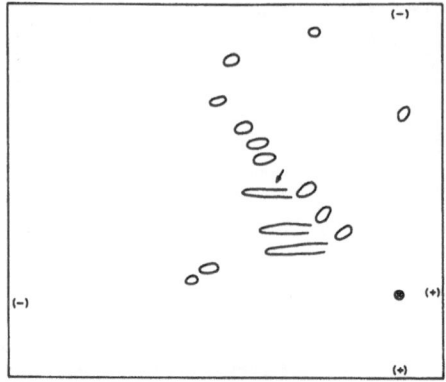

Fig. 10. Thin layer "diagonal" maps of pepsin digests of (left) non-irradiated and (right) irradiated ficin, -S-carboxy-methylated. Electrophoresis at pH 4.4 in both dimensions. Oxidation by performic acid.

apparent between the irradiated and the non-irradiated samples (Fig. 10).

CONCLUSION

The results discussed above show that cross-linking proteins to a variety of agents can modify the secondary effects of ionizing radiation on aqueous suspensions or solutions of the protein complexes formed. Radiolysis of proteins and peptides in aqueous solutions can also produce cross-linking - either by formation of dimers and higher polymers, or by the novel formation of intra-molecular sulfur-sulfur bonds.

REFERENCES

Anbar, M. and Neta, P. (1967). A compilation of specific bio-molecular rate constants for the reactions of hydrated electrons, hydrogen atoms and hydroxyl radicals with inorganic and organic compounds in aqueous solutions. Int. J. Appl. Radiat. Isotop., 18, 493-523.

Brown, J.R. and Hartley, B.S. (1966). Location of disulphide bridges by diagonal paper electrophoresis. The disulphide bridges of bovine chymotrypsinogen A. Biochem. J., 101, 214-231.

Cifionelli, J.A., Montgomery, R. and Smith, F. (1956). The reaction between concanavalin A and glycogen. J. Amer. Chem. Soc., 78, 2485-2488.

Delincee, H. and Radola, B.J. (1975). Structural damage of gamma-irradiated ribonuclease revealed by thin-layer isoelectric focusing. Int. J. Radiat. Biol., 28, 565-579.

Haskill, J.S. and Hunt, J.W. (1967). Radiation damage to crystalline ribonuclease; identification of the physical alterations by gel filtration on Sephadex. Radiat. Res., 31, 327-342.

Haynes, R. and Feeney, R.E. (1967). Fractionation and properties of trypsin and chymotrypsin inhibitor from lima beans. J. Biol. Chem., 242, 5378-5385.

Haynes, R. and Walsh, K.A. (1969). Enzyme envelopes on colloidal particles. Biochem. Biophys. Res. Comm., 36, 235-252.

Hofstee, B.H.J. (1962). Soluble complexes of nucleic acids with α-chymotrypsin and its derivatives. Biochem. Biophys. Acta, 124, 440-454.

Lichtin, W.W., Ogdan, J. and Stern, G. (1972). Fast consecutive radical processes within the ribonuclease molecule in aqueous solution. II. Reaction with OH radicals and hydrated electrons. Biochem. Biophys. Acta, 276, 424-442.

Liebster, J. and Kopoldova, J. (1964). Radiation chemistry of amino acids. In "Advances in Radiation Biology" (L.F. Augenstein, R. Mason and H. Quasther, Eds.) Vol. 1, 157-236. Academic Press, New York.

Lynn, K.R. (1970). X- and γ-irradiation of dilute solutions of chymotrypsin in atmospheres of air and nitrogen: effects on activity towards various substrates. Radiation Res., 43, 525-533.

Lynn, K.R. (1971). The effects of γ-radiation on dilute solutions of trypsin. Radiation Res., 46, 268-278.

Lynn, K.R. (1972a). Effects of γ-radiation on concentrated solutions of α-chymotrypsin. Radiation Res., 49, 51-62.

Lynn, K.R. (1972). Effects of γ-radiation on a silica-trypsin compound suspended in aqueous buffer. Radiation Res., 51, 265-271.

Lynn, K.R. (1974). γ-Radiolysis of trypsin, chymotrypsin and chymotrypsinogen when associated with DNA. Radiation Res., 57, 395-402.

Lynn, K.R. (1976). γ-Radiolysis of ribonuclease in association with DNA. Radiation Res., 66, 644-648.

Lynn, K.R. and Louis, D. (1973). The effects of γ-radiolysis on solutions of papain. Int. J. Radiat. Biol., 23, 477-485.

Lynn, K.R. and Orpen, Gail (1969a). X- and γ-irradiation of dilute solutions of chymotrypsin: the active intermediate. Int. J. Radiat. Biol., 14, 363-371.

Lynn, K.R. and Orpen, Gail (1969). X- and γ-irradiation of dilute solutions of chymotrypsin: Effects on the enzyme. Radiation Res., 39, 15-25.

Lynn, K.R. and Raoult, A.P.D. (1976). γ-irradiation of lima bean protease inhibitor in dilute aqueous solutions. Radiation Res., 65, 41-49.

Marciani, D.J. and Tolbert, B.M. (1972). Analytical studies of
     fractions from lysozyme. <u>Biochim</u>. <u>Biophys</u>. <u>Acta</u>, <u>271</u>, 262-
     273.
Purdie, J.W. and Lynn, K.R. (1973). The influence of gamma-
     irradiation on the structure and biological activity of
     oxytocin, a cyclic oligopeptide. <u>Int</u>. <u>J</u>. <u>Rad</u>. <u>Biol</u>., <u>23</u>,
     583-589.

N.R.C.C. No. 15500

34

ISOLATION AND CHARACTERIZATION OF STABLE PROTEIN-DNA
ADDUCTS INDUCED IN CHROMATIN BY ULTRAVIOLET LIGHT

Gary F. Strniste, Julia M. Hardin, and S. Carlton Rall

Cellular and Molecular Biology Group, Los Alamos
Scientific Laboratory, University of California, Los
Alamos, New Mexico 87545

ABSTRACT

The induction of protein-DNA adducts mediated by ultraviolet
(uv) light was analyzed in two forms of chromatin isolated from
cultured Chinese hamster cells: sheared, soluble chromatin, and
defined, chromatin subunits (nucleosomes). Four methods of analysis were employed to quantify and qualify this photochemical reaction of stabilizing protein to DNA: (1) a membrane filter assay
which retains both protein and protein-DNA complexes; (2) CsCl
equilibrium density gradients in which stable complexes of protein
and DNA band at densities other than their native buoyant densities; (3) gel filtration which allows separation of protein linked
to DNA from the bulk, nonlinked protein; and (4) SDS-polyacrylamide
gel electrophoretic analysis of the chromatin proteins. The following observations have been made: (1) the rate of linkage of
protein-to-DNA is linear with uv light fluence in both forms of
chromatin studied; (2) in sheared, soluble chromatin, the rate of
linkage of nonhistone proteins exceeds the rate of linkage of
histone proteins by two-fold (mass:mass); (3) in sheared, soluble
chromatin, factors which increase condensation of chromatin (divalent metal ions, ionic strength, pH) enhance the photochemical
addition reaction; (4) in both forms of chromatin, it appears as if
all four core histones (H2A, H2B, H3, and H4) participate (to varying degrees) in the uv light-induced crosslinking reaction; and
(5) uv light of 254-nm wavelength induces other photoproducts
(protein-protein adducts and/or aggregates) besides protein-DNA
adducts.

INTRODUCTION

Recent concerns about the potential depleting effects of certain technological waste products on the upper stratospheric ozone layer which acts as a protective shield against incoming, solar-originating, short wavelength, ultraviolet (uv) radiations have stimulated resurgent interests on the effects of uv light on biological systems. It is known that uv light can induce a variety of photoproducts in cells, including the classic and well-documented pyrimidine dimer in nucleic acids. Another rather unique photoproduct discovered simultaneously in uv-irradiated bacteria and mammalian cells is the protein-DNA adduct or complexes of protein and DNA stabilized evidently by a covalent linkage photochemically induced by the radiation (Alexander and Moroson, 1962; Smith, 1962). Although under certain conditions there exists a correlation between the occurrence of uv light-induced protein-DNA adducts and cell lethality in bacteria and mammalian cells (Ashwood-Smith et al., 1965; Smith and O'Leary, 1967; Habazin and Han, 1970; Han et al., 1975), the sublethal biological effects of this kind of photoproduct are not understood.

Several studies have been reported in an attempt to define the photochemical reaction involved in uv light-induced protein-DNA adducts. Various groups have reported the covalent linkage of various amino acids to DNA bases, to synthetic polynucleotides, and to native DNA (Smith and Aplin, 1966; Smith and Meun, 1968; Varghese, 1973; Schott and Shetlar, 1974). Furthermore, it has been shown that a variety of protein molecules which interact with nucleic acids can be photochemically induced to form covalent linkages with nucleic acids, including DNA polymerase (Markovitz, 1972), RNA polymerase (Strniste and Smith, 1974), aminoacyl tRNA synthetases (Schoemaker and Schimmel, 1974), ribosomal proteins (Gorelic, 1975a,b; Baca and Bodley, 1976), and chromatin proteins (Strniste and Rall, 1976a; Todd and Han, 1976). In some of these studies, uv light was used as a stabilizing agent to fix protein to DNA in order to determine the structural relationships or interacting sites of unique proteins and their nucleic acid counterparts.

The DNA of eucaryotic cells is heterogeneously complexed as chromatin with a variety of basic (histone) and acidic (nonhistone) proteins. Since protein-DNA adducts have been observed in uv-irradiated mammalian cells, we undertook this investigation to determine the role of chromosomal proteins in this particular photochemical process. In this report, we review some of our recent published studies concerning uv light-induced protein-DNA adducts in sheared, soluble chromatin (see Strniste and Rall, 1976a) and present some recent experimental findings concerning adduct formation in uv-irradiated, defined chromatin subunits or nucleosomes (for a detailed analysis and characterization of Chinese hamster cell nucleosomes, see Strniste et al., 1976b). In addition to various kinetic studies on the induction of protein–DNA adducts in chromatin and nucleosomes, we present data in an attempt to define qualitatively and quantitatively the role of both histones and nonhistones in the photochemical addition reaction between protein and DNA.

METHODS AND MATERIALS

Tissue Culture

Chinese hamster cells (line CHO) were grown in suspension and labeled with either (methyl-^{3}H)thymidine, (^{14}C)lysine, or (^{14}C)-tryptophan, as described previously (Strniste and Rall, 1976a).

Chromatin Isolation

Sheared, soluble chromatin was prepared from purified CHO nuclei by a decreasing ionic strength procedure, as described previously (Strniste and Rall, 1976a).

Nucleosome Isolation

The nucleosome classes, monomers through tetramers, were isolated from nuclease-treated, (methyl-^{3}H)thymidine- or (^{14}C)lysine-labeled CHO nuclei and were purified via sedimentation through isokinetic sucrose gradients, as recently described by Strniste et al., 1976b.

Irradiation with UV Light

Irradiation was performed at room temperature using an unfiltered germicidal lamp. Samples of chromatin or nucleosomes ($\leq$ 1 ml at 5–60 μg DNA/ml) were exposed to an incident fluence at 254 nm of 6.0 Jm^{-2} s^{-1}.

Filter Assay

A membrane filter assay developed for analysis of protein–DNA complexes has been described in detail elsewhere by Strniste and Smith (1974) and by Strniste and Rall (1976a). Using this technique for the analysis of uv light-induced protein–DNA adducts in nucleosomes, only high salt concentrations (3 $\underline{M}$ NH$_4$C$_2$H$_3$O$_2$) were employed in the dissociating medium.

CsCl Density Gradient Assay

The procedure used for the analysis of uv light-induced protein–DNA adducts has been described previously (Strniste and Smith, 1974; Strniste and Rall, 1976a).

Column Assays

The procedure used for analyzing protein-DNA adducts induced in sheared, soluble chromatin by uv light utilizing the method of gel filtration (Sepharose 4B) has been described recently (Strniste and Rall, 1976a). For analysis of uv light-induced adducts in nucleosomes, gel filtration using Bio-Gel A-1.5m (Bio-Rad Laboratories) was employed. Control or uv-irradiated samples of nucleosomes ($\leq$ 0.3 ml) at a final concentration of 3 $\underline{M}$ $NH_4C_2H_3O_2$ were applied to salt-equilibrated columns (13 x 0.7 cm) and eluted with 3 $\underline{M}$ $NH_4C_2H_3O_2$ into 36 0.2-ml fractions directly into scintillation vials at a flow rate of 9 ml/h. Samples were diluted with 1 ml H_2O and counted after the addition of 10 ml PCS scintillation fluid (Amersham/Searle).

SDS[+]-Polyacrylamide Gel Electrophoresis

Polyacrylamide gel slabs (15%, 0.75 x 160 x 280 mm) using the SDS-Tris-glycine system of Laemmli (1970), as previously described (Strniste et al., 1976b), were employed to analyze the binding patterns of proteins from control and uv-irradiated sheared, soluble chromatin and nucleosomes. The slabs were electrophoresed 10.5 h at 20 mamp constant current. After staining, the slabs were trimmed and then sliced into 16 1.5-cm strips and prepared for counting, as previously described (Strniste et al., 1976b).

RESULTS AND DISCUSSION

Chromatin Isolation and Characteristics

Since two different forms of chromatin which were prepared in different fashions and exhibit different physical and chemical characteristics have been used in our studies, it is necessary to define more precisely these two forms which will be discussed throughout this paper. Sheared, soluble chromatin was isolated from purified nuclei which were hypotonically disrupted. The resultant chromatin was separated from other nuclear contaminants by differential centrifugation. The chromatin pellet was swollen in low ionic strength buffer and solubilized by mechanical shearing. The protein-to-DNA mass ratio was 2.2:1.0, and the protein composition consisted of the five major histone bands and at least 25 distinct nonhistone bands when analyzed on SDS-polyacrylamide gels. The mass ratio of histones-to-nonhistones was about 2.0 (for further details, see Strniste and Rall, 1976a).

[+]Abbreviations used throughout the text: SDS, sodium dodecyl sulfate; SLS, sodium lauroyl sarcosinate; Tris, tris(hydroxymethyl)-aminomethane; CHO, Chinese hamster cells; HSEtOH, 2-mercaptoethanol; cpm, counts per minute.

Nucleosome classes, monomers through tetramers, were isolated from micrococcal DNase-treated nuclei. The nucleosome classes were purified via sedimentation through isokinetic sucrose gradients. The protein-to-DNA mass ratio was substantially reduced (1.2 to 1.4:1.0) primarily due to the considerable reduction in nonhistone proteins associated with the complexes. All five major histones were present in each class, although histone H1 was present in greater amounts as the size of the nucleosome class increased (see Strniste et al., 1976b, for further details).

Methods for Analyzing Protein-DNA Adducts

A rapid assay for detection of protein-DNA adducts utilizes a membrane filter technique. This assay is based on the facts that DNA molecules (double-stranded) readily pass through the filter, whereas protein molecules remain associated or bound to the filter. Therefore, any DNA which is linked to protein, even after treatment with high concentrations of salt and detergent to disrupt non-covalent interactions, will remain bound to the filter along with the protein. Samples of chromatin or nucleosomes were irradiated with uv light for various fluences, dissociated with either 3 $\underline{M}$ NaCl, 0.5% SLS (ss-chromatin)* or 3 $\underline{M}$ $NH_4C_2H_3O_2$ (nucleosomes) and passed through Millipore filters. In Fig. 1, the nonlinear increase in percent of the total cpm for (^{3}H)TdR ss-chromatin¶ remaining on the filter as a function of uv light fluence is shown. The stability of ss-chromatin in low salt (1 m$\underline{M}$ Tris, 1 m$\underline{M}$ EDTA) even after extensive irradiation is also shown (the low salt wash curve).

In Fig. 2, the results of a similar experiment are shown in which (^{3}H)TdR nucleosome classes, monomers through tetramers, were uv-irradiated. Samples of nucleosomes at a constant DNA concentration (5 µg/ml) were irradiated, dissociated in 3 $\underline{M}$ $NH_4C_2H_3O_2$, and applied to Millipore filters under gentle suction. It is interesting to note that the rate of induction of protein-DNA adducts in nucleosome classes by uv light, as detected by entrapment onto membrane filters, progresses from the monomer through the tetramer. Ignoring the initial portion of each induction curve, the rates of percent DNA bound to the filter per 1000 Jm^{-2} are 0.92, 1.75, 2.67, and 3.17 for the monomers through tetramers, respectively. Since,

---

*Throughout the text, sheared, soluble chromatin will be referred to as ss-chromatin.

¶Throughout the text, ss-chromatin labeled with (^{3}H)thymidine, (^{14}C)lysine, or (^{14}C)tryptophan will be referred to as (^{3}H)TdR, (^{14}C)Lys, or (^{14}C)Trp chromatin, respectively; nucleosomes labeled with (^{3}H)thymidine or (^{14}C)lysine are noted as (^{3}H)TdR or (^{14}C)Lys nucleosomes.

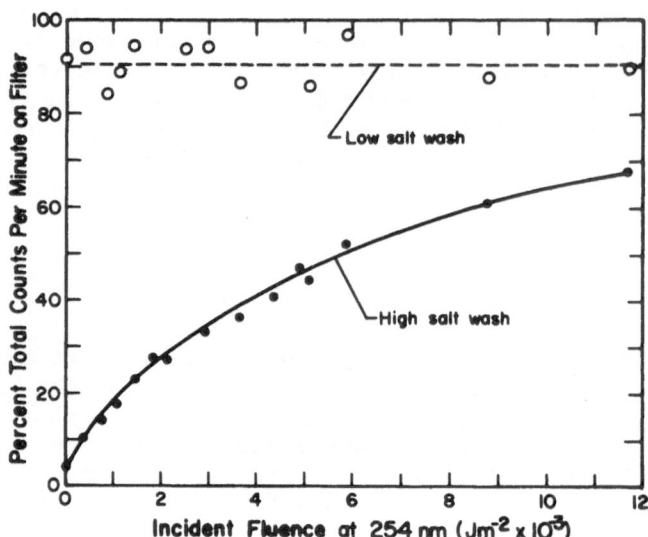

Figure 1. Detection of uv light-induced protein-DNA adducts in (^{3}H)TdR-labeled, sheared, soluble chromatin by a membrane filter assay. The solid line (——●——) shows the fluence-dependent increase in DNA remaining on the filter after dissociation of the chromatin in high salt and detergent. The dashed line (--O--) shows the stability of chromatin in low salt. For experimental details, see Strniste and Rall (1976a).

on the average, the number of protein molecules associated with the unit DNA increases in a ratio of 1:2:3:4 for monomers:dimers:tri-mers:tetramers, it would seem reasonable to expect the above result if only one molecule of protein linked per unit DNA would be neces-sary to entrap the complex to the filter. The amount of DNA involved in the crosslinking process is much less for all nucleo-some classes than for ss-chromatin. A possible explanation for this may be that ss-chromatin has much more protein originally associated with it, but this may be only one of a number of con-tributing factors.

Another method for analyzing protein-DNA adducts is CsCl density equilibrium gradients. Native CHO DNA has a buoyant den-sity in CsCl of about 1.70 g/cc; however, if proteins are linked to the DNA through stable, nondissociable covalent bonds, there will be a shift in the resultant buoyant densities of the complexes to lighter densities. In Fig. 3a, CsCl gradients are shown for (^{3}H)TdR ss-chromatin (control and uv-irradiated). Note the skew-ing in banding pattern of the DNA toward lighter densities, includ-ing material which floats on top of the gradient ($\rho \leq 1.63$ g/cc).

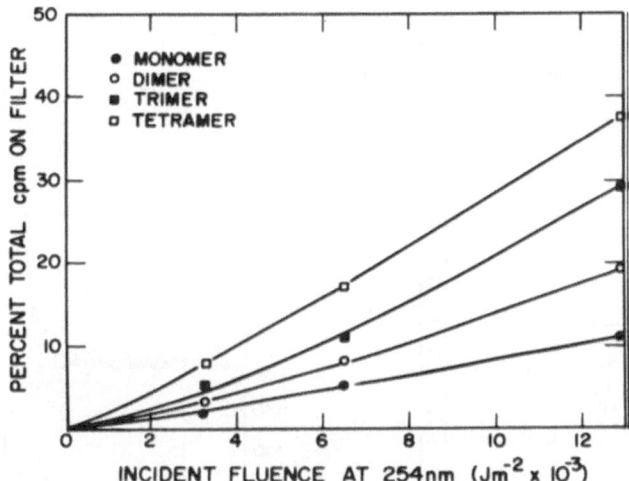

Figure 2. Detection of uv light-induced protein-DNA adducts in (^{3}H)TdR-labeled nucleosomes by a membrane filter assay. Nucleosome classes, monomer through tetramers, were irradiated with uv light at 5-10 μg DNA/ml and assayed as described in Methods and Materials (see also Strniste and Smith, 1974; Strniste and Rall, 1976a).

In Fig. 3b, a similar set of experiments are shown except, in this case, the ss-chromatin was labeled with (^{14}C)Lys. In the control, nonirradiated sample, the (^{14}C) label (protein) essentially floats on top of the gradient, whereas ss-chromatin samples exposed to uv light show an accumulation of counts near the banding position of the DNA (1.70 g/cc) and in the region between the banding positions of native DNA and protein. There is an uv light-fluence-dependent linear increase in the amount of (^{14}C)Lys material linked to the DNA (i.e., ^{14}C cpm banding at positions in the gradient at densities greater than the buoyant density of free protein): for example, at 5856 Jm^{-2}, 3.1% of the total cpm; at 11,712 Jm^{-2}, 6.3% of the total cpm.

Gel filtration is another useful method to study protein-DNA adducts. A proper choice of gel medium enables one to exclude the DNA and protein-DNA adducts while sieving the bulk, nonlinked proteins even under extreme dissociating conditions. For analysis of uv light-induced protein-DNA adducts in ss-chromatin, a column containing Sepharose 4B was chosen. Samples of (^{14}C)Lys ss-chromatin were irradiated with uv light, dissociated with and eluted through the column with 2 M NaCl, 0.1% SLS. In Fig. 4 are shown several chromatographic profiles for (^{14}C)Lys ss-chromatin

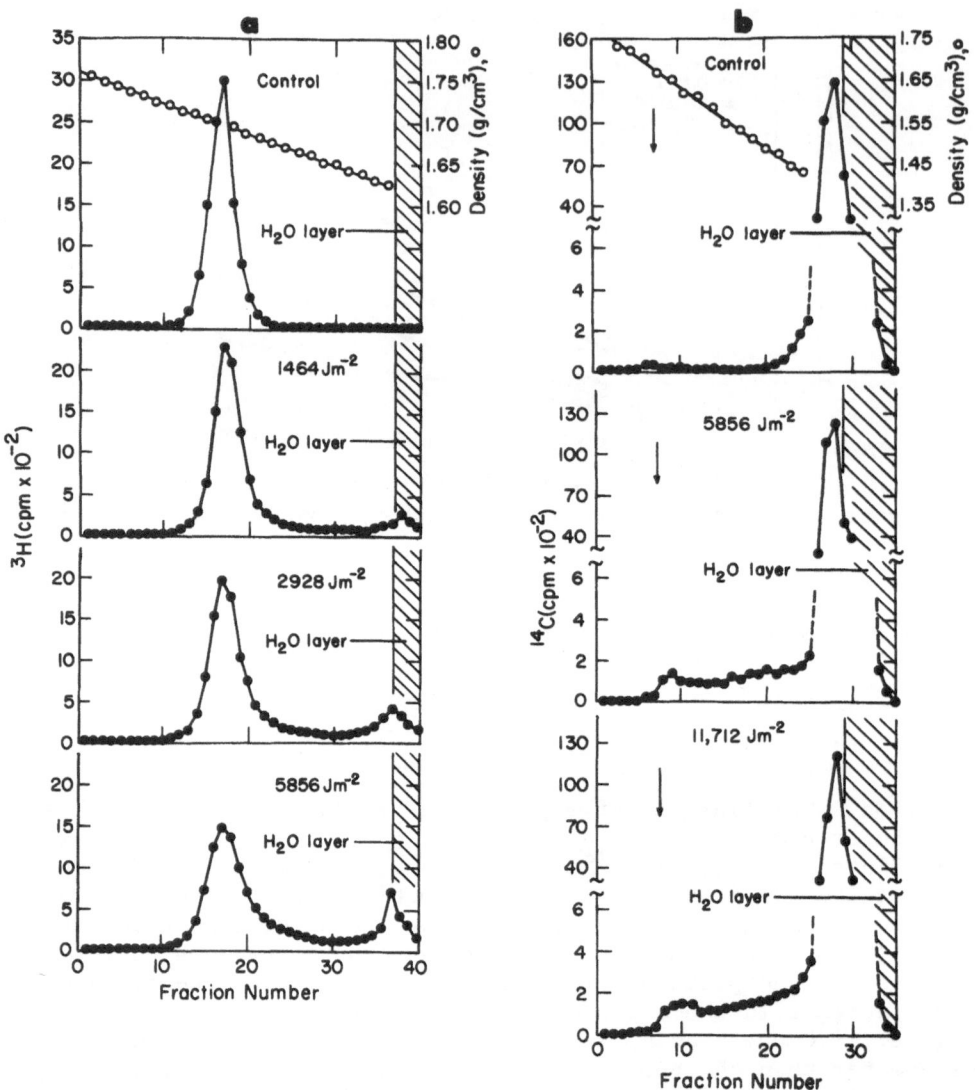

Figure 3.   (a) CsCl density equilibrium gradient profiles of control and uv-irradiated, (^{3}H)TdR-labeled, sheared, soluble chromatin; and (b) CsCl density equilibrium gradient profiles of control and uv-irradiated, (^{14}C)Lys-labeled, sheared, soluble chromatin. For complete details, see Strniste and Rall (1976a).

as a function of uv light fluence.   The insert graphs show the percent of ^{14}C cpm associated with the DNA (hatched areas) for increasing uv light fluences.   It was calculated that, for every 1830 Jm^{-2}, 1% of the total (^{14}C)Lys cpm are linked to the DNA.

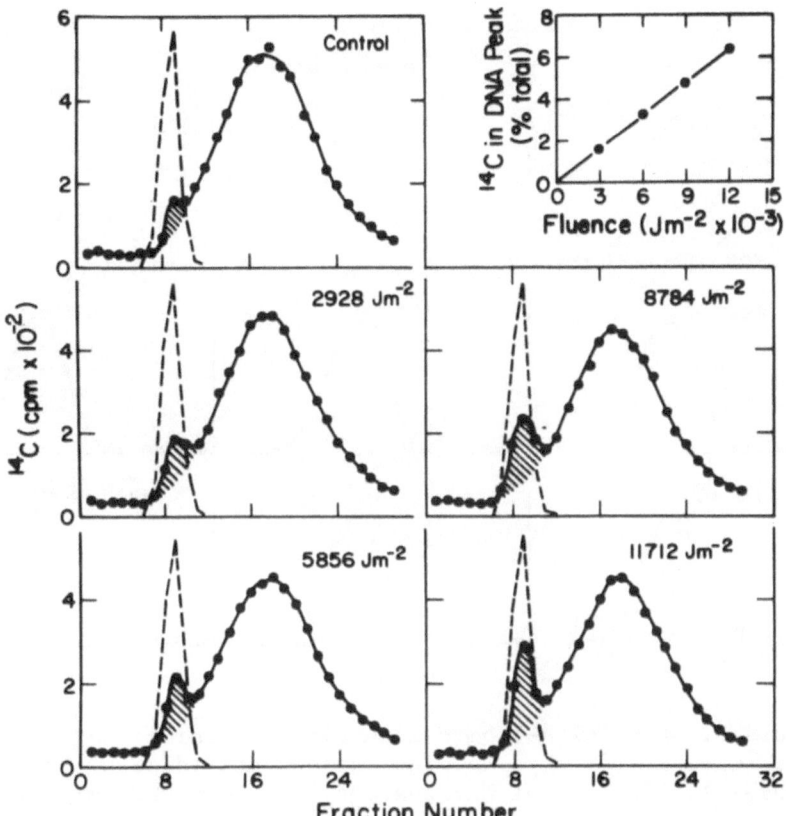

Figure 4. Sepharose 4B column elution profiles of control and uv-irradiated, (^{14}C)Lys-labeled, sheared, soluble chromatin dissociated and eluted with 2 $\underline{M}$ NaCl, 0.1% SLS. The dashed line (-----) reflects the elution pattern of unirradiated DNA [(^{3}H)TdR chromatin]. The solid line (——•——) reflects the ^{14}C cpm (protein profile). The hatched areas represent those ^{14}C cpm (protein) associated with the DNA, plotted as a function of uv light fluence in the insert graph (from Strniste and Rall, 1976a).

In Fig. 5, a similar set of experiments is shown for the formation of uv light-induced protein-DNA adducts in nucleosomes (dimers, trimers, and tetramers). In this set of experiments, the gel medium chosen was Bio-Gel A-1.5m. Tetramer and trimer DNA [shown by the dotted line using (^{3}H)TdR nucleosomes] was essentially excluded in this medium, whereas the dimer DNA was partially excluded. Essentially all monomeric DNA entered the gel (data not shown) and, therefore, did not allow sufficient separation from the bulk proteins. Therefore, monomer studies using this technique were not continued. As was evident for the ss-chromatin, there is

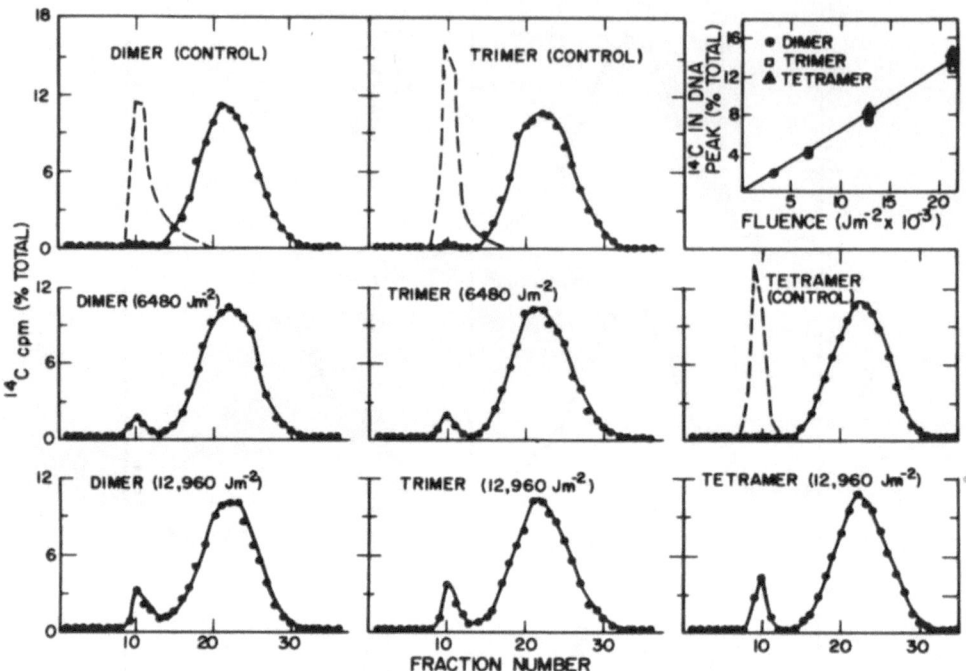

Figure 5. Bio-Gel A-1.5m column elution profiles of control
and uv-irradiated, (^{14}C)Lys-labeled nucleosomes (dimers, trimers,
and tetramers). Samples of nucleosomes at 5-10 μg DNA/ml were irradiated with uv light, dissociated, and eluted through the column
with 3 $\underline{M}$ NH$_4$C$_2$H$_3$O$_2$ (see Methods and Materials). The dashed line
(-----) reflects the elution pattern of unirradiated DNA [(^{3}H)TdR-
labeled dimers, trimers, and tetramers]. The solid line (——•——)
reflects the ^{14}C cpm (protein profile). Those ^{14}C cpm co-eluting
with the DNA were summed together and plotted vs uv light fluence,
as shown in the insert graph.

an uv light-fluence-dependent increase in the amount of (^{14}C)Lys
(protein) associated with the DNA peak region after uv-irradiated
nucleosomes were dissociated in high salt and eluted through the
column. The insert graph shows a linear increase in the amount of
(^{14}C)Lys cpm associated with the DNA over the range of uv light
fluences examined: 1% of the total cpm was linked to the DNA per
1560 Jm^{-2} for dimers, trimers, and tetramers. The fact that
dimers, trimers, and tetramers all show essentially the same rate
of protein linkage to DNA is not surprising, since the protein
composition per mass DNA is similar for all three (see Strniste et
al., 1976b). The difference in the rate for ss-chromatin is
influenced by two factors: (1) ss-chromatin has more nonhistone
protein involved (see following section); and (2) histone linkage

is less in ss-chromatin than in nucleosomes per unit uv light
fluence (later discussion and Fig. 10).

Linkage of Histones and Nonhistones in SS-Chromatin

As stated previously, ss-chromatin isolated from CHO cells
contains the five basic histone proteins and at least 25 nonhistone
proteins. Utilizing the Sepharose 4B column assay, samples of ss-
chromatin which were labeled with either (^{14}C)Lys or (^{14}C)Trp were
uv-irradiated for various fluences, dissociated in high salt-
detergent, and eluted through the column. That portion of the ^{14}C
cpm which eluted coincidentally with the DNA is plotted vs uv light
fluence in Fig. 6 for both (^{14}C)Lys- and (^{14}C)Trp-labeled chromatin. As was noted before, 1% of the (^{14}C)Lys cpm is associated
with the DNA per 1830 Jm^{-2}, whereas 1% of the (^{14}C)Trp cpm is
associated with the DNA per 1200 Jm^{-2}. If it is assumed that
(1) there is no tryptophan in histone protein; (2) the percent of
total CHO chromatin protein which is histone and nonhistone is 66
and 33, respectively (Strniste and Rall, 1976a); and (3) the mole
percent of lysine in histone and nonhistone is 12.7 and 6.3 (on
the average), respectively (see Gurley and Hardin, 1968; Marushige
et al., 1968), it can be shown that, on a mass basis, the rate of
uv light-induced linkage of nonhistone-to-DNA is about twice that
of histone.

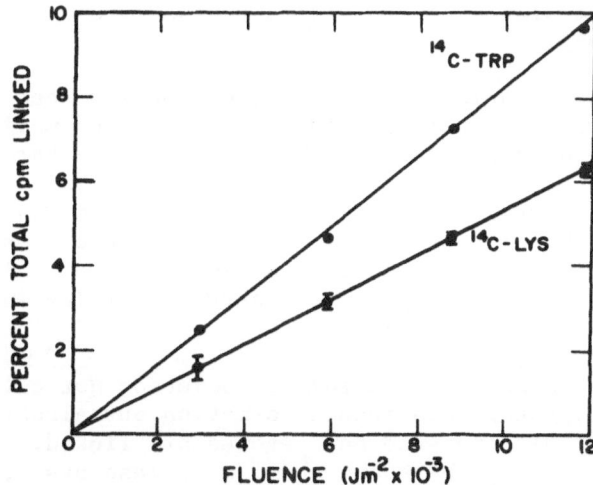

Figure 6. Results from the Sepharose 4B column analysis of
control and uv-irradiated, (^{14}C)Lys- and (^{14}C)Trp-labeled,
sheared, soluble chromatin. Analysis of protein associated with
DNA as a function of uv light fluence was performed as outlined
in Fig. 4 (see also Strniste and Rall, 1976a), and the percent of
total ^{14}C cpm associated with the DNA as a function of uv light
fluence is plotted.

Effects of Ionic Strength, pH, and Divalent

Metals on Protein-DNA Adduct Formation

We have previously demonstrated that alterations in the physical state of ss-chromatin drastically influences the rate of
protein-DNA adducts induced by uv light (Strniste and Rall, 1976a).
In general, it was concluded that variances in condensation state
of the chromatin influenced by changes in ionic strength or pH of
the medium or the addition of divalent metal ions could partially
account for the uv light-induced response (see Davies and Walker,
1974). Other contributing factors which could also aid or depress
the photochemical reaction would be protein migration, redistribution, and exchange, all of which are known to occur in chromatin
under a variety of conditions (see Clark and Felsenfeld, 1971; Van
and Ansevin, 1973; Varshavsky and Ilyin, 1974). In Fig. 7 we present a few representative Sepharose 4B column elution patterns for
(^{14}C)Lys ss-chromatin uv-irradiated under conditions of varying pH
or the addition of NaCl. Over the pH range tested (7.5 to 1.5),
there is a ten-fold change in the amount of ^{14}C cpm (protein) associated with the DNA at the particular fluence of uv light used
(2828 Jm^{-2}). Furthermore, adduct formation in ss-chromatin increased with salt concentration, peaking at twice the level at 0.2 $\underline{M}$
compared to no salt addition, and then rapidly decreased in yield
until at 2 $\underline{M}$ NaCl the amount of counts associated with the DNA peak
equaled the unirradiated control (data for NaCl concentration
> 0.8 $\underline{M}$ not shown).

A similar stimulation in adduct formation was seen with the
inclusion of divalent metal ions in the irradiation medium, as is
evident from the data presented in Table 1. Of the three metal
ions tested (Mg^{2+}, Mn^{2+}, and Zn^{2+}), there was a stimulation of
about eight-fold in the amount of protein linked to DNA at the
particular uv light fluence used. Furthermore, there was a noticeable difference in the concentration of metal ion at which maximum
enhancement in adduct formation was observed, varying in the order
Zn^{2+}< Mn^{2+}< Mg^{2+} (see Table 1).

Alterations in pH of the chromatin solution not only influence the rate of protein-DNA adduct formation but also influence
the rate at which histones and nonhistones are linked. As was
noted previously, it can be shown that, on a mass basis, the rate
of linkage of nonhistones exceeds the rate of linkage of histones
at neutral pH by about two-fold. This difference in rate is maintained to a pH of about 2.3, whereby the rate of linkage of nonhistones-to-histones increases such that, at a pH of 1.5, the rate
of linkage of nonhistones-to-histones is greater than 2.5:1.0
[see Table 2 for actual data concerning the amount of ^{14}C cpm co-
eluting with DNA on a Sepharose 4B column for both uv-irradiated
(^{14}C)Lys and (^{14}C)Trp ss-chromatin as a function of pH].

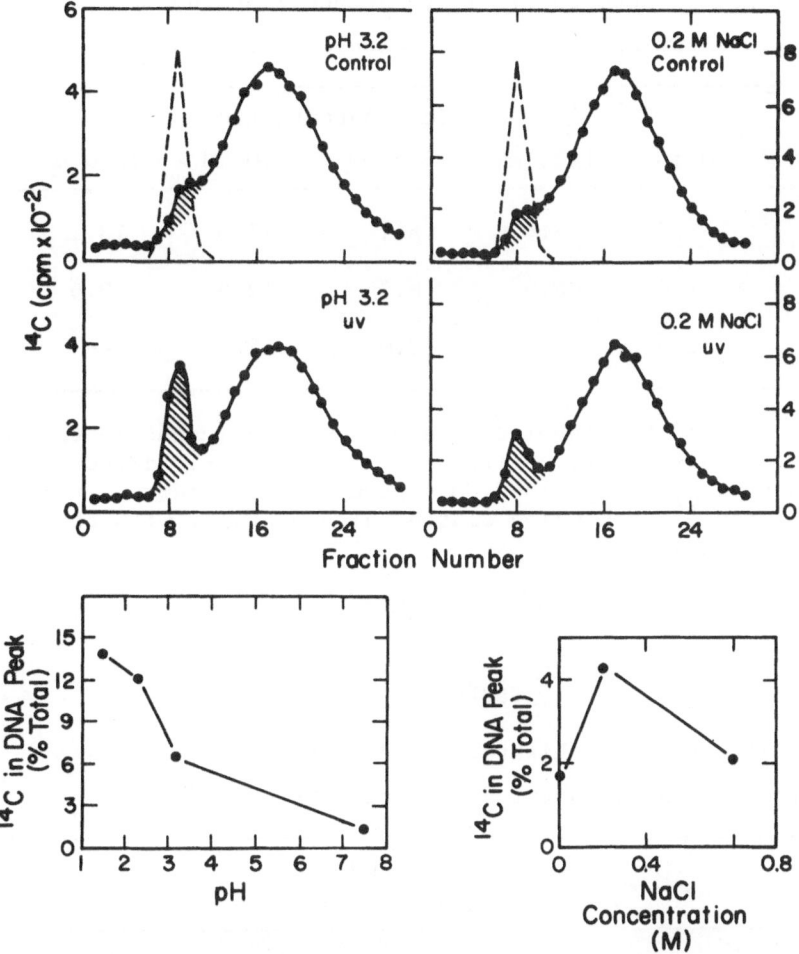

Figure 7. Sepharose 4B column assay of uv-irradiated, (^{14}C)-Lys-labeled, sheared, soluble chromatin as a function of pH or ionic strength of the irradiation medium. Details of the column assay are the same as in Fig. 4: (-----) chromatin DNA profile; and (——●——) ^{14}C cpm (protein profile). The protein associated with the DNA (hatched areas) at a uv light fluence of 2928 Jm^{-2} for various pHs and NaCl concentrations is plotted in the lower graphs (from Strniste and Rall, 1976a).

TABLE 1

Metal Ion Effects on UV Light-Induced Protein-DNA
Crosslinking in Sheared, Soluble Chromatin[a]

	Metal Ion Concentration				
	0 mM	1.25 mM	2.5 mM	5.0 mM	10.0 mM
Mg^{2+}	1.6	3.4	5.1	11.8	13.1
Mn^{2+}	1.6	6.9	8.7	13.6	9.6
Zn^{2+}	1.6	9.6	13.0	12.1	--

[a]Solutions of (^{14}C)Lys-labeled chromatin were adjusted to the noted
metal ion concentrations, uv-irradiated (2928 Jm^{-2}), dissociated,
and eluted through a Sepharose 4B column. The numbers in the table
reflect the net ^{14}C cpm which eluted coincidentally with the DNA
(percent of total).

TABLE 2

pH Effects on UV Light-Induced Protein-DNA
Crosslinking in Sheared, Soluble Chromatin[a]

	pH			
	7.5	3.2	2.3	1.5
(^{14}C)Lys	1.6	6.5	12.2	14.3
(^{14}C)Trp	2.5	10.4	19.0	27.5

[a]Solutions of (^{14}C)Lys- or (^{14}C)Trp-labeled chromatin were
adjusted to the noted pHs, uv-irradiated (2928 Jm^{-2}), dissociated,
and eluted through a Sepharose 4B column. The numbers in the table
reflect the net ^{14}C cpm eluted with the DNA (percent total).

Protein-DNA Adducts Analyzed on SDS-Polyacrylamide Gels

To examine qualitatively the role of various chromosomal proteins in uv light-mediated protein-DNA adduct formation, samples
of both (^{14}C)Lys and (^{14}C)Trp ss-chromatin were uv-irradiated,
lyophilized, resuspended in a buffer containing SDS and HSEtOH,
boiled, and electrophoresed in a slab gel apparatus. Completed
slabs were stained, sliced into 16 1.5-cm strips, and counted. In
Fig. 8, the radioactive distribution of ^{14}C cpm is shown for control and uv-irradiated ss-chromatin. For (^{14}C)Lys ss-chromatin
in which both histone and nonhistone proteins are labeled, note
the disappearance of ^{14}C cpm from the H3, H2A + H2B, and H4 histone banding positions as a function of increasing uv light fluence. It is also apparent that nonhistones are being rapidly
removed from their normal banding positions in the gel, especially
those proteins that migrate in the region of 1.5 to 9.0 cm. It
should be noted that, in experiments containing only (^{3}H)TdR ss-
chromatin, greater than 98% of the ^{3}H cpm was confined to the
first 4.5 cm of the gel (almost 80% remained in the stacking gel
itself), whereas for (^{3}H)TdR ss-chromatin irradiated with
21,600 Jm^{-2}, about 91% of the ^{3}H cpm was confined to the top
4.5 cm of the gel (about 71% in the stacking gel) (data not shown).

There is a fairly good correlation between the ^{14}C cpm remaining in the stacking gel (i.e., protein associated with DNA) and uv
light fluence and the results obtained from the Sepharose 4B column.
For example, at 10,800 Jm^{-2}, there is 5.9% of the ^{14}C cpm in the
stacking gel compared to 5.8% of the ^{14}C cpm which co-elutes with
the DNA in the column assay. Furthermore, it is evident from the
first slice of the gel (0 to 1.5 cm) that there is a fluence-
dependent accumulation of cpm, possibly the result of additional
protein-DNA adducts and also protein-protein adducts and aggregates
(see Martinson et al., 1976). The role that histone H1 plays in
the photochemical induction of protein-DNA adducts is difficult to
monitor using this particular technique due to possible co-electro-
phoresis of protein-protein adducts in the H1 position (see Martinson et al., 1976). Therefore, no attempt was made to include
histone H1 in any further analysis made on the photochemical crosslinking process. The loss of nonhistone proteins in uv-irradiated
ss-chromatin due to protein-DNA adducts and protein-protein adducts
and aggregates is more easily seen in the following experiment.
(^{14}C)Trp ss-chromatin was uv-irradiated and analyzed on SDS-polyacrylamide gels. These results are also shown in Fig. 8. Essentially all regions of the gel pattern are reduced in ^{14}C cpm
(excluding the first fraction from 0 to 1.5 cm and the stacking gel)
with increasing uv fluences. There is good agreement between the
amount of (^{14}C)Trp cpm associated with DNA in the stacking gel and
the results from the Sepharose column assay; at 10,800 Jm^{-2}, there
is 10% of the ^{14}C cpm in the stacking gel compared to 9% of the

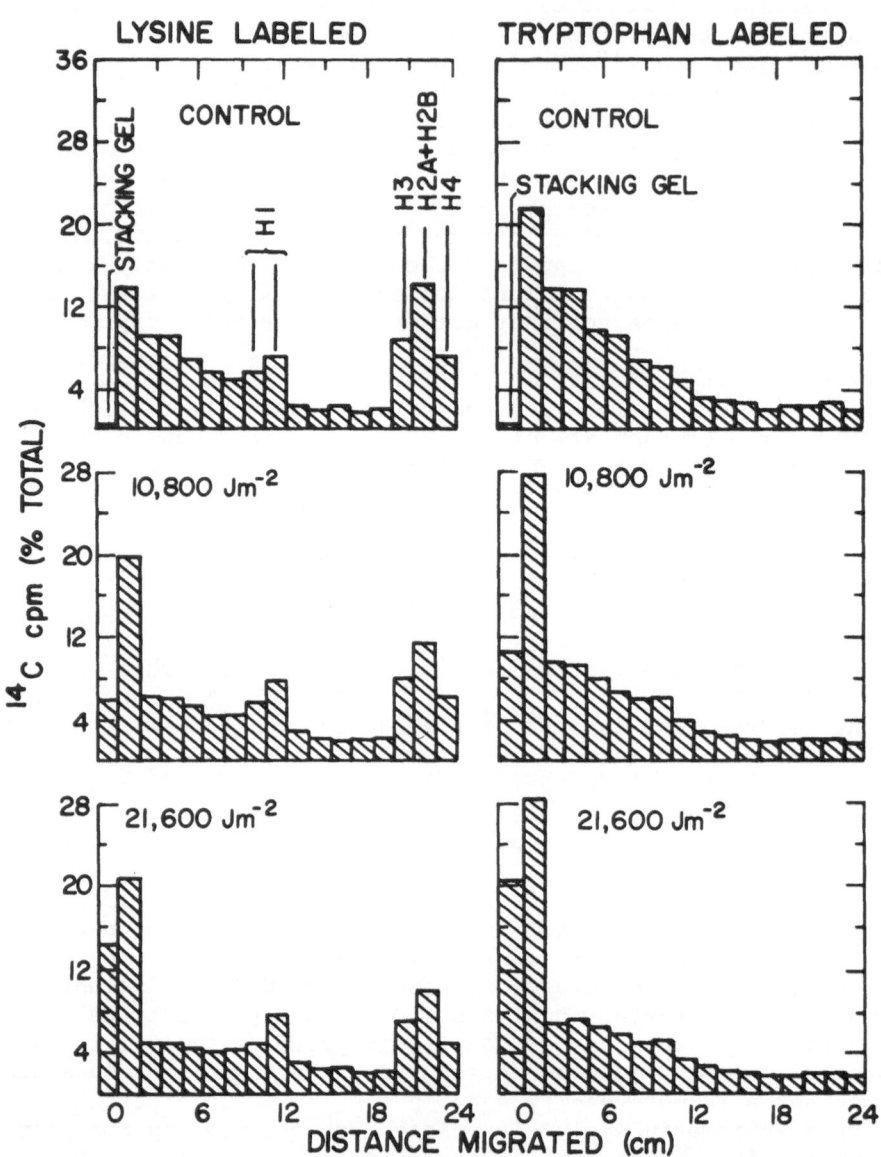

Figure 8. SDS-polyacrylamide gel electrophoretic patterns of control and uv-irradiated, (^{14}C)Lys- and (^{14}C)Trp-labeled, sheared, soluble chromatin (see Methods and Materials for details).

total ^{14}C cpm which co-elutes with the DNA. There is also an accumulation of cpm in the first gel slice (0 to 1.5 cm), possibly indicating additional protein–DNA adducts along with protein–protein adducts and aggregates.

The results of a similar set of experiments using (^{14}C)Lys nucleosomes (dimers and trimers) are shown in Fig. 9. CHO nucleosomes have little nonhistone protein associated with them (see Strniste et al., 1976b); the majority of the protein consists of the five basic histones which contribute to the structure of the nucleosome. UV-irradiated nucleosomes, when analyzed on SDS-polyacrylamide gels, show dramatic losses of the core histones (H3, H2A, H2B, and H4). Since the DNA of nucleosome classes used is relatively small in size, the majority of it migrates through the stacking gel and into the slab itself [for dimers and trimers, a sizable amount of the DNA is found in the first slice, 0 to 1.5 cm (data not shown)]. Spreading of the cpm in the gel (fractions from 1.5 to 15 cm) is probably due not only to protein–DNA adducts but also to histone aggregates and histone–histone adducts.

Figure 10 summarizes the data from SDS-polyacrylamide gels and shows the percent of each histone recovered at its normal gel banding position (compared to the control, nonirradiated values) for the nucleosome classes, dimers, trimers, and tetramers, and for ss-chromatin as a function of uv light fluence. For all the nucleosome classes examined, loss of the core histones from their normal gel banding position proceeds as the following: H4 < H3 < H2A + H2B. Realizing the mole percent lysine for each histone from CHO (see Gurley and Hardin, 1968) and the mole ratio of each histone in the various nucleosome classes (see Strniste et al., 1976b), it can be shown that the decreases observed in Fig. 10 for the four core histones (excluding any possible loss of histone H1 in the calculation) would account for 15% and 30% of the total (^{14}C)Lys cpm for 10,800 and 21,600 Jm^{-2}, respectively. However, as shown in Fig. 5, only 7% and 14% of the total (^{14}C)Lys cpm eluted coincidentally with the DNA at these uv light fluences. Therefore, it is estimated that, of the total amount of histone loss observed by gel analysis, only a maximum of 50% can be attributed to protein–DNA adducts and the remaining loss due possibly to both protein–protein adducts and protein aggregates which would migrate to positions in the gel other than the normal banding regions. At this time, we cannot distinguish between the latter two possibilities.

The loss of histones in ss-chromatin reveals a different pattern from that observed for the nucleosomes in that histone H4 appears to be lost at a greater rate than histone H3. However, since it is known that mechanical shearing destroys much of the original structure of chromatin (see Noll et al., 1975), then the

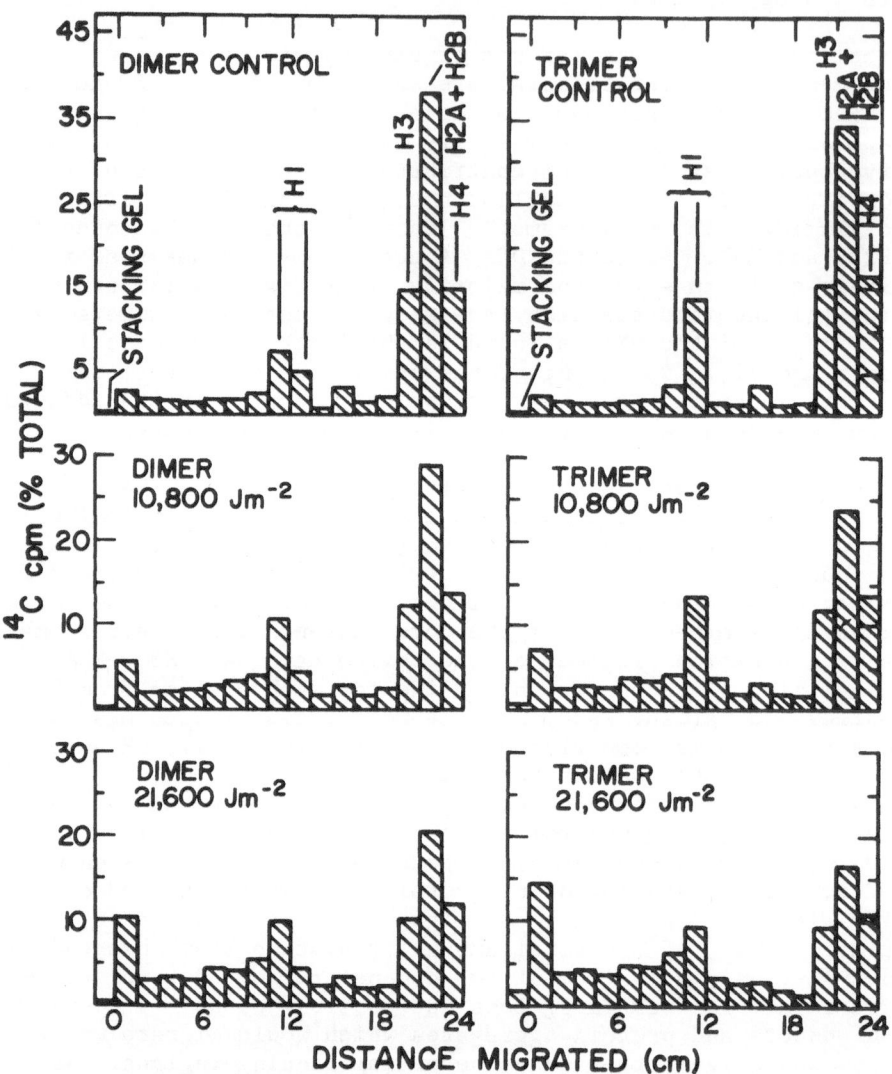

Figure 9. SDS-polyacrylamide gel electrophoretic patterns of
control and uv-irradiated, (^{14}C)Lys-labeled nucleosomes (dimers
and trimers) [see Methods and Materials and Strniste et al. (1976b)
for details].

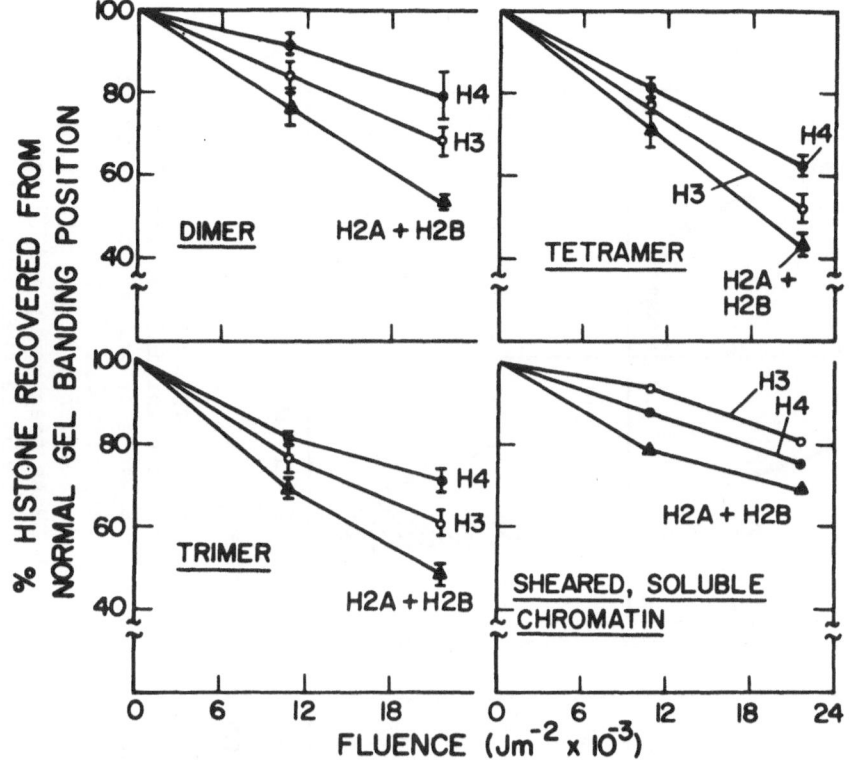

Figure 10. Kinetics of loss of various histones from their
normal banding positions in SDS-polyacrylamide gels from uv-
irradiated nucleosomes (dimers, trimers, and tetramers) and sheared,
soluble chromatin. Gel analysis was performed as outlined in
Methods and Materials, and the results were similar to those shown
in Figs. 8 and 9. The percent total of each histone (H3, H4, and
H2A + H2B) at its normal banding position compared to its control,
nonirradiated value was then plotted vs uv light fluence.

arrangement of the various proteins on the DNA of ss-chromatin is
probably different when compared to the protein-DNA arrangement in
nucleosomes. Furthermore, the presence of nonhistone protein in
the ss-chromatin most likely is influencing the various protein-to-
DNA and protein-to-protein interactions existing in the chromatin
structure. Therefore, it is not surprising that differences in
the pattern of uv light-induced linkages are seen in comparing the
results of ss-chromatin to nucleosomes.

Protein-DNA Adducts vs Protein-Protein Adducts and Aggregates

In the following experiments using the column assay (Sepharose
4B for the analysis of ss-chromatin and Bio-Gel A-1.5m for the

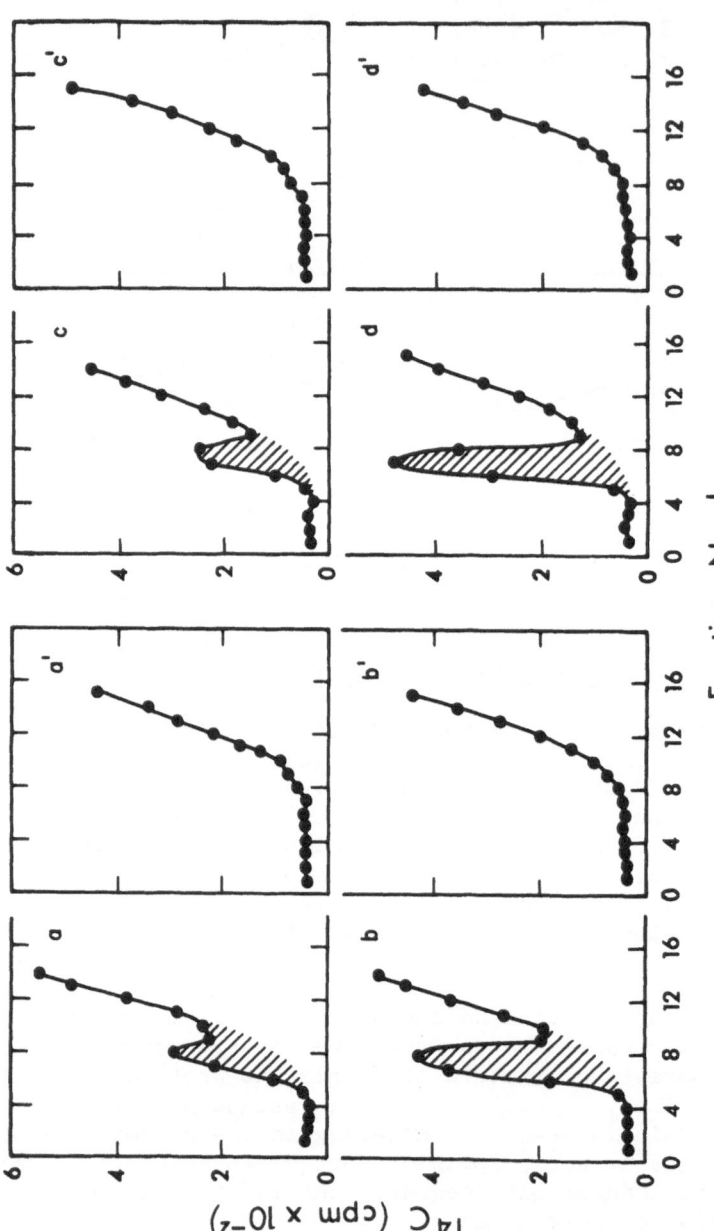

Figure 11. Effects of DNase I on the elution patterns from a Sepharose 4B column of uv-irradiated, (^{14}C)Lys-labeled, sheared, soluble chromatin. Only the initial portion of each profile (DNA elution region) is shown. The various graphs represent (a and a') chromatin ir-radiated with 11,712 Jm^{-2} without and with DNase treatment, respectively; (b and b') chromatin at pH 3.2 uv-irradiated with 2928 Jm^{-2} without and with DNase treatment; (c and c') chromatin in 0.2 $\underline{M}$ NaCl uv-irradiated with 2928 Jm^{-2} without and with DNase treatment; and (d and d') chro-matin in 5 $\underline{mM}$ $MnCl_2$ uv-irradiated with 2928 Jm^{-2} without and with DNase treatment (from Strniste and Rall, 1976a).

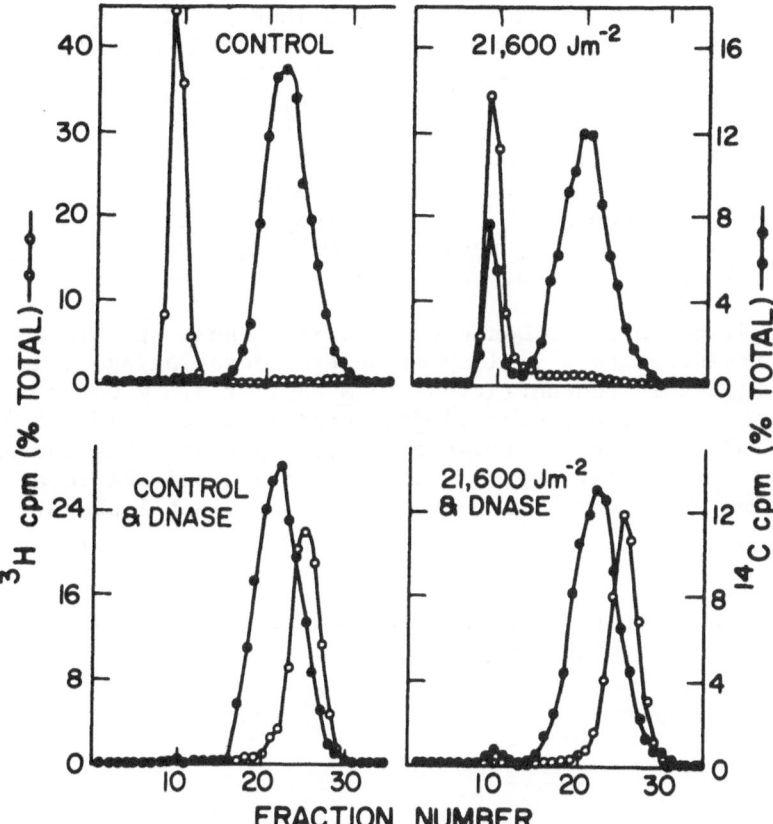

Figure 12. Effects of micrococcal DNase and DNase I on the Bio-Gel A-1.5m column elution patterns of control and uv-irradiated nucleosomes. Mixtures of (^{3}H)TdR- and (^{14}C)Lys-labeled tetramers (total of 3.7 μg DNA) were digested with six units of micrococcal DNase and two units of DNase I, dissociated in 3 $\underline{M}$ NH$_4$C$_2$H$_3$O$_2$, and eluted through the column (see Methods and Materials and Fig. 5).

analysis of nucleosomes), we show that the ^{14}C-labeled material which elutes coincidentally with the DNA in uv-irradiated complexes is protein-DNA adducts and not protein-protein adducts or protein aggregates. In Fig. 11, the results are shown for the elution patterns of (^{14}C)Lys ss-chromatin uv-irradiated at pH 7.5, at pH 3.2, in 0.2 $\underline{M}$ NaCl, or in 5 m$\underline{M}$ MnCl$_2$, without or with DNase I treatment. As is seen in all cases, that material which is associated with the DNA after irradiation (fractions 6-10) is completely removed after nuclease treatment.

A similar DNase experiment is shown in Fig. 12 for control and uv-irradiated mixtures of (^{14}C)Lys and (^{3}H)TdR nucleosomes (tetramers). Again, the vast majority (> 95%) of the ^{14}C-labeled material which co-elutes with the DNA [(3H)TdR profile] after uv-irradiation is sensitive to DNase treatment, thus indicating a protein-DNA complex. It should be noted that protein-protein ad-ducts and aggregates which probably account for at least 50% of the histone loss, as shown in the gel patterns (see Figs. 9 and 10), would probably be indistinguishable from the general elution pat-tern of the bulk, nonlinked protein.

In addition to their DNase sensitivity, these protein-DNA ad-ducts are resistant to a variety of severe dissociating conditions, including high salt concentrations (3 $\underline{M}$ NaCl or 5.5 $\underline{M}$ CsCl), deter-gent (1% SLS or 1% SDS), urea (6 $\underline{M}$), guanidinium chloride (5 $\underline{M}$), and heat (93°C). These observations strongly imply that the asso-ciation photochemically induced between protein and DNA in ss-chromatin and nucleosomes is most likely a covalent linkage.

ACKNOWLEDGMENT

This work was performed under the auspices of the U. S. Energy Research and Development Administration.

REFERENCES

Alexander, P. and Moroson, H. (1962). Cross-linking of deoxyribonucleic acid to protein following ultraviolet irradiation of different cells. Nature, 194, 882-883.

Ashwood-Smith, M. J., Bridges, B. A. and Munson, R. J. (1965). Ultraviolet damage to bacteria and bacteriophage at low temperatures. Science, 149, 1103-1105.

Baca, O. G. and Bodley, J. W. (1976). U. V. induced crosslinking of E. coli ribosomal RNA to specific proteins. Biochem. Biophys. Res. Commun., 70, 1091-1096.

Clark, R. J. and Felsenfeld, G. (1971). Structure of chromatin. Nature New Biol., 229, 101-106.

Davies, K. E. and Walker, I. O. (1974). The solubility of calf thymus chromatin in sodium chloride. Nucleic Acids Res., 1, 129-139.

Gorelic, L. (1975a). Photoinduced covalent crosslinkage, in situ, of Escherichia coli 50S ribosomal proteins to rRNA. Biochim. Biophys. Acta, 390, 209-225.

Gorelic, L. (1975b). Evidence for photoinduced cross-linkage, in situ, of 30S ribosomal proteins to 16S rRNA. Biochemistry, 14, 4627-4633.

Gurley, L. R. and Hardin, J. M. (1968). The metabolism of histone fractions. I. Synthesis of histone fractions during the life cycle of mammalian cells. Arch. Biochem. Biophys., 128, 285-292.

Habazin, V. and Han, A. (1970). Ultraviolet light-induced DNA-to-protein cross-linking in HeLa Cells. Int. J. Radiat. Biol., 17, 569-575.

Han, A., Korbelik, M. and Ban, J. (1975). DNA-to-protein cross-linking in synchronized HeLa cells exposed to ultraviolet light. Int. J. Radiat. Biol., 27, 63-74.

Laemmli, U. K. (1970). Cleavage of structural proteins during the assembly of the head of bacteriophage T4. Nature, 227, 680-685.

Markovitz, A. (1972). Ultraviolet light-induced stable complexes of DNA and DNA polymerase. Biochim. Biophys. Acta, 281, 522-534.

Martinson, H. G., Shetlar, M. D. and McCarthy, B. J. (1976). Histone-histone interactions within chromatin. Crosslinking studies using ultraviolet light. Biochemistry, 15, 2002-2007.

Marushige, K., Brutlag, D. and Bonner, J. (1968). Properties of chromosomal nonhistone protein from rat liver. Biochemistry, 7, 3149-3155.

Noll, M., Thomas, J. O. and Kornberg, R. D. (1975). Preparation of native chromatin and damage caused by shearing. Science, 187, 1203-1206.

Schoemaker, H. J. P. and Schimmel, P. R. (1974). Photo-induced joining of a transfer RNA with its cognate aminoacyl-transfer RNA synthetase. J. Mol. Biol., 84, 503-513.

Schott, H. N. and Shetlar, M. D. (1974). Photochemical addition of amino acids to thymine. Biochem. Biophys. Res. Commun., 59, 1112-1116.

Smith, K. C. (1962). Dose dependent decrease in extractability of DNA from bacteria following irradiation with ultraviolet light or with visible light plus dye. Biochem. Biophys. Res. Commun., 8, 157-163.

Smith, K. C. and Aplin, R. T. (1966). A mixed photoproduct of uracil and cysteine (5-S-cysteine-6-hydrouracil). A possible model for the in vitro cross-linking of deoxyribonucleic acid and protein by ultraviolet light. Biochemistry, 5, 2125-2130.

Smith, K. C. and O'Leary, M. E. (1967). Photoinduced DNA-protein cross-links and bacterial killing: A correlation at low temperature. Science, 155, 1024-1026.

Smith, K. C. and Meun, D. H. C. (1968). Kinetics of the photochemical addition of (^{35}S)cysteine to polynucleotides and nucleic acids. Biochemistry, 7, 1033-1037.

Strniste, G. F. and Smith, D. A. (1974). Induction of stable linkage between the deoxyribonucleic acid dependent ribonucleic acid polymerase and $d(A-T)_n \cdot d(A-T)_n$ by ultraviolet light. Biochemistry, 13, 485-493.

Strniste, G. F. and Rall, S. C. (1976a). Induction of stable protein-deoxyribonucleic acid adducts in Chinese hamster cell chromatin by ultraviolet light. Biochemistry, 15, 1712-1719.

Strniste, G. F., Okinaka, R. T. and Rall, S. C. (1976b). Characterization of chromatin substructure from cultured Chinese hamster cells (manuscript submitted).

Todd, P. and Han, A. (1976). UV-induced DNA to protein crosslinking in mammalian cells. "Aging, Carcinogenesis, and Radiation Biology. The Role of Nucleic Acid Addition Reactions," K. Smith, ed., Plenum Press, New York and London, pp. 83-104.

Van, N. T. and Ansevin, A. T. (1973). Ion-induced splitting in thermal denaturation profiles of F1 nucleohistone. Biochim. Biophys. Acta, 299, 367-373.

Varghese, A. J. (1973). Alpha-S-cysteinyl-5,6-dihydrothymine: A possible model for radiation-induced cross-linking of DNA and protein. Biochem. Biophys. Res. Commun., 51, 858-862.

Varshavsky, A. J. and Ilyin, Yu. V. (1974). Salt treatment of chromatin induces redistribution of histones. Biochim. Biophys. Acta, 340, 207-217.

35

IDENTIFICATION OF BINDING SITES ON THE E. COLI RIBOSOME BY AFFINITY LABELING

Barry S. Cooperman

Department of Chemistry, University of Pennsylvania,

Philadelphia, Pennsylvania   19174

ABSTRACT

Both electrophilic and photolabile derivatives of several different types of ribosomal ligands have been used in affinity labeling studies on the Escherichia coli ribosome. These studies have resulted in the localization of the peptidyl transferase center within a region of the 50S subunit, and the localization of the mRNA binding site within one of two regions on the 30S particle. In addition, labeling data have been obtained for GTP and strepto- mycin affinity labels. The affinity labeling results are discussed along with the results of other studies, and procedures are suggest- ed for improving the resolving power of the affinity labeling technique as applied to ribosomes.

INTRODUCTION

In recent years biochemists have increasingly turned their attention to the study of complex macroscopic structures. Work on E. coli ribosomes is at a particularly interesting juncture (Wittmann, 1976; Nomura et al., 1974), because it now appears that as a result of the vast array of experimental approaches which have and are being directed toward the study of this particle, a detailed model of its structure and function will be forthcoming. The affinity labeling technique (Singer, 1967; Knowles, 1972; Cooperman, 1976) in which one utilizes radioactive, chemically labile derivatives of native ligands to form covalent bonds with, and subsequently identify, specific receptor sites in such structures, has provided important information toward this goal.

Elsewhere we have presented a detailed critical analysis of these
studies (Cooperman, 1977). We here would like to present a
briefer account, emphasizing the general ideas which have emerged
from this analysis.

RIBOSOME STRUCTURE

The E. coli ribosome is designated by its sedimentation
coefficient as a 70S particle (molecular weight, 2.7×10^6 daltons)
and is made up of two dissociable particles. The 50S particle
(molecular weight, 1.8×10^6 daltons) is composed of two RNA
chains, 23S RNA (molecular weight, 1.1×10^6 daltons) and 5S RNA
(molecular weight, 4×10^4 daltons) and approximately 34 different
proteins, designated L1-L34. The peptidyl transferase center is
located on this subunit. The 30S particle is composed of one RNA
chain, 16S RNA (molecular weight, 0.55×10^6 daltons) and 21
different proteins designated S1-S21, and provides the site for
mRNA binding.

TARGET SITES AND REAGENTS

The ligands whose sites have been studied so far are: the
3'-terminus of aminoacyl tRNA (Bispink and Matthaei, 1973;
Bochkareva et al., 1973; Czernilofsky et al., 1974; Eilat et al.,
1974a, b; Girshovich et al., 1974; Hsiung and Cantor, 1974;
Hsiung et al., 1974; Oen et al., 1974; Pellegrini et al., 1974;
Sopori et al., 1974; Barta et al., 1975; Sonenberg et al., 1975;
Yukioka et al., 1975; Breitmeyer and Noller, 1976; Collatz et al.,
1976), the 8-thioU site on tRNA^{Val} (Schwartz and Ofengard, 1974;
Schwartz et al., 1975), mRNA (Budker et al., 1973; Fiser et al.,
1975a, b; Pongs et al., 1975a, b, 1976; Wagner and Gassen, 1975;
Luhrmann et al., 1976; Pongs and Rossner, 1976), chloramphenicol
(Pongs et al., 1973a; Sonenberg et al., 1973, 1974; Pongs and
Messer, 1976), puromycin (Pongs et al., 1973b; Greenwell et al.,
1974; Cooperman et al., 1975; Jaynes et al., 1977), streptomycin
(Pongs and Erdmann, 1973; Girshovich et al., 1976a), and guanosine
nucleotides (Maassen and Möller, 1974, 1975; Girshovich et al.,
1976b).

The kinds of affinity labels used so far are summarized in
Table 1. For tRNA, virtually all of the chemically reactive
groups have been attached via acylation at the α-amino position
of an aminoacyl group attached to the 3'-terminus of a charged
tRNA. An exception is that an aryl azide has been introduced into
an internal tRNA position via alkylation of the 4-thiouridine at
the 8 position of tRNA^{Val}. Both polynucleotides and oligonucleotides
have been used to examine the mRNA binding site. Poly U, poly
4-thiouridine, and poly 5-bromouridine have all been cross-linked

TABLE 1

Affinity Labeling Reagents

Ligand Type / Reagent	Electrophilic				Photolabile					
	haloacetyl	pNO$_2$φcarbamate	nitrogen mustard	α-dicarbonyl	aryl azide	α-diazocarbonyl	aromatic ketone	4-thiouridine	5-bromouridine	intrinsic
tRNA	+[a]	+[d]	+[e]		+[g]	+[j]	+[l]			
mRNA	+[b]		+[e]	+[f]				+[m]	+[n]	+[o]
antibiotics	+[c]				+[h]	+[k]				+[k,p]
GDP, GTP					+[i]					

[a]Eilat et al., 1974a, b; Oen et al., 1974; Pellegrini et al., 1974; Sopori et al., 1974; Yukioka et al., 1975; Breitmeyer and Noller, 1976.

[b]Luhrmann and Gassen, 1976; Pongs and Rossner, 1976; Pongs et al., 1975b, 1976.

[c]Pongs et al., 1973a, b; Pongs and Erdmann, 1973; Sonenberg et al., 1973; Greenwell et al., 1974; Pongs and Messer, 1976.

[d]Czernilofsky et al., 1974; Collatz et al., 1976.

[e]Bochkareva et al., 1973; Budker et al., 1973.

[f]Wagner and Gassen, 1975.

[g]Girshovich et al., 1974; Hsiung and Cantor, 1974; Hsuing et al., 1974; Schwartz and Ofengard, 1974; Schwartz et al., 1975; Sonenberg et al., 1975.

[h]Girshovich et al., 1976a.

[i]Maassen and Möller, 1974, 1975; Girshovich et al., 1976b.

[j]Bispink and Matthaei, 1973.

[k]Cooperman et al., 1975.

[l]Barta et al., 1975.

[m]Fiser et al., 1975b.

[n]Pongs et al., 1975a.

[o]Fiser et al., 1975a.

[p]Sonenberg et al., 1974; Jaynes et al., 1977.

to ribosomes in photo-induced reactions. Electrophilic derivatives
of tri and tetranucleotides have been synthesized via esterification
of the 5' phosphate, via acylation of a modified base (5-amino-
uridine) and via acylation of a modified ribose (2'-deoxy-2'-amino-
ribose). Electrophilic derivatives of chloramphenicol have been
made by replacing the $-NHCOCHCl_2$ moiety of the native molecule
with both the bromo- and the iodoacetamidyl groups. Chloramphenicol
is also sufficiently photolabile to incorporate into ribosomes on
direct photolysis, although with apparently little specificity.
Photolabile and electrophilic derivatives of puromycin have been
synthesized via derivatization at both its 5'-hydroxyl and α-amino
positions. Like chloramphenicol, puromycin itself has been shown
to photoincorporate into ribosomes, with the difference that a
major portion of the incorporation occurs at a puromycin-specific
site. Reactive derivatives of streptomycin have been made through
condensation with its aldehyde group. Photolabile derivatives of
both GDP and GTP have been made both by esterification of the
terminal phosphate and by periodate oxidation of the ribose and
derivatization of the resulting dialdehyde with a photolabile
hydrazine.

AFFINITY LABELING RESULTS

In discussing the results of affinity labeling studies on
ribosomes we will confine our remarks to studies of the two
ribosomal sites examined in greatest detail, the peptidyl transferase center and the m-RNA binding site. In most of the work
discussed at least two of the minimal criteria for affinity
labeling have been met, i.e., that the non-covalent binding of
the affinity labeling reagent mimics that of the native ligand
and that non-covalent binding is a necessary prerequisite for
covalent incorporation. In several of the studies cited, the
additional, very important, criterion has been met that the
covalently bound affinity label retains the function of the non-
covalently bound native ligand. Thus several of the covalently
bound acylaminoacyltRNA's have been found to be active as
peptidyl donors (Hsuing and Cantor, 1974; Hsuing et al., 1974;
Pellegrini et al., 1974; Barta et al., 1975) and all of the
covalently bound oligonucleotides have been found to code for
specific tRNA binding.

The reader will note that our discussion of ribosomal sites
is limited to those proteins which have been implicated at these
sites, and this despite the fact that the ribosome is two-thirds
RNA by weight and that several affinity-labeling reagents have
been shown to incorporate partially or totally into RNA. The
reason for this is that studies to localize sites of RNA labeling
are not yet sufficiently far along to allow detailed topological
conclusions to be reached.

Affinity labeling reagents directed toward the peptidyl
transferase center, acylaminoacyl tRNA's and two inhibitors of
peptidyl transferase, puromycin and chloramphenicol, incorporate
principally into the 50S particle, as expected. The major
proteins labeled are listed in Table 2. Despite the multiplicity
of labeled proteins these results can be seen to have a strong
internal coherence when considered together with results obtained
from other approaches. That is, the proteins which are affinity
labeled appear to constitute a region on the 50S particle, as
determined from cross-linking, immune electron microscopy, and
RNA binding studies, as summarized in Table 3. The results with
the mRNA affinity labels are analogous. Incorporation occurs
principally into the 30S particle, with five proteins, S1, S4,
S12, S18, and S21, found to be highly labeled. Again, as summarized
in Table 4, these proteins appear to be in proximity of each other
within the 30S particle.

IMMUNE ELECTRON MICROSCOPY

Immune electron microscopy results allow, in principle, the
three-dimensional localization of ribosomal proteins within the
30S and 50S proteins. The groups of both Stöffler (Tischendorf
et al., 1975) and Lake (1976) are currently utilizing this
approach. In comparing their findings with those obtained from
affinity labeling, we note that the two groups have significant
differences in their respective models for ribosome structure.
Figure 1 presents the two structural models proposed for the 50S
particles. For this particle, protein localization results are
only available from the Stöffler group (Tischendorf et al., 1975).
Their results suggest that the peptidyl transferase center is
located within the shaded area of Figure 1a. Thus, of the nine
proteins listed in Table 2, all or parts of four of them, L11,
L16, L18 and L23, fall within this area, with the remaining five
having not as yet been localized. This suggestion is in accord
with a proposal of Lake's (1976) who placed the binding site for
the 3'-end of tRNA in an area of his model for the 50S particle
(Figure 1b) which corresponds closely to the shaded area of the
Stöffler model.

The structural models for the 30S particle are presented in
Figure 2. Although (or perhaps, because?) both groups have more
information with respect to protein localization in this particle
than is available for the 50S particle, localization of the mRNA
site is less certain than localization of the peptidyl transferase
center. Of the five proteins implicated in the mRNA site
Tischendorf et al. (1975) have localized four, S4, S12, S18 and
S21. According to these authors, three of these proteins, S4,
S12 and S18 have elongated conformations and are close neighbors

TABLE 2

Affinity Labeled 50S Proteins

L proteins	Affinity Label Reagent Type		
	3'-acylaminoacyl tRNA	Chloramphenicol	Puromycin
2	++[a,c]	++[f]	
5	+[b]		
11	++[b]		
16	++[d]	++[g]	
18	++[b]		
23			++[i]
24	++[e]	++[h]	
27	++[a,c]	++[f]	
32/33	++[e]		

[a]Pellegrini et al., 1974.

[b]Hsuing et al., 1974; Hsuing and Cantor, 1974.

[c]Czernilofsky et al., 1974.

[d]Eilat et al., 1974a.

[e]Eilat et al., 1974b.

[f]Sonenberg et al., 1973.

[g]Pongs et al., 1973a.

[h]Pongs and Messer, 1976.

[i]Cooperman et al., 1975.

TABLE 3

Correlation of Affinity Labeled Proteins
at the Peptidyl Transferase Center
with Results from Other Studies

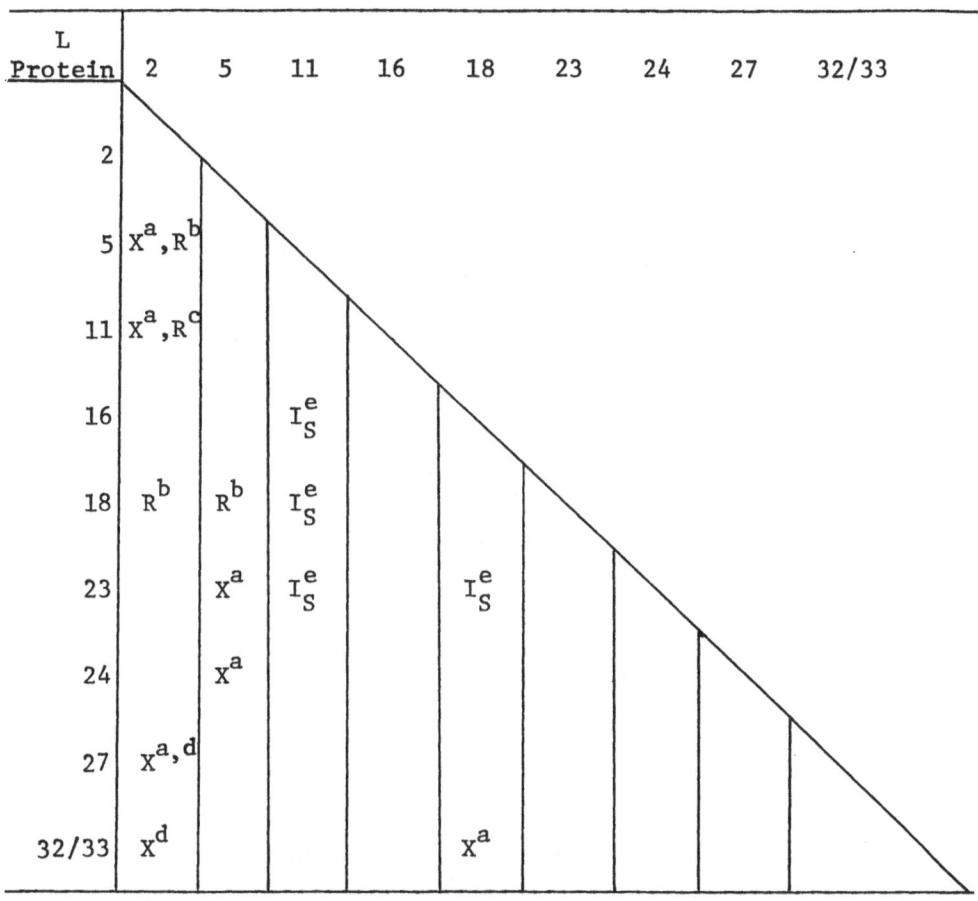

L Protein	2	5	11	16	18	23	24	27	32/33
2									
5	X^a, R^b								
11	X^a, R^c								
16			I_S^e						
18	R^b	R^b	I_S^e						
23		X^a	I_S^e		I_S^e				
24		X^a							
27	$X^{a,d}$								
32/33	X^d				X^a				

X – chemical cross linking.
R – binding to same, limited RNA region or cooperative binding to RNA.
I_S– immune electron microscopy – Stöffler's group.

[a]Traut and Kenney, 1976.

[b]Horne and Erdmann, 1972.

[c]Gray and Monier, 1972.

[d]Barritault et al., 1975.

[e]Tischendorf et al., 1975.

TABLE 4

Correlation of Affinity Labeled Proteins
at the mRNA Binding Site with
Results from Other Studies

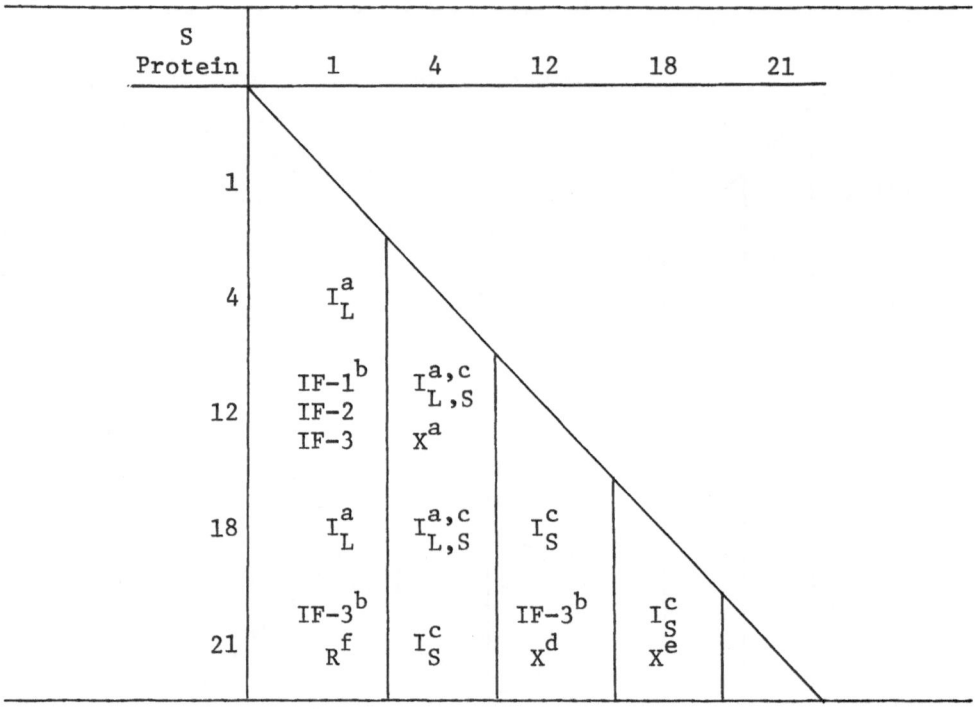

S Protein	1	4	12	18	21
1					
4	I_L^a				
12	IF-1[b] IF-2 IF-3	$I_{L,S}^{a,c}$ X^a			
18	I_L^a	$I_{L,S}^{a,c}$	I_S^c		
21	IF-3[b] R^f	I_S^c	IF-3[b] X^d	I_S^c X^e	

IF-1, IF-2, IF-3 – mutually cross-linked to one or more of these
    initiation factor proteins.
$I_{L,S}$ – immune electron microscopy – Stöffler's and/or Lake's group.
R, X – defined as in Table 3.

[a]Lake, 1976.

[b]Traut, 1976.

[c]Tischendorf et al., 1975.

[d]Sommer and Traut, 1976.

[e]Lutter et al., 1972.

[f]Czernilofsky et al., 1975.

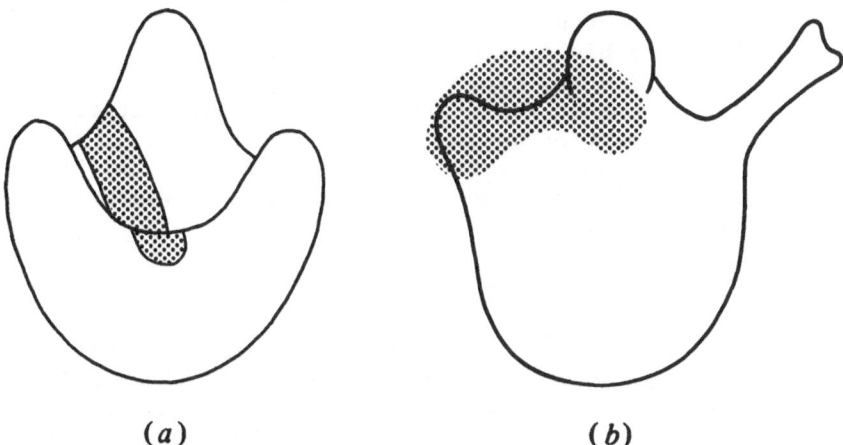

(a) (b)

Figure 1.   Structural models for the 50S particle proposed
from immune electron microscopy studies by Tischendorf et al.
(1975) (a) and by Lake (1976) (b).   In (a), the shaded area
contains determinants for proteins implicated at the peptidyl
transferase center.   In (b) the shaded area is proposed by Lake
(1976) as the binding site for the 3'-end of tRNA.

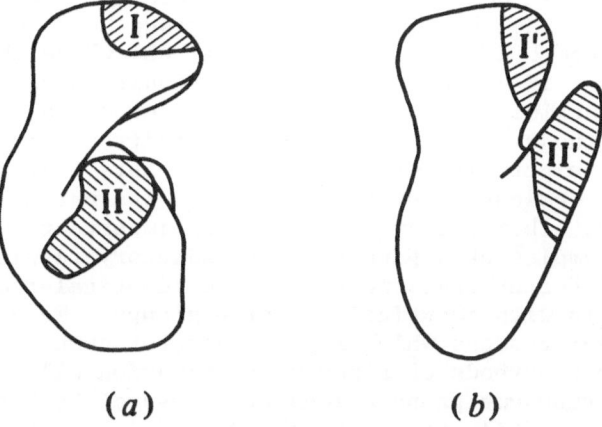

(a) (b)

Figure 2.   Structural models for the 30S particle proposed
from immune electron microscopy studies by Tischendorf et al. (1975)
(a) and by Lake (1976) (b).   The shaded areas contain determinants
for proteins implicated at the mRNA binding site.

in two different regions of the 30S particle, designated I and II
(Figure 2a). The fourth, S21, appears to be localized only in
region I. Lake and his coworkers have also localized four
proteins, S1, S4, S12 and S18. Of these, a part of S4 and S12
fall in region I'(Figure 2b), corresponding roughly to region I,
while S1, another part of S4, and a part of S18 fall in region
II' (Figure 2b), which may correspond to region II. At present
it seems certain only that the m-RNA binding site will fall within
either site I/I' or II/II'.

CONCLUSIONS

     From these results it seems fair to say that affinity labeling
studies on ribosomes have been successful, at least as a low
resolution technique to define what appear to be rather extended
regions surrounding a ligand binding site. However, the affinity
labeling technique is in principle capable of considerably higher
resolution (here we may recall that affinity labeling was originally
introduced to study active sites within single proteins). We are
thus left with the question of why different investigators using
similar reagents have reported disparate results, disagreeing in
some cases on the identity of the major labeled proteins (as
discussed above)(Pongs et al., 1973a; Sonenberg et al., 1973) and
in others as to whether the protein or RNA fractions of the
ribosomes present the principal targets for affinity labeling
(Girshovich et al., 1974; Hsiung and Cantor, 1974; Hsiung et al.,
1974; Pellegrini et al., 1974; Sonenberg et al., 1975; Breitmeyer
and Noller, 1976). Some of these differences are undoubtedly due
to the nature of the chemically reactive group used in affinity
labeling. As has been discussed in detail elsewhere (Knowles,
1972; Cooperman, 1976), it is clear that results obtained with
electrophilic affinity labels must be considered inherently less
reliable than those obtained with photoaffinity labels, since the
high chemical selectivity of the former reagents increases the
likelihood of covalent insertion outside of the native binding
site. However, other factors must also be important, since there
are several examples of apparently contradictory results being
obtained by different investigators using identical reagents or
similar reagents with identical reactive groups. We believe there
to be four major reasons which may account for such results.
Firstly, current methods of ribosome preparation all yield
heterogeneous ribosome preparations with respect both to activity
and composition. While there is no simple solution to this
problem, since purification methods that rigorously remove non-
ribosomal ligands such as tRNA and protein factors also remove
some of the more loosely bound ribosomal proteins, the result is
that different laboratories may be working with ribosomes whose
binding sites for a particular ligand may differ significantly
one from the other. Clearly, in the future more attention will

have to be paid to studies of labeling pattern obtained as a
function of the method of ribosomal preparation. Secondly,
ribosomes are conformationally labile and are quite sensitive to
changes in, for example, Mg^{2+} and monocation concentration and
in pH. In the experiments cited above the reaction media used
have often been significantly different in different laboratories,
raising the possibility that the differences observed in the
labeling results reflect conformational changes in the ribosomal
binding site for the affinity label. It should be emphasized
that as a consequence a systematic study of the variation in
labeling pattern as a function of reaction medium could provide
valuable insight into the dynamic functioning of the ribosome.
Thirdly, comparatively little attention has been paid to examination of labeling pattern as a function of affinity label concentration, raising the possibility that in some instances, particularly
with weakly bound ligands such as oligonucleotides and antibiotics,
the observed labeling patterns reflect covalent incorporation
into both tight (specific) and weak (non-specific) binding sites.
Finally, because of the finding that at least some ribosomal
proteins have elongated conformations within the ribosome, it is
possible that portions of several proteins may be close to the
reactive group of the affinity label in the affinity label·ribosome
complex, i.e., that the multiplicity of labeled proteins are an
accurate reflection of the ligand binding site. This possibility
brings up the interesting and somewhat unanticipated point that
for ribosomal proteins having elongated conformations it probably
will be necessary for both affinity labeling and immune electron
microscopy experiments to be localized to the level of protein
fragments in order to be certain of obtaining significant topological information. If the effort is made both to more stringently
control the experimental variables discussed above and to localize
sites of covalent incorporation into protein and RNA with greater
precision, then it should be possible for affinity labeling experiments to yield higher resolution models for ribosomal ligand site
structure than has been obtained heretofore. Hopefully such
efforts will be forthcoming.

REFERENCES

Barritault, D., Expert-Bezancon, A., and Milet, M. (1975). Etude
    des relations de voisinage entre des proteines de la sous-
    unite ribosomique 50S d'E. coli. C.R. Acad. Sc. Paris, 281, 1043.
Barta, A., Kuechler, E., Branlant, C., Sriwadada, J., Krol, A.,
    and Ebel, J. P. (1975). Photoaffinity labelling of 23S RNA
    at the donor-site of the Escherichia coli ribosome. FEBS
    Lett., 56, 170.
Bispink, L., and Matthaei, H. (1973). Photoaffinity labeling of
    23S rRNA in Escherichia coli ribosomes with poly(U)-coded
    ethyl 2-diazomalonyl-Phe-tRNA. FEBS Lett., 37, 291.

Bochkareva, E. S., Budker, V. G., Girshovich, A. C., Knorre, D. G., and Teplova, N. M. (1973). Specific chemical modification of the ribosome close to the peptidyl transferase center. Molekul. Biol., 7, 278.

Breitmeyer, J. B., and Noller, H. F. (1976). Affinity labeling of specific regions of 23S RNA by reaction of N-bromoacetyl-phenylalanyl-transfer RNA with Escherichia coli ribosomes. J. Mol. Biol., 101, 297.

Budker, V. G., Girshovich, A. S., Grineva, N. I., Karpova, G. G., Knorre, D. G., Kobets, N. D. (1973). Specific chemical modification of ribosomes near the mRNA-binding center. Dokl. Akad. Nauk. SSSR, 211, 725.

Collatz, E., Kuechler, E., Stöffler, G., and Czernilofsky, A. P. (1976). The site of reaction on ribosomal protein L27 with an affinity label derivative of tRNA$_f^{met}$. FEBS Lett., 63, 283.

Cooperman, B. S. (1976). Photoaffinity labeling of proteins and more complex receptors, in "Aging, carcinogenesis and radiation biology. The role of nucleic acid addition reactions." K. C. Smith, ed., Plenum Press, N.Y., p. 315.

Cooperman, B. S. (1977). Affinity labeling studies on Escherichia coli ribosomes. Bioorg. Chem., in press.

Cooperman, B. S., Jaynes, E. N., Brunswick, D. J., and Luddy, M. A. (1975). Photoincorporation of puromycin and N-(ethyl-2-diazomalonyl)puromycin into Escherichia coli ribosomes. Proc. Nat. Acad. Sci. USA, 72, 2974.

Czernilofsky, A. P., Collatz, E. E., Stöffler, G., and Kuechler, E. (1974). Proteins at the tRNA binding sites of Escherichia coli ribosomes. Proc. Nat. Acad. Sci. USA, 71, 230.

Czernilofsky, A. P., Kurland, C. G., and Stöffler, G. (1975). 30S ribosomal proteins associated with the 3'-terminus of 16S RNA. FEBS Lett., 58, 281.

Eilat, D., Pellegrini, M., Oen, H., de Groot, N., Lapidot, Y., and Cantor, C. R. (1974a). Affinity labelling the acceptor site of the peptidyl transferase centre of the Escherichia coli ribosome. Nature, 250, 514.

Eilat, D., Pellegrini, M., Oen, H., Lapidot, Y., and Cantor, C. R. (1974b). A chemical mapping technique for exploring the location of proteins along the ribosome-bound peptide chain. J. Mol. Biol., 88, 831.

Fiser, I., Margaritella, P., and Kuechler, E. (1975a). Photoaffinity reaction between polyuridylic acid and protein S1 on the Escherichia coli ribosome. FEBS Lett., 52, 281.

Fiser, I., Scheit, K. H., Stöffler, G., and Kuechler, E. (1975b). Proteins at the mRNA binding site of the Escherichia coli ribosome. FEBS Lett., 56, 226.

Girshovich, A. S., Bochkareva, E. S., Kramarov, V. M., and Ovchinnikov, Y. A. (1974). E. coli 30S and 50S ribosomal subparticle components in the localization region of the tRNA acceptor terminus. FEBS Lett., 45, 213.

Girshovich, A. S., Bochkareva, E. S., and Ovchinnikov, Y. A. (1976a).
    Identification of components of the streptomycin-binding
    center of E. coli MRE 600 ribosomes by photo-affinity labelling.
    Mol. Gen. Genet., 144, 205.
Girshovich, A. S., Pozdnyahov, V. A., and Ovchinnikov, Y. A. (1976b).
    Localization of the GTP-binding site in the ribosome•elonga-
    tion factor-G•GTP complex. Eur. J. Biochem., 69, 321.
Gray, P. N., and Monier, R. (1972). Partial localization of the 5S
    RNA binding site on 23S RNA. Biochimie, 54, 41.
Greenwell, P., Harris, R. J., and Symons, R. H. (1974). Affinity
    labelling of 23-S ribosomal RNA in the active centre of
    Escherichia coli peptidyl transferase. Eur. J. Biochem., 49,
    539.
Horne, J. R., and Erdmann, V. A. (1972). Isolation and characteri-
    zation of 5S RNA-protein complexes from bacillus stearothermo-
    philus and Escherichia coli ribosomes. Mol. Gen. Genet., 119,
    337.
Hsiung, N., and Cantor, C. R. (1974). A new simpler photoaffinity
    analogue of peptidyl tRNA. Nucleic Acids Res., 1, 1753.
Hsiung, N., Reines, S. A., and Cantor, C. R. (1974). Investigation
    of the ribosomal peptidyl transferase center using a photo-
    affinity label. J. Mol. Biol., 88, 841.
Jaynes, E. N., Jr., Grant, P. G., Wieder, R., Giangrande, G., and
    Cooperman, B. S. Photo-induced affinity labeling of the
    Escherichia coli ribosome puromycin site. Submitted for
    publication.
Knowles, J. R. (1972). Photogenerated reagents for biological
    receptor-site labeling. Accounts Chem. Res., 5, 155.
Lake, J. (1976). Ribosome structure determined by electron micro-
    scopy of Escherichia coli small subunits, large subunits, and
    monomeric ribosomes. J. Mol. Biol., 105, 131.
Luhrmann, R., and Gassen, H. G. (1976). Identification of the
    30-S ribosomal proteins at the decoding site by affinity
    labelling with a reactive oligonucleotide. Eur. J. Biochem.,
    66, 1.
Lutter, L. C., Zeichhardt, H., Kurland, C. G., and Stöffler, G.
    (1972). Ribosomal protein neighborhoods. I. S18 and S21
    as well as S5 and S8 are neighbors. Mol. Gen. Genet., 119,
    357.
Maassen, J. A., and Moller, W. (1974). Identification by photo-
    affinity labeling of the proteins in Escherichia coli
    ribosomes involved in elongation factor G-dependent GDP
    binding. Proc. Nat. Acad. Sci. USA, 71, 1277.
Maassen, J. A., and Moller, W. (1975). Comparison by photoaffinity
    labeling of the proteins involved in GTP hydrolysis from 70S
    ribosomes and a 5S RNA-protein complex from bacillus
    stearothermophilus. Biochem. Biophys. Res. Commun., 64, 1175.
Nomura, M., Tissieres, A., and Lengyel, P. (1974). Ribosomes.
    Monograph Series, Cold Spring Harbor Laboratory.

Oen, H., Pellegrini, M., and Cantor, C. R. (1974). Peptidyl
    transferase inhibitors alter the covalent reaction of BrAcPhe-
    tRNA with the E. coli ribosome. FEBS Lett., 45, 218.
Pellegrini, M., Oen, H., Eilat, D., and Cantor, C. R. (1974). The
    mechanism of covalent reaction of bromoacetyl-phenylalanyl-
    transfer RNA with the peptidyl-transfer RNA binding site of
    the Escherichia coli ribosome. J. Mol. Biol., 88, 809.
Pongs, O., Bald, R., and Erdmann, V. A. (1973a). Identification of
    chloramphenicol-binding protein in Escherichia coli ribosomes
    by affinity labeling. Proc. Nat. Acad. Sci. USA, 70, 2229.
Pongs, O., Bald, R., Wagner, T., and Erdmann, V. A. (1973).
    Irreversible binding of N-iodoacetylpuromycin to E. coli
    ribosomes. FEBS Lett., 35, 137.
Pongs, O., and Erdmann, V. A. (1973). Affinity labeling of E. coli
    ribosomes with a streptomycin-analogue. FEBS Lett., 37, 47.
Pongs, O., Lanka, E., and Bald, R. (1975a). 10th FEBS Meeting
    Abstr. No. 447.

Pongs, O., Stöffler, G., and Lanka, E. (1975b). The codon binding
    site of the Escherichia coli ribosome as studied with a chemi-
    cally reactive A-U-G analog. J. Mol. Biol., 99, 301.
Pongs, O., and Messer, W. (1976). The chloramphenicol receptor
    site in vivo affinity labeling by monoiodoamphenicol. J. Mol.
    Biol., 101, 171.
Pongs, O., and Rossner, E. (1976). Comparison of the reactions of
    chemically reactive analogs of U-G-A and of A-U-G with ribosomes
    of Escherichia coli. Nuc. Acids Res., 3, 1625.
Pongs, O., Stöffler, G., and Bald, R. W. (1976). Location of protein
    S1 of Escherichia coli ribosomes at the 'A'-site of the codon
    binding site. Affinity labeling studies with a 3'-modified
    A-U-G analog. Nuc. Acids Res., 3, 1635.
Schwartz, I., Gordon, E., and Ofengand, J. (1975). Photoaffinity
    labeling of the ribosomal a site with S-(p-azidophenacyl)valyl-
    tRNA. Biochem., 14, 2907.
Schwartz, I., and Ofengand, J. (1974). Photo-affinity labeling of
    tRNA binding sites in macromolecules. I. Linking of the
    phenacyl-p-azide of 4-thiouridine in (Escherichia coli) valyl-
    tRNA to 16S RNA at the ribosomal P site. Proc. Nat. Acad. Sci.
    USA, 71, 3951.
Singer, S. J. (1967). Covalent labeling of active sites. Adv.
    Prot. Chem., 22, 1.
Sommer, A., and Traut, R. (1976). Identification of neighboring
    protein pairs in E. Coli in 30S ribosome subunits by cross-
    linking with methyl-4-mercaptoethyl-imidate.J.Mol.Biol.106,995-100
Sonenberg, N., Wilchek, M., and Zamir, A. (1973). Mapping of
    Escherichia coli ribosomal components involved in peptidyl
    transferase activity. Proc. Nat. Acad. Sci. USA, 70, 1423.

Sonenberg, N., Zamir, A., and Wilchek, M. (1974). A photo-induced reaction of chloramphenicol with E. coli ribosomes: Covalent binding of the antibiotic and inactivation of peptidyl transferase. Biochem. Biophys. Res. Commun., 59, 693.

Sonenberg, N., Wilchek, M., and Zamir, A. (1975). Identification of a region in 23S rRNA located at the peptidyl transferase center. Proc. Nat. Acad. Sci. USA, 72, 4332.

Sopori, M., Pellegrini, M., Lengyel, P., and Cantor, C. R. (1974). Affinity labeling of Escherichia coli ribosomal proteins with an analog of the natural initiator tRNA. Biochemistry, 13, 5432.

Tischendorf, G. W., Zeichhardt, H., and Stöffler, G. (1975). Architecture of the Escherichia coli ribosome as determined by immune electron microscopy. Proc. Nat. Acad. Sci. USA, 72, 4820.

Traut, R. (1976). Private communication.

Traut, R., and Kenney, J. (1976). Private communication.

Wagner, R., and Gassen, H. G. (1975). On the covalent binding of mRNA models to the part of the 16S RNA which is located in the mRNA binding site of the 30S ribosome. Biochem. Biophys. Res. Commun., 65, 519.

Wittmann, H. G. (1976). Structure, function, and evolution of ribosomes. Eur. J. Biochem., 61, 1.

Yukioka, M., Hatayama, T., and Morisawa, S. (1975). Affinity labeling of the ribonucleic acid component adjacent to the peptidyl recognition center of peptidyl transferase in Escherichia coli ribosomes. Biochem. Biophys. Acta, 390, 192.

36

PHOTOINDUCED NUCLEIC ACID-PROTEIN CROSSLINKAGE

IN RIBOSOMES AND RIBOSOME COMPLEXES

Lester Gorelic

Department of Chemistry, Wayne State University

Detroit, Michigan   48202

I. ABSTRACT

Exposure of aqueous buffered solutions of E. coli 30S and 50S ribosomal subunits and of E. coli 70S ribosomes to ultraviolet radiation results in changes in a number of the initial physical properties of the ribosome components consistent with the formation of covalent crosslinks between the rRNA and protein components of the ribosomal subunits. These changes in physical properties include a dose-dependent decrease in the separability of the rRNA and protein components of the ribosome subunits under conditions normally denaturing for the ribosome structure, co-elution from gel filtration media of the rRNA and protein components of irradiated ribosomes under conditions where the native structure of the ribosomes is completely disrupted, enhanced mobilities of specific ribosomal proteins towards the anodic (+) electrodes in an applied electrical field, and radioactive labelling of specific ribosomal proteins in irradiated ribosomal subunits and 70S ribosomes initially labelled with tritium only in their rRNA components.

Irradiation of 70S poly(x) complexes with 254 nm radiation effects changes in the separabilities of the poly(x) and 30S ribosome components of the complex, and irradiation of 70S ^3H - poly(x) complexes results in the radioactive labelling of a small number of ribosomal proteins in the 30S subunit. It has therefore been concluded that covalent crosslinks are also formed between the poly(x) components of 70S poly(x) complexes and 30S ribosomal proteins.

Application of the photoinduced formation of covalent crosslinks between the nucleic acid and protein components of the above
ribosome systems has resulted in the identification of the mRNA-
binding proteins on the 30S ribosomal subunit and determination of
the relative proximities of the ribosomal proteins to the rRNA
bases in the intact E. coli 30S and 50S ribosomal subunits. In
addition, a molecular basis has been developed for relating the
reactivity of specific ribosomal proteins in photoinduced crosslinkage to the rRNA components in their corresponding subunits to
rRNA-protein interactions in the native topographical states of
the ribosomal subunits.

II.   INTRODUCTION

The ribosomes of E. coli are ribonucleoprotein complexes that
sediment at a velocity of 70S in an applied centrifugal field and
that dissociate into a smaller and a larger subunit at low magnesium
concentrations. The smaller subunit sediments at a velocity of 30S
in an applied centrifugal field and is constituted of a 16S rRNA
molecule and 21 proteins (Kaltschmidt and Wittmann, 1970). The
larger of the subunits sediments at a velocity of 50S in an applied
centrifugal field and is constituted of 34 proteins (Kaltschmidt
and Wittmann, 1970), and a 5S rRNA and a 23S rRNA molecule.

The 30S and 50S ribosomes of E. coli constitute the major components of the protein synthetic machinery of this organism. In
this capacity, the 30S and 50S ribosomal subunits associate with
mRNA and tRNA molecules to form the high molecular weight complexes -- i.e. polysomes -- upon which the nascent protein chains
are assembled.

Photochemical studies of a number of nucleoprotein complexes
have indicated that exposure of such complexes to ultraviolet
radiation results in the formation of stable covalent crosslinks
between their nucleic acid and protein components (for a review
of these studies see Smith, 1975). Since the ribosomes are also
nucleoprotein complexes, it would be reasonable to expect that
irradiation of E. coli ribosomes and of ribosome-mRNA (tRNA) complexes would result in the formation of covalent crosslinks between
the nucleic acid and protein components of these ribosome systems.
Recent reports of the effects of ultraviolet radiation on ribosome
systems have, in fact, afforded evidence consistent with the conclusions that covalent nucleic acid-protein crosslinks are present
in irradiated ribosomes and ribosome complexes. The results presented in some of these studies have also indicated that the photoinduced formation of nucleic acid-protein crosslinks between the
ribosome somponents and/or the components of ribosome complexes

could be used to study rRNA-protein interactions in the ribosomes and mRNA-ribosome interactions in ribosome complexes.

The primary objectives of this article are to review the published reports of photoinduced nucleic acid-protein crosslinkage in irradiated ribosomes and ribosome complexes and of the applications of this photoreaction to studies of ribosome interactions; and to offer a critical analysis of the data contained in these reports.

III. EVIDENCE THAT COVALENT RNA-PROTEIN CROSSLINKS ARE PRESENT IN IRRADIATED E. COLI RIBOSOMES AND RIBOSOME-mRNA (tRNA) COMPLEXES

A. Evidence for RNA-Protein Crosslinks in Irradiated E. Coli 30S and 50S Ribosomal Subunits. The possibility that covalent RNA-protein crosslinks are present in irradiated E. coli 30S and 50S ribosomal subunits has been examined by studying the effects of irradiation of the intact ribosomal subunits with 254 nm radiation on a number of initial physical properties of the ribosomal components (Gorelic, 1975a, 1975b). The ribosomal subunits used in these studies were prepared from E. coli 70S ribosomes that had been repeatedly centrifuged through high magnesium-buffered sucrose solutions. Solutions of the ribosomal subunits in a phosphate-based buffer (final ribosome concentration one A_{260} unit per ml) were irradiated under anaerobic conditions in a photochemical reactor using a Hanau low pressure mercury lamp as the light source. The flux of incident 254 nm radiation was 1.5×10^{18} quanta S^{-1} (2.9×10^{2} ergs $mm^{-2} S^{-1}$), of which ca. 30% was actually absorbed by the ribosome solutions.

The results of a study of the effects of 254 nm radiation on the distribution of the rRNA components of E. coli 30S and 50S ribosomes between soluble and insoluble phases in 4M urea-2M LiCl are represented by the data in Figure 1. The data in this figure indicate that the rRNA components of unirradiated ribosomal subunits are quantitatively fractionated into the insoluble phase in 4M urea-2M LiCl. The protein components of the ribosomal subunits were found to be fractionated by 4M urea-2M LiCl treatment exclusively into the soluble phase (data not shown). Exposure of the ribosomal subunits to increasing doses of 254 nm radiation results in a decrease in the amount of rRNA-derived material detected in the insoluble phase and a concomitant increase in the amount of rRNA-derived material detected in the soluble phase. Increasing doses of 254 nm radiation did not effect a change in the initial fractionation of the protein components of the ribosomal subunits exclusively into the soluble phase in 4M urea-2M LiCl (data not shown). The results of control studies indicated that exposure of the free rRNA

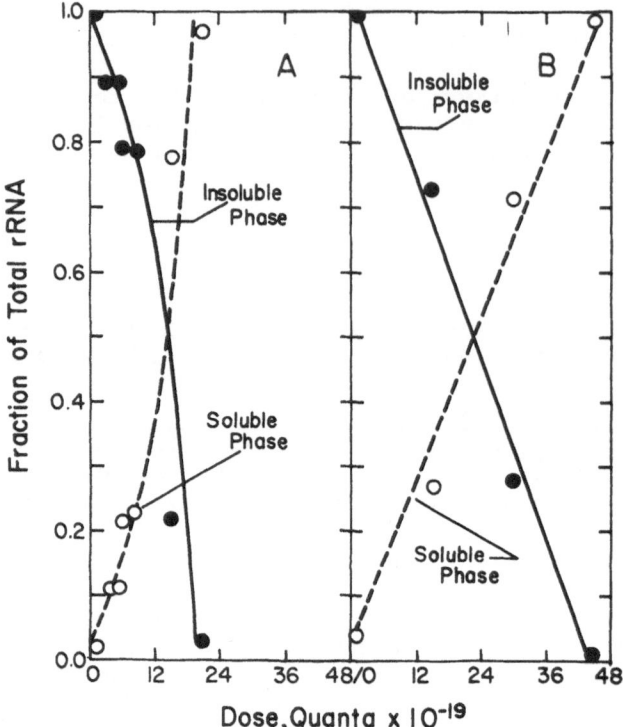

Figure 1.   Distribution of rRNA components of irradiated
ribosomes between soluble and insoluble phases in 4M urea-2M LiCl:
Irradiated ribosomes were treated in 4M urea-2M LiCl.   The resultant mixtures were centrifuged, and the RNA contents in the
resultant supernatant and pellet fractions determined by the Orcinol
method (Mejbaum, 1939).   A.   30S Ribosomes.   B.   50S Ribosomes
(Redrawn from Gorelic, 1975a, 1975b).

components of the ribosomal subunits to doses of 254 nm radiation
used in the studies represented by the data in Figure 1 did not
alter their initial fractionation exclusively into the insoluble
phase in 4M urea-2M LiCl, even in the presence of added total
ribosomal proteins from the corresponding ribosomal subunit.   It
was concluded from these findings that exposure of E. coli ribosomal
subunits to 254 nm radiation results in a reduction in the separability of the ribosome components in 4M urea-2M LiCl; and that the
basis for this effect does not reside in protein-independent
modifications in the primary structures of the rRNA components that
enhance the inherent solubilities of the rRNA components in 4M
urea-2M LiCl or that increase the strength of the non-covalent
associations of the rRNA components of the ribosomes with their
corresponding proteins.

The results of a study of the effects of 254 nm radiation on the gel filtration properties of the ribosome components are represented by the data in Figure 2. The samples used in these

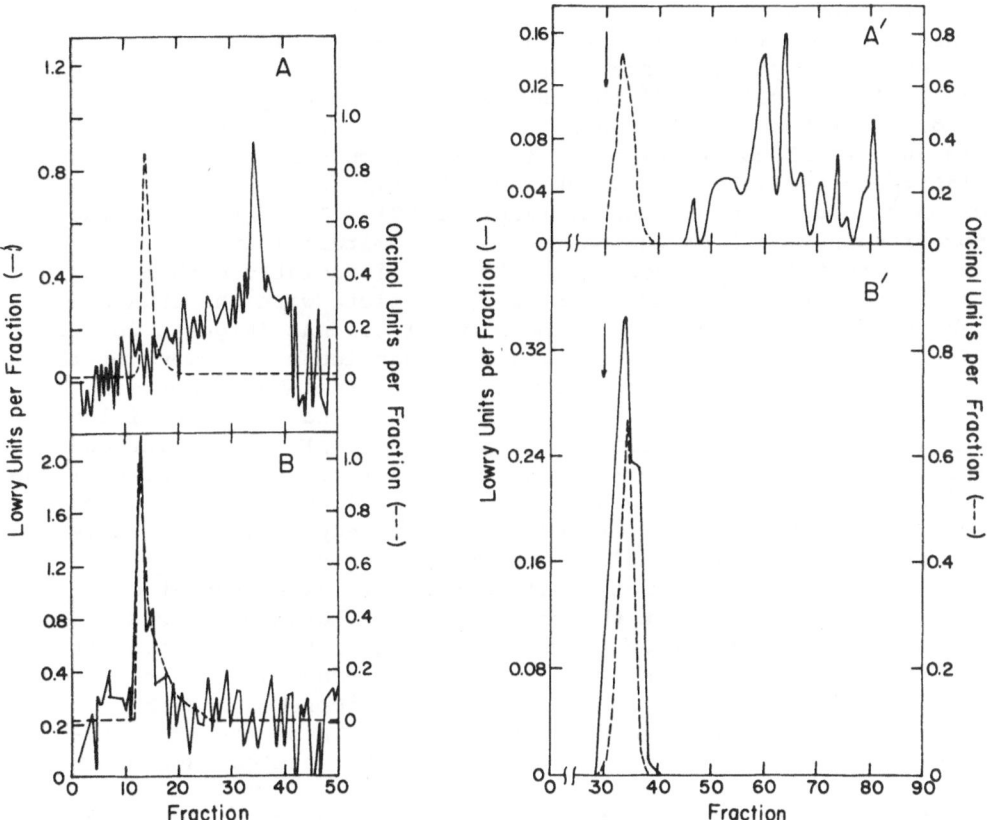

Figure 2. Gel filtration properties of ribosome components: The free rRNA and protein components of unirradiated ribosomal sub-units, and urea-LiCl soluble fractions prepared by low-speed centrifugation of irradiated ribosomes were subjected to gel filtration on Bio-Gel columns. The samples were eluted with a Tris-6M urea buffer. The protein contents of the fractions were determined by the Lowry method (Lowry, et.al., 1951) and the RNA contents by the Orcinol reaction (Mejbaum, 1939). A, B: elution profiles, on Bio-Gel A-5.0 m, of the rRNA and protein components of samples prepared from unirradiated and irradiated 30S ribosomes, respectively; A', B': elution profiles, on Bio-Gel A-5.0 m, of the rRNA and protein components of samples prepared from unirradiated and irradiated ribosomes, respectively. (Redrawn from Gorelic, 1975a, 1975b).

studies were either the free rRNA and total protein components of
unirradiated ribosomes obtained by treatment of the ribosomal subunits in 4M urea-2M LiCl, or were the soluble phases of ribosomal
subunits irradiated with a dose of 3×10^{20} quanta of 254 nm
radiation and subsequently treated in 4M urea-2M LiCl. (It should
be noted that 80-100% of the total rRNA components of ribosomal
subunits exposed to 3×10^{20} quanta of 254 nm radiation is fractionated into the soluble phase in 4M urea-2M LiCl.) The data in
Figures 2A and 2A' indicate that there is a substantial difference
in the elution volumes of the rRNA and protein components of unirradiated ribosomes, findings that are consistent with the 10-20
fold larger molecular weight of the rRNA components of the ribosomes
relative to the molecular weights of the ribosomal proteins
(Dzionara, Kaltschmidt, and Wittmann, 1970). In contrast to the
above findings, all of the Lowry-reactive material in urea-LiCl
soluble fractions obtained from ribosomes irradiated with a dose of
3×10^{20} quanta (1.7×10^{4} ergs mm^{-2}) co-elute with the rRNA components of the ribosomes from the Bio-Gel columns (Figures 2B, 2B').
Control studies indicated that prior ribonuclease treatment of the
urea-LiCl soluble fractions prepared from irradiated ribosomes
resulted in the removal of most of the Lowry-reactive material
from the rRNA (or high molecular weight regions) of the elution
profiles. Based upon the findings cited here and the results of the
control studies associated with the data in Figure 1, it was concluded that a high molecular weight ribonucleoprotein complex is
present in urea-LiCl soluble fractions prepared from E. coli
ribosomal subunits irradiated with a dose of 254 nm radiation of
3×10^{20} quanta; and that the basis for the photoinduced changes in
the initial gel filtration properties of the ribosome components
does not reside in photochemically enhanced non-covalent interactions between the rRNA components of the ribosomes and the
individual ribosomal proteins or the co-elution of the rRNA components with high molecular weight crosslinked protein aggregates.

Finally, the effects of 254 nm radiation on the electrophoretic
mobilities of the ribosomal proteins were investigated. The samples
used in these studies were the soluble phases obtained by treatment
in 4M urea-2M LiCl of unirradiated ribosomal subunits or ribosomal
subunits irradiated with a dose of 1.2×10^{20} quanta (7×10^{3} ergs
mm^{-2}) of 254 nm radiation. The 2D-gel system used in these studies
was the Howard and Traut modification of the original Kaltschmidt
and Wittmann procedure (Howard and Traut, 1973; Kaltschmidt and
Wittmann, 1970) further modified so that only 10-20% of the total
ribosomal proteins applied to the first dimension of the 2D-gel
system did not penetrate into the second dimension. The ribosomal
proteins were visualized by staining the gels with Coomassie
Brilliant Blue. The 2D-electropherograms in Figures 3A and 4A
indicate that 19 of the 21-30S ribosomal proteins and 30 of the

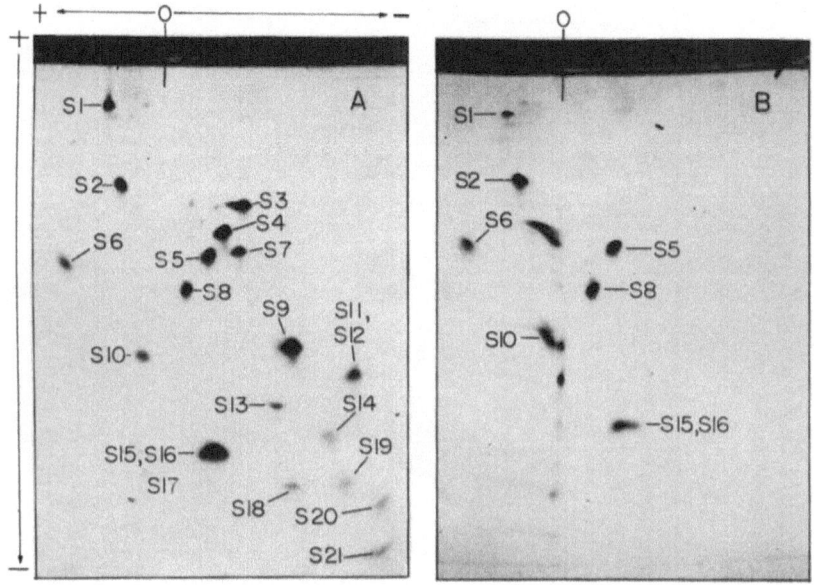

Figure 3. Two-dimensional electropherograms of urea-LiCl Soluble fractions prepared from unirradiated and irradiated E. coli 30S ribosomal subunits: the urea-LiCl soluble fractions were prepared from unirradiated and irradiated E. coli 30S ribosomes. Equivalent amounts, in terms of protein content, of the urea-LiCl soluble fractions were analyzed by 2D-gel electrophoresis. The proteins were visualized by staining the gels with Coomassie Brillant Blue. A. Unirradiated ribosomes. B. Irradiated ribosomes. (Redrawn from Gorelic, 1976a)

34-50S ribosomal proteins can be resolved into discrete spots. In contrast to this situation, only 7 of the 30S ribosomal proteins and half of the 50S ribosomal proteins can be detected as discrete spots on 2D-electropherograms of urea-LiCl soluble fractions obtained from irradiated ribosomal subunits; and many of the latter spots are detected at stained intensities substantially reduced relative to their initial values. The photoinduced crosslinkage of ribosomal protein to the rRNA components of its corresponding ribosomal subunit would result in the formation of a negatively-charged ribonucleo-protein complex. Since electrophoresis in the second dimension of the 2D-gel system is from anode (+) to cathode (-), the photoinduced crosslinkage of a ribosomal protein to the rRNA molecule should result in the formation of a complex with an electrophoretic mobility of zero in the second dimension. It was therefore concluded that the observed effects of 254 nm radiation on 2D-electropherograms of urea-LiCl soluble fractions prepared from the E. coli 30S and 50S

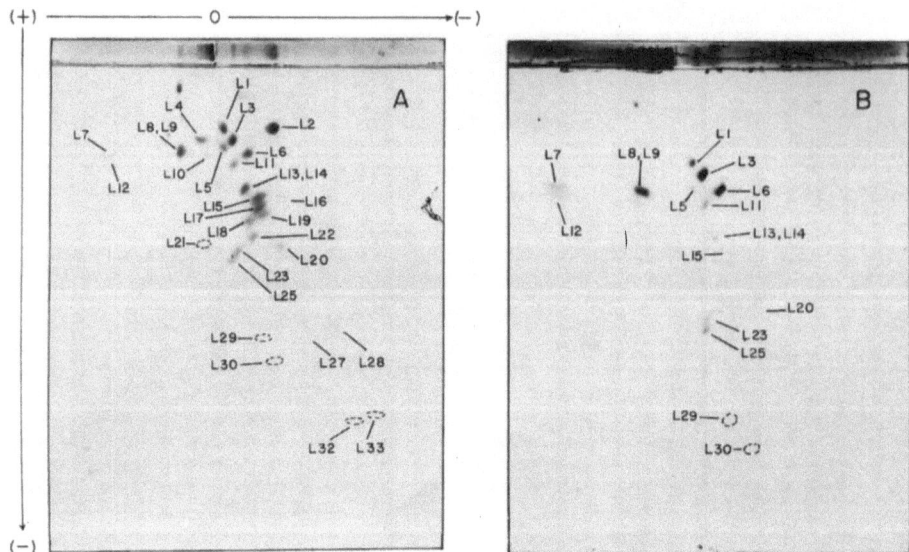

Figure 4. Two-dimensional electropherograms of urea-LiCl
soluble fractions prepared from unirradiated and irradiated E. coli
50S ribosomes: Urea-LiCl soluble fractions were prepared from
unirradiated and irradiated E. coli 50S ribosomes. Equivalent
amounts, in terms of proteins content, of the urea-LiCl soluble
fractions were analyzed by 2D-gel electrophoresis. The proteins
were visualized by staining with Coomassie Brillant Blue.
A. Unirradiated ribosomes.   B. Irradiated ribosomes (Redrawn
from Gorelic, 1976b)

ribosomal subunits was a result of the covalent crosslinkage of the
components of the ribosomes to the rRNA components. The results of
one control study for these conclusions indicated that ribonuclease
pre-treatment of the urea-LiCl soluble fractions prepared from
irradiated ribosomal subunits completely, or substantially, restored
the initial stained intensities on 2D-electropherograms of most of
the ribosomal proteins. It was not possible, however, to ascertain
whether the positions of the restored spots were shifted towards
the anodes in the first and second dimensions relative to the spots
on the control 2D-electropherograms. The results of a second control
for these conclusions indicated that irradiation of the ribosomal
subunits with a dose of 1.2×10^{20} quanta did not substantially
alter the initial molecular weight distributions of the ribosomal
proteins. Therefore, it also was concluded that photoinduced modifications in the primary structures of the ribosomal proteins by
reactions such as protein-protein crosslinkage, peptide bond cleavage

or the loss of positively charged side-chain groups could not account for the dose-dependent changes observed in the 2D-electropherograms in Figures 4 and 5.

Based upon the results of the above studies and the associated controls, it was concluded that covalent RNA-protein crosslinks are present in irradiated E. coli ribosomal subunits, and that the presence of such crosslinks accounts for the observed effects of 254 nm radiation on the separability of the ribosome components and on the initial gel filtrations and electrophoretic properties of the ribosomal proteins.

The possibility that covalent RNA-protein crosslinks are present in irradiated ribosomal subunits has also been examined by determining the radioactive content before and after irradiation of the ribosomal proteins in ribosomal subunits initially radioactively labelled in their rRNA components (Moller and Brimacombe, 1975). The ribosomal subunits used in these studies were prepared from 70S ribosomes that were purified by high salt washings and that were radioactively labelled in their rRNA components with $^{32}PO_4^{3-}$. Solutions of the ribosomal subunits in a Tris-based buffer (final ribosome concentration, 5 A_{260} units per ml) were irradiated in plastic containers with a Sylvania germicidal lamp. The output of the lamp was not disclosed. The irradiated ribosomal subunits were treated with ribonucleases and a mixture of concentrated formic acid and 60% acetic acid in order to release the ribosomal proteins from the irradiated ribosomes and to destroy non-covalently attached rRNA fragments. Analysis, by the technique of 2D-gel electrophoresis, of the protein mixtures obtained from irradiated ribosomal subunits revealed the presence of two stained spots not detected on 2D-electropherograms of protein mixtures obtained from unirradiated ribosomal subunits. The two new spots were detected at positions northwest to the positions of the spots corresponding to proteins S7 and L4. Autoradiograpraphy of the 2D-electropherograms of protein mixtures prepared from irradiated ribosomal subunits indicated that all of the radioactivity detected in the second dimension of the 2D-electropherograms co-electrophoresed with the two new spots.

Moller and Brimacombe concluded from the results of the autoradiographic analysis of the 2D-electropherograms cited above that at least two of the protein components of the irradiated E. coli ribosomal subunits had been covalently crosslinked to the rRNA components of their corresponding subunit(s). The photoinduced modification of a ribosomal protein by the covalent attachment of negatively-charged nucleotides should increase the electrophoretic mobility of the protein on 2D-polyacrylamide gels towards the anodic (+) electrodes (located, by convention, west in the first dimension of the 2D-gels and north in the second dimension), these investigators concluded that the ribosomal proteins that had been covalently

crosslinked to the rRNA components were proteins S7 and L4. These
latter assignments were subsequently confirmed by immunochemical
methods using the radioactively labelled ribosomal proteins released from the irradiated ribosomal subunits as a source of antigen
and anti-sera specific against each of the ribosomal proteins.
Since a weak reaction was observed between the mixture of labelled
ribosomal proteins from the E. coli 50S ribosomes and anti-sera-L2,
it was concluded that protein L2 had also been crosslinked to the
rRNA components of the irradiated E. coli 50S ribosomal subunit.

   B. Evidence for Covalent Nucleic Acid-Protein Crosslinks in
Irradiated E. coli 70S Ribosomes and 70S-mRNA Complexes. The
possibility that covalent RNA-protein crosslinks are present in
irradiated 70S ribosomes has been investigated by Baca and Bodley
(1976). The 70S ribosomes used in their investigation were radioactively labelled in their rRNA components with $^{32}PO_4^{3-}$. Solutions
of 70S ribosomes (final ribosome concentration, 47 A_{260} units per ml)
in a Tris-based buffer were irradiated in a Rayonet RPR-100 photochemical reactor. The flux of incident radiation at 254 nm was
calculated to be 254 quanta min^{-1} per ribosome, which corresponds
to 2×10^{17} quanta min^{-1}. The radioactively labelled 70S ribosomes
were irradiated with doses of 254 nm radiation of $1.2-2.4 \times 10^{19}$
quanta, and the ribosomal proteins separated from non-covalently
attached rRNA fragments by ribonuclease treatment under denaturing
conditions followed by precipitation with 20% trichloroacetic
acid.

   Two-dimensional electropherograms of the protein mixtures prepared in the above manner from irradiated 70S ribosomes were significantly different from 2D-electropherograms of the corresponding
protein mixtures prepared from unirradiated ribosomes. The most
obvious differences that were detected by staining of the 2D-gels
were a new spot between the stained spots corresponding to proteins
S2 and L4 and a heavily stained tail radiating outward in a north-
westerly direction from the spot corresponding to the protein L2.
Autoradiography of the stained 2D-electropherogram indicated that
most of the radioactivity detected in the second dimension of the
2D-gel co-electrophoresed with the new spot between S2 and L4, and
with the tail on the spot corresponding to protein L2. In addition,
a radioactive spot was detected near the stained spot corresponding to protein S8; and substantial radioactivity was detected in
the one-dimensional gel comprising the electropheoretic origin of
the electrophoresis in the second dimension. Baca and Bodley concluded from these findings that exposure of the E. coli 70S
ribosomes to 254 nm radiation had resulted in the covalent crosslinkage of ribosomal proteins L2 and L4, and possibly S8, to the
rRNA components of their corresponding subunits.

The data obtained by Baca and Bodley clearly indicate that co-valent crosslinks are formed between the radioactivily-labelled rRNA components of the irradiated 70S ribosomes used in their studies and certain ribosomal proteins. However, the complexity of 2D-electro-pheorgrams of total ribosomal proteins from 70S ribosomes -- i.e. 55 resolvable spots under the best of conditions -- precludes, in this reviewer's opinion, definitive assignments of the ribosomal proteins actually crosslinked to their corresponding rRNA molecules based solely upon electropheoretic data. In addition, since substantial amounts of radioactivity were not electropheoresed from the lD-gels into the second dimension of the 2D-gels, it is conceivable that a substantially larger number of ribosomal proteins had been radio-actively labelled than was indicated by autoradiography of the second dimension.

The possibility that the covalent crosslinks are formed bet-ween the ribosomal proteins in irradiated 70S polynucleotide com-plexes and the polynucleotide components has also been investigated. Schenkman and co-workers (1974) have reported the effects of ultra-violet radiation on the nature of the interactions between the ribosome components and polynucleotide constituent of an E. coli 70S ribosome [3H]-poly(U) complex. Solutions of the complexes (final ribosome concentration, 10 A_{260} units per ml) in a Tris-based buffer were irradiated in glass petri dishes with a Sylvania germicidal lamp. The flux of incident radiation was 10 ergs mm^{-2} S^{-1}; the fraction of incident radiation actually absorbed by the solutions of the complex was not determined. Sedimentation analysis, on low magnesium-sucrose gradients, of an unirradiated 70S[3H]poly (U) complex indicated that none of the poly(U)-associated radio-activity co-sedimented with the 30S or 50S ribosomal subunits. On the other hand, a significant portion of the poly(U)-associated radioactivity co-sedimented with the 30S subunit of 70S-[3H]poly(U) samples irradiated with a total dose of 1.2 x 10^4 ergs mm^{-2} of 254 nm radiation. Isolation of the co-sedimenting 30S-[3H]poly(U) material from the sucrose gradients of the irradiated 70S-poly(U) samples, followed by treatment with 6M urea, was shown by sedmen-tation techniques similar to those cited above to result in the release of ca. 50% of the initially bound poly(U)-associated radioactivity from the 30S ribosomes. [The results of an earlier study had indicated that treatment of 30S ribosomes with 6M urea results in the release of ca. 6-7 of the 30S ribosomal proteins. Included in this released group of proteins was the 20S ribosomal protein Sl, a ribosomal protein that had been shown by reconstitu-tion methods to be required for poly(U) binding to the E.coli 30S ri-bosomal subunit(Kurland, 1974). In addition to the noted effects of 6M urea on the 30S-[3H]poly(U) co-sedimentry material, treatment of this putative 30S-[3H]poly(U) complex in 1% SDS-0.1M LiCl was shown by sedimentation techniques to result in the release of ca. 80% of

the initially bound poly(U) from the complex and the release of all
of the ribosomal proteins from the 16S rRNA component from the 30S
ribosomes. The latter sedimentation studies also indicated that the
remaining 20% of initially-bound [3H]-poly(U) co-sedimented with the
protein-free 16S rRNA molecule after treatment of the putative 30S-
[3H]poly(U) complex with 1% SDS-0.1M LiCl.

Schenkman and co-workers concluded from the above findings that
ca. 80% of the total poly(U) co-sedimenting with the 30S ribosomal
subunit of the irradiated 70S poly(U) complex had been covalently
crosslinked to some of the protein components of the 30S ribosomal
subunit; and that the remaining 20% had been covalently crosslinked
to the 16S rRNA components. However, it should be noted that the
noncovalent binding affinities of irradiated 30S ribosomes for the
poly(U) template or of irradiated poly(U) for the ribosome components
were not determined in these studies. Consequently, it is not
possible to rule out an alternative interpretation of the data that
the co-sedimentation of the 30S ribosome components of an irradiated
70S poly(U) complex with the poly(U) component is a result of photoinduced modifications that primarily affect the strengths of the
non-covalent associations between the ribosome components and the
polynucleotide.

A less ambiguous demonstration of the formation of covalent
crosslinks between the ribosome and polynucleotide components of an
irradiated 70S-poly(x) complex has been presented by Fiser and
co-workers (1974, 1975). The polynucleotide component used in these
studies was [3H]-poly(^{4}SU). Solutions of the 70S-[3H]poly(^{4}SU) complex were prepared in a Tris-based buffer (final ribosome concentration, 5 A260 units per ml) and were irradiated with ultraviolet
light in the spectral region of maximum absorption by the 4-thio-
uridine residues of the poly(^{4}SU) component. Sedimentation analysis,
on low magnesium-sucrose gradients, of the irradiated 70S-[3H]poly-
(^{4}SU) complex indicated that a significant fraction of the poly-
(^{4}SU)-associated radioactivity co-sedimented with the 30S subunit of
the complex. The 30S-[3H]poly(^{4}SU) co-sedimenting material was
isolated from the gradients, treated with ribonuclease and 0.1M
alkali in order to release the ribosomal proteins from the 30S subunit and to remove non-covalently attached rRNA fragments, and the
resultant protein mixture prepared for analysis on SDS-polyacrylamide gels by heating the protein samples in the presence of SDS to
100°C. The SDS-electropherogram of the treated protein mixture
obtained from irradiated 70S-[3H]poly(^{4}SU) complexes indicated that
most of the poly[^{4}SU]-derived radioactivity exhibited an electro-
pherotic mobility comparable to that of the largest 30S protein S1.
The remaining radioactivity was distributed throughout the lower
molecular weight regions of the electropherogram. A control study
indicated that none of the poly(^{4}SU) associated radioactivity

could be detected on SDS-electropherograms of a mixture of 30S ribosomes and [3H]-poly(^{6}SU) that had been treated in the same manner as the 30S-[3H]-poly(^{4}SU) complex isolated from an irradiated 70S-[3H]-poly(^{4}SU) complex. It was concluded from these findings that irradiation of a 70S-poly(^{4}SU) complex results in the covalent crosslinkage of the the polynucleotide to at least protein S1 of the 30S ribosomal subunit (Fiser, et. al, 1974). A subsequent immunochemical analysis of the radioactivly-labelled ribosomal proteins released by ribonuclease treatment from an irradiated 70S-poly(^{4}SU) complex confirmed this assignment; and in addition, indicated that proteins S18 and S21 had also been covalently crosslinked to the poly(^{4}SU) template (Fiser, et. al, 1975).

IV.   APPLICATIONS

The photoinduced formation of covalent crosslinks between the nucleic acid and protein components in E. coli ribosomes and ribosome-poly(x) complexes has been used to identify specific nucleic acid-protein interactions within these complexes. Fiser and co-workers have concluded from the pattern of radioactive labelling of ribosomal proteins in an irradiated 70S-[3H]-poly(^{4}SU) complex that only proteins S1, S18, and S21 constitute the mRNA-binding site of the E. coli 30S ribosomes (Fiser, et. al, 1974, 1975). Moller and Brimacombe have concluded from studies of radioactive labelling of the ribosomal proteins in irradiated [^{32}P-rRNA]-E. coli 30S and 50S ribosomal subunits that only protein S7 of the 30S subunit and proteins L2 and L4 of the 50S subunit are sufficiently close to the rRNA components of their respective subunits to be reactive in photoinduced RNA-protein crosslinkage (Moller and Brimacombe, 1975). Finally, a study of the photoinduced radioactive labelling patterns of ribosomal proteins in [^{32}P-rRNA]-E. coli 70S ribosomes has led Baca and Bodley to conclude that proteins S8, and L2 and L4 are the only ribosomal proteins sufficiently close to the rRNA components of their respective subunits to be able to participate in the formation of covalent RNA-protein crosslinks.

The above assignments of nucleic acid-binding ribosomal proteins in irradiated ribosome and ribosome-poly(x) complexes were based upon the observations that the cited proteins were the only ribosomal proteins that were radioactivly labelled at specific doses of ultraviolet radiation. The molecular basis for this rationale is that those ribosomal proteins in actual physical contact with the rRNA (poly(x)) bases in the native topographical states of the tested ribosome systems should crosslink to the nucleic acid components at lower doses of ultraviolet radiation than ribosomal proteins distal to the nucleic acid bases (the latter type of which would require for crosslinkage to the nucleic acid components prior photoinduced

changes in ribosome topography). The validity of an analysis of
ribosome interactions based upon the rationale described here is
contingent upon the validities of a number of assumptions. These
assumptions are: (1) Photoinduced changes in ribosome topography
do not precede the photoinduced crosslinkage of the "low-dose
reactive" ribosomal proteins to the nucleic acid molceules; (2) The
differences between the initial rates of crosslinkage to the
nucleic acid molecules of the "low-dose reactive" molecules are not
so large that doses sufficient to reslt in the crosslinkage to the
nucleic acid molecules of a small fraction of the more reactive
of the "low-dose reactive" proteins would not be sufficient to result in the crosslinkage of a large enough fraction of the less
reactive proteins to the nucleic acid molecules to be detected
experimentally; (3) All ribosomal proteins interacting directly
with the bases of the nucleic acid molecules in the native topographical states of the ribosome systems are capable of participating
in the photoinduced formation of at least one stable nucleic acid-
protein crosslink.

The results of studies by Moller and Brimacombe and in our own
laboratory have indicated that photoinduced changes in the tertiary
structure of the ribosomal subunits do not occur at doses lower
than those required to result in the covalent crosslinkage of most
of the ribosomal proteins to the rRNA components of their respective
ribosomal subunits (Moller and Brimacombe, 1975; Gorelic, 1975a,
1976b). Therefore, it can be concluded that a photochemical analysis
of the type described above of ribosome interactions is probably not
distorted by photoinduced changes in overall ribosome conformation.
However, since the methods used in the above studies of ribosome
conformation -- i.e. sedimentation and electropheoretic techniques
-- are not sufficiently sensitive to permit the detection of small
changes in ribosome conformation, the possibility that subtle photoinduced changes in ribosome topography precede the photoinduced
crosslinkage of the "low-dose reactive" ribosomal proteins to their
respective nucleic acid molecules cannot be eliminated from consideration. Furthermore, since detailed analyses of the kinetics of
photoinduced formation of nucleic acid-protein crosslinks in
irradiated ribosomes or ribosome-poly(x) complexes have not been
reported, it is not possible to determine whether ribosomal proteins
in addition to those already identified as "low-dose reactive,"
might belong to this group of ribosomal proteins. Finally, since the
ribosome samples used in the analyses of ribosome interactions were
prepared under highly acidic or basic conditions or at elevated
temperatures, yet numerous studies have shown that pyrimidine
photoadducts of amino acid and amino acid analogues rapidly revert
to starting materials under these conditions (Wang, 1962; Smith and
Aplin, 1966; Smith and Muen, 1968; Gorelic and Yang, 1972), it is
conceivable that not all of the nucleic acid-binding ribosomal proteins have been identified in the tested ribosome systems.

Based upon the discussion in the above sections, it seems reasonable to conclude that most of the attempted applications of photochemical methods to studies of ribosome interactions have not provided an unequivocal answer to the question of which ribosomal proteins are able to interact directly with the rRNA or poly(x) molecules in the native topographical states of ribosomes or ribosomal poly(x) systems.

An alternative approach to a photochemical analysis of ribosome interactions has been proposed in our laboratory and applied to an analysis of rRNA-protein interactions in intact E. coli 30S and 50S ribosomes (Gorelic, 1976a, 1976b). In the selected approach, the kinetics of crosslinkage of the individual ribosomal proteins to the rRNA components in their respective subunits is determined by quantitatively monitoring by 1D- and 2D-polyacrylamide gel electrophoresis the stained intensities of the individual proteins in mixtures obtained at neutral pH and low temperature from unirradiated and ribosomes irradiated with different doses of 254 nm radiation. The resultant kinetic data are analyzed by target theory, assuming that a single-hit multitarget model applies to the crosslinkage reaction. The back-extrapolates to zero dose of the exponential portions of the resultant curves provide a measure of the extrapolation number or the complexity of the photoprocesses required for rRNA-protein crosslinkage. The principal advantages of this approach are that, with the exception of the actual electrophoresis studies, the ribosomal protein mixtures obtained from irradiated ribosomes are not exposed to conditions capable of destroying covalent crosslinks between the rRNA and protein components of the ribosomal; and the complexities of the photoprocesses required for the crosslinkage of specific ribosomal proteins to their respective rRNA molecules can be obtained directly from the kinetic data.

The results of a photochemical analysis of the type outlined above of ribosome interactions are presented in Figures 5 and 6. The data in these figures indicate that photoinduced crosslinkage of the protein components in the intact ribosomal subunits to the rRNA components occurs in a sequential manner with certain proteins exhibiting detectable crosslinkage to the rRNA components at lower doses than other proteins The data in Figures 5 and 6 also indicate that there are statistically significant differences in the magnitudes of the extrapolation numbers associated with the photoinduced crosslinkage of the various ribosomal subunits to the rRNA components in their respective ribosomal subunits. The values of these extrapolation numbers vary from unity for the 30S ribosomal proteins S3, S7-S9, S15-S17, and for the 50S ribosomal proteins L2, L4, L5, L16, L21, and L29; to values of 2 to 4 for the 30S ribosomal proteins S11, S18, and S19; and for the 50S ribosomal proteins L1, L3, L10,

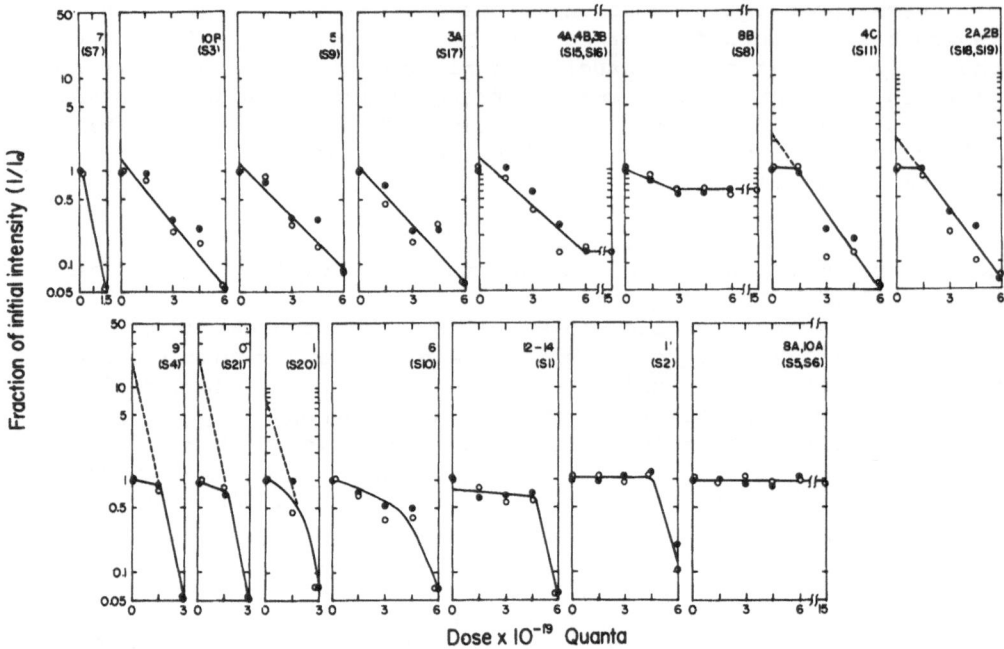

Figure 5. Kinetics of crosslinkage of protein components of
E. coli 30S ribosomes to 16S rRNA component: Urea-LiCl soluble
fractions were prepared from unirradiated and irradiated E. coli
30S ribosomes. The fractions were analyzed by 1D-gel electrophoresis according to the procedure of Traut, et. al (1969). The
proteins were visualized by staining the electropherograms with
coomassive Brilliant Blue. The kinetics of crosslinkage were
determined by monitoring dose-dependent reductions in the initial
stained intensities of the bands on the 1D-electropherograms
according to the procedure described by Gorelic (1976a). (Redrawn
from Gorelic, 1976a).

L17–L19, L22, and L28; to values of greater than or equal to 10 for
the remaining detectable 30S and 50S ribosomal proteins. Supple-
mentation of the presented kinetic data for the ribosomal proteins
of the 30S subunit with data obtained from a similar kinetic
analysis made by 2D-gel electrophoresis indicated that the extrapolation numbers associated with the photoinduced crosslinkage of
proteins S12, S13 and S14 to the 16S rRNA were 2, greater than 10
and greater than 10, respectively (Gorelic, 1976a). Finally, the data
in Figures 5 and 6 indicate that it cannot be assumed, a priori,

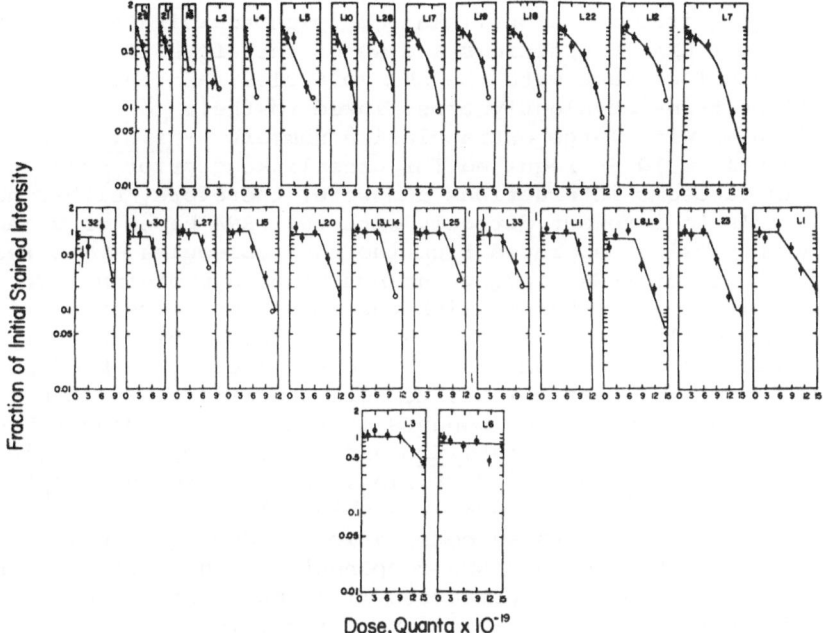

Figure 6. Kinetics of crosslinkage of protein components of
E. coli 50S ribosomes to the rRNA components: urea-LiCl soluble
fractions were prepared from unirradiated and irradiated E. coli
50S ribosomes. The fractions were analyzed by 2D-gel electrophoresis according to the procedure of Howard and Traut (1973).
The proteins were visualized by staining the 2D-electropherograms
with coomassive Brilliant Blue. The kinetics of crosslinkage was
determined by monitoring dose-dependent reductions in the initial
stained intensities of the spots on the 2D-electropherograms
according to the procedure described by Gorelic (1976b). (Redrawn
by Gorelic, 1976b).

that the initial rates of photoinduced crosslinkage to the rRNA of
the low-dose reactive ribosomal proteins are similar, an assumption
that is inherent in an analysis of the ribosomal interactions in
terms of low-dose reactivity in RNA-protein crosslink formation.

It was concluded from the above functions that those ribosomal
proteins covalently crosslinked to their corresponding rRNA molecules by photoprocesses with low extrapolation numbers -- i.e. one
to four -- required for crosslinkage only photoexcitation of the
bases at their RNA-interactions and were, therefore, normally in
direct physical contact with the RNA molecules; and that those
ribosomal proteins covalently crosslinked to their corresponding
rRNA molecules with large extrapolation numbers -- i.e. greater
than or equal to 10 -- required for crosslinkage prior photoinduced
changes in ribosome interactions and were, therefore, either not
normally able to interact directly with the rRNA bases or were even
possibly physically separated from the rRNA molecules. The resulting assignments of rRNA-binding and non-binding proteins were, in
fact, in quite good agreement with the rRNA-binding characteristics
of the individual ribosomal proteins determined from non-photochemical studies of ribosome interactions. These rRNA-binding
characteristics included ability to form specific and stable complexes with the free rRNA components of the corresponding ribosomal subunit, presence in in vitro and in vitro protein-deficient
rRNA-containing subribosomal particles, and requirement for the
early steps in in vitro ribosome assembly. There were a limited
number of exceptions to these correlations. Certain ribosomal
proteins crosslinked to the rRNA components of the corresponding
ribosomal subunits by photoprocesses with small extrapolation
numbers did not exhibit rRNA-binding characteristics; and other
ribosomal proteins crosslinked to the rRNA components of the
corresponding ribosomal subunits by photoprocesses with large extrapolation numbers exhibited rRNA-binding characteristics. The
first type of exception to the observed correlation was attributed
to the inability of the non-photochemical methods to detect RNA-
binding proteins with rRNA-binding constants too small to permit the
formation of stable non-covalent protein-rRNA complexes or to
result in their inclusion in in vitro and in vivo subribosomal
particles. The second type of exception to the observed correlations was tentatively attributed to the inability of certain
ribosomal proteins to participate in the formation of covalent RNA-
protein crosslinks stable under the irradiation condition or under
both the acidic and basic conditions used in the characterization of
the irradiated ribosomal subunits by electrophoretic techniques;
and to steric factors preventing access of the reactive groups in
certain rRNA-binding proteins to the photoexcited bases in the
rRNA molecules. Previously published reports indicating that some
of the rRNA-binding ribosomal proteins do, in fact, interact with
the rRNA molecules in regions of complex secondary structure, and

a number of photoadducts of pyrimidines and purines with amino
acids and amino acid analogues readily revert to starting materials
under acidic or basic conditions were cited in support of these
latter proposals. Based upon the observed correlation between
extrapolation numbers and rRNA-binding characteristics of the
ribosomal proteins, and the associated exceptions, it was finally
concluded that a low extrapolation number associated with the
photoinduced crosslinkage of a ribosomal protein to the rRNA components in the corresponding ribosomal subunit can confidently be
accepted as evidence that the ribosomal protein interacts directly
with the rRNA bases in the native topographical state of the
ribosome. However, a comparison of the RNA-binding characteristics
of a ribosomal protein crosslinked to the rRNA components with large
extrapolation number with the photochemical data is absolutely
required for an unequivocal assignment of the relative spatial
orientations of the ribosomal protein and the rRNA molecules in the
native topographical state of the ribosome.

REFERENCES

Baca, O. G. and Bodley, J. W. (1976). UV Induced Crosslinking of
    E. coli Ribosomal RNA to Specific Proteins. Biochem. Biophys.
    Res. Commun., 70, 1091.

Dzionara, M., Kaltschmidt, E. and Wittmann, H. G. (1970). Ribosomal
    Proteins, XIII. Molecular Weights of Isolated Ribosomal Pro-
    teins of E. coli. Proc. Nat'l Acad. Sci., USA, 67, 1909.

Fiser, I., Scheit, K. H., Stoffler, G. and Kuechler, E. (1974).
    Identification of Protein S1 at the Messenger RNA Binding
    Site of the E. coli Ribosome. Biochem. Biophys. Res. Commun.,
    60, 1112.

Fiser, I., Scheit, K. H., Stoffler, G. and Kuechler, E. (1975).
    Proteins at the RNA Binding Site of the E. coli Ribosome.
    FEBS Lett., 56, 226.

Gorelic, L. (1975a). Photoinduced Covalent Crosslinkage, in situ,
    of E. coli 50S Ribosomal Proteins to rRNA. Biochem. Biophys.
    Acta, 390, 209.

Gorelic, L. (1975b). Evidence for Photoinduced Cross-Linkage, in
    situ, of 30S Ribosomal Proteins to 16S rRNA. Biochemistry,
    14, 4627.

Gorelic, L. (1976a). Photoinduced Cross-Linkage, in situ, of E.
    coli 30S Ribosomal Proteins to 16S rRNA: Identification of
    Cross-Linked Proteins and Relationships Between Reactivity
    and Ribosome Structure. Biochemistry, 15, 3599.

Gorelic, L. (1976b). Analysis of the Kinetics of Photoinduced
    Crosslinkage of the Protein and RNA Components of the E. coli
    50S Ribosomal Subunit. Biochem. Biophys. Acta (in press).

Gorelic, L., Lisagor, P. and Yang, N. C. (1972). The Photochemical
    Reactions of 1,3-Dimethyluracil with 1-Aminopropane and Poly-
    lysine. Photochem. Photobiol., 16, 465-80.

Howard, G. A. and Traut, R. R. (1973). Separation and Autoradio-
    graphy of Microgram Quantities of Ribosomal Proteins by Two-
    Dimensional Polyacrylamide Gel Electrophoresis. FEBS Lett.,
    29, 177.

Kaltschmidt, G. and Wittmann, H. G. (1970). Ribosomal Proteins,
    XII. Number of Proteins in Small and Large Ribosomal Subunits
    of E. coli as Determined by Two-Dimensional Gel Electro-
    phoresis. Proc. Nat'l Acad. Sci., USA, 67, 1276.

Kaltschmidt, G. and Wittmann, H. G. (1970). Ribosomal Proteins
    VII. Two-Dimensional Polyacrylamide Gel Electrophoresis for
    Finger Printing of Ribosomal Proteins. Anal. Biochem., 36,
    401.

Kurland, C. G. (1972). Structure and Function of Bacterial
    Ribosomes, Ann. Rev. Biochem., 41, 377.

Lowry, O. H., Rosebrough, N. J., Farr, A. L. and Randall, R. J.
    (1951). Protein Measurement with the Folin-Phenol Reagent.
    J. Biol. Chem., 193, 265.

Mejbaum, E. Z. (1939). Estimation of Small Amounts of Pentose,
    Especially in Derivatives of Adenylic Acid. Z. Physiol. Chem.,
    258, 117.

Moller. K. and Brimacombe, R. (1975). Specific Crosslinking of
    Proteins S7 and L14 to Ribosomal RNA by UV Irradiation of E.
    coli Ribosomal Subunits. Mol. Gen. Genetics, 141, 343.

Schenkman, M. I., Ward, D. C. and Moore, P. B. (1974). Covalent
    Attachment of a Messenger RNA to the Eschericia coli Ribosome.
    Biochim. Biophys. Acta, 353, 503.

Smith, D. C. and Aplin, R. T. (1966). A Mixed Photoproduct of
    Uracil and Cysteine (5-S-Cysteine-6-dihydrouracil). A
    Possible Model for the in vivo Cross-Linking of Deoxyribo-
    nucleic Acid and Protein by Ultraviolet Light. Biochemistry,
    5, 2125-30.

Smith, K. C., ed (1975). Aging, Carcinogenesis, and Radiation Biology: The Role of Nucleic Acid Addition Reactions. Plenum Press, New York.

Smith, K. C. and Muen, D. H. C. (1968). Kinetics of the Photo-chemical Addition of 5S-Cysteine to Polynucleotides and Nucleic Acids. Biochemistry, 7, 1033-37.

Traub, P., Mizushima, P., Lowry, C. and Nomura, M. (1971). Re-constitution of Ribosomes from Subribosomal Components. Methods of Enzymology, 20C, 391.

Wang, W. Y., Apicalla, M. and Stone, B.R. (1956). Ultraviolet Irradiation of 1,3-Dimethyluracil. J. Am. Chem. Soc., 78, 4180.

37

CROSSLINKING OF NUCLEIC ACIDS AND PROTEINS BY BISULFITE

Robert Shapiro and Aviv Gazit

Department of Chemistry, New York University

New York, New York 10003

ABSTRACT

Sodium bisulfite is a food and beverage additive. It is also a salt of the urban air pollutant, sulfur dioxide. Bisulfite catalyzes, at neutral pH and physiological temperatures, the transamination reactions of cytosine derivatives with amines. The products of the reactions are N^4-substituted cytosines. Both the α- and ε-amino groups of L-lysine react with cytosine, and its nucleosides, in the presence of bisulfite. Bisulfite catalyzes the binding of cytosine to polylysine, lysine to polycytidylic acid, and polylysine to polycytidylic acid. Polylysine crosslinks with heat-denatured, but not native, calf thymus DNA, in the presence of bisulfite. Other workers have demonstrated crosslinking of viral RNA with maturation and coat protein, after treatment of bacteriophage MS2 with bisulfite. Nucleic acid-protein crosslinking reactions may contribute to the adverse effects of sulfur dioxide and bisulfite upon health.

INTRODUCTION

Sulfur dioxide and its salts, bisulfites and sulfites, are commonly used additives to foods and beverages. The terms bisulfite, or sulfur dioxide, will be used in this article to refer to any of these readily interconverted substances. The antimicrobial, antioxidant, and decolorizing properties of these substances have led to their wide use in wines, beer, preserved fruits and vegetables, fruit juices and syrups, meats and fish (Chichester and Tanner, Jr., 1972). The U.S. Food and Drug Administration includes bisulfite on its GRAS (generally recognized as safe) list. The principle source of bisulfite ion in the diet is in wines, in which it may occur at a level of several hundred ppm (Institute of Food Technologists, 1976).

Sulfur dioxide is also an important urban air pollutant, which is produced primarily by the burning of fossil fuels in electrical power plants. Because the presence of SO_2 in the air is statistically correlated with respiratory disease and other adverse health aspects (U.S. Dept. of Health, Education, and Welfare, 1969), its level in the air is strictly regulated by the Environmental Protection Agency.

TRANSAMINATION REACTIONS OF CYTOSINE DERIVATIVES

It was observed by Janion and Shugar (1967) that dihydrocytosine derivatives react with glycine to give a product of type III, Figure 1 (X = R = H, R' = CH_2CO_2H). Similar reactions were observed with dihydrocytosine and several other amino acids and dipeptides, methylamine and semicarbazide. At the same time, it was noted in our laboratory that cytosine derivatives would undergo transamination when heated with aromatic amines in an acidic carboxylate buffer (Shapiro and Klein, 1967). The function of the carboxylate buffer is to produce a small concentration of a reactive dihydrocytosine intermediate by addition to the 5,6-double bond of cytosine (I $\longrightarrow$ II). A side reaction is the formation of the uracil derivative IV, via the intermediate V.

We subsequently found sodium bisulfite to be a far better catalyst for transamination than carboxylate buffers (Shapiro and

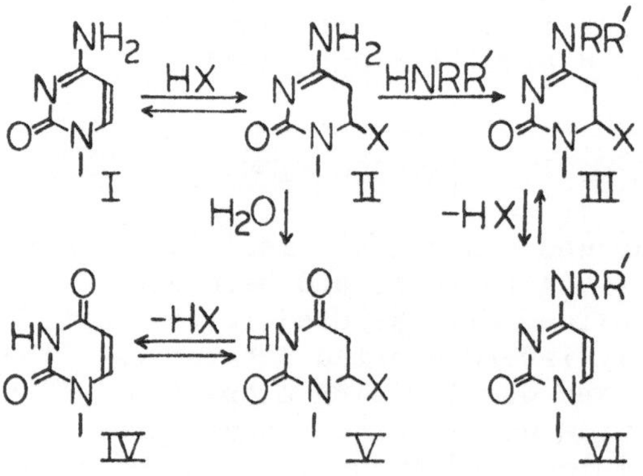

Figure 1   Transamination and deamination of cytosine derivatives

Weisgras, 1970). The route is the same one illustrated in Figure 1, with X = SO3. The reaction was best run at neutral pH, where deamination was minimal. The following amines were used success-fully: methylamine, dimethylamine, pyrrolidine, aniline, o-amino-phenol, βnaphthylamine, and glycine. The success of the glycine reaction led us to predict that bisulfite could catalyze the crosslinking of nucleic acids and proteins.

In an extension of this work, Boni and Budowsky (1973) de-monstrated the reaction of a dipeptide, glycylglycine, with cyti-dine 5'-phosphate, under bisulfite catalysis. The side reaction, deamination was further limited by restricting the bisulfite con-centration to 0.2 M. The same workers incubated cytidine, polyly-sine, and bisulfite. A portion of the ultraviolet absorption of cytidine remained with the polymer upon dialysis, but was released by O-methylhydroxylamine treatment (this reagent removes N^4-substi-tuents from cytosine). Thus, indirect evidence was provided for transamination of a polypeptide.

Transamination at the polynucleotide level was carried out by us using yeast RNA with bisulfite and methylamine or aniline. Cy-tosines in the RNA were specifically converted to their N^4-methyl or N^4-phenyl derivatives. The reaction was single-strand specific, as cytosines in a poly(I)-poly(C) complex did not react (Shapiro, et al., 1972).

REACTIONS WITH LYSINE AND POLYLYSINE

In more recent, unpublished, work we have attempted to explore the chemistry of the transamination reaction of cytosine and its nucleosides with lysine. Cytosine, cytidine, and deoxycytidine were each allowed to react with excess lysine, in the presence of 0.5 M or 1.0 M sodium bisulfite, 60°, for 2-7 days. The workup in each case involved an adsorbtion-desorbtion process on charcoal, and preparative thin-layer chromatography on silica in ethanol-conc. ammonia, 7:3 (solvent A). In each case, starting material (VII), as well as a faster-moving (VIII) and slower-moving (IX) transamination product was observed. The assignment of structure to VIII and IX is discussed below. The R_f values observed were: VII a, 0.80; b, 0.65; c. 0.90; VIII a, 0.75; b, 0.50; c, 0.85; IX a, 0.40; b, 0.20; c, 0.45. Each substance was also homogenous in a second thin-layer chromatography system on cellulose: isopropanol-HCl-H_2O, 17:4:4 (solvent B). The R_f values in this system were: VII a, 0.60; b, 0.60; c, 0.65; VIII a, 0.45; b, 0.45; c, 0.50; IX a, 0.20; b, 0.15; c, 0.20. The ultraviolet spectra of VIII and IX were those expected of N^4-substituted cytosines, with λ max (0.1 N HCl) 279-282 nm and λ max (H_2O) 270-272 nm. The nmr spectra of the products, in D_2O, showed cytosine (or cytosine nucleoside) and ly-sine C-H peaks in a 1:1 ratio.

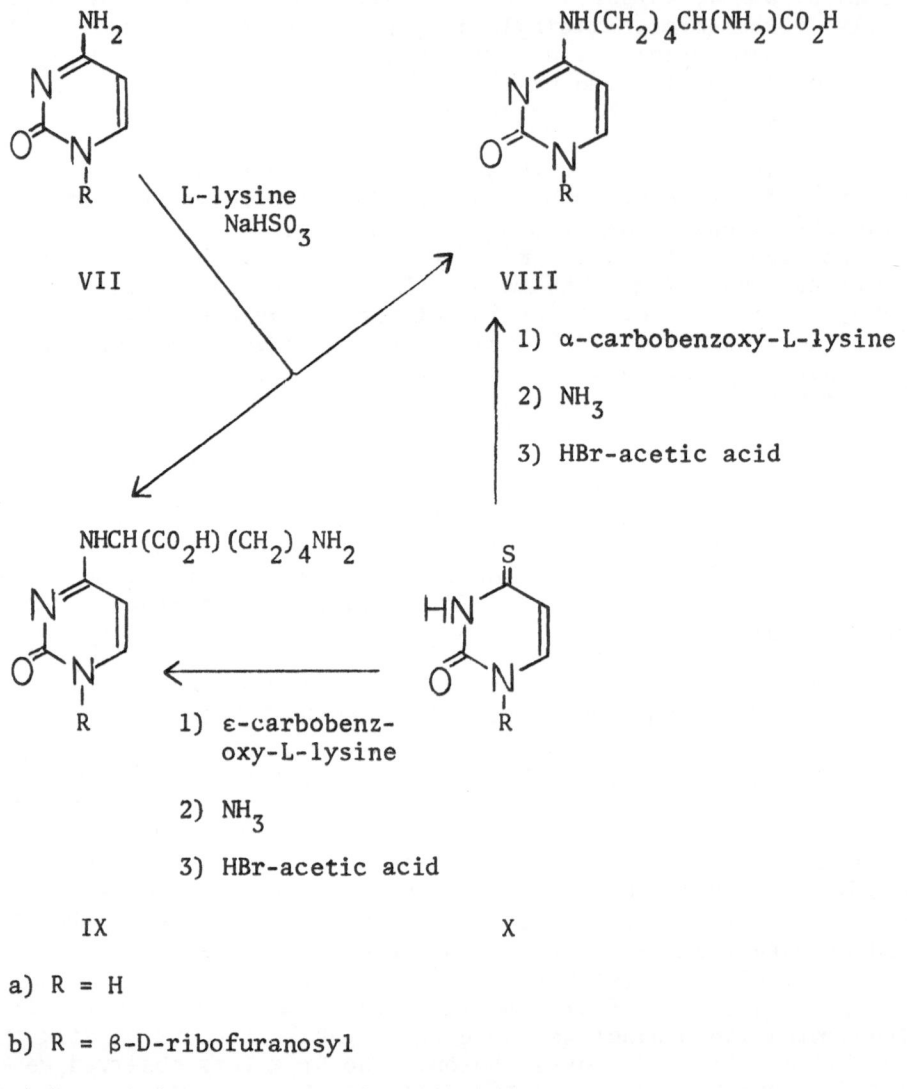

a) R = H

b) R = β-D-ribofuranosyl

c) R = 2'-deoxy-β-D-ribofuranosyl

d) R = 2', 3', 5'-tri-O-benzoyl-β-D-ribofuranosyl

Figure 2. Preparation of the transamination products of cytosine derivatives with L-lysine

In order to identify these products, VIII b and IX b were prepared in an alternative manner. (Figure 2). 2',3',5'-tris-0-
benzoyl-4-thiouridine (Xd) was condensed with α- and ε-carbobenz-
oxy-L-lysine in refluxing 50:50 ethanol-water, with added triethylamine. The blocking groups were then removed by successive treatments with NH_3 and HBr in acetic acid. The reaction mixtures
were then worked up in a manner similar to the bisulfite-catalyzed reactions. The reaction with α-carbobenzoxy-L-lysine gave
only the fast moving band (VIII b) on t.l.c. in solvent A, while
the reaction with ε-carbobenzoxyl-L-lysine afforded only the slow
moving band. Products VIII b and IX b were converted to VIII a
and IX a, respectively, by treatment with periodate and methylamine
at pH 8.1. Derivatives VIII c and IX c were converted to the bases
VIII a and IX a, by hydrolysis with 70% $HClO_4$ at 100°.

Crosslinking between monomers and polymers by bisulfite has
been studied by us. Polylysine (0.33 M in monomer) was allowed
to react with 0.4 M cytidine in 1 M $NaHSO_3$ at 37°, pH 7.45. Ultraviolet absorbtion remained with the polylysine fraction upon dialysis. The yield (based on polylysine) of presumed cytosine-ly-
sine cross-links rose to 22% over 14 days, and then leveled off.
Under the same conditions, Boni and Budowsky (1973) had obtained
about twice as many cross-links. Our preparation was hydrolyzed
by 70% $HClO_4$ at 100° for 2 hours, and worked up by preparative
t.l.c. in solvent B. Product VIII b was isolated in 87% of the
amount expected on the basis of the ultraviolet assay. We also
observed crosslinking using lower concentration of reactants. A
yield of 3% of VIII b was obtained with 0.003 M cytosine, 0.0017 M
(in monomer) polysine, and 0.25 M $NaHSO_3$ at pH 6.9, 37°, after 21
days.

The crosslinking of polycytidylic acid (0.0026 M in monomer)
with L-lysine (0.325 M) and 0.25 M bisulfite, 37 °, pH 7.0, was now
examined. The workup involved quenching with ammonia, dialysis,
70% $HClO_4$ hydrolysis (100°, 2 hours) and preparative thin layer
chromatography in solvent B. Products VIII a, and IX a were obtained, in 4.3 and 2.8% yields, respectively, by this procedure.

CROSSLINKING OF POLYPEPTIDES AND PROTEINS WITH POLYNUCLEOTIDES AND NUCLEIC ACID.

The bisulfite (0.5 M)-catalyzed crosslinking of polylysine
(0.0015 M in monomer) and polycytidylic acid (0.0027 M in monomer
at 37 °, pH 7.3 was studied. Aliquots were withdrawn, hydrolyzed
with 2N NaOH at 37°, for 16 hours, and then dialyzed. The extent
of crosslinking was measured by the residual non-dialyzable ultraviolet absorbtion remaining with the polylysine fraction. The
yield of cross-links (based on polylysine) varied with reaction

time as follows: 7 days, 38%; 10 days, 19%; 15 days, 11%; 18 days, 8.6%, 30 days, 7.7%. An initial rapid crosslinking reaction was observed, which decreased in amount with time.

When a more dilute $NaHSO_3$ solution (0.15 M) was used, the variation of crosslinking with time was different: 5 days, 4.7%; 12 days, 6.9%; 19 days, 7.8%; 26 days, 8.2%. The identity of the cross-links were confirmed in these studies by demonstrating their lability upon treatment for 5 hours with 1 M CH_3ONH_2, pH 6.0, 37°.

Crosslinking of polylysine (0.025 M in monomer) with a two-fold excess of heat-denatured calf thymus DNA was now explored (0.5 M $NaHSO_3$, pH 7.3, 37°). The final reaction mixture was hydrolyzed first with deoxyribonuclease I and venom phosphodiesterase at pH 8.5, and then with spleen phosphodiesterase and alkaline phosphatase at pH 7.7. Finally, perchloric acid hydrolysis and thin layer chromatography in solvent B, as described above, were performed. The results were similar to those with polycytidylic acid and polylysine: crosslinking rose and then fell, with time. The yields, based on lysine monomer, were: 8 days, 0.8%; 11 days, 1.4%; 18 days, 0.6%; 30 days, 0.15%. When 0.15 M $NaHSO_3$ was substituted in the above procedure, the yields were: 6 days, 0.5%; 12 days, 0.8%; 18 days, 1.2%; 30 days, 1.3%. In the more concentrated bisulfite solution, the decrease in the yield of cross-links with longer reaction time was probably due to deaminations by the route: VI $\longrightarrow$ III $\longrightarrow$ V (Figure 1).

Attempts to crosslink double stranded, native, calf thymus DNA with polylysine under the above conditions failed. A preliminary experiment using calf thymus histones and DNA appeared more successful. Equal weights of the two substances were dialyzed against 4 M NaCl at 25°, and the salt concentration was reduced stepwise over 8 days. A final dialysis was performed against distilled water, and histones and DNA were separated by hydroxyapatite chromatography (Levina and Mirzabekov, 1975). No histones were detected in the DNA peak, using the Lowry method. When this procedure was repeated using sodium bisulfite instead of sodium chloride, the DNA peak was found to contain 6% by weight of histones.

Crosslinking of proteins and nucleic acids by bisulfite has successfully been demonstrated in the RNA bacteriophage, MS2 (Turchinsky, et al., 1974). Modification was performed with 1 M bisulfite, pH 7, for 1-4 hours. Following this treatment, the phage protein was labeled by reaction with N-[^{14}C]acetoxysuccinimide, and RNA was separated from protein by either gel filtration or phenol extraction. Between 1 and 2% of the protein remained with the RNA fraction. After pancreatic ribonuclease treatment of the RNA, the presence of both maturation and coat proteins in the RNA fraction was demonstrated by gel electrophoresis. When the bisulfite treatment was conducted at a more acidic pH, the amount of protein

bound to RNA reached a peak within an hour, and then fell. This
result was attributed to deamination of the initially formed transamination products, as discussed above. The loss of cross-links
was not observed at pH 7.4, as deamination is slow at that pH.

In more recent work by the same group (Budowsky, et al., 1974)
MS2 phage was prepared with ^{14}C-pyrimidines in the RNA. After
treatment with 1 M bisulfite, pH 7.2, 30°, for 4 hours, the phage
was separated from bisulfite by gel filtration. The RNA was hydrolyzed by heating in 1 M HCl 66% acetic acid, and protein separated
from the RNA hydrolysate by a second gel filtration step. The
proteins were in turn hydrolyzed, using 6 N HCl, 110°, for 48 hours.
The presence of ε-N-(2-ketopyrimidyl-4)-lysine (VIII a) was demonstrated by thin layer chromatography, using the non radioactive
material as a marker (the preparation and characterization of this
marker was not described, however).

A brief statement has appeared (Tikchonenko, et al., 1973)
that proteins and nucleic acid of the double-stranded DNA phage,
Sd, can be crosslinked by bisulfite. No details have been provided
up to this time.

CONCLUSIONS

The ability of bisulfite to crosslink nucleic acids and proteins has been demonstrated with synthetic polymers, DNA in vitro,
and with viruses. Single-stranded nucleic acids react readily.
Double-stranded DNA does not react with amines or polylysine, but
may react in nucleohistone or other nucleoprotein complexes. Human
beings are continually taking in bisulfite by ingestion of foods
and beverages, and by inhalation of polluted air containing sulfur dioxide. The steady-state level of bisulfite in human tissues
is unknown, but certainly much lower than those used in the experiments described here. By their very nature, however, crosslinking reactions have the capacity for causing a great deal of
damage at a low level of incidence. Crosslinking reactions by bisulfite in such vital cellular targets as chromosomes and ribosomes may be an important cause of the adverse effects of sulfur
dioxide and bisulfite on human health.

ACKNOWLEDGEMENTS

The work from our laboratory described here has been supported
by a grant from the National Institute of Environmental Health
Sciences, N.I.H. (ES-01033). Robert Shapiro is a Research Career
Development Awardee of the National Institute of General Medical
Sciences, N.I.H. (GM-50188).

REFERENCES

Boni, I.V. and Budowsky, E.I. (1973). Transformation of non-
    covalent interactions in nucleoproteins into covalent bonds
    induced by nucleophilic reagents. I. The preparation and pro-
    perties of the products of bisulfite ion-catalyzed reaction
    amino acids and peptides with cytosine derivatives. J. Bio-
    chem. (Tokyo), 73, 821-830.

Budowsky, E.I., Simukova, N.A., Turchinsky, M.F., Boni, I.V. and
    Skoblov, Yu. M. (1976). Induced formation of covalent bonds
    between nucleoprotein components. V. UV or bisulfite induced
    polynucleotide-protein crosslinkage in bacteriophage MS2.
    Nucleic Acids Res., 3, 261-276.

Chichester, D.F. and Tanner, F.W. Jr. (1972). Antimicrobial food
    additives, T.E. Furia ed. Handbook of Food Additives, 2nd Ed.,
    CRC Press, Cleveland, pp 115-184.

Institute of Food Technologists (1976). Sulfites as food additives.
    Nutrition Revs., 34, 58-62.

Janion, C. and Shugar, D. (1967). Reactions of amines with dihydro-
    cytosine analogs and formation of amino acid and peptidyl de-
    rivatives of dihydropyrimidines. Acta Biochim. Polon., 14,
    293-302.

Levina, C.S. and Mirzabekov, A.D. (1975). Covalent bonding of
    proteins to DNA in chromatin. Doklady Biochem., 221, 172-
    175.

Shapiro, R. and Klein, R.S. (1967). Reactions of cytosine deri-
    vatives with acidic buffer solutions. II. Studies on trans-
    amination, deamination, and deuterium exchange. Biochemistry,
    7, 3576-3582.

Shapiro, R., Law, D.C.F. and Weisgras, J.M. (1972). A new chemi-
    cal probe for single-stranded RNA. Biochem. Biophys. Res.
    Commun., 49, 358-363.

Shapiro, R. and Weisgras, J.M. (1970). Bisulfite-catalyzed trans-
    amination of cytosine and cytidine. Biochem. Biophys. Res.
    Commun., 40, 839-843.

Tikchonenko, T.I., Kisseleva, N.P., Zintshenko, A.I., Ulanov, B.P.
    and Budowsky, E.I. (1973). Peculiarities of the secondary
    structure of bacteriophage DNA in situ. IV. Covalent cross-
    links between DNA and protein that arise in the reaction of
    S_d phage with O-methylhydroxylamine, J. Mol. Biol., 73, 109-
    119.

U.S. Department of Health, Education and Welfare (1969). Air
    Air Quality Criteria for Sulfur Oxides. National Air Pollu-
    tion Control Administration Publication No. AP-50, Washing-
    ton, D.C.

38

CROSS-LINKING OF AMINO ACIDS BY FORMALDEHYDE. PREPARATION AND ^{13}C NMR SPECTRA OF MODEL COMPOUNDS

David P. Kelly, M.K. Dewar, R.B. Johns,
Shao Wei-Let and J.F. Yates

Department of Organic Chemistry
University of Melbourne
Parkville, Victoria 3052, Australia

ABSTRACT

Model cross-linked systems have been prepared by reacting amino acids or alkylamines with formaldehyde and various amino acid model compounds such as 2,4-dimethylphenol (tyrosine), 3-methylindole (tryptophan) and alkylamides (glutamine, asparagine. ^{13}C NMR spectra of the products show the resonances of the formaldehyde-derived methylene carbons in the region 45-60 ppm. Interferences occur from resonances of the α-amino acid methine carbons. From the data for these products and other model compounds it has been possible to predict the shifts of the residual methylene carbons in a variety of cross-linked systems. This NMR technique shows promise as a rapid non-degradative method for identification of cross-linking sites.

INTRODUCTION

The identification of the sites of cross-linking of peptide chains by formaldehyde is of importance because of the profound changes in the physiological properties of proteins after treatment with formaldehyde. The methods used in attempts to identify these sites include (1) estimation of bound formaldehyde (Fraenkel-Conrat et al. 1945), (2) ^{14}C tracer techniques (Bowes et al. 1965; Caldwell and Milligan, 1972), (3) chemical degradation (Blass et al. 1968; Feairheller et al. 1967) and (4) enzymic degradation (Caldwell and Milligan, 1972).

In the least equivocal method of mild enzymic hydrolysis, Caldwell and Milligan (1972) tentatively identified N^ε, $N^{\varepsilon'}$-methylene dilysine and N^δ-methylolglutamine from the chromatographic behaviour of the hydrolysate of ^{14}C formaldehyde treated wool. A more recent report (Trézl et al. 1976) indicates that one of the major products of formaldehyde treated wool is N^ε-methyllysine, which may result either from direct methylation of the protein or from hydrolysis of the cross-link. Thus none of these methods provide unequivocal proof of the exact site of the formaldehyde cross-link in the proteins.

The difficulty of working with the natural materials has resulted in the use of compounds of lower molecular weight, for example, protected amino acids (Blass et al. 1968) or mixtures of model compounds and amino acids (Fraenkel-Conrat et al. 1948; Bowes et al. 1965; Blass et al. 1967). We have followed these examples in an attempt to obtain low molecular weight, easily identifiable, cross-linked compounds in which the position of the formaldehyde residue could be identified beyond doubt. In order to avoid any ambiguity in these identifications it was necessary to use a non-degradative method; thus we chose NMR spectroscopy. Although proton spectroscopy has been used successfully in the study of peptides (Roberts and Jardetzky, 1970) we considered that the problem of overlapping resonances of both other methylene groups and aqueous solvents would be much greater than in the case of ^{13}C NMR spectroscopy.

We have therefore prepared a number of cross-linked products by reacting amino acids and model compounds with aqueous formaldehyde at pH < 7 and 37° for several days, and analysed their ^{13}C spectra.

DISCUSSION

Intramolecular Cross-linking

The treatment of single amino acids such as cysteine, asparagine, histidine and tryptophan with formaldehyde solution yields crystalline, cyclized products 1 to 4 respectively (Ratner and Clarke, 1937; Stammer, 1961; Neuberger, 1944; Jacobs and Craig, 1936) the carbon spectra of which show the formaldehyde-derived methylene carbon (*) at 54-56 ppm (1, 2) and 42 ppm (3, 4). However, it is unlikely that such products form in the reaction of proteins with formaldehyde unless these amino acids occur at the ends of the peptide chains.

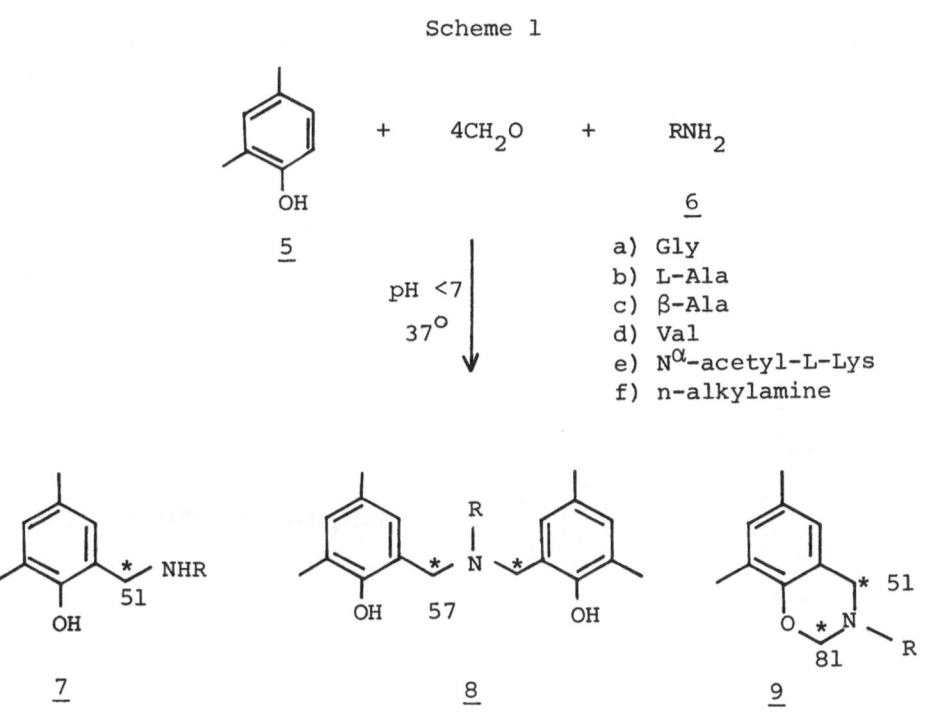

Intermolecular Cross-linking

Tyrosine models. When 2,4-dimethylphenol (5) was treated with glycine (6a), L-alanine (6b) or β-alanine (6c) and excess formaldehyde two major products were isolated, 7 and 8 (Scheme 1). The use of (more sterically demanding) L-valine (6d) however, resulted in only one isolatable product 7, whilst the use of N^{α}-acetyllysine yielded only 8. The reactions between methyl-N^{α}-acetyltyrosine and either N^{α}-acetyllysine, valine or glycine all produced complex mixtures which we were unable to separate by thin layer, paper or ion exchange chromatography.

Since the participation of lysine had been confirmed by Caldwell and Milligan (1972) we were anxious to obtain a model for the tyrosine-CH_2-N^ε-lysine residue and we thus replaced the amino acid in the above reaction with various n-alkylamines (6f). With excess formaldehyde the major product was the cyclized compound 3-alkyl-3,4-dihydro-1,3-2H-benzoxazine (9) (Dewar et al. 1974). All three compounds 7, 8 and 9 were present in all the reactions with n-alkylamines (CH_3NH_2 to $C_5H_{11}NH_2$), although the proportion of 9 could be reduced by lowering the formaldehyde/phenol ratio (Burke, 1949; Gaines and Swanson, 1971). We did not observe any dihydro-benzoxazines in the reactions with the amino acids (6a to 6c).

The carbon shifts of the residual formaldehyde carbons in 7, 8 and 9 all occur in the range 50-60 ppm downfield from $(CH_3)_4Si$, except in the case of those between nitrogen and oxygen atoms in 9 which occur at lower field, 80-82 ppm. From a study of carbon shifts of model compounds, the presence of an amide carbonyl group β to the methylene carbon will cause an upfield shift of this carbon of ca. 8 ppm (Dewar et al. 1974). We expect therefore, that the residual methylene carbon resonance in tyrosine-CH_2-N^δ-glutamine or tyrosine-CH_2-N^γ-asparagine cross-links will occur at ca. 45 ppm.

Glutamine and lysine models. As a model for asparagine, Fraenkel-Conrat (1948) used acetamide and prepared 10 with L-alanine. We have prepared the bisamides 11 (n = 1,2) by the reaction between the appropriate amides and formaldehyde. The cross-linking carbons resonate at 45 ppm.

$\overset{*}{CH_3CONHCH}_2NHCH(CH_3)COOH$ $(CH_3(CH_2)_nCONH)_2\overset{*}{CH}_2$

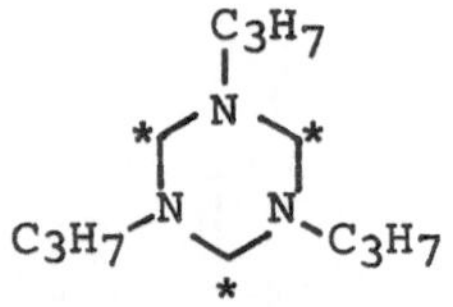

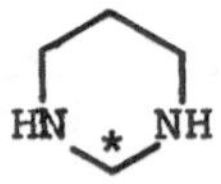

51   45

10   11

75   61   (64)

12   13   14

Attempts to prepare lysine-N^ε-CH_2-N^ε-lysine models were frustrated by the well recognized triazine (12) formation when n-alkylamines are treated with formaldehyde (Walker, 1964). We thus prepared the hexahydropyrimidine 13 (Evans 1967) which shows the residual carbon resonance at 61 ppm. From published data for alkylamines (Eggert and Djerassi, 1973) and cyclic, saturated hetero-cycles (Johnson and Jankowski, 1972) we are able to estimate the residual methylene carbon shift in the ideal model 14 as 64 ppm.

Tryptophan models. The participation of tryptophan in the formaldehyde cross-linking of wool was implicated by the work of Gruen and Nicholls (1969). Following the example of others (Fraenkel-Conrat et al. 1947) we used 3-methylindole as a model for tryptophan, the product obtained being substituted at the 1-position, 15. However, under prolonged reaction with formaldehyde in acidic conditions, the product isolated was the 2-substituted-3-methylindole, 16. Our studies indicate the operation of two different mechanisms. Compound 15 is formed under kinetic control by direct condensation at the indole nitrogen, but on further exposure to acid it is hydrolysed to the indole and a reactive imine (Shao 1970).

Substitution of the imine at C_3 followed by a 1,2 shift ($C_3 \rightarrow C_2$) of the methylene amino acid side chain yields 16. This mechanism is supported by our observation of the unsymmetrically coupled product 17 from the formaldehyde treatment of 1,3-dimethyl-indole and glycine.

From all of the above results we are able to predict the chemical shifts of the formaldehyde derived methylene carbons in a variety of cross-linked systems, as given in Table 1. However, there is the problem of interfering resonances. From the chemical shifts of a range of phenols, benzylamines and amino acids we have been able to show that the only interfering resonances in the region 50-60 ppm (apart from the benzylic carbon of tyrosine) are those from the α-amino acid methine carbons.

60   *CH_2
      NHCH(CH_3)COOH
      57

15

$R = NHCH(CH_3)COOH$

16

17

TABLE 1

Predicted ^{13}C Chemical Shifts of Formaldehyde Derived Methylene
Carbons of Cross-linked Systems

System	δ^a	System	δ
Lys N^ϵ-CH_2-N^ϵ Lys	64	Glu N^δ-CH_2-N^α Ala	51
Try N_1-CH_2-N Ala	60	Tyr C_3-CH_2-N^δ Glu	45
Tyr C_3-CH_2-C_3 Tyr	58-55	Glu N^δ-CH_2-N^δ Glu	45
cyc-Asp NH_2[b]	56	cyc-His[b]	42
cyc-Cys SH[b]	54	cyc-Try[b]	42
Tyr C_3-CH_2-N^ϵ Lys	52-50		

[a] ppm downfield from $(CH_3)_4Si$. [b] The product formed by cyclization
of the amino acid with CH_2O, 1-4.

In low molecular weight systems such as those studied here,
these methine carbons can be readily distinguished by single
frequency off-resonance decoupling experiments. However, in higher
molecular weight systems these resonances may cause significant
interference. In the region 40-50 ppm, interference will arise
from lysine side chain resonances particularly the methylene carbons
adjacent to nitrogen atoms, δ 45-48. However, the use of ^{13}C
enriched formaldehyde would overcome these problems.

As an aid to the identification of the residual methylene
carbons we have started to measure spin lattice relaxation times
of some of the model compounds. Preliminary results show that the
methylene carbons attached to tyrosine residues as in 7 have
relatively short T_1 (ca. 0.6 sec.) indicating restriction to their
movement when attached to the rigid planar aryl system, whereas
those in the middle of flexible cyclic (1) or acyclic (14) chains
have relatively long T_1 values (ca. 3 sec).

^{13}C NMR spectroscopy thus shows promise as a rapid non-
degradative method for the identification of cross-linking sites
in formaldehyde treated peptides.

ACKNOWLEDGEMENTS

We are grateful to Dr. B. Milligan of C.S.I.R.O. for helpful discussions, to Miss E. Cochrane for preparation of some of the model compounds and to the Australian Research Grants Committee for financial support.

REFERENCES

Blass, J., Bizzini, B., and Raynaud, M. (1967). Bull.Soc.Chim.Fr., 3957.

Blass, J., Bizzini, B., and Raynaud, M. (1968). Ann.Inst.Pasteur, Paris, 115, 881.

Bowes, J.H., Cater, C.W., and Ellis, M.J. (1965). J.Amer.Leather Chem.Ass., 60, 275.

Burke, W.J. (1949). J.Amer.Chem.Soc., 71, 609.

Caldwell, J.B. and Milligan, B. (1972). Text.Res.J., 42, 122.

Dewar, M.K., Johns, R.B., Kelly, D.P., and Yates, J.F. (1974). Aust.J.Chem., 28, 917.

Eggert, H., and Djerassi C. (1973). J.Amer.Chem.Soc., 95, 3710.

Evans, R.F. (1967). Aust.J.Chem., 20, 1643.

Feairheller, S.H., Taylor, M.M., Gruber, H.A., Mellon, E.F., and Filachione, E.M. (1967). Amer.Chem.Soc., Div.Polym.Chem. Prepr., 8, 775.

Fraenkel-Conrat, H., Cooper, M., and Olcott, H.S. (1945). J.Amer. Chem.Soc., 67, 950.

Fraenkel-Conrat, H., Brandon, B.A. and Olcott, H.S. (1947). J.Biol. Chem., 168, 99.

Fraenkel-Conrat, H., and Olcott, H.S. (1948). J.Biol.Chem., 174, 827.

Gaines, J.R., and Swanson, A.W. (1971). J.Heterocycl.Chem., 8, 249.

Gruen, L.C., and Nicholls, P.W. (1969). Aust.J.Chem., 22, 2137.

Jacobs, W.A., and Craig, L.C. (1936). J.Biol.Chem., 113, 759.

Johnson, L.F., and Jankowski, W.C. (1972). "Carbon-13 N.M.R. Spectra", Wiley-Interscience, New York, N.Y.

Neuberger, A. (1944). Biochem.J., 38, 309.

Ratner, S., and Clarke, M.T. (1937). J.Amer.Chem.Soc., 59, 210.

Roberts, G.C.K., and Jardetzky, O. (1970). Advan.Protein Chem., 24, 449.

Shao Wei-Let (1970). Ph.D. thesis, University of Melbourne.

Stammer, C.H. (1961). J.Org.Chem., 26, 2556.

Trézl, L., Heiszman, J., and Tyihak, E. (1976). 'Proc.Int.Wool Text.Res.Conf., 5th, Aachen 1975' in "Schriftenr.Dtsch. Wollforsch.Inst.Tech.Hochschule Aachen", in press.

Walker, J.F. (1964). "Formaldehyde", Reinhold, New York, N.Y.

39

ELECTRON MICROSCOPY OF AN OLIGOMERIC PROTEIN STABILIZED

BY POLYFUNCTIONAL CROSS-LINKING

C. N. Gordon

Department of Molecular Biology and Biochemistry

University of California, Irvine, California 92717

ABSTRACT

Oligomeric proteins can be intramolecularly cross-linked with polylysine in a reaction in which a water soluble carbodiimide mediates an amide linkage between the protein carboxyl groups and the ε-amino groups of polylysine. Studies carried out with a cytochrome P-450 indicate that a small number of molecules in a population which has been cross-linked in this way retain important features of their tertiary and quaternary structure when negatively stained and examined in the electron microscope. Use of the method in determining the subunit geometry of oligomeric proteins is discussed.

INTRODUCTION

The subunit geometry of oligomeric proteins can, in favorable circumstances, be resolved in the transmission electron microscope. In almost all examples reported to date, the molecules have been visualized by a technique called "negative staining". Two recent, excellent reviews (Haschemeyer and Myers, 1972; Haschemeyer and de Harven, 1974) describe the negative staining technique in detail. In brief, a heavy metal salt solution (stain) dries around the protein, the latter having been mounted on a thin membrane covering an electron microscope mesh grid (e.g. Fig. 3.1, Haschemeyer and Myers, 1972). If the protein is to retain its native three dimensional structure it must withstand the disruptive effects of surface tension and increase in

salt concentration as the stain dries around it. Reference to
the reviews cited above indicates that the literature contains
successful examples of proteins which can withstand the hostile
environment created during specimen preparation and yield aesthet-
ically pleasing, symmetrical images, as well as unsuccessful cases
where obvious collapse and distortion are evident. In addition,
cases of "intermediate" distortion can be recognized in which
some elements of native structure may be retained (Haschemeyer
and de Harven, 1974). Examples of proteins which are at opposite
ends of the spectrum in this regard are Escherichia coli glutamine synthetase and rabbit muscle pyruvate kinase, shown in Figures 1 and 2.

Figure 1 is an electron micrograph of E. coli glutamine synthetase, a protein which shows good structural preservation. The
enzyme is known to contain 12 subunits and the dodecamer was
interpreted by Valentine et al. (1968) as having D_6 symmetry
i.e., 2 hexagonal rings stacked upon each other. This protein is
remarkably stable when stained and the symmetry is obvious upon
inspection.

Figure 2 is a micrograph of rabbit muscle pyruvate kinase.
The oligomer has a molecular weight of 240,000 and contains four
subunits which are highly similar, if not identical (Cottam, et
al., 1969). The symmetry of the oligomer has not been determined
but four identical subunits can be arranged with either C_4 (square
planar) or D_2 (tetrahedral) symmetry and have all subunits in
equivalent environments (Klotz, et al., 1970). Examination of
the micrograph leaves little basis for choice between these alternatives. It is clear that the staining process has severely disrupted the structural organization of the molecule.

POLYFUNCTIONAL CROSS-LINKING

Our interest in developing cross-linking methods which might
stabilize the structure of proteins originated during a study of
the subunit structure of a cytochrome P-450 from bovine adrenocortical mitochondria (Gordon et al., 1974). Prior physical-
chemical data indicated that this protein was composed of equi-
molecular weight subunits which could be arranged in alternative
oligomeric forms of four, eight or sixteen subunits (Shikita and
Hall, 1973). When the protein was stained, the images were unin-
terpretable, indicating that obvious distortion was occurring
during the staining process. The protein was then cross-linked
with glutaraldehyde, formaldehyde and dimethyl suberimidate; reagents which are known to introduce bifunctional cross-links into
proteins (Korn, et al., 1972; Myers and Hardman, 1972; Davies and
Stark, 1970). However, in all three cases the appearance of the

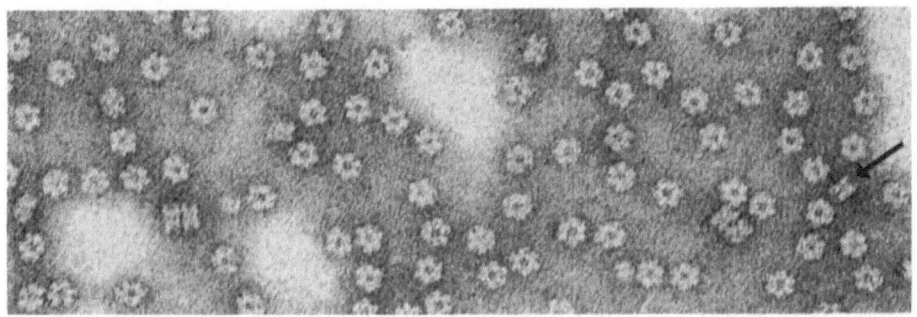

Fig. 1. Electron micrograph of E. coli glutamine synthetase negatively stained with lithium tungstate. Some of the molecules are
lying on their side (e.g., arrow) showing the double layer of two
hexagonal rings. Magnification: 308,000

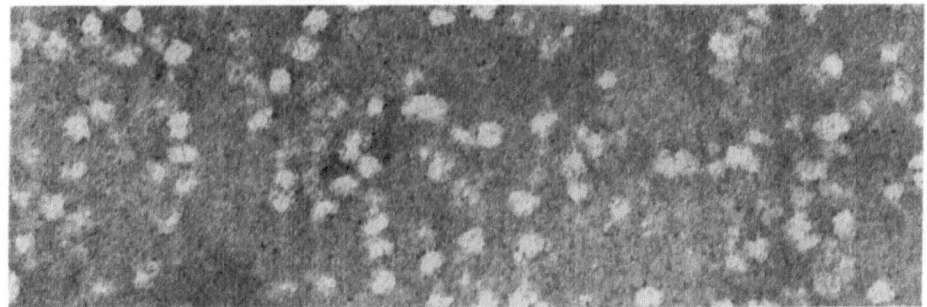

Fig. 2. Electron micrograph of rabbit muscle pyruvate kinase negatively stained with uranyl acetate. The irregularity of shapes
and sizes indicates artifactual disorganization induced by the
staining process. Magnification: 308,000

cross-linked protein in the electron microscope was essentially
indistinguishable from that of the unlinked molecule. "Fixation"
of cells and tissues prior to dehydration and embedding for electron microscopy is a well-established procedure and indeed, all
three of these reagents have been used in this way (Hayat, 1970;
Hassell and Hand, 1974). However, the literature contains little
indication that these reagents, in general, produce marked effects
on protein structure in cases where negative staining has caused
obvious distortion and this lack of effect was borne out in our
studies with the cytochrome P-450.

The failure of bifunctional reagents to stabilize the subunit
structure of the cytochrome P-450 led us to explore the use of
polyfunctional cross-linking agents for this purpose. Figure 3
is a schematic diagram illustrating how a polyfunctional crosslinking agent might stabilize a folded polypeptide chain (protomer) against loss of native three-dimensional structure. It is
assumed that linkage of the three points indicated by "X" is
required in order to maintain the tertiary structure of the protomer against surface tension forces which arise during drying of
the negative stain around the molecule. Because of the multiplicity of reactive groups and its flexible conformation, a single
polyfunctional linker can simultaneously bridge all three sites.
By contrast, a bifunctional reagent can link together only two of
the three sites.

Figure 4 shows the cross-linking system used in the study of
the subunit structure of the cytochrome P-450. The carboxyl
groups of the protein are joined in amide linkage with the ε-
amino groups of polylysine in a reaction mediated by a water-
soluble carbodiimide (Hoare and Koshland, 1967). The cytochrome
was allowed to react with polylysine in the presence of carbodiimide and then was stained and examined in the electron microscope.
A small number of the molecules showed obvious symmetry and it
was assumed that these molecules had acquired enhanced stability
as a consequence of the cross-linking. The symmetrical images
observed were used to construct a model of the native, quaternary
structure of the protein (Gordon, et al., 1974).

The mechanism of the carbodiimide mediated coupling reaction
between a carboxyl group and an appropriate nucleophile has been
discussed by Hoare and Koshland (1967). The carbodiimide reacts
with the carboxyl to form an O-acylisourea. Then, attack by a
nucleophile yields the adduct, which is an N-substituted amide if
the nucleophile is a primary amine. However, the O-acylisourea
can also rearrange irreversibly to an N-acylurea via an intramolecular acyl transfer. If this happens, the carboxyl group is per-
manently prevented from forming an adduct with the nucleophile.
Kinetic studies indicated that the rearrangement was slow compared
to nucleophilic attack provided the concentration of nucleophile
was sufficiently high, e.g. 1 M (Hoare and Koshland, 1967). However, in the polylysine cross-linking system, it would be desirable
to have a low concentration of nucleophile (polylysine) so that
polyvalent rather than monovalent attachment to the protein would
be favored (Wold, 1972). In such circumstances, one might expect
that the reaction would proceed slowly and that the rearrangement
would compete seriously with nucleophile attack and remove a significant fraction of the protein's carboxyl groups from participation in the cross-linking reaction.

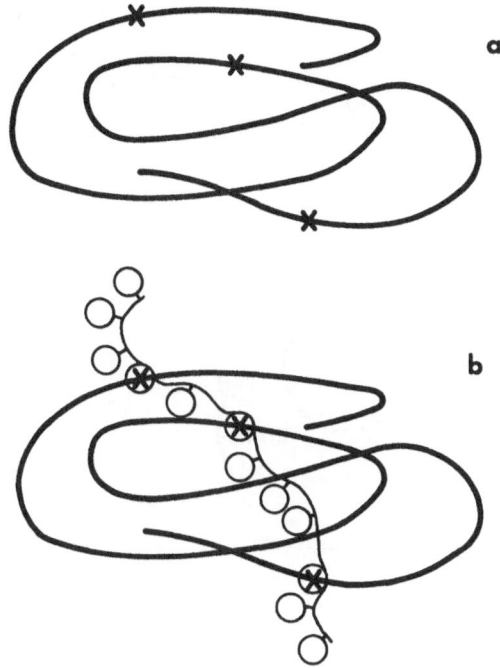

Fig. 3. Schematic diagram showing linkage of three points of a
protomer (X) by a polyfunctional linking agent. (a) Unlinked
protomer. (b) Cross-linked protomer.

To gain an indication of how serious a problem this might be,
bovine catalase was cross-linked with a low concentration of polylysine (about 10 polylysine linkers per subunit of catalase) and
the extent of cross-linking occurring after 2 and 11 hours was
determined. Polylysine (molecular weight of 3,400) was methylated
with HCl-methanol (Wilcox, 1967). The reaction mixture at pH 6
contained bovine catalase (100 µg/ml), methylated polylysine (50
µg/ml) and 0.05 M 1-ethyl-3-(3-dimethylaminopropyl) carbodiimide.
During the course of the reaction the pH was maintained at 6 by
addition of HCl and the carbodiimide concentration maintained at
0.05 M by periodic addition of solid reagent. Portions were
removed after 2 hours and 11 hours, excess 1 M sodium acetate was
added to quench the reaction and the solutions were dialyzed at
4° against 0.01 M NaCl. After concentration in a Diaflo membrane
ultrafiltration apparatus, the solutions were examined on 5% dode-
cylsulfate-acrylamide gels (Weber et al., 1972). The gel patterns
are shown in Figure 5. After 2 hours, most of the catalase is

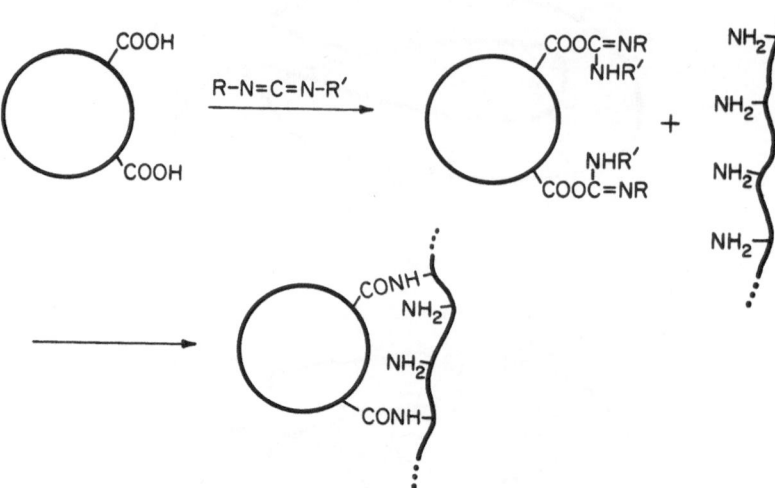

Fig. 4. Intra-protomer linkage of carboxyl groups with polylysine, mediated by a water soluble carbodiimide (Hoare and Koshland, 1967; Gordon et al., 1974).

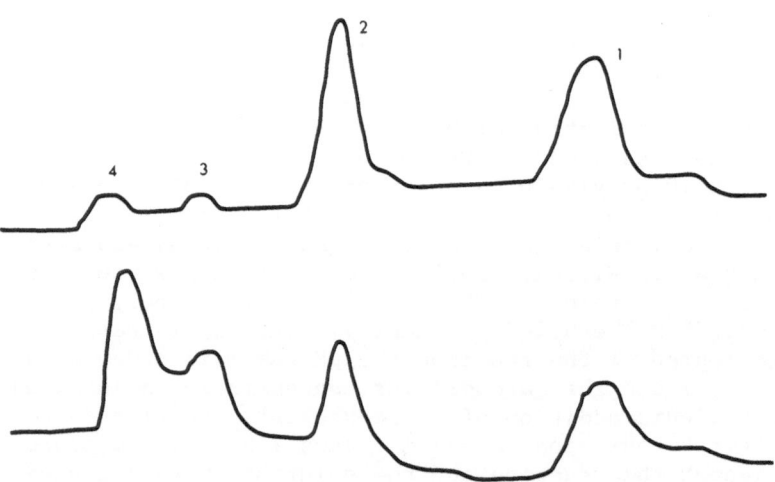

Fig. 5. Dodecyl sulfate-acrylamide gel electrophoresis of catalase cross-linked with polylysine for 2 hr. (top) and 11 hr.
(bottom). See text for details of reaction. 1, 2, 3, 4: monomer,
dimer, trimer, tetramer. Migration is from left to right.

present as the monomer and dimer with small amounts of trimer and
tetramer. After 11 hours, there is further conversion to the
trimer and tetramer but still appreciable amounts of monomer and
dimer. It is evident that the reaction is proceeding slowly and
is still incomplete after 11 hours. The slowness of the reaction
is probably a consequence of the low concentration of polylysine
nucleophile used. It is likely that a number of carboxyl groups
of the protein have been converted to N-acylurea groups during
the course of the extended reaction and are thus removed from
participating in cross-linking.

A disturbing feature of the cytochrome P-450 work (Gordon,
et al., 1974) was the very small number of symmetrical images
observed in any given field. The explanation offered at the time
was that the cross-linking proceeded in quasi-random fashion and
only a small fraction of the molecules might have key strategic
regions linked in such a way that collapse during staining was
prevented. While this may partially account for the low yield of
symmetrical oligomers, it is now felt that loss of available carboxyl groups due to the rearrangement was also a contributing
factor. The reaction proceeded for a total of 66 hours, at which
time dodecyl sulfate-gel electrophoresis showed the absence of
monomer. The electrophoresis, of course, only indicated whether
or not subunits were linked to one another and did not gauge the
extent of cross-linking within subunits. This intra-subunit crosslinking is postulated to be the key element in stabilizing subunits
against loss of tertiary structure during negative staining (Figure 3). Thus, in spite of the fact that all subunits within an
oligomer were linked to each other, the extent of intra-subunit
linkage may have been minimal because of carboxyl group rearrangement and this may account in part for the low percentage of symmetrical images.

Work in progress is aimed at increasing the efficiency of
cross-linking while maintaining a low ratio of linker to protomer.
Among the variables which might cause a faster rate of coupling
are pH and the use of a polymer other than polylysine in which the
nucleophile group has a lower pK. A low efficiency is not a serious drawback in cases where the symmetry of the oligomer is apparent, such as a square planar (C_4) arrangement, since one can
readily identify the exceptional, stabilized images. However, in
cases of more complex symmetry where a number of alternative pro-
jections are possible (Haschemeyer and de Harven, 1974) a high
percentage of stabilized molecules increases the likelihood of
correctly identifying the subunit geometry.

REFERENCES

Cottam, G. L., Hollenberg, P. F. and Coon, M. J. (1969). Subunit structure of rabbit muscle pyruvate kinase. J. Biol. Chem., 244, 1481.

Davies, G. E. and Stark, G. R. (1970). Use of dimethyl suberimidate, a cross-linking reagent in studying the subunit structure of oligomeric proteins. Proc. Nat. Acad. Sci., U.S.A., 66, 651.

Gordon, C. N., Shikita, M. and Hall, P. F. (1974). The use of a novel cross-linking method in demonstrating the subunit structure of an oligomeric protein by negative staining. J. Ultrastruc. Res., 47, 285.

Haschemeyer, R. H. and de Harven. (1974). Electron microscopy of enzymes. Ann. Rev. Biochem., 43, 279.

Haschemeyer, R. H. and Myers, R. J. (1972). Negative staining. In "Principles and Techniques of Electron Microscopy", Vol. 2, Van Nostrand Reinhold Co., New York, N. Y.

Hassel, J. and Hand, A. R. (1974). Tissue fixation with diimidoesters as an alternative to aldehydes. J. Histochem. Cytochem., 22, 223.

Hayat, M. A. (1970). "Principles and Techniques of Electron Microscopy", Vol. 1, Van Nostrand Reinhold Co., New York, N. Y.

Hoare, D. G. and Koshland, D. E., Jr. (1967). A method for the quantitative modification and estimation of carboxylic acid groups in proteins. J. Biol. Chem., 242, 2447.

Klotz, I. M. and Langerman, N. R. (1970). Quaternary structure of proteins. Ann. Rev. Biochem., 39, 25.

Korn, A. H., Feairheller, S. H. and Filachione, E. M. (1972). Glutaraldehyde: nature of the reagent. J. Mol. Biol. 65, 525.

Myers, J. S. and Hardman, J. K. (1971). Formaldehyde-induced cross-linkages in the α subunit of the Escherichia coli tryptophan synthetase. J. Biol. Chem., 246, 3863.

Shikita, M. and Hall, P. F. (1973). Cytochrome P-450 from bovine adrenocortical mitochondria: an enzyme for the side chain cleavage of cholesterol. II. Subunit structure. J. Biol. Chem., 248, 5605.

Valentine, R. C., Shapiro, B. M. and Stadtman, E. R. (1968). Regulation of glutamine synthetase. XII. Electron microscopy of the enzyme from Escherichia coli. Biochemistry, 7, 2143.

Weber, K., Pringle, J. R. and Osborn, M. (1972). Measurement of molecular weights by electrophoresis on SDS-acrylamide gel. Methods Enzymol., 26, 3.

Wilcox, P.E. (1967). Esterification. Methods Enzymol., 11, 605.

Wold, F. (1972). Bifunctional reagents. Methods Enzymol., 25, 623.

40

FISH MYOFIBRILLAR PROTEIN AND LIPID INTERACTION IN AQUEOUS MEDIA AS DETECTED BY ISOTOPE LABELING, SUCROSE GRADIENT CENTRIFUGATION, POLYACRYLAMIDE ELECTROPHORESIS AND ELECTRON PARAMAGNETIC RESONANCE

Soliman Y.K. Shenouda* and George M. Pigott**

*National Marine Fisheries Service, Gloucester, MA 01930
**Institute for Food Science and Technology, College of
  Fisheries, University of Washington, Seattle, Washington
  98195

INTRODUCTION

The integrity and function of various essential natural systems depends on the existence of stable complexes of lipid and protein. Examples of such systems are numerous, such as the fluid mosaic cell membranes, the blood coagulation process, various lipase reactions, actomyosin ATPase activity, blood serum lipoproteins, etc. In food systems, on the other hand, complexes of lipid and protein coexisting naturally, or artificially formed, show an important role in food manufacturing (dough mixing, bread making, dairy products, meat sausage, food emulsions, etc.) and in the instability of processed foods, particularly fishery products.

Quality deterioration, especially in texture, of frozen fish blocks (particularly in minced form) is partially attributed to be a consequence of lipid-protein complex formation. Instability of fish protein concentrates (FPC) is another example where residual lipids complexed to protein are charged as being the cause. Protein-lipid complexes could be a faster route for the oxidation of the lipid moiety; however, loss in nutritional value and protein quality are, in general, imputed to the interaction of oxidized lipids, or their products, with food protein. Such interaction results in damage to certain amino-acid residues (Lys., His., Tyr., Met., and Cys.) which also causes inactivation of various enzyme proteins such as RNAase, trypsin, and pepsin. Toxic products are another outcome of the interaction between protein and oxidized lipids in fish meal. The toxic effect was not due to the direct toxicity of the fat itself, but it was found to be associated with the non-fat fraction.

Myofibrillar proteins constitute the majority of the proteins (between 65-75 percent) of whole fish muscle, and their concentration increases as water-washing is included in the manufacturing steps which eliminate most of the sarcoplasmic proteins, as is the case with FPC or washed minced fish.

The list of isolated myofibrillar proteins (Table 1) has grown to include at least half a dozen proteins whose function and location in the sarcomere are reasonably defined. It was reported (Lowey, 1972) that with a few exceptions, notably actin and the newly discovered C-protein, the majority of fibrous proteins consist of more than one polypeptide chain.

In our study, myosin and actin were chosen from the myofibrillar proteins to study this interaction with fish lipid as these two proteins possess quite different characteristics when considering their shape, configuration, molecular weight, number of peptide chains/molecule, enzymatic activity, sensitivity to denaturation, and affinity for combination with each other to form actomyosin.

In the following section, the composition, properties, and characteristics of myosin, actin, and fish lipids will be briefly reviewed.

Table 1

Contractile Proteins, Types, and Properties

Localization in myofibril	% Total protein	Protein	Molecular weight	Sub-unit molecular weight	% α-Helix
Thick filament	55	Myosin	470,000	200,000 20,000	57
Thick filament	2	C-Protein	140,000	140,000	10
Thin filament	25	G-Actin	46,000	46,000	26
Thin filament	5	Tropomyosin	64,000	32,000	90
Thin filament	5	Troponin	80,000	37,000 24,000 21,000	35
Z-line	tr.	α-Actinin	180,000	90,000	60
M-line	tr.	M-Protein(s)	180,000 - 88,000	90,000 140,000 43,000	40 <10 26

Source: Reproduced from S. Lowey (1972).

Myosin

Myosin is one of the major proteins which is responsible, at the molecular level, for the conversion of the chemical energy of ATP into mechanical energy of contractile or motile processes. Myosin is a very asymmetrical protein with an axial ratio of more than 40:1. It is a long molecule (about 1620 A^O with 26 A^O thickness)with a globular head. It is believed that through most of the length of the myosin molecule each of the large polypeptide chains exists in helical conformation, and the two chains are supercoiled to form a double helical structure. At the end of the myosin molecule, both of the polypeptide chains are folded into globular structure. In addition, bound to the globular head with secondary forces are the other short polypeptide chains.

Disagreement about the precise molecular weight of myosin is found in the literature. This arises in part from its highly asymmetrical structure and the tendency of myosin and other myofilament proteins to associate tightly with each other. Fish myosin preparations, moreover, are very labile and not easy to prepare (Connell, 1964). The difficulty of preparation stems initially from the fact that the rate of extraction of actin during the normal extraction of myosin is very rapid in the case of fish, so that procedures which with mammalian muscle yield almost pure myosin, invariably give preparations heavily contaminated with actin or actomyosin; and hence extra purification steps are needed. Affinity chromatography technique was recently (Trayer and Trayer, 1975) employed for obtaining pure myosin preparations.

Dreizen and Gershman (1970) indicated that myosin is dissociated in concentrated salt solution at pH 7.0 and at 4^OC., into a heavy-chain core of 420,000 daltons molecular weight and light components that are comprised of 12 percent of myosin and contain 2.7 ($\pm$0.3) light polypeptide chains of average molecular weight 20,000. Tsuchiya and Matsumoto (1975) found only one light chain of 17.5 x 10^3 daltons associated with carp myosin. Weeds (1967) reported that the light chains showed similarity in peptide maps and identical thiol sequences. Dreizen and Gershman (1970) suggested that the heavy chain contained the hydrolytic site (for ATPase) and the adjacent sub-unit (two light chains) stabilized the unique conformation required. The third light chain (fast electrophoretic band) probably played a regulatory role in the reaction.

More information regarding the structure of myosin and its activity has come from enzymatic fragmentation studies. Trypsin or chymotrypsin cleave a specific peptide linkage near the center of the tail to yield heavy meromyosin and light meromyosin. Papain, on the other hand, gives different types of fragmentation.

It was concluded from such studies that the ATPase activity of myosin resided entirely in its head, and there are two catalytic sites per myosin molecule. Fish myosin was split by trypsin in the same way, but the instability of the meromyosins formed has slowed much further progress toward their characterization (Stainier-Lambrecht, 1962).

Myosin activity is characteristically dependent on two classes of sulfhydryl groups, which differ in their susceptibility to alkylation or to mercaptide formation (Lehninger, 1970). When the more susceptible SH groups were blocked, the ATPase activity of the myosin was increased suggesting that this class of sulfhydryl group was normally inhibitory to the enzyme. When the second class of sulfhydryl group was then titrated, the ATPase activity was completely abolished suggesting that the presence of the second class is required for the hydrolytic process (Quinlivan et al., 1969). Myosin from different parts of chicken muscle, legs, or breast was reported to have different properties, i.e., ATPase activity, extractability, and sensitivity to tryptic digestion (Chung Wu, 1969).

On the basis of their work, Huszar and Elzinga (1969) stated that at least three amino acids, methylated derivatives of histidine and lysine are present in the acid hydrolysate of myosin. Both 3-methylhistidine and methylated lysine have been localized in subfragment -1, which represents the globular part of the myosin molecule and contains both the ATPase and actin binding sites of myosin. The amount of 3-methylhistidine has been found to vary depending upon the source of myosin (Huszar and Elzinga, 1971). It was also noticed that the appearance of the 3-methylhistidine in the myosin molecule (during the first months of the rabbit's life) coincides with an increase in ATPase activity. The same authors reported that the amino acid sequence around the single 3-methylhistidine in actin and myosin was found to be different. One common feature in both peptides was the presence of an acidic residue adjacent to the 3-methylhistidine and therefore the side chain of 3-methylhistidine probably lies at or near the surface of actin and myosin. Kuehl and Adelstein (1970) found that myosin prepared from white fibers contains two residues of 3-methylhistide/molecule (one residue/heavy chain) whereas myosin from red fibers contains no such residue.

Effect of Some Processing Steps on Myosin

The effect of different processing steps on myosin denaturation measured as degree of solubility, intrinsic viscosity, iso-electric point, ATPase activity, UV absorption spectra, and O.R.D. was studied by several investigators, among them Hao-Chu and

Sterling (1970). These authors stated that the denaturation of myosin increased in the order of fresh, frozen, frozen stored, dehydrated, dehydrated stored, and cooked. Mao and Sterling (1970) noticed a decrease in the free SH group content of fish myosin, a little in freezing, and more in frozen storage; and none were present after dehydration or cooking.

In 1962, Connell stated that 70-80 percent of cod myosin became nonextractable, at a rate similar to that at which the total myofibrillar proteins of the flesh became nonextractable, during storage at -14°C. Takama et al. (1972) related the decrease in fish myofibrillar extractability after frozen storage at -20°C. to the complex formation with short chain free fatty acids (mainly caproic) and aliphatic aldehydes (mainly propanal), while Babbitt et al. (1972) related the decrease in protein extractability of stored hake at -2°C. to the formation of dimethylamine and formaldehydes. Buttkus (1970) reported that trout myosin monomers aggregated to dimers or trimers during freezing, and that the aggregation process reached maximum rate near the eutectic point of the myosin-potassium chloride-water solution (-11°C.). The ultra-structural studies showed that the aggregation proceeded in a side-to-side monomer.

Wu and Sayre (1971) showed that aging of white and red chicken muscles did not affect the number of SH groups of myosin.

In 1971, Suzuki showed that the maximum denaturation rate of myofibrillar proteins during dehydration occurred when dehydration exceeded a critical point which coincided with that of an abrupt change in the line plotting the residual moisture vs. half value of NMR spectra in dehydrated sea bass muscle (calculated as 28-20 percent moisture).

The effect of heating on muscle proteins (Chrystall, 1971) showed that myosin between 40-60°C. broke down into smaller components which were isolated on DEAE-cellulose chromatography.

Actin

Actin can exist in both globular (G-actin) and fibrous (F-actin) forms. The actin monomers are 55 A° in diameter, arranged in a right-handed helix whose pitch is about 2 x 360 A° in the filaments. Tropomyosin appears to lie in the two grooves of the actin helix where it positions a troponin about every 400 A° along the thin filament (Lowey, 1972). There are approximately 13 to 15 globular sub-units per turn of the helix.

Similar to the other myofibrillar proteins (Shenouda and Pigott, 1975a; Briskey and Fukazawa, 1971), actin is high in dicarboxylic amino acids, though not as high as tropomyosin or myosin, with glutamic, aspartic, alanine, threonine, and glycine being the most abundant. Actin possesses only a few more acidic amino acids than basic amino acids giving an iso-electric point of 4.7-4.8.

Gillibrand (1972) reported the formation of an intermediate form of actin accompanied by structural conformational changes in the G-actin during the formation of F-actin. Rizzino et al. (1970) cited that actin binds to heavy meromyosin in a 1:1 ratio.

Comparatively little work has been done on fish actin; moreover, it was recognized recently that most of the fish actin preparations were probably seriously contaminated with tropomyosin. Ebashi and Maruyama (1965) detected the presence of α- and B -actinin as contaminants in actin preparations. α-Actinin was found to evoke the gelation of F-actin (Maruyama and Ebashi, 1965). Drabikowski et al. (1968) showed that part of the tropomyosin was very tightly bound with actin, and the loosely bound part was easily dissociable in the absence of troponin and became much more tightly bound in the presence of troponines. Shenouda and Pigott (1975b) proposed a method of isolation and purification of fish actin as checked with sodium dodecyl sulfate-polyacrylamide gel electrophoresis (SDS-PAGE).

Elzinga (1970) showed that rabbit actin is composed of 420 amino acids including one 3-methylhistidine, 5 cysteines, 5 tryptophanes, and 9 histidines. Lusty and Fasold (1969) and Bridgen (1972) reported that the three surface sulfhydryl groups were not directly involved in the actin polymerization reaction, and the other two sulfhydryl groups were covered in the course of polymerization.

Johnson and Perry (1970) noticed that while the 3-methylhistidine content varied in myosin according to the type of muscles (red or white) it was constant in actin in all types of rabbit muscles. They also stated that this residue is not essential for actin-myosin interaction.

Fish Lipids

Fish are known to consist of groups of lipids rather different from the warm-blooded animals and vegetable lipid family. Many research works have been published covering the composition of fish lipids and the factors which affect variation in both their content and composition (Borgstrom, 1962). Dominova (1970), using literature data, cited that the aliphatic fish lipids contained considerable quantities of unsaturated fatty acids with 4 and 5 and 6 double bonds, and the carbon number of the fish lipid acids ranged between C_3 to C_{26} and an odd number of carbons (11-23) and double bonds in unusual positions occurred. He also reported that apart from triglycerides, the unsaponifiable fraction contained: etherglycol, C_3- and C_4- poly alcohols, ether of glycerin and higher alcohols (stearyl and palmityl), lecithin, cephalin, sphingomylin, sterols (cholesterol), coloring matters (carotenoids), hydrocarbons (squalene, pristane), and vitamins A, D, and E.

Generally speaking, polar lipids contain the highest proportion of polyunsaturated fatty acids, while neutral lipids are much less unsaturated and contain the highest proportion of monounsaturated acids. Lovern (1956), working with haddock, stated that each class of lipids appeared to have a characteristic fatty acid (FA) composition. He found that the fatty acids of cholesterol esters were of an unusually high average unsaturation owing mainly to a high content of C_{20} and C_{22} unsaturated acids and to a lesser extent a rather low content of saturated FA. The triglycerides contained major proportions of hexadecaenoic acids with certain unsaturated FA which were rich in C_{16} and C_{18}. Stansby (1967) reported that the majority of lipids in fish species which have low fat contents occur as phospholipids and non-glyceride forms such as lecithin, waxes, alcohols, cholesterol, free FA, phosphatidylethanolamine (PE), phosphatidylinositol (PI), cholesterol esters, hydrocarbons, plasmalogens, phosphatidylserine (PS), sphingomylin, and other unidentified components. Wessels and Spark (1973) noticed a constant high level of unsaturated skin lipids relative to muscle lipids in various hake species. The major saturated fatty acids were C_{16} and monoenoic C_{18}, while $C_{22:6}$ dominated and $C_{20:5}$ was second in polyenoic fatty acids. Furthermore, the phospholipids in general were particularly rich in fatty acids with $C_{16:0}$ and $C_{22:6}$.

According to Wood and Hintz (1971), during storage of the fish at ice temperature, there was a selective loss in polyunsaturated acids where most of the loss was from the PE fraction even though it contained only 25 percent of the total polyunsaturated acids. Small amounts were lost from neutral lipids and virtually none from lecithin. The decrease in these groups was accompanied by an equivalent increase in free fatty acids.

Olley et al. (1969) correlated the preferential breakdown of phosphatidylcholine (PC) and PE (containing $C_{16:0}$, $C_{18:1}$, and $C_{20:5}$) in stored haddock and lemon sole at -7 to -29°C. with protein denaturation and taste panel assessment of texture. Furthermore, it was reported (El-Bastavizi and Smirnova, 1972a) that freezing itself does not have any influence on muscle lipids of carp; however, during storage at 0 to -4°C., the accumulated free fatty acids from phospholipids caused reduction in the solubility of the myofibrial proteins. The sarcoplasmic proteins were relatively stable.

LIPID-PROTEIN COMPLEXES

Definitions and Terms

Lipid protein complexes were defined as complexes of proteins and lipids which migrate as discrete units and resist reseparation into proteins and lipids by physical means. They were described by different terms such as lipoproteins, phospholipoproteins, and proteolipids, according to their solubility properties in aqueous solvents. Other terms were based on their relative composition (i.e., the ratio between neutral, phospholipid, and protein) such as low-density and high-density lipoproteins.

Forces

A variety of chemical bonds and other bonds are responsible for the interaction between the building units of the protein and lipid molecules to form and stabilize the lipid-protein complexes.

The types of bonds or interactions usually considered to contribute in lipid-protein complexes are:

A. Covalent Bonds. This is an electron sharing mechanism with an energy of 30-100 KCal/mole. It is proposed that the existence of this type of interaction is uncommon in biological systems (Burley, 1971). However, various models of artificially formed lipid-protein complexes showed difficulty when trying to re-extract the lipid, even after treatment with urea (Marinette and Pettit, 1968) suggesting the existence of covalent bonding. Also, interaction between protein sulfhydryl groups and lipid double bonds may occur through the addition reaction of thiol to olefin (Robinson, 1966).

B. Ionic Binding. This is a Coulomb attraction force between charged groups of opposite signs with an energy range of 10-20 KCal/mole. In addition to the charged groups on proteins and lipids, groups with permanent or inducible dipoles (i.e., -OH, -CO) would also be expected to interact with the ionic groups.

Ionic binding is also reported to be existent between phospholipid and protein via a mixed chelation of divalent metals such as calcium or magnesium (Fullington, 1969).

The stability of ionic binding depends on the factors which affect the state of the charge of the functional groups of the protein and lipid such as pH, ionic strength, or the dielectric constant of the media.

C. Weak Secondary Forces. Various weak forces contribute in the interaction of lipid and protein, among them the attraction between apolar groups by Van der Waals forces with an energy of 1-3 KCal/mole. Although Van der Waals forces are individually weak, collectively they confer great strength provided that close steric proximity is possible.

Hydrogen bonding was reported to be of less importance for the binding of lipid and protein because most lipids have a low proportion of labile hydrogen atoms. However, evidence recently from I.R. studies (Kimm and Grewther, 1968) confirm its participation.

D. Hydrophobic Interaction. These types of forces which exist only in aqueous environment are highly significant in stabilizing complexes formed between lipids and proteins. Evidence of such interaction has been repeatedly reported in the literature.

Specificity of Binding

Binding between lipids and proteins, as tested in various model systems, is not very specific as compared to the "enzyme-substrate" systems but is rather a preferential specificity. Binding specificity is presumed to reside primarily in the proteins as they are more complex and more sensitive to their microenvironment. However, specificity in binding as attributed to lipids has been reported, such as (a) the effect of their steric configuration and dielectric properties (Colacicco and Rapport, 1966); (b) the type and class of lipids (Camejo et al., 1968); and (c) the type of liposomes they form, i.e., lamellar or micellar liposomes and the different micellar size and properties as affected by degree of unsaturation and charges (Pitlick and Nemerson, 1970). Although not studied in detail, lipid-lipid and protein-protein interaction is another area which logically should signify the amount and specificity of lipid-protein interaction.

MYOSIN-LIPID INTERACTION

Detection of Lipid in Purified Myosin Preparation

On applying the direct thin layer chromatography (D-TLC) method as described by Cherayil and Scaria (1970), traces of lipids

were detected in the purified myosin preparation eluted from the
Sephadex G-200 Column. Thereafter, a large size preparation of
myosin (Table 2) was extracted by chloroform-methanol-water (Folch
et al. method, 1956), and the contaminating lipids in the myosin
preparation totaled about 1.3 percent on a myosin dry weight basis.
When these myosin-contaminating lipids were further separated on
TLC, they contained basically the same subgroups present in the
original fish-lipid extracts (Shenouda, 1974).

                    Interaction of Myosin with Lipid

        The sucrose gradient centrifugation method was applied for
separating pre-incubated (overnight at 4oC) mixtures of myosin with
various carbon-14 labeled fish lipids. A typical profile after the
centrifugation period (at 150,000 x g/24 hrs.) of myosin and polar
fish lipids is presented in Fig. 1.

        The Carbon-14 fish lipids were obtained by injecting a live
fish with C-14 labeled acetate, fatty acids, and glycerols. After
three days, the lipids were extracted from the injected fish,
purified and fractionated into polar and neutral lipids.

                              Table 2

        Quantitative Determination of Lipids in Myosin Preparation

| MYOSIN PREPARATION USED | | | Weight of lipid extracted** (mg) | Percent lipids (myosin dry wt.) |
O.D. @ 279 nm	Volume (ml)	Myosin*(mg) dry-weight		
0.59 x 10	100	1,060	0.0135	1.27%
0.60 x 10	76	813	0.0121	1.48%
				1.37%

 * 10 mg myosin/1 ml corresponds to 5.6 absorbancy at 279 nm.

** Lipid extracted by chloroform-methanol-water from myosin
   preparation after Sephadex-G200 column purification.

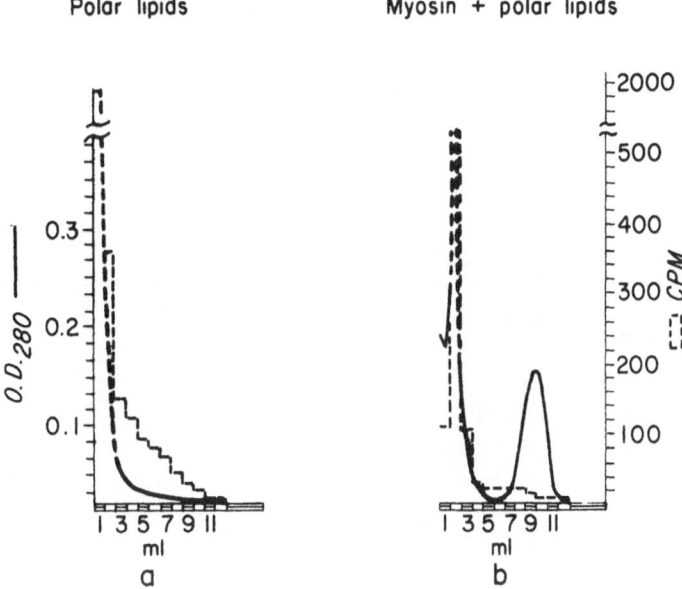

Fig. 1. Sucrose gradient centrifugation profile of myosin and C-14 polar lipids. The sucrose gradient was 5-35%. 1 ml myosin (1%) was incubated with 1 ml lipid (0.3 -2.0 mg sonicated in 1 ml buffer) at 4°C. overnight. The mixture was layered on top of the gradient, centrifugation was applied at 35,000 rpm for 20 hrs. Myosin was traced by absorbancy at 280 nm; fractions of 1 ml were collected in scintillation vials for tracing the radioactive lipids. (a) C-14 polar lipid alone; (b) myosin and polar lipid mixture; ---- cpm; ——— O.D. @ 280 nm.

Table 3 presents the results of this part indicating that, in general, the myosin preparation did not have the ability to bind with either pure triglycerides (triolein or tripalmitin) or fish polar or neutral lipids.

The electron paramagnetic resonance studies (Shenouda and Pigott, 1974) of myosin and fish lipid labeled with nitroxide spin labels (12-doxyl methyl stearate) (Fig. 2) confirmed the fact that no interaction occurred between freshly prepared, undenatured myosin, and polar or neutral lipids.

Table 3

Effect of Various Treatments on the Percentage
of Lipids Bound to Myosin*

Treatments	% of lipid bound to myosin
I. Incubation at 4°C.:	
Triolein + myosin	Traces
Tripalmitin + myosin	Traces
Neutral lipids + myosin	0.00
Polar lipids + myosin	Traces
II. Other treatments of myosin:	
Heating with triolein at 70°C./30 sec.	70.4
Heating with triolein at 50°C./15 min.	32.3
Heating with polar lipids at 50°C./15 min.	18.2
Aged myosin + triolein at 4°C./overnight	29.8
Agitating (myosin + triolein) for 30 sec.	53.8
Agitating (myosin + neutral lipids) for 3 min.	62.2

* 0.3 - 2.0 mg lipids were sonicated in buffer solution and added
  to 10 mg myosin preparation. The mixtures were subjected to the
  different treatments before fractionation on the sucrose
  gradients. Values are averages % of lipids migrated (bound)
  with the myosin fractions on the gradient.

Sounders (1968), Chapman et al. (1968), and Fulk et al. (1969)
showed that the class of lipid governs the type of liposome
aggregates in aqueous dispersion. For example, lysophosphatidyl
choline forms micellar liposomes while phosphatidyl choline (PC)
and phosphatidyl serine (PS) form lamellar aggregates. Based on
that, Berger (1971) related the degree of lipid-protein interaction
to the type of liposome. They stated that micellar lipids bind
more strongly to apoproteins from human erythrocyte membranes than

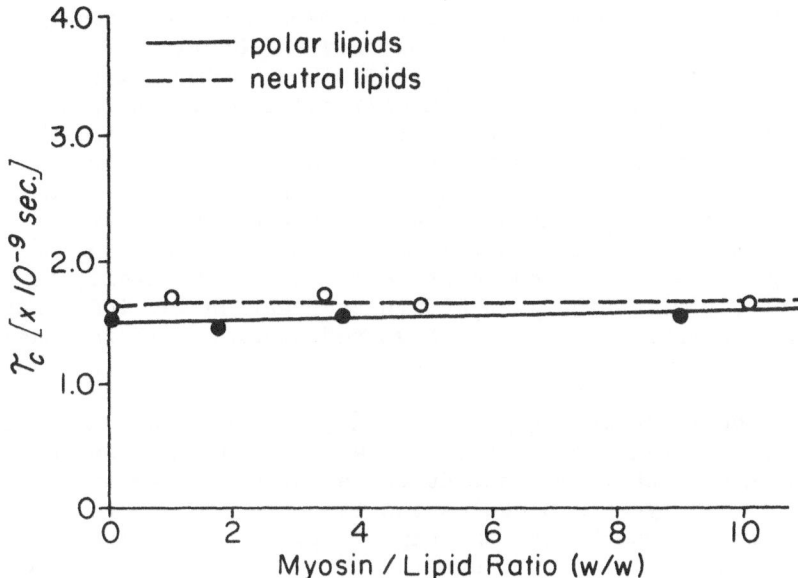

Fig. 2. A plot of reorientation correlation time (T_c) calculated
from the EPR spectra of fish lipids, labeled with 12-doxyl methyl
stearate, as a function of added myosin. The spin label
concentration was 2 x 10^{-4} M.

do lamellar lipids. Our findings from the EPR and sucrose gradient
studies indicated that the state of myosin, and not the lipid, is
the important factor in forming lipid-myosin complexes.

Inducing Myosin-Lipid Interaction by Denaturing the Myosin Molecules

Several factors caused myosin-lipid complex formation through
their effect on the myosin state. The factors tested were myosin
aging, myosin agitation (foam formation), and the effect of heating.

A. Aging of Myosin. Storing the myosin preparation for several
weeks at 4°C. before mixing it with lipids caused a noticeable
change in the capability of myosin to bind with fish lipids so that
one-third of the added lipids bind to the myosin fraction (Table 3).
Moreover, scanning the sucrose gradient at 280 nm showed more than
one distinctive peak for myosin.

Investigators showed that aging of myosin caused configurational changes in the molecules. Godfrey and Harrington (1970)
reported that there was a rapidly reversible monomer-dimer
equilibrium of myosin molecules in salt solutions, and this
equilibrium was pH and ionic strength dependent. Dreizen and
Gershman (1970) stated that pure myosin tended to dissociate into
sub-units, a heavy-chain core and light component in various salt
solutions. These light chains underwent aggregation during prolonged
salt treatment, especially in the absence of thiol protection, and
heavy-chain cores aggregated to at least the dimer level. Lowey
and Holtzer (1959) stated that myosin aggregation was independent
of protein concentration, and it involved local configurational
change.

Since the fresh myosin preparations, as indicated earlier,
have a very weak tendency to interact with lipid, the configurational changes which take place during aging may cause either
unfolding of the myosin molecules and exposure of hindered
functional groups which have the capability to react with lipids,
or may cause an activating or untying of the surface functional
groups on the myosin molecules and free them to react with other
molecules (lipids or proteins).

B. Agitation and Foam Formation of Myosin. Myosin is very
sensitive to physical treatments such as agitation. Aggregates of
myosin were observed even after short agitation treatments (30
seconds on a test tube shaker) indicating a rapid protein-protein
interaction. After such treatment, about 50 percent of the added
lipids accompanied the myosin peaks on the sucrose gradient.
Agitation was apparently accompanied by coagulation indicating
remarkable configurational changes and significant increases in
lipid-myosin interaction. Agitation for longer periods (3 min.)
formed visible aggregates which bound to 62.2 percent of the added
lipids (Table 3). The forces that bound the lipid to the aggregated
myosin were apparently strong enough to resist dissociation under
the high centrifugal force applied (about 150,000 x g for 24 hrs.).

C. Effect of Heating. It was noted that even mild heat
treatment markedly changed the myosin properties. Four factors were
investigated for their effect on myosin-lipid interaction:

1. Effect of temperature
2. Effect of lipid-protein ratio during heating
3. Effect of changing pH after heating
4. Effect of the presence of detergents during heating

1. Effect of temperature. A mild heat treatment (50°C.
for 15 min.) showed that the majority of the myosin molecules still
retained their distinct monomer peak on the sucrose gradient (Fig. 3).

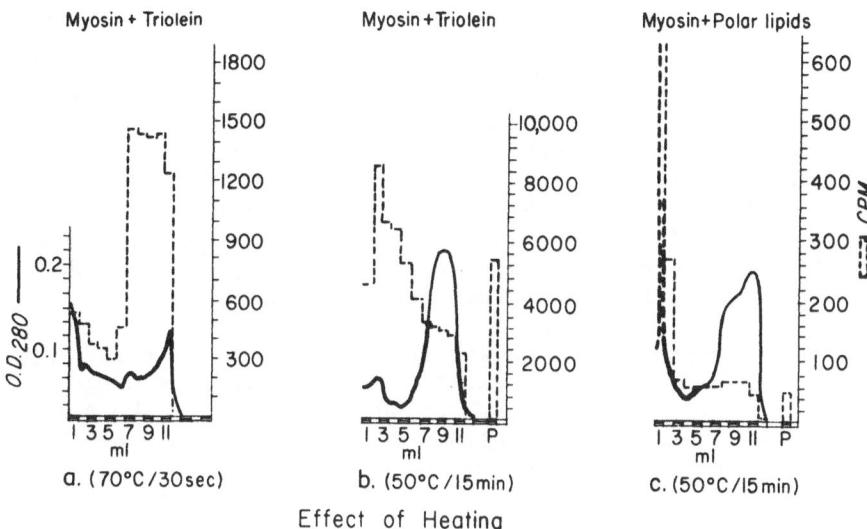

a. (70°C/30sec)   b. (50°C/15min)   c. (50°C/15min)

Effect of Heating

Fig. 3. Sucrose gradient profile of heated mixtures of myosin and lipid. Condition of fractionation as cited in Fig. 1. (a) myosin-triolein mixture heated at 70°C for 30 sec.; (b) myosin-triolein mixture heated at 50°C. for 15 min.; (c) myosin-polar lipid mixture heated at 50°C. for 15 min. P = precipitated pellets.

Polymers of higher molecular weight also resulted from this treatment and were observed as a pellet at the bottom of the tubes after the centrifugation period (150,000 x g/24 hrs.). Between 1/5 and 1/3 of the added lipids were bound to myosin fractions (monomers and pellets) (Table 3).

Increasing the temperature to 70°C. for extremely short times (30 seconds) caused drastic effects on the myosin behavior. The myosin monomer peak was completely abolished from the sucrose gradient, and about 3/4 of the added lipids were complexed with myosin fractions. Chrystall (1971) reported along the lines that myosin molecules were broken down into smaller components between 40-60°C. Apparently, this process was accompanied by another process where protein-protein interaction took place forming high molecular weight molecules.

Increasing the heating temperature of myosin-lipid solutions to 100°C. for 15 minutes, the majority of the myosin formed a visible precipitate which was collected at low rpm centrifugation. It was reported (Mao and Stilling, 1970) that heating at cooking temperature caused destruction of the free SH and aldehyde groups on myosin molecules and was postulated that sulphydryl groups were oxidized in the formation of cross-links during the heating process. Other changes in myofibrillar proteins occurring at higher temperatures were also reported (Hamm, 1966), such as alteration in pH and the iso-electric precipitation of protein. However, the nature of lipid-protein interaction upon heating has not been fully studied, and the cross-linkage of protein's SH groups and lipid may be one of the possible routes.

Our studies (1974) showed that the amount of bound lipids retained in the heated myosin pellet (not including the lipid bound to soluble fractions) was about 40-43 percent of the added lipid (Table 4). Washing the pellets twice with H_2O did not reduce the amount of bound lipids, indicating that the lipids were tightly bound and not just occluded on the surface of the myosin aggregates.

2. Effect of Lipid Myosin Ratio During Heating. A linear increase in bound polar and neutral lipid to myosin (after heating at 100°C. for 15 min.) as a function of increasing the lipid/myosin ratio was observed (Fig. 4). This trend indicates that the interaction is dependent on the lipid availability, i.e., the saturation sites on myosin. The rate of increase of bound polar lipids was higher than neutral lipids, suggesting the importance of the ionic interaction between heated myosin and lipids.

3. Effect of Changes in pH After Heating. Changing the pH from 7.0 to 4.5 after the heat treatment (100°C. for 15 min.) caused a 9 percent decrease in the bound neutral lipid. No effect on the amount of bound polar lipids was observed (Table 4). Thus, lowering the pH to the iso-electric point, i.e.p., of myosin may have caused an increase in the total charges of the myosin molecules if the majority of the functional groups have pK's around this i.e.p. value. Hence, the presence of these extra charges will decrease the chances of hydrophobic interaction between neutral lipid groups and the heavily charged myosin. In another way, the results of this experiment confirm, indirectly, the participation of hydrophobic forces between lipid and myosin.

4. Effect of the Presence of Non-ionic Detergents (Triton X-100) During Heating. Addition of Triton X-100 decreased significantly the amount of neutral lipid retained in the myosin pellets (from 43 to 8.5 percent) indicating the role of hydrophobic interaction between myosin and lipid. Triton X-100 was reported (Helenius and Simon, 1972) to bind primarily via hydrophobic inter-

Table 4

Effect of Heat Denaturation* and Subsequent pH Change
on Myosin-Lipid Complex Formation

	mg lipid added to 136 mg myosin	Pellet pH	mg lipid bound to the pellet	% lipid retained
C-14 polar lipid	5.668	7.5	2.443	43.1
	5.668	7.5 + Wash**	2.267	39.9
	5.668	4.5	2.443	43.1
	5.668	4.5 + Wash**	2.358	41.6
C-14 neutral lipid	8.455	7.5	2.827	33.5
	8.171	7.5 + Wash**	2.787	34.1
	7.995	4.5	2.043	25.6
	7.869	4.5 + Wash**	1.879	23.9
C-14 neutral lipid plus Triton X-100 (16 mg)	8.876	7.5	0.754	8.5

* Myosin lipid mixtures were heated at 100°C for 15 min.; pellets were collected by centrifugation. pH adjustment after the heat treatment.

** The pellets were re-washed in 40 ml. H_2O and recentrifuged.

action to the hydrophobic regions on the protein. The same authors reported that the removal of lipid by detergent treatments would thus be an exchange of bound lipid for bound detergent. Connell (1969) reached the same conclusion with ionic detergents when SDS replaced the residual lipid in FPC.

ACTIN—LIPID INTERACTION

Pure actin preparations, differing from myosin preparations, interact easily with lipids. The amount of lipid bound to actin (Table 5) as detected on sucrose gradients was between 1/3 to 1/2

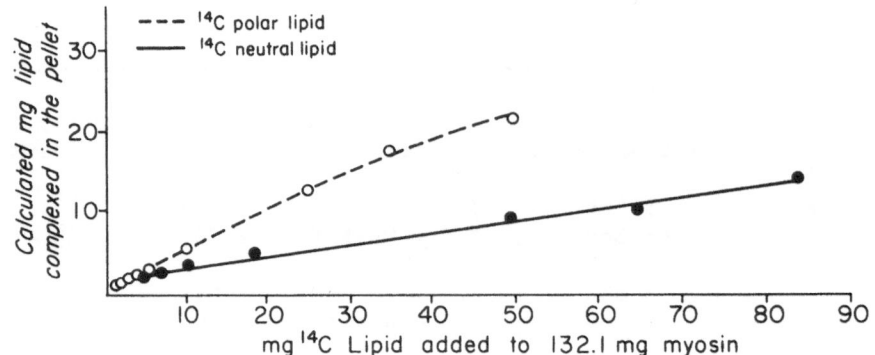

Fig. 4. A plot of mg C-14 fish lipids complexed in the myosin pellets, after heating at 100°C. for 15 min., as a function of lipid/myosin ratio before heating.

of the added lipid (400 ugm/7mg lipid), depending on the incubation temperature and type of lipid. In general, increasing the incubation (reaction) temperature from 4°C. to 25°C. (room temperature) caused 9 percent and 13 percent increases in the bound polar and neutral lipids respectively.

For the effect of lipid types, actin preparation bound 11 to 15 percent less neutral lipid than polar lipid. On the other hand, the EPR results (Fig. 5) showed that the degree of binding between neutral-lipid and actin was stronger than polar-lipid and actin as indicated by the rate of change of T_c (tumbling rate of the nitroxide spin label attached to the lipid) as a function of the protein/lipid ratio (Shenouda and Pigott, 1976).

Based on this information, it seems that the participation of "ionic interaction" manifolds the attraction between the actin and lipid moeity, but the hydrophobic interaction plays an important role in stabilizing the lipoprotein complexes. Consequently,

Table 5

Effect of Various Treatments on the Percentage of Lipid Bound to Actin

Treatments*	Percent of bound lipid**
Incubation at 4°C. with neutral lipids	33
Incubation at 4°C. with polar lipids	48
Incubation at room temp. with neutral lipids	46
Incubation at room temp. with polar lipids	57
Addition of Ca++ to actin-neutral lipids	93
Addition of Mg++ to actin-neutral lipids	89
Addition of Ca++ to actin-polar lipids	39
Addition of Mg++ to actin polar lipids	33
Heating (75°C./3 min.) actin-neutral lipids	49
Heating (75°C./3 min.) actin-polar lipids	26
Agitation (3 min.) actin-neutral lipids	63
Agitation (3 min.) actin-polar lipids	32

* 400 ugm lipids were added to 7 mg actin.

** Total bound lipids to all forms of actin separated on the sucrose gradients (monomers, soluble polymer and precipated actin polymers).

difficulties existed in extracting lipids and purifying actin preparations without using agents that would disturb the hydrophobic interaction.

The interaction between actin-polar and actin-neutral lipids was also tested on SDS-polyacrylamide gel electrophoresis SDS-PAGE (Fig. 6). The actin used in these runs was taken before the final Sephadex purification and therefore the preparation contained about 12 percent tropomyosin as an impurity (Shenouda and Pigott, 1975). The actin-lipid complexes migrated as a unit under the influence of the electrical current. It was observed

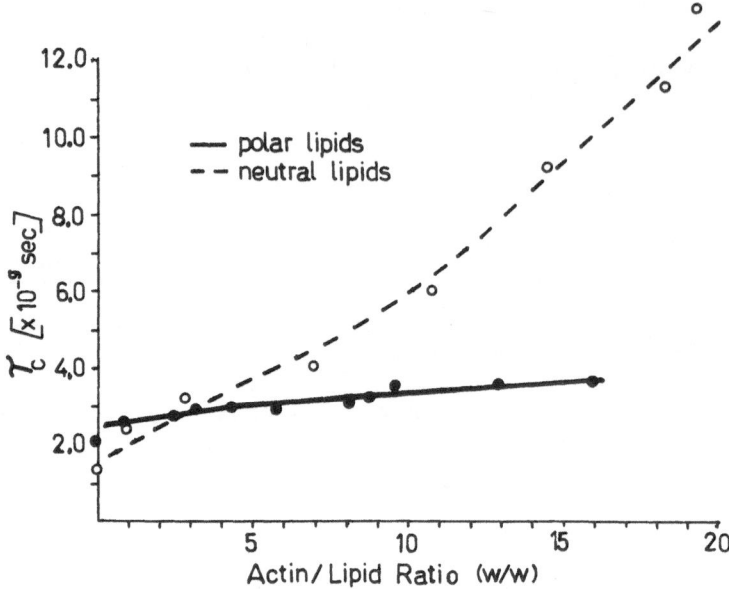

Fig. 5. A plot of reorientation correlation time (T_c) calculated from the EPR spectra of fish lipids labeled with 12-doxyl stearic acid, dispersed in water, as a function of added actin. The spin label concentration was 2×10^{-4}M.

that both polar and neutral lipids caused formation of lipoprotein complexes of high molecular weight. These lipoprotein complexes also showed resistance to dissociation when treated by SDS and urea, indicating that hydrogen bonding is not primarily responsible for the binding of lipid to actin in the lipoprotein. Neutral lipids showed less effect on forming actin polymers of higher molecular weight on SDS PAGE than polar lipids.

Actin Form (G versus F)

The G-form and F-form of actin as fractionated on sucrose gradients (Shenouda and Pigott, 1975) showed different binding capacities. As a consistent rule, the polymer form (F actin) either soluble or precipitated exhibited a greater binding capability to fish lipid than the monomer form (G actin).

The increase in bound lipid (polar or neutral) to the polymerized form of actin suggested that polymerization of actin was associated with conformational changes which caused a considerable increase in the interacting sites on the actin molecule.

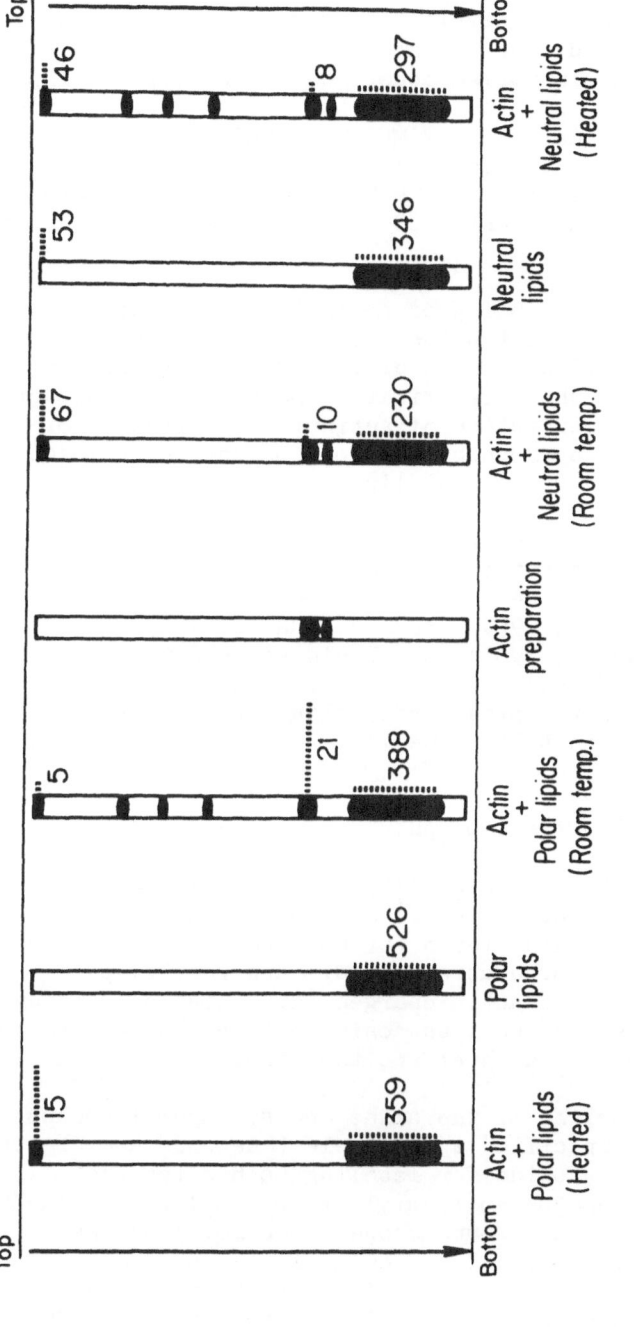

S.D.S. Polyacrylamide Gel Electrophoresis

········· CPM

Fig. 6. Schematic presentation of the main actin bands seen on SDS polyacrylamide gel electrophoresis (5% acrylamide) with the corresponding carbon-14 counts. Samples and gels were prepared according to Weber and Osborn. Actin (1 mg/gel), C-14-neutral lipids (1.3 mg/gel), C-14-polar lipid (0.9 mg/gel). Actin-lipid mixtures were incubated at room temperature or heated at 100°C for 15 min. prior to electrophoresis.

Divalent Cations

G-actin polymerized to form F-actin by increasing the ionic strength or by the presence of divalent cations such as calcium or magnesium. It was found that the amount of lipid bound to the divalently polymerized actin varied according to the type of lipid, i.e., 89-93 percent of the neutral lipid was complexed with actin in the presence of approximately $0.1M$ $MgCl_2$ or $CaCl_2$, respectively, while only 33-39 percent of the polar lipid was bound. In comparison, increasing the ionic strength with KCl resulted in a linear increase in the bound lipid reaching 80 and 99 percent of the polar and neutral lipid at 3.0μ.

It is obvious that the binding of actin to lipid, especially the polar group, has a different pattern according to the presence or absence of divalent cations. This difference in binding can probably be attributed to the polymerization mechanism of the actin itself. Apparently, polymerization of actin in the presence of Ca^{++} or Mg^{++} exposes more hydrophobic regions on the actin molecule which suppress the binding to neutral lipids.

Unlike the other system described for lipid-protein interaction in the presence of divalent cations where calcium or magnesium ions act as bridges between the charged groups on the protein and lipid molecules (Fullington, 1969; Braun and Radin, 1969; Henderickson and Fullington, 1965), the lipid-actin interaction in the presence of Ca^{++} or Mg^{++}, on the contrary, caused a significant decrease in the bound polar lipids, indicating that the bridge theory is not applicable in this system.

Effect of pH

pH greatly affects the actin ability for lipid interaction. Results from sucrose gradients (Fig. 7) showed that the minimum lipid binding occurred at pH's over 6 for both neutral and polar lipids and also at pH 3.0 for neutral lipid. Apparently, pH changes caused changes in the actin charges (dissociation constants of the functional groups), as well as conformational changes in the structure of the molecules, which affect the binding pattern.

By differential labeling of the amino and sulfhydryl groups of the actin with "site specific" spin labels (Shenouda and Pigott, 1976), using maleimide nitroxide for labeling both sites and iodoacetamide for labeling the sulfhydryl groups, all the sulfhydryl and the majority of the amino groups showed free mobility near neutral pH's, 7-7.5, (Fig. 8). This free mobility suggests that these sites are located on the surface of the actin molecules, and the environment around them is quite fluid. The absence of EPR

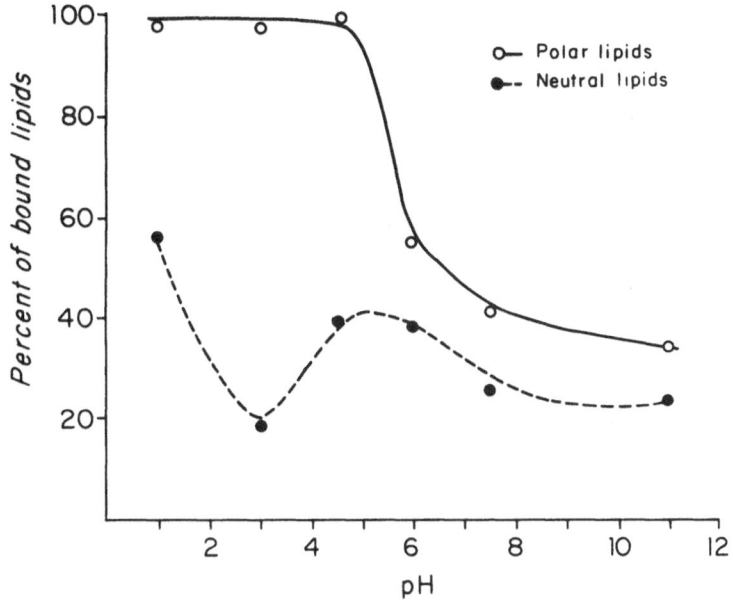

Fig. 7. Percent of bound lipid to actin as a function of pH. 600 ugm of C-14 polar or neutral lipids were incubated with 12 mg actin for 3 hrs. at room temperature at different pH's. The mixtures were fractionated by density gradient centrifugation (sucrose gradient 5-20%, centrifugation at 40,000 rpm for 48 hrs. at 4°C.). Fractionation procedure was as cited in Fig. 1. Percent bound lipid = % (bound lipid/total added lipid).

spectral changes on the addition of lipids (neutral or polar) should at least indicate that the labeled sites have retained their structural integrity during actin-lipid interaction, and no conformational changes, which affect the mobility of the spin labels, occurred near these sites.

At pH's close to the iso-electric precipitation, pH 4.5-6.0 (Fig. 9), a considerable restriction in the mobility of the label sites was noticed, indicating either folding of the molecule or excessive protein-protein interaction. The mobility of these sites was restored again by further lowering the pH to 3.0. At higher pH (10-11), the partial restriction of amino group sites on the actin molecules noticed at lower pH's was completely abolished. Furthermore, from the sucrose gradient studies, actin molecules were less dense at these pH's (10-11). Consequently, at very high pH, actin molecules expanded accompanied by an unfolding and exposure of the hidden amino group sites to the surface of the molecule.

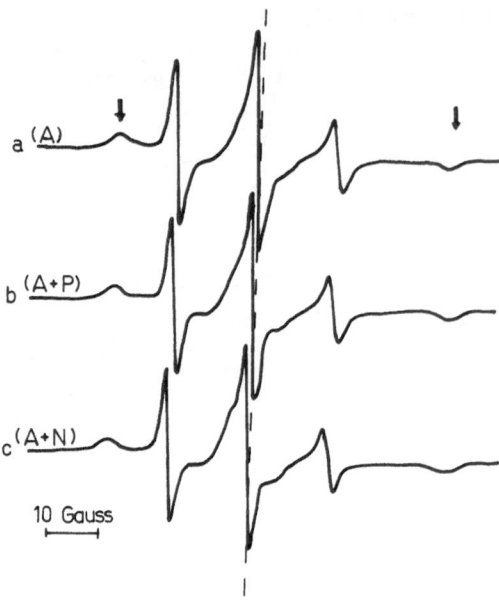

Fig. 8. EPR spectra of fish actin covalently labeled with N(1-oxyl
2,2,5,5 tetramethyl pyrrolidyl) maleimide. 5 mg of the spin label
was stirred with 100 mg actin dissolved in Tris-ATP buffer, pH 7.5,
at 4°C. for 3 days. The unreacted spin label was removed by
exhaustive dialysis and Sephadex G-200 column elution. The spectra
were recorded before and after the addition of lipids. (a) Labeled
actin in 7.5 pH buffer. Vertical arrows indicate bands arising from
strongly immobilized spin label. (b) Polar lipid added to actin in
a 1:1 ratio (w/w). (c) Neutral lipid added to actin in a 1:1 ratio.
Labeling the actin with N(1-oxyl 2,2,5,5 tetramethyl pyrrolidyl)
iodoacetamide (for labeling SH groups) gave identical EPR spectra
lacking the immobilized peaks.

Other Factors Tested

A. Effect of Ionic Strength. Increasing the ionic strength
caused a gradual increase in the formation of higher molecular
actin polymers, accompanied by a linear increase in the bound lipids
(Fig. 10). Increasing the ionic strength over 1.0 u caused a
steeper increase in the bound neutral lipid but no further increase

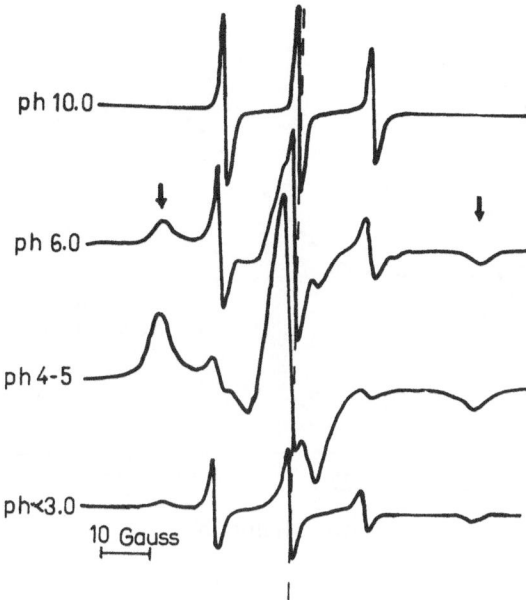

Fig. 9.   pH effect on the EPR spectra of fish actin label with
N(1-oxyl 2,2,5,5 tetramethyl pyrrolidyl) maleimide. Labeling
as cited in Fig. 8. The arrows indicate the immobilized spin
labels at neutral or acidic pH's. These immobilized bands completely
disappeared at higher pH's.

in the bound polar lipid was observed over 0.5 ionic strength.
This experiment demonstrated the effect of salt concentration on
the actin, where the molecules probably underwent configurational
changes resulting in vast interacting sites to bind fish lipids
hydrophilically and hydrophobically.

    B. Effect of Heating. Heating actin and lipid mixtures at
75°C. for 15 minutes caused a small increase in the bound neutral
lipids as fractionated on the sucrose gradient (Table 5). A reverse
effect was observed in the case of polar lipids where about one-
half of the bound polar lipid (at room temperature) was released
after the heat treatment.

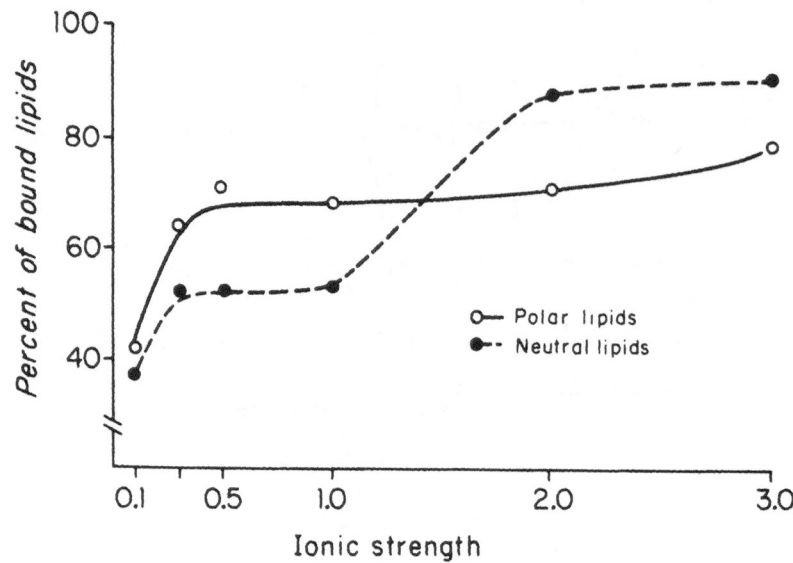

Fig. 10. Percent of bound lipid to actin as a function of
increasing the ionic strength. 600 ugm C-14 polar or neutral lipids
were incubated with 12 mg actin at different KCl concentrations,
pH 7.5, at room temperature for 3 hours. The mixtures were
fractionated on sucrose gradients as cited in Fig. 7, and the per-
cents of bound lipids were calculated.

     C. Effect of Agitation. Agitation of lipid-actin mixtures
produced similar effects on lipid binding as the heat treatment
(Table 5). Probably such treatments caused exposure of more
hydrophobic regions on the actin molecules causing an increase in
the bound neutral lipid and a decrease in the bound polar lipid.

REFERENCES

Babbitt, J., Crawford, D.L. and Law, D.K. (1972). Decomposition of TMAO and changes in protein extractability during frozen storage of minced and intact hake muscle. J. Agric. Food Chem., 20:1052.

Berger, K.U., Barratt, M.D. and Kamat, V.B. (1971). Magnetic resonance studies on the components of human erythrocyte membranes. Chem. Phys. Lipids, 6:351.

Borgstrom, G. (1962). Fish as food. Vol. II, Academic Press, N.Y.

Braun, P.E. and Radin, N. (1969). Interaction of lipids with a membrane structural protein. Biochemistry, 8:4310.

Bridgen, J. (1972). The reactivity and function of thiol groups in trout actin. Biochem. J., 126:21.

Briskey, E.J. and Fukazawa, T. (1971). Myofibrillar proteins of skeletal muscle. Adv. Food Res., 19:279.

Burley, R.W. (1971). Lipoproteins "Biochemistry and methodology of lipids." Eds. Johnson, A.R. and Davenport, S.P. Wiley. Intersci., N.Y.

Buttkus, H. (1970). Accelerated denaturation of myosin in frozen solution. J. Food Sci., 35:558.

Camejo, G, Colacicco, G. and Rapport, M.M. (1968). Lipid monolayers: interaction with the apoprotein of high density plasma lipoprotein. J. Lipid Res., 9:562.

Chapman, D., Fluck, D.J., Penkett, S.A. and Shipley, G.G. (1968). Physical studies of phospholipids. 10. The effect of sonication on aqueous dispersion of egg yolk lecithin. Biochem. Biophys., Acta, 163:255.

Cherayil, G.D. and Scaria, K.S. (1970). Thin layer chromatography of tissue lipids without extraction. J. Lipid Res., 11:378.

Chrystall, B.B. (1971). Macroscopic, microscopic and physical studies on the influence of heating on muscle tissues and protein. Dissertation Abst., Intl. Sec. B. The Science of Eng., 31:6050. c.f. Fd. Sci. Tech. Abst. 4(2):25189.

Chung Wu, C.S. (1969). Comparative studies on myosin from breast and leg muscles of chicken. Biochemistry, 8:29.

Colacicco, G. and Rapport, M.M. (1966). Lipid monolayers: action of phospholipase-A of Crotalus artox and Naja naja venoms on phosphatidylcholine and phosphatidalcholine. J. Lipid Res., 7:258.

Connell, J.J. (1969). The FPC story VIII on the use of detergents in F.P.C. production. Food Technol., 23:206.

Connell, J.J. (1964). Fish muscle proteins and some effects on them of processing. "Symposium on Foods: Proteins and their reaction." Eds. Schultz, H.W. and Anglemier, A.F. The AVI Publishing Co., Inc., Conn.

Connell, J.J. (1962). Changes in amount of myosin extractable from cod flesh during storage at -14°C. J. Sci. Food Agric., 13:607.

Dominova, S.R. (1970). New data on the composition of fish lipids.
    Food Sci. Technol. Abst., 3(3):3R88.
Drabikowski, W., Kaming, D.R. and Maruyama, K. (1968). Effect of
    troponin on the reversibility of tropomyosin binding to F-actin.
    J. Biochem., 63:802.
Dreizen, P. and Gershman, L.C. (1970). Relationship of structure
    to function in myosin. 2. Salt denaturation and recombination
    experiments. Biochemistry, 9:1970.
Ebashi, S. and Maruyama, L. (1965). Preparation and some properties
    of ⍺-actinin-free actin. J. Biochem., 58:20.
El-Bastavizi, A.M. and Smirnova, G.A. (1972). Changes of lipid
    and protein in carp muscle during freezing and frozen storage.
    c/f Food Sci. Technol. Abst., 5:1R39.
Elzinga, M. (1970). Amino acid sequence studies on rabbit skeletal
    muscle actin. Cyanogen bromide cleavage of the protein and
    determination of the sequence of seven of the resulting
    peptides. Biochemistry, 9:1365.
Folch, J., Lees, M. and Sloanestanley, H.G. (1956). A simple
    method for the isolation and purification of total lipids
    from animal tissues. J. Biol. Chem., 214:497.
Fulk, D.J., Henson, A.F., and Chapman, D. (1969). The structure
    of dilute lecithin-water system revealed by freeze-etching
    and electron microscope. J. Ultrastoric. Res., 29:416.
Fullington, J.G. (1969). Lipid-protein interaction. Bakers
    Digest, 43:34.
Gillibrand, J.M. (1976). A study of aggregation phenomena occurring
    in actin solution during polymerization. Biochem. J., 127:737.
Hamm, R. (1966). Heating of muscle systems. "Muscle as a food."
    Eds. Briskey, E.J.; Cassens, R.G. and Trautman, J.C. The
    Univ. of Wisconsin Press, Madison, Wis.
Hao-Chu, G. and Sterling, C. (1970). Parameters of texture changes
    in processed fish: myosin denaturation. J. Texture Studies,
    1:214.
Helenius, A. and Simon, K. (1972). The binding of detergents to
    lipophilic and hydrophobic proteins. J. Biol. Chem., 247:3656.
Hendrickson, H.S. and Fullington, J.G. (1965). Stabilities of
    metal complexes of phospholipids Ca++, Mg++ and Ni++ complex
    of phosphatidylserine and triphosphoinositide. Biochemistry,
    4:1599.
Huszar, G. and Elzinga, M. (1971). Amino acid sequence around the
    single 3-methylhistidine residue in rabbit skeletal muscle
    myosin. Biochemistry, 10:229.
Huszar, G. and Elzinga, M. (1969). E-N-Methyl lysine in myosin.
    Nature, 223:834.
Johnson, P., Perry, S.V. (1970). Biological activity and 3-methyl-
    histidine content of actin and myosin. Biochem. J., 119:293.
Kimm, S. and Grewther, W.G. (1968). "Symposium on fibrous proteins."
    Butterworth, Sydney.

Kuehl, W.M. and Adelstein, R.S. (1970). The absence of 3-methyl-histidine in red cardiac and fetal myosin. Biochim. Biophys. Res. Comm., 39:956.

Lehninger, A.L. (1970). "Biochemistry." Worth Publishers, Inc., N.Y.

Lovern, J.A. (1956). The lipids of fish. 8. The triglycerides and cholesterol of haddock flesh. Biochem. J., 63:373.

Lowey, S. (1972). Protein assemblies in muscle. "Protein-protein interaction." Eds. Jaenicke, R. and Helmreich, H., Springer-Verlag, N.Y.

Lowey, S. and Holtzer, A. (1959). The aggregation of myosin. J. Am. Chem. Soc., 81:1378.

Mao, W.W. and Sterling, C. (1970). Parameters of texture changes in processed fish: cross-linkage of proteins. J. Texture Studies, 1:484.

Marinette, G.V. and Pettit, D. (1968). The interaction of γ-globulin with lipids. Chem. Phys. Lipids, 2:17.

Olley, J., Farmer, J. and Stephen, E. (1969). The rate of phospholipid hydrolysis in frozen fish. J. Food Technology, 4:27.

Pitlick, F.A. and Nemerson, Y. (1970). Some technical problems in egg research. Proc. Soc. Anal. Chem., 7:63.

Quinlivan, J., McConnell, H.M., Stowring, L., Cooke, R. and Morales, M.F. (1969). Myosin modification as studied by spin labeling. Biochemistry, 8:3644.

Rizzino, A.A., Barouch, W.W., Eisenberg, E., and Moos, C. (1970). Actin-heavy meromyosin binding. Determination of binding stoichiometry from ATP-ase kinetic measurements. Biochemistry, 9:2404.

Robinson, J.D. (1966). Interaction between protein sulphydryl groups and lipid double bonds in biological membranes. Nature, 212:199.

Shenouda, S.Y.K. and Pigott, G.M. (1976). Electron paramagnetic resonance studies of actin-lipid interaction in aqueous media. J. Agric. Food Chem., 24:11.

Shenouda, S.Y.K. and Pigott, G.M. (1975a). Lipid-protein interaction during aqueous extraction of fish-protein: Fish actin preparation and purification. J. Food Sci., 40:520.

Shenouda, S.Y.K. and Pigott, G.M. (1975b). Lipid-protein interaction during aqueous extraction of fish protein: Actin-lipid interaction. J. Food Sci., 40:523.

Shenouda, S.Y.K. and Pigott, G.M. (1974). Lipid protein interaction during aqueous extraction of fish protein. Myosin-lipid interaction. J. Food Sci., 39:726.

Sounders, G. (1968). Molecular aggregation in aqueous dispersion of phosphatidyl and lysophosphatidyl choline. Biochim. Biophys. Acta, 125:70.

Stainier-Lambrecht, A. (1962). Heterogenity of carp L-meromyosin. Arch. Intn. Physid. Biochem., 70:682.

Stansby, M.E. (1967). "Industrial fishery technology." Reinhol
    Publishing Corp., Conn.
Suzuki, A. (1971). Denaturation of fish muscle proteins during
    dehydration. J. Food Sci. Technol. (Japan), 18:167.
Takama, K., Zama, K. and Igarashi, H. (1972). Changes in the flesh
    lipids of fish during frozen storage. III. Relation between
    rancidity in fish flesh and protein extractability. Bull.
    Jap. Soc. Sci. Fish., 38:607.
Trayer, H.R. and Trayer, I.P. (1975). A new and rapid method for
    the isolation of myosin from small amounts of muscle and
    non-muscle tissue by affinity chromatography. FEBS Letters,
    54:291.
Tsuchiya, T. and Matsumoto, J.J. (1975). Isolation, purification
    and structure of carp myosin, HMM, and LMM. Bull. Jap. Soc.
    Sci. Fish., 41:1319.
Weber, K. and Osborn, M. (1969). The reliability of molecular
    weight determination by SDS-polyacrylamide gel electrophoresis.
    J. Biol. Chem., 244:4406.
Weeds, A.G. (1967). Small sub-units of myosin. Biochem. J.,
    105:25C.
Wessels, J.P.H. and Spark, A.A. (1973). The fatty acid composition
    of the lipids from two species of hake. J. Sci. Food Agric.,
    24:1939.
Wood, G. and Hintz, L. (1971). Lipid changes associated with the
    degradation of fish tissue. J. Assoc. Off. Anal. Chem.,
    54:1019.
Wu, C.S. and Sayre, R.N. (1971). Myosin stability in intact
    chicken muscle and a protein component released after aging.
    J. Food Sci., 36:133.

ACKNOWLEDGMENTS

    The authors express their thanks to Valerie E. Shenouda for
reviewing this manuscript.

    This work was supported by the National Sea Grant Program.

41

GAS-LIQUID CHROMATOGRAPHY AND MASS SPECTROMETRY OF LANTHIONINE, LYSINOALANINE, AND S-CARBOXYETHYLCYSTEINE

Munenori Sakamoto, Fumitaka Nakayama and Koh-Ichi Kajiyama
Department of Textile and Polymeric Materials, Tokyo Institute of Technology, O-okayama, Meguro-ku, Tokyo, Japan

ABSTRACT

A programmed-temperature gas-liquid chromatographic method of analysis of amino acids as their n-butyl esters of N(O)-trifluoroacetyl derivatives with a dual column system using OV-17 and Dexsil 300 GC as stationary phase is discussed. The method is particularly suitable for analysis of chemically modified amino acids in the hydrolyzates of chemically modified wool and other protein fibers. Combined gas-liquid chromatography-mass spectrometry is used to identify unequivocally the molecular structure of modified amino acids. Lanthionine, lysinoalanine, and S-(2-carboxyethyl)-cysteine in the hydrolyzates of chemically modified wool fibers can be quantitatively analyzed by the gas-liquid chromatographic method on Dexsil 300 GC. N(ε)-(2-Carboxyethyl)lysine was detected by the combined gas-liquid chromatography-mass spectrometry in the hydrolyzate of reduced wool treated with acrylonitrile.

INTRODUCTION

For the studies on the chemical treatment of wool and other protein fibers, a convenient and reliable method for the analysis of chemically modified amino acids produced by reactions of reactive side groups of protein molecules during the treatment, in the presence of a number of protein amino acids, is a necessity. The analysis of such a complex mixture cannot be accomplished without use of one of chromatographic techniques. Ion-exchange chromatography (IEC) has been utilized as fully automated amino acid

analyzers and is the most widely used technique for the analysis
of amino acids. The main drawback of the automated amino acid
analyzers is their low sensitivity and relatively long operating
time. The analyzers are expensive and their versatility is
limited. Chemically modified amino acids with reduced solubility
in aqueous media cannot be eluted. For mixtures containing both
protein amino acids and non-protein amino acids, IEC elution data
are generally not sufficient enough to identify unequivocally the
amino acids present, and there is no definite method for identification of the molecular structures of the amino acids.

Much work has been done in recent years for the development
of dependable gas-liquid chromatographic (GLC) methods for the
separation and quantification of common protein amino acids as
their volatile and thermally stable derivatives. A recent review
(Hušek and Macek, 1975) summarizes all the papers on this subject
published during the period 1956 - mid 1974. Lamkin and Gehrke
(1965) compared various amino acid derivatives and concluded that
the most promising derivative with respect to appropriate volatility for the GLC quantitative analysis of natural protein amino
acids was n-butyl ester of N(O)-trifluoroacetyl (BTFA) derivative.
Since then, Gehrke and his coworkers have been engaging in the
development of a reliable, sensitive, and rapid GLC method for the
quantitative analysis of amino acids. The experimental details of
their BTFA method established at macro, semimicro, and micro
levels were presented in a monograph (Gehrke et al., 1968). The
GLC operation was carried out on a dual-column system with EGA as
stationary phase for the separation of most amino acids and OV-17
for the analysis of arginine, histidine, and cystine which were
unstable on the polar EGA column. Later refinements of the method
with respect to sensitivity and speed and its applications to
complex biological mixtures such as blood plasma and geochemically
important Lunar and other samples were reviewed by Gehrke et al.
(1971) and by Gehrke (1972). Experimental details of their
refined method was described by Kaiser et al. (1974).

Through the continued work of Gehrke and his coworkers, the
usefulness of the GLC method of amino acid analysis has been
widely realized and now, in many laboratories, the Gehrke's GLC
method is routinely used with or without modification. Raulin et
al. (1972) extended the Gehrke's method to the analysis of various
non-protein amino acids in the presence of protein amino acids on
EGA, in relation to their study on prebiological organic chemistry.
Amico et al. (1976) also reported the application of the Gehrke's
method to the analysis of protein and non-protein amino acids for
their study on the chemical constituents of marine algae.

Gehrke and Takeda (1973) reported the single column separation of the BTFA derivatives with Apiezon M, however, according to

March (1975) this method did not provide good resolution for the peaks with short retention times. Amino acid derivatives other than BTFA derivatives are preferred by other workers because the GLC analysis of common protein amino acids can be made on a single column. Islam and Darbre (1972) and Cliffe *et al.* (1973) developed refined GLC methods for the analysis of amino acids as their methyl esters of N(O)-trifluoroacetyl derivatives on a mixed phase (XE-60, QF-1, and MS-200). Some of these methyl esters are highly volatile so that it is difficult to concentrate before injection without substantial loss of the more volatile components. Moss *et al.* (1971) and March (1975) separated protein amino acids as their n-propyl esters of N(O)-heptafluorobutyryl derivatives on OV-1 and on a mixed phase with OV-101 and OV-7, respectively. Zanetta and Vincendon (1973) considered that even BTFA derivatives of some amino acids were too volatile and recommended use of isoamyl esters of N(O)-heptafluorobutyryl derivatives which were well separated on SE-30. Felker and Bandurski (1975) used the isoamyl esters on SP 2100 and noted that histidine, cysteine, cystine, and tryptophane could not be quantified by the method. Mackenzie and Tanaschuk (1974) separated isobutyl esters of N(O)-heptafluorobutyryl derivatives of protein amino acids on SE-30 and reported that the method was suitable for the analysis of plant seed proteins and was also applicable to glycoproteins. Methods to synthesize volatile amino acid derivatives suitable for the GLC analysis in a single reaction step have been looked for but until now no satisfactory method has been developed.

We have considered that the GLC method is particularly suitable for the study of chemical modification of wool keratin and other proteins. The hydrolyzates of chemically modified protein samples may contain not only usual protein amino acids but also unusual amino acids with chemically modified substituents and compounds other than amino acids all of which being produced by the chemical modification of reactive side groups of the proteins followed by acid hydrolysis. The advantage of the GLC method over the conventional IEC method includes: (1) compounds other than amino acids may be detected, (2) identification and characterization of unknown substances can easily be made by comparison of chromatograms taken on a number of columns of different nature, (3) identification of unknown substances can definitely be made by application of the technique of combined gas-liquid chromatography and mass spectrometry (GC-MS) when necessary, and (4) the response to a flame ionization detector (FID) of a compound can be estimated from its chemical structure, therefore, semiquantitative analysis can be made without tedious preparation of an authentic sample.

In this article we present a GLC method applicable to the study of chemical modifications (some of them are related to crosslinking) of wool and other protein fibers (Sakamoto *et al.*,

1974, 1975a - 1975d, 1976a, 1976b) and discuss particularly the GLC method of analysis of lanthionine and lysinoalanine as constituents in chemically modified wool fibers. Both α,α'-diamino dicarboxylic acids are produced during alkaline and other treatments of wool fibers (Asquith and García-Domínguez, 1968; Asquith and Carthew, 1973; Cuthbertson and Phillips, 1945; Horn *et al*., 1941; Miró and García-Domínguez, 1967, 1968, 1973; Robson *et al*., 1969; Ziegler, 1964, 1965) and quantitative determinations of these diamino di-carboxylic acids as well as natural diamino dicarboxylic acid, cystine, are important in the elucidation of physical properties of chemically modified wool fibers. We also describe the GLC and GC-MS analysis of the reduced wool fibers treated with acrylo-nitrile. The hydrolyzate of the sample contained S-(2-carboxy-ethyl)cysteine and N(ε)-(2-carboxyethyl)lysine.

METHODS

Gas-Liquid Chromatography (GLC)

Reagents.   Analytical grade, constant-boiling hydrochloric acid obtained from Wako Junyaku Kogyo Co. Ltd., Tokyo, Japan was used for hydrolysis of wool and, after dilution, for the prepa-ration of amino acid stock solutions. Methanol and methylene chloride of GR grade were dried over molecular sieves. *n*-Butanol and trifluoroacetic anhydride of GR grade were used without further purification. A mixture of hydrochloric acid and methanol was prepared as follows. Hydrogen chloride gas, generated by the slow addition of conc. hydrochloric acid to conc. sulfuric acid, was dried over conc. sulfuric acid, washed with anhydrous methanol and absorbed in anhydrous methanol to give a 5.25 w/w% HCl-methanol. A 5.25 w/w% HCl-*n*-butanol was prepared in a similar manner. *n*-Butyl stearate of EP grade was purchased from Tokyo Kasei Co. Ltd., Tokyo, Japan. A 2.5 mmole/1. *n*-butyl stearate in *n*-butanol was used as a standard solution.

Wool hydrolyzates. About 50 mg of a wool sample was weighed exactly and subjected to hydrolysis without drying. The hydrolysis was carried out with 5 ml of constant-boiling hydrochloric acid for various times (usually 20 - 24 h) at 100°C in a sealed tube under nitrogen atmosphere. The hydrolyzate was filtered with a sintered glass funnel of G 3 under reduced pressure. One ml of the aliquot (*ca.* 10 mg of total amino acids) of the hydrolyzate was transfered into a 100-ml, round-bottomed flask and evaporated on a rotary evaporator at 60°C under reduced pressure. About 5 ml of water was added to the flask and evaporated. After this procedure was per-formed four or five times repeatedly, the residue was dried azeotropically with added methylene chloride (10 ml). The drying procedure was repeated. The dry weight of the wool sample was calculated from the moisture content determined separately with another 30 mg of the sample.

Derivatization. The derivatization of amino acids to n-butyl esters of N-trifluoroacetyl amino acids was carried out at the macro level according to Gehrke et al. (1968): to the dried mixture of amino acids or the dried wool hydrolyzate was added 10 ml of 5.25 w/w% HCl-methanol, and the container was stoppered and kept at room temperature for 30 min with magnetic stirring. The reaction mixture was evaporated to dryness on a rotary evaporator at 60°C under reduced pressure. To the residue, 10 ml of 5.25 w/w% HCl-n-butanol and 1 ml of the standard n-butyl stearate solution were added, and the flask was fitted with CaSO$_4$ drying tube. The mixture was allowed to react at 100°C for 2.5 h with magnetic stirring. The mixture was evaporated to dryness on a rotary evaporator at 60°C under reduced pressure. The mixture of n-butyl esters of amino acids was treated with a mixture of 1 ml of trifluoroacetic anhydride and 7 ml of methylene chloride at room temperature for 15 min with magnetic stirring. To achieve complete acylation, about 3 ml of the aliquot of the reaction mixture was transfered into an acylation tube with a Teflon-lined screw cap and heated at 100°C for 1 h.

Chromatographic conditions. A Shimadzu Model GC-4BMPF dual column gas chromatograph equipped with two separable hydrogen flame ionization detectors (FID) was used extensively. One of the two glass columns (1 m long x 3 mm I. D.) was packed with 1.5 w/w% of OV-17 on 80 - 100 mesh acid-washed and heat-treated High Performance Chromosorb G, and the other with 1.5 w/w% of Dexsil 300 GC on the same quality of Chromosorb G, both column packings being purchased from Nishio Kogyo Co. Ltd., Tokyo, Japan. The columns were set parallel to each other in the dual-column system. The temperature program most frequently used was an initial temperature of 100°C at 8°C/min or 6°C/min, and a final temperature of 270°C. The injection port and detector temperatures were 280°C. Flow rates employed were 70 ml/min for carrier nitrogen, 50 ml/min for hydrogen, and 1100 ml/min for air. Usually 5 μl of the acylated mixture was injected into the glass column directly. Chart speed employed was 4 cm/min for the quantitative analysis. The sensitivity was usually set at 10^{3}MΩ and the range at 128 or 64 x 0.01V.

Determination of retention indices. The n-paraffins for the determination of retention indices were pairs of hydrocarbons selected from C$_{10}$, C$_{12}$, C$_{14}$, C$_{16}$, C$_{18}$, C$_{20}$, C$_{24}$, C$_{28}$, C$_{32}$, and C$_{36}$ n-paraffins obtained from Applied Science Labs., State College, Pa., U. S. A. The temperature program used was an initial temperature of 100°C at 8°C/min.

Determination of FID relative molar response (RMR). FID molar

responses of amino acids as BTFA derivatives were determined from
the gas chromatogram obtained from a mixture of glutamic acid
(reference) and amino acids whose peaks did not overlap. The peak
area ratio was obtained from weights of pieces of chromatogram
paper of the peaks cut. Three independent derivatizations were
made for the mixture of amino acids, and the RMR values determined
for each samples were averaged.

Quantitative analysis. Usually two independent derivatizations were made for each hydrolyzate, and the peak area ratio
between the amino acid and n-butyl stearate peaks was determined.

Combined Gas-Liquid Chromatography and Mass Spectrometry (GC-MS)

The derivatization of the amino acid mixture was carried out
in a usual manner for the GLC analysis. The solvent was removed
at a reduced pressure at 60°C. The residue was acylated again with
0.2 ml of trifluoroacetic anhydride and 1.4 ml of methylene
chloride in an acylation tube with a Teflon-screw cap at 100°C for
1 h to obtain a BTFA mixture of a five-fold concentration over that
of the usual mixture for the GLC analysis.

The GC-MS analysis was made on an Shimadzu-LKB Gas Chromato-
graph-Mass Spectrometer Model GC-MS 9000 with the glass column
(1 m x 3 mm I. D.) packed with 2.0 w/w% of OV-17 on 60 - 80 mesh
acid washed (DMCS) Chromosorb W. The temperature program used was
an initial temperature of 100°C at 5°C/min. The flow rate of
carrier He was 30 ml/min. The ionization energy to obtain mass
spectra was 70 eV.

THE STANDARD GLC METHOD

The BTFA derivative which has been extensively studied is
chosen in our work as the derivative for the GLC analysis of amino
acids. Gehrke's macro method (Gehrke et al., 1968) was employed
for the preparation of BTFA amino acids. The derivatization
consists of the following steps:
(a) Removal of water from an acid hydrolyzate or a mixture of amino
acids in hydrochloric acid.
(b) Esterification of the dry mixture to form methyl ester hydrochlorides.
(c) Interesterification of the methyl ester hydrochlorides to form
n-butyl ester hydrochlorides.
(d) Acylation of the butyl ester hydrochlorides with trifluoroacetic
anhydride to form BTFA derivatives in solution.
n-Butyl ester hydrochlorides may be produced directly from amino

acid hydrochlorides under ultrasonification (Kaiser *et al.*, 1974).

GLC operations were carried out with dual all-glass columns with dual hydrogen flame ionization detectors so that chromatograms taken on two different columns could always be compared for identification and characterization of unknown substances. Wool samples contain much cystine, therefore, the stationary phase such as EGA which makes BTFA cystine decompose is undesirable. Chemically modified amino acids obtained from the chemically modified wool and other protein fibers usually have molecular weights higher than those of unmodified ones and the BTFA derivatives from the modified amino acids have higher boiling points. The stationary phases chosen were thermally stable OV-17 and Dexsil 300 GC. Rohrschneider constants of the two are compared with other OV stationary phases in Table 1. Dexsil 300 GC is less polar than OV-17, and usually the BTFA amino acid was eluted at a lower temperature on OV-17 than on Dexsil 300 GC. The relative molar response to FID usually does not differ much on either columns and this is useful for peak identification when one compares two chromatograms taken on OV-17 and Dexsil 300 GC.

The GLC analysis was made by the linear programmed-temperature method. For characterization of the peak position, retention time, retention temperature and relative retention are often used. However, these retention values fluctuate from one determination to another considerably. Therefore, it appears better to express the peak position as the retention index of Kováts (1958). The retention index of the BTFA amino acid determined under isothermal conditions was found to change linearly with temperature and temperature coefficient of the retention index differs much, depending on amino acid and stationary phase, as seen in Table 2. Therefore,

TABLE 1

Rohrschneider Constants of Various Stationary Phases

Stationary phase	Phenyl substitution, %	X	Y	Z	U	S
OV-1	0	0.16	0.20	0.50	0.85	0.48
OV-3	10	0.42	0.81	0.85	1.52	0.89
OV-17	50	1.30	1.66	1.79	2.83	2.47
Dexsil 300 GC	0	0.43	0.64	1.11	1.51	1.01

TABLE 2

Increase per 10°C in the Retention Index Determined under
Isothermal Conditions for BTFA Amino Acids

Amino acid	Stationary phase			
	OV-1	Dexsil 300 GC	OV-17	OV-210
Alanine	- 6.2	- 3.5	- 7.4	2.0
Aspartic acid	- 1.9	- 1.6	- 5.9	5.0
Ornithine	- 7.9	- 1.0	- 9.6	0.7
Serine	- 5.7	- 6.1	- 9.4	2.1
Phenylalanine	0.7	1.5	0.7	7.8
Tyrosine	- 2.2	- 2.9	- 6.7	- 2.3
Methionine	0.1	0.6	- 3.0	5.7

the retention index should be taken on the same temperature program
for comparison. In our study, the retention indices of BTFA amino
acids were determined on the linear programmed-temperature method
with an initial temperature of 100°C at 8°C/min. For quantitative
analysis, other temperature programs such as that with an initial
temperature of 100°C at 6°C/min were also used.

The retention index taken with the fixed program changes
slightly, depending on column preparation and column aging. Table
3 lists retention indices of BTFA derivatives of common protein
amino acids and n-butyl stearate which is often used as an internal
standard for quantification. Table 4 lists retention indices of
S-substituted cysteines and α,α'-diamino dicarboxylic acids which
are often encountered in the study of wool chemistry. For a number
of amino acids, two set of data are shown in Tables 3 and 4 to show
how much the retention index changes over a long period of time.
These set of the data were taken on different column preparations
by different operators. We have been using commercial column
packing materials. It was observed that the property of the OV-17
column packing was different slightly from one preparation to
another. Thus, lysine (LYS) peak appeared as a shoulder of the
glutamic acid (GLU) peak on one OV-17 column while the two peaks
overlapped completely on another one.

The retention index can be used to estimate the chemical

TABLE 3

Retention Indices of BTFA Derivatives of Protein Amino Acids

Amino acid (Abreviation)	Retention indices		Δ I
	OV-17	Dexsil 300 GC	
Alanine (ALA)	1261	1218	43
Glycine (GLY)	1330	1238	92
Threonine (THR)	1260	1260	0
Serine (SER)	1317	1300	17
	1300	1291	9
Valine (VAL)	1357	1321	36
Leucine (LEU)	1429	1400	29
Isoleucine (ILEU)	1435	1410	25
Cysteine (CYSH)	1499	1470	29
Proline (PRO)	1678	1563	115
Methionine (MET)	1747	1670	77
Aspartic acid (ASP)	1835	1734	101
Histidine (HIS) (diacyl)	1832	1746	86
Phenylalanine (PHE)	1846	1748	98
Tyrosine (TYR)	1928	1864	64
	1910	1870	40
Glutamic acid (GLU)	1977	1873	104
	1977	1878	99
Lysine (LYS)	2000	1930	70
Arginine (ARG)	2063	2057	6
Tryptophane (TRY)	2243	2192	51
n-Butyl stearate (BS)	2480	2400	80
Cystine (CYS)	2612	2524	88
	2611	2529	82

TABLE 4

Retention Indices of BTFA Derivatives of
Chemically Modified Amino Acids

| Amino acid | Retention indices | | Δ I |
(Abreviation)	OV-17	Dexsil 300 GC	
S-Substituted cysteines			
S-Methylcysteine	1627	1530	97
S-Ethylcysteine	1680	1595	85
S-n-Propylcysteine	1752	1670	82
S-n-Butylcysteine	1835	1757	78
S-Carboxymethylcysteine	2225	2092	133
	2228	2088	140
S-(2-Carboxyethyl)cysteine (SCEC)	2334	2200	134
S-(2-Carboxypropyl)cysteine	2320	2200	120
S-(3-Carboxypropyl)cysteine	2436	2307	129
S-(1,2-Dicarboxyethyl)cysteine	2744	2594	150
S-(2-Aminoethyl)cysteine	2117	2000	117
S-(2-Aminopropyl)cysteine	2065	2000	65
S-(1-Ethyl-2,5-dioxopyrrolidin-3-yl)cysteine			
	2542	2321	221
S-(2-Pyrid-2-ylethyl)cysteine	2454	2267	187
α,α'-Diamino dicarboxylic acids			
Lanthionine (LAN)	2374	2307	67
S,S'-Methylenebiscysteine	2788	2667	121
S,S'-Ethylenebiscysteine	2902	2800	102
Lysinoalanine (LAL)	2608	2597	11

structure of the amino acid in some cases (Sakamoto *et al.*, 1974, 1975a, 1976a). The differences (Δ I) between the retention indices on OV-17 and Dexsil 300 GC are also shown in Tables 3 and 4, which are valuable for identification purpose because BTFA amino acids of very similar structure have the Δ I values close to each other. The change in the Δ I from one determination to another is more pronounced than the change in the retention indices.

The GLC method discussed aims at analysis of unusual, chemically modified amino acids. We do not intend to analyze most of protein amino acids quantitatively by the present method. Tyrosine (TYR), lysine (LYS), and cystine (CYS) which form most abundant reactive amino acid residues in wool or silk proteins can be quantitatively determined on OV-17 or on Dexsil 300 GC.

The synthesis of pure non-protein amino acids to obtain authentic samples usually involves tedious and time-consuming work. If the relative molar response (RMR) can be estimated from the chemical formula of a given amino acid, then, the quantitative determination of the amino acid can be greatly facilitated. It is important in our work to use FID as detector not only because it is sensitive and non-specific towards organic compounds but also because the FID molar response is an additive property of the structural features. The contribution of a structural unit to the molar response is usually expressed as the effective carbon number (ECN). The ECN of a molecule is a sum of the ECN values of structural units in the molecule. The ECN values we have used for calculation of RMR values of amino acids (Sakamoto *et al.*, 1974, 1975b, 1976a) are recorded in Table 5. The calculated RMR values for a number of amino acids compare well with the observed ones (within ± 20% range, mostly within ± 15% range) except for homocystine; therefore,

TABLE 5

Effective Carbon Number (ECN)

Atom or group	ECN
C in alkyl and aryl group	1.0
S in thioether	- 0.9
O in ether	- 1.0
-COO- in ester	- 0.5
$-N(H)COCF_3$	- 0.7
$-OCOCF_3$	0.4

the RMR value calculated by the ECN method can be used for semi-quantitative analysis of chemically modified amino acid when an authentic amino acid sample is not available.

ANALYSIS OF LANTHIONINE AND LYSINOALANINE

The GLC method of analysis of lanthionine (LAN) and lysino-alanine (LAL) in the presence of protein amino acids, S-carboxy-ethylcysteine (SCEC) and *n*-butyl stearate (BS) was studied (Sakamoto *et al.*, 1975a). SCEC is produced by the reaction of cysteine (CYSH) residues in proteins with acrylonitrile followed by acid hydrolysis. Treatment of wool samples with acrylonitrile was employed for CYSH analysis by Robson *et al.* (1969).

The GLC method discussed in the previous section was examined for the analysis of LAN, LAL, and SCEC. A mixture of DL- and *meso*-LAN gave a single peak on either stationary phase, OV-17 or Dexsil 300 GC. We have experienced in a number of cases that diastereomers of an amino acid with two asymmetric carbons, in which the second asymmetric carbon is introduced as a consequence of reaction of a reactive side chain of a protein amino acid residue, are not resolved by the present GLC method. This makes the chromatogram simpler. It should be noted, however, that Raulin *et al.* (1972) and Amico *et al.* (1976) separated several pairs of diastereomers (such as threonine and allothreonine) of amino acids of simpler structure on EGA.

The LAN peak overlapped with the SCEC peak, and the CYS peak was not resolved from the LAL peak on OV-17. The resolution of CYS and LAL was satisfactory on Dexsil 300 GC. SCEC and trypto-phane (TRY) peaks overlapped each other. However, as TRY in wool was almost completely destroyed during acid hydrolysis, TRY did not interfere with the determination of SCEC in wool hydrolyzates. The BS peak appeared between LAN and CYS peaks and was suitable as an internal standard. The LYS peak was separated from others on Dexsil 300 GC. Thus, LAN, LAL, and SCEC as well as LYS and CYS could be quantified on Dexsil 300 GC. Proline (PRO), methionine (MET), and arginine (ARG) may be determined as well on Dexsil 300 GC. Since PRO residues in proteins are chemically inert, it is expected to use PRO as an inner standard for the analysis of amino acids in the hydrolyzates of chemically modified protein fibers. It is advisable from our experience that care should be taken for each particular sample if any unknown peak overlaps with the PRO peak or not.

Dexsil 300 GC was extremely stable at high temperatures, and little base line bleed was observed. For the analysis of LAN, LAL, and SCEC as well as LYS and CYS, the programmed-temperature GLC op-eration with a single Dexsil 300 GC column should give satisfactory

results. It is preferable, however, to use the dual-column system
with OV-17 and Dexsil 300 GC and to compare chromatograms on either
stationary phases in order to confirm the peak characterization,
especially for analysis of new types of samples which may contain
other chemically modified amino acids. Tyrosine (TYR) could be
determined on OV-17 under normal operating conditions.

Quantitative Analysis

Three independently derivatized samples made from a stock
solution of PRO, PHE, LYS, LAN, CYS, and LAL and another stock
solution of PRO, PHE, TYR, LAN, and CYS were analyzed, after BS was
added, on the same Dexsil 300 GC and OV-17 columns, respectively.
The results are shown as RMR values to GLU in Table 6. The deviation observed was very small, indicating that the derivatization
was carried out with good reproducibility. The RMR values on Dexsil
300 GC and OV-17 agree fairly well with each other. The RMR values
calculated by the ECN method are also shown in Table 6. The calculated RMR values compare well with the observed ones. Changes in
RMR values during prolonged usage are not negligibly small in some
cases, and occasional checks of the values are highly recommended
in order to get accurate values in routine analysis. Table 7 shows
the range of the RMR values observed for selected amino acids.
These data were collected over a period of four years on different

TABLE 6

Relative Molar Responses (RMR) to GLU of Various Amino Acids

| Amino | RMR on OV-17 | | | | RMR on Dexsil 300 GC | | | | RMR |
acid	1	2	3	Av	1	2	3	Av	Calcd
PRO	0.67	0.67	0.66	0.67	0.72	0.73	0.74	0.73	0.65
PHE	1.07	1.08	1.08	1.08	1.24	1.20	1.20	1.21	1.16
TYR	1.06	1.06	1.06	1.06					1.20
LYS					0.84	0.84	0.82	0.83	0.76
LAN	0.90	0.83	0.88	0.87	0.88	0.91	0.91	0.90	0.94
CYS	0.75	0.75	0.78	0.76	0.82	0.85	0.83	0.83	0.84
LAL					1.16	1.23	1.20	1.20	1.35
BS	1.97	1.96	2.07	2.00	2.13	2.12	2.12	2.12	1.97

TABLE 7

Variations of RMR (to GLU) of Various Amino Acids

Amino acid	RMR	
	OV-17	Dexsil 300 GC
SCEC	0.96 - 1.00	1.00 - 1.04
CYS	0.74 - 0.81	0.65 - 0.83
LAN	0.77 - 0.90	0.88 - 0.91
LAL		1.14 - 1.20
BS	1.67 - 2.09	1.78 - 2.12

TABLE 8

Analysis of Merino Wool Hydrolyzates Containing
Added Lanthionine and Lysinoalanine

LAN			LAL			CYS
Added	Observed	Recovery	Added	Observed	Recovery	Observed
µmoles/g		%	µmoles/g		%	µmoles/g
0	13		0	0		411
31.5	49	114	40.8	46	113	396
94.5	110	103	122.4	130	106	406
157.5	165	97	204.0	211	103	393

columns by different operators. RMR values of CYS and BS were most
variable. RMR values for CYS in the literature varied considerably
from one report to another. RMR values to BS calculated from RMR
to GLU are used to obtain the amino acid contents by comparison of
their peak areas with that of internal standard, BS, in the chromatogram.

In order to see whether the interference from other components
in wool hydrolyzates occurs, known amounts of LAN and LAL were
added to hydrolyzates of untreated wool fibers and the mixtures
were analyzed by the method developed. The hydrolyzate of untreated
wool contained 13 µmoles/g (dry wool) of LAN; therefore the recovered LAN value was corrected by substraction of this value. The
results shown in Table 8 indicate that the observed and theoretical

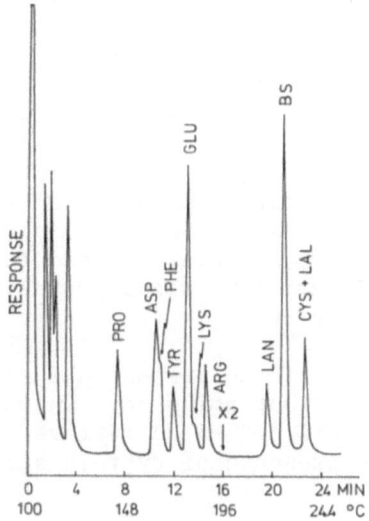

Figure 1.   A gas chromatogram of alkali-treated Lincoln wool on OV-17.   (Sakamoto *et al.*, 1975a).

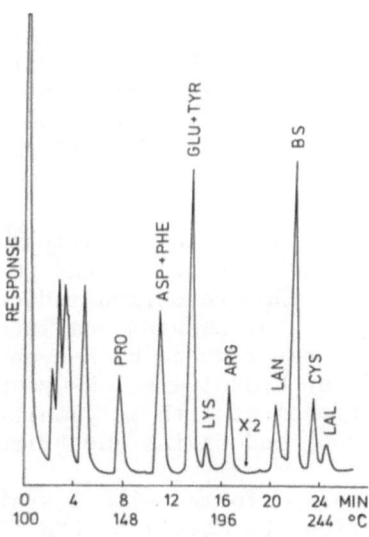

Figure 2.   A gas chromatogram of alkali-treated Lincoln wool on Dexsil 300 GC.   (Sakamoto *et al.*, 1975a).

contents of LAN and LAL are in sufficient agreement. Table 8 also
lists CYS contents observed, indicating that the agreement of the
values for different samples is satisfactory.

The GLC method of anlysis of LAN and LAL was applied successfully to alkali-treated wool fibers (Sakamoto *et al.*, 1975a) and
wool fibers treated with aqueous KCN (Sakamoto *et al.*, 1975c).
Figures 1 and 2 are the chromatograms of an alkali-treated wool
sample which contains both LAN and LAL. It is important that the
hydrolyzate is filtered with a sintered glass funnel before derivatization, otherwise, interfering small peaks may appear at the
later part of the chromatogram.

The main disadvantage of the present GLC method is that
cysteic acid, another important amino acid in wool chemistry, cannot be converted to a volatile BTFA derivative and, therefore,
cannot be analyzed. The presence of cysteic acid did not interfere
in routine analysis.

Reduced Wool Treated with Acrylonitrile

Wool was completely reduced with 1% tri-*n*-butylphosphine in a
9:1 mixture of methanol and water for 3 h, then, treated with 1%
acrylonitrile in a 2:1 mixture of a borate-acetate buffer of pH 8
and *n*-propanol at room temperature for 24 h. Figures 3 and 4 are
the chromatograms of the reduced wool treated with acrylonitrile.
The CYS peak disappeared but no CYSH peak appeared. A large SCEC
peak appeared on both columns as expected. The molar content (837
μmoles/g) of SCEC in the chemically modified wool sample compared
well with twice the molar content of CYS (419 μmoles/g) in the
untreated wool sample.

The LYS peak was decreased by the treatment with acrylonitrile, and a new small peak appeared at the position near to
that of CYS in the chromatogram (retention indices: 2604 on OV-17
and 2553 on Dexsil 300 GC). The unknown peak seemed to be that of
a N(ε)-(2-carboxyethyl)lysine produced by N-cyanoethylation of LYS
residues in wool followed by hydrolysis. N-Cyanoethylation of LYS
residues in proteins by treatment with acrylonitrile was reported
by Riehm and Scheraga (1966) and Cavins and Friedman (1967).

A model experiment was performed with a synthetic polypeptide
containing a high proportion (26 mole%) of LYS residues. The
polypeptide was synthesized by copolymerization of N-carboxy
anhydrides of N(ε)-carbobenzyloxy-L-lysine and γ-methyl L-glutamate
(molar ratio, 1:3) with triethylamine as initiator (*A/I*, 40) in
dioxane (concentration of total anhydrides, 3 g/dl) at 25°C for
6 h, followed by decarboxybenzylation with 36% HBr in glacial

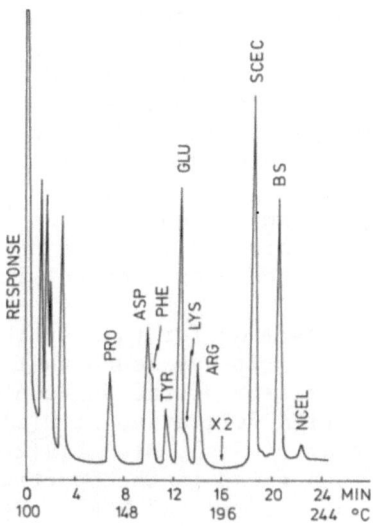

Figure 3.   A gas chromatogram on OV-17 of reduced Merino wool treated with acrylonitrile.   (Sakamoto *et al.*, 1975a).

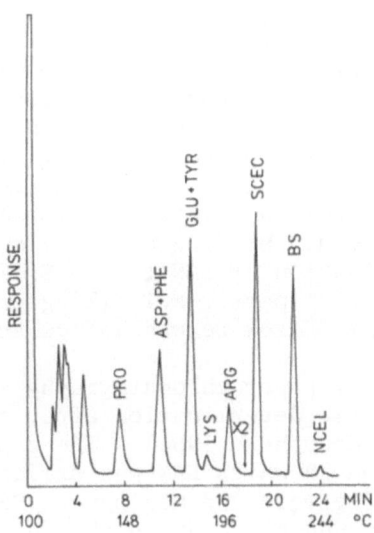

Figure 4.   A gas chromatogram on Dexsil 300 GC of reduced Merino wool treated with acrylonitrile.   (Sakamoto *et al.*, 1975a).

acetic acid. This synthetic polypeptide (cast film, 0.16 g) was treated with 5.0 ml of acrylonitrile in 60 ml of methanol containing 0.6 g of NaOH at room temperature for 24 h. The sample (10 mg) was hydrolyzed with a mixture of 2.0 ml of trifluoroacetic acid and 5.0 ml of constant-boiling hydrochloric acid at 100°C for 36 h in a sealed tube. The GLC analysis of the BTFA derivatives of the hydrolyzate was made. The LYS peak was almost completely lost and a large new peak (retention indices: 2600 on OV-17 and 2558 on Dexsil 300 GC) appeared at the same position of the unidentified peak of the reduced wool treated with acrylonitrile.

The GC-MS analysis was made for this synthetic polypeptide treated with acrylonitrile. It was found that the parent peak of m/e of 522 in the mass spectrum of the unknown peak corresponded to the molecular ion of BTFA derivative of N(ε)-(2-carboxyethyl)-lysine (NCEL) and not that of N(ε), N(ε)-bis(2-carboxyethyl)-lysine. The fragment ions found in the spectrum also supported the structure of the peak material as NCEL. Details of the mass spectrum of NCEL will be shown in the later section. The RMR (to GLU) of NCEL estimated by the ECN method is 1.37. The molar content of NCEL of the acrylonitrile-treated polypeptide obtained by use of the calculated RMR value accounted for *ca*. 60% of the LYS lost by the treatment.

Other Methods of Analysis of LAN and LAL

LAN gave two peaks on a Spackman-Moore-Stein type automated amino acid analyzer under normal running conditions. The slower eluting one was due to *meso*-LAN and was not resolved from GLU (Zacharius and Talley, 1962). Moore and Stein type manual ion-exchange chromatography was applied to LAN by Blackburn and Lee (1954). LAN could be determined with a Technicon Autoanalyzer using Hamilton's single-column method (Asquith and García-Domínguez, 1968). LAN could be determined by an automated amino acid analyzer after oxidation of LAN to the S-dioxide which gave a single peak just before aspartic acid (Ziegler, 1965). Reported conversion of LAN to the S-dioxide was as low as 77%.

LAN was separated by paper chromatography and analyzed by extraction and photometric determination after the reaction with ninhydrin (Dowling and Crewther, 1964). The paper chromatographic method was also reported in which LAN was oxidized to the S-oxide before chromatography (Decroix and Mazingue, 1958). Miró and García-Domínguez (1966a and b) reported the determination of LAN and LAL by use of paper electrophoresis. LAL can be determined on an automated amino acid analyzer with a 15-cm column, although its peak may not be resolved from that of TRY. The methods of Dowling and Crewther (1964) and Miró and García-Domínguez (1966a)

TABLE 9

Important Fragment Ions of Low m/e for BTFA Amino Acids

Ion	m/e	Ion	m/e
Common fragment ions			
C_2H_5	29	CF_3	69
C_3H_5	41	CF_3CO	97
For lysine and N-substituted lysines			
C_5H_7	67	$CF_3CONHC_2H_4$	140
$CF_3CONHCH_2$	126	$CF_3CONHC_5H_8$	180
For S-substituted cysteines			
$CF_3CONHCHCOOH$ $\mid$ CH_2	184		
For O-substituted tyrosines			
$CH_2C_6H_5$	91	$CHCHC_6H_4OH$	119
$CH_2C_6H_4OH$	107	$CH_2CHC_6H_4OH$	120

seem to be most widely used for the analysis of LAN in wool chemistry.

THE GC-MS ANALYSIS

The GC-MS technique is the best method for unequivocal identification of molecular structures of unknown peak materials
separated by GLC. The GC-MS technique is also useful to check if
a GLC peak is an overlapped one or not. The GC-MS analysis of
BTFA derivatives of nineteen protein amino acids was studied by
Gelpi et al. (1969). Only 20 eV mass spectra were discussed in
their work. Lawless and Chadha (1971) studied the 70 eV fragmentation patterns of BTFA derivatives of some non-protein amino
acids such as sarcosine. Felker and Bandurski (1975) reported
the GC-MS analysis of isoamyl esters of N-heptafluorobutyryl
derivatives of 21 protein and non-protein amino acids. The ionization voltage used was 70 eV. Collection of the 70 eV mass
fragmentation patterns of BTFA derivatives of amino acids related
to the study of the chemical reactions of wool and other protein
fibers is in progress in our laboratory and the GC-MS technique
was successfully applied to a number of chemically modified wool

and other protein fibers in order to identify the reaction products
of wool and other protein fibers with various reagents (Sakamoto *et
al.*, 1975d, 1976b).

In general, the confirmation of the anticipated molecular
structures of BTFA amino acids from their 70 eV mass spectra can be
made with ease. Table 9 lists the most common ion fragments in the
low *m/e* region. Usually the molecular ion (M) is detected even for
chemically modified amino acids of high molecular weights and its
intensity being variable with the molecular structure. In a few
cases, the "M + 1" ion is found to predominate over the "M" ion.

Some characteristic fragments of the high *m/e* region are listed
in Table 10. The relative intensities of the fragment ions depend
much on the molecular structure, and, in many cases, the ions listed
in Table 10 are not necessarily found.

Tables 11 - 14 list fragment patterns of SCEC, LAN, LAL, and
NCEL. Molecular ions and most abundant fragment ions (relative
abundance > 18%) of *m/e* higher than 115 are shown in the Tables.
The base peak chosen is the most intense peak in the region of *m/e*
higher than 115. The fragment ions of low *m/e* such as those shown
in Table 9 are often more abundant than the base peak ion chosen.

TABLE 10

Possible Fragment Ions of High *m/e* for BTFA Amino Acids

---

m/e   [Eliminating group]

M - 55 [C_4H_7]   M - 100 [C_4H_8OCO]
M - 56 [C_4H_8]   M - 101 [C_4H_9OCO]
M - 69 [CF_3]   M - 102 [C_4H_9OCOH]
M - 73 [C_4H_9O]   M - 113 [CF_3CONH_2 or CF_3COO]
M - 74 [C_4H_9OH]   M - 114 [CF_3COOH]
M - 97 [CF_3CO]   M - 226 [$CF_3CONHCHCOOC_4H_9$]

Ions formed from M by double elimination such as

    M - 129 [$C_4H_9O + C_4H_8$]
    M - 169 [$CF_3CONH_2 + C_4H_8$ or $CF_3COO + C_4H_8$]

TABLE 11

Fragmentation Pattern of BTFA Derivative of SCEC

m/e	[Structure]	(Relative abundance)
401	[M]	(0.7)
288	[M - CF_3CONH_2]	(26.0)
214	[M - CF_3CONH_2 - C_4H_9OH]	(40.6)
158	[M - CF_3CONH_2 - C_4H_9OH - C_4H_8]	(67.8)
119	[$CH_2SCH_2CH_2COOH$]	(100)

TABLE 12

Fragmentation Pattern of BTFA Derivative of LAN

m/e	[Structure]	(Relative abundance)
512	[M]	(0.4)
399	[M - CF_3CONH_2]	(18.1)
286	[M - $CF_3CONHCHCOOC_4H_9$]	(62.5)
230	[$CF_3CONHCH(CH_2SCH_2)COOH$]	(28.3)
183	[$CF_3CONHC(=CH_2)COOH$]	(100.0)
182	[$CF_3CONHC(=CH)COOH$]	(41.7)
172	[$CF_3CONHCHCH_2SH$]	(18.5)
170	[$CF_3CONHC=CHSH$]	(21.7)
138	[$CF_3CONHC_2H_2$]	(33.5)
117	[$HOOCCHCHSCH_2$]	(33.0)

SCEC gave a fragment ion of m/e of 184 with relative abundance of 13.4%. The fragment ion of m/e of 184 has been found in the 70 eV mass spectra of CYSH derivatives. The fragment ion was the base peak ion for CYS and the fragment ion of m/e of 183 was an abundant ion for CYS. A similar pair of fragment ions was found for LAN but one mass unit smaller, respectively, that is, ions of m/e of 183 and 182. LAL and NCEL gave fragment ions typical of N-substituted lysines.

ACKNOWLEDGMENTS

The advice and encouragement of Professor H. Tonami of Tokyo Institute of Technology and Dr. F. Bekku of Japan Branch, International Wool Secretariat is greatly appreciated. The GC-MS operations were carried out by Mr. T. Kitsuwa, Application Lab., Shimadzu Seisakusho, Ltd., Japan. Parts of the present work were made by Dr. T. Teshirogi, Mr. Y. Sato, Mr. N. Imamura, Mr. J. Sekine, and Miss N. Ojima of Tokyo Institute of Technology.

TABLE 13

Fragmentation Pattern of BTFA Derivative of LAL

m/e	[Structure]	(Relative abundance)
633	[M]	(0.1)
351	[M - $CF_3CONHCHCOOC_4H_9$ - C_4H_8]	(71.8)
333	[M - $CF_3CONHCHCOOC_4H_9$ - C_4H_9OH]	(24.1)
209	[$CF_3CONHC(C_4H_9)CO$]	(62.7)
180	[$CF_3CONHC_5H_8$]	(100.0)
178	[$CF_3CONC_5H_7$?]	(41.3)
140	[$CF_3CONHC_2H_4$]	(80.9)
126	[$CF_3CONHCH_2$]	(18.4)

TABLE 14

Fragmentation Pattern of BTFA Derivative of NCEL

m/e	[Structure] (Relative abundance)
522	[M] (4.1)
449	[M - C_4H_9O] (29.8)
448	[M - C_4H_9OH] (53.4)
425	[M - CF_3CO] (66.1)
374	[M - 2 C_4H_9OH] (45.0)
351	[M - CF_3CO - C_4H_9OH] (36.2)
347	[M - C_4H_9OCO - C_4H_9OH] (94.8)
227	[$CF_3CONHCH_2COOC_4H_9$] (23.4)
222	[M - $CF_3CONHCOOC_4H_9$ - C_4H_9OH] (37.1)
209	[$CF_3CONHC(C_4H_9)CO$] (19.0)
207	[$C_5H_9N(COCF_3)CHCH_2$] (21.7)
198	[$CH_2N(COCF_3)CH_2CH_2COOH$] (92.8)
194	[$C_4H_8N(COCF_3)CHCH_2$] (30.0)
192	[$C_4H_6N(COCF_3)CHCH_2$] (46.6)
184	[$CF_3CONCH_2CH_2COOH$] (25.0)
180	[$CF_3CONHC_5H_8$] (100.0)
171	[$CF_3CONHCH_2COOH$] (24.5)
168	[$CF_3CONHC_4H_8$] (48.2)
166	[$CF_3CONHC_4H_6$] (20.0)
153	[$CF_3CONHC_3H_5$] (21.8)
152	[$CF_3CONHC_3H_4$] (26.4)
140	[$CF_3CONHC_2H_4$] (51.5)
126	[$CF_3CONHCH_2$] (33.2)

REFERENCES

Amico, V., Oriente, G. and Tringali, C. (1976). Quantitative Gas-Liquid Chromatography of Non-Protein Amino Acids. J. Chromatogr., 116, 439-444.

Asquith, R. S. and García-Domínguez, J. J. (1968). Crosslinking Reactions Occurring in Keratin under Alkaline Conditions. J. Soc. Dyers Colourists, 84, 211-216.

Asquith, R. S. and Carthew, P. (1973). The Competitive Addition Reaction of Dehydroalanine Residues Formed during the Alkaline Degradation of Wool Cystine. J. Text. Inst., 64, 10-20.

Blackburn, S. and Lee, G. R. (1954). Chromatographic Separation of the Diastereoisomerides of Lanthionine. Chemistry and Industry, 1252.

Cavins, J. F. and Friedman, M. (1967). New Amino Acids Derived from Reactions of ε-Amino Groups in Proteins with α,β-Unsaturated Compounds. Biochem., 6, 3766-3770.

Cliffe, A. J., Berridge, N. J. and Westgarth, D. R. (1973). Determination of Some Amino Acids by Gas Chromatography of Derivatives. J. Chromatogr., 78, 333-341.

Cuthbertson, W. R. and Phillips, H. (1945). The Action of Alkalis on Wool. 1. The Subdivision of the Combined Cystine into Two Fractions Differing in Their Rate and Mode of Reaction with Alkalis. Biochem. J., 39, 7-17.

Decroix, G. and Mazingue, G. (1958). Séparation Chromatographique et Dosage dans la Laine de la Lanthionine à l'Étal de Lanthionine Sulfoxyd. Bull. Inst. Text. France, 73, 41-52.

Dowling, L. M. and Crewther, W. G. (1964). Determination of Lanthionine in Protein Hydrolysates. Anal. Biochem., 8, 244-256.

Felker, P. and Bandursky, R. S. (1975). Quantitative Gas-Liquid Chromatography and Mass Spectrometry of the N(O)-Perfluorobutyryl-O-Isoamyl Derivatives of Amino Acids. Anal. Biochem., 67, 245-262.

Gehrke, C. W., Roach, D., Zumwalt, R. W., Stalling, D. L. and Wall, L. L. (1968). "Quantitative Gas-Liquid Chromatography of Amino Acids in Proteins and Biological Substances." Analytical Biochemistry Laboratories Inc., Columbia, Mo., U. S. A.

Gehrke, C. W., Zumwalt, R. W. and Kuo, K. (1971). Quantitative Amino Acid Analysis by Gas Chromatography. J. Agr. Food Chem., 19, 605-618.

Gehrke, C. W. (1972). Quantitative-Micro Gas-Liquid Chromatography. J. Assoc. Off. Anal. Chemists, 55, 449-457.

Gehrke, C. W. and Takeda, H. (1973). Gas Liquid Chromatographic Studies on the Twenty Protein Amino Acids: A Single Column Separation. J. Chromatogr., 76, 63-75.

Gelpi, E., Koenig, W. A., Gilbert, J. and Oró, J. (1969). Combined Gas Chromatography-Mass Spectrometry of Amino Acid Derivatives. J. Chromatogr. Sci., 7, 607-613.

Horn, M. J., Jones, D. B. and Ringel, S. J. (1941). Isolation of

a New Sulfur-Containing Amino Acid (Lanthionine) from Sodium
    Carbonate-Treated Wool. J. Biochem., 138, 141-149.
Husek, P. and Kacek, K. (1975). Gas Chromatography of Amino Acids.
    J. Chromatogr., 113, 139-230.
Islam, A. and Darbre, A. (1972). Gas-Liquid Chromatography of
    Trifluoroacetylated Amino Acid Methyl Esters: Determination
    of Their Molar Responses with the Flame Ionization Detector.
    J. Chromatogr., 71, 223-232.
Kaiser, F. E., Gehrke, C. W., Zumwalt, R. W. and Kuo, K. C. (1974).
    Amino Acid Analysis. Hydrolysis, Ion-Exchange Clean-Up,
    Derivatization, and Quantitation by Gas-Liquid Chromatography.
    J. Chromatogr., 94, 113-133.
Kovats, E. (1958). Gas-Chromatographische Charakterisierung
    Organischer Verbindungen. Teil 1: Retentionsindices
    Aliphatischer Halogenide, Alkohole, Aldehyde und Ketone.
    Helv. Chim. Acta., 41, 1915-1932.
Lamkin, W. M. and Gehrke, C. W. (1965). Quantitative Gas
    Chromatography of Amino Acids: Preparation of n-Butyl N-
    Trifluoroacetyl Esters. Anal. Chem., 37, 383-389.
Lawless, J. G. and Chadha, M. S. (1971). Mass Spectral Analysis
    of C_3 and C_4 Aliphatic Amino Acid Derivatives. Anal.
    Biochem., 44, 473-485.
Mackenzie, S. L. and Tenaschuk, D. (1974). Gas-Liquid Chromato-
    graphy of N-Heptabutyryl Isobutyl Esters of Amino Acids.
    J. Chromatogr., 97, 19-24.
March, J. F. (1975). A Modified Technique for the Quantitative
    Analysis of Amino Acids by Gas Chromatography Using Hepta-
    fluorobutyric n-Propyl Derivatives. Anal. Biochem., 69,
    420-442.
Miró, P. and García-Domínguez, J. J. (1966a). Bestimmung von
    Lanthionin in Wollhydrolysaten. Melliand Textilber., 47,
    68-72.
Miró, P. and García-Domínguez, J. J. (1966b). Bestimmung von
    Lysinoalanine in Hydrolysaten von Wolle nach Hitze- und
    Alkalinenbehandlung. Melliand Textilber., 47, 676-680.
Miró, P. and García-Domínguez, J. J. (1967). Action of Nucleo-
    phylic Reagents on Wool. J. Soc. Dyers Colourists, 83, 91-95.
Miró, P. and García-Domínguez, J. J. (1968). Action of Nucleo-
    phylic Reagents on Wool II. Action of Sodium Sulfite at pH
    8.6. J. Soc. Dyers Colourists, 84, 310-313.
Miró, P. and García-Domínguez, J. J. (1973). Action of Ammonium
    and Sodium Hydroxides on Keratin Fibers in Relation to Their
    Morphological Structure. J. Soc. Dyers Colourists, 89, 137-140.
Moss, C. W., Lambert, M. A. and Diaz, F. J. (1971). Gas-Liquid
    Chromatography of Twenty Protein Amino Acids on a Single
    Column. J. Chromatogr., 60, 134-136.
Raulin, F., Shapshak, P. and Khare, B. N. (1972). Quantitative
    Gas-Liquid Chromatography of Non-Protein Amino Acids in the
    Presence of the Twenty Protein Amino Acids. J. Chromatogr.,
    73, 35-41.

Riehm, J. P. and Scheraga, H. A. (1966). Structural Studies of
    Ribonuclease. XX. Acrylonitrile. A Reagent for Blocking Amino
    Groups of Lysine Residues in Ribonuclease. Biochem., 5, 93-99.
Robson, A., Williams, M. J. and Woodhouse, J. M. (1969). The
    Formation of Lysinoalanine and Lanthionine in Wool Fibers
    Stretched in Boiling Water, and Their Relation to Permanent
    Set. J. Text. Inst., 60, 140-151.
Sakamoto, M. Kajiyama, K.-I. and Tonami, H. (1974). Gas-Liquid
    Chromatographic Behaviours of N-Trifluoroacetyl n-Butyl Esters
    of Various S-Substituted Cysteines. J. Chromatogr., 94,
    189-207.
Sakamoto, M., Kajiyama, K.-I., Teshirogi, T. and Tonami, H.
    (1975a). Determination of Lanthionine and Lysinoalanine as
    N-Trifluoroacetyl n-Butyl Esters by Gas-Liquid Chromatography.
    Text. Res. J., 45, 145-154.
Sakamoto, M., Kajiyama, K.-I., Shiozaki, H. and Tanaka, Y. (1975b).
    Gas Chromatographic Analysis of Artifact Amino Acids in Silk
    and Wool Treated with Alkylene Oxides. (in Japanese). Sen-i
    Gakkaishi, 31, T158-T168.
Sakamoto, M., Kajiyama, K.-I., Iwata, M. and Tonami, H. (1975c).
    Reaction of Wool with Potassium Cyanide. Presented at the
    5th Internat. Wool Text. Res. Conf. Aachen, 1975; to be
    published as special issues of 'Schriftenreihe Deutsches
    Wollforschungsinstitut an der Technischen Hochschule Aachen',
    1976.
Sakamoto, M., Kajiyama, K.-I., Sato, Y. and Nakayama, F. (1975d).
    Gas Chromatography of Artifact Amino Acids for Chemically
    Modified Wool Fibers. Presented at the 5th Internat. Wool
    Text. Res. Conf. Aachen, 1975; to be published as special
    issues of 'Schriftenreihe Deutsches Wollforschunsinstitut
    an der Technischen Hochschule Aachen', 1976.
Sakamoto, M., Kajiyama, K.-I., Shiozaki, H. and Tanaka, Y. (1976a).
    Gas Chromatographic Analysis of Silk and Wool Treated with
    Aryl Glycidyl Ethers. (in Japanese). Sen-i Gakkaishi, 32,
    T335-T339.
Sakamoto, M. and Nakayama, F. (1976b). Gas Chromatography-Mass
    Spectrometry of Epoxide-Treated Wool. (in Japanese).
    Preprint, Annual Meeting of Soc. Fiber Sci. Technol., Japan,
    June 16-18, Tokyo, pp 36.
Zacharius, R. M. and Talley, E. A. (1962). Elution Behavior of
    Naturally Occurring Ninhydrin-Positive Compounds during Ion
    Exchange Chromatography. Anal. Chem., 34, 1551-1556.
Zanetta, J. P. and Vincendon, G. (1973). Gas-Liquid Chromato-
    graphy of N(O)-Heptafluorobutyrates of Isoamyl Esters of
    Amino Acids. J. Chromatogr., 76, 91-99.
Ziegler, K. (1964). New Cross-Links in Alkali-Treated Wool.
    J. Biol. Chem., 239, 2713-2714.
Ziegler, K. (1965). The Influence of Alkali Treatment on Wool.
    Proc. 3rd Internat. Wool Text. Res. Conf. Paris (CIETEL),
    2, 403-471.

42

MASS SPECTRA OF CYSTEINE DERIVATIVES

Mendel Friedman

Western Regional Research Laboratory, Agricultural
Research Service, U.S. Department of Agriculture,
Berkeley, California 94710

ABSTRACT

The mass spectra of a series of cysteine derivatives of
structure $X-CH_2CH_2SCH_2CH(NH_2)COOH$ were examined to assess the influence of the electron-withdrawing functional group X on the mass
spectral fragmentation patterns. Measurable molecular ions were
present in most of the spectra although in some cases such peaks
had relative abundances below a few per cent. More useful information on the nature of the substituent could be obtained from the
M - 74 peak corresponding to cleavage $C_2H_4O_2N$ at the sulfur atom.
The results show that mass spectroscopy is valuable for identifying the S-alkyl side chain in S-alkyl cysteine derivatives, a
process frequently required in studies on chemical modification of
sulfhydryl groups and in determining disulfide bonds in proteins.
The observed fragmentation patterns are discussed in terms of
localization of positive charges of ionic species on either sul-
ful, nitrogen, or heterocyclic rings and in terms of substituent
effects on available decomposition pathways.

INTRODUCTION

Chemical modification of the sulfhydryl group in amino acids,
peptides, and proteins by mono- and bifunctional reagents occurs
very often in protein chemistry (Friedman, 1973). To establish
the extent and stoichiometry of reaction, it is usually necessary
to obtain and characterize the expected cysteine derivatives by
several analytical techniques which may include mass spectrometry
(Kiryushkin et al., 1968; Polan et al., 1970; Toubiana et al.,
1970; Harpp and Gleason, 1971; Shemyakin et al., 1971; Nishimura
et al., 1972; Nishimura and Mizutani, 1975; Tsang and Harrison,

1976). Generally the observed fragmentation pathways of cysteine derivatives and of cystine appear to differ significantly from those of non-sulfur amino acids, apparently because sulfur amino acids undergo desulfurization (elimination) reactions to dehydro-alanine intermediates that rearrange and fragment further. Thus, Kiryushkin et al. (1968) examined mass spectra of cystine and cys-teine-containing peptides and drew the following conclusions: (a) molecular ion peaks are formed only from dipeptides; (b) high-er molecular weight peptides undergo disulfide bond rupture with concurrent transfer of a hydrogen atom from the neutral half to the charged half of the cleaved molecule; (c) carbon-sulfur bonds are also cleaved; and (d) S-β-aminoethylcysteine side chains undergo elimination to form a dehydroalanine derivative.

To establish the specificity of reaction between SH groups and various vinyl compounds, Friedman and collaborators (Friedman et al., 1965; Cavins and Friedman, 1970; Friedman and Noma, 1970; Friedman and Tillin, 1970; Krull et al., 1971; Friedman et al., 1973) synthesized a series of S-derivatives of cysteine (Figure 1). In this paper I describe the mass spectra of some of these deriva-tives. The results show how mass spectrometry can be used to es-tablish the identity of S-alkyl side chains in cysteine deriva-tives.

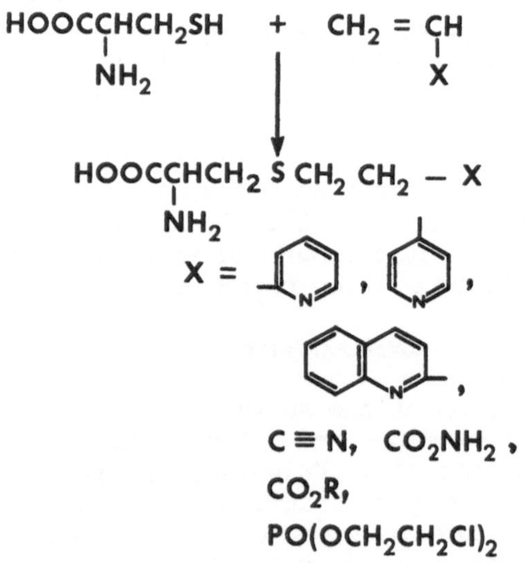

Fig. 1. Preparation of cysteine derivatives.

EXPERIMENTAL

The spectra were obtained on a Nuclide mass spectrometer Model 1290G using conventional electron impact techniques. All spectra reported are at 70 electron volts ionizing voltage. The direct introduction probe of the mass spectrometer was necessary for all the compounds. Volatilization temperatures were between 160 and 185°C for all compounds except cysteine-S-propionamide, indicating that the volatility is determined principally by the cysteine portion. Possible thermal decomposition of the 2-pyridyl and 2-quinolyl derivatives was indicated by the presence of major odd-electron ions corresponding in molecular weight and elemental composition to the original vinyl compounds used to make these cysteine derivatives. In both of these cases, however, much of the compound volatilized without thermal decomposition and produced the expected molecular ion.

RESULTS AND DISCUSSION

Decomposition Pathways. The decomposition pathways in the mass spectrometer for molecular and secondary ions of the cysteine derivatives include: (a) homolytic cleavage of the bond between sulfur and carbon (α cleavage in Figure 2) followed by formation of a new bond to sulfur (b) cleavage of the bond β to sulfur (β cleavage in Figure 2) followed by formation of a new bond to sulfur (these cleavages may be accompanied by hydrogen migration (Figure 3); (c) homolytic cleavage with charge on the heterocyclic ring (Figure 4); and (d) loss of HCN (Figure 5).

A major decomposition pathway for molecular ions of these cysteine derivatives is cleavage of the bond β to the sulfur atom followed by formation of a new bond to sulfur. When it occurs, such cleavage should indicate localization of positive charge at the sulfur atom. Because all of the compounds have identical structure near the sulfur atom, I examined possible effects of the β-ethyl substituent on the degree to which such cleavage occurs. Beta cleavage can actually occur at two points in these cysteine derivatives, but none of the compounds gave peaks corresponding to loss of 134 mass units; therefore the amino acid end of the molecule must destabilize the charge on the sulfur atom. In contrast, the illustrated β cleavage within the cysteine portion of the molecule occurred for all derivatives.

Cysteine. For comparison, the mass spectrum of cysteine is shown in Figure 6. Note the presence of the molecular ion peak, the characteristic peaks due to M − COOH and β cleavage, a mechanism for which is shown in the Figure, and the formation of positively charged species including radical cations.

Heterocyclic Derivatives. The mass spectrum of the 2-pyridyl derivative contained a molecular ion of 0.6% relative abundance.

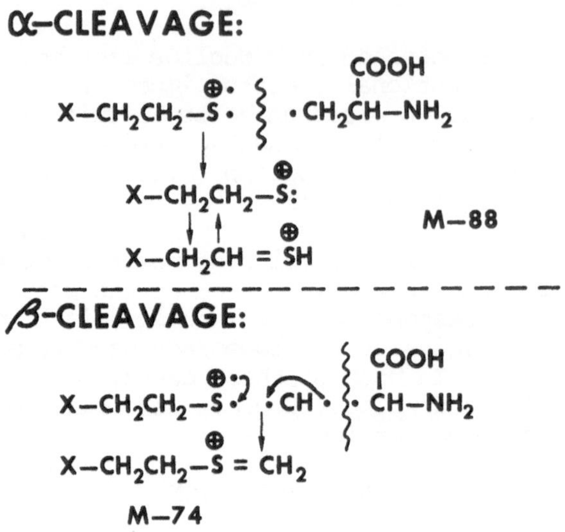

α–CLEAVAGE:

$$X-CH_2CH_2-\overset{\oplus}{S}\cdot \quad \cdot CH_2\overset{\text{COOH}}{\underset{|}{CH}}-NH_2$$

$$X-CH_2CH_2-\overset{\oplus}{S}:$$

$$X-CH_2CH=\overset{\oplus}{S}H \qquad M-88$$

- -

β–CLEAVAGE:

$$X-CH_2CH_2-\overset{\oplus}{S}\cdot \quad \cdot CH\cdot \quad \overset{\text{COOH}}{\underset{|}{CH}}-NH_2$$

$$X-CH_2CH_2-\overset{\oplus}{S}=CH_2$$

$$M-74$$

Fig. 2. α and β Cleavage of cysteine derivatives.

α–CLEAVAGE WITH REARRANGEMENT

$$\text{[pyridine]}-CH_2CH_2-\overset{\oplus}{S}\cdot \quad CH_2-\overset{\text{COOH}}{\underset{|}{CH}}-NH_2$$

$$\text{[pyridine]}-CH_2CH_2-\overset{\oplus}{S}H \qquad m/e\ 139$$

$$M-87$$

- - - - - - - - - - - - - - - - - - - -

β–CLEAVAGE WITH REARRANGEMENT

$$\text{[pyridine]}-CH_2CH_2-\overset{\oplus}{S}-CH_2 \quad \overset{\text{COOH}}{\underset{|}{CH}}-NH_2$$

$$\text{[pyridine]}-CH_2CH_2-\overset{\oplus}{S}-\underset{H}{CH_2}\cdot$$

$$m/e\ 139$$
$$M-73$$

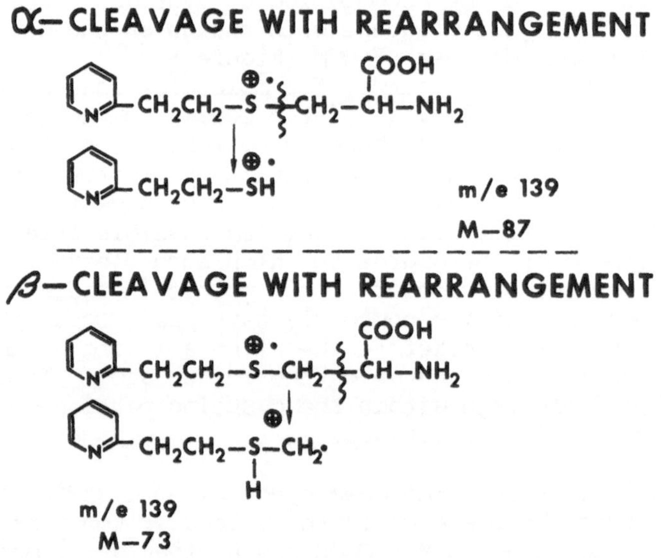

Fig. 3. α and β Cleavage of cysteine derivatives accompanied by hydrogen migration.

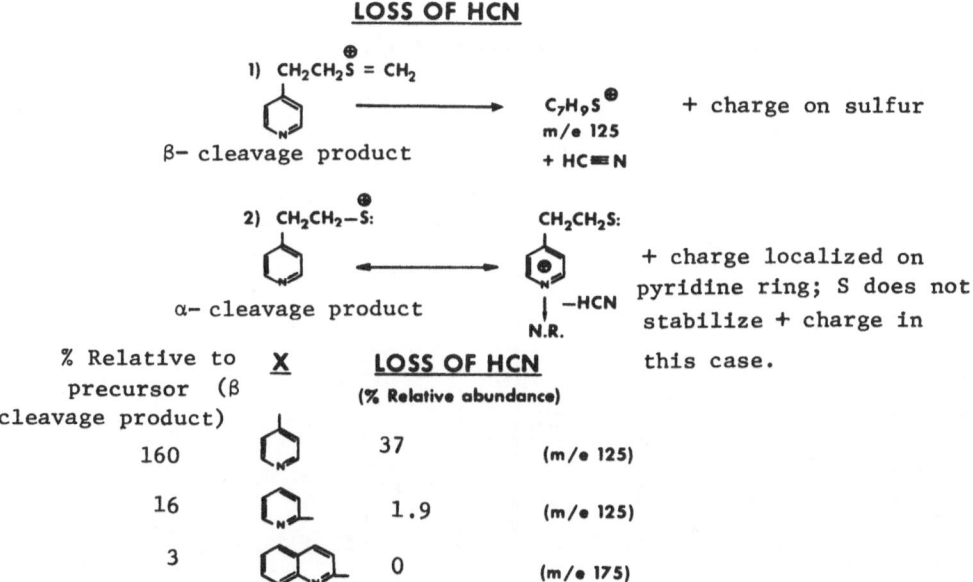

Fig. 4. Homolytic cleavage of carbon-sulfur bond.

LOSS OF HCN

1) $CH_2CH_2\overset{\oplus}{S} = CH_2$

β- cleavage product

$\longrightarrow$ $C_7H_9S^{\oplus}$
m/e 125

+ HC≡N

+ charge on sulfur

2) $CH_2CH_2-\overset{\oplus}{S}:$

α- cleavage product

$\longleftarrow$ $CH_2CH_2S:$

−HCN

N.R.

+ charge localized on
pyridine ring; S does not
stabilize + charge in
this case.

% Relative to precursor (β cleavage product)	X	LOSS OF HCN (% Relative abundance)	
160		37	(m/e 125)
16		1.9	(m/e 125)
3		0	(m/e 175)

Fig. 5. Loss of HCN during fragmentation of heterocyclic cysteine derivatives.

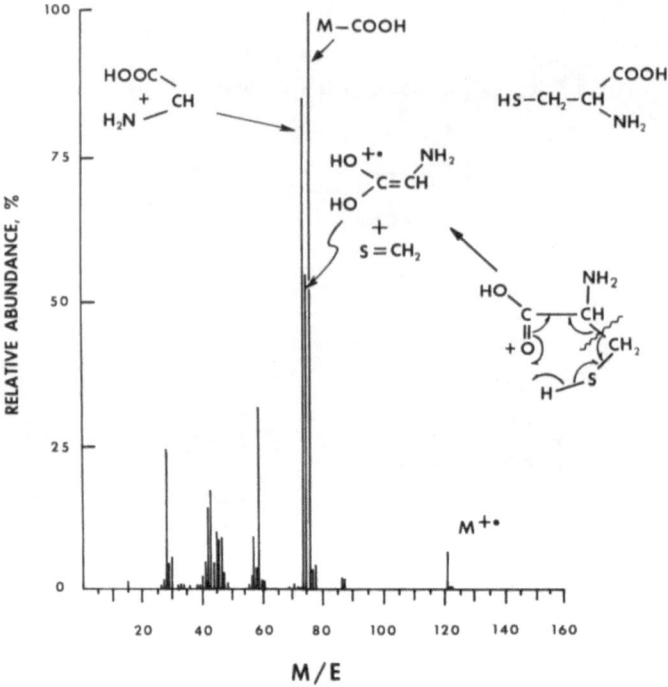

Fig. 6.   Mass spectrum of L-cysteine.

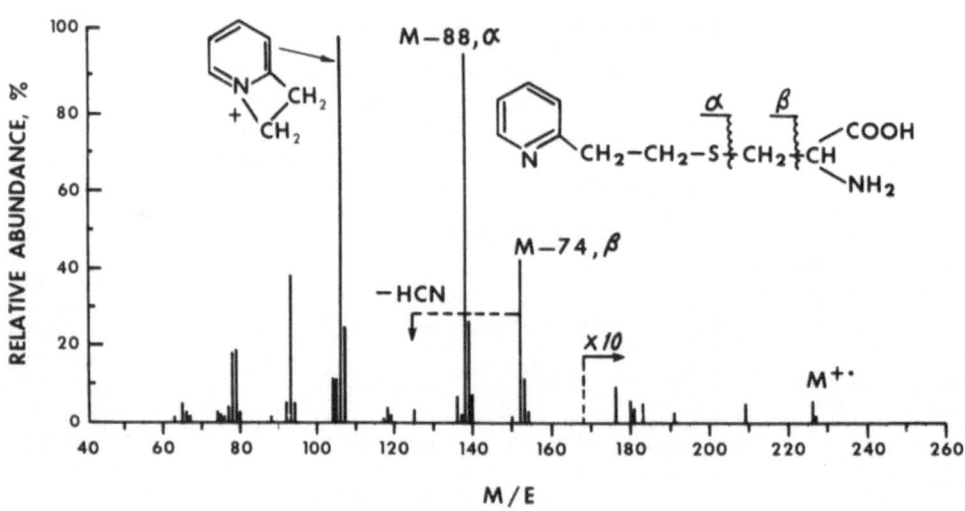

Fig. 7.   Mass spectrum of S-β-(2-pyridylethyl)-L-cysteine.

The presence of the pyridine ring appears to enhance the stability of ions having the charge localized at the sulfur atom, accounting for the high abundance of M − 74 (β cleavage) and M − 88 (α cleavage). Both of these ions are diagnostically useful for characterizing the substituent. Rearrangement of ions are present at m/e 139 and 153 (19% and 8% relative abundance after subtraction of C_{13} isotope). The base peak in the spectrum at m/e 106, possibly formed as illustrated in Figure 5, is as expected for pyridine compounds (Brown and Moser, 1971; Elliott and Waller, 1972).

There is one essential difference between the mass spectrum of the 2-pyridyl derivative (Figure 7) and the 4-pyridyl analogue (Figure 8). There is a major ion fragment at m/e 125 formed by loss of HCN from the ion at mass 152 arising from β-cleavage, amounting to 37% relative abundance. The corresponding values for the 2-pyridyl and 2-quinolyl derivatives are 1.9% and 0%, respectively (Figure 5). No obvious explanation is apparent to rationalize these large differences.

A molecular ion is present in the mass spectrum of the 2-quinoline derivative (Figure 9) with a relative abundance of 0.7%. The major high mass ion is at M − 74 (202) corresponding to cleavage β to sulfur. Such cleavage is also useful in this case for characterizing the substituent. Rearrangement of one hydrogen with α-cleavage gives rise to the most important high-mass ion at mass 189. Rearrangement predominates by a factor of 3 over simple α-cleavage.

Straight-chain Derivatives. The nature of the side group of S-cyanoethylcysteine can be deduced from the elemental composition of the M − 74 ion, as with other derivatives (Figure 10). Rearrangement ions at m/e 87 and 101 are important, relative to simple α and β cleavage, respectively. The base peak of the spectrum is at mass 74, corresponding to the amino acid end of the molecule. The major high mass ion for S-carboethoxyethylcysteine arises from cleavage β to sulfur (Figure 11). The major peak in the spectrum at mass 59 is probably C_2H_3O. Fragmentation reactions of S-carbobutoxyethylcysteine (Figure 12) differ dramatically from those of the other derivatives, apparently because of the proximity of the sulfur and carboxyl groups. The expected ester cleavage reactions for this compound (at the acyl position) are surprisingly absent. For example, there is no peak at mass 101 nor at mass 194 (double hydrogen rearrangement), and the abundance of the M − 73 peak is very low. Rearrangement of hydrogen to the sulfur atom is relatively less important than in the other derivatives. The identity of the mass 130 peak for both ethyl and butyl ester derivatives cannot be accounted for by simple cleavage reactions or hydrogen rearrangments.

All major peaks in the spectrum for the S-propionamide derivative (Figure 13) can be accounted for by retention of the charge on the amide portion of the molecule. This was the only derivative of the series for which β cleavage did not produce an easily

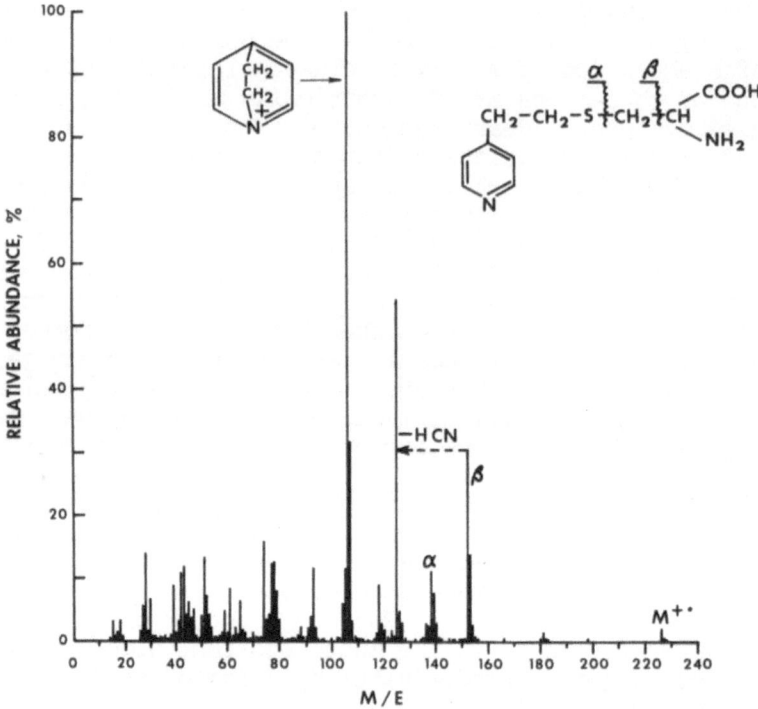

Fig. 8.   Mass spectrum of S-β-(4-pyridylethyl)-L-cysteine.

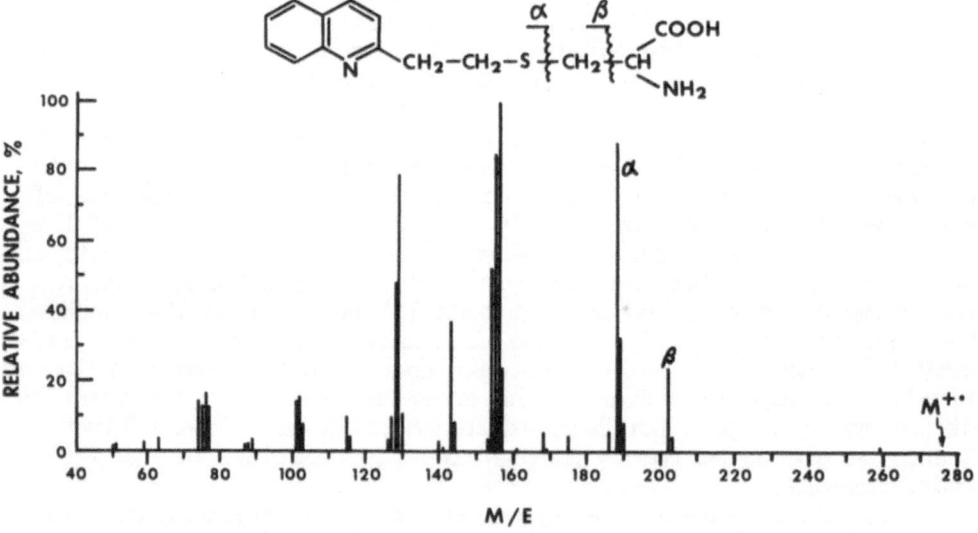

Fig. 9.   Mass spectrum of S-β-(2-quinolylethyl-L-cysteine.

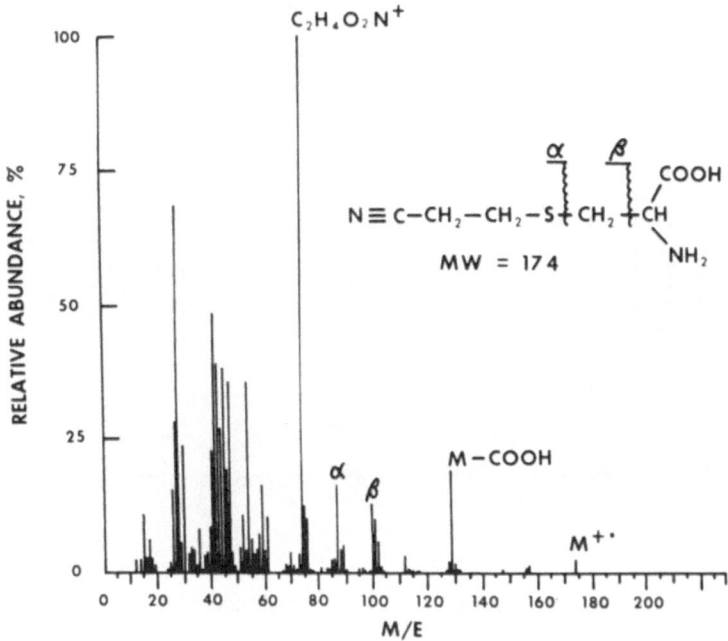

Fig. 10.   Mass spectrum of S-cyanoethyl-L-cysteine.

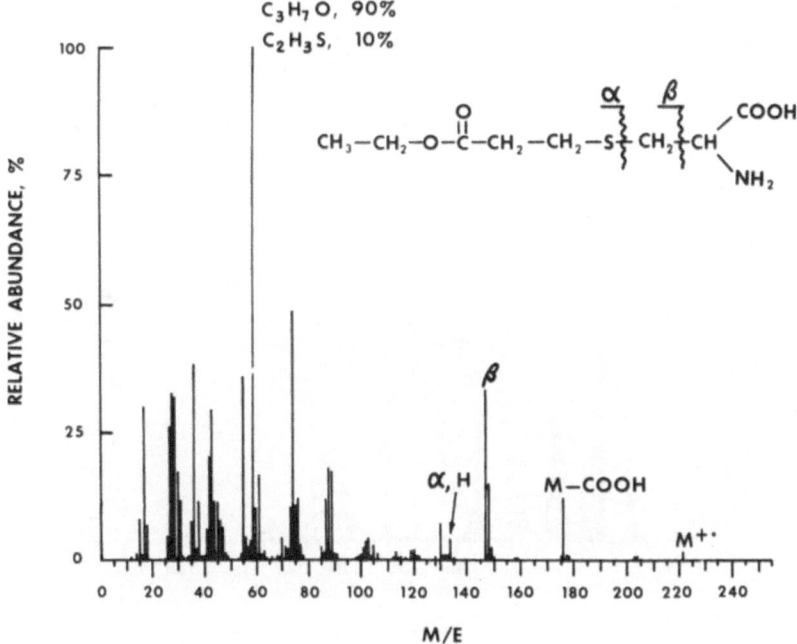

Fig. 11.   Mass spectrum of S-carboethoxyethyl-L-cysteine.

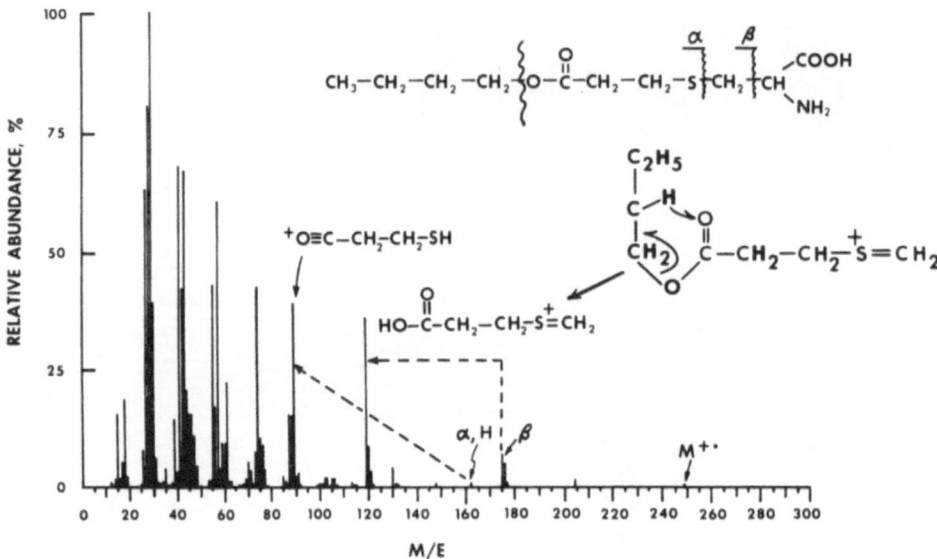

Fig. 12.   Mass spectrum of S-carbobutoxyethyl-L-cysteine.

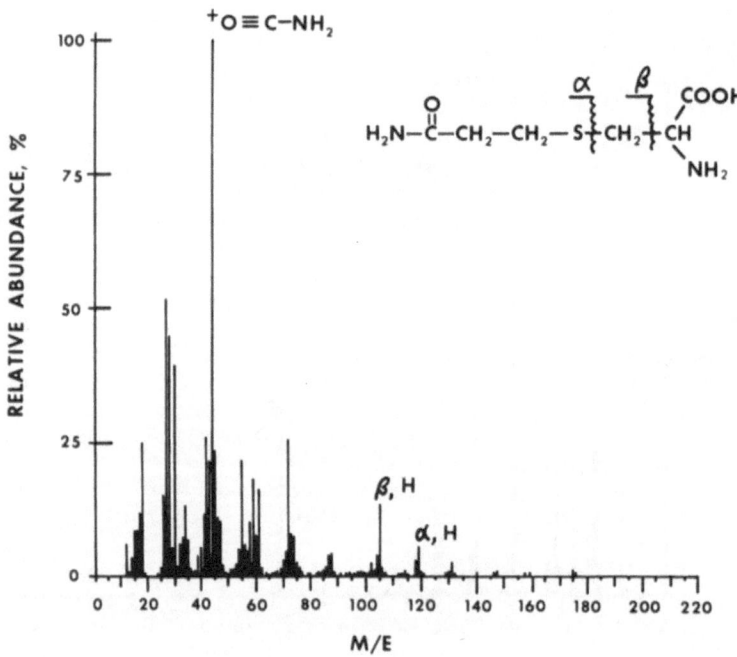

Fig. 13.   Mass spectrum of cysteine-S-propionamide.

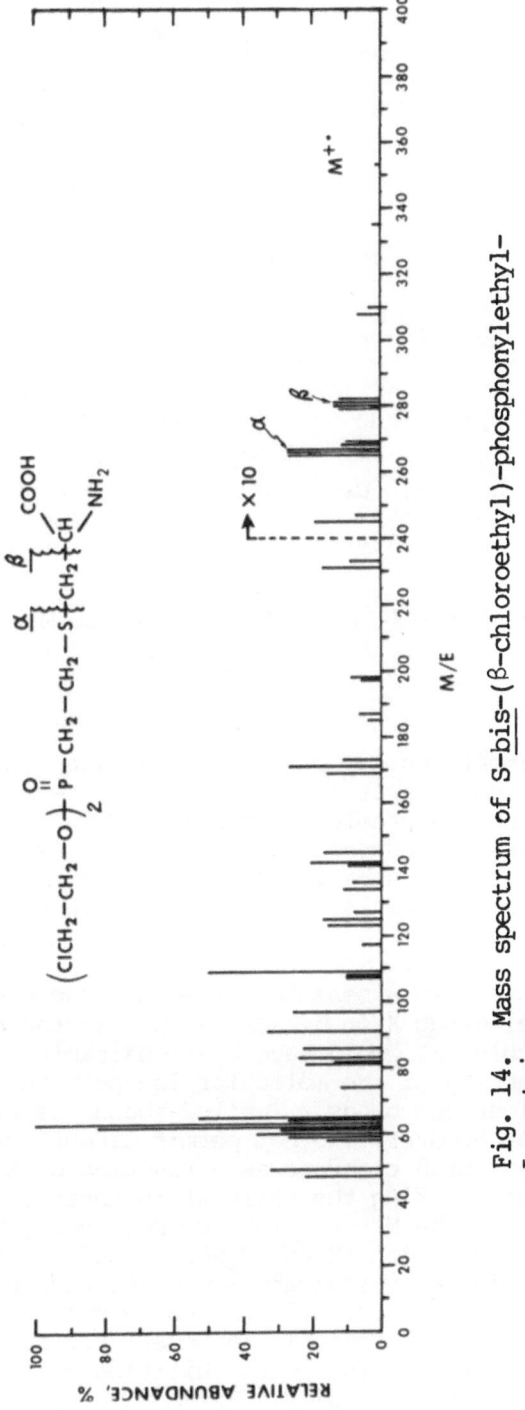

Fig. 14. Mass spectrum of S-bis-(β-chloroethyl)-phosphonylethyl-L-cysteine.

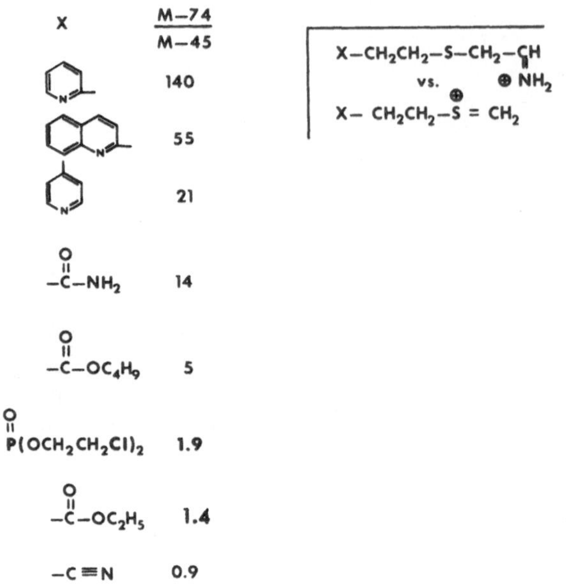

X	$\dfrac{M-74}{M-45}$
(2-pyridyl)	140
(quinolyl)	55
(4-pyridyl)	21
$-\overset{\overset{\textstyle O}{\|}}{C}-NH_2$	14
$-\overset{\overset{\textstyle O}{\|}}{C}-OC_4H_9$	5
$\overset{\overset{\textstyle O}{\|}}{P}(OCH_2CH_2Cl)_2$	1.9
$-\overset{\overset{\textstyle O}{\|}}{C}-OC_2H_5$	1.4
$-C\equiv N$	0.9

$$X-CH_2CH_2-S-CH_2-\overset{\textstyle CH}{\underset{\textstyle \overset{\oplus}{N}H_2}{}}$$

vs.

$$X-CH_2CH_2-\overset{\oplus}{S}=CH_2$$

Fig. 15.   Ratio of M – 74 peak (β–cleavage) to M – 45 peak (loss
of COOH).

recognizable fragment ion.   The α and β cleavage with hydrogen rearrangement produces ions at mass 105 and 119, respectively.   This
compound yielded no measurable molecular ion.

Fragment peaks corresponding to β cleavage (M – 74) can be
used to characterize the substituent in the S–dichloroethyl phosphonate derivative (Figure 14).   A molecular ion peak was also observed.

Substituent Effects.   To assess the effect of the electron–withdrawing functional group X on β cleavage within the cysteine portion of the molecule, it would have been desirable to compare the
ratio of the intensity of the molecular ion peak to the M – 74
peak for the various compounds.   The low abundance or absence of
the molecular ion, however, did not permit direct correlation of
the relative extent of β cleavage as a function of X.   We used an
alternate procedure, taking the ratio of intensities of the M – 74
peak (β cleavage) to the M – 45 peak, corresponding to the loss of
COOH from the molecular ion ($X-CH_2CH_2SCH_2CH=\overset{+}{N}H_2$ vs. $X-CH_2CH_2S^+=CH_2$).

The data in Figure 15 indicate a wide spread in this ratio,
from 140 to 0.9 for the 2-pyridyl and cyanoethyl derivatives, respectively.   The observed ratio is, however, not directly related
to the relative electron–withdrawing abilities of X as determined
from rate and equilibrium studies.   Evidently, free energy parame-

ters that govern relative reactivities of X–CH=CH$_2$ with amino acids and ionization equilibria of the protonated amino group in C$_6$H$_5$–CH$_2$CH(COOH)NH$_2$ –CH$_2$CH$_2$–X in solution (Friedman and Romersberger, 1968) do not influence gas-phase reactions of analogous amino acid derivatives in the mass spectrometer in the same way.

ACKNOWLEDGEMENTS

It is a pleasure to thank R. E. Rohwedder and W. F. Haddon for obtaining and helping to interpret the mass spectra.
Reference to a company or product name does not imply approval or recommendation of that product by the U.S. Department of Agriculture to the exclusion of others that may be suitable.

REFERENCES

Brown, E. V. and Moser, R. J. (1971). Mass spectra of some 2-, 3-, and 4-pyridine carboxylic acids. Further evidence of a N–H interaction from loss of carbon dioxide. J. Heterocyclic Chem., 8, 189–192.

Cavins, J. F. and Friedman, M. (1970). Preparation and evaluation of S-β-(4-pyridylethyl)-L-cysteine as an internal standard for amino acid analyses. Anal. Biochem., 35, 489–493.

Elliott, W. H. and Waller, G. R. (1972). Vitamins and cofactors. In "Biochemical Applications of Mass Spectrometry," G. R. Waller, Ed., Wiley-Interscience, New York, New York, Chapter 18.

Friedman, M. (1973). "The Chemistry and Biochemistry of the Sulfhydryl Group in Amino Acids, Peptides, and Proteins," Pergamon Press, Oxford, England and Elmsford, New York, 485 + viii p.

Friedman, M. and Noma, A. T. (1970). Cystine content of wool. Textile Res. J., 40, 1073–1078.

Friedman, M. and Romersberger, J. A. (1968). Relative influences of electron-withdrawing functional groups on basicities of amino acid derivatives. J. Org. Chem., 33, 154–157.

Friedman, M. and Tillin, S. J. (1970). Flame-resistant wool. Textile Res. J., 40, 1045–1047.

Friedman, M., Cavins, J. E., and Wall, J. S. (1965). Relative nucleophilic reactivities of amino groups and mercaptide ions in addition reactions with α,β-unsaturated compounds. J. Amer. Chem. Soc., 87, 3672–3682.

Friedman, M., Noma, A. T., and Masri, M. S. (1973). New internal standards for basic amino acid analyses. Anal. Biochem., 51, 280–287.

Harpp, D. N. and Gleason, J. G. (1971). Preparation and mass spectral properties of cystine and lanthionine derivatives. A novel synthesis of L-lanthionine by selective desulfurization. J. Org. Chem., 36, 73–80.

Kiryushkin, A. A., Gorlenko, V. A., Agadzhanyan, Ts.E., Rosinov,
    B. V., Ovchinnikov, Yu. A., and Shemyakin, M. M. (1968).
    Mass spectrometric determination of the amino acid sequence
    in cystine and cysteine-containing peptides. Experientia,
    24, 883-885.
Krull, L. H., Gibbs, D. E., and Friedman, M. (1971). 2-Vinyl-
    quinoline, a reagent to determine protein sulfhydryl groups
    spectrophotometrically. Anal. Biochem., 49, 80-85.
Nishimura, H. and Mizutani, J. (1975). Photochemistry and radia-
    tion chemistry of sulfur-containing amino acids. A novel re-
    action of 1-propenylthiyl radicals. J. Org. Chem., 40, 1567-
    1575.
Nishimura, H., Tahara, S., Okuyama, H., and Mizutani, J. (1972).
    Mass spectra of sulphur-containing amino acids and peptides.
    Tetrahedron, 28, 4503-4513.
Polan, N. L., McMurray, W. J., Lipsky, S. R., and Lande, S.
    (1970). Mass spectroscopy of cysteine-containing peptides.
    Biochem. Biophys. Res. Commun., 38, 1127-1133.
Shemyakin, M. M., Ovchinnikov, Yu. A., and Kiryushkin, A. A.
    (1971). Mass spectrometry of amino acids and peptides. In
    "Mass Spectrometry: Techniques and Applications," G. W. A.
    Milne, Ed., Wiley-Interscience, New York, New York, pp. 289-
    325.
Toubiana, R., Barnett, J. E. G., Sach, E., Das, B. C., and
    Lederer, E. (1970). Determination of amino acid sequences
    in peptides by mass spectrometry. Desulfurization of sulfur-
    containing peptides. FEBS Letters, 8, 207-209.
Tsang, C. W. and Harrison, A. G. (1976). Chemical ionization of
    amino acids. J. Amer. Chem. Soc., 98, 1301-1308.
Tschesche, H., Schneider, M., and Wachter, E. (1972). Mass spec-
    tral identification and quantification of phenylthiohydantoin
    derivatives from Edman degradation of proteins: cysteine de-
    rivatives. FEBS Letters, 23, 367-372.

Presented at the 159th National Meeting of the American Chemical
    Society, Houston, Texas, Feb. 22-27, 1970, Abstracts p. ANAL
    91.

43

A NUCLEAR MAGNETIC DOUBLE RESONANCE STUDY OF N-β-BIS-

(β'-CHLOROETHYL)PHOSPHONYLETHYL-DL-PHENYLALANINE

Mendel Friedman[*] and Walter A. Boyd

Northern Regional Research Laboratory, Agricultural
Research Service, U. S. Department of Agriculture,
Peoria, Illinois 61604

ABSTRACT

Studies were carried out on the effect of decoupling, deuterium labeling, concentration, temperature, and solvent media on the NMR parameters of the vinyl phosphonate adduct of phenylalanine, $C_6H_5CH_2CH(COO^-)NH_2^+CH_2CH_2PO(OCH_2CH_2Cl)_2$. The results permit assignments of chemical shifts and coupling constants to the various protons of this molecule which contains unique structural features. The NH_2^+-$\underline{CH_2}$-protons are deshielded by more than 1 ppm than the $\underline{CH_2}$-PO-protons. The -OCH_2-protons are nonequivalent exhibiting a fine split. Possible sources of the fine split include NH...O=P hydrogen bonding. The deuterium-labeling method should be applicable for synthesizing deuterium- and tritium-labeled crosslinked amino acids such as lysinoalanine and lanthionine and demonstrating analgous dehydroalanine-α-amino group-crosslinking.

INTRODUCTION

Nuclear magnetic resonance (NMR) spectroscopy is a useful technique in studies of addition reactions of COOH, NH_2, and SH groups with vinyl compounds and of SH oxidations (Weisleder and Friedman, 1968; Friedman et al., 1965; Sharples and Flavin, 1966; Snow et al., 1976; 1975). A series of N-substituted derivatives of structure $RCH(COO^-)NH_2^+CH_2CH_2X$ were previously prepared to determine the influence of X on acid-base equilibria of the amino groups (Friedman and Romersberger, 1968). This paper discusses the NMR spectra of one of these derivatives and the corresponding deuterium-labeled analogue. The results are explained in terms of N-H···O=P and N-H···O=C internal hydrogen bonding giving rise to conformational isomers. Insofar as we know, possible physicochemical consquences of N-H···O=P hydrogen bonding in proteins have not yet been studied.

EXPERIMENTAL

Bis(β–chloroethyl)vinyl phosphonate was obtained from Stauffer Chemical Company and was redistilled before use. N–β–bis(β'–chloroethyl)phosphonylethyl-DL–phenylalanine (1) was prepared as described by Friedman and Romersberger (1968). The deuterated analogue 2 was prepared in the same manner except that H_2O was replaced by D_2O in all operations. m.p. 169–170°C.

Anal. Calcd. for $C_{15}H_{19}D_3NCl_2PO_3$ (401.25): N, 3.49; Cl, 17.68.
Found: N, 3.49; Cl, 17.64.

All spectra were obtained in D_2O, trifluoroacetic acid (TFA), d_1–TFA, and NaOD on a Varian Associates HA–100 Spectrometer operating at a frequency of 100 MHz, and on an A–60 spectrometer operating at 60 MHz. Decoupling experiments were performed at 30°C in the frequency sweep mode with a Hewlett Packard Model 200 A.B. audio oscillator. All solutions were w/v, and τ values were obtained from internal tetramethylsilane (TMS) for TFA and d_1–TFA, external TMS for D_2O, and DDS for NaOD.

RESULTS AND DISCUSSION

DL–Phenylalanine. The NMR spectra of this amino acid in TFA, d_1–TFA, and D_2O are shown in Figure 1. The assigned chemical shifts and coupling constants are summarized in Table 1. Integration of the spectrum obtained in TFA indicates that the signals associated with $-NH_3^+$ protons overlap with those from aromatic protons. The spectrum in d_1–TFA confirms this conclusion. In this solvent, the NH_3^+ groups are converted to ND_3^+ as a result of rapid exchange, and integration shows only the five–proton signal of the aromatic ring.

The multiplet centered at 5.42τ (TFA) is assigned to the methine hydrogen resonance which is split as a result of unequal coupling to the benzyl protons and coupling with the ammonium protons. Several experiments support this assignment. Coupling with NH_3^+ protons was removed when the spectrum was determined in either D_2O or d_1–TFA. In these solvents the complex multiplet is changed to a symmetrical four–peak pattern. The non–equivalency of the benzyl protons is shown by the eight–line AB–pattern in the region 6–7τ characteristic of an ABX system. Decoupling of the methine from the benzyl protons (Figure 1d) leaves an AB–pattern for the benzyl protons. Analysis of this pattern according to eq. 1 (Bible, 1965) locates the benzyl proton chemical shifts at 6.48τ and 6.23τ, respectively, where $(\nu_1 - \nu_3)$ equals the separation of peaks 1 and 3 of the quartet.

$$\nu^2_{AB} = (\nu_1 - \nu_3)^2 - J^2_{AB} \qquad (1)$$

Bis(β-chloroethyl) Vinyl Phosphonate. The NMR spectrum of this vinyl compound is shown in Figure 2 and the assigned chemical shifts and coupling constants are summarized in Table 2.

The four protons in the $-OCH_2-$ group appear to be approximately magnetically equivalent and to give rise to a doublet of triplets centered at 5.57τ. Evidently the triplet that results from coupling of adjacent methylene groups is further split into two triplets as a result of long-range coupling to P^{31}. This assignment was confirmed by a decoupling experiment. Irradiation of the spectrum at 6.25τ transformed the two triplets to a doublet resonance (Figure 2b). The doublet in the decoupled spectrum undoubtedly arises as a result of coupling of the $-OCH_2-$ protons to P^{31}.

The triplet centered at 6.25τ is assigned to two equivalent $-CH_2Cl$ groups. The additional multiplicity of the peaks may arise from long-range coupling to P^{31}, second-order splitting, or to the nonequivalent two protons. As expected, irradiation at 5.57τ changed the triplet to a single peak (Figure 2b).

The pattern in the region $3-5\tau$ is assigned to the vinyl protons. The multiplicity and complexity of the peaks are probably caused by vinyl proton coupling with each other, with P^{31}, and to second-order effects. No attempt was made to assign chemical shifts and coupling constants to the three vinyl protons.

N-β-Bis(β'-chloroethyl)phosphonylethyl-DL-phenylalnine (1). The assignment of chemical shifts and coupling constants to 1 will be discussed with reference to the NMR spectra illustrated in Figures 3-7 and to the data summarized in Table 3.

To test the first-order assignments, the parameters were used to calculate a theoretical spectrum of 1 with NMRIT. Several choices were tried with variations of the signs of the coupling constants but we had insufficient resolution to favor one choice over the others. Consequently, all signs were assumed positive. With this decision and the NMRIT results, input data from NMREN were prepared and trial energy levels obtained. The parameters, energy levels, and line frequencies were then entered into NMRIT and ten iterations were performed.

For the purpose of fitting, the spectrum was treated as three problems: two 5-spin systems and a 3-spin system. The first 5-spin system, an A_2B_2X, was $-PCH_2CH_2N$. The second, also an A_2B_2X, was $-POCH_2CH_2Cl$. The 3-spin system was $C_6H_5CH_2CH(COO^-)N$, an ABC.

The results and errors are compared to the first order assignments in Table 3. With the exception of the coupling constants for the methine proton, the theoretical and first order data agree reasonably well.

The benzylic protons appear to be nonequivalent. Evidence for this conclusion was obtained from a decoupling experiment in which the sample was irradiated at the resonance frequency ($\tau5.39$) of the methine proton. Irradiation removes the coupling of the methine with the benzyl protons and simplifies the spectrum. The two benzyl

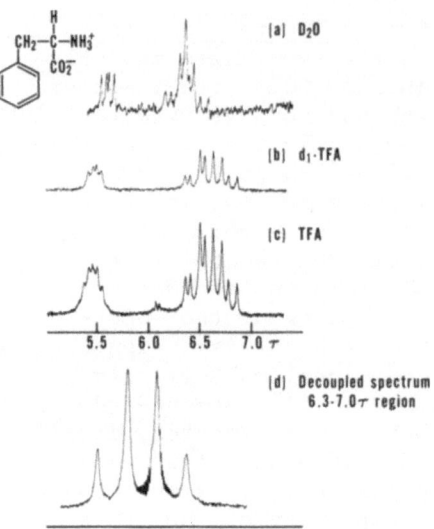

Figure 1. NMR spectra of DL–phenylalanine at 100 MHz. (a) in D_2O;
(b) in $\underline{d_1}$–TFA (trifluoracetic acid); (c) in TFA; (d) decoupled
spectrum (irradiated at 5.42τ) of methylene protons in TFA.

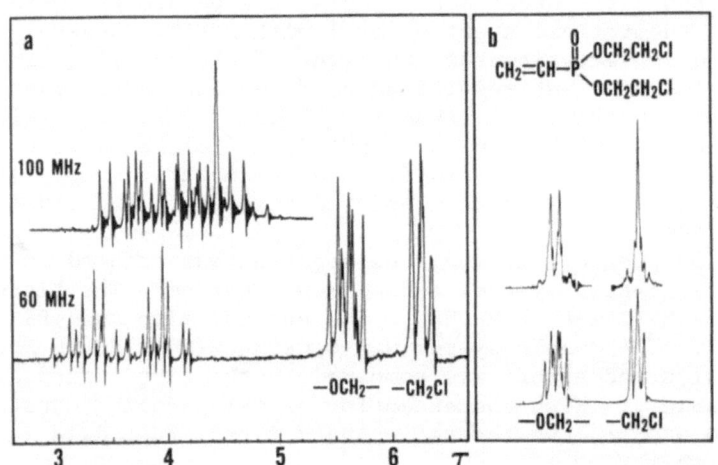

Figure 2. NMR spectra of bis(β–chloroethyl) vinyl phosphonate
in TFA at 60 MHz; upper curve in (a), vinylic region at 100 MHz;
(b) decoupled spectra (irradiated at 6.25 and 5.57τ).

Table 1

NMR Assignments for $C_6H_5CH_2CH(NH_3^+)COO^-$

Protons	Solvent	Chemical Shift (τ units)	Remarks
C_6H_4 and NH_3^+	TFA	2.7	Center of unsymmetrical multiplet
	$\underline{d_1}$-TFA	2.7	Center of broad absorption; NH_3^+ exchanged to ND_3^+
-CH-	TFA	5.42	Center of unsymmetrical multiplet
	$\underline{d_1}$-TFA	5.42	Quartet
	NaOD	6.50	Quartet
-CH$_2$-	TFA and $\underline{d_1}$-TFA	6.23 and 6.48	Calculated chemical shift for nonequivalent protons $\underline{J}_{vic}$ = 8.5 and 5.0 cps $\underline{J}_{gem}$ = 15.0 cps
	NaOD	6.99 and 7.12	$\underline{J}_{vic}$ = 7.9 and 5.1 cps J_{gem} = 13.5 cps

Table 2

NMR Assignments for $CH_2{=}CH{-}\underset{\underset{O}{\|}}{P}\overset{OCH_2CH_2Cl}{\underset{OCH_2CH_2Cl}{\diagdown}}$
(10% w/v TFA solutions)

Protons	Chemical Shift (τ units)	Remarks
-OCH$_2$-	5.57	Center of two triplets due to coupling to -CH$_2$Cl protons ($<\underline{J}>$ = 5.5 cps) and to P^{31} ($\underline{J}$ = 7.4 cps)
-CH$_2$Cl	6.25	Center of triplet $<\underline{J}>$ = 5.5 cps due to splitting by adjacent CH$_2$. Additional multiplicity due to long-range coupling to P^{31}, second-order splitting, or nonequivalence of these protons.
-CH$_2{=}$CH-	3.5	Complex pattern

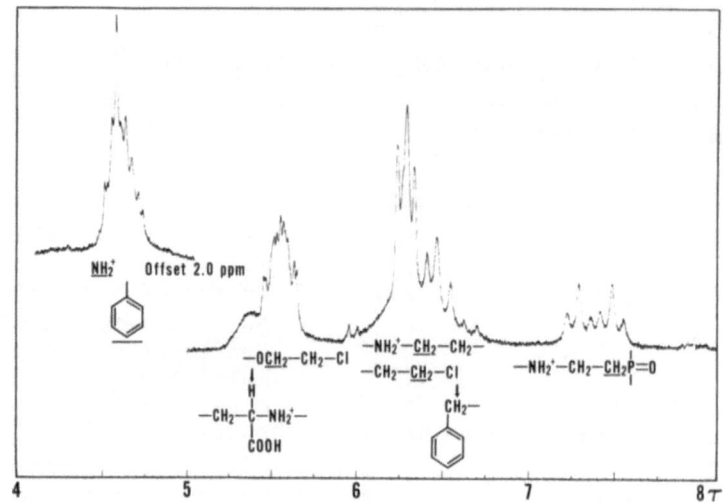

Figure 3.   NMR spectra of N–β–bis(β'–chloroethyl)phosphonylethyl–
DL–phenylalanine (1) in TFA at 100 MHz.

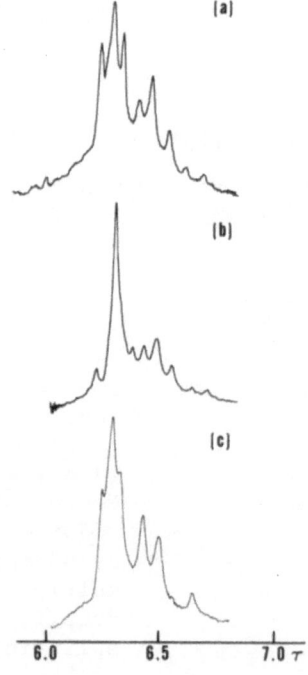

Figure 4.   (a) NMR spectrum of N–β–bis(β'–chloroethyl)phosphonyl–
ethyl–DL–phenylalanine in TFA at 100 MHz in the region 6–7τ; (b)
spectrum irradiated at 5.55τ; (c) spectrum irradiated at 5.39τ.

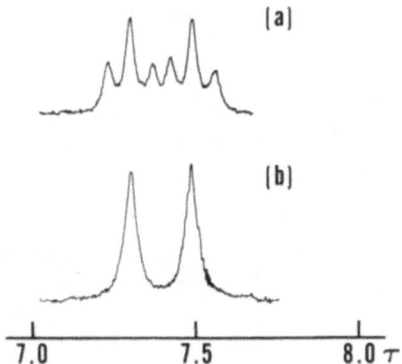

Figure 5. (a) NMR spectrum of N–β–bis(β'–chloroethyl)phosphonyl–ethyl–DL–phenylalanine in TFA at 100 MHz in the region 7–8τ; (b) decoupled spectrum (irradiated at 6.32τ).

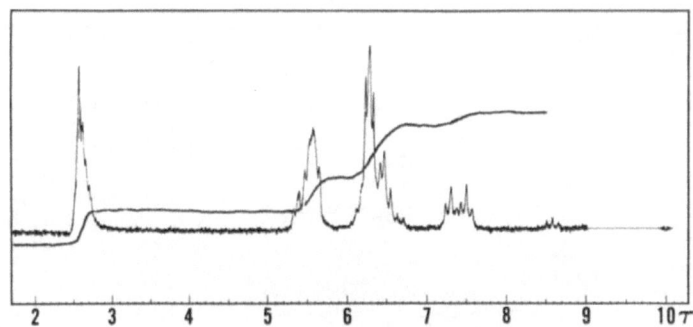

Figure 6. NMR spectrum of N–β–bis(β'–chloroethyl)phosphonylethyl–DL–phenylalanine in d₁–TFA at 100 MHz.

Table 3

NMR Assignments* for $C_6H_5CH_2CH(COO^-)NH_2{}^+CH_2CH_2\overset{\underset{\|}{O}}{P}(OCH_2CH_2Cl)_2$

| Protons | Chemical Shifts | | | |
	First Order**		Calculated***	
$CH_2P=O$	7.39 (7.75)	J_{PH} = 19.7 (19.9)	7.41 (0.005)	J_{PH} =19.6 (0.7)
$N\underline{CH}_2CH_2P=O$	6.32 (7.10)	J_{PH} = 0.0	6.32 (0.005)	J_{PH} =0.0 (0.7)
(A_2B_2X)		J_{HH} = 7.0 (8.0)		J_{HH} =6.0 (0.5)
$O=POCH_2$	5.55 (5.66)	J_{PH} = 7.6 (7.6)	5.57 (0.004)	J_{PH} =8.3 (0.7)
$O=POCH_2\underline{CH}_2Cl$	6.29 (6.22)	J_{PH} = 0.0	6.31 (0.005)	J_{PH} =0.0 (0.7)
(A_2B_2X)		J_{HH} = 5.0 (5.0)		J_{HH} =5.0 (0.5)
$C_6H_5CH_{a,b}CH_c$ a	6.56 (6.98)	J_{gem}= 15.0	6.62 (0.002)	J_{gem}=14.9 (0.4)
(ABC) b	6.39		6.36 (0.002)	
$C_6H_5CH_2\underline{CH}(COO^-)$	5.39 (6.43)	J_{HAa}= 7.9	5.38 (0.002)	J_{HHa}=9.2 (0.4)
		J_{HHb}= 6.0		J_{HHb}=4.6 (0.4)
C_6H_5, $NH_2{}^+$	2.55 (2.67)			

*Chemical shifts in τ and coupling constants in cps.

**In TFA and NaOD; values obtained in NaOD are shown in parenthesis.

***In TFA. Values in parenthesis are errors.

protons now appear as an AB-pattern with the most downfield line obscured. However, it was still possible to analyze this pattern with eq. 1. The values for the nonequivalent chemical shifts and coupling constants are shown in Figure 4 and Table 3.

The complex, unresolved multiplet centered at $\tau 5.39$ (Fig. 3) is assigned to the methine proton. The multiplet arises as a result of coupling to the benzyl protons, and is broadened by $-NH_2^+$ coupling. The broadening disappears in $\underline{d}_1$-TFA as a result of rapid exchange of $-NH_2^+$ protons for deuterium. Also, the pattern of peaks associated with the methine protons is sharpened in going from TFA to NaOD.

Integration of the peak centered at $\tau 2.6$ indicates that the resonance due to the $-NH_2^+-$ protons is masked by a signal arising from the aromatic protons. The spectrum of 1 in $\underline{d}_1$-TFA confirms this conclusion. In addition, the changed pattern of peaks centered at $\tau 5.4$ in this solvent suggests that the methine proton is coupled to the adjacent $-NH_2^+$ protons. The doublet of triplets centered at $\tau 7.39$ is assigned to the CH_2 group directly linked to the phosphorus atom (Fig. 5a). The pattern arises through coupling with the adjacent methylene protons ($\underline{J}_{AB}$=7.0 cps) and P^{31}($\underline{J}_{P}$-H = 19.7 cps).

The signals of $-NH_2^+CH_2-$ protons are not openly visible but are masked by signals from the $-CH_2Cl$ and benzyl protons. Because $-NH_2^+CH_2-$protons are coupled to $-CH_2P$ protons, a spin decoupling technique should locate the center of the $-NH_2^+CH_2-$resonance.

Irradiating at $\tau 6.32$ to collapse partially the structure of the resonance at $\tau 7.39$ provides the appropriate spectral data for the $-NH_2^+CH_2-$protons. Decoupling left a doublet centered at $\tau 7.39$ resulting from P^{31}-H coupling (Fig. 5b). In the absence of P^{31}-H coupling the signal should have been a singlet ($\underline{Cf}$. Kohler and Klein, 1976).

The double resonance experiment that decoupled $-CH_2P$ from $-NH_2^+CH_2-$protons did, in fact, give a value of $\tau 6.32$ for the center of the $-NH_2^+-CH_2-$resonance. It was not, however, possible to establish whether $-NH_2^+CH_2-$protons are coupled to the $-NH_2^+-$protons because (a) the $-NH_2^+-$resonance is masked by the signal arising from the aromatic protons and (b) the signal from the $-NH_2^+CH_2-$protons is masked by those arising from $-CH_2Cl$ and $C_6H_5CH_2-$protons.

The question arises as to whether the $-PCH_2-$ and $-NH_2^+CH_2-$assignments are reversed since such a structure has apparently not been previously analyzed. The broadening of the signals of the doublet (Fig. 5b) might have been due to coupling with $-NH_2^+-$protons. Furthermore, if the assignment is correct, no sharpening of the line should occur in $\underline{d}_1$-TFA. This was exactly the case since broadening of the signals does not disappear in this solvent. (Line width and half-heights remain the same in the two solvents.)

Unequivocal evidence for the assignments to the two methylene groups comes from analysis of the NMR spectrum of the deuterated analogue of 1. The labeled compound (2) was prepared in D_2O:

Deuterium Exchange:

$$C_6H_5CH_2CH(COO^-)NH_3^+ + D_2O \xrightarrow{\text{Et}_3N} C_6H_5CH_2CH(COO^-)ND_2$$

Phenylalanine

Nucleophilic Addition:

$$C_6H_5CH_2CH(COO^-)\ddot{N}D_2 + CH_2=CH-PO(OCH_2CH_2Cl)_2$$

$$\xrightarrow{D_2O} C_6H_5CH_2CH(COO^-)ND_2^+CH_2\underset{D}{CH}-PO(OCH_2CH_2Cl)_2$$

Deuterium labeled phenylalanine
derivative, 2

Equilibration of phenylalanine in D_2O in the presence of triethyl-amine results in exchange of the hydrogens on the amino group for deuterium atoms. The labeled phenylalanine then adds to the double bond of the vinyl compound to give a derivative that has one non-exchangeable deuterium atom on the C-atom adjacent to phosporus.

The NMR spectrum of 2 in TFA has a signal at $\tau 7.39$ with an intensity one-half the value for the corresponding signal in the NMR spectrum of 1. The NMR spectra for the two compounds were otherwise identical. These results confirm the assignments for the two methylene groups. Evidently, the $-NH_2^+\underline{CH_2}$-protons are more deshielded (by about 1 ppm) than the $-\underline{CH_2}P$ protons.

[+]Insofar as we know, this is the first reported synthesis of a deuterium-labeled amino acid derivative by adding its amino group to the double bond of a vinyl compound. This method is evidently applicable for synthesizing deuterium and tritium labeled amino acid, peptide, and protein derivatives, including labeled lysino-alanine, ornithinoalanine, and lanthionine, as illustrated:

$$HOOCCH(NHCOCH_3)(CH_2)_3CH_2NH_2 + CH_2=C(NHCOCH_3)COOCH_3$$

α-N-acetyllysine   N-Acetyldehydroalanine methyl
                                                    ester

1. X_2O/Et_3N (X = D or T)

2. Hydrolysis

$$HOOCCH(NH_2)(CH_2)_3CH_2NH-CH_2\underset{X}{C}(NH_2)COOH$$

Deuterium or tritium labeled lysinoalanine

Such amino acids should be useful in studying the metabolism of lysinoalanine, ornithinolanine, etc. For example, if lysino-alanine exerts its pharmacological effect by undergoing a reverse-Michael reaction in vivo to generate reactive dehydroalanine that

---

[+]The discussion on amino acid crosslinking complements that given on pp. 1–27 of Part A of this volume (Friedman, 1977a).

then alkylates sensitive tissue sites (Friedman, 1977a; Gould and MacGregor, 1977), the deuterium or tritium label will be lost. If, on the other hand, lysinoalanine is cleaved by nucleophilic displacement to generate lysine and serine, as illustrated, the label will not be lost.

Reverse-Michael Reaction:

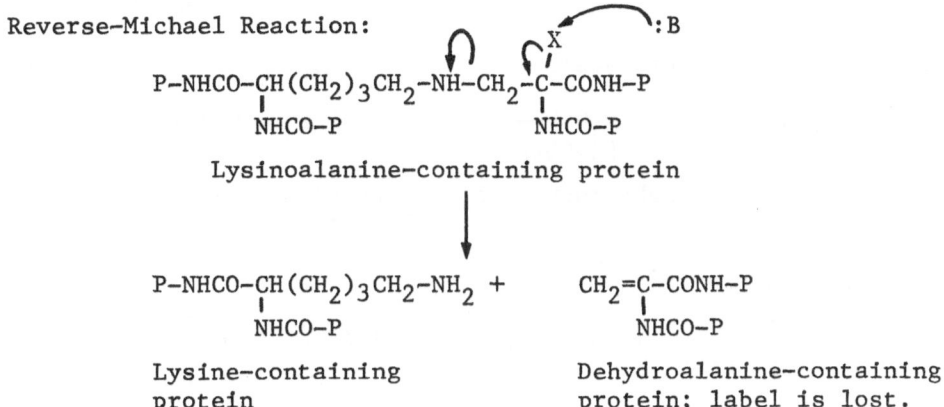

Lysinoalanine-containing protein

P–NHCO–CH(CH$_2$)$_3$CH$_2$–NH$_2$ +
        |
      NHCO–P

CH$_2$=C–CONH–P
        |
      NHCO–P

Lysine-containing
protein

Dehydroalanine-containing
protein; label is lost.

Lysinoalanine Cleavage by Nucleophilic Displacement:

B:

P–NHCO–CH(CH$_2$)$_3$CH$_2$–NH–CH$_2$–C–CONH–P
                                    |
                                  NHCO–P
                                  X

P–NHCO–CH(CH$_2$)$_3$CH$_2$–NH$_2$ +
        |
      NHCO–P

                X
                |
HO–CH$_2$–C–CONH–P
                |
              NHCOP

Lysine-containing protein

Serine-containing protein;
label retained

Since the described deuterium and tritium labeling method results in introducing one label for each crosslink formed, it could, in principle, also be used to determine the total number of crosslinks formed during alkali-treatment of a protein and to characterize α-amino acid-crosslinked amino acids. The total number of crosslinks includes not only those derived from reaction of dehydroalanine side chains with free protein functional groups such as ε-amino groups of lysine, imidazole groups of histidine, and sulfhydryl groups of cysteine side chains as discussed in detail elsewhere (Friedman, 1977a), but also those formed by adding α-amino groups to dehydroalanine. These α-amino groups are present in N-terminal amino acids and can arise also from peptide bond cleavage that may occur during the alkaline treatment, as illustrated:

Deuterium or Tritium Exchange of Peptide Bonds:

$$P-CO-NH-CH(R)-P \quad + \quad X_2O/NaOX \rightleftharpoons P-CO-NX-CH(R)-P$$

Partial Hydrolysis of Labeled Protein During Alkali Treatment:

$$P-CO-NX-CH(R)-P \quad + \quad X_2O/NaOX \longrightarrow P-COOX \quad + \quad X_2N-CH(R)-P$$

Nucleophilic Addition of Labeled α-Amino Group:

$$P-CH(R)-NX_2 \quad + \quad \underset{NX-CO-P}{CH_2=C-P} \longrightarrow P-CH(R)-NX-CH_2-\overset{X}{\underset{NX-CO-P}{C}}-P$$

                            Dehydroprotein   Labeled crosslinked
                                                      protein

Protein Acid Hydrolysis:

$$\underset{NX-CO-P}{P-CH(R)-NX-CH_2-\overset{X}{C}-P} \quad + \quad 6N\ HCl \longrightarrow \underset{\underset{NH_2}{NH-CH_2-CH-COOH}}{R-\overset{X}{CH}-C-COOH}$$

                                      Labeled crosslinked α-amino acid
                                      = glycinoalanine for R = H, etc.

   P = protein side chain
   X = D or T
   R = H for glycine, CH_3 for alanine, $C_6H_5CH_2$ for phenylalanine,
       $HOCH_2$ for serine, etc.

   It is quite possible that some of the new, unidentified peaks
we noted on amino acid chromatograms of acid hydrolysates of
alkali-treated soy protein and polyamino acids (Finley and Friedman,
1977; Friedman, 1977a) may be due to such α-amino acid-crosslinked
amino acids.

   Returning to our analysis of the phenylalanine derivative, we
note the presence of a strong triplet in the NMR spectrum centered
at τ6.29 which is assigned to the -CH_2Cl-protons. The triplet
arises from equal coupling to the adjacent methylene groups.
Irradiation of the spectrum at τ5.55 collapses the triplet to a
single peak centered at τ6.29.
   The complex multiplet at τ5.55 was assigned to the -OCH_2-protons.
Part of the complexity is removed by decoupling the -OCH_2-protons
from the adjacent -CH_2Cl-protons (Fig. 7). The decoupled spectrum
(irradiated at τ6.29) shows that the -OCH_2- protons appear as two
peaks (J^{31}_{P-H} =7.5cps) and that each of these is further split into
two additional peaks (Fig. 7b). The large splitting is attributed
to long-range coupling of the -OCH_2-protons to P^{31}. The smaller
splitting may be due to the nonequivalence of the -OCH_2 groups in 1.

The fine split of each multiplet is visible at 100 MHz and not at 60 MHz. The splitting is apparently not caused by a spin-splitting interaction since in that case the fine split would be identical at both frequencies.

The fine split is observed unchanged at 28, 40, 60, and 70°C with 5 and 10% solutions. A study of the effect of concentration on the NMR spectra indicates that the fine split remains at concentrations (w/v) of 5, 10, 20 (room temperature), and 25% (50°C). Since the fine split is neither affected by changes in temperature nor concentration, apparently it is not due to hindered rotation or intermolecular effects.

A possible explanation for the observed spectrum is that the two protons within each $-OCH_2-$ group are nonequivalent, the fine split being the separation of the two center lines of an AB quartet with the outer lines being weak and not observable at the present amplification. An alternate possibility is that the two $-OCH_2-$ groups are nonequivalent.

It should be emphasized that because of the presence of an asymmetric center in 1, it was expected that certain protons would be intrinsically nonequivalent (Shafer et al., 1961). However, model studies do not permit the prediction that the $-OCH_2-$protons should be the only ones that would fall under this category. In view of the distance separating the two $-OCH_2-$ groups from the asymmetric center, the explanation that this is another case of "acetal-type" nonequivalence (Mislow, 1966; Kondo and Mislow, 1967; Rattet, 1967) is also not warranted. Furthermore, since the fine split is not observed in the decoupled spectrum of the vinyl compound where there is no possibility for hydrogen-bonding, evidently any intrinsic asymmetry of the phosphorus atom is not responsible for the fine split.

Examination of molecular models indicates that in certain conformations of 1 the oxygen is conveniently situated for internal hydrogen bonding to the NH_2 protons forming a six-membered ring (Figure 8). Such hydrogen-bonding interactions render the two $-OCH_2-$ groups magnetically nonequivalent. At any instant these groups are in a different spatial relationship with respect to ring current effects of the phenyl and other multiple-bonded groups in the molecule. This explanation requires strong N-H····O=P bonding where the exchange is slower than the NMR measurments.

The following observations also support the postulated conformation of the phenylalanine derivative. The decoupled spectrum of the $-OCH_2-$group in NaOD consists of the expected doublet due to coupling to P^{31} but shows not fine split (Fig. 7c). The disappearance of the fine split in NaOD may be due to a solvent effect which renders the two $-OCH_2-$protons magnetically equivalent. Alternately, the fine split would also disappear on disruption of the internal hydrogen-bonding interaction by the NaOD. In a related preliminary study it was noted that the NMR spectrum of the related phenylalanine-methyl vinyl ketone adduct has two methyl peaks.

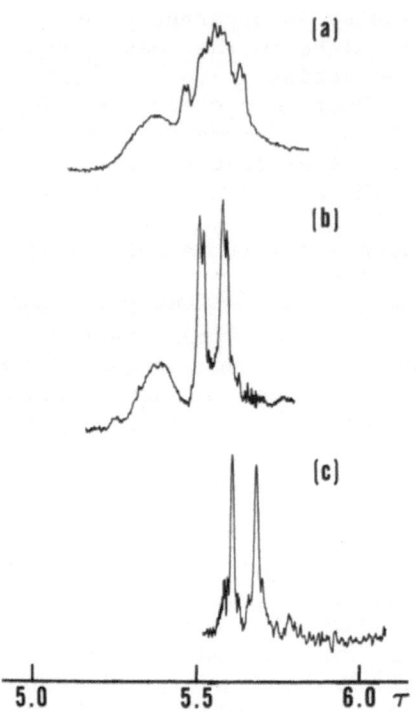

Figure 7.   (a) NMR spectrum of N-β-bis(β'-chloroethyl)phosphonyl-
ethyl-DL-phenylalanine in TFA at 100 MHz in the 5-6τ region;
(b) decoupled spectrum (irradiated at 6.29τ); (c) decoupled
spectrum in NaOD.

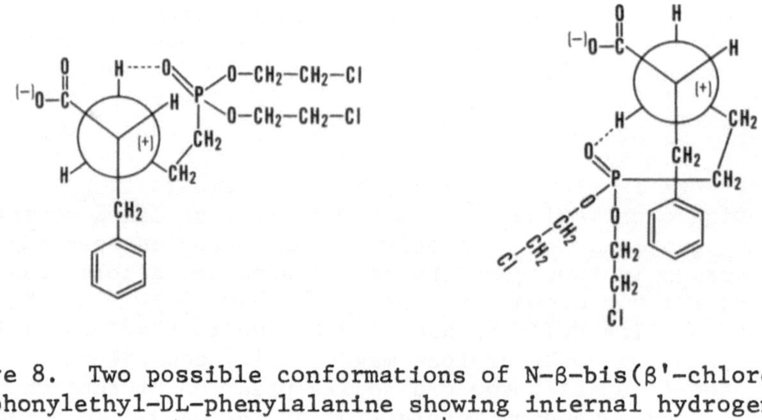

Figure 8.   Two possible conformations of N-β-bis(β'-chloroethyl)
phosphonylethyl-DL-phenylalanine showing internal hydrogen-
bonding between -P=O oxygen and NH_2^+-hydrogen.

A possible explanation is that this compound is also strongly hydrogen-bonded to give rise to a <u>spiro</u>-type bicyclic system, analogous to that shown in Fig. 8, in which hydrogen-exchange is slower than the NMR measurment. Such intramolecular N–H···O=C hydrogen bonding would give rise to two substances with the starred hydrogen <u>cis</u> in one and <u>trans</u> in the other, as illustrated, and would account for the two methyl resonances (Cf. Friedman and Wall, 1966).

It is also noteworthy that strong N–H···O=P hydrogen bonding appears to take place in an aziridine derivative (Berlin and Rengaraju, 1972). Possible effects of such hydrogen bonding on protein properties merit investigation.

Finally, the best way to distinguish between the two possible sources of the "fine split" of the methylenes would be to look at the spectrum at even higher field strength. The split at 100 MHz is about 1 cps and at 220 MHz would be about 2.2 cps. Since the peak areas of the outer peaks in an AB quartet equal the areas of the inner peaks times the ratio of the distance between the outer lines to the distance between the inner lines, the size of the outer peaks jumps from 1/29 of the inner peaks at 100 MHz to about 1/14 of the inner peaks at 220 MHz. These peaks should be visible at 220 MHz if the nonequivalence is of protons in the same methylenic group (J_{AB} was assumed to be -14 cps in the calculation of these fractions).

Reactions of the vinyl phosphonate derivative with wool and reduced wool and with the sulfhydryl group of cysteine are described elsewhere (Friedman and Tillin, 1970), as are toxicological studies of the phenylalanine, tyrosine, and cysteine vinyl phosphonate adducts (Friedman, 1977b), and the mass spectrum of the cysteine adduct (Friedman, 1977c).

ACKNOWLEDGMENTS

It is a pleasure to thank R. B. Bates and R. W. Lundin for valuable suggestions and C. A. Glass for the computer·calculations.

*Present Address: Western Regional Research Laboratory, Agricultural Research Service, USDA, Berkeley, California 94710.

REFERENCES

Berlin, D. K. and Rengaraju, S. (1972). A case of slow nitrogen
    inversion due to intramolecular hydrogen bonding. Study of
    the slow nitrogen inversion in diethyl 2-aziridinylphospho-
    nate from paramagnetic-induced shifts in the PMR spectra
    using tris(dipivalomethanato) europium (III). 163rd American
    Chemical Society Meeting, Boston, Massachusetts, April 9-14,
    1972. Abstracts p. ORGN 33.

Bible, R. N., Jr. (1965). "Interpretation of NMR Spectra",
    Plenum Press, New York, p. 82.

Finley, J. W. and Friedman, M. (1977). New amino acids formed by
    alkaline treatment of proteins. In "Protein Crosslinking:
    Nutritional and Medical Consequences", edited by M. Friedman,
    Advances in Experimental Medicine and Biology, Volume 86B,
    Plenum Press, New York, pp. 123-130.

Friedman, M. (1977a). Crosslinking amino acids -- stereochemistry
    and nomenclature. In "Protein Crosslinking: Nutritional and
    Medical Consequences", edited by M. Friedman, Advances in
    Experimental Medicine and Biology, Volume 86B, Plenum Press,
    New York, pp. 1-27.

Friedman, M. (1977b). Chemical basis for pharmacological and
    therapeutic actions of penicillamine. In "Protein Crosslin-
    king: Nutritional and Medical Consequences", edited by
    M. Friedman, Advances in Experimental Medicine and Biology,
    Volume 86B, Plenum Press, New York, pp. 649-673.

Friedman, M. (1977c). Mass spectra of cysteine derivatives. In
    "Protein Crosslinking: Biochemical and Molecular Aspects",
    edited by M. Friedman, Advances in Experimental Medicine
    and Biology, Volume 86A, Plenum Press, New York, pp. 713-726.

Friedman, M. and Romersberger, J. A. (1968). Relative influences
    of electron-withdrawing functional groups on basicities of
    amino acid derivatives. J. Org. Chem., 33, 154-157.

Friedman, M. and Tillin, S. (1970). Flame-resistant wool.
    Text. Res. J., 40, 1045-1047.

Friedman, M. and Wall, J. S. (1966). Additive linear free energy
    relationships in reaction kinetics of amino groups with α,β-
    unsaturated compounds. J. Org. Chem., 31, 2888-2894.

Friedman, M., Cavins, J. E., and Wall, J. S. (1965). Relative
    nucleophilic reactivities of amino groups and mercaptide ions
    in addition reactions with α,β-unsaturated compounds.
    J. Amer. Chem. Soc., 87, 3672-3682.

Gould, D. G. and MacGregor, J. T. (1977). Nutritional and biologi-
    cal consequences of protein crosslinking: an overview.
    In "Protein Crosslinking: Nutritional and Medical Consequen-
    ces", edited by M. Friedman, Advances in Experimental Medicine
    and Biology, Volume 86B, Plenum Press, New York, pp. 29-48.

Kondo, K. and Mislow, K. (1967). Chemical shift nonequivalence of diastereotropic groups in sulfonium salts. Tetrahed. Lett., No 14, pp. 1325-1328.

Kohler, S. J. and Klein, M. P. (1976). ^{31}P NMR chemical shielding tensors of phosphorylethanolamine, lecithin, and related compounds: application to head-group motion in model membranes. Biochemistry, 15, 967-973.

Mislow, K. (1966). "Introduction to Stereochemistry", W. A. Benjamin, Inc., New York, Chapter 2.

Rattet, L. S. Mandell, L., and Goldstein, J. H. (1967). The ^{13}C-H satellite NMR spectrum of nonequivalent protons in acetal. J. Amer. Chem. Soc., 89, 2253-2255.

Sharples, N. E. and Flavin, M. (1966). The reaction of amines and amino acids with maleimides. Structure of the reaction products deduced from infrared and NMR spectroscopy. Biochemistry, 5, 2963-2971.

Snow, J. T., Finley, J. W., and Friedman, M. (1976). Relative reactivities of sulfhydryl groups with N-acetyldehydroalanine and N-acetyldehydroalanine methyl ester. Int. J. Peptide Protein Res., 8, 57-64.

Snow, J. T., Finley, J. W., and Friedman, M. (1975). Oxidation of sulfhydryl groups to disulfides by sulfoxides. Biochem. Biophys. Res. Commun., 441-447.

Weisleder, D. and Friedman, M. (1968). The addition of halogenated acetic acids to alkyl vinyl ketones. An NMR study of the kinetics. J. Org. Chem., 33, 3542-3543.

Presented at the 154th American Chemical Society Meeting, Chicago, Illinois, September 10-15, 1967. Abstracts p. S-88.

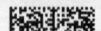